Rover 3 & 3.5 Litre Parts Catalogue Saloon and Coupe

Rover-British Leyland UK Limited
Solihull, Warwickshire, England

1st Edition
Issued December 1972

Service Department
Solihull, Warwickshire
Telephone: 021-743 4242
Telegrams : Rovrepair Solihull
Telex : 33-156

Parts Department
P.O. Box 79
Cardiff
Great Britain
Telephone: Cardiff 33681
Telegrams : Rovparts, Cardiff
Telex : 49-359

Please note that prices and specifications are subject to alteration without notice

Always quote type and number of vehicles when ordering

By Appointment to Her Majesty Queen Elizabeth II  Manufacturers of Motor Cars and Land-Rovers

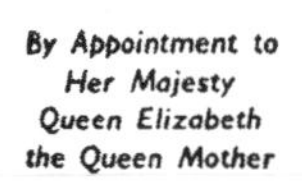

By Appointment to Her Majesty Queen Elizabeth the Queen Mother  Suppliers of Motor Cars and Land-Rovers

Part No. 608264

The Rover P5 Series, is a group of large Saloon and Coupe automobiles that were produced by Rover from 1958 until 1973.

Brooklands Books Ltd., PO Box 146, Cobham,
Surrey KT11 1LG, England.
E-mail: sales@brooklands-books.com www.brooklands-books.com

Part Number: 608264

ISBN 9781855202375 Ref: RV40PH 10T6/2438

CONTENTS

ROVER 3 LITRE
COMMENCING CAR AND UNIT NUMBERS

ROVER 3-LITRE Mk I SALOON

1959

Model	Car and chassis commencing numbers	Engine commencing number	Gearbox commencing number	Rear axle commencing number
4-speed, Home, RH Stg	625900001	625900001 Basic 8.75:1 626900001 to 626930000 with Overdrive 8.75:1 626930001 to 626960000 Basic 7.5:1 626960001 onwards with Overdrive 7.5:1	625900001	625900001 Basic 626900001 Overdrive
4-speed, Export, RH Stg	626900001			
4-speed, CKD, RH Stg	627900001			
4-speed, Export, LH Stg	628900001			
4-speed, CKD, LH Stg	629900001			
Automatic transmission, Home, RH Stg	630900001	630900001 8.75:1	Borg-Warner R5B 1000 Violet	625900001
Automatic transmission Export, RH Stg	631900001			
Automatic Transmission, CKD, RH Stg	632900001			
Automatic transmission, Export, LH Stg	633900001	631900001 7.5:1		
Automatic transmission, CKD LH Stg	634900001			

1960

Model	Car and chassis commencing numbers	Engine commencing number	Gearbox commencing number	Rear axle commencing number
4-speed, Home, RH Stg	625000001	625000001 Basic 8.75:1 626000001 to 626030000 with Overdrive 8.75:1 626030001 to 626060000 Basic 7.5:1 626060001 onwards with Overdrive 7.5:1	625000001	625000001 Basic 626000001 Overdrive
4-speed, Export, RH Stg	626000001			
4-speed, CKD, RH Stg	627000001			
4-speed, Export, LH Stg	628000001			
4-speed, CKD, LH Stg	629000001			
Automatic transmission, Home, RH Stg	630000001	630000001 8.75:1	Borg-Warner R5B 1000 Violet plate	625000001
Automatic transmission, Export, RH Stg	631000001			
Automatic transmission, CKD, RH Stg	632000001			
Automatic transmission, Export, LH Stg	633000001	631000001 7.5:1		
Automatic transmission, CKD, LH Stg	634000001			

NOTE: On 1959-60 3-Litre 4-speed models the fifth digit in the 6269 and 6260 range of engine numbers is also used to indicate the various types of units used. See above for details.

ROVER 3 LITRE
COMMENCING CAR AND UNIT NUMBERS - continued

1961

Model	Car and chassis commencing numbers	Engine commencing number	Gearbox commencing number	Rear axle commencing number
4-speed, Home RH Stg	625100001	625100001 Basic 8.75:1 626100001 Overdirve 8.75:1 627100001 Basic 7.5:1 628100001 Overdrive 7.5:1	625100001	625100001 Basic 626100001 Overdrive
4-speed, Export, RH Stg	626100001			
4-speed, CKD, RH Stg	627100001			
4-speed, Export, LH Stg	628100001			
4-speed, CKD, LH Stg	629100001			
Automatic transmission Home, RH Stg	630100001	630100001 8.75:1	Borg-Warner R5B 1000K Violet plate	625100001
Automatic transmission, Export, RH Stg	631100001			
Automatic transmission, CKD, RH Stg	63210001			
Automatic transmission, Export, LH Stg	633100001	631100001 7.5:1		
Automatic transmission, CKD, LH Stg	634100001			

ROVER 3- LITRE MK IA SALOON

Model	Car and chassis commencing numbers	Engine commencing number	Gearbox commencing number	Rear axle commencing number
4-speed, Home, RH Stg	72500001a	72500001a8.75:1without Overdrive 72600001a 8.75:1 with Overdrive 72700001A 7.5:1 with Overdrive 72800001a 7.5:1 with Overdrive	72500001a	72500001a without Overdrive 72600001a with Overdrive
4-speed, Export RH Stg	72600001a			
4-speed, CKD, RH Stg	72700001a			
4-speed, Export, LH Stg	72800001a			
4-speed, CKD, LH Stg	72900001a			
Automatic transmission, Home RH Stg	73000001a	73000001a 8.75:1	Borg-Warner R5B 1000P	72500001a
Automatic transmission, Export RH Stg	73100001a			
Automatic transmission, CKD, RH Stg	73200001a			
Automatic transmission, Export, LH Stg	73300001a	73100001a 7.5:1		
Automatic transmission, CKD, LH Stg	73400001a			

ROVER 3 LITRE
COMMENCING CAR AND UNIT NUMBERS

ROVER 3-LITRE Mk II SALOON

Model	Car and chassis commencing numbers	Engine commencing number	Gearbox commencing number	Rear axle commencing number
4-speed, Home, RH Stg	77000001a			
4-speed, Export, RH Stg	77100001a	77000001a 8.75:1		
4-speed, CKD, RH Stg	77200001a		77000001a	77000001a
4-speed, Export, LH Stg	77300001a	77100001a 8:1		
4-speed, CKD, LH Stg	77400001a			
Automatic transmission, Home, RH Stg	77500001a			
Automatic transmission, Export, RH Stg	77600001a			
Automatic transmission, CKD, RH Stg	77700001a	77500001a	Borg-Warner R5B 1000KP	77500001a
Automatic transmission, Export, LH Stg	77800001a			
Automatic transmission CKD, LH Stg	77900001a			

ROVER 3-LITRE MK II COUPE

Model	Car and chassis commencing numbers	Engine commencing number	Gearbox commencing number	Rear axle commencing number
4-speed, Home, RH Stg	73500001a			
4-speed, Export, RH Stg	73600001a	77000001a8.75:1		
4-speed, CKD, RH Stg	73700001a		77000001a	77000001a
4-speed, Export, LH Stg	73800001a	77100001a8:1		
4-speed, CKD, LH Stg	73900001a			
Automatic transmission, Home RH Stg	74000001a			
Automatic transmission, Export, RH Stg	74100001a			
Automatic transmission, CKD, RH Stg	74200001a	77500001a	Borg-Warner R5B 1000KP	77500001a
Automatic transmission, Export, LH Stg	74300001a			
Automatic transmission, CKD, LH Stg	74400001a			

ROVER 3 LITRE
COMMENCING CAR AND UNIT NUMBERS - continued

ROVER 3-LITRE MK III SALOON

Model	Car and chassis commencing numbers	Engine commencing number	Gearbox commencing number	Rear axle commencing number
4-speed, Home RH Stg	79500001a			
4-speed,Export, RH Stg	79600001a	79500001a 8.75:1		
4-speed, CKD, RH Stg	79700001a		77000001a	79500001a
4-speed, Export, LH Stg	79800001a	79600001a 8:1		
4-speed, CKD, LH Stg	79900001a			
Automatic transmission, Home RH Stg	80000001a			
Automatic transmission, Export, RH Stg	80100001a	80000001a 8.75:1		
Automatic transmission, CKD, RH Stg	80200001a		3EU 1001	80000001a
Automatic transmission, Export, LH Stg	80300001a	80100001a 8:1		
Automatic transmission, CKD, LH Stg	80400001a			

ROVER 3- LITRE Mk III COUPE

Model	Car and chassis commencing numbers	Engine commencing number	Gearbox commencing number	Rear axle commencing number
4-speed, Home, RH Stg	80500001a			
4-speed, Export, RH Stg	80600001a	79500001a 8.75:1		
4-speed, CKD, RH Stg	80700001a		77000001a	79500001a
4-speed, Export, LH Stg	80800001a	79600001a 8:1		
4-speed, CKD, LH Stg	80900001a			
Automatic transmission, Home, RH Stg	81000001a			
Automatic transmission, Export, RH Stg	81100001a	80000001a 8.75:1		
Automatic transmission CKD, RH Stg	81200001a		3EU 1001	80000001a
Automatic transmission, Export, LH Stg	81300001a	80100001a 8:1		
Automatic transmission, CKD, LH Stg	81400001a			

ROVER 3½ LITRE SALOON AND COUPE
COMMENCING CAR AND UNIT NUMBERS

ROVER 3½ LITRE SALOON

Model	Car and chassis commencing numbers	Engine commencing number	Gearbox commencing number	Rear axle commencing number
Automatic transmission, Home, RH Stg	84000001a			
Automatic transmission, Export, RH Stg	84100001a			
Automatic transmission, CKD, RH Stg	84200001a	84000001a	Borg-Warner	84000001a
Automatic transmission, Export, LH Stg	84300001a			
Automatic transmission, CKD, LH Stg	84400001a			

ROVER 3½ LITRE COUPE

Model	Car and chassis commencing numbers	Engine commencing number	Gearbox commencing number	Rear axle commencing number
Automatic transmission Home, RH Stg	84500001a			
Automatic transmission, Export, RH Stg	84600001a			
Automatic transmission, CKD, RH Stg	84700001a	84000001a	Borg-Warner	84000001a
Automatic transmission Export, LH Stg	84800001a			
Automatic transmission CKD, LH Stg	84900001a			

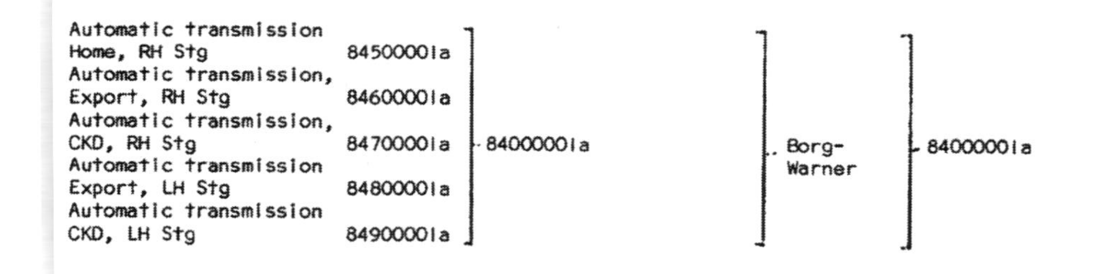

GENERAL EXPLANATION

Some of the information in this Publication is based on illustration and script type of presentation and the remainder is part-numbered illustrations with no script.

As the list covers both Home and Export models, reference is made throughout to the 'left-hand' (LH) and 'right-hand' (RH) sides of the vehicle, rather than to 'near-side' and 'off-side'. The 'left-hand side' is that to the left hand when the vehicle is viewed from the rear, similarly, 'left-hand steering' (LH Stg) models are those having the driving controls on the left-hand side, again when the vehicle is viewed from the rear.

The part numbers quoted are common to all 3 Litre and 3½ Litre models, unless otherwise stated, either against the part number or in the remarks section on the page or illustration, as applicable.

Part numbers marked with an asterisk * indicate new parts which have not been used on any previous Rover model.

Script and illustrations

Assemblies are printed in capitals, and all the parts comprising that assembly are shown by indenting the alignment of the description column, the indentation reverting to the original alignment when the assembly is complete. The indentation is clearly shown by the numbers 1, 2, 3, 4 at the top of each page.

It will be seen that the 'ROCKER COVER, TOP, ASSEMBLY' 11-49, includes the 'tappet clearance plate' and 'drive screw fixing plate' (indentation 1), while the 'joint washer', by reverting to alignment with 'ROCKER', is excluded from the assembly.

In all cases where the entire assembly is required, quote only the assembly number.

Part-numbered illustrations

A new presentation of parts information is used for certain parts of this 3 Litre and 3½ Litre Publication. That is, part numbered illustrations with no description, this gives a quicker look-up time and obviates the need for Parts Catalogue translation.

Information about assemblies and the components which comprise an assembly together with part numbers of gasket sets etc. will be found at the bottom of the illustration.

Figures in brackets represent the quantity used per vehicle.

OBSERVATIONS GENERALES

Pour une partie de ce catalogue nous utilisons des illustrations accompagnées d'un texte, tandis que le reste est constitué par des illustrations auxquelles les reference ont été incorporées.

Vu que cette fiche est valable bien pour les véhicules destinés au marché intérieur que pour ceux prévus pour l'exportation, nous avons recours à la spécification gauche et droite, droite et gauche étant définis par rapport au vehicule, lorsque celui-ci est examiné de l'arrière.

Sauf avis contraire, toutes les références sont valables pour tous les modèles 3 Litres et 3½ Litres. Toute exception sera indiquée soit au bas de la page, soit sur la gravure même.

Les pièces repérées par un astérisque sont des pièces utilisées pour la prémiere fois sur un véhicule Rover.

Texte et illustrations

Les ensembles sont imprimés en lettres majuscules et toutes les pièces comprises dans un ensemble sont indiquées en deplaçant la désignation de ces pièces d'une colonne vers la droite par rapport à la position qu'occupe la désignation de l'ensemble. La désignation de la pièce suivante non comprise dans l'ensemble est ramenée a l'alignement des désignations et clairement montrée par les numéros 1, 2, 3 et 4 en haut de chaque page.

Par example, l'on remarquera que l'ensemble 'ROCKER COVER, TOP, ASSEMBLY' 11-49, comprend la 'tappet clearance plate' et la 'drive screw fixing plate' (colonne 2), tandis que le 'joint washer' est exclu de l'ensemble, la désignation de celle-ci étant ramenée a la colonne 1, en ligne avec 'ROCKER'.

Lorsqu'un ensemble complet est requis, il suffira d'indiquer le numéro de l'ensemble.

Illustrations numerotées

Certaines parties de catalogue sont présentées sous une nouvelle forme. C'est-à-dire, les référence sont incorporées aux illustrations sans en utiliser une désignation, ce qui permet de rechercher une pièce plus rapidement. Ainsi, aucune traduction du catalogue de pièces detachées s'imposera.

Tous les détails au sujet des ensembles et de leurs pièces constituantes sont donnés en bas de la page avec les références des pochettes de joints.

Le chiffres entre parenthèses indiquent la quantité requise par véhicule.

ALLGEMEINE ERKLAERUNGEN

Ein Teil dieses Kataloges ist in der bisherigen Aufmachung gehalten, d.h. die Abbildung auf der einen und der Text auf der gegenuberliegenden Seite, d.h. die Textseite fallt weg und die Ersatzteilnummern sind direkt in der Abbildung enthalten.

Das Transparent hat fur Fahrzeuge fur den Binnen- und Exportmarkt Geltung. Es wird deshalb stets auf die linke oder rechte Fahrzeugseite Bezug genommen. Unter der linken oder rechten Seite ist diejenige Seite zu verstehen, wenn das Fahrzeug von hinten her betrachtet wird. Bei einem linksgesteuerten Fahrzeug ist das Lenkrad demnach auf der linken Fahrzeugseite angeordnet, wenn dieses von hinten betrachtet wird.

Falls dies nicht anderweitig angegeben ist, haben alle aufgefuhrten Teilnummern fur alle 3 Liter und 3½ Liter Modelle Geltung. Allfallige Abweichungen sind entweder am Fuss der Seite oder in der Abbildung selbst angegeben.

Bei Teilen, die mit einem Stern (*) bezeichnet sind, handelt es sich um solche, welche bisher noch bei keinem Rover Modell Verwendung gefunden haben.

Text und Abbildung

Die Baugruppen sind in Grossbuchstaben gedruckt. Alle in der Baugruppe enthaltenen Teile sind eingeruckt. Der Schluss der Baugruppe wird durch das Vorrucken an den Rand angegeben. Die Einruckung wird am oberen Seitenrand durch die Zahlen 1, 2, 3 und 4 angezeigt.

So umfasst die Baugruppe 'ROCKER COVER, TOP, ASSEMBLY' 11-49, die 'tappet clearance plate' und die 'drive screw' (eingeruckt), wahrend der 'joint washer' in der Baugruppe nicht eingeschlossen ist, und wieder an den Rand vorgeruckt wird.

Wo die ganze Baugruppe benotigt wird, ist ausschliesslich die Teilnummer dieser Baugruppe aufzufuhren.

Abbildungen mit Teilnummern

Ein gewisser Teil dieses Kataloges ist in der neuen Aufmachung dargestellt. Bei dieser Darstellung sind die Teilnummern in der Abbildung eingetragen. Auf eine Bezeichnung der abgebildeten Teile wird verzichtet, um ein rascheres Auffinden der Teile zu ermoglichen. Mit dieser Methode erubrigt sich auch das Uebersetzen der Ersatzteilliste.

Die entsprechenden Angaben uber Baugruppen und die darin enthaltenen Teile sind jeweils mit den Teilnummern fur Dichtungssatze, etc. am Fuss der Seite aufgefuhrt.

Die in Klammern aufgefuhrten Zahlen geben die pro Fahrzeug benotigte Anzahl an.

EXPLICACAO GERAL

Alguns dos dados nêste texto estão baseados sobre a presentação do tipo de ilustração e títulos, sendo o resto representado por ilustrações sem títulos.

Como a lista compreende os modelos para o mercado británico e para exportação, refere-se, em todas partes, aos lados esquerdos e direitos (left-hand LH) e (right-hand RH) do veículo, em lugar de 'lado mais perto do pavimento' (near-side) e 'lado mais distante do pavimento' (off-side). O 'lado esquerdo' e o lado esquerdo visto quando ve-se o veícolo desde a parte traseira; similarmente, os modelos com volante à esquerda são os modelos com volante à esquerda, os modelos com os comandos de condução situados no lado esquerdo, também quando ve-se o veículo desde a parte traseira.

Os Peca Nos. marcados com um asterisco (*) indicam peças novas que não teem sido empregadas em qualquer modelo Rover anterior.

Títulos e ilustrações

Os conjuntos impressos em letras maíusculas e todas as peças que formam êste conjunto, são indicados com recortes dentados do alinhamento da coluna de descrição, o recorte revertendo ao alinhamento primitivo quando o conjunto está completo. O recorte é indicado claramente pelos numeros 1, 2, 3, 4 na parte superior da página.

E possível observar que a 'ROCKER COVER, TOP, ASSEMBLY' 11-49 (TAPA SUPERIOR DE BALANCEIRO' CONJUNTO DE', compreende a 'tappet clearance plate' (chapa da folga de tucho) e a 'drive screw fixing plate' (chapa de seguranca de parafuso regulador (recorte 1), mas a 'joint washer' (arruela de junta), revertendo ao alinhamento com o 'ROCKER' (BALANCEIRO), é excluido do conjunto.

Em todos casos onde precisa-se do conjunto completo, citar sòmente o número do conjunto.

Ilustrações com Peca Nos.

Emprega-se uma presentação nova de dados de peças, para certas secções dêste texto do Land-Rover, quer dizer, ilustrações com Peça Nos. sem descrição, o que proporciona tempo de investigação mais breve, assim evitando a necessidade de tradução do Catalógo de Peças.

Dados acerca de conjuntos e componentes, que compreendem um conjunto, juntamente com Peça Nos. de jôgos de juntas, etc., acharam-se ao pe da ilustração.

As cifras em parenteses representam a quantidade empregada por veículo.

EXPLICACIONES GENERALES

Algunos de los datos en la presente son basados sobre la presentación tipo de ilustración y texto; por lo que se refiere al resto, esto consta de ilustraciones con número de pieza sin título.

Como la lista comprende modelos para la exportación y para el mercado britanico, se refiere, en todas partes, al lado izquierdo (LH) y al lado derecho (RH) del vehículo, en vez de 'near side' (lado más cercano al solado) y de 'off side' (lado mas lejos del solado). El lado izquierdo es el lado izuierdo visto desde la parte trasera del vehiculo ? igualmente, los modelos con volante a la izquierda, son los modelos que tienen el volante de direccion la izquierda, visto también desde la parte trasera del vehiculo.

Los Pieza Nos. citados son comunes para todos los modelos 3 Litros y 3½ Litros, salvo indicaciones contrarias, que están ubicados sea contra el Pieze No., sea en la sección de observaciones en la página o ilustración, según aplicable.

Los Pieza Nos. marcados con asterisco (*) indican piezas nuevas que no han sido utilizadas en cualquier modelo Rover anterior.

Titulos e ilustraciones

Los conjuntos están imprimidos en letras mayúsculas y todas las piezas que comprenden dicho conjunto vienen indicadas sangrando (haciendo entrar) la alineación de la columna de descripción, volviendo el sangrado a la alieneación primitiva cuando el conjunto está competo. El sangrado esta indicado con claridad por los numeros 1, 2, 3, 4 en la parte superior de cado página.

Se verá que 'ROCKER COVER, TOP, ASSEMBLY' 11-49, (CONJUNTO DE TAPA DE BALANCIN SUPERIOR), incluye la 'place de huelgo de taqué' (tappet clearance plate) y 'drive screw fixing plate' (placa de sujeción de tornillo), mientras que la 'joint washer' (arandela de junta), volviendo a alinearse con el 'ROCKER' (BALANCIN), se excluye del conjunto.

En todos casos donde se requiere el conjunto completo, sólo se citará el número del conjunto.

Illustraciones con Pieza Nos.

Se usa una presentación nueva de datos pertinentes a piezas para ciertas partes del presente texto, es decir, ilustraciones con Pieza Nos., sin descripción, lo que proporciona tiempo de investigación mas breve y evita la necesidad de traducir el Catalogo de Piezas.

Datos pertinentes a conjuntos y a componentes que comprenden un conjunto, conjuntamente con pieza nos. de juegos de juntas, etc., se encontrarán al pie de la ilustración.

Las cifras entre parentesis representan la cantidad usada por vehículo.

GROSS INDEX. INDEKS GRUPPER. INHOUD-HOOFDSTUKKEN. RYHMAHAKEMISTO.
TABLE DES SECTIONS. GRUPPENVERZEICHNIS. INDICE DEI GRUPPI.
NUMBERACAO DE GRUPO. INDICE DEL GRUPO. GRUPPINDEX.

3 LITRE 1116 to 1185	1205 to 1216 1244 to 1273	3 LITRE 1217 to 1243 3½ LITRE 1331 to 1350
3½ LITRE 1274 to 1285 1305 to 1327	1352 to 1359	1360 to 1373
1421 to 1430	MANUAL 1431 to 1440 1460 to 1465 POWER 1441 to 1459 1466 to 1467	3 LITRE 1468 to 1477 3½ LITRE 1478 to 1483
3 LITRE 1505 to 1518 3½ LITRE 1519 to 1522	3 LITRE 1523 to 1534 3½ LITRE 1535 to 1538	3 LITRE 1539 to 1542 3½ LITRE 1543 to 1546
1549 to 1578	1579 to 1584	1605 to 1614

GROSS INDEX. INDEKS GRUPPER. INHOUD-HOOFDSTUKKEN. RYHMAHAKEMISTO.
TABLE DES SECTIONS. GRUPPENVERZEICHNIS. INDICE DEI GRUPPI.
NUMBERACAOE DE GRUPO. INDICE DEL GRUPPO. GRUPPINDEX.

3 LITRE 1624 to 1653 3½ LITRE 1654 to 1661	3 LITRE 1662 to 1683 1711 to 1726 3½ LITRE 1705 to 1710 1727 to 1732	1733 to 1752
1760 to 1769	1770 to 1783	1805 to 1826
1827 to 1834	1835 to 1859	1860 to 1869 1905 to 1922
1870 to 1885 1923 to 1949	1950 to 1957	1958 to 1959
1961	1962 to 1967	1968 to 1971

PAGE INDEX. INDEKS SIDER. INHOUD-BLADZIJDEN. SIVUHAKEMISTO. INDEX. SEITENVERZEICHNIS.
INDICE DELLE PAGINE. NUMERACAO DE PAGINA. INDICE DE LA PAGINA SIDOINDEX

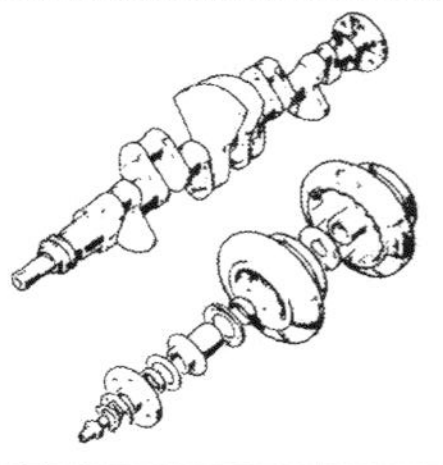

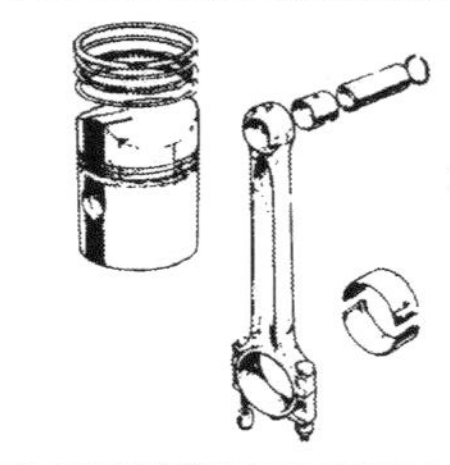

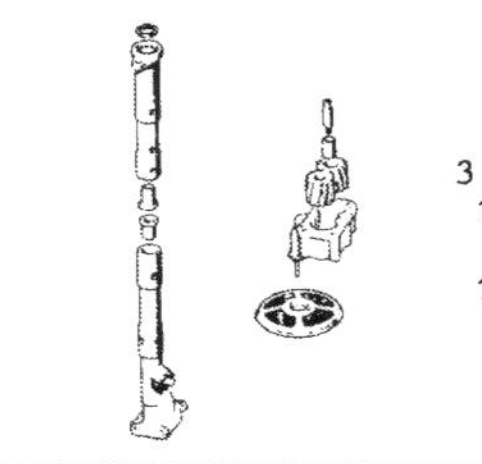

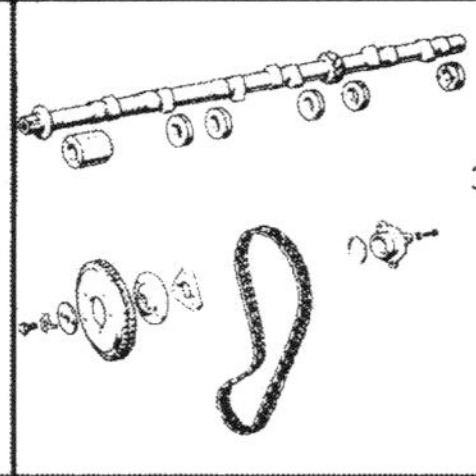

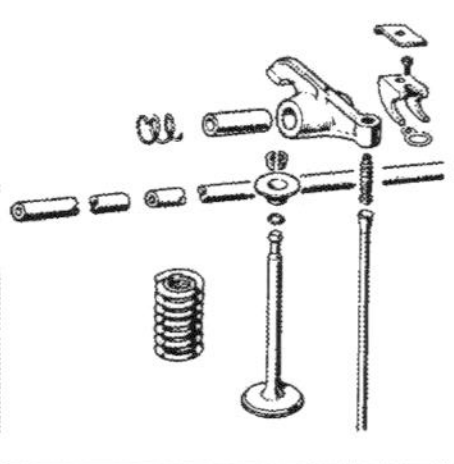

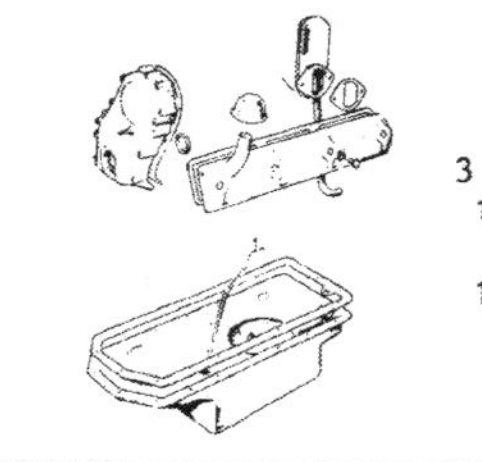

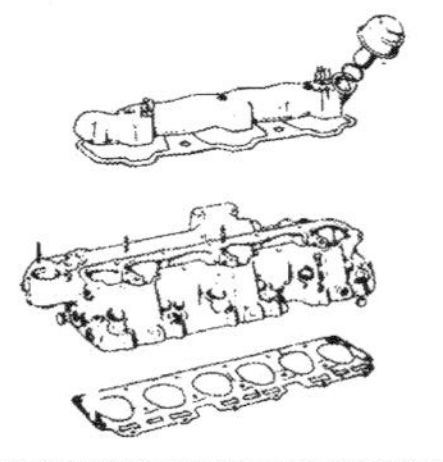

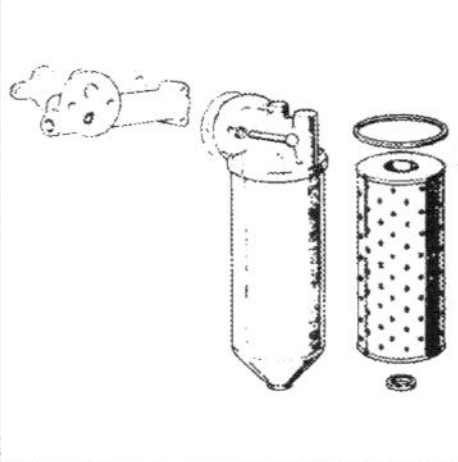

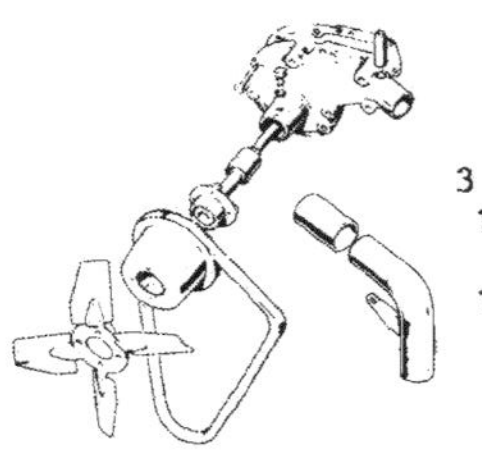

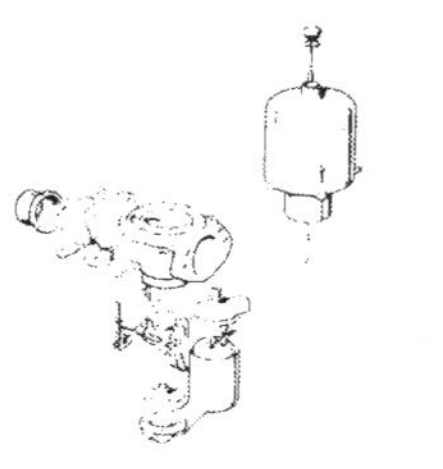

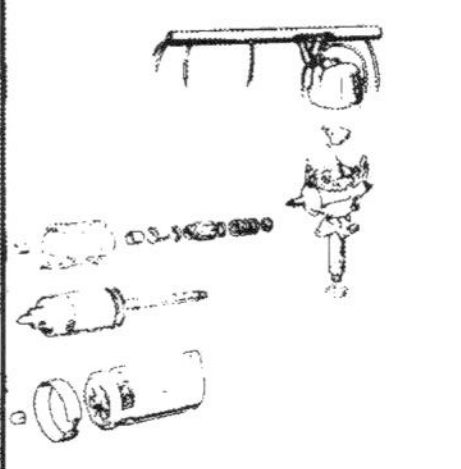

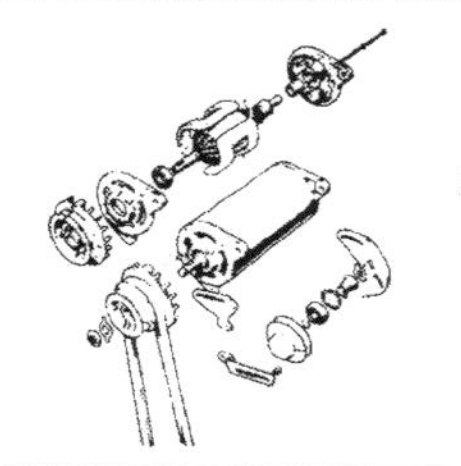

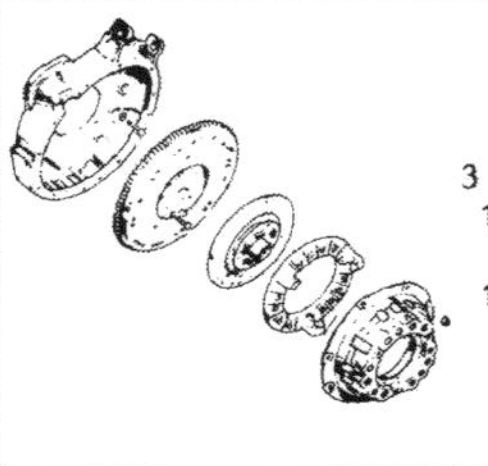

PAGE INDEX. INDEKS SIDER. INHOUD-BLADZIJDEN. SIVUHAKEMISTO. INDEX. SEITENVERZEICHNIS.
INDICE DELLE PAGINE. NUMERACAO DE PAGINA. INDICE DE LA PAGINA SIDOINDEX

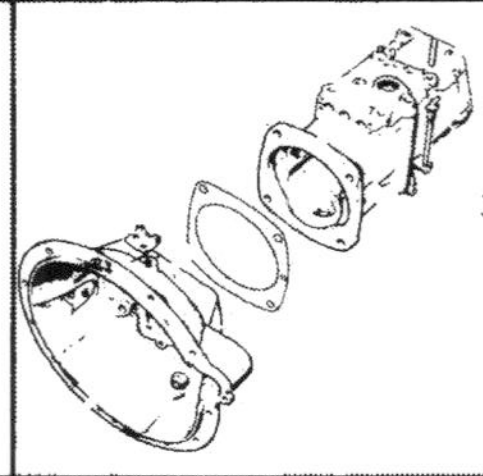

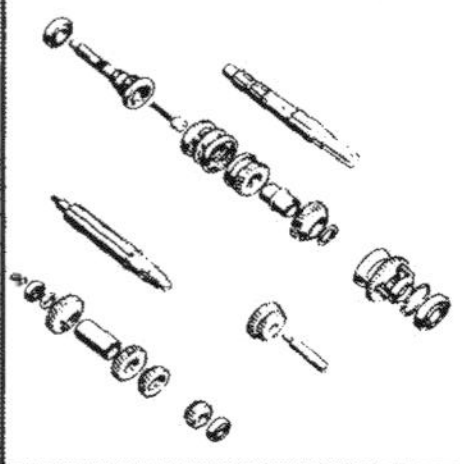

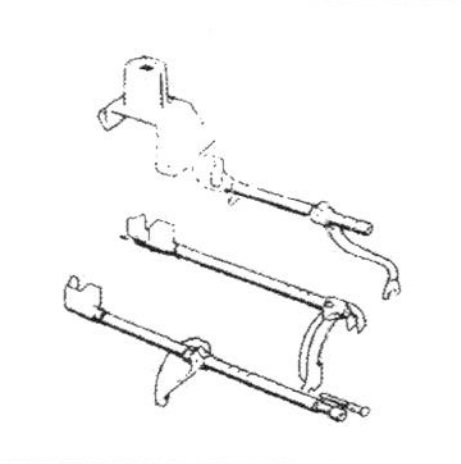

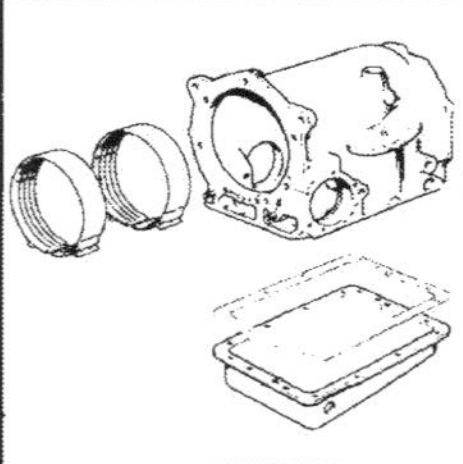

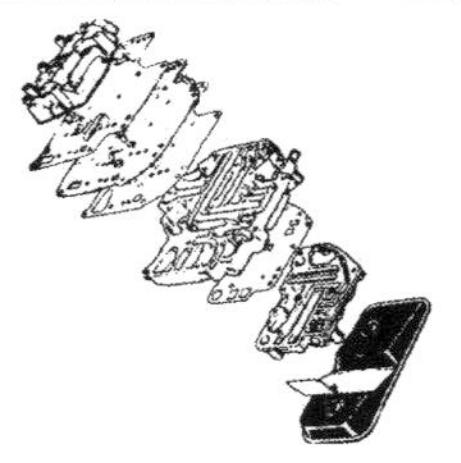

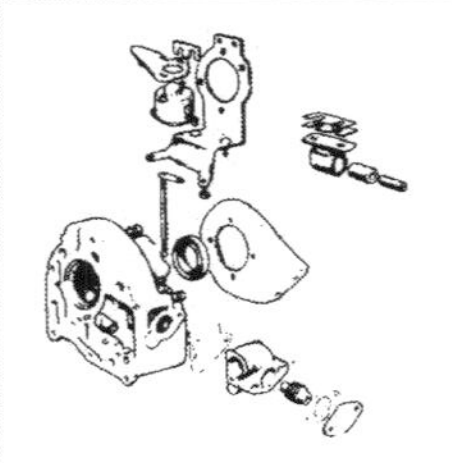

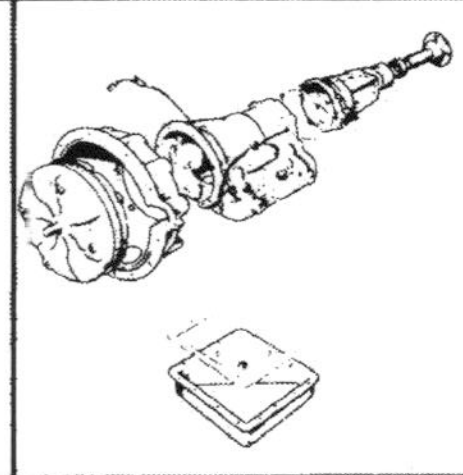

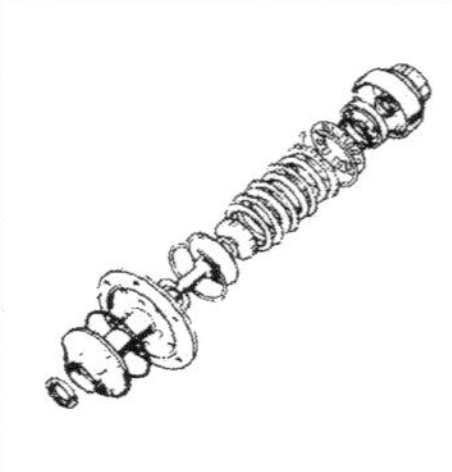

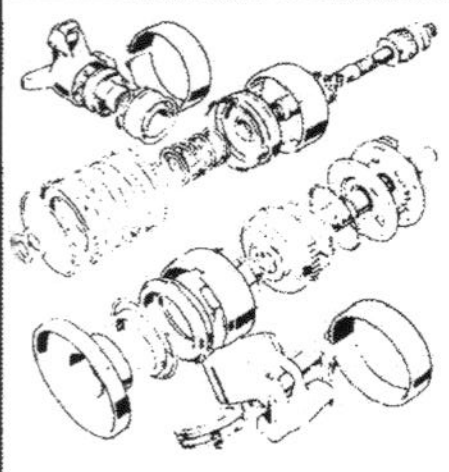

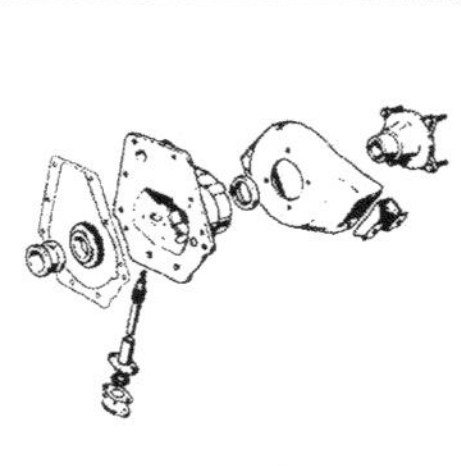

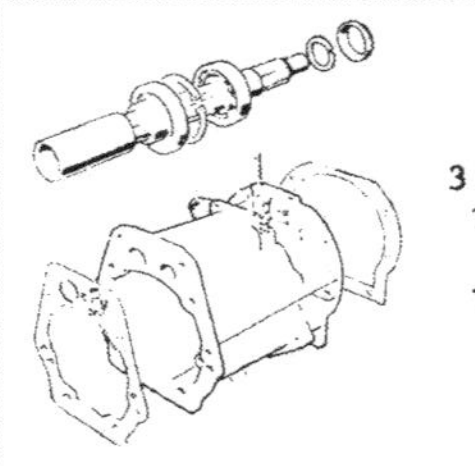

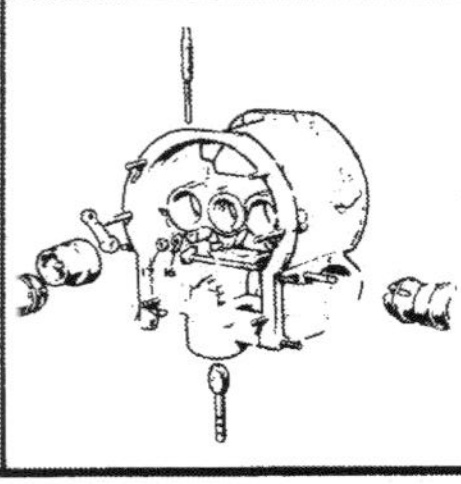

PAGE INDEX. INDEKS SIDER. INHOUD-BLADZIJDEN. SIVUHAKEMISTO. INDEX. SEITENVERZEICHNIS.
INDICE DELLE PAGINE. NUMERACAO DE PAGINA. INDICE DE LA PAGINA SIDOINDEX

PAGE INDEX. INDEKS SIDER. INHOUD-BLADZIJDEN. SIVUHAKEMISTO. INDEX. SEITENVERZEICHNIS.
INDICE DELLE PAGINE. NUMERACAO DE PAGINA. INDICE DE LA PAGINA SIDOINDEX

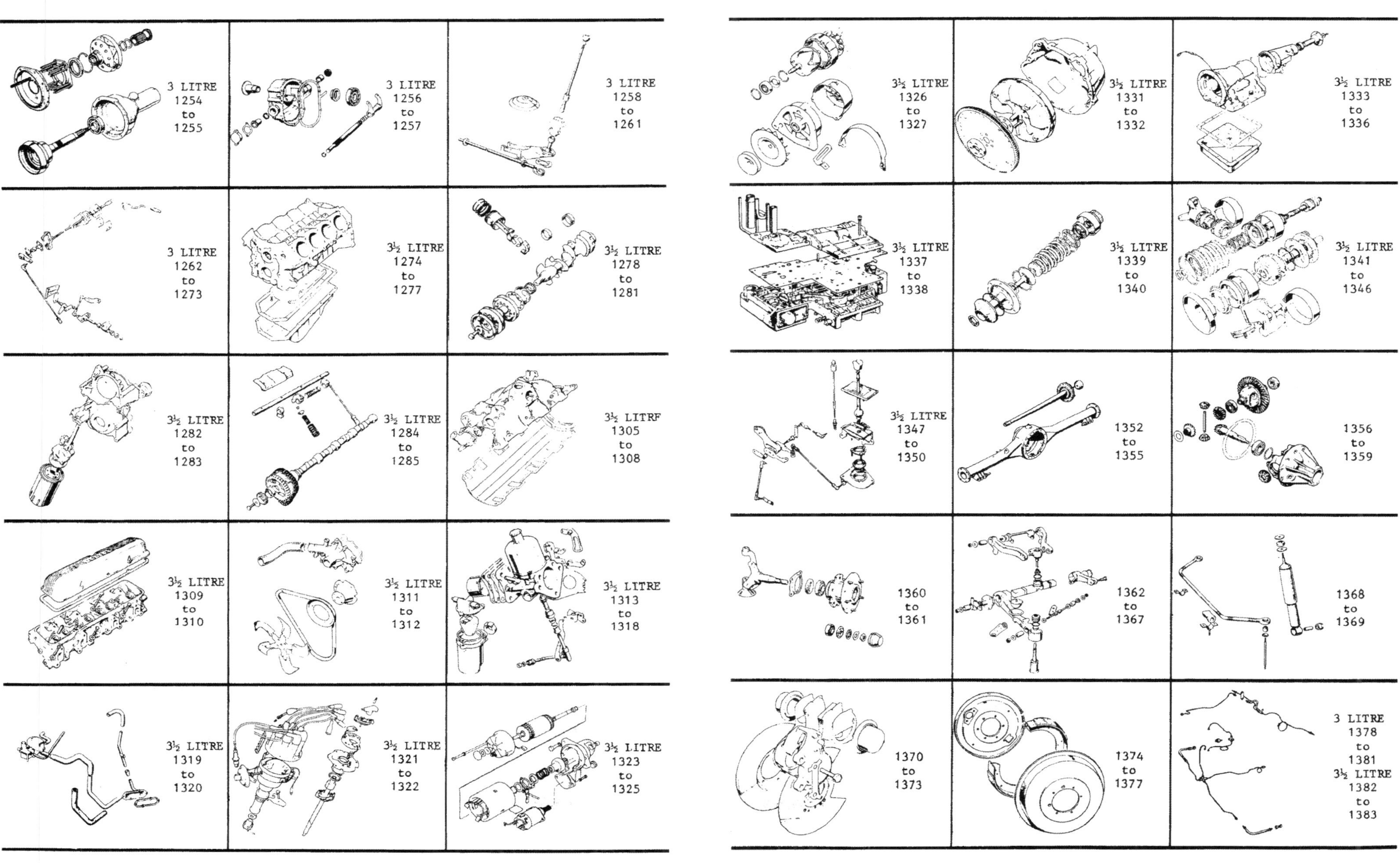

PAGE INDEX. INDEKS SIDER. INHOUD-BLADZIJDEN. SIVUHAKEMISTO. INDEX. SEITENVERZEICHNIS.
INDICE DELLE PAGINE. NUMERACAO DE PAGINA. INDICE DE LA PAGINA SIDOINDEX

PAGE INDEX. INDEKS SIDER. INHOUD-BLADZIJDEN. SIVUHAKEMISTO. INDEX. SEITENVERZEICHNIS.
INDICE DELLE PAGINE. NUMERACAO DE PAGINA. INDICE DE LA PAGINA SIDOINDEX

PAGE INDEX. INDEKS SIDER. INHOUD-BLADZIJDEN. SIVUHAKEMISTO. INDEX. SEITENVERZEICHNIS.
INDICE DELLE PAGINE. NUMERACAO DE PAGINA. INDICE DE LA PAGINA SIDOINDEX

PAGE INDEX. INDEKS SIDER. INHOUD-BLADZIJDEN. SIVUHAKEMISTO. INDEX. SEITENVERZEICHNIS.
INDICE DELLE PAGINE. NUMERACAO DE PAGINA. INDICE DE LA PAGINA SIDOINDEX

CYLINDER BLOCK

H731

Plate Ref.	1 2 3 4	Description	Qty.	Part No.	Remarks
		ENGINE AND CLUTCH ASSEMBLY	1	525048	Mk I and Mk IA 4-speed. 8.75:1 compression ratio
		ENGINE AND CLUTCH ASSEMBLY	1	525050	Mk I and Mk IA 4-speed. 7.5:1 compression ratio
		ENGINE ASSEMBLY	1	525052	Mk I and Mk IA Automatic. 8.75:1 compression ratio
		ENGINE ASSEMBLY	1	525054	Mk and Mk IA Automatic. 7.5:1 compression ratio
		ENGINE AND CLUTCH ASSEMBLY	1	600371	Mk II 4-speed. 8.75:1 compression ratio
		ENGINE AND CLUTCH ASSEMBLY	1	600377	Mk II 4-speed. 8:1 compression ratio
		ENGINE ASSEMBLY	1	516951	Mk II Automatic. 8:1 compression ratio with engine suffix 'A' and 'B'
		ENGINE ASSEMBLY	1	600373	Mk II Automatic. 8:1 compression ratio with engine suffix 'C'
		ENGINE AND CLUTCH ASSEMBLY	1	600379	Mk II 4-speed NADA. 8.75:1 compression ratio
		ENGINE ASSEMBLY	1	600235	Mk II Automatic NADA. 8:1 compression ratio with engine suffix 'A' and 'B'
		ENGINE ASSEMBLY	1	600380	Mk II Automatic NADA. 8:1 compression ratio with engine suffix 'C'
		ENGINE AND CLUTCH ASSEMBLY	1	601455	Mk III 4-speed. 8.75:1 compression ratio
		ENGINE AND CLUTCH ASSEMBLY	1	601454	Mk III 4-speed. 8:1 compression ratio
		ENGINE ASSEMBLY	1	601453	Mk III Automatic. 8.75:1 compression ratio
		ENGINE ASSEMBLY	1	601452	Mk III Automatic. 8:1 compression ratio
		ENGINE AND CLUTCH ASSEMBLY	1	601456	Mk III 4-speed NADA and France. 8.75:1 compression ratio
		ENGINE ASSEMBLY	1	601457	Mk III Automatic NADA and France. 8.75:1 compression ratio

NOTE: For engine compression ratio identification numbers see
NADA indicates parts peculiar to cars exported to the North American dollar area.

*Asterisk indicates a new part which has not been used on any previous model.

CYLINDER BLOCK

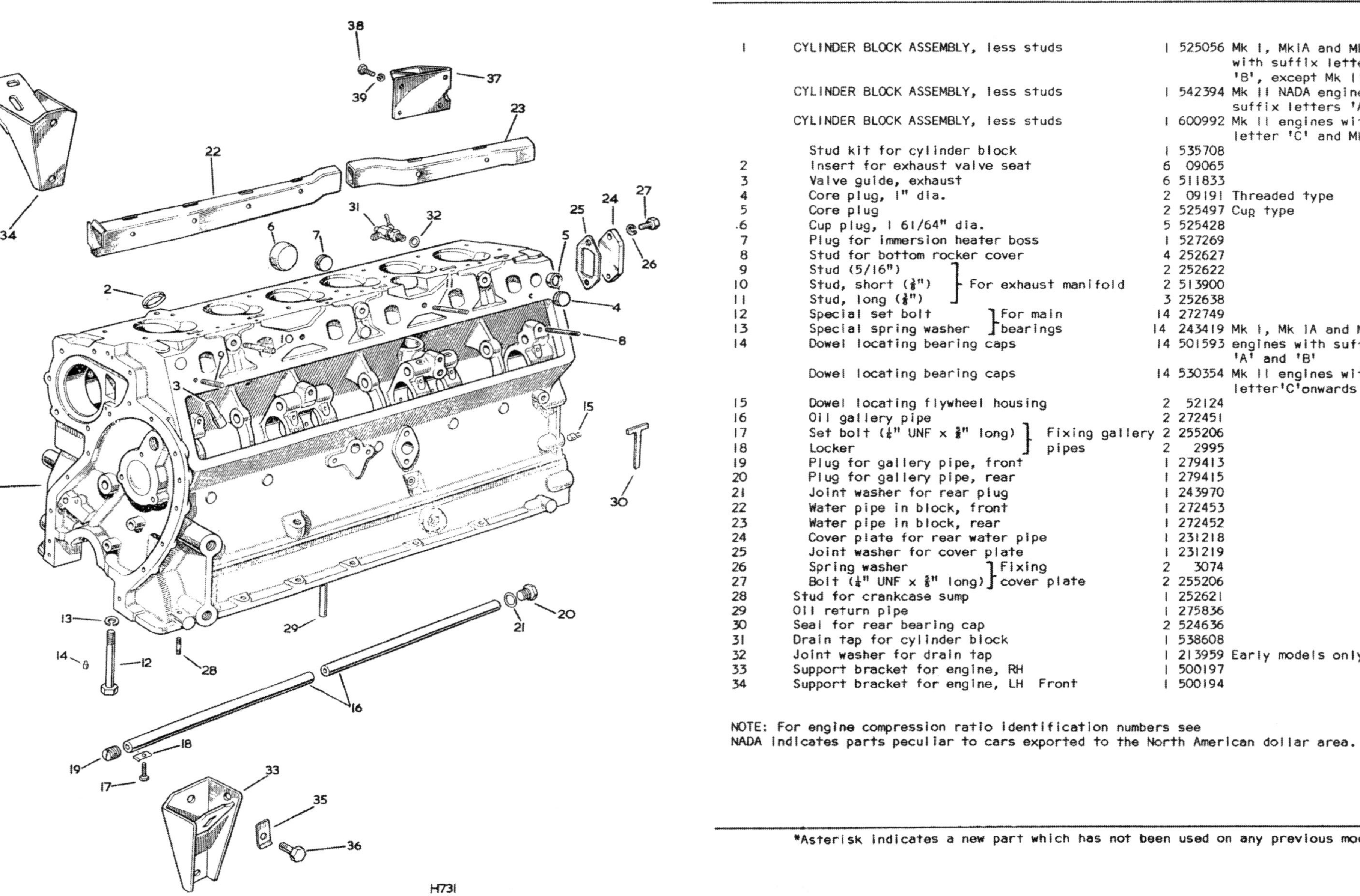

Plate Ref.	1 2 3 4	Description	Qty.	Part No.	Remarks
1		CYLINDER BLOCK ASSEMBLY, less studs	1	525056	Mk I, MkIA and Mk IIengines with suffix letters 'A' and 'B', except Mk II NADA
		CYLINDER BLOCK ASSEMBLY, less studs	1	542394	Mk II NADA engines with suffix letters 'A' and 'B'
		CYLINDER BLOCK ASSEMBLY, less studs	1	600992	Mk II engines with suffix letter 'C' and Mk III
		Stud kit for cylinder block	1	535708	
2		Insert for exhaust valve seat	6	09065	
3		Valve guide, exhaust	6	511833	
4		Core plug, 1" dia.	2	09191	Threaded type
5		Core plug	2	525497	Cup type
6		Cup plug, 1 61/64" dia.	5	525428	
7		Plug for immersion heater boss	1	527269	
8		Stud for bottom rocker cover	4	252627	
9		Stud (5/16") } For exhaust manifold	2	252622	
10		Stud, short (⅜") } For exhaust manifold	2	513900	
11		Stud, long (⅜") } For exhaust manifold	3	252638	
12		Special set bolt } For main bearings	14	272749	
13		Special spring washer } For main bearings	14	243419	Mk I, Mk IA and Mk II
14		Dowel locating bearing caps	14	501593	engines with suffix letters 'A' and 'B'
		Dowel locating bearing caps	14	530354	Mk II engines with suffix letter'C'onwards and Mk III
15		Dowel locating flywheel housing	2	52124	
16		Oil gallery pipe	2	272451	
17		Set bolt (¼" UNF x ⅜" long) } Fixing gallery pipes	2	255206	
18		Locker } Fixing gallery pipes	2	2995	
19		Plug for gallery pipe, front	1	279413	
20		Plug for gallery pipe, rear	1	279415	
21		Joint washer for rear plug	1	243970	
22		Water pipe in block, front	1	272453	
23		Water pipe in block, rear	1	272452	
24		Cover plate for rear water pipe	1	231218	
25		Joint washer for cover plate	1	231219	
26		Spring washer } Fixing cover plate	2	3074	
27		Bolt (¼" UNF x ⅜" long) } Fixing cover plate	2	255206	
28		Stud for crankcase sump	1	252621	
29		Oil return pipe	1	275836	
30		Seal for rear bearing cap	2	524636	
31		Drain tap for cylinder block	1	538608	
32		Joint washer for drain tap	1	213959	Early models only
33		Support bracket for engine, RH	1	500197	
34		Support bracket for engine, LH Front	1	500194	

NOTE: For engine compression ratio identification numbers see
NADA indicates parts peculiar to cars exported to the North American dollar area.

*Asterisk indicates a new part which has not been used on any previous model.

CYLINDER BLOCK

Plate Ref.	1 2 3 4	Description		Qty.	Part No.	Remarks
35	Lock washer		Fixing front support brackets to cylinder block	4	212430	
36	Set bolt ($\frac{1}{2}$" UNF x $\frac{7}{8}$" long)			4	255084	
37	Support bracket for engine, RH rear			1	518821	@@
38	Set bolt (5/16" UNF x $\frac{3}{4}$" long)		Fixing rear support bracket to cylinder block	4	255226	@@
39	Spring washer			4	3075	@@

@@ Cars numbered from 625101046,626100263, 628100110, 630100841, 631100354, 633100079 onwards

*Asterisk indicates a new part which has not been used on any previous model.

CRANKSHAFT, MK I, MK IA AND MK II ENGINES WITH SUFFIX LETTERS 'A' AND 'B'

Plate Ref.	1 2 3 4 Description	Qty.	Part No.	Remarks
1	CRANKSHAFT ASSEMBLY, STD	1	500548	
	CRANKSHAFT ASSEMBLY, .010" US	1	505990	
	CRANKSHAFT ASSEMBLY, .020" US (Not Export)	1	505991	4-Speed
	CRANKSHAFT ASSEMBLY, .030" US	1	505992	
	CRANKSHAFT ASSEMBLY, .040" US	1	505993	
	CRANKSHAFT ASSEMBLY, STD	1	279685	
	CRANKSHAFT ASSEMBLY, .010" US	1	503205	
	CRANKSHAFT ASSEMBLY, .020" US (Not Export)	1	503206	Mk and Mk IA Automatic
	CRANKSHAFT ASSEMBLY, .030" US	1	503207	
	CRANKSHAFT ASSEMBLY, .040" US	1	503208	
	CRANKSHAFT ASSEMBLY, STD	1	531929	
	CRANKSHAFT ASSEMBLY, .010" US	1	536534	Mk II Automatic
	CRANKSHAFT ASSEMBLY, .020" US (Not Export)	1	536535	engines with suffix
	CRANKSHAFT ASSEMBLY, .030" US	1	536536	letters 'A' and 'B'
	CRANKSHAFT ASSEMBLY, .040" US	1	536537	
2	Dowel for flywheel or flexible drive plate	1	265779	
3	Main bearing, std	7	536394	
	Main bearing, .010" US	7	536395	
	Main bearing, .020" US	7	536396	
	Main bearing, .030" US	7	536397	
	Main bearing, .040" US	7	536398	
4	Thrust washer for crankshaft, std (Pairs)	1	523296	
	Thrust washer for crankshaft, .0025" OS	1	523297	
	Thrust washer for crankshaft, . 005" OS	1	523298	
	Thrust washer for crankshaft, .0075" OS	1	523299	
	Thrust washer for crankshaft, . 010" OS	1	523300	
	Thrust washer for crankshaft, .0125" OS	1	523353	
	BEARING SET FOR CRANKSHAFT, STD	1	542193	
	BEARING SET FOR CRANKSHAFT, .010" US	1	542194	Sets include main bearings,
	BEARING SET FOR CRANKSHAFT, .020" US	1	542195	thrust washers and
	BEARING SET FOR CRANKSHAFT, .030" US	1	542196	connecting rod bearings
	BEARING SET FOR CRANKSHAFT, .040" US	1	542197	
5	Chainwheel on crankshaft	1	03013	
6	Key locating chainwheel	1	52015	
7	Oil thrower on crankshaft	1	276455	
	VIBRATION DAMPER ASSEMBLY	1	503118	1959-60
	VIBRATION DAMPER ASSEMBLY	1	522880	1961 Mk I and Mk IA
8	DAMPER FLYWHEEL AND BUSH ASSEMBLY	1	500198	1959-60 (Not
	DAMPER FLYWHEEL AND BUSH ASSEMBLY	1	522879	1961 Mk I and Mk IA (part of
9	Bush for flywheel	2	236289	1959-60 (engine
	Bush	2	522928	1961 Mk I and Mk IA (assem-
10	Driving flange	1	503119	1959-60 (bly)
	Driving flange for damper	1	522672	1961 Mk I and Mk IA
11	Rubber disc for damper	2	03017	
12	Lock wahser	3	247647	Mk I and Mk IA
13	Set bolt (¼" UNF x ⅜" long) fixing washer	6	255206	
14	VIBRATION DAMPER ASSEMBLY	1	531908	Mk II engines with suffix letters 'A' and 'B'
	VIBRATION DAMPER ASSEMBLY	1	546834	Alternative to 531908 for Mk II engines with suffix letters 'A'and'B' 4 speed models only

*Asterisk indicates a new part which has not been used on any previous model.

CRANKSHAFT, MK I, MK IA AND MK II ENGINES WITH SUFFIX LETTERS 'A' AND 'B'

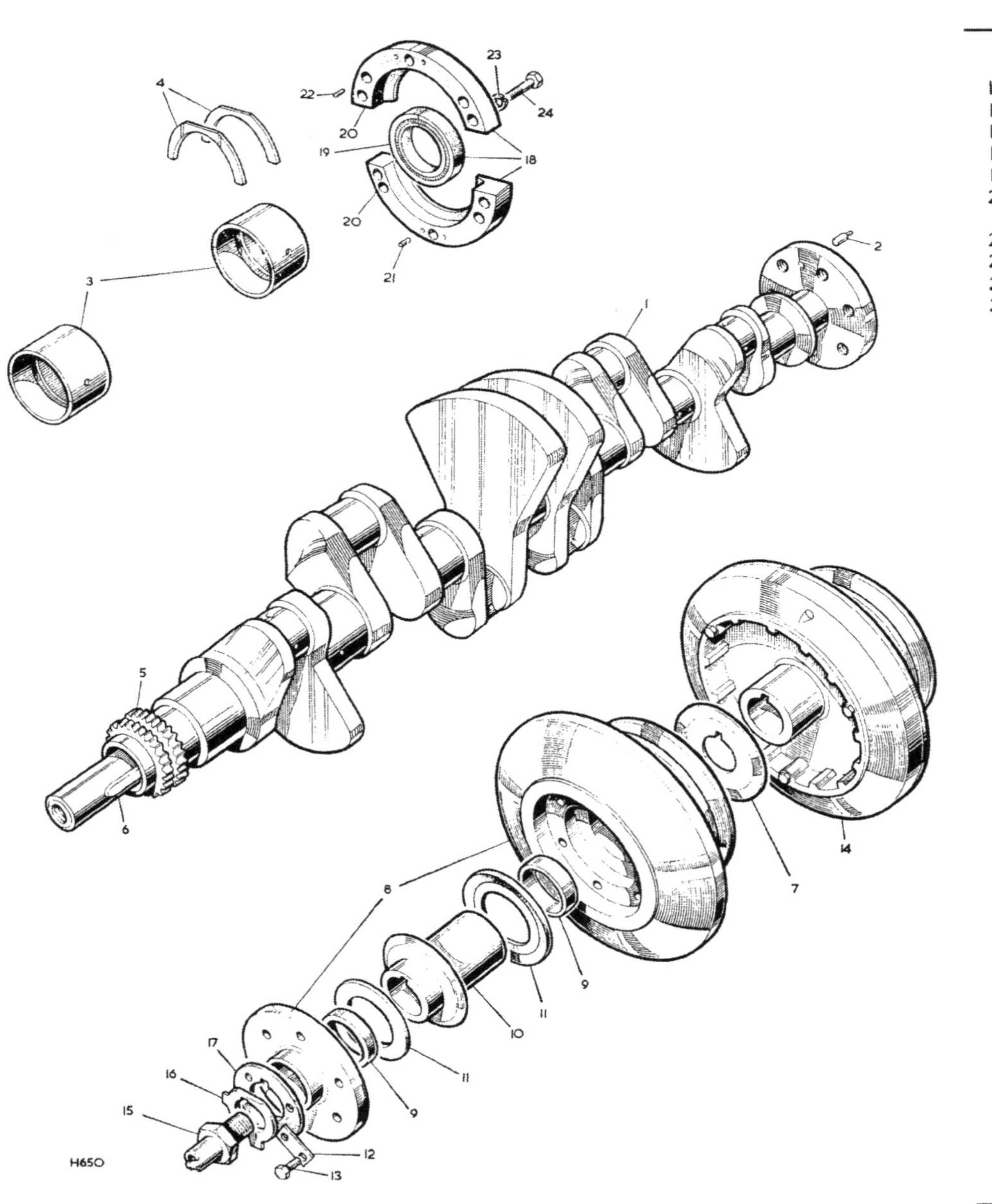

CRANKSHAFT MK I, MK IA AND MK II 3 LITRE ENGINES WITH SUFFIX LETTERS 'A' AND 'B'

Plate Ref.	1 2 3 4 Description		Qty.	Part No.	Remarks
15	Starting dog		1	503120	
16	Lock washer for starting dog		1	546333	
17	Coupling disc at front of crankshaft		1	503124	Automatic
18	CRANKSHAFT OIL RETAINER AND SEAL ASSEMBLY		1	542497	
19	Oil seal assembly		1	542495	
20	Retainer halves for oil seal		2	523240	
	Silicone grease		1	270656	
21	Dowel, lower	Fixing retainer	2	519064	
22	Dowel, upper	to cylinder block	2	246464	
23	Spring washer	and rear	10	3074	
24	Bolt ($\frac{1}{4}$"UNFx$\frac{7}{8}$" long)	bearing cap	10	255208	

*Asterisk indicates a new part which has not been used on any previous model.

CRANKSHAFT, MK II ENGINES WITH SUFFIX LETTER 'C' AND MK III

Plate Ref.	1 2 3 4 Description		Qty.	Part No.	Remarks
	CRANKSHAFT ASSEMBLY, STD		1	541904	4-speed; Mk II engines with suffix letter 'C'
	CRANKSHAFT ASSEMBLY, .010" US	Not Export	1	600157	4-speed; Mk II engines with suffix letter 'C'
	CRANKSHAFT ASSEMBLY, .020" US	Not Export	1	600158	4-speed; Mk II engines with suffix letter 'C'
	CRANKSHAFT ASSEMBLY, .030" US	Not Export	1	600159	4-speed; Mk II engines with suffix letter 'C'
	CRANKSHAFT ASSEMBLY, .040" US	Not Export	1	600160	4-speed; Mk II engines with suffix letter 'C'
	CRANKSHAFT ASSEMBLY, STD		1	541907	Automatic; Mk II engines with suffix letter 'C'
	CRANKSHAFT ASSEMBLY, .010" US	Not Export	1	600166	Automatic; Mk II engines with suffix letter 'C'
	CRANKSHAFT ASSEMBLY, .020" US	Not Export	1	600167	Automatic; Mk II engines with suffix letter 'C'
	CRANKSHAFT ASSEMBLY, .030" US	Not Export	1	600168	Automatic; Mk II engines with suffix letter 'C'
	CRANKSHAFT ASSEMBLY, .040" US	Not Export	1	600169	Automatic; Mk II engines with suffix letter 'C'
	CRANKSHAFT ASSEMBLY, STD		1	541904	Mk III 4-speed and Automatic
1	CRANKSHAFT ASSEMBLY, .010" US	Not Export	1	600157	Mk III 4-speed and Automatic
	CRANKSHAFT ASSEMBLY, .020" US	Not Export	1	600158	Mk III 4-speed and Automatic
	CRANKSHAFT ASSEMBLY, .030" US	Not Export	1	600159	Mk III 4-speed and Automatic
	CRANKSHAFT ASSEMBLY, .040" US	Not Export	1	600160	Mk III 4-speed and Automatic
2	Dowel for flywheel or flexible drive plate		1	265779	
3	Main bearing, std	Pairs	7	600161	
	Main bearing, .010" US	Pairs	7	600162	
	Main bearing, .020" US	Pairs	7	600163	
	Main bearing, .030" US	Pairs	7	600164	
	Main bearing, .040" US	Pairs	7	600165	
4	Thrust washer for crankshaft, std	Pairs	1	600177	
	Thrust washer for crankshaft, .0025" OS	Pairs	1	600174	
	Thrust washer for crankshaft, . 005" OS	Pairs	1	600175	
	Thrust washer for crankshaft, .0075" OS	Pairs	1	600176	
	Thrust washer for crankshaft, .010" OS	Pairs	1	600178	
	Thrust washer for crankshaft, .0125" OS	Pairs	1	600179	
	BEARING SET FOR CRANKSHAFT, STD		1	600181	Sets include main bearings, thrust washers and connecting rod bearings
	BEARING SET FRO CRANKSHAFT, . 010" US		1	600182	Sets include main bearings, thrust washers and connecting rod bearings
	BEARING SET FOR CRANKSHAFT, . 020" US		1	600183	Sets include main bearings, thrust washers and connecting rod bearings
	BEARING SET FOR CRANKSHAFT, . 030" US		1	600184	Sets include main bearings, thrust washers and connecting rod bearings
	BEARING SET FOR CRANKSHAFT, . 040" US		1	600185	Sets include main bearings, thrust washers and connecting rod bearings
5	Chainwheel on crankshaft		1	541882	
6	Key, front, locating vibration damper		1	542622	
7	Key, rear, locating chainwheel		1	542623	
8	Oil thrower on crankshaft		1	530343	
9	Oil seal ring		1	541921	
10	VIBRATION DAMPER ASSEMBLY		1	554469	
11	Starting dog		1	546324	
12	Lock washer for starting dog		1	546333	
13	Coupling disc at front of crankshaft		1	503124	Automatic
14	CRANKSHAFT OIL RETAINER AND SEAL ASSEMBLY		1	542494	
15	Oil seal assembly		1	542492	
16	Retainer halves for oil seal		2	523240	
	Silicone grease		1	270656	
17	Dowel, lower	Fixing retainer to cylinder block and rear bearing cap	2	519064	
18	Dowel, upper	Fixing retainer to cylinder block and rear bearing cap	2	246464	
19	Spring washer	Fixing retainer to cylinder block and rear bearing cap	10	3074	
20	Bolt (¼" UNF x ⅞" long)	Fixing retainer to cylinder block and rear bearing cap	10	255208	

*Asterisk indicates a new part which has not been used on any previous model.

CONNECTING ROD AND PISTON, MK I, MK IA AND MK II ENGINES

Plate Ref.	1 2 3 4 Description	Qty.	Part No.	Remarks
1	CONNECTING ROD ASSEMBLY	6	524490	
2	Gudgeon pin bush	6	273163	
3	Special bolt	12	518100	
4	Self-locking nut	12	272317	
	Complete set of nuts and bolts for connecting rod	1	525069	
5	Connecting rod bearing, std (Pairs)	6	523341	
	Connecting rod bearing, .010" US (Pairs)	6	523342	
	Connecting rod bearing, .020" US (Pairs)	6	523343	
	Connecting rod bearing, .030" US (Pairs)	6	523344	
	Connecting rod bearing, .040" US (Pairs)	6	523345	
	PISTON ASSEMBLY, STD	6	500630	Mk I and Mk IA engines with 7.5:1 compression ratio. Check before ordering
	PISTON ASSEMBLY, .010" OS	6	506008	
	PISTON ASSEMBLY, .020" OS	6	506009	
	PISTON ASSEMBLY, .030" OS	6	506010	
	PISTON ASSEMBLY, .040" OS	6	506011	
6	PISTON ASSEMBLY, STD, GRADE 'Z'	6	537263	Mk I, Mk IA and Mk II engines with 8.75:1 compression ratio
	PISTON ASSEMBLY, STD, GRADE 'A'	6	537264	
	PISTON ASSEMBLY, STD, GRADE 'B'	6	537265	
	PISTON ASSEMBLY, STD, GRADE 'C'	6	537266	
	PISTON ASSEMBLY, STD, grade 'D'	6	537267	
	PISTON ASSEMBLY, .010" OS	6	537268	
	PISTON ASSEMBLY, .020" OS	6	537269	
	PISTON ASSEMBLY, .030" OS	6	537270	
	PISTON ASSEMBLY, .040" OS	6	537271	
	PISTON ASSEMBLY, STD, GRADE 'Z'	6	536267	Mk II 4-speed 8:1 compression ratio and all Mk II Automatic models
	PISTON ASSEMBLY, STD, GRADE 'A'	6	536268	
	PISTON ASSEMBLY, STD, GRADE 'B'	6	536269	
	PISTON ASSEMBLY, STD, GRADE 'C'	6	536270	
	PISTON ASSEMBLY, STD, GRADE 'D'	6	536271	
	PISTON ASSEMBLY, .010" OS	6	536272	
	PISTON ASSEMBLY, .020" OS	6	536273	
	PISTON ASSEMBLY, .030" OS	6	536274	
	PISTON ASSEMBLY, .040" OS	6	536275	
7	Piston ring, compression, std	12	231155	
	Piston ring, compression, .010" OS	12	236173	
	Piston ring, compression, .020" OS	12	236174	
	Piston ring, compression, .030" OS	12	236175	
	Piston ring, compression, .040" OS	12	236176	
8	Duaflex scraper ring, std	6	554620	
	Duaflex scraper ring, .010" OS	6	554789	
	Duaflex scraper ring, .020" OS	6	554790	
	Duaflex scraper ring, .030" OS	6	554791	
	Duaflex scraper ring, .040" OS	6	554792	
9	Gudgeon pin, std	6	264569	
	Gudgeon pin, .001" OS	6	267257	
	Gudgeon pin, .003" OS	6	267258	
10	Circlip for gudgeon pin	12	235100	

*Asterisk indicates a new part which has not been used on any previous model.

CONNECTING ROD AND PISTON, MK III MODELS

G627

CONNECTING ROD AND PISTON, MK III MODELS

Plate Ref.	1 2 3 4 DESCRIPTION	Qty	Part No.	REMARKS
1	CONNECTING ROD ASSEMBLY	6	524490	
2	Gudgeon pin bush	6	273163	
3	Special bolt	12	518100	
4	Self-locking nut	12	272317	
	Complete set of nuts and bolts for connecting rod	1	525069	
5	Connecting rod bearing, std (Pairs)	6	523341	
	Connecting rod bearing, .010" US (Pairs)	6	523342	
	Connecting rod bearing, .020" US (Pairs)	6	523343	
	Connecting rod bearing, .030" US (Pairs)	6	523344	
	Connecting rod bearing, .040" US (Pairs)	6	523345	
6	PISTON ASSEMBLY, STD, GRADE 'Z'	6	537263	Mk III 4-speed and automatic engines with 8.75:1 compression ratio
	PISTON ASSEMBLY, STD, GRADE 'A'	6	537264	Mk III 4-speed and automatic engines with 8.75:1 compression ratio
	PISTON ASSEMBLY, STD, GRADE 'B'	6	537265	Mk III 4-speed and automatic engines with 8.75:1 compression ratio
	PISTON ASSEMBLY, STD, GRADE 'C'	6	537266	Mk III 4-speed and automatic engines with 8.75:1 compression ratio
	PISTON ASSEMBLY, STD, GRADE 'D'	6	537267	Mk III 4-speed and automatic engines with 8.75:1 compression ratio
	PISTON ASSEMBLY, .010" OS	6	537268	Mk III 4-speed and automatic engines with 8.75:1 compression ratio
	PISTON ASSEMBLY, .020" OS	6	537269	Mk III 4-speed and automatic engines with 8.75:1 compression ratio
	PISTON ASSEMBLY, .030" OS	6	537270	Mk III 4-speed and automatic engines with 8.75:1 compression ratio
	PISTON ASSEMBLY, .040" OS	6	537271	Mk III 4-speed and automatic engines with 8.75:1 compression ratio
	PISTON ASSEMBLY, STD, GRADE 'Z'	6	536267	Mk III 4-speed and automatic engines with 8:1 compression ratio
	PISTON ASSEMBLY, STD, GRADE 'A'	6	536268	Mk III 4-speed and automatic engines with 8:1 compression ratio
	PISTON ASSEMBLY, STD, GRADE 'B'	6	536269	Mk III 4-speed and automatic engines with 8:1 compression ratio
	PISTON ASSEMBLY, STD, GRADE 'C'	6	536270	Mk III 4-speed and automatic engines with 8:1 compression ratio
	PISTON ASSEMBLY, STD, GRADE 'D'	6	536271	Mk III 4-speed and automatic engines with 8:1 compression ratio
	PISTON ASSEMBLY, .010" OS	6	536272	Mk III 4-speed and automatic engines with 8:1 compression ratio
	PISTON ASSEMBLY, .020" OS	6	536273	Mk III 4-speed and automatic engines with 8:1 compression ratio
	PISTON ASSEMBLY, .030" OS	6	536274	Mk III 4-speed and automatic engines with 8:1 compression ratio
	PISTON ASSEMBLY, .040" OS	6	536275	Mk III 4-speed and automatic engines with 8:1 compression ratio
7	Piston ring, compression, std	12	231155	
	Piston ring, compression, .010" OS	12	236173	
	Piston ring, compression, .020" OS	12	236174	
	Piston ring, compression, .030" OS	12	236175	
	Piston ring, compression, .040" OS	12	236176	
8	Duaflex scraper ring, std	6	554620	
	Duaflex scraper ring, .010" OS	6	554789	
	Duaflex scraper ring, .020" OS	6	554790	
	Duaflex scraper ring, .030" OS	6	554791	
	Duaflex scraper ring, .040" OS	6	554792	
9	Gudgeon pin, std	6	264569	
	Gudgeon pin, .001" OS	6	267257	
	Gudgeon pin, .003" OS	6	267258	
10	Circlip for gudgeon pin	12	235100	

* Asterisk indicates a new part which has not been used on any previous Rover model

OIL PUMP AND OIL PIPES

F 581

COLLINS-JONES

Plate Ref.	1 2 3 4 Description		Qty.	Part No.	Remarks
	OIL PUMP ASSEMBLY		1	542697	
1	OIL PUMP BODY ASSEMBLY		1	542396	
2	Bush for drive shaft		1	212309	
3	Oil pump gear, driver		1	240555	
4	OIL PUMP COVER ASSEMBLY		1	502214	Mk I, Mk IA and Mk II engines with suffix letters 'A' and 'B'
	OIL PUMP COVER ASSEMBLY		1	542696	Mk II engines with suffix letter 'C' and Mk III
5	Dowel locating body		2	52710	
6	Spindle for idler wheel		1	502209	
7	Stud for oil strainer		1	274804	
8	OIL PUMP IDLER GEAR ASSEMBLY		1	278109	
9	Bush for idler gear		1	214995	
10	Oil pump shield		1	09225	
11	Bolt (5/16" UNF x 2¼" long)	Fixing cover to body	4	256229	Mk I, Mk IA and Mk II engines with suffix letters 'A' and 'B'
12	Self-locking nut (5/16" UNF)	Fixing cover to body	4	252211	Mk I, Mk IA and Mk II engines with suffix letters 'A' and 'B'
	Set bolt (5/16" UNF x ⅞" long)	Fixing cover to body	4	255227	Mk II engines with suffix letter 'C' and Mk III
	Spring washer	Fixing cover to body	4	3075	Mk II engines with suffix letter 'C' and Mk III
13	Oil strainer		1	266900	
14	Castle nut	Fixing strainer to pump cover	1	254950	
15	Split pin	Fixing strainer to pump cover	1	2556	
16	DISTRIBUTOR HOUSING ASSEMBLY		1	274084	
17	Bush for drive shaft		1	521583	
18	Cork washer for distributor housing		1	52278	
19	Oil pump drive shaft		1	278980 @	
20	Oil pump and distributor gear and shaft assembly (phosphor-bronze gear)		1	515969 @@	
21	Drive shaft for distributor		1	267829	
22	Oil pump driving gear		1	212308 @	
23	Taper pin fixing gear to shaft		1	3005 @	
24	Steel ball	For oil pressure release valve	1	01035	
25	Plunger	For oil pressure release valve	1	245940	
26	Spring	For oil pressure release valve	1	504997	
27	Retaining cap	For oil pressure release valve	1	504995	
28	Joint washer	For oil pressure release valve	1	243971	
29	Special set screw	Fixing oil pump to cylinder block	1	274086	
30	Locknut	Fixing oil pump to cylinder block	1	254851	
31	Oil feed bolt locating distributor housing		1	274928	
32	Locker for bolt		1	2504	
	Oil pipe complete to cylinder head		1	267980	Mk I and Mk IA
	Oil pipe complete to cylinder head		1	520208	Mk II engines with suffix letters 'A' and 'B'

@ Engines numbered up to 625100017, 626100169, 627100025, 628100016, 630100138, 631100055.

@@ Engines numbered from 625100018, 626100170, 627100026, 628100017, 63 0100139, 631100056 onwards

*Asterisk indicates a new part which has not been used on any previous model.

OIL PUMP AND OIL PIPES

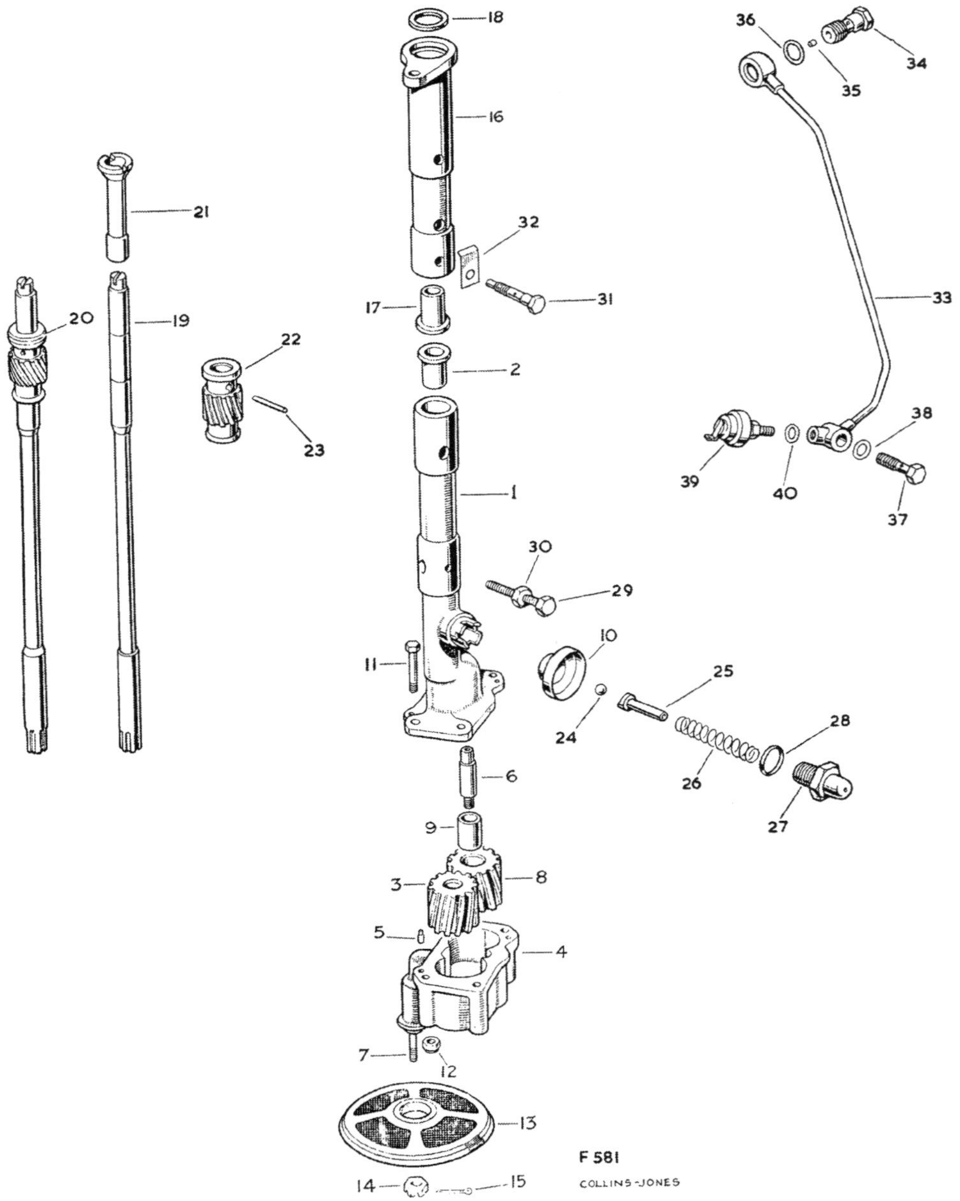

F 581

COLLINS-JONES

Plate Ref.	1 2 3 4	Description	Qty.	Part No.	Remarks
		Oil pipe complete to cylinder head	1	546347	Mk II engines with suffix letter 'C'
33		Oil pipe complete to cylinder head	1	554965	Mk III
34		Banjo bolt fixing oil pipe to cylinder head	1	265038	
36		Joint washer for banjo bolt	2	231577	
37		Banjo bolt fixing oil pipe to cylinder block	1	233520	
38		Joint washer for banjo bolt	2	232039	
39		Oil pressure switch complete	1	519864	
40		Joint washer for switch	1	232039	

*Asterisk indicates a new part which has not been used on any previous model.

CAMSHAFT AND TENSIONER MECHANISM

H648

Plate Ref.	1 2 3 4 Description		Qty.	Part No.	Remarks
	Camshaft		1	277064	Mk I and Mk IA
1	Camshaft		1	523139	Mk II and Mk III
2	Bearing, front	For camshaft	1	274116	
3	Bearing, intermediate		4	274115	
4	Bearing, rear		1	274117	
5	Spring washer	Fixing camshaft bearings to block	6	3075	
6	Special set screw		6	274118	
7	Thrust plate for camshaft		1	502266	
8	Locker	Fixing thrust plate to block	3	2500	
9	Set bolt (1/4" UNF x 5/8" long)		3	255206	
10	Hub for camshaft chainwheel		1	502965	For Early type chainwheel with separate hub
11	Sealing plate for camshaft		1	530481	
12	Joint washer for sealing plate		1	276541	
13	Set bolt (1/4" UNF x 5/8" long)	Fixing sealing plate to cylinder block	3	255206	
14	Spring washer		3	3074	
15	Key locating chainwheel		1	230313	
16	Retaining washer	Fixing chainwheel hub to camshaft	1	09093	
17	Locker		1	09210	
18	Set bolt (3/8" UNFx3/4"long)		1	255245	Alternatives
	Set bolt (3/8" UNFx7/8"long)		1	255046	Check before ordering
19	Chainwheel for camshaft		1	502966	Early type with separate hub
	Chainwheel for camshaft		1	563145	Late type with integral hub
20	Locker	Fixing chainwheel to hub	3	2499	For early type chainwheel
21	Special bolt		3	502265	with separate hub
22	Camshaft chain		1	266662	
23	Vibration damper for timing chain		1	275234	
24	Set bolt (1/4" UNFx 1/2" long)	Fixing damper to cylinder block	2	255204	
25	Locker		2	557523	
26	TENSIONER FOR TIMING CHAIN		1	266661	
27	Tab washer for tensioner		1	504443	
28	Set bolt (1/4" UNF x 1 3/8" long)	Fixing tensioner to cylinder block	2	256202	
29	Spring washer		2	3074	

*Asterisk indicates a new part which has not been used on any previous model.

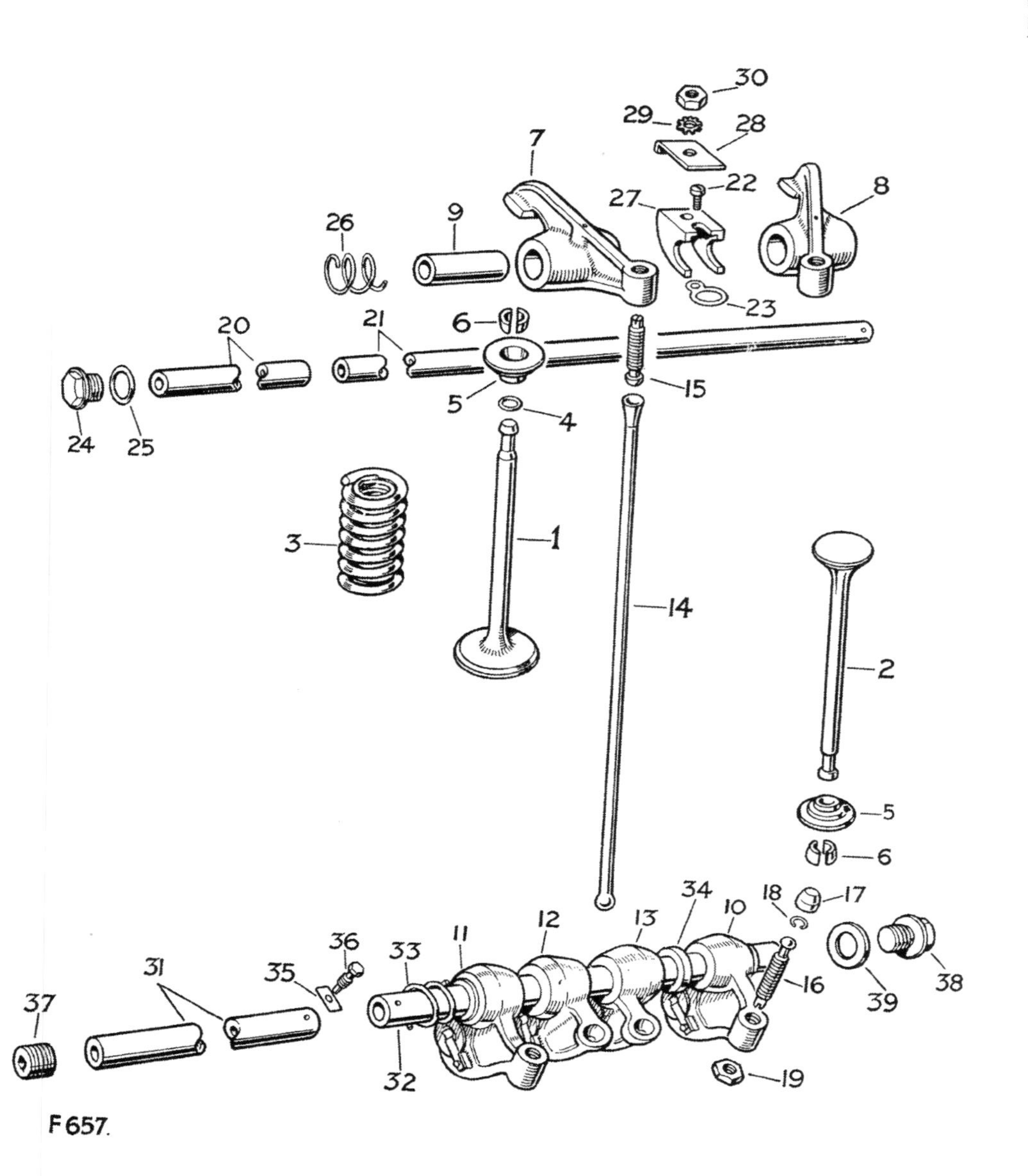

Plate Ref.	1 2 3 4 DESCRIPTION	Qty	Part No.	REMARKS
1	Inlet valve	6	525124	
2	Exhaust valve	6	525308	
3	Valve spring, inner and outer	12	502904	
4	Rubber ring for valve guide	6	233419	
5	Valve spring cup	12	268292	
6	Split cone halves for valves	24	268293	
7	VALVE ROCKER ASSEMBLY, INLET, LH	3	525911	
8	VALVE ROCKER ASSEMBLY, INLET, RH	3	525910	
9	Bush for inlet valve rocker	6	230034	
10	VALVE ROCKER, EXHAUST, LH	3	531874	
11	VALVE ROCKER, EXHAUST, RH	3	531873	
	Roller follower for exhaust valve rockers	6	517429	
12	INLET CAM FOLLOWER, LH	3	273292	Engines numbered up to 72500020, 72600770, 72700032 72800255, 73000997, 73100233
13	INLET CAM FOLLOWER, RH	3	273291	
14	Tappet push rod, inlet	6	279604	
	INLET CAM FOLLOWER, LH	3	536873	Engines numbered from 72500021, 72600771, 72700033 72800256, 73000998, 73100234 onwards
	INLET CAM FOLLOWER, RH	3	536872	
	Roller follower for inlet cam followers	6	517429	
	Tappet push rod, inlet	6	526236	
	Shim for cam follower	2	505597	
15	Tappet adjusting screw, inlet	6	506816	
16	TAPPET ADJUSTING SCREW ASSEMBLY, EXHAUST	6	506817	
17	Cap for screw	6	212160	
18	Circlip fixing cap	6	212161	
19	Locknut for tappet adjusting screws	12	254881	
20	Top valve rocker shaft, front	1	231348	
21	Top valve rocker shaft, rear	1	231349	
22	Special set screw } Fixing rocker shafts	3	231343	
23	Special washer } Fixing rocker shafts	3	231352	
24	Plug ($\frac{3}{8}$" BSP) } For front end of rocker shaft	1	250978	
25	Joint washer for plug } For front end of rocker shaft	1	243959	
26	Spring for top rockers	6	500609	
27	Spacer for top rockers	3	231344	
28	Retaining plate for rocker shaft set screw	3	238871	
29	Shakeproof washer } Fixing top rocker spacers	3	72614	
30	Nut ($\frac{5}{16}$" UNF) } Fixing top rocker spacers	3	254851	
31	Bottom valve rocker shaft, front	1	273306	
32	Bottom valve rocker shaft, rear	1	273307	
33	Spring for bottom rocker shaft	6	500610	
34	Spacing washer, thick } For bottom rocker shaft	2	273308	
	Spacing washer, thin } For bottom rocker shaft	2	273309	
35	Locker } Fixing rocker shafts to block	2	09005	
36	Locating screw } Fixing rocker shafts to block	2	09004	
	End plug, front, with slot } For bottom rocker shaft	1	247127	Alternative types } Check before ordering
37	End plug, front, with square recess } For bottom rocker shaft	1	12455	Alternative types } Check before ordering
38	End plug, rear } For bottom rocker shaft	1	3088	
39	Joint washer for rear plug } For bottom rocker shaft	1	231577	

* Asterisk indicates a new part which has not been used on any previous Rover model

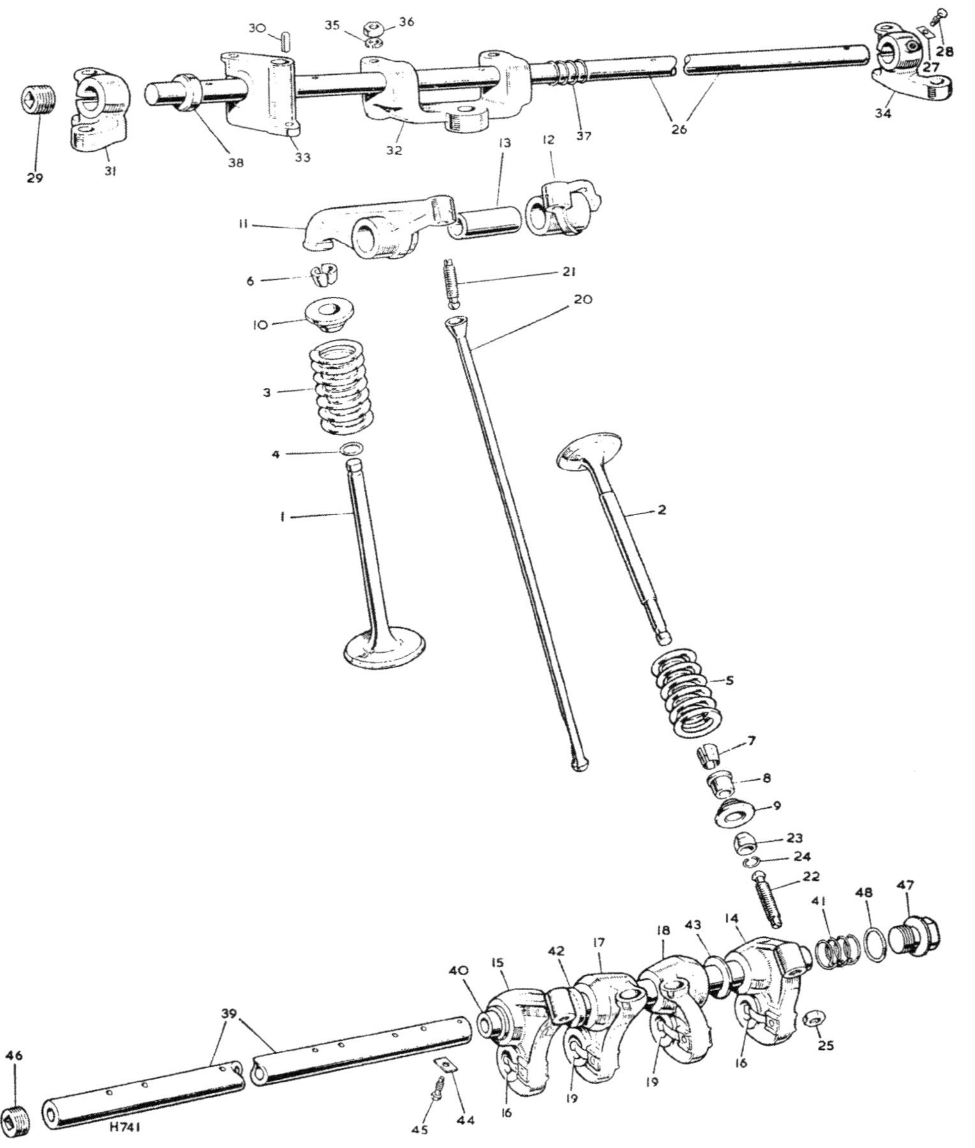

Plate Ref.	1 2 3 4 DESCRIPTION	Qty	Part No.	REMARKS
1	Inlet valve	6	526231	
2	Exhaust valve	6	512140	
3	Valve spring, inlet, inner and outer	6	535989	
4	Rubber ring for valve guide	6	247186	
5	Valve spring, exhaust, inner and outer	6	502904	
6	Split cone halves for inlet valves	12	268293	
7	Valve collet for exhaust valves, halves	12	512142	
8	Valve locking sleeve, exhaust	6	512143	
9	Valve spring cap for exhaust valves	6	512144	
10	Valve spring cap for inlet valves	6	268292	
11	VALVE ROCKER ASSEMBLY, INLET, LH	3	525907	
12	VALVE ROCKER ASSEMBLY, INLET, RH	3	525906	
13	Bush for inlet valve rocker	6	501267	
14	VALVE ROCKER, EXHAUST, LH	3	531874	
15	VALVE ROCKER, EXHAUST, RH	3	531873	
16	Roller follower for exhaust valve rockers	6	517429	
17	INLET CAM FOLLOWER, LH	3	536873	
18	INLET CAM FOLLOWER, RH	3	536872	
19	Roller follower for inlet cam followers	6	517429	
20	Tappet push rod, inlet	6	524737	
21	Tappet adjusting screw, inlet	6	506816	
22	TAPPET ADJUSTING SCREW ASSEMBLY, EXHAUST	6	506817	
23	Cap for screw	6	212160	
24	Circlip fixing cap	6	212161	
25	Locknut for tappet adjusting screws	12	254881	
26	Top valve rocker shaft	1	534131	
27	Locking plate } Fixing top rocker shaft	1	09005	
28	Locating screw } Fixing top rocker shaft	1	09004	
29	Plug for top rocker shaft	2	12012	
30	Dowel for rocker brackets	5	59059	
31	Rocker shaft bracket, front	1	520202	
32	Rocker shaft bracket, intermediate, long	2	512982	
33	Rocker shaft bracket, intermediate, short	3	520203	
34	Rocker shaft bracket, rear, oil feed	1	520204	
35	Spring washer } Fixing brackets to cylinder head	9	3075	
36	Nut ($\frac{5}{16}$" UNF) } Fixing brackets to cylinder head	9	254811	
37	Spring for top rockers	6	500609	
38	Spacer for top rockers	3	501147	
39	Bottom valve rocker shaft, front	1	273306	
40	Bottom valve rocker shaft, rear	1	273307	
41	Spring for bottom rocker shaft	6	500610	
42	Spacing washer, thick } For bottom rocker shaft	2	273308	
43	Spacing washer, thin } For bottom rocker shaft	2	273309	
44	Locker } Fixing rocker shafts	2	09005	
45	Locating screw } Fixing rocker shafts	2	09004	
46	End plug, front } For bottom rocker shaft	1	12455	
47	End plug, rear } For bottom rocker shaft	1	536577	
48	Joint washer for rear plug } For bottom rocker shaft	1	231577	

* Asterisk indicates a new part which has not been used on any previous Rover model

FRONT COVER, SIDE COVER AND SUMP

FRONT COVER, SIDE COVER AND SUMP, 3 LITRE MODELS

Plate Ref.	1 2 3 4 Description		Qty.	Part No.	Remarks
1	FRONT COVER ASSEMBLY		1	522859	
2	Dowel locating front cover		2	6395	
3	Oil seal for front cover		1	213744	
4	Joint washer for front cover		1	272835	
5	Spring washer	Fixing front cover to cylinder block	11	3075	
6	Set bolt (5/16" UNFx1$\frac{7}{8}$"long)	Fixing front cover to cylinder block	11	256025	10 off on Mk III
7	Nut (5/16" UNF)		1	254811	Mk III
8	Timing pointer for front cover		1	554647	Mk III
9	Rocker cover, side, and oil filler		1	500785	Except Mk II and Mk III NADA
	Rocker cover, side, and oil filler		1	542425	Mk II and Mk III NADA
10	Joint washer for rocker cover		1	GEG433	
11	Special nut	Fixing side cover	4	274091	
12	Rubber sealing washer	Fixing side cover	4	267828	Not part of engine assembly (items 12–22)
13	Breather filter and oil filler cap		1	236406	Except Mk II and Mk III NADA
	Oil filler cap		1	542427	Mk II and Mk III NADA
14	Breather pipe for engine		1	533995	Except Mk II and Mk III NADA
	Breather pipe for engine		1	546280	Mk II and Mk III NADA
15	Joint washer for breather pipe		1	214058	
16	Spring washer	Fixing breather pipe to cylinder block	2	3075	
17	Set bolt (5/16" UNF x$\frac{3}{4}$" long)	Fixing breather pipe to cylinder block	2	255026	
18	Clip for breather pipe		1	243321	
19	Bracket for breather pipe clip		1	274130	Mk I and Mk IA
	Bracket for breather pipe clip		1	534023	Mk II and Mk III
20	Bolt ($\frac{1}{4}$" UNF x $\frac{3}{8}$" long)	Fixing clip to breather pipe and bracket	1	255206	
21	Plain washer	Fixing clip to breather pipe and bracket	1	3840	
22	Spring washer	Fixing clip to breather pipe and bracket	1	3074	
23	Nut ($\frac{1}{4}$" UNF)	Fixing clip to breather pipe and bracket	1	254810	
24	CRANKCASE SUMP		1	529825	Mk I, Mk IA and Mk II
	CRANKCASE S MP		1	554257	Mk III
25	Drain plug for sump		1	536577	
26	Joint washer for plug		1	243959	
27	Joint washer for sump		1	GEG538	
28	Set bolt (5/16" UNC x $\frac{7}{8}$" long)	Fixing sump to cylinder block	2	253027	
29	Spring washer	Fixing sump to cylinder block	19	3075	
	Nut (5/16" UNF)	Fixing sump to cylinder block	1	254811	
30	Set bolt (5/16" UNF x $\frac{7}{8}$" long)	Fixing sump to cylinder block	16	255027	
31	SUMP UNIT FOR OIL LEVEL GAUGE		1	503601	Mk I and Mk IA
	SUMP UNIT FOR OIL LEVEL GAUGE		1	545227	Mk II

NADA indicates parts peculiar to cars exported to the North American dollar area.

*Asterisk indicates a new part which has not been used on any previous model.

FRONT COVER, SIDE COVER AND SUMP

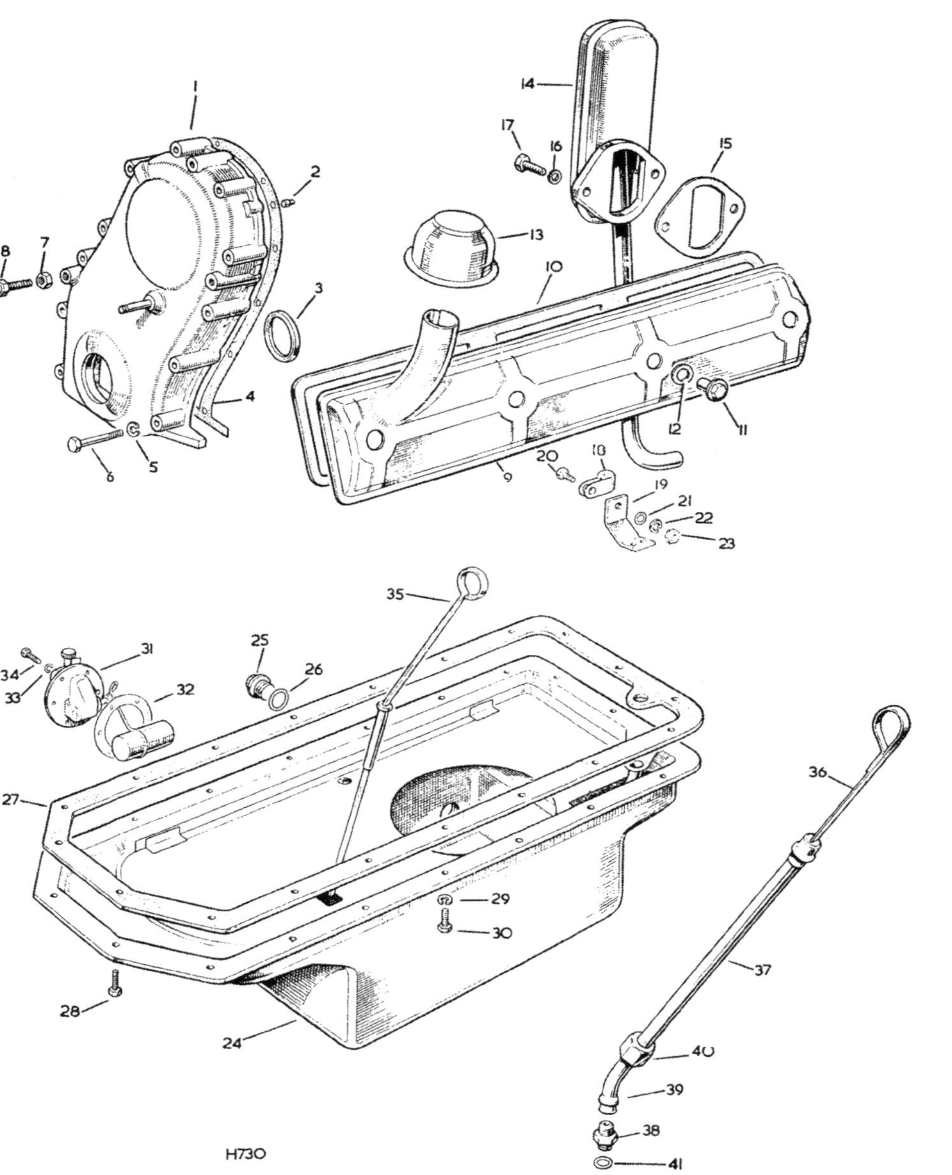

Plate Ref.	1 2 3 4	Description	Qty.	Part No.	Remarks
32		Joint washer for sump unit	1	546488	Mk I, Mk IA and Mk II
33		Spring washer] Fixing unit	6	3101	
34		Set screw (3 BA x ½" long)] to sump	6	3100	
35		Oil level rod	1	275627	Mk I, Mk IA and Mk II engines with suffix letters 'A' and 'B'
36		Oil level rod	1	542422	Mk II engines with suffix letter 'C' and Mk III
37		Tube for oil level rod	1	546718	
38		Adaptor for oil level rod tube	1	541266	
39		Olive] Fixing tube	1	236408	
40		Union nut] to adaptor	1	236407	
41		Sealing ring for oil level rod	1	532387	

*Asterisk indicates a new part which has not been used on any previous model.

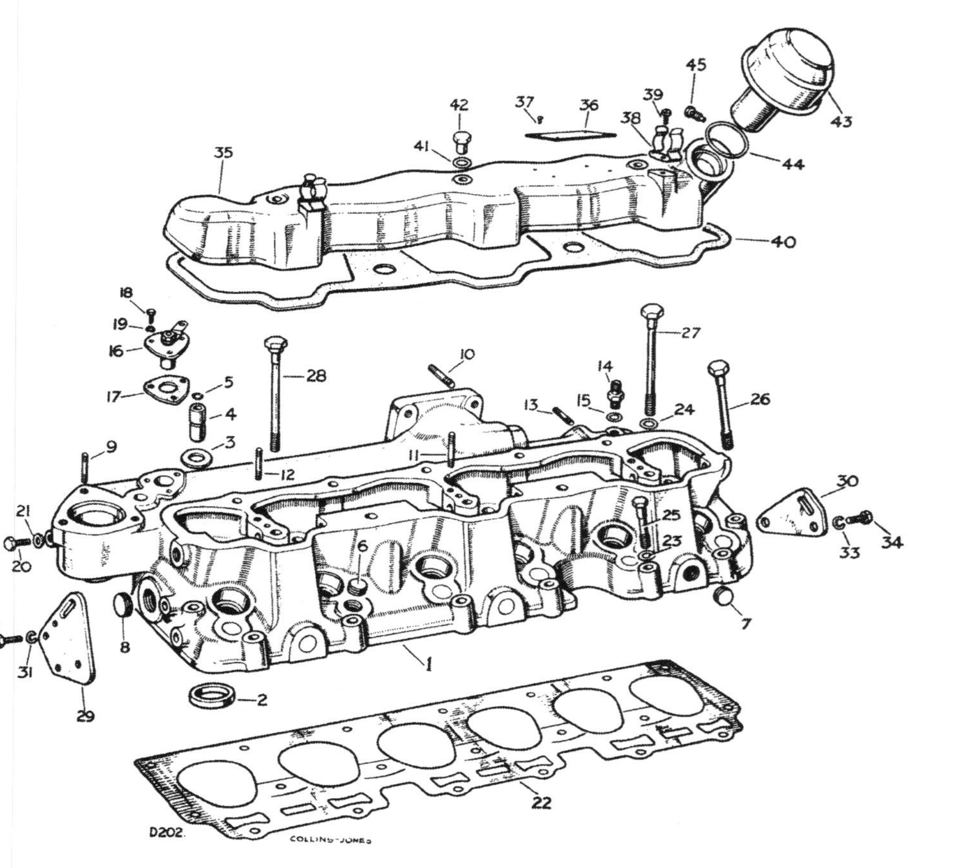

Plate Ref.	1 2 3 4 Description		Qty.	Part No.	Remarks
1	CYLINDER HEAD ASSEMBLY, less studs		1	542591	
2	Insert for inlet valve seat		6	266321	
3	Washer for valve spring		6	230062	
4	Valve guide, inlet		6	504169	
5	Rubber ring for valve guide, inlet		6	233419	
6	Core plug, 11/16" dia.		7	230251	
7	Core plug, $\frac{7}{8}$" dia.		4	210492	
	Core plug, 1" dia.		1	09191	
8	Core plug, 1$\frac{3}{8}$" dia.		2	230250	
9	Stud for water outlet pipe		3	252501	
10	Stud for carburetter		4	500668	
11	Stud for rocker cover, long		2	503047	
12	Stud for rocker cover, short		1	506046	
13	Stud for accelerator shaft bracket		3	252497	
14	Double-ended union for cylinder head		1	2336	@
	Double-ended union for cylinder head		1	513171	@@
15	Joint washer for union		1	243958	
16	Thermostat switch for choke warning light		1	545010	
17	Joint washer for thermostat switch		1	236022	
18	Set bolt (2 BA x 7/16" long)	Fixing	3	251002	
19	Spring washer	switch	3	3073	
20	Set bolt ($\frac{1}{4}$" UNC x 7/16" long)	For hole in	1	253003	
21	Joint washer for bolt	thermostat housing	1	232037	
22	Cylinder head gasket		1	542351	
23	Plain washer, small		6	2210	
24	Plain washer, large	Fixing	11	3843	
25	Special set bolt ($\frac{3}{8}$" UNF x 1.906" long)	cylinder head	6	274093	
26	Special set bolt (7/16" UNF x 2.531" long)	to block	3	574103	
27	Special set bolt (7/16" UNF x 4.906" long)		8	574104	
28	Special set bolt (7/16" UNF x 5.156" long)		3	587338*	
29	Front lifting bracket and coil bracket		1	519313	
30	Rear engine lifting bracket		1	500628	
31	Spring washer	Fixing	2	3075	
32	Set bolt (5/16" UNC x$\frac{3}{4}$" long)	front bracket	2	253026	
33	Spring washer	Fixing	2	3076	
34	Set bolt ($\frac{3}{8}$" UNC x $\frac{3}{4}$" long)	rear bracket	2	253045	

@ Engines numbered up to 625000378, 626000119, 628000046, 630000162, 631000040, 633000094

@@ Engines numbered from 625000379, 626000120, 628000047, 630000163, 631000041, 633000095 onwards

*Asterisk indicates a new part which has not been used on any previous model.

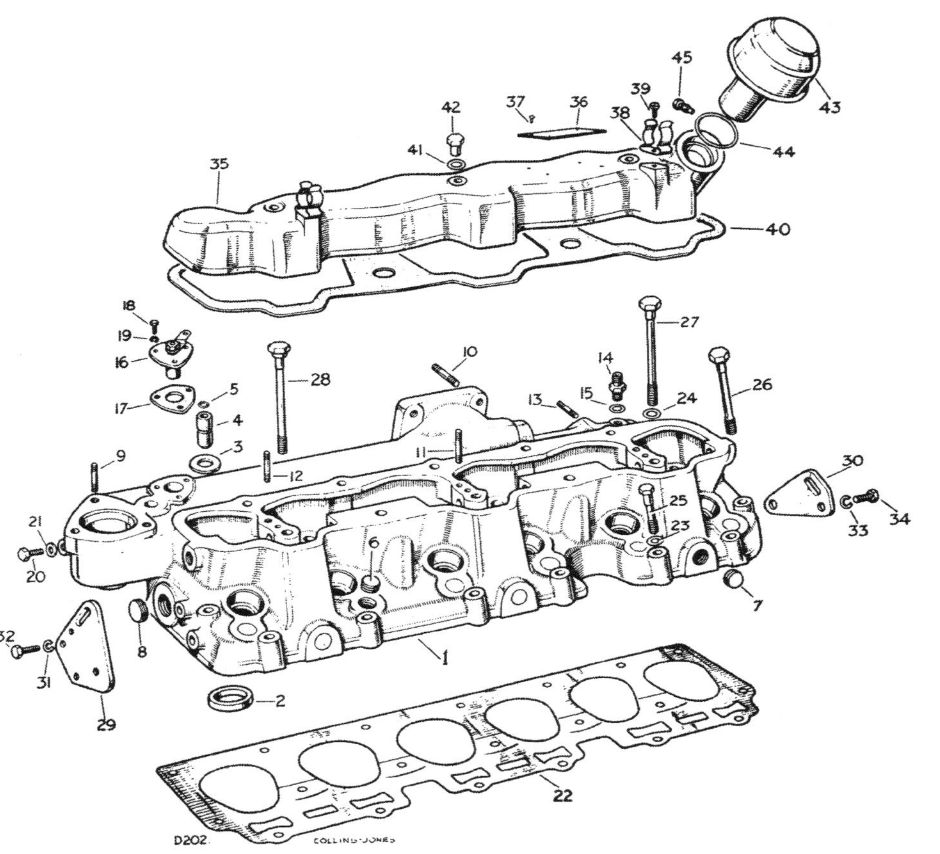

Plate Ref.	1 2 3 4	Description	Qty.	Part No.	Remarks
35		ROCKER COVER ASSEMBLY, TOP	1	274097	
36		Tappet clearance plate	1	274100	
37		Hammer drive screw for plate	4	3767	
38		Clip for ignition wire carrier	2	566813	
39		Drive screw fixing clip	2	77919	
40		Joint washer for rocker cover	1	GEG432	
41		Sealing washer } Fixing top	3	231576	
42		Special nut } rocker cover	3	274089	
43		Filter for engine breather	1	530702	Not part of engine assembly
44		Sealing ring for breather filter	1	268887	Not part of engine assembly
45		Special set screw fixing filter	1	515291	Not part of engine assembly
		Sealing washer for set screw	1	232037	Not part of engine assembly
		Gasket set, decarbonising	1	GEG163	

*Asterisk indicates a new part which has not been used on any previous model.

CYLINDER HEAD, MK II AND MK III MODELS

H733

Plate Ref.	1 2 3 4 Description	Qty.	Part No.	Remarks
	CYLINDER HEAD ASSEMBLY, less studs	1	522669	Engines with suffix letters 'A' and 'B'
1	CYLINDER HEAD ASSEMBLY, less studs	1	542349	Engines with suffix letter 'C' and Mk III
2	Insert for inlet valve seat	6	501269	
3	Washer for valve spring	6	230062	
4	Valve guide inlet	6	516934	
5	Rubber ring for valve guide, inlet	6	247186	
6	Core plug, 7/8" dia.	3	210492	
7	Core plug, 1 3/8" dia.	1	230250	
8	Stud for inlet manifold	14	506046	
9	Stud for water outlet pipe	3	252501	
10	Stud, short, for rocker bracket	4	252517	
11	Stud, long, for rocker bracket	5	252720	
12	Plain washer	14	2257	
13	Special set (7/16" UNF x 3.968" long)	7	501144	
14	Special set bolt (7/16" UNF x 4.539" long)	2	512989	
15	Special set bolt (7/16" UNF x 5.406" long)	3	501143	
16	Special set bolt (7/16" UNF x 4.531" long)	2	525074	
17	Special set bolt (3/8" UNF x 1.906" long)	6	274093	
18	Plain washer	6	2210	
19	Front lifting bracket	1	518767	
20	Set bolt (5/16" UNC x 5/8" long) } Fixing front lifting bracket	2	253025	
21	Spring washer } Fixing front lifting bracket	2	3075	
	Rear lifting bracket	1	500628	Mk II engines with suffix letters 'A' and 'B'
22	Rear lifting bracket	1	547771	Mk II engines with suffix letter 'C' and Mk III
23	Set bolt (3/8" UNC x 5/8" long) } Fixing rear lifting bracket to head	2	253044	
24	Spring washer } Fixing rear lifting bracket to head	2	3075	
25	'O' ring for cylinder head	1	532319	Engines with suffix letter 'C' onwards and Mk III
	Cylinder head gasket	1	542351	Engines with suffix letters 'A' and 'B'
26	Cylinder head gasket	1	GEG339	Engines with suffix letter 'C' onwards and Mk III
27	INLET MANIFOLD ASSEMBLY	1	547810	Except NADA
	INLET MANIFOLD ASSEMBLY	1	547812	NADA
28	Welch plug	2	250830	
29	Core plug	1	504854	
30	Dowel for inlet manifold	2	213700	
31	Plain washer } Fixing inlet manifold to cylinder head	14	3830	
32	Spring washer } Fixing inlet manifold to cylinder head	14	3075	
33	Nut (5/16" UNF) } Fixing inlet manifold to cylinder head	14	254811	
34	Thermostat switch (choke light)	1	545010	
35	Joint washer for switch	1	236022	
36	Spring washer } Fixing switch to manifold	3	3073	
37	Set bolt (2 BA x 7/16" long) } Fixing switch to manifold	3	251002	

NADA indicates parts peculiar to cars exported to the North American dollar area

*Asterisk indicates a new part which has not been used on any previous model.

CYLINDER HEAD, MK II AND MK III MODELS

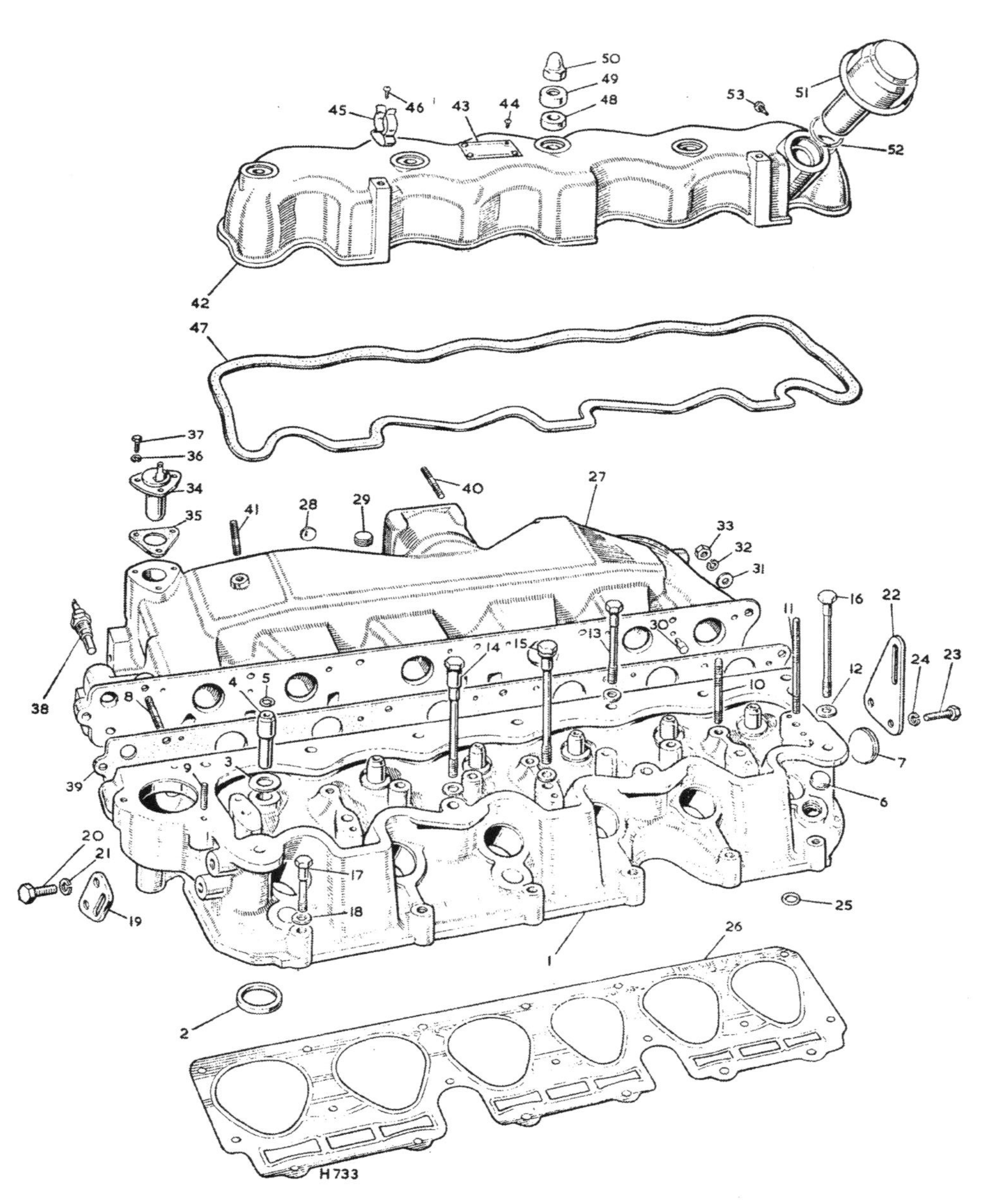

Plate Ref.	1 2 3 4	Description		Qty.	Part No.	Remarks
38		Water temperature transmitter		1	559259	
39		Joint washer for inlet manifold		1	GEG641	
40		Stud for carburetter		4	506046	
41		Stud fixing air silencer to inlet manifold		1	512700	
		Adaptor for heater water pipe (½" BSP)		1	530115	
		Joint washer for adaptor		1	530110	
42		ROCKER COVER ASSEMBLY, TOP		1	515553	
43		Tappet clearance plate		1	274100	
44		Hammer drive screw fixing plate		4	3767	
45		Clip for ignition wire carrier		2	566813	
46		Drive screw fixing clip		2	77919	
47		Joint washer for rocker cover		1	GEG434	
48		Rubber washer	Fixing rocker cover	5	506069	
49		Cover for rubber washer	Fixing rocker cover	5	538359	
50		Nut (5/16" UNF)	Fixing rocker cover	5	538360	
51		Filter for engine breather		1	530702	Except NADA
		Filter for engine breather		1	542420	NADA
52		Sealing ring for breather filter		1	268887	
53		Special set screw fixing filter		1	515291	
		Gasket set, decarbonising		1	GEG164	

*Asterisk indicates a new part which has not been used on any previous model.

ENGINE OIL FILTER AND ADAPTOR 3 LITRE MODELS

Plate Ref.	1 2 3 4 Description		Qty.	Part No.	Remarks
1	OIL FILTER FOR ENGINE		1	513591	Not part of engine assembly
2	Element	For oil filter	1	GFE111	
3	Gasket, large		1	246261	
4	Gasket, small		2	516370	
	Rubber washer for centre bolt		1	269889	
5	Adaptor for oil filter		1	274103	Saloon
6	Adaptor assembly for oil filter		1	536520	Coupe
7	Oil pressure transmitter		1	537138	
8	Shim washer, transmitter to oil filter adaptor		1	512387	
9	Joint washer, front	For oil filter adaptor	1	274609	
10	Joint washer, rear		1	274104	
11	Set bolt (5/16" UNF x 1¾" long)	Fixing adaptor to cylinder block	2	256024	
12	Spring washer		2	3075	
13	Joint washer	Fixing adaptor and oil filter to cylinder block	1	272839	Not part of engine assembly
14	Set bolt (7/16"UNF x2½"long)		2	256066	
15	Spring washer		2	3077	
16	Plug	For adaptor oil way	1	536577	Also part of 536520
17	Joint washer		1	231577	

*Asterisk indicates a new part which has not been used on any previous model.

Plate Ref.	1 2 3 4 Description	Qty.	Part No.	Remarks
	WATER PUMP ASSEMBLY	1	500619	1959-60 — Not part of engine assembly
	WATER PUMP ASSEMBLY	1	529472	1961 and Mk IA — Not part of engine assembly
	WATER PUMP ASSEMBLY	1	530602	Mk II and Mk III — Not part of engine assembly
1	Casing for water pump	1	500614	Mk I and Mk IA
	Casing for water pump	1	515784	Mk II and Mk III
2	Spindle and bearing complete	1	523354	
	Hub for fan blade	1	274738	1959-60
3	Hub for fan blade	1	524624	1961 onwards
4	Spring washer } Locating bearing in casing	1	3074	
5	Special set bolt } Locating bearing in casing	1	247078	
6	Carbon ring and seal	1	557626	
7	Impeller for pump	1	272613	
	Adaptor for thermostat by-pass (¼" BSP)	1	231734	Mk I and Mk IA
8	Adaptor for thermostat by-pass (¼" BSP)	1	515946	Mk II and Mk III
9	Adaptor for heater pipe (⅜" BSP)	1	529473	
10	Dowel for water pump casing	2	6395	Not part of engine assembly
11	Joint washer for water pump	1	503629	
12	Sealing washer, water pump to cylinder head	1	09170	
13	Spring washer } Fixing water pump to block	5	3074	
14	Set bolt (¼" UNF X 1" long) } Fixing water pump to block	6	255009	
15	Set bolt (¼" UNF x 1⅜" long) } Fixing water pump to block	1	256004	
16	Set bolt (¼" UNF x 1¾" long) } Fixing water pump to block	1	256005	
17	Set bolt (¼" UNF x 2¼" long) } Fixing water pump to block	2	256208	
18	Inlet pipe for water pump	1	500861	Not part of engine assembly
19	Hose, water inlet pipe to pump	1	500860	Not part of engine assembly
20	Clip for hose	2	50321	Not part of engine assembly
21	Thermostat stamped '170'	1	515962	
	Outlet pipe to radiator	1	276354	Mk I and Mk IA
22	Outlet pipe to radiator	1	547592	Mk II and Mk III
23	Joint washer for water outlet pipe	1	276510	
24	Spring washer } Fixing pipe to cylinder head	3	3074	
25	Nut (¼" UNF) } Fixing pipe to cylinder head	3	254810	
	Adaptor for thermostat by-pass	1	241797	Mk I and Mk IA
	Rubber by-pass pipe for thermostat	1	500697	Mk I and Mk IA
	Clip for pipe	2	50305	Mk I and Mk IA
26	Pulley for fan	1	272615	Not part of engine assembly
27	Fan blade	1	522920	Not part of engine assembly
	Packing for fan blade	1	274737	1959-60 — Not part of engine assembly
28	Spring washer } Fixing pulley and blade to hub	4	3074	Not part of engine assembly
	Set bolt (¼"UNFx1⅛" long) } Fixing pulley and blade to hub	4	255010	1959-60 — Not part of engine assembly
29	Set bolt (¼" UNFx¾" long) } Fixing pulley and blade to hub	4	255227	1961 onwards — Not part of engine assembly
	Fan belt	1	GFB167	Engines without jockey pulley assembly — Not part of engine assembly
30	Fan belt	1	527288	Engines with jockey pulley assembly — Not part of engine assembly

*Asterisk indicates a new part which has not been used on any previous model.

H701

Plate Ref.	1 2 3 4 Description	Qty.	Part No.	Remarks
	CARBURETTER	1	277517	@ Mk I and
	CARBURETTER	1	511651	@@ Mk IA
	CARBURETTER	1	546250	Mk II and Mk III except NADA — Not part of engine assembly
	CARBURETTER	1	542429	Mk II NADA with engine suffix letters 'A' and 'B'
	CARBURETTER, less throttle lever	1	546542	Mk II and Mk III NADA engines with suffix letter 'C'
	Carburetter body	1	504092	@
	Carburetter body	1	541330	@@
1	Carburetter body	1	601518	Mk II and Mk III
2	Piston lift pin	1	542316	
	Spring for lift pin	1	600113	
	Circlip fixing lift pin	1	542318	
3	Adaptor, ignition and weakening device	1	504082	
4	Gasket for adaptor	1	274950	
5	Shakeproof washer } Fixing adaptor	2	274951	
6	Screw } Fixing adaptor	2	262493	
7	Union for ignition pipe	1	274953	
8	Union for economiser pipe	1	245295	
9	Suction chamber and piston complete	1	504097	
10	Special screw fixing suction chamber	4	262493	
11	Spring for piston (red and green)	1	504084	
12	Thrust washer for suction chamber	1	504085	
	Needle, UF	1	507087	Alternative to UK needle on Mk IA
	Needle UK	1	538650	Alternative to UF needle on Mk IA
13	Needle, UR	1	542321	Mk II and Mk III
14	Special screw fixing needle	1	605865	
15	Oil cap complete	1	511061	
16	Jet complete	1	504086	
17	Jet bearing	1	274957	
18	Jet screw	1	274958	
19	Jet spring	1	274959	
	Jet housing complete	1	504087	@
20	Jet housing complete	1	536860	@@
21	Throttle spindle	1	504089	
22	Throttle butterfly	1	504090	Except Mk II and Mk III NADA
	Throttle butterfly(plain,no hole)	1	546176	Mk II and Mk III NADA
23	Screw for throttle butterfly	2	262481	
	Throttle stop	1	274963	
24	Throttle stop	1	601519	
	Gland washer for throttle spindle, brass	2	274964	Mk I and Mk IA

@ Engines numbered up to 625900655, 626900802, 626930156,626960036,630900793, 631900121
@@Engines numbered from 625900656, 626900803,626930157,626960037,630900794,631900122 onwards
NADA indicates parts peculiar to cars exported to the North American dollar area

*Asterisk indicates a new part which has not been used on any previous model.

Plate Ref.	1 2 3 4	Description	Qty.	Part No.	Remarks
		Spring for throttle spindle gland	2	274965	Mk and Mk IA
		Gland washer for throttle spindle, langite	2	274966	Mk and Mk IA
		Retainer cap for gland washer	2	274967	Mk and Mk IA
25		Nylon bush for throttle spindle	2	601515	Mk II and Mk III
26		Clip retaining nylon bush,	2	601516	Mk II and Mk III
27		Sleeve retaining nylon bush	2	601517	Mk II and Mk III
28		Slow running adjusting valve	1	274968	
29		Gland spring for slow running	1	274969	
30		Gland washer, for slow running, rubber	1	274970	
31		Brass washer for slow running	1	274971	
32		Float chamber	1	274972	
33		Bolt } Fixing float chamber	4	274973	
34		Shakeproof washer } Fixing float chamber	4	274951	
35		Float	1	41710	
		Lid for float chamber	1	274974	Mk I and Mk IA
36		Lid for float chamber	1	601524	Mk II and Mk III
37		Joint washer for float chamber lid	1	261980	
38		Needle valve and seat	1	601600	
		Lever for float	1	262425	Mk I and Mk IA
39		Lever for float	1	601522	Mk II and Mk III
40		Pin for lever	1	267909	
41		Banjo	1	245279	
42		Fibre washer for banjo } On float chamber	1	261981	
43		Alum. washer for banjo } On float chamber	1	262427	
44		Cap nut fixing banjo } On float chamber	1	245277	
45		Double-ended union for carburetter	1	245278	
46		Washer for union	1	232006	
47		Filter and spring for carburetter body	1	262446	
48		Economiser union for rubber tube	1	245258	
49		Pipe for economiser	1	504098	
50		Union for economiser pipe	1	245266	
		Bracket for choke cable	1	274976	@
51		Bracket for choke cable	1	513990	@@
52		Clip for cable bracket	1	274977	
53		Screw (2 BA x ½" long) } Fixing clip to choke bracket	1	71164	
54		Spring washer } Fixing clip to choke bracket	1	3073	
55		Nut (2 BA) } Fixing clip to choke bracket	1	2247	

@ Engines numbered up to 625900655, 626900802, 626930156, 626960036, 630900793, 631900121

@@Engines numbered from 625900656, 626900803, 626930157, 626960037, 630900794, 631900122 onwards

*Asterisk indicates a new part which has not been used on any previous model.

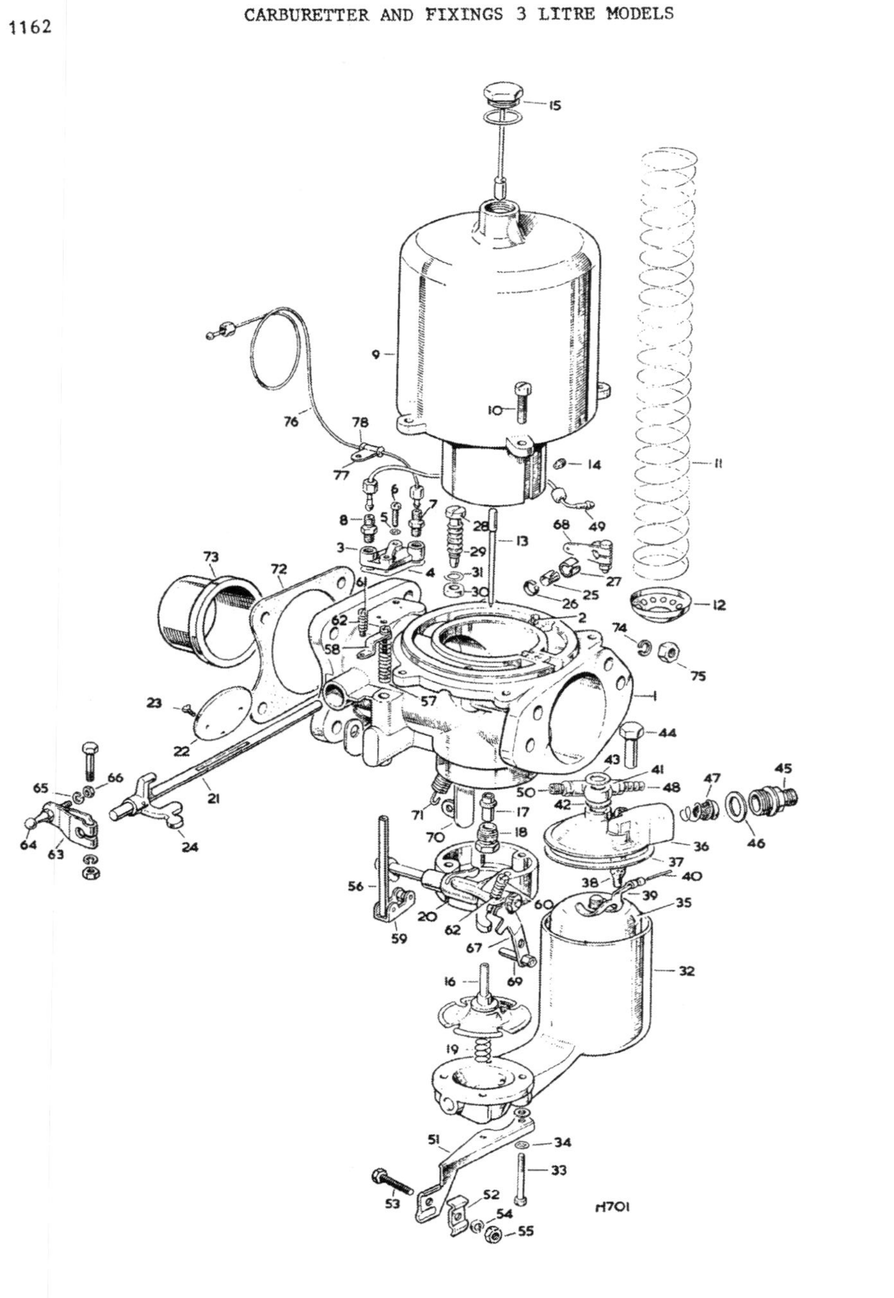

Plate Ref.	1 2 3 4 Description	Qty.	Part No.	Remarks
	Sliding rod	1	514523	@
56	Sliding rod and roller	1	536797	@@
57	Spring for sliding rod	1	274979	
58	Top plate	1	274980	
59	Gam shoe	1	511332	
60	Stop screw, bottom	1	262488	
61	Stop screw, top	1	601521	
62	Spring for stop screw	2	262489	
63	Throttle lever for carburetter	1	505796	Mk I, Mk IA and Mk II engines with suffix letters 'A' and 'B'
64	Ball end for carburetter lever	1	273964	
65	Spring washer } Fixing ball end	1	3073	
66	Nut (2 BA) } Fixing ball end	1	2247	
	Cold start lever	1	505797	@
67	Cold start lever	1	513991	@@
68	Lever for throttle return spring	1	501374	
69	Swivel for cold start control	1	566902	
70	Bracket for throttle return spring	1	505787	
71	Throttle return spring	1	505786	
	Joint washer for carburetter	2	279111	@
72	Joint washer for carburetter	2	511652	@@ 1 off on Mk II and MkIII
73	Liner for manifold	1	277076	} Not part of engine assembly
	Distance piece for carburetter	1	511653	Mk I and Mk IA } Not part of engine assembly
74	Spring washer } Fixing carburetter to cylinder head	4	3075	
75	Nut (5/16" UNF) } Fixing carburetter to cylinder head	4	254831	
	Suction pipe complete	1	501769	Mk I and Mk IA } Not part of engine assembly
76	Suction pipe complete	1	515797	Mk II and Mk III } Not part of engine assembly
77	Clip for suction pipe	1	275037	
78	Rubber grommet for clip	1	214229	

@ Engines numbered up to 625900655, 626900802, 626930156, 626960036, 630900793, 631900121

@@ Engines numbered from 625900656, 626900803, 62630157, 626960037, 630900794, 631900122 onwards

*Asterisk indicates a new part which has not been used on any previous model.

Plate Ref. 1 2 3 4	Description	Qty.	Part No.	Remarks	
	DISTRIBUTOR COMPLETE	1	546420	Mk I and Mk IA engines with 8.75:1 compression ratio	
	DISTRIBUTOR COMPLETE	1	517196	Mk I and Mk IA engines with 7.5:1 compression ratio	
	DISTRIBUTOR CAP	1	GDC106	Except distributor 546420	
	DISTRIBUTOR CAP	1	GDC101	For distributor 546420	
	Brush and spring for cap	1	262703		
	Rotor arm	1	262704		
	Contact points	1	GCS102	Except distributor 546420	
	Contact points	1	GCS109	For distributor 546420	
	Condenser	1	GSC101		
	Auto advance spring, set	1	276788	8.75:1 compression ratio. For distributor LU 40721	
	Auto advance spring, set	1	600533	For distributor 546420	
	Auto advance spring, set	1	264992	7.5:1 compression ratio	
	Auto advance weight	2	262708		Mk I and Mk IA
	Vacuum unit	1	245864	Except distributor 546420	
	Vacuum unit	1	539573	For distributor546420	Not part of engine assembly
	Nylon pad and spring for distributor	1	268241		
	Clamping plate	1	245857		
	Base plate for contact breaker	1	245860	Except distributor546420	
	Base plate for contact breaker	1	502282	For distributor 546420	
	Cam,	1	245861	Except distributor 546420	
	Cam	1	600532	For distributor 546420	
	Shaft	1	245863	Except distributor 546420	
	Shaft and action plate	1	539575	For distributor 546420	
	Bearing	1	245865		
	Bush	1	245866		
	Clip for cover	2	502287	8.75:1 compression ration. For distributor LU 40721	
	Clip for cover	2	539576	For distributor 546420	

NOTE: For engine compression ratio identification see pages (i) and(ii)

*Asterisk indicates a new part which has not been used on any previous model.

Plate Ref. 1 2 3 4	Description	Qty.	Part No.	Remarks	
	Clip for cover	2	245013	7.5:1 compression ratio	
	Driving dog	1	254014	Except distributor 546420	Mk I and Mk IA
	Driving dog	1	600534	For distributor 546420	
	Sealing ring	1	539577	For distributor 546420	Not part of engine assembly
	Sundry parts kit	1	245868		
	Cable nut (For distributor with top)	7	214278	Mk I	
	Washer for cable nut outlet cap (For distributor with top)	7	214279	Mk I	

NOTE: For engine compression ratio identification numbers see pages (i) and (ii)

*Asterisk indicates a new part which has not been used on any previous model.

F399

COLLING-JONES

Plate Ref.	1 2 3 4 Description	Qty.	Part No.	Remarks	
	DISTRIBUTOR COMPLETE	1	538101	8.75:1 compression ratio	Mk II and Mk III
1	DISTRIBUTOR COMPLETE	1	538102	8:1 compression ratio	
2	DISTRIBUTOR CAP	1	GDC101		
3	Brush and spring	1	262703		
4	Rotor arm	1	262704		
5	Contact points	1	GSC109		
6	Condenser	1	GSC101		
	Auto advance spring, set	1	539571	8.75:1compression ratio	
7	Auto advance spring, set	1	539579	8:1 compression ratio	
8	Auto advance weight	2	539572		Not part of engine assembly
	Vacuum unit	1	539573	8.75:1 compression ratio	
9	Vacuum unit	1	502286	8:1 compression ratio	
10	Clamping plate	1	245857		
11	Base plate for contact breaker	1	502282		
	Cam	1	539578	8.75:1 compression ratio	
12	Cam	1	539578	8:1 compression ratio	
13	Shaft and action plate	1	539575		
14	Clip for cover	2	539576		
15	Driving dog	1	600534		
	Sealing ring	1	539577		
	Sundry parts kit	1	245868		
16	Cork washer for distributor housing	1	52278		
17	Set bolt ($\frac{1}{4}$" UNF x 9/16"long) } Fixing distributor lever	1	253005		
18	Spring washer } Fixing distributor lever	1	3074		
19	Plain washer } Fixing distributor lever	1	3911		
20	Heat shield for distributor	1	531937		
21	Clip for heat shield	2	243321		
22	Bolt ($\frac{1}{4}$" UNF x $\frac{3}{8}$" long) } Fixing heat shield to water pipe	2	255208		
23	Spring washer } Fixing heat shield to water pipe	2	3074		
24	Nut ($\frac{1}{4}$" UNF) } Fixing heat shield to water pipe	2	254810		
25	Ignition wire carrier	1	231234	Not part of engine assembly	
26	Sparking plug, Champion N5	6	512445		
27	Washer for plug	6	40441		
28	Suppressor for sparking plug	6	240138		
29	Cover for sparking plug	6	214262	Not part of engine assembly	
30	Rubber sealing ring for plug cover	6	213172		
31	Cable nut	6	214278		
32	Washer for cable nut	6	214279		
	HT Wire black rubber	As reqd	80603	Alternative to silk cord suppressed leads	

NOTE: For engine compression ratio identification numbers see pages (i) and(ii)

*Asterisk indicates a new part which has not been used on any previous model.

F399

COLLINS-JONES

Plate Ref. 1 2 3 4	Description	Qty.	Part No.	Remarks
33	COMPLETE SET OF PLUG LEADS (Silk cord suppressed leads)	1	534309	Alternative to rubber HT leads and suppressors except France (Not part of engine assembly)
	Sparking plug lead No.1 (Silk cord suppressed leads)	1	532643	
	Sparking plug lead No.2 (Silk cord suppressed leads)	1	532644	
	Sparking plug lead No.3 (Silk cord suppressed leads)	1	532645	
	Sparking plug lead No.4 (Silk cord suppressed leads)	1	532646	
	Sparking plug lead No.5 (Silk cord suppressed leads)	1	532647	
	Sparking plug lead No.6 (Silk cord suppressed leads)	1	532648	
	Coil to distributor lead	1	532649	
34	Connector for screw cap nut on leads	13	600247	
35	Staple pin for leads	13	600246	
	Boot for coil lead	1	542126	
	COMPLETE SET OF PLUG LEADS	1	546183	France Mk II and Mk III with screened ignition system (Not part of engine assembly)
	Sparking plug lead No. 1	1	546185	
	Sparking plug lead No. 2	1	546186	
	Sparking plug lead No. 3	1	546187	
	Sparking plug lead No. 4	1	546188	
	Sparking plug lead No. 5	1	546189	
	Sparking plug lead No. 6	1	546190	
	Coil to distributor lead	1	546184	
	Ignition wire carrier	1	231234	
	Top plate (For distributor shroud)	1	546241	
	Side plate (For distributor shroud)	1	546244	
	Side plate with cut-out for distributor shroud	1	546245	
	Bolt (1/4" UNF x 9/16" long) (Fixing side plate to distributor clamp plate)	1	255205	
	Plain washer (Fixing side plate to distributor clamp plate)	1	3946	
	Spring washer (Fixing side plate to distributor clamp plate)	1	3074	
	Nut (1/4" UNF) (Fixing side plate to distributor clamp plate)	1	254810	
	Bolt (10 UNF x 3/8" long) (Fixing top plate to side plates)	2	257309	
	Plain washer (Fixing top plate to side plates)	2	4555	
36	Cleat for plug leads	1	240429	
37	STARTER MOTOR COMPLETE	1	510236	
38	Bracket for starter, commutator end	1	516095	
39	Bracket, drive end	1	516096	
40	Armature,	1	516097	
41	Bush, commutator end	1	242958	
42	Bush, pinion end	1	244714	
43	Pinion and sleeve	1	538351	
44	Spring for pinion	1	244712	
45	Main spring for pinion	1	244710	
46	Nut for pinion	1	244709	Alternative to circlip and cup fixing

*Asterisk indicates a new part which has not been used on any previous model.

DISTRIBUTOR AND STARTER

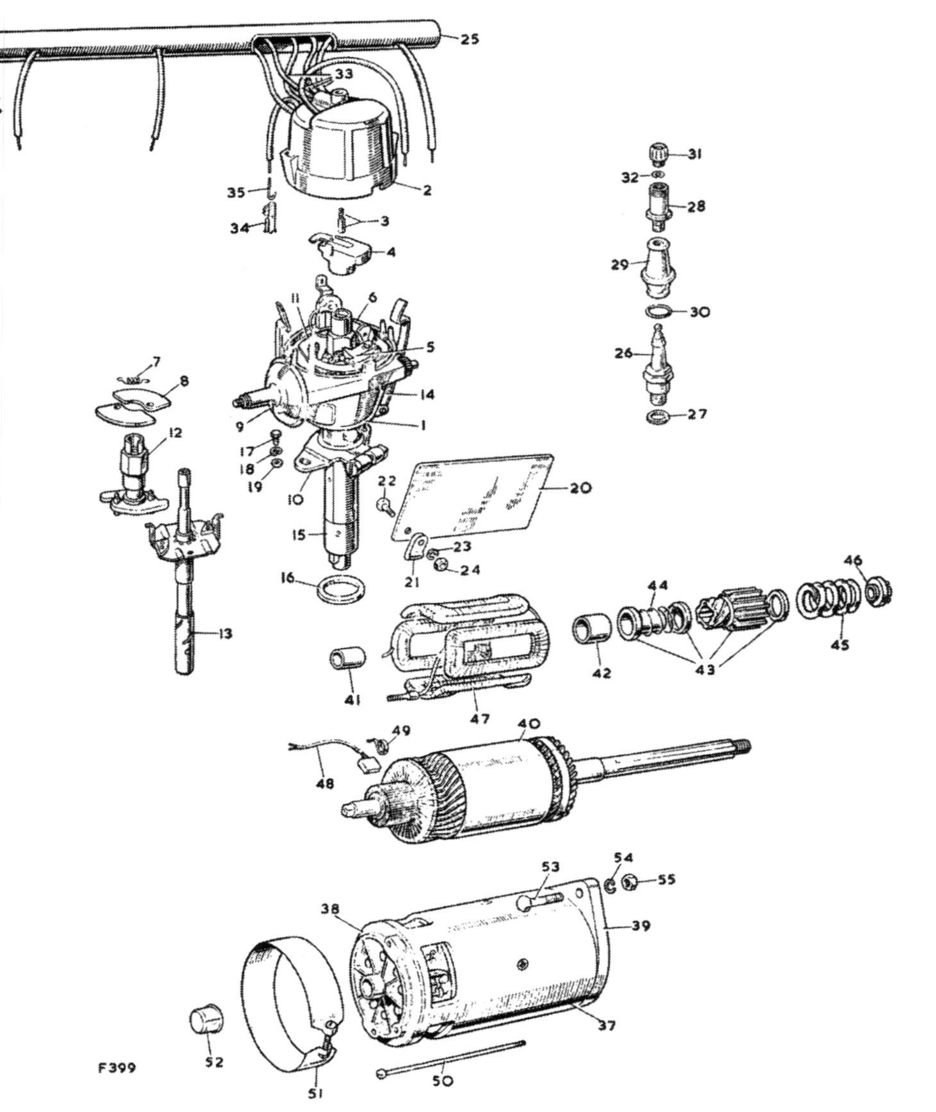

Plate Ref.	1 2 3 4 Description	Qty.	Part No.	Remarks
	Circlip } For armature	1	606220*	} Alternative to
	Cup } shaft, drive end	1	606219*	} nut fixing
47	Field coil for starter	1	516098	For early-type starter LU 26164
	Field coil for starter	1	605065	
48	Brushes for starter motor, set	2	GSB103	
49	Spring set for brushes	1	261239	For early-type starter LU 26164
	Spring set for brushes	1	601754	
50	Bolt for bracket	2	244717	
51	Cover band	1	516094	
52	Grease cap	1	243095	
	Sundry parts kit	1	244718	
	Insulating sleeve for spare terminal on starter motor	1	519870	
	Set bolt (3/8" UNC x 1" long) } Fixing	1	253047	
53	Bolt (3/8" UNF x 1½" long) } starter motor	1	256041	
54	Spring washer } to flywheel	2	3076	
55	Nut (3/8" UNF) } housing	1	254812	

(Items Circlip to Nut (3/8" UNF): Not part of engine assembly)

NOTE: For engine compression ratio identification numbers see page (i)

*Asterisk indicates a new part which has not been used on any previous model.

DYNAMO AND FIXINGS, MK I AND MK IA MODELS

DYNAMO AND FIXINGS, MK I AND MK IA 3 LITRE MODELS

Plate Ref.	Description	Qty.	Part No.	Remarks
1	DYNAMO COMPLETE	1	512248	
2	Armature	1	512800	
	Brushes for dynamo, set	1	GGB107@	
3	Brushes for dynamo, set	1	GGB103@@	
	Spring set for brushes	1	244708@	
4	Spring set for brushes	1	512799@@	
5	Bush, commutator end	1	263395	
6	Ball bearing, front	1	260026	Mk I and Mk IA with manual
	Support bracket, commutator end	1	507604@	steering
7	Support bracket, commutator end	1	512801@@	Not part of engine
8	Support bracket, drive end	1	264433	assembly
9	Oiler for dynamo	1	264431	
10	Field coil	1	264435	
11	Terminal	1	514188	
12	Bolt for bracket	2	279019	
	Sundry parts kit	1	264438	
	DYNAMO COMPLETE,	1	277629	
	Armature,	1	523269	
	Brushes for dynamo, set	1	GGB103	
	Spring set for brushes	1	512799	
	Bearing, commutator end	1	523265	
	Bearing, front	1	523267	Mk I and Mk IA with power
	Support bracket, commutator end	1	523266	steering
	Support bracket, drive end	1	523268	Not part of engine assembl
	Field coil	1	264435	
	Terminal	1	539568	
	Bolt for bracket	2	523270	
	Sundry parts kit	1	264438	
13	Pulley for dynamo	1	510992	1959-60
14	Pulley for dynamo	1	522705	1961 and Mk IA
15	Woodruf key } Fixing pulley to dynamo	1	1664	
16	Lock washer } Fixing pulley to dynamo	1	03748	
17	Special nut } Fixing pulley to dynamo	1	3466	
18	Support bracket for dynamo	1	269248	1959-60
19	Support bracket for dynamo	1	530572	1961 and Mk IA
20	Distance piece for dynamo mounting	1	521569	Mk IA
21	Spring washer } Fixing bracket to cylinder block	2	3075	
22	Set bolt (5/16"UNF×¾"long) } Fixing bracket to cylinder block	2	255426	
23	Bolt (5/16" UNF ×1"long) } Fixing dynamo to bracket	1	255428	
24	Bolt (5/16" UNF ×1¼"long) } Fixing dynamo to bracket	1	256420	
25	Spring washer } Fixing dynamo to bracket	2	3075	
26	Nut (5/16" UNF) } Fixing dynamo to bracket	2	254811	
27	Adjusting link for dynamo	1	279571	1959-60
28	Adjusting link for dynamo	1	524852	1961 and Mk IA

@ Engines numbered up to 625900493, 626930104, 626900632, 626960025, 630900512, 631900063

@@ Engines numbered from 625900494, 62630105, 6269000633, 626960026, 630900513, 631900064 onwards

*Asterisk indicates a new part which has not been used on any previous model.

DYNAMO AND FIXINGS, MK I AND MK IA MODELS

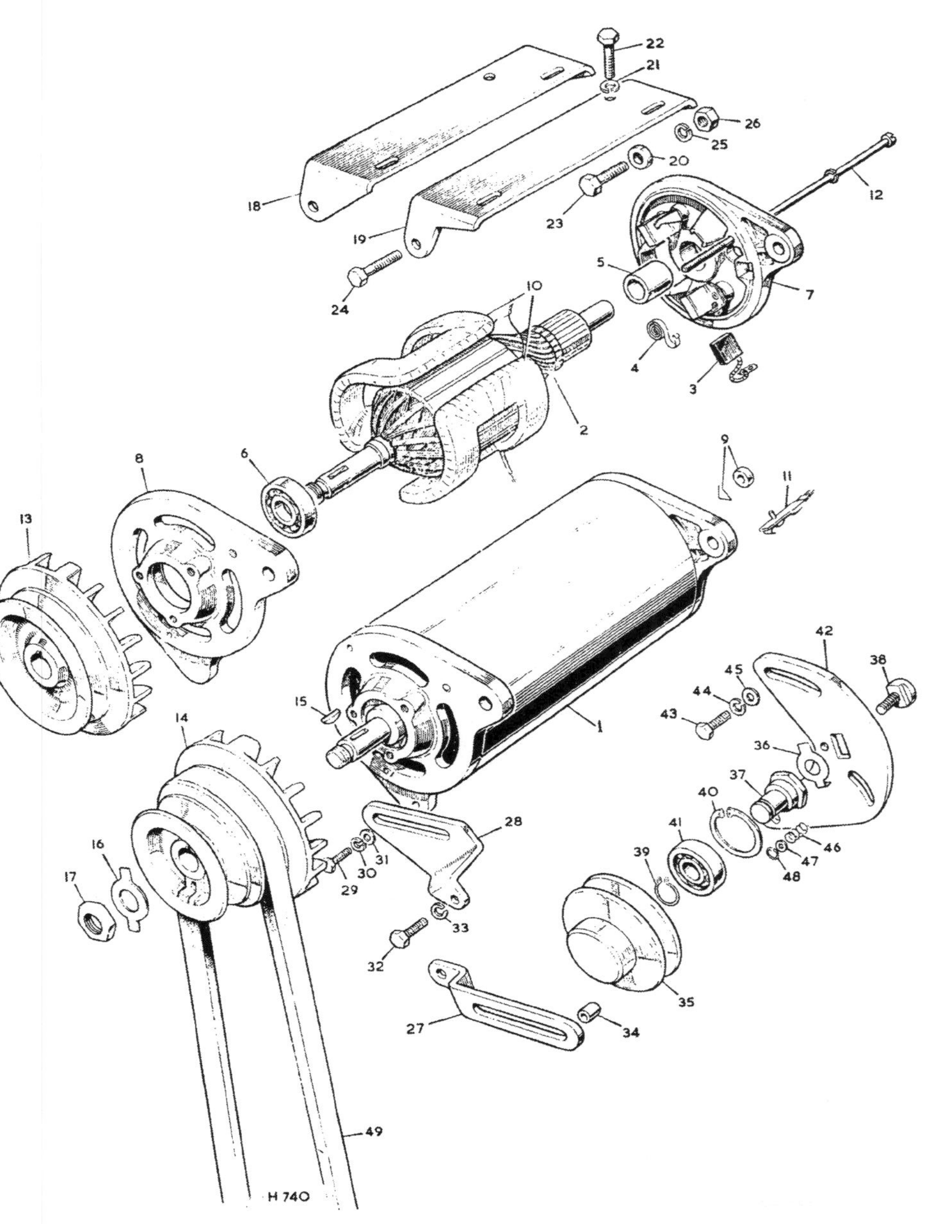

Plate Ref.	1 2 3 4 Description		Qty.	Part No.	Remarks
29	Special bolt		1	253027	
30	Spring washer	Fixing adjusting link to dynamo	1	3075	
31	Plain washer		1	3868	
32	Set bolt (5/16" UNF x 2¼" long)	Fixing adjusting link to cylinder block	1	255237	1959-60
	Set bolt (5/16" UNF x ¾" long)		1	255226	1961 and Mk IA
33	Spring washer		1	3075	
34	Distance piece		1	210566	1959-60
	JOCKEY PULLEY AND ADJUSTING LINK ASSEMBLY		1	522941	Cars with power steering
35	Jockey pulley		1	520224	
36	Lock plate for jockey pulley shaft		1	523074	
37	Shaft for jockey pulley		1	520191	
38	Special bolt, shaft to adjusting link		1	522835	
39	Retaining ring for bearing, small		1	520223	
40	Retaining ring for bearing, large		1	217500	
41	Bearing for shaft		1	520278	
42	Adjusting lnk for jockey pulley		1	522813	
43	Set bolt (5/16" UNF x ¾" long)	Fixing jockey pulley adjusting link to front cover	2	255226	
44	Spring washer		2	70822	
45	Plain washer		2	4148	
46	Spring	For swivel pin on front cover	1	2355	
47	Plain washer		1	2208	
48	Circlip		1	522838	
49	Dynamo driving belt		1	GFB138	

*Asterisk indicates a new part which has not been used on any previous model.

DYNAMO AND FIXINGS, MK II AND MK III MODELS

H683

Plate Ref.	1 2 3 4 Description	Qty.	Part No.	Remarks
	DYNAMO COMPLETE	1	514858	Mk II with manual steering. Not part of engine assembly
	Armature	1	538630	
	Brushes for dynamo, set	1	GGB101	
	Spring set for brushes	1	514192	
	Bush, commutator end	1	538628	
	Ball bearing, front	1	242672	
	Support bracket, commutator end	1	538627	
	Support bracket, drive end	1	538629	
	Oiler for dynamo	1	514189	
	Field coil	1	538631	
	Bolt for bracket	2	538632	
	Sundry parts kit	1	532567	
	DYNAMO COMPLETE	1	601833*	Mk II — With power steering
1	DYNAMO COMPLETE, negative earth	1	551499	Mk III — With power steering
2	Armature	1	539566	With power steering
3	Field coil	1	549567	With power steering
4	Support bracket, commutator end	1	539564	With power steering
5	Brushes for dynamo, set	1	GGB101	For dynamos 545183 and 551499 — With power steering
	Brushes for dynamo, set	1	601902*	For dynamo 601833 — With power steering
6	Spring set for brushes	1	514192	With power steering
7	Ball bearing, commutator end	1	523265	With power steering
8	Support bracket, drive end	1	539565	For dynamos 545183 and 551499 — With power steering
	Support bracket, drive end	1	601903*	For dynamo 601833 — With power steering
9	Bearing, drive end	1	242672	With power steering
	Collar for drive end bearing	1	601904*	For dynamo 601833 — With power steering
10	Bolt for bracket	2	538632	With power steering
11	Connector, lucar, 35-amp	1	600431	With power steering
12	Connector, lucar, 117.5-amp	1	600430	With power steering
	Terminal screw	1	539568	With power steering
	Sundry parts kit	1	539569	
	Clip fixing harness to dynamo	1	523203	
	Pulley for dynamo	1	532618	Mk II engines with suffix letter 'A'
13	Pulley for dynamo	1	546287	Mk II engines with suffix letter 'B' onwards and Mk III
14	Fan for dynamo	1	554055	
15	Distance piece for dynamo pulley	1	533860	
16	Woodruff key — Fixing pulley to dynamo	1	1664	
	Lock washer — Fixing pulley to dynamo	1	217781	Alternatives
17	Spring washer — Fixing pulley to dynamo	1	547604	Alternatives
18	Special nut — Fixing pulley to dynamo	1	3466	
19	Support bracket for dynamo	1	532552	
20	Set bolt (5/16" UNF × ¾" long) — Fixing bracket to cylinder block	2	255426	
21	Spring washer — Fixing bracket to cylinder block	2	3075	
22	Plain washer — Fixing bracket to cylinder block	2	2550	
	Bolt (5/16" UNF × 1" long) — Fixing dynamo to bracket	1	255428	
23	Bolt (5/16" UNF × 1¼" long) — Fixing dynamo to bracket	1	256420	
24	Spring washer — Fixing dynamo to bracket	2	3075	
25	Nut (5/16" UNF) — Fixing dynamo to bracket	2	254811	

*Asterisk indicates a new part which has not been used on any previous model.

DYNAMO AND FIXINGS, MK II AND MK III MODELS

Plate Ref.	1 2 3 4 Description	Qty.	Part No.	Remarks
	Adjusting link for dynamo	1	538398	Mk II engines with suffix letter 'A'
26	Adjusting link for dynamo	1	546354	Mk II engines with suffix letter 'B' onwards and Mk III
27	Special bolt } Fixing adjusting link to dynamo	1	253027	
28	Spring washer } Fixing adjusting link to dynamo	1	3075	
29	Plain washer } Fixing adjusting link to dynamo	1	3868	
30	Set bolt (5/16" UNF x ¾" long) } Fixing adjusting link to cylinder block	1	255226	
31	Spring washer } Fixing adjusting link to cylinder block	1	3075	
	JOCKEY PULLEY AND ADJUSTING LINK ASSEMBLY	1	522941	} Mk II engines with suffix letter 'B' onwards and Mk III, also all earlier cars with power steering
32	Jockey pulley	1	520224	
33	Lock plate for jockey pulley shaft	1	523074	
34	Shaft for jockey pulley	1	520191	
35	Special bolt, shaft to adjusting link	1	522835	
36	Retaining ring for bearing, small	1	520223	
37	Retaining ring for bearing, large	1	217500	
38	Bearing for shaft	1	520278	
39	Adjusting link for jockey pulley	1	522813	
40	Set bolt (5/16"UNFx¾"long) } Fixing jockey pulley adjusting link to front cover	2	255226	
41	Spring washer } Fixing jockey pulley adjusting link to front cover	2	70822	
42	Plain washer } Fixing jockey pulley adjusting link to front cover	2	4148	
43	Spring } For swivel pin on front cover	1	2355	
44	Plain washer } For swivel pin on front cover	1	2208	
45	Circlip } For swivel pin on front cover	1	522838	
46	Dynamo driving belt	1	GFB138	

H683

*Asterisk indicates a new part which has not been used on any previous model.

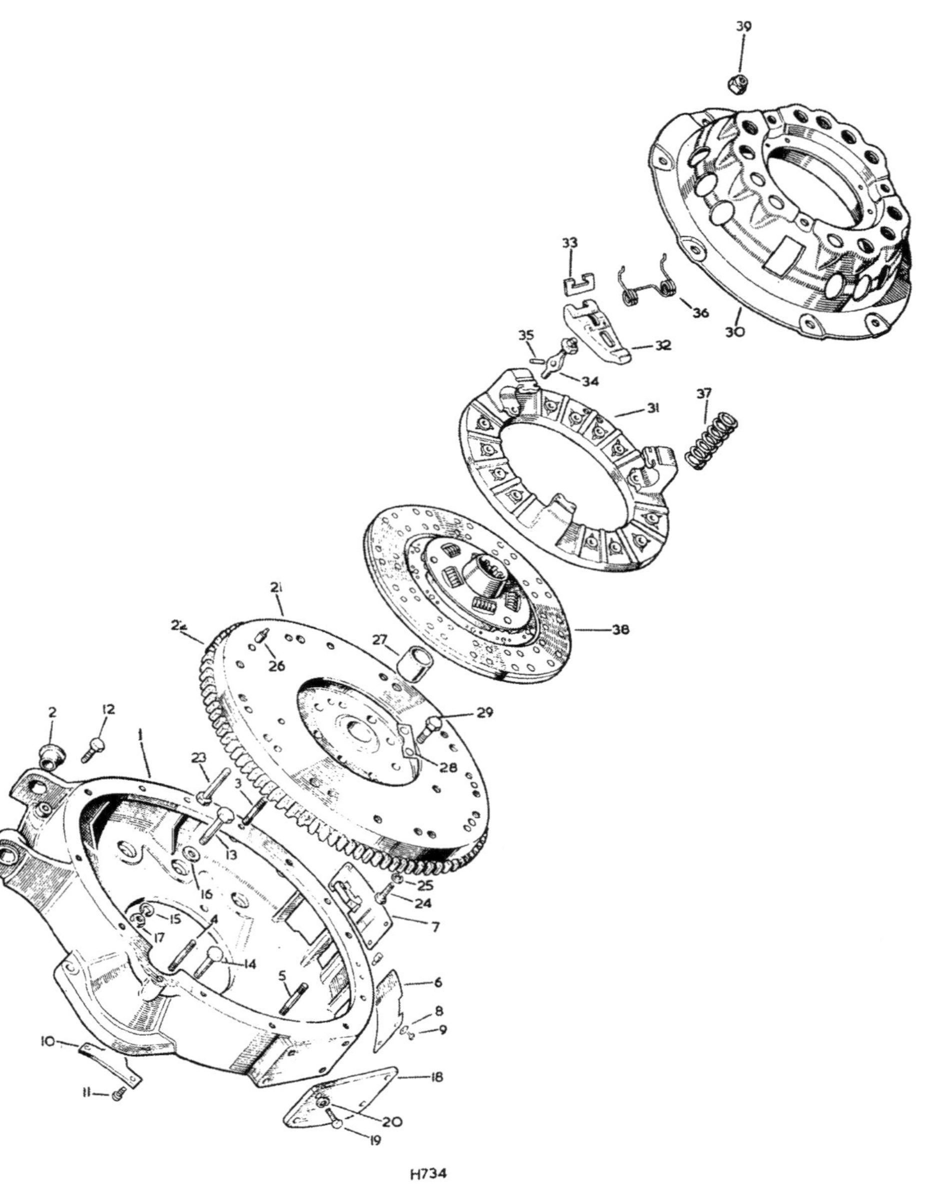

Plate Ref.	1 2 3 4 Description		Qty.	Part No.	Remarks
	HOUSING FOR FLYWHEEL		1	600483	Mk I, Mk IA and Mk II
1	HOUSING FOR FLYWHEEL		1	550071	Mk III
2	Bush		2	511127	Mk II with suffix letter 'C' and Mk III
	Stud, short (3/8"), 1½" long	Fixing flywheel Housing to bell housing	12	277458@	
3	Stud, short (3/8"), 1½" long		8	277458@@	
4	Stud (3/8"), 2" long		3	252722@@	
5	Stud (3/8"), 2⅛" long)		1	252529@@	
	Stud (5/16""		1	277459	
	Stud (¼") for inspection cover		2	252496	Except Mk III
	Inspection cover plate		1	56140	
	Indicator for engine timing		1	500691	
	Nut (¼" UNF) fixing cover plate		2	254810	
6	Cover for lower holes in flywheel housing		2	279563	1 off on models @@
7	Cover plate for flywheel housing		1	524166@@	
8	Plain washer	Fixing cover to flywheel housing	8	2256	
9	Drive screw		8	77704	
10	Cover plate for starter motor aperture		1	554532	Mk III
11	Drive screw fixing cover plate		2	77704	
12	Bolt (3/8" UNF x 1⅛" long)	Fixing flywheel housing to cylinder block	2	256040	
13	Bolt (3/8" UNF x 1½" long)		3	256041	
14	Bolt (3/8" UNF x 1⅛" long)		3	255048	
15	Spring washer		8	3076	
16	Plain washer		6	2219	
17	Nut (3/8" UNF)		3	254812	
18	Engine mounting plate, LH rear		1	521480@@	
19	Set bolt (5/16" UNF x 1" long)	Fixing mounting plate to flywheel housing	2	255228@@	
20	Spring washer		2	3075@@	
21	FLYWHEEL ASSEMBLY		1	554681	
	Ring gear for flywheel		1	279145	Mk I, Mk IA and Mk II
22	Ring gear for flywheel		1	550068	Mk III
23	Special fitting bolt for ring gear		6	596965*	
24	Set bolt (¼" UNF x 1" long)	Fixing ring gear to flywheel	2	255009	
25	Spring washer		2	3074	
26	Dowel for clutch cover plate		2	6395	
27	Bush for primary pinion		1	08566	
28	Locker	Fixing flywheel to crankshaft	4	526161	
29	Special set bolt		8	247135	
	CLUTCH ASSEMBLY		1	279050	
30	Cover plate for clutch		1	504954	
31	Pressure plate for clutch		1	504951	
32	Release lever		3	504952	
33	Strut for release lever		3	231884	
34	Special eyebolt and nut, cover to eyebolt		3	242996	
35	Pin for release lever		3	231885	
36	Anti-rattle spring		3	504953	
37	Clutch spring, cream		12	231881	
38	CLUTCH PLATE COMPLETE		1	539751	
	Lining package for clutch plate		1	504955	
39	Self-locking nut (3/8" UNF)fixing clutch to flywheel		6	252162	
@	Cars numbered up to 625101045, 626100262, 628100109				
@@	Cars numbered from 625101046, 626100263, 628100110 onwards				

*Asterisk indicates a new part which has not been used on any previous model.

TORQUE CONVERTER, FLYWHEEL AND BELL HOUSING MK I, MK IA AND MK II AUTOMATIC MODELS

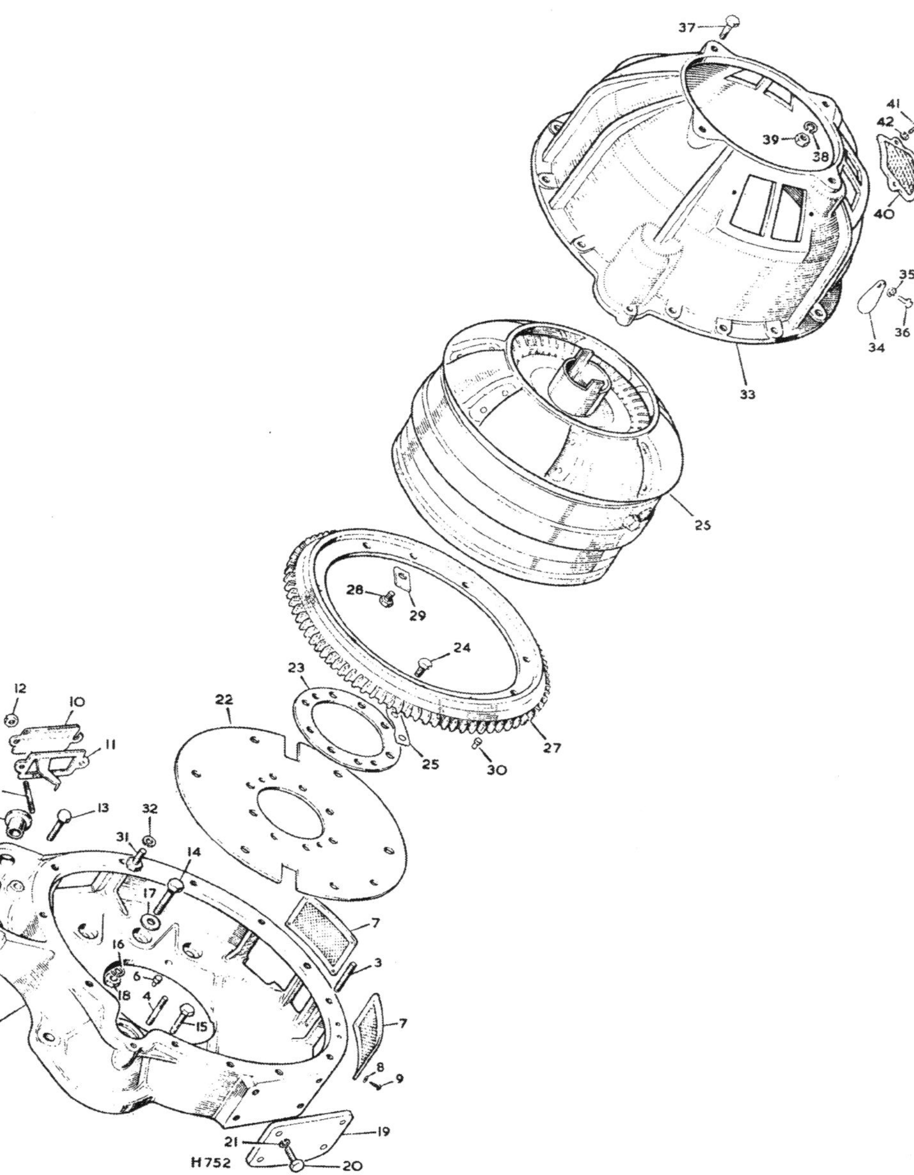

TORQUE CONVERTER, FLYWHEEL AND BELL HOUSING MK I, MK IA AND MK II AUTOMATIC MODELS

Plate Ref.	1 2 3 4 DESCRIPTION	Qty	Part No.	REMARKS
	HOUSING FOR FLYWHEEL	1	600483	
2	Bush for accelerator cross-shaft	2	511127	Mk II with suffix letter 'C'
3	Stud (3/8") } Fixing flywheel housing to bell housing	12	277458	
4	Stud (5/16") } Fixing flywheel housing to bell housing	1	277459	
5	Stud (1/4") for inspection cover	2	252496	
6	Dowel for flywheel housing	2	52124	
7	Cover for flywheel housing lower holes	2	279564	
8	Plain washer } Fixing cover	8	2256	
9	Drive screw } Fixing cover	8	77704	
10	Inspection cover plate	1	56140	
11	Indicator for engine timing	1	500683	
12	Nut (1/4" UNF) fixing cover plate	2	254810	
13	Bolt (3/8" UNF x 1 3/8" long) } Fixing flywheel housing to cylinder block	2	256040	
14	Bolt (3/8" UNF x 1 1/2" long) } Fixing flywheel housing to cylinder block	3	256041	
15	Bolt (3/8" UNF x 1 1/8" long) } Fixing flywheel housing to cylinder block	3	255048	
16	Spring washer } Fixing flywheel housing to cylinder block	8	3076	
17	Plain washer } Fixing flywheel housing to cylinder block	6	2219	
18	Nut (3/8" UNF) } Fixing flywheel housing to cylinder block	3	254812	
19	Engine mounting plate, LH rear	1	521480††	
20	Set bolt (5/16" UNF x 1" long)	2	255228††	
21	Spring washer	2	3075††	
22	Flexible drive plate	1	279637	
23	Reinforcing plate for drive plate	1	279186	
24	Special set bolt } Fixing flexible drive plate to crankshaft	8	279647	
25	Locker } Fixing flexible drive plate to crankshaft	4	526161	
	TORQUE CONVERTER UNIT AND DIRECT DRIVE CLUTCH	1	527498†	Not part of engine assembly
26	TORQUE CONVERTER UNIT AND DIRECT DRIVE CLUTCH	1	538212††	Not part of engine assembly
	Drain plug for torque converter unit (3/8")	1	518830	Alternatives; Not part of engine assembly
	Joint washer for drain plug (3/8")	1	518831	Alternatives; Not part of engine assembly
	Drain plug for torque converter unit (7/16")	1	521391	Alternatives; Not part of engine assembly
	Joint washer for drain plug (7/16")	1	521392	Alternatives; Not part of engine assembly
27	Starter ring	1	524886	Not part of engine assembly
28	Special set bolt (3/8" UNF) } Fixing starter ring	2	269388	Not part of engine assembly
29	Locker } Fixing starter ring	2	2819	Not part of engine assembly
30	Dowel } Fixing torque converter unit to flexible drive plate	1	269453	Not part of engine assembly
31	Special set bolt (3/8" UNF) } Fixing torque converter unit to flexible drive plate	6	269389	Not part of engine assembly
32	Spring washer } Fixing torque converter unit to flexible drive plate	6	501598	Not part of engine assembly
33	Bell housing	1	500558	
34	Cover plate, bottom, for bell housing	1	507523	
35	Spring washer } Fixing cover plate	1	3074	
36	Set bolt (1/4" UNC x 3/8" long) } Fixing cover plate	1	253002	
37	Set bolt (7/16" UNC x 1 1/2" long) } Fixing bell housing to gearbox	2	254060	
38	Spring washer } Fixing bell housing to gearbox	4	3077	
39	Nut (7/16" UNF) } Fixing bell housing to gearbox	2	254853	
40	Air inlet cover	3	269353	
41	Set bolt (1/4" UNC x 1/2" long) } Fixing cover to bell housing	6	253004	
42	Spring washer } Fixing cover to bell housing	6	3074	

* Asterisk indicates a new part which has not been used on any previous Rover model
† Cars numbered up to 630100840, 631100353, 633100078
†† Cars numbered from 630100841, 631100354, 633100079 onwards

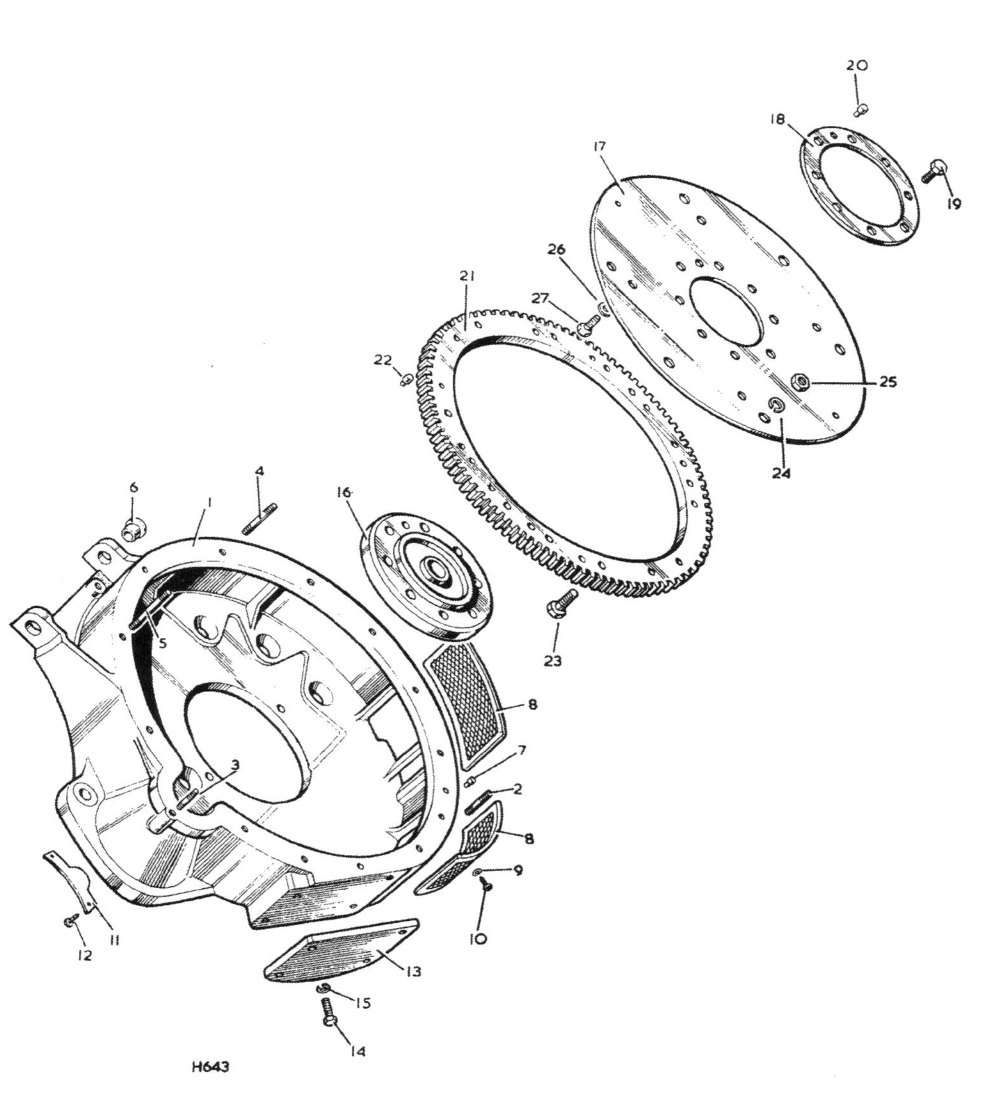

Plate Ref.	1 2 3 4 DESCRIPTION	Qty	Part No.	REMARKS
1	HOUSING FOR FLYWHEEL	1	550071	
2	Stud ($\frac{3}{8}$" UNC/UNF)	8	277458	
3	Stud ($\frac{5}{16}$" UNC/UNF)	1	277459	
4	Stud ($\frac{3}{8}$" UNC/UNF x 2" long)	3	252722	
5	Stud ($\frac{3}{8}$" UNC/UNF x 2$\frac{1}{8}$" long)	1	252529	
6	Bush for accelerator cross-shaft	2	511127	
7	Dowel for flywheel housing	2	52124	
8	Cover for flywheel housing lower holes	2	279564	
9	Plain washer } Fixing cover plate	8	2256	
10	Drive screw } Fixing cover plate	8	77704	
11	Cover plate for starter motor aperture	1	554532	
12	Drive screw fixing cover plate	2	77704	
13	Engine mounting plate, LH rear	1	521480	
14	Set bolt ($\frac{5}{16}$" UNF x 1" long) } Fixing mounting plate to housing	2	255228	
15	Spring washer } Fixing mounting plate to housing	2	3075	
16	Centre plate for flexible drive plate	1	550038	
17	Flexible drive plate	1	550037	
18	Reinforcing plate for flexible drive plate	1	279186	
19	Set bolt ($\frac{7}{16}$" UNF x $\frac{7}{8}$" long) } Fixing flexible drive plates to crankshaft	8	247135	
20	Dowel } Fixing flexible drive plates to crankshaft	1	265779	
21	Starter ring	1	550068	
22	Dowel } Fixing starter ring to flexible drive plate	2	529364	
23	Bolt ($\frac{3}{8}$" UNF x 1" long) } Fixing starter ring to flexible drive plate	6	255247	
24	Spring washer } Fixing starter ring to flexible drive plate	6	3076	
25	Nut ($\frac{3}{8}$" UNF) } Fixing starter ring to flexible drive plate	6	254812	
26	Plain washer } Fixing torque converter to drive plate	4	2204	
27	Set bolt ($\frac{3}{8}$" UNF x $\frac{5}{8}$" long) } Fixing torque converter to drive plate	4	255444	

* Asterisk indicates a new part which has not been used on any previous Rover model

GEARBOX CASING AND BELL HOUSING, 4-SPEED MODELS

Plate Ref.	1 2 3 4 Description	Qty.	Part No.	Remarks
	GEARBOX COMPLETE ASSEMBLY	1	279055@	Models without
	GEARBOX COMPLETE ASSEMBLY	1	520999@@	overdrive
	GEARBOX ASSEMBLY, LESS EXTENSION CASING AND OVERDRIVE UNIT ASSEMBLY	1	514381	Models with overdrive Gearboxes numbered up to 625101417
	GEARBOX ASSEMBLY, LESS EXTENSION CASING AND OVERDRIVE ASSEMBLY	1	525065	Models with overdrive Gearboxes numbered from 625101418 and Mk IA models
	GEARBOX ASSEMBLY, LESS EXTENSION CASING AND OVERDRIVE ASSEMBLY	1	536813	Mk II and Mk III
	GEARBOX CASING ASSEMBLY	1	248711@	
1	GEARBOX CASING ASSEMBLY	1	515687@@	
2	Stud (¼")] Fixing top cover to gearbox	2	216034	
3	Spring washer] Fixing top cover to gearbox	2	3074	
4	Nut (¼" BSF)] Fixing top cover to gearbox	2	2823	
5	Stud, long (⅜")] Fixing top cover to gearbox	2	241772	
6	Spring washer] Fixing top cover to gearbox	4	3076	
7	Nut (⅜" BSF)] Fixing top cover to gearbox	4	2827	2 off from @@
	Stud for cover and gear change	2	241743@	
8	Bolt (⅜"Whit x 1¾" long) fixing top cover	2	561484@@	
	Nut (⅜" BSF) fixing top cover	2	2827@@	
9	Dowel for cover	2	07289	
10	Stud for housing, short	6	3238	
11	Dowel for housing	2	55636	
	Oil level rod	1	243451	Models without overdrive
12	Oil level rod	1	277318	Models with overdrive
13	Oil filler cap	1	268564	
14	Washer for oil filler cap	1	11932	
15	Retaining spring for oil filler	1	243449	
16	Plain washer, small] Fixing retaining spring to gearbox cover	1	3946	
17	Plain washer, large] Fixing retaining spring to gearbox cover	1	3821	
18	Set bolt(¼"whit x 9/16"long)] Fixing retaining spring to gearbox cover	1	215769	
19	Drain plug	1	241110	
20	Joint washer for plug	1	515599	
	BELL HOUSING ASSEMBLY	1	521000	Mk I and Mk IA
21	BELL HOUSING ASSEMBLY	1	528696	Mk II and Mk III
22	Stud for withdrawal race housing	3	269413	
	Stud gear selector housing	1	241580@	
23	Set bolt (¼" BSF x 1" long)] For second speed stop	1	250696	
24	Locknut (¼" BSF)] For second speed stop	1	2823	
25	Bracket for reverse stop	1	242331	
26	Spring washer] Fixing bracket to bell housing	2	3074	
27	Set bolt (¼" Whitx½"long)] Fixing bracket to bell housing	2	215758	

@ Gearboxes numbered up to 625101668
@@ Gearboxes numbered from 625101669 onwards
NB: Gearbox assemblies include selector shafts, clutch withdrawal housing, top support assembly and clutch slave cylinder

*Asterisk indicates a new part which has not been used on any previous model.

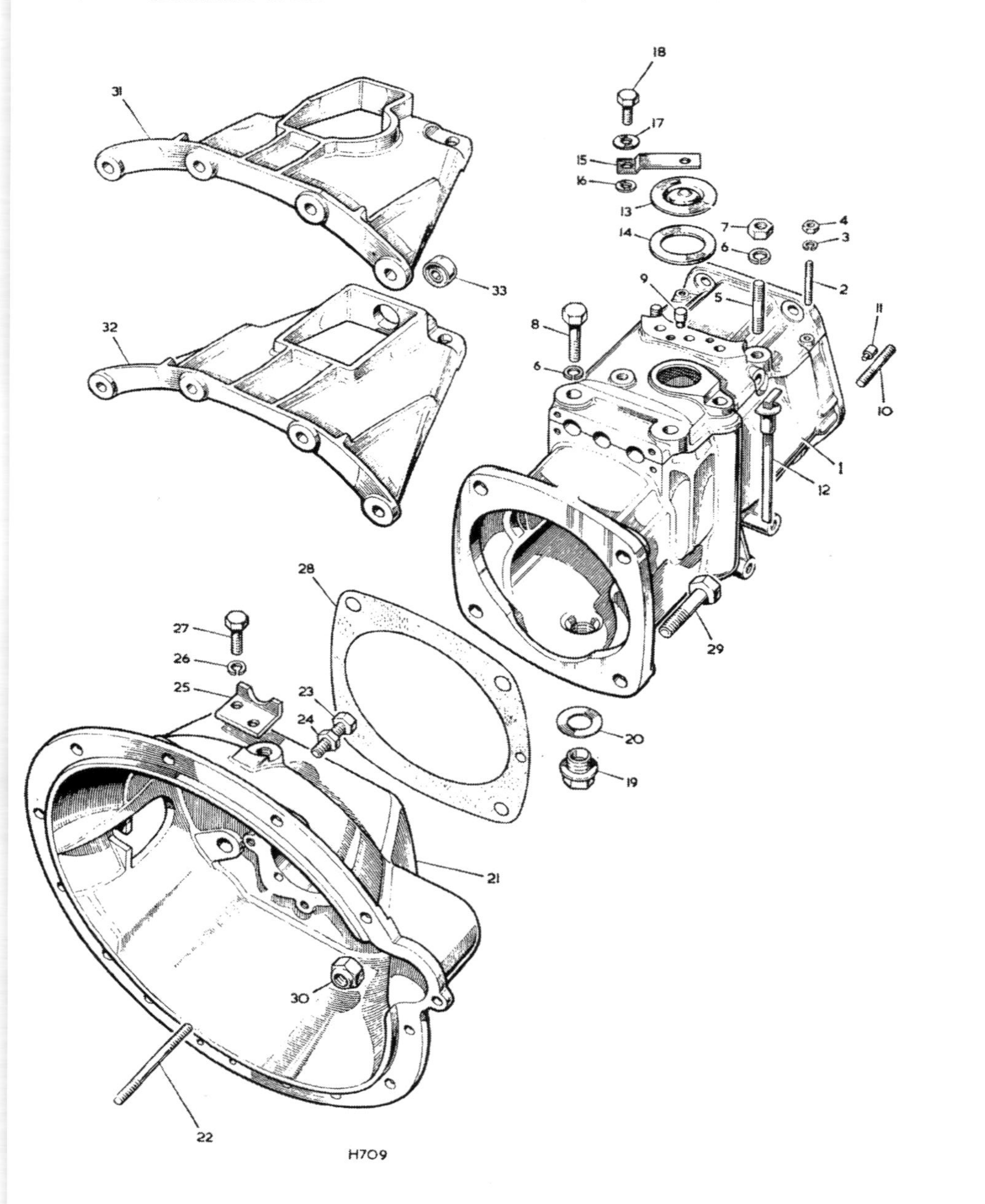

Plate Ref.	1 2 3 4 DESCRIPTION	Qty	Part No.	REMARKS
28	Joint washer, bell housing to gearbox	1	241590	
29	Special bolt (2″ long) } Fixing bell housing	4	248720	
30	Self-locking nut ($\frac{1}{2}$″ BSF) } to gearbox casing	4	251324	
	Inspection cover for bell housing	1	241606	Mk I and Mk IA
	Rubber seal for inspection cover } Fixing inspection cover	1	241607	Mk I and Mk IA
	Spring washer } Fixing inspection cover	2	3074	Mk I and Mk IA
	Set bolt ($\frac{1}{4}$″ BSF x $\frac{5}{8}$″ long) } Fixing inspection cover	2	237140	Mk I and Mk IA
31	Top support stiffener	1	518547††	
32	TOP SUPPORT AND BUSH ASSEMBLY ..	1	533002	Mk II and Mk III
33	Bush, self-aligning	1	532958	Mk II and Mk III

* Asterisk indicates a new part which has not been used on any previous Rover model
†† Gearboxes numbered from 625101669 and Mk IA 4-speed models

H735

Plate Ref.	1 2 3 4 Description		Qty.	Part No.	Remarks
1	PRIMARY PINION AND CONSTANT GEAR		1	518473	
2	Shield for primary pinion		1	01017	
3	Spring clutch plate decelerator		1	515360	
4	Button	Retaining spring in	1	515391	
5	Pin	primary pinion	1	515392	
6	Ball bearing for primary pinion		1	55714	
7	Lock washer	Fixing bearing to	1	08250	
8	Lock nut	primary pinion	1	213416	
9	Retaining plate	Fixing bearing	1	213666	
10	Extension stud	to bell	4	214090	
11	Self-locking nut	housing	4	251321	
	Layshaft		1	265096	Mk I and Mk IA
12	Layshaft		1	528686	Mk II and Mk III
	Mainshaft		1	515433	
	Special washer	For	1	513454	Models without overdrive.
	Self-locking nut (⅜" UNF)	mainshaft	1	513188	
13	Mainshaft		1	515499	Models with overdrive
14	Peg for second gear thrust washer		1	06405	
15	Peg for mainshaft distabce sleeve		1	09561	
	Thrust washer, .125"		1	267572	
16	Thrust washer, .128"	For second	1	267573	As required
	Thrust washer, .130"	speed gear	1	267574	
	Thrust washer, .135"		1	267575	
17	First speed layshaft gear		1	522375	
	Second speed layshaft and mainshaft gear		1	269521	Mk I and Mk IA
18	Second speed layshaft and mainshaft gear		1	536811	Mk II and Mk III
	Third speed layshaft and mainshaft gear		1	245767	Mk I and Mk IA
19	Third speed layshaft and mainshaft gear		1	536812	Mk II and Mk III
20	Distance sleeve for mainshaft		1	571218	
	Thrust washer, .125"		1	08188	
21	Thrust washer, .128"	For third speed	1	50702	As required
	Thrust washer, .130"	mainshaft gear	1	50703	
	Thrust washer, .135"		1	231737	
22	Spring fixing second and third mainshaft gears		1	06402	
23	Sleeve for layshaft 3.329" long		1	245743	As required
	Sleeve for layshaft 3.321" long		1	245741	As required
	Bearing for layshaft, front		1	576207	
	Plain washer		1	08185	Mk I and Mk IA
	Slotted Nut	Fixing bearing to layshaft	1	09932	
	Split pin		1	2980	
24	Bearing for layshaft, front		1	528701	
25	Plain washer		1	528692	Mk II and Mk III
26	Slotted nut	Fixing bearing to layshaft	1	528691	
	Split pin		1	2766	

NOTE: 4-speed models numbered from 625003710 are fitted with clutch plate decelerator

*Asterisk indicates a new part which has not been used on any previous model.

GEARBOX SHAFTS AND GEARS, 4-SPEED MODELS

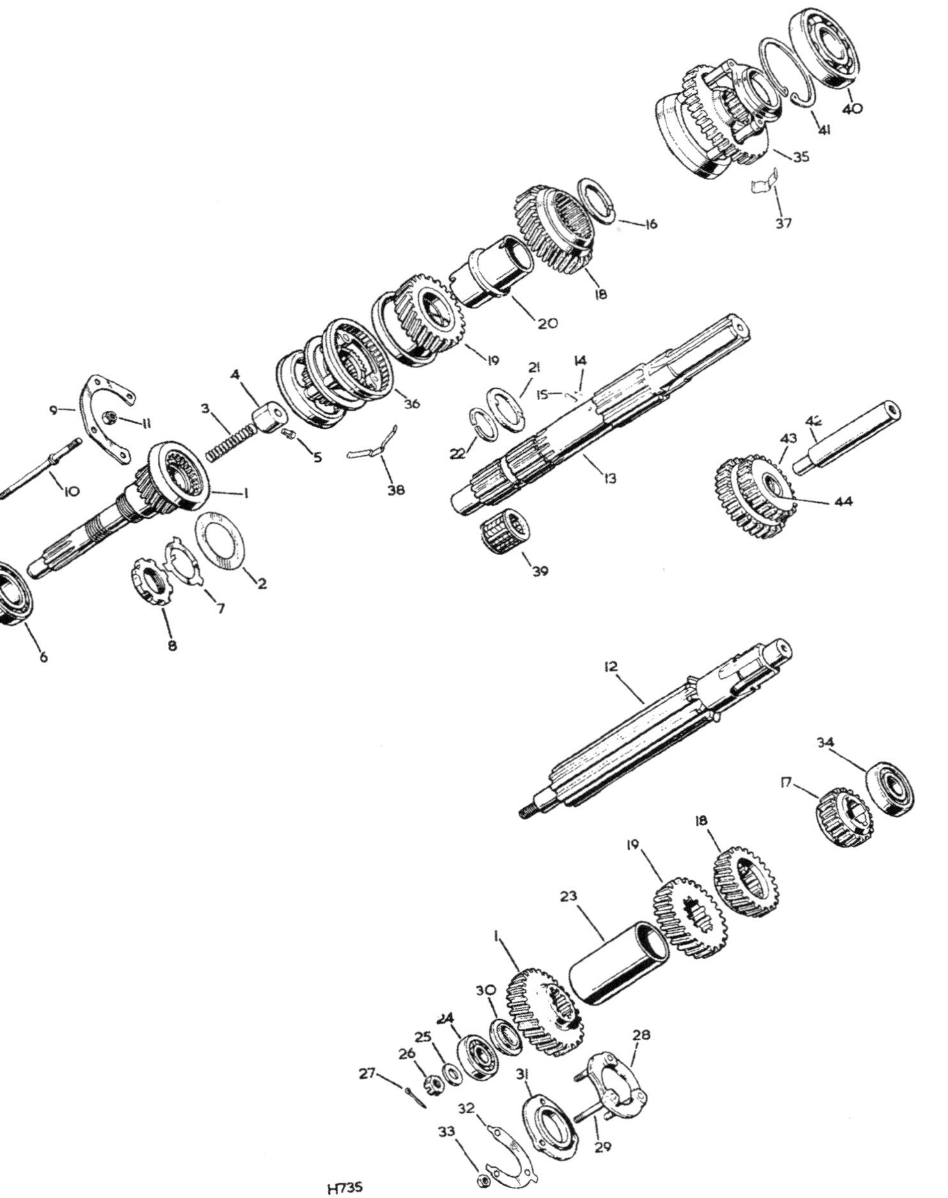

Plate Ref.	1 2 3 4 Description	Qty.	Part No.	Remarks
	BEARING PLATE ASSEMBLY FOR LAYSHAFT	1	214792	Mk I and Mk IA
28	BEARING PLATE ASSEMBLY FOR LAYSHAFT	1	528685	Mk II and Mk III
29	Stud for bearing plate	3	213417	
	Distance piece, .372" } For layshaft	1	245736	Mk I and Mk IA. As required
	Distance piece, .392" } For layshaft	1	245737	Mk I and Mk IA. As required
	Distance piece, .412" } For layshaft	1	245738	Mk I and Mk IA. As required
30	Distance piece, std } For layshaft	1	528687	Mk II and Mk III. As required
	Distance piece, .020" } For layshaft	1	528688	Mk II and Mk III. As required
	Distance piece, .040" } For layshaft	1	528689	Mk II and Mk III. As required
	Retaining plate for layshaft front bearing	1	213419	Mk I and Mk IA
	Lock washer } Fixing plate and bearing to bell housing	1	09931	Mk I and Mk IA
	Nut (5/16" BSF) } Fixing plate and bearing to bell housing	3	2828	Mk I and Mk IA
31	Retaining plate for layshaft front bearing	1	528690	Mk II and Mk III
32	Lock washer } Fixing plate and bearing to bell housing	1	528683	Mk II and Mk III
33	Nut (5/16" UNF) } Fixing plate and bearing to bell housing	3	254851	Mk II and Mk III
34	Bearing for layshaft, rear	1	55715	
	SYNCHRONISING CLUTCH, SECOND SPEED AND FIRST SPEED MAINSHAFT GEAR	1	246557	Mk I and Mk IA
35	SYNCHRONISING CLUTCH, SECOND SPEED AND FIRST SPEED MAINSHAFT GEAR	1	522373	Mk II and Mk III
36	SYNCHRONISING CLUTCH, THIRD AND FORTH SPEED	1	513521	
37	Detent spring for clutch, second speed	3	246556	
38	Detent spring for clutch, third and fourth speed	3	06395	
39	Roller bearing for mainshaft	1	06397	
40	Ball bearing for mainshaft	1	1645	
41	Circlip fixing bearing	1	09960	
42	Shaft for reverse gear	1	241589	
43	REVERSE GEAR ASSEMBLY	1	56936	
44	Bush for reverse gear	1	561954	

*Asterisk indicates a new part which has not been used on any previous model.

F599.

Plate Ref.	1 2 3 4 DESCRIPTION	Qty	Part No.	REMARKS
	GEAR CHANGE SELECTOR ASSEMBLY ..	1	528409	Mk I and Mk IA (items to 14)
1	Gear change selector lever	1	246815†	
	Gear change selector lever	1	521886§	
	Gear change selector lever	1	528411§§	
2	Retainer for selector lever	1	528354	
3	Insulator housing for gear change lever	1	246931	Cars numbered up to §§
	Insulator housing for gear change lever	1	532247§§	
4	Self-locking nut (¼" BSF) fixing insulator to lever	1	251320	
5	Spring } Fixing retainer on gear lever	3	07190	
6	Special screw } to selector housing	3	07189	
7	HOUSING ASSEMBLY FOR GEAR SELECTOR ..	1	278137†	
8	HOUSING ASSEMBLY FOR GEAR SELECTOR ..	1	532971††	
9	Insulating pad for housing	1	244043†	
10	Rubber bush } For gear selector	6	243615†	
11	Distance sleeve } housing	3	243616†	
12	Plain washer } Fixing selector	3	2219†	
13	Spring washer } housing to	3	3076†	
14	Nut (⅜" BSF) } bell housing	3	2827†	
15	Gear selector lever, lower	1	532996	Mk II and Mk III
16	Bolt (⅜" UNF x 1½" long) } Clamping lever to	1	256441	Mk II and Mk III
17	Self-locking nut } gear change shaft	1	252162	Mk II and Mk III
	SELECTOR SHAFT ASSEMBLY, FIRST AND SECOND SPEED	1	515886†	Mk I and Mk IA
	SELECTOR SHAFT ASSEMBLY, FIRST AND SECOND SPEED	1	532969††	Mk I and Mk IA
18	SELECTOR SHAFT ASSEMBLY, FIRST AND SECOND SPEED	1	533338	Mk II and Mk III
19	Interlocking pin	1	55697	
20	Peg fixing interlocking pin	1	55775	
21	Selector fork, first and second speed	1	241593	
22	Plug for second speed selector shaft	1	515888††	
23	Bolt (5/16" BSF x 1" long) fixing fork to shaft ..	1	237160	
24	Locating pin for second gear fork	1	241588	
	Selector shaft, third and fourth speed	1	241591†	Models without overdrive
	Selector shaft, third and fourth speed	1	528885††	Models without overdrive
	Selector shaft, third and fourth speed	1	273602†	Models with overdrive
	Selector shaft, third and fourth speed	1	528884††	
25	Selector shaft, third and fourth speed	1	533340	Mk II and Mk III
26	Selector fork, third and fourth speed	1	06422	
27	Bolt (5/16" BSF x 1" long) fixing fork to shaft ..	1	237160	
	Selector shaft, reverse	1	533341†	Mk I and Mk IA
	Selector shaft, reverse	1	521007††	Mk I and Mk IA
28	Selector shaft, reverse	1	533341	Mk II and Mk III
29	Selector fork, reverse	1	241594	
30	Bolt (5/16" BSF x 1" long) fixing fork to shaft ..	1	237160	
31	Sealing ring, forward selector shaft	2	272596	

* Asterisk indicates a new part which has not been used on any previous Rover model
† Cars numbered up to 625101045, 626100262, 628100109
†† Cars numbered from 625101046, 626100263, 628100110 and Mk IA, Mk II and Mk III 4-speed models
§ Cars numbered from 625101046, 626100263, 628100110 up to cars numbered 72500721, 72600112, 72800136
§§ Mk IA cars numbered from 72500722, 72600113, 72800137

F599.

Plate Ref.	1 2 3 4 DESCRIPTION		Qty	Part No.	REMARKS
32	Sealing ring, reverse selector shaft		1	272597	
33	Retaining plate for sealing rings		2	241598	
34	Spring washer	Fixing retaining plate	4	3074	
	Bolt ($\frac{1}{4}$" Whit x $\frac{1}{2}$" long)	Fixing retaining plate	2	215758	
35	Bolt ($\frac{1}{4}$" Whit x $\frac{7}{16}$" long)	Fixing retaining plate	2	215769	
36	Interlocking plunger	For selectors	2	55638	
37	Steel ball ($\frac{3}{8}$")	For selectors	3	1643	
38	Selector spring, forward		2	243465	
39	Selector spring, reverse		1	243465	
40	Retaining plate for selector spring and plug		1	248580	
41	Joint washer for retaining plate		1	248581	
42	Bolt fixing retaining plate		2	215593	
43	Plain washer	Fixing retaining plate	2	3840	
44	Spring washer	Fixing retaining plate	2	3074	
45	Plug for interlocker hole		1	248582	
46	Hinge plate for reverse stop		1	245306	
47	Set bolt ($\frac{1}{4}$" BSF x $\frac{5}{8}$" long)	Fixing hinge plate to reverse selector shaft	2	250693	
48	Spring washer	Fixing hinge plate to reverse selector shaft	2	3074	
49	Spring for reverse stop		1	231116	
50	Switch for reverse lamp, LU 31807A		1	503347	
51	Set bolt (2 BA x $\frac{3}{8}$" long)	Fixing switch to bracket	2	237119	
52	Spring washer	Fixing switch to bracket	2	3073	
53	Nut (2 BA)	Fixing switch to bracket	2	2247	
54	Bracket for reverse light switch		1	516101	
	Striker arm for reverse switch		1	243906†	
	Striker arm for reverse switch		1	521665††	
55	Striker arm for reverse switch		1	243906	Mk II and Mk III

* Asterisk indicates a new part which has not been used on any previous Rover model
† Cars numbered up to 625101045, 626100262, 628100109
†† Cars numbered from 625101046, 626100263, 628100110 and Mk IA 4-speed models

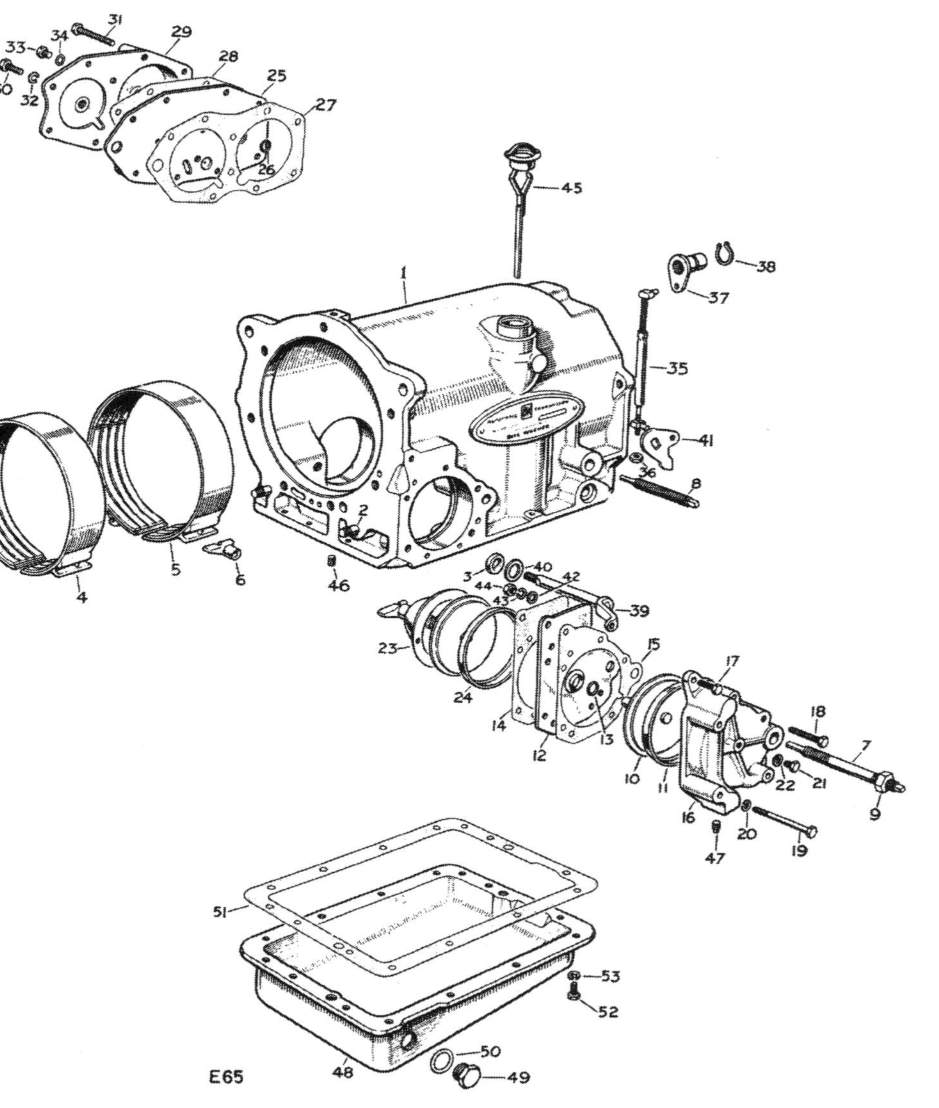

AUTOMATIC TRANSMISSION MAIN CASING, MK I, MK IA AND MK II MODELS

Plate Ref.	1 2 3 4 DESCRIPTION	Qty	Part No.	REMARKS
	AUTOMATIC TRANSMISSION COMPLETE ASSEMBLY, FIRST SPEED START	1	505788	1959–60
	AUTOMATIC TRANSMISSION COMPLETE ASSEMBLY, SECOND SPEED START, three-point engine mounting	1	522624	1961 cars numbered up to 630100840, 631100353, 633100078
	AUTOMATIC TRANSMISSION COMPLETE ASSEMBLY with bracket for mounting solenoid	1	529502	1961 cars numbered from 630100841, 631100354, 633100079 and Mk IA
	AUTOMATIC TRANSMISSION COMPLETE ASSEMBLY	1	535583	Mk II
	Switch for reverse lamp, LU 31807A	1	503347	
	Set bolt (2 BA x $\frac{3}{8}$" long) } Fixing switch to bracket	2	237119	
	Spring washer } Fixing switch to bracket	2	3073	
	Nut (2 BA) } Fixing switch to bracket	2	2247	
	Bracket for reverse light switch	1	516101	
	Striker arm for reverse switch	1	243906	
1	MAIN CASING ASSEMBLY	1	505388	
2	Stud ($\frac{7}{16}$" UNF) fixing main casing to bell housing	2	505389	
3	Oil seal for selector control shaft	1	505390	
4	Reverse brake band	1	505431	
5	Forward and low brake band	2	505430	
6	Side strut for brake band	3	505418	
7	Adjusting screw for forward and reverse brake band	2	505416	
8	Adjusting screw for low brake band	1	505415	
9	Locknut for adjusting screw	3	505417	
10	Piston complete, outer, for reverse and forward band cylinder	2	505410	
11	Rubber sealing ring for outer reverse and forward band pistons, lipped, $3\frac{1}{4}$" dia.	2	505404	
12	Plate for reverse band brake cylinder	1	505406	
13	Rubber ring for plate	1	505462	
14	Joint washer, inner } For reverse band brake cylinder plate	1	505405	
15	Joint washer, outer } For reverse band brake cylinder plate	1	505409	
16	Brake cylinder for reverse band	1	505414	
17	Set bolt ($\frac{5}{16}$" UNC x 1" long) } Fixing brake cylinder and plate to gearbox casing	1	505411	
18	Set bolt ($\frac{5}{16}$" UNC x 2" long) } Fixing brake cylinder and plate to gearbox casing	2	505413	
19	Set bolt ($\frac{5}{16}$" UNC x $2\frac{3}{4}$" long) } Fixing brake cylinder and plate to gearbox casing	3	505412	
20	Spring washer } Fixing brake cylinder and plate to gearbox casing	6	3075	
21	Plug } In reverse brake cylinder for pressure take-off ..	1	505429	
22	Copper washer } In reverse brake cylinder for pressure take-off	1	505428	
23	Piston complete for inner forward, inner reverse and low band cylinder	3	505403	
24	Rubber sealing ring for inner forward, inner reverse and low band pistons, lipped, $3\frac{1}{4}$" dia.	3	505404	
25	Plate for forward band brake cylinder	1	505425	
26	Rubber ring for plate	1	505408	
27	Joint washer, inner } For forward and low band brake cylinder plate	1	505427	
28	Joint washer, outer } For forward and low band brake cylinder plate	1	505423	
29	Brake cylinder for forward band	1	505422	

* Asterisk indicates a new part which has not been used on any previous Rover model

E65

Plate Ref.	1 2 3 4 DESCRIPTION	Qty	Part No.	REMARKS
30	Set bolt ($\frac{5}{16}$" UNC x 1" long) } Fixing brake cylinder and plate to gearbox casing	4	505411	
31	Set bolt ($\frac{5}{16}$" UNC x 2" long) } Fixing brake cylinder and plate to gearbox casing	5	505413	
32	Spring washer } Fixing brake cylinder and plate to gearbox casing	9	3075	
33	Plug } In forward and low brake cylinder for pressure take-off	2	505429	
34	Copper washer } In forward and low brake cylinder for pressure take-off	2	505428	
35	PARKING BRAKE ACTUATING ROD COMPLETE	1	505398	
36	Nut ($\frac{1}{4}$" UNF) fixing bottom swivel to actuating rod	1	505399	
37	Toggle arm lever complete	1	505401	
38	Circlip fixing arm to toggle shaft	1	505402	
39	Selector control shaft	1	505392†	
	Selector control shaft	1	541923††	
40	Plain washer for control shaft	1	505393	
41	Selector lever for control shaft	1	505394	
42	Plain washer } Fixing selector lever to control shaft	1	505395	
43	Spring washer } Fixing selector lever to control shaft	1	3075	
44	Nut ($\frac{5}{16}$" UNF) } Fixing selector lever to control shaft	1	505397	
45	Oil level rod for gearbox	1	279601	
46	Allen screw for direct drive clutch and front pump pressure take-off	2	507020	
47	Special plug for converter pressure take-off ..	2	507021	
48	FLUID SUMP	1	505273	
49	Drain plug for sump	1	505277	
50	Joint washer for plug	1	505274	
51	Joint washer for sump	1	505278	
52	Set bolt ($\frac{5}{16}$" UNC x $\frac{5}{8}$" long) } Fixing sump to gearbox casing	14	505279	
53	Spring washer } Fixing sump to gearbox casing	14	3075	

* Asterisk indicates a new part which has not been used on any previous Rover model
† Mk I, Mk IA and Mk II cars numbered up to 77500123, 77600051, 77800002
†† Mk II cars numbered from 77500124, 77600052, 77800003

E66

AUTOMATIC TRANSMISSION SHAFTS AND GEARS, MK I, MK IA AND MK II 3 LITRE MODELS

Plate Ref.	1 2 3 4 Description		Qty.	Part No.	Remarks
1	Mainshaft complete		1	505312	
2	Oil transfer tube for mainshaft		1	279260	
	FRONT OIL PUMP AND COLLECTOR RING ASSEMBLY		1	505300	
3	Collector ring complete		1	505301	
4	FRONT PUMP BODY AND GEARS		1	505306	
5	Oil seal		1	505307	
6	Rubber sealing ring, large	For collector ring	1	505308	
7	Rubber sealing ring, small	For collector ring	1	505462	
8	Set bolt (¼" UNC x 1¾" long)	Fixing collector ring to oil pump body	1	505302	
9	Set bolt(5/16" UNC x⅞"long)	Fixing collector ring to oil pump body	2	505304	
10	Set bolt (5/16" UNC x 1½"long)	Fixing collector ring to oil pump body	2	505305	
11	Spring washer	Fixing collector ring to oil pump body	1	3074	
12	Spring washer	Fixing collector ring to oil pump body	4	3075	
13	Joint washer, collector ring to main casing		1	505282	
14	Set bolt (5/16" UNC x ⅞"long)	Fixing collector ring to main casing	4	505283	
15	Set bolt (5/16" UNC x 1¼"long)	Fixing collector ring to main casing	3	505284	
16	Spring washer	Fixing collector ring to main casing	7	3075	
17	Thrust washer, .052" - .054"	For rear of collector ring	As reqd	505280	
	Thrust washer, .061" - .063"	For rear of collector ring	As reqd	505524	
	Thrust washer, .072" - .074"	For rear of collector ring	As reqd	505525	
	Thrust washer, .082" - .084"	For rear of collector ring	As reqd	505526	
18	Front ring gear complete		1	505362	
19	Piston ring for front ring gear, 1⅜" dia.		3	505320	
20	Thrust washer for rear of front ring gear		1	505359	
21	Front planet carrier complete		1	505360	
22	Washer for front sun gear		1	505356	
23	Reverse brake drum		1	505357	
24	Snap ring for brake drum		1	505358	
25	Front sun gear and low drum complete		1	505342	
26	Piston for multi-disc clutch		1	505345	
27	Rubber sealing ring for piston, lipped, 3" dia.		1	505343	
28	Piston ring for clutch		1	505344	
29	Retractor plate complete for clutch		1	505347	
30	Spacer plate for clutch		1	505348	Mk I and Mk IA
31	Steel clutch plate		4	505349	Mk I
32	Bronze clutch plate		3	505351@	Mk I
33	Spring for clutch		12	505352	Mk I
34	Retainer for clutch		1	505354	Mk I
	Steel clutch plate		5	505349@@	Mk I and Mk IA
	Bronze clutch plate		4	505351	Mk I and Mk IA
	Spring for clutch		21	505352	Mk I and Mk IA
	Steel clutch plate		4	505349	Mk II
	Friction clutch plate		3	538386	Mk II
	Spring for clutch		27	505352	Mk II
	Retainer for clutch (alloy)		1	505354	Mk II
35	Circlip fixing clutch		1	505355	

@ Automatic transmissions numbered up to R5B 2401

@@ Automatic transmissions numbered from R5B 2402 onwards

*Asterisk indicates a new part which has not been used on any previous model.

AUTOMATIC TRANSMISSION SHAFTS AND GEARS, MK1, MK1A and MK11 3 LITRE

E66

COLLINS-JONES

AUTOMATIC TRANSMISSION SHAFTS AND GEARS, MK I, MK IA AND MK II MODELS

Plate Ref.	1 2 3 4 DESCRIPTION	Qty	Part No.	REMARKS
36	REAR RING GEAR COMPLETE	1	505322	
37	Rubber sealing ring, small	1	505323	
38	Circlip for front planet carrier	1	505341	
39	Piston ring for rear ring gear, 2$\frac{3}{8}$" dia.	1	505339	
40	Rubber sealing ring, large	1	505340	
41	Thrust washer for rear planet carrier	1	505321	
42	Oil ring retainer	1	505318	
43	Circlip fixing oil ring retainer	1	505319	
44	Piston ring for retainer, 1$\frac{3}{8}$" dia.	1	505320	
45	Rear planet carrier complete	1	505313	
46	Washer for rear planet carrier	1	505317	
47	Thrust washer for rear sun gear	1	505316	
48	Piston ring on mainshaft, 1" dia.	3	505364	
49	Rear sun gear complete	1	505365	
50	Washer } For rear	1	505380	
51	Spacer } sun gear	1	505381	
52	One-way clutch lock-up plate complete	1	505367	
53	Roller for lock-up plate	2	505369	
54	Circlip fixing lock-up plate	1	505370	
55	One-way clutch complete, narrow	1	505368	
56	Forward brake drum complete	1	505373	
57	One-way clutch complete, wide	1	505374	
58	Circlip for forward brake drum	2	505375	
59	Spacer for one-way clutch	1	505378	
60	Pin fixing spacer	1	505377	
61	Spacer for mainshaft rear bearing	1	505382	
62	Rear bearing for mainshaft	1	505383	
63	Circlip fixing bearing	1	505285	
64	Spacer for parking brake and governor drive gear	1	505384	
65	Parking brake gear	1	505385	
66	Governor drive gear	1	505386	
67	Spacer for propeller shaft flange	1	505286	
68	FLANGE FOR PROPELLER SHAFT, square	1	505756	} Mk I and
69	Special bolt in flange	4	507023	} Mk IA
	FLANGE FOR PROPELLER SHAFT	1	535799	} Mk II
	Special bolt in flange	4	237341	} Mk II
70	Special washer	1	505299	
71	Self-locking nut ($\frac{3}{4}$" UNF)	1	505387	

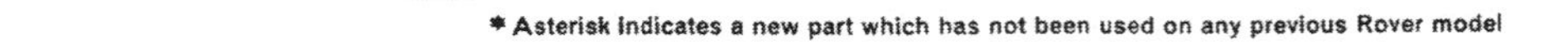

* Asterisk indicates a new part which has not been used on any previous Rover model

VALVE BLOCK, AUTOMATIC TRANSMISSION, MK I, MK IA AND MK II MODELS

D216

COLLINS-JONES

VALVE BLOCK, AUTOMATIC TRANSMISSION, MK I, MK IA AND MK II MODELS

Plate Ref.	1 2 3 4 DESCRIPTION	Qty	Part No.	REMARKS
	VALVE BLOCK COMPLETE ASSEMBLY	1	505433†	
	VALVE BLOCK COMPLETE ASSEMBLY (with modified shuttle valve)	1	530099††	
1	Manifold plate	1	505434	
	SELECTOR VALVE FOR VALVE BLOCK ASSEMBLY	1	542088	
2	Spring } For selector valve detent	1	537285	
3	Ball, $\frac{1}{4}$" dia. } For selector valve detent	1	505437	
4	Spring } For selector valve detent	1	505436	Mk I
	Spring } For selector valve detent	1	537285	Mk IA and Mk II
5	Ball, $\frac{1}{4}$" dia. } For selector valve detent	1	505437	
6	Joint washer, manifold to base plate	1	505438	
7	Base plate	1	505441	
8	RELIEF VALVE BODY COMPLETE	1	505442	
9	Spring } For main relief valve	1	505443	
10	Retainer } For main relief valve	1	505445	
11	Retainer for main relief valve plug	1	505444	
12	Reverse interlock valve spring retainer	1	505447	
13	Accumulator valve spring retainer	1	505446	
14	Joint washer, base plate to relief valve body	1	505448	
15	Ball, $\frac{3}{8}$" dia.	2	505449	
16	CONVERTER VALVE BODY COMPLETE	1	505450†	
	CONVERTER VALVE BODY COMPLETE (with modified shuttle valve)	1	529390††	
17	Retainer for shuttle valve sleeve	1	505453	
	Shuttle valve	1	529405††	
18	Joint washer, manifold to converter valve body	1	505455	
19	Set bolt ($\frac{1}{4}$" UNC x 2$\frac{3}{8}$" long) } Fixing valve block together	8	505456	
20	Spring washer } Fixing valve block together	8	3074	
21	Set bolt ($\frac{5}{16}$" UNC x 1$\frac{1}{8}$" long) } Fixing valve block to gearbox casing	7	505275	
22	Spring washer } Fixing valve block to gearbox casing	7	3075	
23	Oil filter	1	505272	
24	Retainer for oil filter	1	505457	
25	Screw ($\frac{1}{4}$" UNC x $\frac{3}{8}$" long) } Fixing retainer to converter valve body	1	505759	
26	Shakeproof washer } Fixing retainer to converter valve body	1	505458	

* Asterisk indicates a new part which has not been used on any previous Rover model
† Cars numbered up to 630100840, 631100353, 633100078
†† Cars numbered from 630100841, 631100354, 633100079 onwards

1227

AUTOMATIC TRANSMISSION EXTENSION CASING MK1, MK1A AND MK 11 MODELS

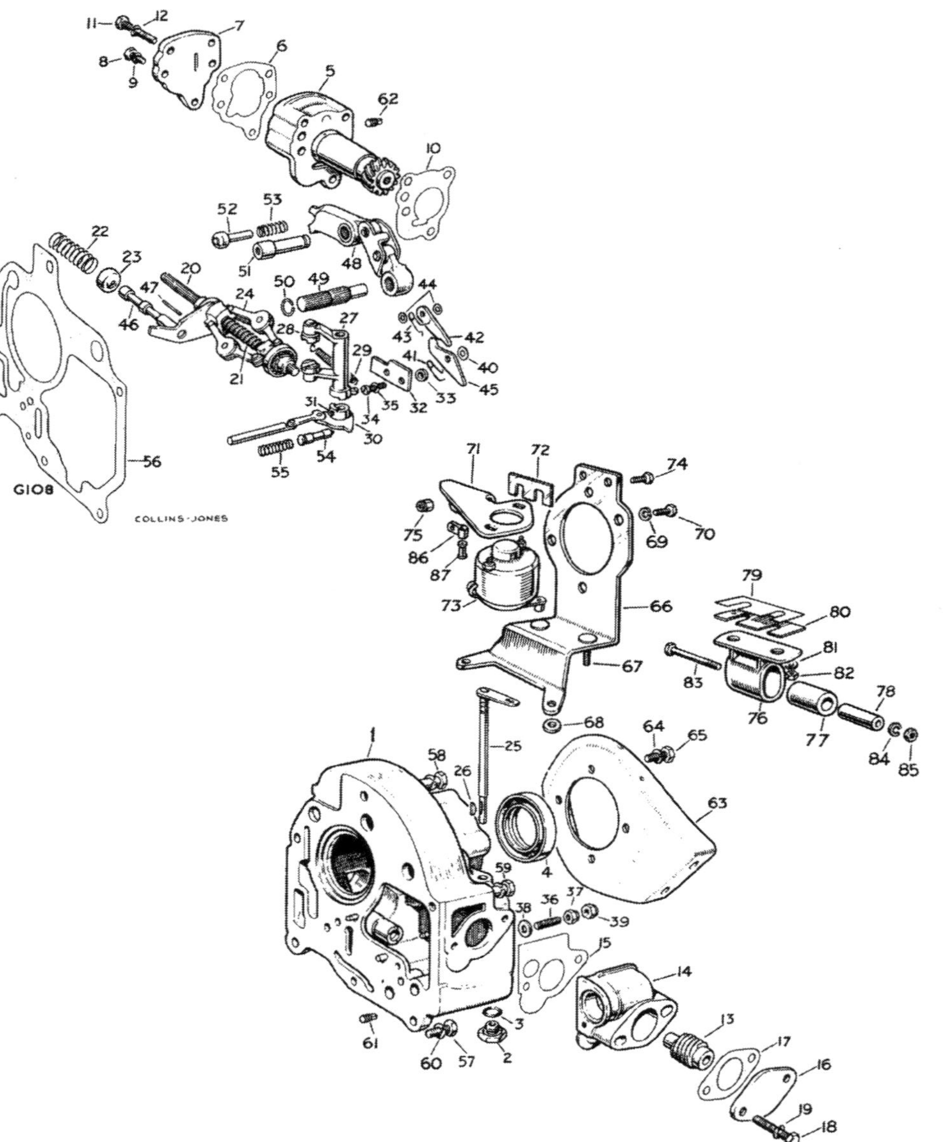

1228

AUTOMATIC TRANSMISSION EXTENSION CASING, MK I, MK IA AND MK II MODELS

Plate Ref.	1 2 3 4 DESCRIPTION	Qty	Part No.	REMARKS
1	EXTENSION CASING ASSEMBLY	1	505460	1959–60 (first speed start)
	EXTENSION CASING ASSEMBLY	1	522537	1961 onwards (second speed start)
2	Bush } For governor control shaft	1	505461	
3	Rubber ring } For governor control shaft	1	505462	
4	Oil seal for mainshaft, rear	1	505463	
5	REAR OIL PUMP COMPLETE	1	505490	
6	Joint washer for oil pump cover, 0.002"	As reqd	505492	
	Joint washer for oil pump cover, 0.003"	As reqd	505493	
	Joint washer for oil pump cover, 0.005"	As reqd	505494	
	Joint washer for oil pump cover, 0.006"	As reqd	505495	
	Joint washer for oil pump cover, 0.008"	As reqd	505496	
7	Cover for rear oil pump	1	505497	
8	Set bolt ($\frac{1}{4}$" UNC x $\frac{3}{4}$" long) } Fixing cover to pump	2	505498	
9	Spring washer } Fixing cover to pump	2	3074	
10	Joint washer, rear pump to extension casing ..	1	505499	
11	Set bolt ($\frac{5}{16}$" UNC x $2\frac{1}{4}$" long) } Fixing oil pump to extension casing	3	505500	
12	Spring washer } Fixing oil pump to extension casing	3	3075	
13	Gear for speedometer drive	1	505502	
14	Housing for speedometer gear	1	505501	
15	Joint washer, housing to extension casing ..	1	505504	
16	Cover plate for speedometer gear housing ..	1	505522	
17	Joint washer for cover	1	505521	
18	Set bolt ($\frac{5}{16}$" UNC x 2" long) } Fixing cover and housing to extension casing	2	505413	
19	Spring washer } Fixing cover and housing to extension casing	2	3075	
20	Shaft for governor	1	505503	
21	Spring for governor	1	505518	Mk I and Mk IA
	Spring for governor	1	538387	Mk II
22	Spring for governor booster valve	1	505520	1959–60
	Spring for governor booster valve	1	519606	1961 onwards
23	Retainer for booster valve spring	1	505519	
24	Governor complete	1	505515	
25	Governor adjusting shaft complete	1	530100	
26	Key for governor shaft	1	505757	
27	Governor adjusting fork	1	505477	
28	Yoke shoe for governor adjusting fork	2	505479	
29	Return spring for governor adjusting fork ..	1	505480	
30	Detent cam with link and plunger	1	505478	1959–60
	Detent cam with link and plunger	1	519610	1961 Mk I and Mk IA
	Detent cam with link and plunger	1	538388	Mk II
31	Allen screw fixing detent cam to governor shaft	1	505761	
32	Stop for governor	1	505472	
33	Spacer for governor stop	1	505475	1959–60
34	Screw ($\frac{1}{4}$" UNC x $\frac{5}{8}$" long) } For governor stop	1	505760	
35	Spring washer } For governor stop	1	3074	
36	Special screw ($\frac{1}{4}$" UNC) } For governor cam stop	1	505762	
37	Self-locking nut ($\frac{1}{4}$" UNC) } For governor cam stop	1	505523	
38	Copper washer } For governor cam stop	1	505763	
39	Special nut ($\frac{1}{4}$" UNC) } For governor cam stop	1	505764	

* Asterisk indicates a new part which has not been used on any previous Rover model

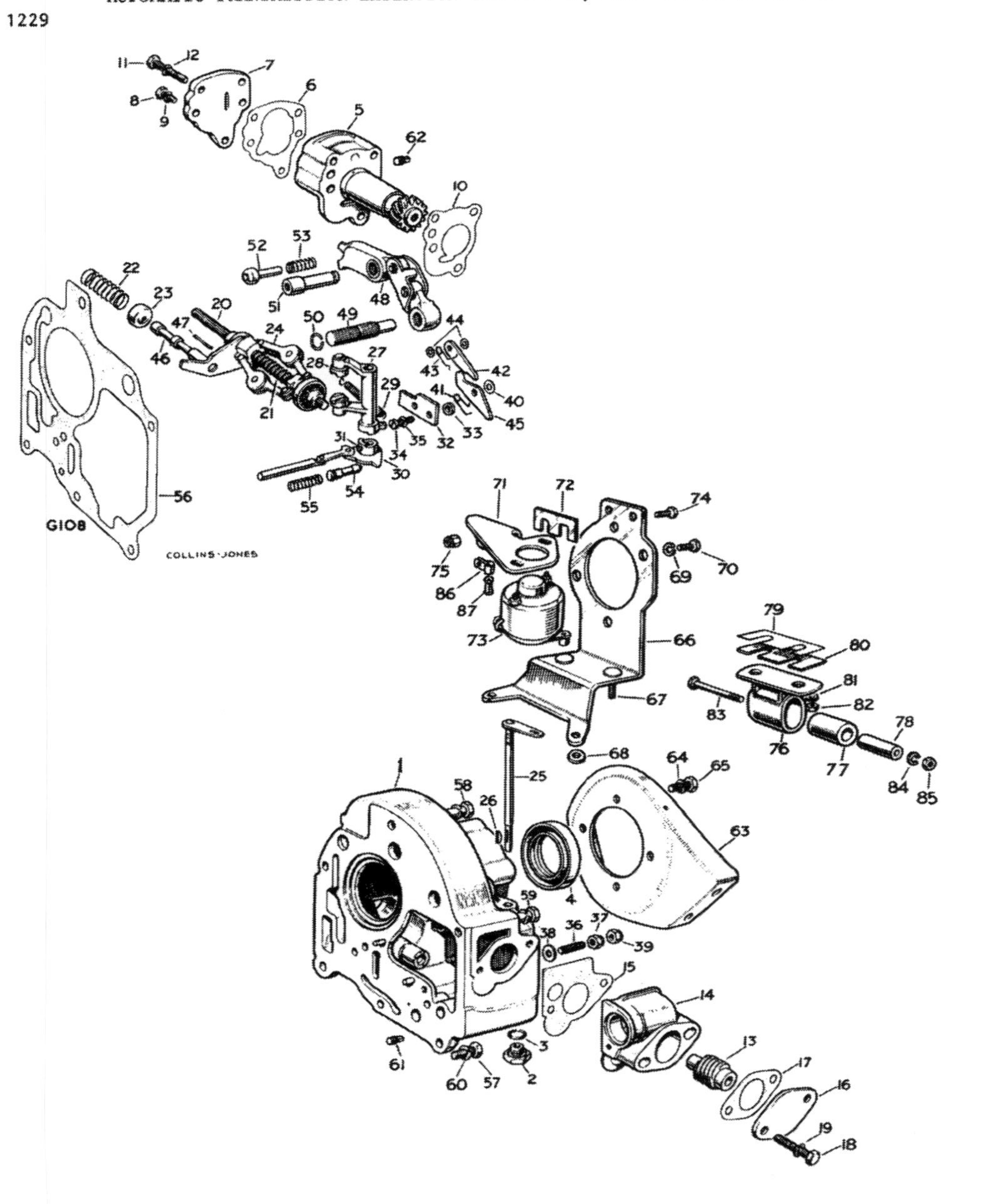

AUTOMATIC TRANSMISSION EXTENSION CASING, MK I, MK IA AND MK II MODELS

Plate Ref.	1 2 3 4 DESCRIPTION	Qty	Part No.	REMARKS
40	Washer for rocker arm control	1	505474	1959–60
41	Spring for rocker arm control	1	507024	1959–60
42	Pawl for second speed drive	1	505464	1959–60
43	Spring for second speed drive pawl	1	507247	1959–60
44	Plain washer for pawl	2	505468	1959–60
45	Rocker arm	1	505471	1959–60
46	Governor control valve complete	1	505481	
47	Detent piston for governor control valve	1	505482	
48	Pawl and toggle for parking brake	1	505483	
49	Shaft for parking brake toggle	1	505485	
50	Circlip fixing pawl and toggle to shaft	1	505488	
51	Pawl pivot pin	1	505489	
52	Parking brake interlock piston	1	505291	
53	Spring for parking brake interlock piston ..	1	505292	
54	Plunger } For accelerator detent	1	505288	1961 onwards
55	Spring } For accelerator detent	1	505289	1961 onwards
56	Joint washer, extension to main casing	1	505290	
57	Set bolt (⅜" UNC x 1" long) } Fixing extension casing to main casing	4	505293	
58	Set bolt (⅜" UNC x 2¼" long) } Fixing extension casing to main casing	1	505295	
59	Set bolt (⅜" UNC x 3" long) } Fixing extension casing to main casing	1	505296	
60	Spring washer } Fixing extension casing to main casing	6	3076	
61	Special plug for multi-disc clutch pressure take-off	1	507021	
62	Special plug for rear pump pressure take-off ..	1	507022	
63	Rear mounting bracket for gearbox	1	500705†	
64	Spring washer } Fixing bracket to extension casing	4	3076†	
65	Set bolt (⅜" UNC x ⅞" long) } Fixing bracket to extension casing	4	253046†	
66	Reaction bracket complete	1	538434††	
	Bolt (5/16" UNF x 1¼" long) for reaction bracket ..	1	256420	
67	Bolt (5/16" UNF x 1⅛" long) in reaction bracket ..	2	255029††	
68	Plain washer, bracket to sump stiffener	2	2220††	
69	Spring washer } Fixing bracket to gearbox	4	3076††	
70	Set bolt (⅜" UNC x ⅞" long) } Fixing bracket to gearbox	4	253046††	
71	Solenoid support bracket	1	521896	Mk IA and Mk II
72	Shim, solenoid to bracket .. .. .. As reqd		305232	Mk IA and Mk II
73	Solenoid for intermediate gear hold	1	526369	Mk IA and Mk II
74	Bolt (¼" UNF x ¾" long) } Fixing support bracket to reaction bracket	2	255207	
75	Self-locking nut (¼" UNF) } Fixing support bracket to reaction bracket	2	252160	
76	Stabiliser bracket at gearbox	1	520915††	
77	Bush for stabiliser	1	521349††	
78	Tube for stabiliser	1	521350††	
79	Shim, 22 SWG } Fixing stabiliser bracket to reaction bracket — As reqd	4	520917††	
80	Shim, 16 SWG } Fixing stabiliser bracket to reaction bracket — As reqd	4	528199††	
81	Spring washer } Fixing stabiliser bracket to reaction bracket	2	3075††	
82	Nut (5/16" UNF) } Fixing stabiliser bracket to reaction bracket	2	254821††	
83	Bolt (5/16" UNF x 2¾" long) } Fixing stabiliser to sub-frame bracket	1	256029††	
84	Spring washer } Fixing stabiliser to sub-frame bracket	1	3075††	
85	Nut (5/16" UNF) } Fixing stabiliser to sub-frame bracket	1	254811††	
86	Cable clip at solenoid support bracket	1	275037	Mk IA and Mk II
87	Grommet for clip	1	214229	Mk IA and Mk II

* Asterisk indicates a new part which has not been used on any previous Rover model
† Cars numbered up to 630100840, 631100363, 633100078 Mk I models
†† Cars numbered from 630100841, 631100364, 633100079 onwards

AUTOMATIC TRANSMISSION MAIN CASING, MK III MODELS

231

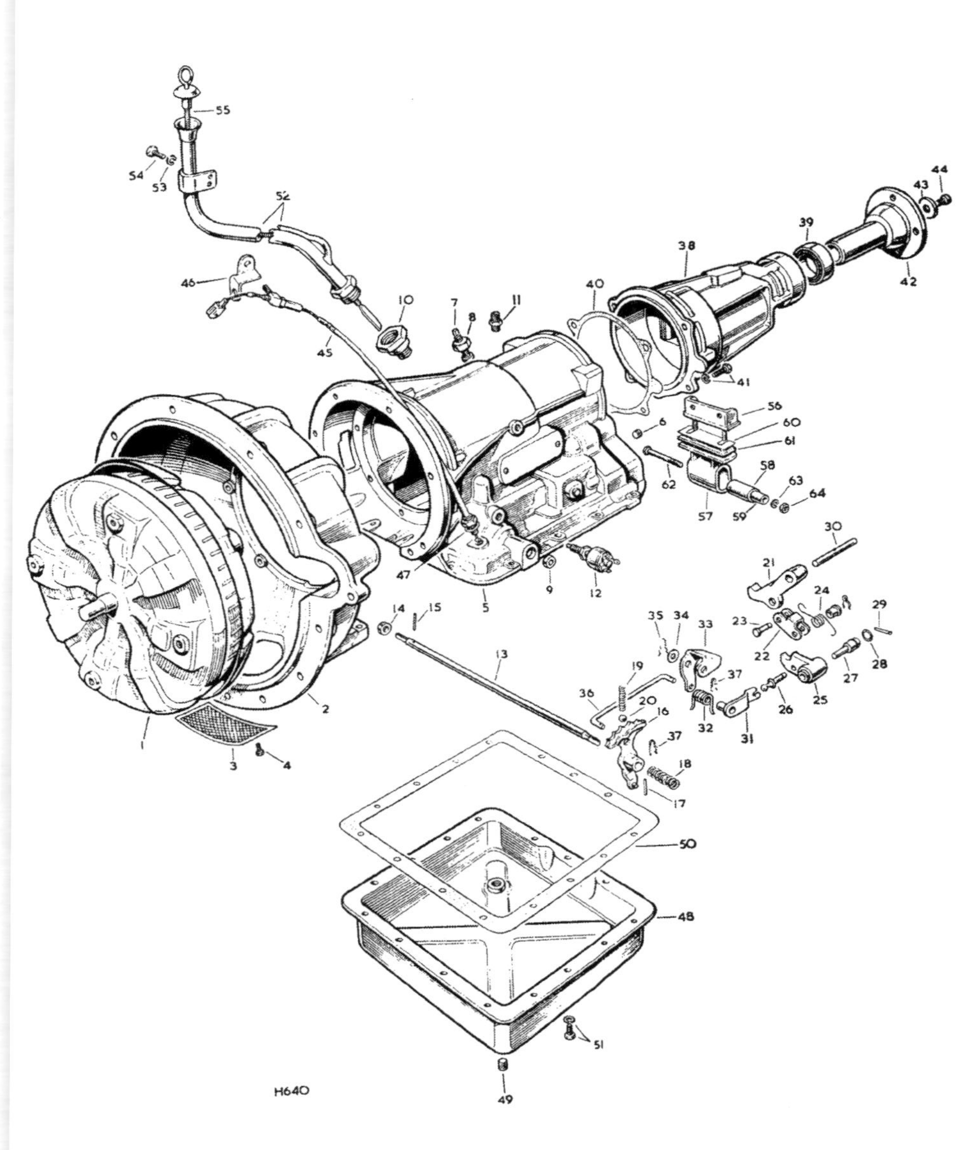

Plate Ref.	1 2 3 4 Description	Qty.	Part No.	Remarks
	AUTOMATIC TRANSMISSION AND TORQUE CONVERTER	1	601367	
1	Torque converter assembly	1	601321	
2	HOUSING FOR TORQUE CONVERTER	1	601392	
3	Stone guard	1	605758	
4	Special bolt fixing stone guard	2	601004	
	Special bolt and washer fixing housing to case	6	601003	
5	MAIN CASE ASSEMBLY	1	601390	
6	Pipe plug	1	601006	
7	Screw, rear servo adjusting	1	601007	
8	Nut, locking adjusting screw	1	601008	
9	Oil seal for manual control shift	2	601009	
10	Adaptor for tube connector	1	601369	
11	Connector, tube outlet, oil cooler	2	601368	
12	INHIBITOR SWITCH	1	601010	
	Locknut for switch	1	601925	
	Shaft and lever kit	1	601744	Transmissions numbered up to 3 EU 2542
13	Shaft for manual valve lever	1	601375	Transmissions numbered from 3 EU 2543
14	Collar for shaft	1	601145	
15	Roll pin for collar	1	601147	
16	Lever for manual valve detent	1	601772	Transmission numbered from 3 EU 2543
17	Roll pin for lever	1	601144	
18	Spring for lever	1	601143	
19	Spring for manual valve detent	1	601906	
20	Ball (⅜" dia.) for manual valve detent	1	601148	
21	Pawl for parking brake	1	601085	
22	Link for parking brake	1	601086	
23	Pin for toggle link, pawl end	1	601087	
24	Spring, parking brake release	1	601088	
25	Toggle lever	1	601089	
26	Pin and spring for toggle lever, ball-ended	1	601695	
27	Pin for toggle lever, shouldered	1	601093	
28	'O' ring for toggle lever	1	601094	
29	Roll pin for toggle lever pin	1	601095	
30	Anchor pin for pawl	1	601096	
31	Toggle left lever, forked	1	601097	
32	Spring for lift lever	1	601098	
33	Torsion lever for parking brake	1	601099	
34	Plain washer for torsion lever	1	601100	
35	Retaining spring for torsion lever	1	601101	
36	Link for parking brake	1	601102	
37	Clip for linkage rod	2	601103	
38	EXTENSION HOUSING ASSEMBLY	1	601380	
39	Oil seal for housing	1	601160	
40	Gasket for extension casing	1	601161	
41	Special bolt and washer fixing extension housing to case	4	601426	
42	Coupling flange	1	601381	
43	Plain washer for flange	1	601163	
44	Special screw fixing coupling flange	1	601164	
45	Downshift cable complete	1	601374	
46	Bracket for downshift cable	1	553812	
47	Sealing washer for cable	1	601946	
48	Oil pan	1	601376	

*Asterisk indicates a new part which has not been used on any previous model.

AUTOMATIC TRANSMISSION MAIN CASING, MK III MODELS

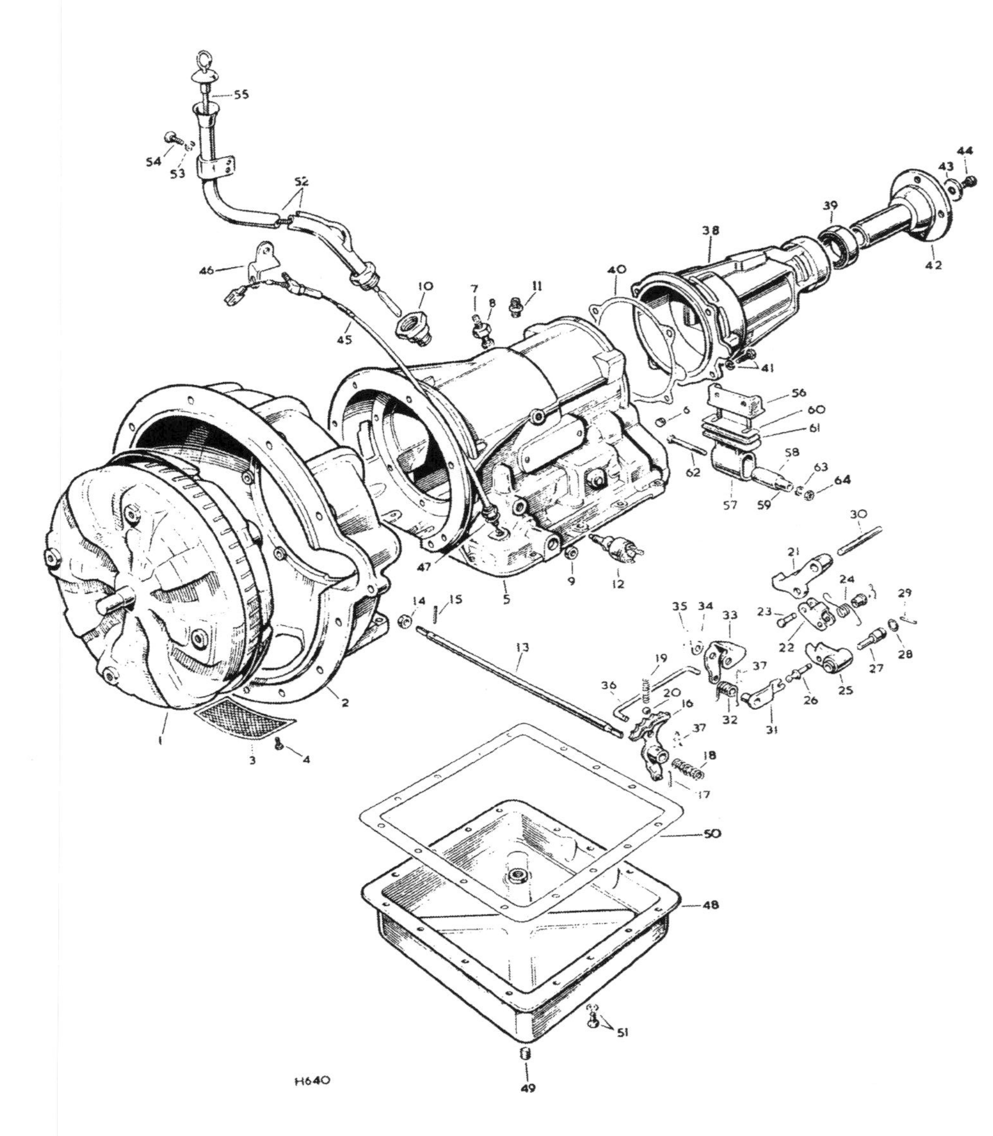

AUTOMATIC TRANSMISSION MAIN CASING, MK III MODELS

Plate Ref.	1 2 3 4 DESCRIPTION		Qty	Part No.	REMARKS
49	Drain plug		1	601377	
40	Gasket for oil pan		1	601150	
41	Special bolt and washer fixing oil pan		15	601151	
52	Oil filler tube		1	553458	
53	Spring washer	Fixing oil filler tube to inlet manifold	2	3074	
54	Set bolt ($\frac{1}{4}$" UNC x $\frac{1}{2}$" long)		2	253004	
55	Dipstick for gearbox		1	553214	
56	Mounting bracket for gearbox		1	553097	
	Spring washer	Fixing bracket to gearbox extension housing	2	3076	
	Set bolt ($\frac{3}{8}$" UNC x 1" long)		2	253047	
57	Stabiliser bracket for gearbox		1	520915	
58	Bush for stabiliser bracket		1	521349	
59	Tube for stabiliser bracket		1	521350	
60	Shim, 22 SWG	Fixing stabiliser bracket to mounting bracket	As reqd 4	520917	
61	Shim, 16 SWG		As reqd 4	528199	
	Spring washer		2	3075	
	Nut ($\frac{5}{16}$" UNF)		2	254811	
62	Bolt ($\frac{5}{16}$" UNF x $2\frac{3}{4}$" long)	Fixing stabiliser bracket to sub-frame bracket	1	256029	
63	Spring washer		1	3075	
64	Nut ($\frac{5}{16}$" UNF)		1	254811	
	Plain washer	Fixing automatic transmission to engine assembly	8	2210	
	Nut ($\frac{3}{8}$" UNF)		12	254812	
	Plain washer		1	2550	
	Nut ($\frac{5}{16}$" UNF)		1	254811	

* Asterisk indicates a new part which has not been used on any previous Rover model

Plate Ref.	1 2 3 4 DESCRIPTION	Qty	Part No.	REMARKS
1	VALVE BODIES ASSEMBLY	1	601372	
2	Check valve, front pump	1	601113	
3	Check valve, rear pump	1	601114	
4	Ball (¼" dia.)	1	601157	
5	Special screw and washer fixing end plate to valve body lower	3	601117	
6	Special screw and washer fixing valve body upper to valve body lower	5	601121	
7	Special screw and washer } Fixing valve body upper to valve body lower	2	601014	
8	Special screw and washer } Fixing valve body upper to valve body lower	1	601122	
9	Special screw and washer fixing plate to valve body	2	601121	
10	Special screw and washer fixing end plates to upper valve bodies	6	601129	
11	Special screw and washer fixing oil tube plate to valve body lower	2	601123	
12	Special screw and washer fixing oil tube plate to valve body lower	6	601124	
13	Cam assembly for downshift valve	1	601125	
14	Special screw and washer fixing downshift valve cam assembly	2	601126	
15	Oil strainer for front pump	1	601127	
16	Oil strainer for rear pump	1	601128	
17	Special screw and washer fixing front pump strainer	4	601129	
18	Special screw and washer fixing rear pump strainer	2	601130	
19	Special bolt and washer, short, fixing valve body	1	601131	
20	Special bolt and washer, long, fixing valve body	2	601132	
21	Tube, front servo apply	1	601133	
22	Tube, front servo release	1	601373	
23	Tube, rear clutch	1	601135	
24	Tube, rear servo	1	601139	
25	Tube, front pump outlet	1	601136	
26	Tube, front pump inlet	1	601137	
27	Tube, converter out and oil cooler in	1	601138	
28	Tube, converter inlet	1	601140	

* Asterisk indicates a new part which has not been used on any previous Rover model

FRONT PUMP AND FRONT CLUTCH, AUTOMATIC TRANSMISSION, MK III MODELS

1237

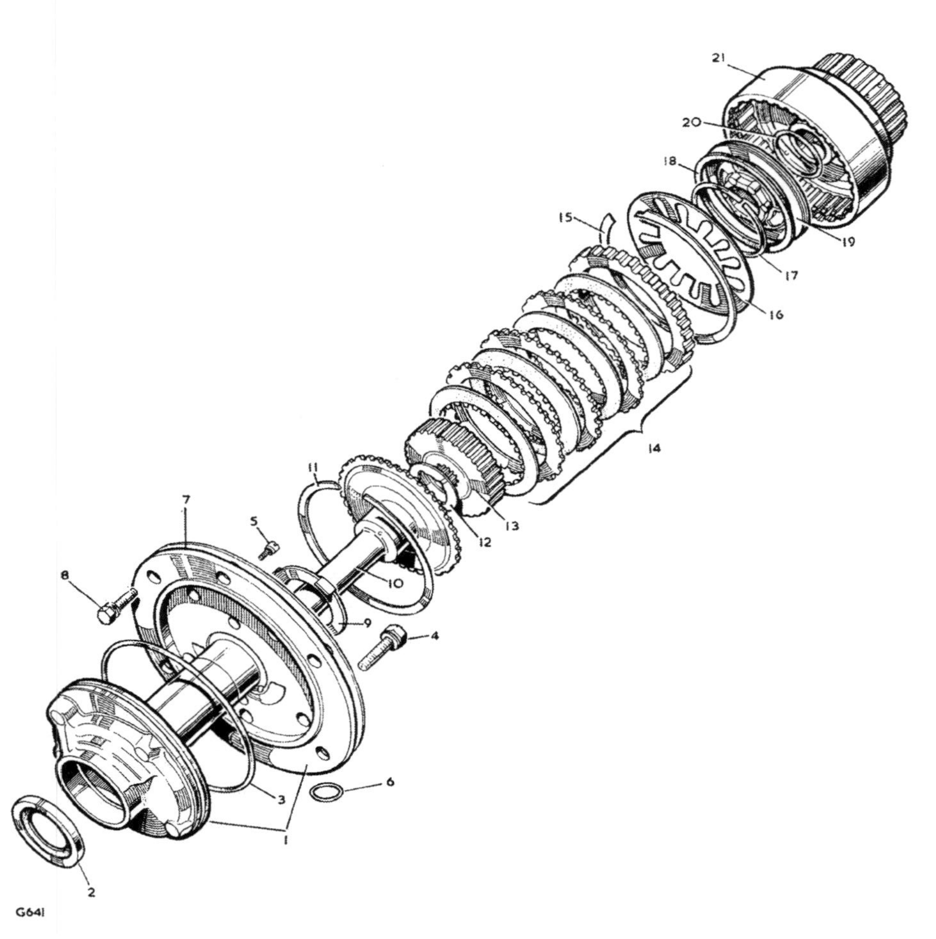

Plate Ref.	1 2 3 4 Description	Qty.	Part No.	Remarks
1	FRONT PUMP ASSEMBLY	1	601297	
2	Seal for pump	1	601012	
3	'O' ring for pump	1	601013	
4	Special bolt and washer (7/8" long)	5	601003	
5	Special screw and washer (3/8" long) fixing pump halves together	1	601014	
6	'O' ring for pump inlet tube	1	601015	
7	Gasket pump to main case	1	601016	
8	Special bolt and washer fixing pump to case	6	601003	
9	Thrust washer	1	601386@	
10	Input shaft complete	1	601028	
11	Snap ring for input shaft	1	601025	
12	Thrust washer	1	601386@	
13	Hub for front clutch	1	601026	
14	Clutch plates(service kit)for front clutch	1	601165	
15	Snap ring for front clutch	1	601025	
16	Diaphragm spring and snap ring for front clutch	1	605973	
17	Bearing ring for clutch spring	1	601022	
18	Sealing ring, large, for piston	1	601023	
19	Piston and reed assembly for front clutch	1	601019	
20	'O' ring, small, for piston	1	601021	
	Cylinder for front clutch and forward sun gear (modification kit)	1	605974*	Transmissions numbered up to 3 EU 5968
21	Cylinder for front clutch	1	605981*	Transmissions numbered from 3 EU 5969

@ Part number 601386 represents a Service kit of all thrust washers used in the transmission
Individual thrust washers are not serviced separately

*Astorisk indicates a new part which has not been used on any previous model.

REAR CLUTCH, REAR PUMP, SERVOS AND GOVERNOR, AUTOMATIC TRANSMISSION, MK III MODELS

1239

G 642

1240

Plate Ref.	1 2 3 4 Description	Qty.	Part No.	Remarks
1	Piston and reed assembly, rear clutch	1	601030	
2	'O' ring, small, for piston rear clutch	1	601029	
3	Sealing ring large for piston	1	601031	
4	Spring for rear clutch	1	601032	
5	Seat for spring	1	601033	
6	Snap ring	1	601034	
7	Clutch plates (service kit)	1	601427	
8	Snap ring, large	1	601025	
9	Thrust washer	1	601386@	
10	Front drum assembly	1	601429	
11	Oil seal ring	3	601035	
12	Forward sun gear assembly	1	605974	Transmissions numbered up to 3 EU 5968
	Forward Sun Gear Assembly	1	605980	Transmission numbered from 3 EU 5969
13	Oil seal ring, nylon	1	601039	
14	Oil seal ring for forward sun gear	2	601041	
15	Thrust needle bearing	1	601386@	
16	Thrust needle bearing and plate	1	601386@	
17	Centre support assembly	1	601042	
18	Special bolt fixing centre support	2	601043	
19	Special lock washer	2	601044	
20	One-way clutch assembly	1	601046	
21	PLANET GEARS AND REAR DRUM ASSEMBLY	1	601432	
22	Outer race assembly for one-way clutch	1	601045	
23	Snap ring for one-way clutch	1	601025	
24	Thrust needle bearing and plate	1	601386@	
25	Ring gear	1	601053	
26	Driven shaft assembly	1	601370	
27	Thrust washer for driven shaft	1	601386@	
28	Snap ring, .055"] Fixing ring	1	601054	] Alternatives
	Snap ring, .057"] gear to driven shaft	1	601055	
29	Oil sealing ring for driven shaft	3	601060	
30	REAR PUMP ASSEMBLY	1	601428	
31	Plate for rear pump	1	601048	
32	Key for rear pump drive	1	601049	
33	Special screw and shakeproof washer	1	601051	
34	Special screw and washer fixing rear pump to case	5	601050	
35	GOVERNOR ASSEMBLY	1	601378	
36	Spring for governor	1	601154	
	Retainer for governor spring	1	605972	
37	Special lock screw fixing cover plate	2	601155	
38	Special screw and washer fixing governor body	2	601156	

@ Part number 601386 represents a Service kit of all thrust washers used in the transmission
Individual thrust washers are not serviced separately

*Asterisk indicates a new part which has not been used on any previous model.

REAR CLUTCH, REAR PUMP, SERVOS AND GOVERNOR, AUTOMATIC TRANSMISSION, MK III MODELS

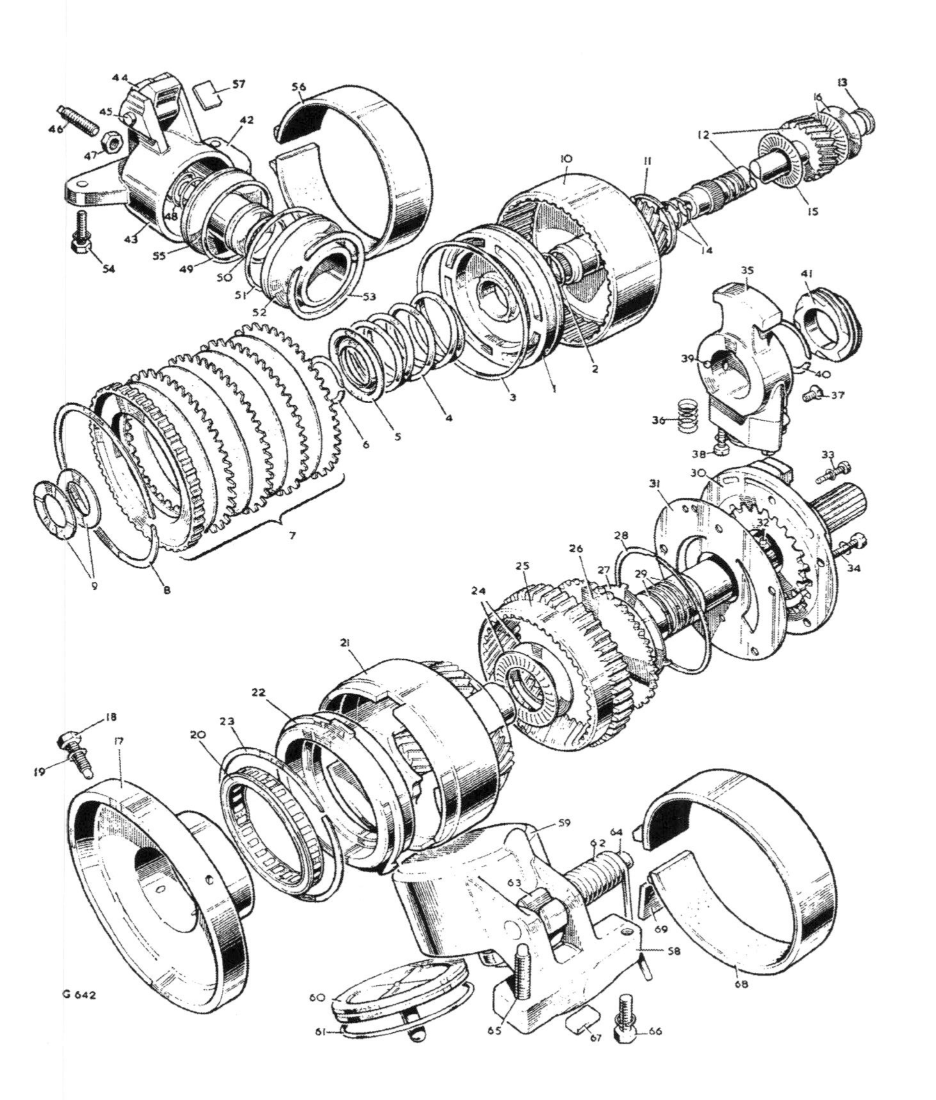

Plate Ref.	1 2 3 4 Description	Qty.	Part No.	Remarks
39	Steel ball (¼" dia.)	1	601157	
40	Snap ring for governor	1	601034	
41	Gear for speedo drive	1	601379	
42	SERVO ASSEMBLY, FRONT	1	601385	
43	Body for front servo	1	601063	
44	Lever for front servo	1	601062	
45	Pivot pin for lever	1	601070	
46	Adjusting screw for front servo	1	601071	
47	Locknut for adjusting screw	1	601072	
48	Spring for front servo	1	601371	
49	'O' ring, large, for front servo piston	1	601064	
50	'O' ring, small, for front servo piston	1	601067	
51	Oil sealing ring for front servo piston sleeve	1	601066	
52	Sleeve for front servo piston	1	601065	
53	Snap ring for front servo piston	1	601068	
54	Special bolt and washer (1⅞" long) fixing servo to case	1	601073	
	Special bolt and washer (1½" long) fixing servo to case	1	601426	
55	Piston assembly (service kit)	1	601430	
56	Brake band for front servo	1	601104	
57	Strut for brake band	1	601105	
58	SERVO ASSEMBLY, REAR	1	601075	
59	Body for rear servo	1	601076	
60	Piston for rear servo	1	601077	
61	'O' ring for piston	1	601078	
62	Spring for rear servo	1	601079	
63	Lever for servo	1	601080	
64	Pivot pin for lever	1	601081	
65	Special bolt, rear, fixing rear servo to case	1	601082	
	Special spring washer	1	601083	
66	Special bolt and washer, rear, fixing rear servo to case	1	601084	
67	Magnet attached to rear bolt of rear servo	1	601152	
68	Brake band for rear servo	1	601104	
69	Strut for brake band	1	601105	

*Asterisk indicates a new part which has not been used on any previous model.

This page is intentionally left bank

SERVICE KITS

Plate Ref.	1 2 3 4	Description	Qty.	Part No.	Remarks
		Front clutch spring and snap ring kit	1	605973*	
		Forward sun gear and front clutch cylinder kit	1	605974*	Transmission numbered up to 3 EU 5968
		Front servo sleeve and 'O' ring kit	1	605595	
		Rubber ring and oil seal kit	1	601389	
		Sealing and piston ring kit	1	601388	
		Thrust washer kit	1	601386	
		Front clutch plates	1	601165	
		Rear clutch plates	1	601427	
		Rear pump kit	1	601428	
		Governor kit	1	601387	
		Springs for valve bodies	1	601384	
		Front drum assembly	1	601429	
		Piston for front servo kit	1	601430	
		Gasket kit	1	601443	
		Planet cover gears and rear drum assembly	1	601432	
		Spring and toggle pin kit	1	601695	
		Manual valve detent lever kit	1	601774	
		Pressure plug salvage kit	1	601431	
		Sealant, Loctite grade 'AV'	1	600303	
		Sealant, Loctite grade 'CV'	1	601168	

*Asterisk indicates a new part which has not been used on any previous model.

Plate Ref.	1 2 3 4 Description	Qty.	Part No.	Remarks
1	HOUSING ASSEMBLY FOR SPEEDOMETER DRIVE	1	500726@	4-speed models without over-drive (items 1–27)
2	HOUSING ASSEMBLY FOR SPEEDOMETER DRIVE	1	519126@@	
3	Insert for housing	1	232846	
4	Stud for rear mounting	4	252524@	
5	Stud for rear stabiliser	2	530613@@	
6	Plug for tapped hole	1	253040@@	
7	Breather pipe complete	1	232855	
8	Joint washer for breather pipe	1	52064	
9	Oil seal for mainshaft	1	217507	
10	FLANGE FOR PROPELLER SHAFT	1	500730	
11	Special bolt in flange	4	500732	
12	Special washer	1	233929	Alternative to 2 lines below. Check before ordering
13	Self-locking nut (½" BSF)	1	251324	
	Special washer	1	513454	Alternative to 2 lines above. Check before ordering
	Self-locking nut (⅝" UNF)	1	513188	
14	Speedometer worm	1	241544	
15	Distance piece for speedometer worm	1	277940	
16	Spindle for speedometer	1	232639	
17	Bush for speedometer spindle	1	232568	
18	Oil seal for spindle	1	218337	
19	Retaining plate for speedometer spindle	1	232565	
20	Joint washer for spindle bush and retaining plate	2	267782	
21	Set screw (¼" Whit x 1"long) fixing plate	2	20361	
22	Joint washer for tailshaft housing	1	242371	
23	Bolt (5/16" BSF x 1⅜"long) — Fixing speedometer drive housing to gearbox	2	250527	
24	Spring washer — Fixing speedometer drive housing to gearbox	2	3075	
25	Nut (5/16" BSF) — Fixing speedometer drive housing to gearbox	2	2828	
26	Spring washer — Fixing speedometer drive housing to gearbox	6	3076	
27	Nut (⅜" BSF) — Fixing speedometer drive housing to gearbox	6	2827	
28	HOUSING ASSEMBLY FOR SPEEDOMETER DRIVE	1	279263	Mk I, Mk IA and Mk II Automatic (items 28–38)
29	Bush for spindle	2	277453	
30	Rubber ring for housing	1	279513	
31	Spindle for speedometer drive	1	279364	
32	Oil seal for spindle	1	277736	
33	Speedometer driven gear	1	279265	
34	Adaptor for speedometer cable	1	277737	
35	Set screw (2 BA x ½" long) fixing adaptor	3	277738	
36	Stud for speedometer spindle retainer	1	279552	
37	Retainer for speedometer spindle	1	279542	
38	Self-locking nut (¼" UNF) fixing retainer	1	252210	

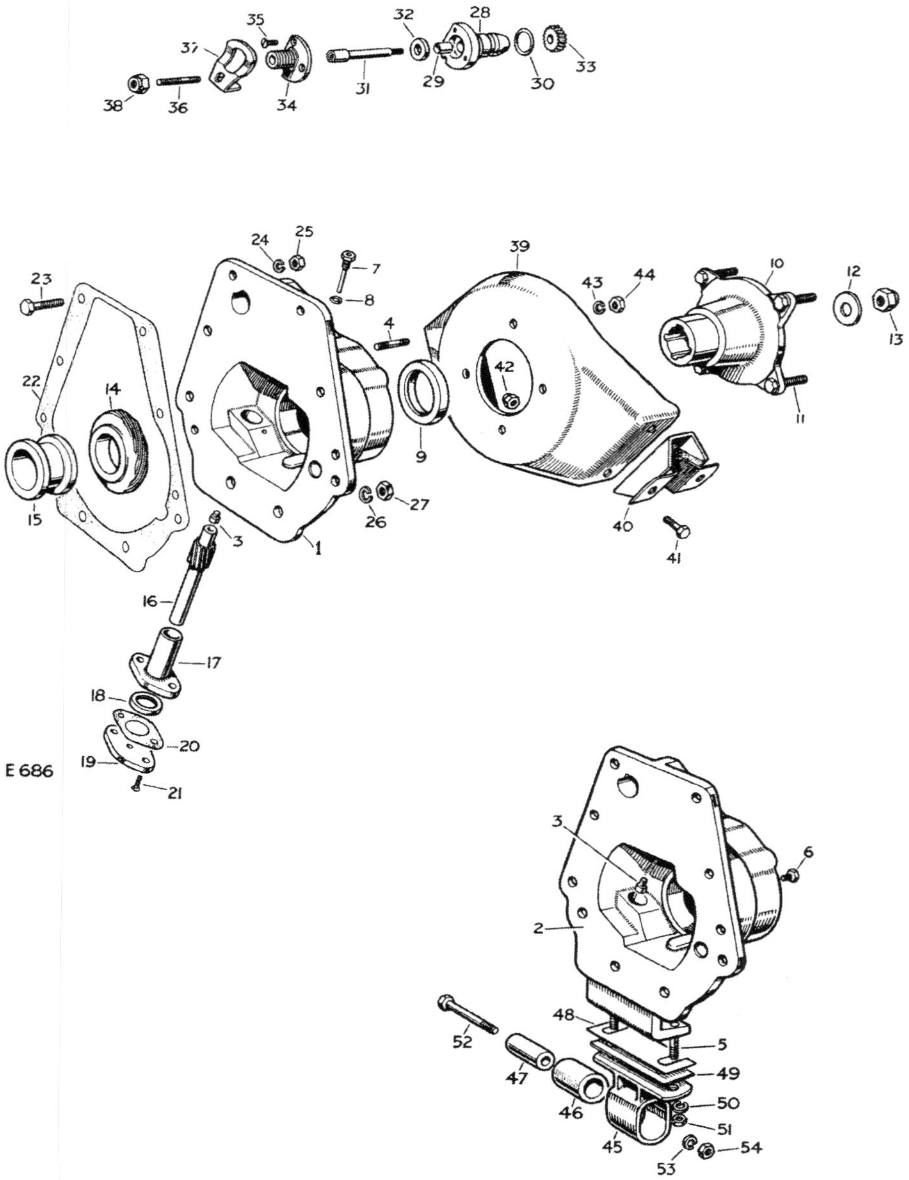

@ Cars numbered up to 625101045, 626100262, 628100109, 630100840, 631100353, 63310078

@@ Cars numbered from 625101046, 626100263, 628100110 onwards

*Asterisk indicates a new part which has not been used on any previous model.

E 686

COLLINS-JONES

SPEEDOMETER DRIVE HOUSING

Plate Ref.	1 2 3 4 DESCRIPTION		Qty	Part No.	REMARKS
	HOUSING ASSEMBLY FOR SPEEDOMETER DRIVE		1	553099	Mk III Automatic
	Adaptor for housing		1	549061	Mk III Automatic
	Oil seal		1	528747	Mk III Automatic
	'O' ring for housing		1	532516	Mk III Automatic
	SPEEDOMETER DRIVEN GEAR AND SPINDLE		1	553102	Mk III Automatic
	Spindle for speedo drive		1	553104	Mk III Automatic
	Spring washer	Fixing speedo drive housing to gearbox extension housing	2	3074	Mk III Automatic
	Set bolt ($\frac{1}{4}$" UNC x 1" long)	Fixing speedo drive housing to gearbox extension housing	2	253409	Mk III Automatic
39	Rear mounting bracket for gearbox		1	500705†	Models without overdrive
40	Rear mounting rubber, identification, blue		2	513586	Cars numbered up to 625001876, 626000570, 628000263
	Rear mounting rubber, identification, brown		2	518699†	Cars numbered from 625001877, 626000571, 628000264 onwards
	Rear mounting rubber, identification, red		2	500708†	Automatic
41	Bolt ($\frac{5}{16}$" UNF x $\frac{7}{8}$" long)	Fixing mounting rubbers to bracket	4	255227†	Models without overdrive
42	Self-locking nut ($\frac{5}{16}$" UNF)	Fixing mounting rubbers to bracket	4	252211†	Models without overdrive
	Set bolt ($\frac{5}{16}$" UNC x $\frac{7}{8}$" long)	Fixing mounting rubbers to bracket	4	253027†	Models with overdrive
	Lock washer	Fixing mounting rubbers to bracket	4	265260†	Models with overdrive
43	Spring washer	Fixing mounting bracket to speedometer housing	4	3076†	4-speed without overdrive
44	Nut ($\frac{3}{8}$" UNF)	Fixing mounting bracket to speedometer housing	4	254852†	4-speed without overdrive
	Spring washer	Fixing bracket to gearbox	4	3076†	Automatic
	Set bolt ($\frac{3}{8}$" UNC x $\frac{7}{8}$" long)	Fixing bracket to gearbox	4	253046†	Automatic
45	Stabiliser bracket at gearbox		1	520915††	
46	Bush for stabiliser		1	521349††	
47	Tube for stabiliser		1	521350††	
48	Shim, 22 SWG	Stabiliser to gearbox	As reqd 4	520917††	
49	Shim, 16 SWG	Stabiliser to gearbox	As reqd 4	528199††	
50	Spring washer	Fixing stabiliser bracket to gearbox	2	3075††	
51	Nut ($\frac{5}{16}$" UNF)	Fixing stabiliser bracket to gearbox	2	254811††	
52	Bolt ($\frac{5}{16}$" UNF x $2\frac{3}{4}$" long)	Fixing stabiliser to sub-frame bracket	1	256029††	
53	Spring washer	Fixing stabiliser to sub-frame bracket	1	3075††	
54	Nut ($\frac{5}{16}$" UNF)	Fixing stabiliser to sub-frame bracket	1	254811††	
	Plain washer, small	Fixing gearbox to engine	1	2550	
	Plain washer, large	Fixing gearbox to engine	11	2210	12 off on Automatic
	Nut ($\frac{5}{16}$" UNF)	Fixing gearbox to engine	1	254801	
	Nut ($\frac{3}{8}$" UNF)	Fixing gearbox to engine	12	254802	

* Asterisk indicates a new part which has not been used on any previous Rover model
† Cars numbered up to 625101045, 626100262, 628100109, 630100840, 631100353, 633100078
†† Cars numbered from 625101046, 626100263, 628100110 onwards

Plate Ref.	1 2 3 4	DESCRIPTION	Qty	Part No.	REMARKS
		EXTENSION CASING ASSEMBLY FOR OVERDRIVE	1	500724	Mk I and Mk IA
1		EXTENSION CASING ASSEMBLY FOR OVERDRIVE	1	532976	Mk II and Mk III
2		Oil seal for mainshaft	1	518140	
3		Bearing for mainshaft	1	1645	
4		Circlip fixing bearing to casing	1	09960	
5		Stud ($\frac{5}{16}$" UNF/UNC x $1\frac{7}{8}$" long) Fixing gear lever support bracket	2	252515	Mk II and Mk III
6		Stud ($\frac{5}{16}$" UNF/UNC x $2\frac{1}{8}$" long)	2	252516	
		Mainshaft for overdrive	1	264884	Mk I and Mk IA
7		Mainshaft for overdrive	1	518452	Mk II and Mk III
8		Split ring fixing bearing to shaft	1	55720	
9		Spacing collar for mainshaft	1	264885	
10		Joint washer for extension casing	1	242371	
11		Dipstick for extension casing	1	533342	Mk II and Mk III
12		Breather pipe for extension casing	1	232855	
13		Washer for breather pipe	1	52064	
14		Insert for isolator switch plunger	1	273593	
15		Plunger for isolator switch	1	273594	
		Isolator switch, LU 31077	1	237539	Mk I and Mk IA
16		Isolator switch, LU 31824	1	514854	Mk II and Mk III
17		Shim for isolator switch	As reqd	273595	
		Shim for isolator switch	As reqd	03609	
18		Joint washer, extension casing to main casing	1	268118	
19		Bolt ($\frac{5}{16}$" BSF x $1\frac{3}{8}$" long) Fixing extension casing to gearbox	2	250527	
20		Spring washer	2	3075	
21		Nut ($\frac{5}{16}$" BSF)	2	2828	
22		Spring washer	6	3076	
		Nut ($\frac{3}{8}$" BSF)	6	2827†	
23		Nut ($\frac{3}{8}$" BSF)	4	2827††	
		Nut ($\frac{3}{8}$" BSF)	2	2323††	
		Plain washer	4	2204††	
24		Locker	1	2499	
25		Reaction bracket complete	1	519222††	
26		Bolt ($\frac{5}{16}$" UNC x 1" long) Fixing reaction bracket to casing	4	253028††	
27		Spring washer	4	3075††	
28		Shim, 22 SWG Reaction bracket to stabiliser bracket	As reqd	520917††	
29		Shim, 16 SWG	As reqd	528199††	
30		Stabiliser bracket	1	520915††	
31		Bush For stabiliser	1	521349††	
32		Tube	1	521350††	
33		Bolt ($\frac{5}{16}$" UNF x $2\frac{3}{4}$" long) Fixing stabiliser bracket to sub-frame bracket	1	256029††	
34		Spring washer	1	3075††	
35		Nut ($\frac{5}{16}$" UNF)	1	254811††	
36		Bolt ($\frac{1}{4}$" UNF x $\frac{3}{4}$" long) Fixing stabiliser bracket to reaction bracket	2	255207††	
37		Spring washer	2	3074††	
38		Nut ($\frac{1}{4}$" UNF)	2	254810††	
39		Spring washer Fixing main casing to extension casing	6	3075	
40		Nut ($\frac{5}{16}$" UNF)	6	254811	

* Asterisk indicates a new part which has not been used on any previous Rover model
† Cars numbered up to 625101045, 626100262, 628100109
†† Cars numbered from 625101046, 626100263, 628100110 onwards

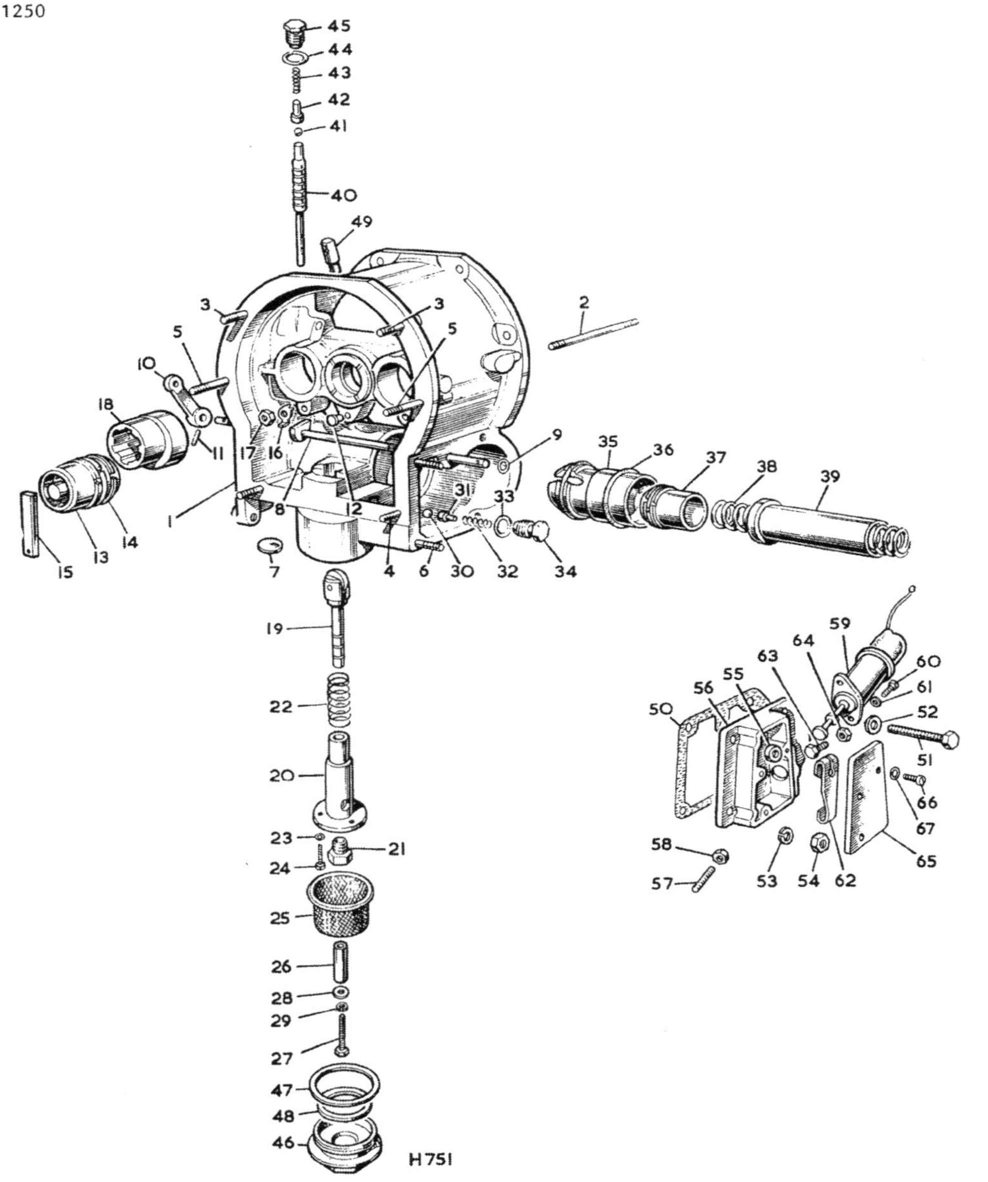

Plate Ref.	1 2 3 4 DESCRIPTION		Qty	Part No.	REMARKS
	OVERDRIVE UNIT ASSEMBLY		1	521919	Mk I and Mk IA
	OVERDRIVE UNIT ASSEMBLY		1	549007	MK II and MK III
	OVERDRIVE MAIN CASING ASSEMBLY ..		1	275688	Mk I and Mk IA
1	OVERDRIVE MAIN CASING ASSEMBLY ..		1	600690	MK II and MK III
2	Stud, $2\frac{15}{16}$" long, bottom rear	For main casing	2	268261	
3	Stud, $1\frac{3}{8}$" long, main to extension		3	268262	
4	Stud, 1" long, main to extension		1	268259	
5	Stud, $2\frac{5}{8}$" long, main to extension		2	268260	
6	Stud, $1\frac{3}{8}$" long, side cover plate		2	268262	
7	Core plug for main casing		1	268093	
	Valve operating shaft		1	275689	Gearboxes numbered up to 77005681
8	Valve operating shaft		1	600709	Gearboxes numbered from 77005682 onwards
9	Oil seal for operating shaft		2	268123	
	Operating valve setting lever		1	268113	Gearboxes numbered up to 77005681
10	Operating valve setting lever		1	600708	Gearboxes numbered from 77005682 onwards
	Taper pin for setting lever		2	268124	Gearboxes numbered up to 77005681
11	Spring dowel for setting lever		1	600710	Gearboxes numbered from 77005682 onwards
12	Pump plunger guide peg		1	268112	
13	OPERATING PISTON ASSEMBLY		2	513891	
14	'O' ring for operating piston		4	513893	
15	Bridge piece for operating pistons		2	268111	
16	Tab washer	Fixing bridge piece	4	268272	
17	Nut ($\frac{1}{4}$" UNF)		4	254810	
18	Cam for oil pump plunger		1	268128	
19	Oil pump plunger		1	541487	
20	Oil pump body		1	268115	
21	Plug for pump body		1	268096	Gearboxes numbered up to 77005681
	Plug for pump body		1	600707	Gearboxes numbered from 77005682 onwards
22	Pump plunger spring		1	268094	
23	Spring washer	Fixing oil pump	2	3073	
24	Special screw		2	268270	
25	Oil filter		1	521815	
26	Distance tube for filter		1	268117	Not required with notched-type drain plug
27	Special bolt	Fixing filter	1	268097	
28	Plain washer		1	2226	
29	Spring washer		1	3073	
30	Steel ball ($\frac{1}{4}$") for valve		1	52459	
31	Ball valve plunger		1	268090	
32	Valve spring		1	268089	
33	Copper washer for valve plug		1	268088	
34	Plug for valve		1	268095	

* Asterisk indicates a new part which has not been used on any previous Rover model

GEARBOX OVERDRIVE UNIT, MAIN CASING

H751

Plate Ref.	1 2 3 4 Description	Qty.	Part No.	Remarks
	ACCUMULATOR PISTON ASSEMBLY	1	511705	
35	Accumulator housing	1	268131	
36	Oil seal for accumulator piston housing	2	268136	Gearboxes numbered up to
37	Accumulator piston ring, set	2	268132	77005681
38	Accumulator spring	1	275695	
	ACCUMULATOR PISTON ASSEMBLY	1	600682	
	Piston ring for accumulator piston	2	600684	Applicable to overdrive
	ACCUMULATOR HOUSING ASSEMBLY	1	600685	assembly 549007. Gearboxes
	Oil seal for housing	1	268136	numbered from 77005682
	Accumulator spring, set	1	600687	onwards
39	Spacing tube for accumulator spring	1	268083	Gearboxes numbered up to 77005681
	Tube for piston	1	600711	Gearboxes numbered from 77005682 onwards
40	Operating valve	1	268102	
41	Steel ball (5/16") for valve	1	3050	
42	Ball valve plunger	1	268090	
43	Valve spring	1	268089	
44	Copper washer for valve plug	1	268088	
45	Plug for valve	1	268095	
46	Drain plug, with provision for magnet	1	521814	
47	Washer for drain plug	1	268092	
48	Magnet for drain plug	1	600639	
49	Breather for main casing	1	268122	
50	Joint washer, solenoid bracket to main casing	1	268114	
51	Special bolt ⎤	2	253434	
52	Plain washer ⎥ Fixing solenoid bracket	2	2223	
53	Spring washer ⎥	2	3075	
54	Nut (5/16" UNF) ⎦	2	254811	
55	Collar for operating shaft	1	271399	
56	BRACKET FOR SOLENOID	1	271402	
	Stop for solenoid	1	271405	
	Dowel for solenoid bracket	1	600509	Gearboxes numbered
57	Adjusting screw for plunger stop	1	600510	from 77004139
58	Locknut for adjusting screw	1	600511	onwards
59	Solenoid	1	547101	
	Joint washer for solenoid	1	600712	
60	Special screw ⎤ Fixing	2	268270	
61	Spring washer ⎦ solenoid	2	3073	
62	Operating lever for solenoid	1	271401	
63	Bolt (¼" BSF x ¾" long) ⎤ Fixing	1	250694	
64	Nut (¼" BSF) ⎦ operating lever	1	2823	
65	Cover plate for solenoid	1	271400	
66	Special screw ⎤ Fixing	3	268270	
67	Spring washer ⎦ cover plate	3	3073	

*Asterisk indicates a new part which has not been used on any previous model.

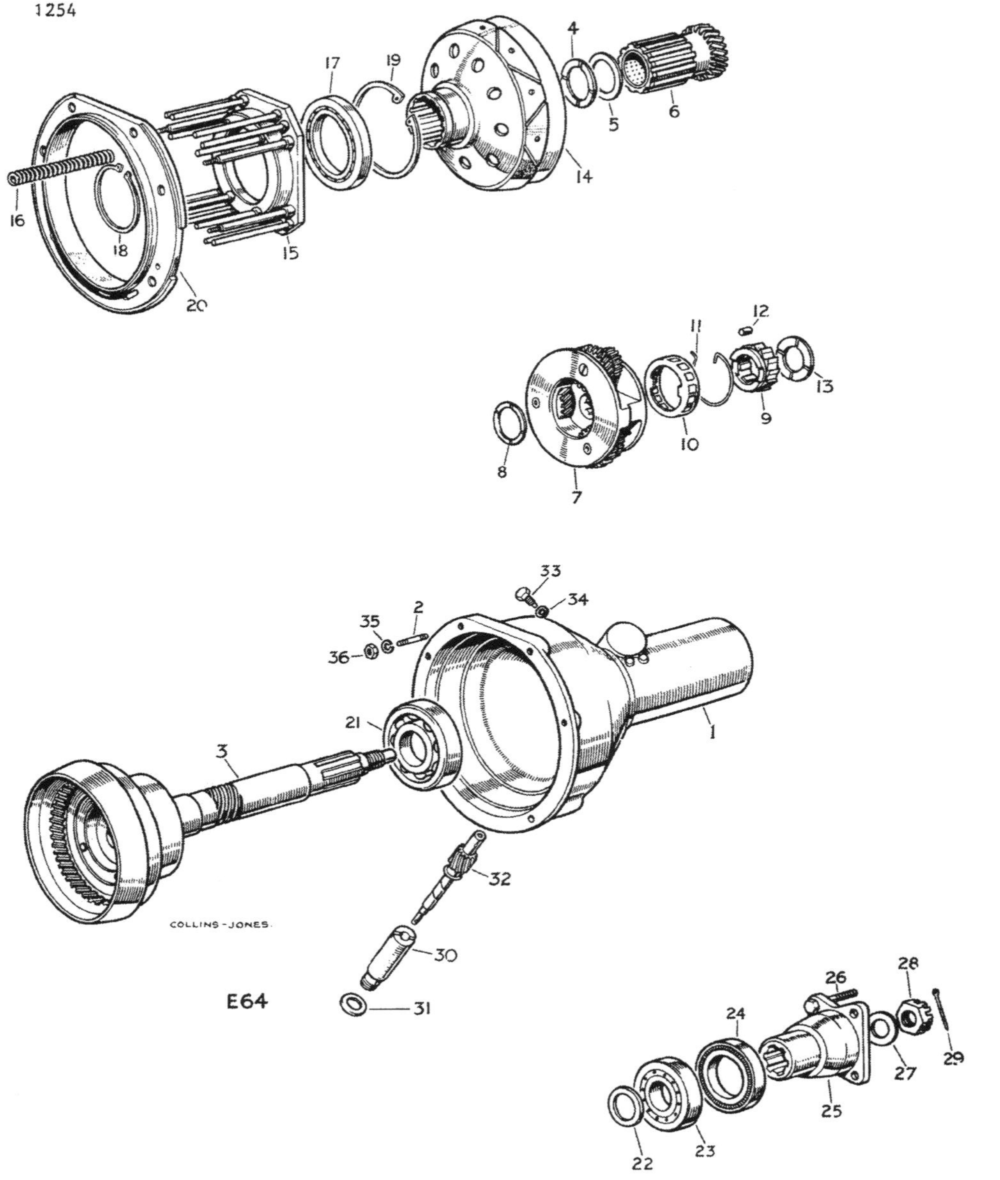

COLLINS-JONES

E64

GEARBOX OVERDRIVE UNIT, TAILSHAFT CASING

Plate Ref.	1 2 3 4 DESCRIPTION	Qty	Part No.	REMARKS
1	OVERDRIVE TAILSHAFT CASING ASSEMBLY	1	268073†	
	OVERDRIVE TAILSHAFT CASING ASSEMBLY	1	528259††	
2	Stud for tailshaft casing	4	268263	
3	Annulus complete	1	268081	
4	Spacing washer, bronze	1	268098	
5	Thrust washer, .077" thick } For sun wheel, front	1	268109	As required
	Thrust washer, .083" thick } For sun wheel, front	1	268108	As required
	Thrust washer, .089" thick } For sun wheel, front	1	268107	As required
	Thrust washer, .095" thick } For sun wheel, front	1	268106	As required
	Thrust washer, .101" thick } For sun wheel, front	1	268105	As required
	Thrust washer, .107" thick } For sun wheel, front	1	268104	As required
	Thrust washer, .113" thick } For sun wheel, front	1	268103	As required
6	Sun wheel complete	1	268079	
7	PLANET CARRIER ASSEMBLY	1	268080	
8	Thrust washer, rear, for planet carrier assembly	1	268100	
9	Uni-directional clutch inner member	1	268119	
10	Uni-directional clutch cage	1	268120	
11	Spring for uni-directional clutch	1	268121	
12	Steel roller for uni-directional clutch .. Set	1	268084	
13	Thrust washer between uni-directional clutch and annulus	1	268101	
14	Cone clutch	1	268077	
15	Cone clutch thrust ring	1	507992	
16	Clutch springs .. Set	1	507007	
17	Thrust bearing for clutch	1	268125	
18	Circlip, small } Fixing clutch thrust ring bearing	1	268268	
19	Circlip, large } Fixing clutch thrust ring bearing	1	268269	
20	Brake ring between main and tailshaft casings	1	268127	
21	Front bearing for tailshaft and annulus	1	217325	
22	Spacing washer, .146" thick } For rear tailshaft bearing	1	268140	As required
	Spacing washer, .151" thick } For rear tailshaft bearing	1	268141	As required
	Spacing washer, .156" thick } For rear tailshaft bearing	1	268142	As required
	Spacing washer, .161" thick } For rear tailshaft bearing	1	268143	As required
	Spacing washer, .166" thick } For rear tailshaft bearing	1	268144	As required
23	Rear bearing for tailshaft and annulus	1	268126	
24	Oil seal in tailshaft housing	1	268135	
25	Flange for tailshaft	1	268139	
26	Bolt for flange	4	237341	
27	Special plain washer } Fixing flange to tailshaft	1	268266	
28	Slotted nut } Fixing flange to tailshaft	1	268267	
29	Split pin } Fixing flange to tailshaft	1	2428	
30	SPEEDOMETER BEARING AND INNER SEAL	1	268085†	
	SPEEDOMETER BEARING AND INNER SEAL	1	528258††	
31	Outer sealing ring for speedometer bearing	1	270290	
32	Speedometer driven gear	1	268138†	
	Speedometer driven gear	1	528262††	
33	Locking screw for speedometer bearing	1	268264	
34	Copper washer for locking screw	1	271460	
35	Spring washer } Fixing main casing to tailshaft casing	6	3075	
36	Nut ($\frac{5}{16}$" UNF) } Fixing main casing to tailshaft casing	6	254811	

* Asterisk indicates a new part which has not been used on any previous Rover model

† Engines numbered up to 625101417

†† Engines numbered from 625101418 onwards

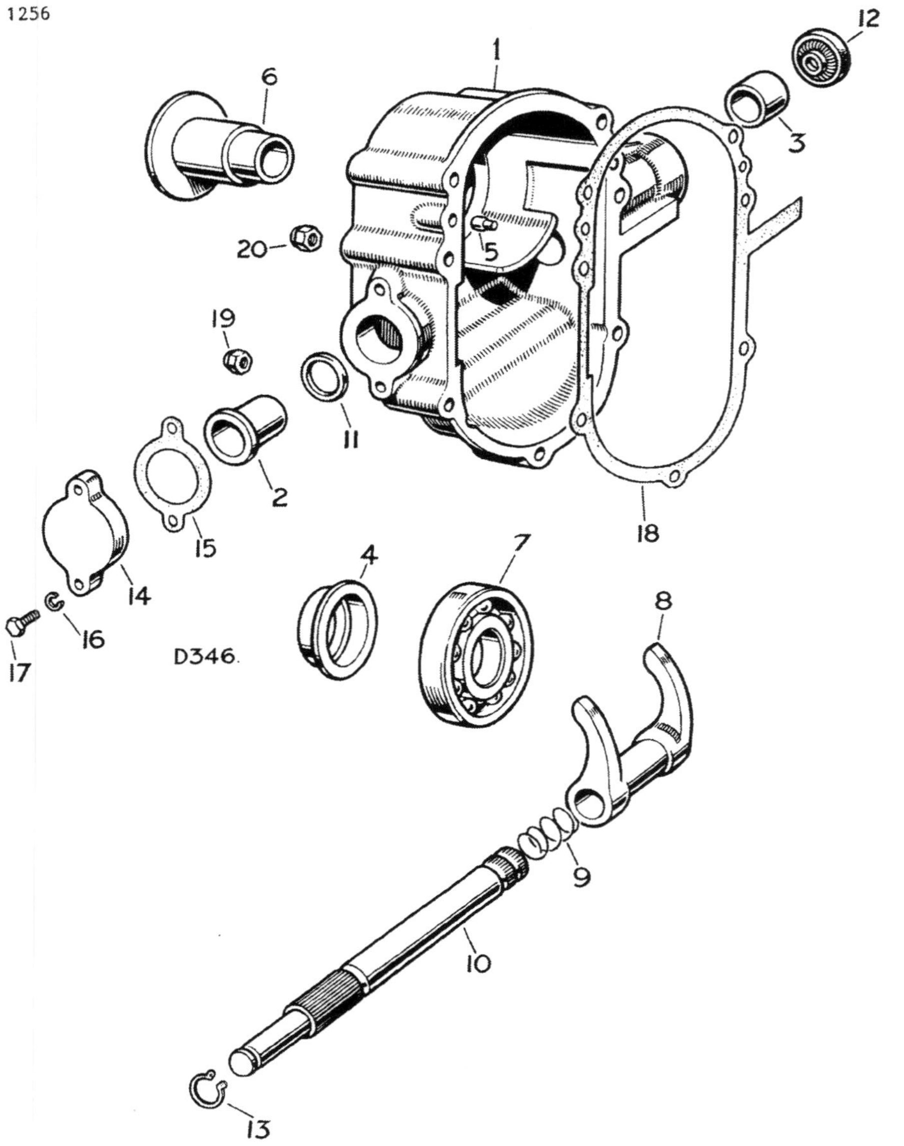

CLUTCH WITHDRAWAL MECHANISM, 4-SPEED MODELS

Plate Ref.	1 2 3 4 DESCRIPTION	Qty	Part No.	REMARKS
1	CLUTCH WITHDRAWAL HOUSING ASSEMBLY	1	501159	Mk I and Mk IA
	CLUTCH WITHDRAWAL HOUSING ASSEMBLY	1	528700	Mk II and Mk III
2	Bush for cross-shaft, flanged	1	501161	
3	Bush for cross-shaft, plain	2	501162	
4	Bush for withdrawal race sleeve	1	231075	
5	Dowel locating housing	2	213700	
6	Clutch withdrawal sleeve	1	231074	
7	Withdrawal race thrust bearing	1	214797	
8	Operating fork for clutch	1	501163	
9	Spring for fork	1	264806	
10	Cross-shaft for clutch operation	1	501165	
11	Thrust washer for cross-shaft	2	501164	
12	Oil seal for cross-shaft	1	501166	
13	Circlip retaining cross-shaft	1	09129	
14	End cover for cross-shaft	1	501168	
15	Joint washer for end cover	1	501167	
16	Spring washer } Fixing end cover	2	3074	
17	Set bolt ($\frac{1}{4}$" Whit x 1$\frac{1}{16}$" long) } Fixing end cover	2	215597	
18	Joint washer for clutch withdrawal housing ..	1	501157	Mk I and Mk IA
	Joint washer for clutch withdrawal housing ..	1	528698	Mk II and Mk III
19	Self-locking nut ($\frac{1}{4}$" BSF) } Fixing housing to bell housing	3	251320	
20	Self-locking nut ($\frac{5}{16}$" BSF) } Fixing housing to bell housing	4	251321	

* Asterisk indicates a new part which has not been used on any previous Rover model

GEAR CHANGE LEVER, MK I AND MK IA 4-SPEED MODELS

E7II.

Plate Ref.	1 2 3 4 Description	Qty.	Part No.	Remarks
	GEAR CHANGE LEVER COMPLETE ASSEMBLY	1	279056@	
	GEAR CHANGE LEVER COMPLETE ASSEMBLY	1	521893@@	
	GEAR CHANGE LEVER AND FORKED ROD ASSEMBLY	1	514557	
1	Gear change lever	1	514554	
2	FORKED ROD ASSEMBLY	1	243542	
3	Bush for forked rod	2	243030	
4	Spring for button	1	241301	
5	Button for forked rod	1	241185	
6	FORK END ASSEMBLY	1	243091	
7	Bush for fork end	2	243030	
8	Joint pin fixing rod and fork ends	2	244026	
9	Belleville washer] Fixing rod	2	244027	
10	Special clip] Fixing rod	2	236961	
11	Support bracket for gear lever pivot	1	278138@	
12	Support bracket for gear lever pivot, threaded internally	1	528351@@	
13	Bolt (5/16" BSF x 1¼" long)] Clamping support bracket to selector housing	1	250526@@	
14	Self-locking nut (5/16" BSF)] Clamping support bracket to selector housing	1	251321@@	
15	External seat for pivot	1	241176	
16	Compression spring for external seat	1	269395	
17	Collar for gear change lever	1	269394	
18	Internal seat for pivot	1	241177	
19	Lock washer Fixing pivot seats	1	241179	
20	Set bolt (¼" BSF x½" long) to support bracket	1	237139	
21	Knob for gear lever	1	242179	
22	Locknut (¼" BSF) for knob	1	522241	
23	Mounting plate for support bracket	1	530302	
24	Plain washer] Fixing support bracket to mounting bracket	4	3840@@	
25	Spring washer] Fixing support bracket to mounting bracket	4	3074@@	
	Nut (¼" BSF)] Fixing support bracket to mounting bracket	4	2823@@	] Alternative threads Check before ordering
26	Nut (¼" UNF)] Fixing support bracket to mounting bracket	4	254810@@	] Alternative threads Check before ordering
	Nut (¼" UNF)] Change support to gearbox cover	4	254810	] Alternative threads Check before ordering
27	Nut (¼" BSF)] Change support to gearbox cover	4	2823	] Alternative threads Check before ordering
28	Spring washer] Change support to gearbox cover	4	3074	
29	Plain washer] Change support to gearbox cover	4	3900	
30	Locker, forked rod to selector	1	243556	
31	Rubber cover for gear change	1	243545@	

@ Cars numbered up to 625101045, 626100262, 628100109

@@ Cars numbered from 625101046, 626100263, 628100110 and Mk IA 4-speed models

*Asterisk indicates a new part which has not been used on any previous model.

GEAR CHANGE LEVER, MK II AND MK III 4-SPEED MODELS

F623.

Plate Ref.	1 2 3 4 Description	Qty.	Part No.	Remarks
1	GEAR LEVER SUPPORT BRACKET ASSEMBLY	1	539621	
2	Bush, self- aligning	1	539615	
3	Spherical seat	1	539685	Part of assembly 600693
4	Locating pin for spherical seat	1	533550	
5	Circlip for spherical seat	1	539618	Early models only
6	Retaining cap, top } For spherical seat	1	539616	
7	Retaining cap, bottom } For spherical seat	1	539617	
8	Set screw } Fixing retaining caps to support bracket	3	257306	
	Plain washer } Fixing retaining caps to support bracket	6	3925	
9	Spring washer } Fixing retaining caps to support bracket	3	3072	
10	Nut (6 UNC) } Fixing retaining caps to support bracket	3	257191	
11	Reverse stop plate	1	532957	
12	Set screw (¼" UNC x 9/16" long) } Fixing reverse stop plate to bracket	2	253005	
13	Spring washer } Fixing reverse stop plate to bracket	2	3074	
14	Shim washer, .005" thick } Gear lever support bracket to extension casing	2	539624	
	Shim washer, .010" thick } Gear lever support bracket to extension casing	2	539625	
	Shim washer, .032" thick } Gear lever support bracket to extension casing	2	539649	
15	Spring washer } Fixing gear lever support assembly to overdrive extension casing	4	3075	
16	Locknut (5/16" UNF) } Fixing gear lever support assembly to overdrive extension casing	4	254861	
17	Gear lever (lower) complete	1	533546	
18	Spring for reverse stop	1	533294	} Part of assembly 600692
	Plunger for reverse stop	1	533082	} Part of assembly 600692
19	Pin for reverse	1	533300	} Part of assembly 600692
20	Gear change shaft assembly	1	539620	
21	Reverse stop washer for gear change shaft	1	539623	
	GEAR CHANGE LEVER ASSEMBLY	1	600693	
22	Gear change lever assembly	1	600692	
	Gear change lever for use with tolerance ring	1	533007@	
	Gear change lever for use with rubber sleeve	1	549377@@	
23	Locknut for base of gear change lever	1	539854	
24	Tolerance ring fixing gear lever knob	1	533081@	
25	Rubber sleeve fixing gear lever knob	1	540043@@	
	Knob for gear lever with tolerance ring	1	533010@	
26	Knob for gear lever with rubber sleeve	1	540039@@	
27	Button for knob	1	540042	
	'O' ring for button	1	247186@@	
28	Control tube for reverse stop	1	539776	
29	Grommet for gear lever, green	1	381094	
	Grommet for gear lever, red	1	381093	
	Grommet for gear lever, blue	1	381098	
	Grommet for gear lever, tan	1	381095	
	Grommet for gear lever, grey	1	381096	
	Grommet for gear lever, dark fawn	1	381097	
	Grommet for gear lever, Toledo red	1	385790	
	Grommet for gear lever, Mortlake brown	1	385791	
	Insulating pad for grommet	1	383266	

@ Cars numbered up to 77000748, 77100082, 77300100, 73500041, 73600000, 73800012

@@ Cars numbered from 77000749, 77100083, 77300101, 73500042, 73600001, 73800013 onwards

*Asterisk indicates a new part which has not been used on any previous model.

SELECTOR LEVER AND RODS, MK I, MK IA AND MK II AUTOMATIC MODELS

Plate Ref. 1 2 3 4	Description	Qty.	Part No.	Remarks
	GEAR SELECTOR LEVER COMPLETE ASSEMBLY	1	521682	
1	GEAR LEVER ASSEMBLY	1	500872	
2	Bush for gear lever	2	243030	
3	Tube for gear lever	1	522023	
4	Spring for gear lever	1	243465	
5	Belleville washer } Fixing gear lever to tube end	1	244027	
6	Joint pin } Fixing gear lever to tube end	1	244026	
7	Special clip } Fixing gear lever to tube end	1	236961	
8	Pointer for gear selection	1	500870@	
	Indicator arrow for gear selection	1	540286@@	
9	Set bolt (2 BA x $\frac{3}{8}$" long) } Fixing pointer to tube end	1	237119@	
10	Spring washer } Fixing pointer to tube end	1	3073@	
	Speed nut fixing indicator to tube	1	556251@@	
11	Knob for gear lever	1	279918	
12	Clip for knob	1	279955	
13	Clamp for gear lever tube	1	279910	
14	Bolt ($\frac{1}{4}$" UNF x $\frac{3}{4}$" long) fixing clamp to tube	1	255207	
15	Finger for gear selector	1	556567	
16	Locker for finger	1	2500	
17	Nylon liner for gear selector gate	1	279909	RH stg
	Nylon liner for gear selector gate	1	501406	LH stg
18	Backing plate for liner	1	279896	RH stg
	Backing plate for liner	1	501407	LH stg
19	Special screw fixing liner and backing plate to upper bracket and gate	4	500218	
20	Housing for spherical bearing } For gear change at lever end	2	217983	
21	Spherical bearing } For gear change at lever end	1	279903	
22	Bolt (5/16" UNF x $\frac{3}{4}$" long) } Fixing spherical bearing to bracket and gate	2	255226	
23	Self-locking nut (5/16" UNF) } Fixing spherical bearing to bracket and gate	2	252211	
24	Upper bracket and gate for gear change	1	536530	RH stg
	Upper bracket and gate for gear change	1	536531	LH stg
25	Sealing washer for gear lever tube at bottom	1	510802	
26	Housing for spherical bearing } For gear change at lower end	2	217983	
27	Felt ring } For gear change at lower end	2	279902	
28	Spherical bearing } For gear change at lower end	1	279903	
29	Self-locking nut (5/16" UNF) fixing spherical bearing to lower bracket	2	252211	
30	Lower bracket for gear change	1	539646	RH stg
	Lower bracket for gear change	1	511962	LH stg
	Gear selection lever, lower	1	279905	
31	Gear selection lever, lower	1	539644	Cars with ball joint type control rod
	Bolt (5/16" UNF x $1\frac{1}{8}$" long) fixing lever to gear lever tube	1	255029	} Alternative fixings
32	Bolt (5/16" UNF x $1\frac{1}{2}$" long)	1	256222	} Alternative fixings
33	Self-locking nut (5/16" UNF)	1	252161	} Alternative fixings
34	Sliding switch complete with fixings	1	279975	
35	Plain washer for switch	2	3851	
37	Bracket for switch	1	279976	

@ Cars numbered up to 630100840, 631100353, 633100078

@@ Cars numbered from 630100841, 631100354, 633100079 onwards

*Asterisk indicates a new part which has not been used on any previous model.

SELECTOR LEVER AND RODS, MK I, MK IA AND MK II AUTOMATIC MODELS

Plate Ref.	1 2 3 4 Description		Qty.	Part No.		Remarks
38	Set bolt (¼" UNF x ½" long)	Fixing switch bracket to dash	2	255204		
39	Shakeproof washer		2	70884		
40	Ball end	On gear selection lever for sliding switch rod	1	273964		
41	Locknut (2 BA)		1	4017		
42	Rod, gear selection lever to sliding switch		1	279977		
43	Ball socket for rod		2	1659		
44	Locknut for socket (2 BA)		2	2247		
45	Rod, lower lever to compensator		1	279904	RH Stg	Plain-type rods Mk I and Mk IA cars numbered up to 73000544, 73100062, 73300106
	Rod, lower lever to compensator		1	521606	LH Stg	
	Split pin	Fixing rod to levers	2	2392		
	Plain washer		2	2208		
	Rod, lever to compensator		1	539779	RH Stg	Ball joint rods (⅜" dia.) Mk IA cars numbered from 73000545, 73100063, 73300107 onwards
	Ball end for lever		2	521604		
	Locknut (⅜" UNF)		2	254852		
	Spring washer	Fixing ball joints to levers	2	3075		
	Nut (5/16" UNF)		2	254811		
	Rod, lever to compensator		1	521606	LH Stg	
	Ball end for lever		1	521604		
	Locknut (⅜" UNF)		1	254852		
	Spring washer	Fixing ball joints to levers	1	3075		
	Nut (5/16" UNF)		1	254811		
	Control rod assembly lever to compensator		1	539779	RH Stg	Ball joint rods (5/16" dia.) with nylon ball joints. Alternative to ⅜" dia. rods listed above
46	Control rod		1	539778		
47	Ball joint, nylon		2	532967		
	Washer, dust excluder, for ball joint		2	549202		
48	Locknut (5/16" UNF)		2	254851		
49	Spring washer		2	3075		
50	Nut (5/16" UNF)		2	254811		
	Control rod assembly lever to compensator		1	540092	LH Stg	
	Control rod		1	540061		
	Ball joint, nylon		1	532967		
	Washer, dust excluder, for ball joint		2	549202		
	Locknut (5/16" UNF)		1	264851		
51	Gear selector compensator		1	279901@	RH Stg	
	Gear selector compensator		1	521668@@	RH Stg	For control rod without ball joints
52	Gear selector compensator		1	521687	RH stg	For control rod with ball joints
	Gear selector compensator		1	501402@	LH Stg	
	Gear selector compensator		1	521669@@	LH Stg	Mk I, Mk IA and Mk II cars with ⅜" dia. control rod
	Gear selector compensator		1	540060	LH Stg	Mk II cars with 5/16" dia. control rod and nylon ball joints

@ Cars numbered up to 630100522, 631100295, 633100024

@@ Cars numbered from 630100523, 631100296, 633100025 and Mk IA and Mk II Automatic models

*Asterisk indicates a new part which has not been used on any previous model.

SELECTOR LEVER AND RODS, MK I, MK IA AND MK II AUTOMATIC MODELS

Plate Ref.	1 2 3 4 Description		Qty.	Part No.	Remarks
53	Support bracket for compensator shaft		1	503519@	RH Stg
54	Support bracket for compensator shaft		1	520966@@	RH Stg
55	Bolt (5/16" UNF x ¾" long)	Fixing bracket to gearbox cover	2	255226	RH Stg
56	Plain washer	Fixing bracket to gearbox cover	2	4148	RH Stg
57	Self-locking nut (5/16" UNF)	Fixing bracket to gearbox cover	2	252211	RH Stg
	Support bracket for compensator shaft		1	503518@	LH Stg
	Support bracket for compensator shaft		1	521667@@	LH Stg
	Bolt (¼" UNF x ¾" long)	Fixing bracket to gearbox cover	2	255207	LH Stg
	Plain washer	Fixing bracket to gearbox cover	2	3467	LH Stg
	Self-locking nut (¼" UNF)	Fixing bracket to gearbox cover	2	252210	LH Stg
58	Housing for spherical bearing		2	217983@	
59	Felt ring	For gear selector compensator	2	279902@	
60	Spherical bearing	For gear selector compensator	2	279903@	
61	Bolt (5/16" UNF x ¾" long)	Fixing spherical bearing to bracket	2	255226@	
62	Self-locking nut (5/16" UNF)	Fixing spherical bearing to bracket	2	252211@	
63	Strap for compensator shaft		1	520968@@	RH Stg
	Strap for compensator shaft		1	520972@@	LH Stg
	Support for strap		1	520973@@	LH Stg
	Plate for 'Vulcollan' strap		1	540906	
	Bolt (¼" UNF x ⅞" long)	(Fixing strap to (bracket	2	255208@@	
	Plain washer	(Fixing strap to (bracket	2	2213@@	
	Self-locking nut	(Fixing strap to (bracket	2	252160@@	
	Plain washer	Fixing strap to compensator	1	3840@@	
	Self-locking nut (¼" UNF)	Fixing strap to compensator	1	252160@@	
64	Gear selector cross-shaft		1	279891%	RH Stg
	Gear selector cross-shaft		1	533047%%	RH Stg
	Gear selector pivot shaft and tube complete		1	533043	LH Stg
65	Bush for compensator in shaft		2	242691	
66	Tube complete for cross-shaft		1	279893	RH Stg
67	Bush for cross-shaft tube		2	274185	
68	Bolt (5/16" UNC x ¾" long)	Fixing cross-shaft tube to gearbox	3	253026	2 off on LH Stg
69	Spring washer	Fixing cross-shaft tube to gearbox	3	3075	
70	Rubber sealing ring for tube		2	267828	
71	Lever for cross-shaft		1	279895%	RH Stg
	Lever for cross-shaft		1	532966%%	RH Stg
72	Rubber sealing ring for cross-shaft spline		1	267077	
73	Plain washer	Fixing lever to shaft	1	2251	
74	Castle nut (⅜" UNF)	Fixing lever to shaft	1	254952	
75	Split pin	Fixing lever to shaft	1	2556	
76	Special bolt	Fixing cross-shaft to compensator	2	273840@	
	Bolt (⅜" UNF x 1⅜" long)	Fixing cross-shaft to compensator	2	256040@@	
77	Self-locking nut (5/16" UNF)	Fixing cross-shaft to compensator	2	252211@	
	Self-locking nut (⅜" UNF)	Fixing cross-shaft to compensator	2	252162@@	

@ Cars numbered up to 630100522, 631100295, 633100024

@@ Cars numbered from 630100523, 631100296, 633100025 and Mk IA and Mk II Automatic models

% Cars numbered up to 77500123, 776000051, 77800002

%% Cars numbered from 77500124, 776000052, 77800003 onwards

*Asterisk indicates a new part which has not been used on any previous model.

SELECTOR LEVER AND RODS, MK I, MK IA AND MK II AUTOMATIC MODELS

1268

F600.

Plate Ref.	1 2 3 4	Description		Qty.	Part No.	Remarks
78	Rod, front	Fixing lever on cross-shaft to selector lever on gearbox		1	506923%	
79	Rod, rear			1	506924%	
80	Turn buckle		For adjusting rods	1	279899%	
81	Locknut, front (⅜" UNF, RH thread			1	254852%	
82	Locknut, rear (⅜" UNF, LH thread)			1	255803%	
83	Plain washer	For rods		2	2208%	
84	Split pin			2	2392%	
85	Operating rod			1	532968%%	
86	Nylon ball joint	Fixing lever on cross-shaft to selector lever on gearbox		2	532967%%	
87	Spring washer			2	3075%%	
88	Nut (5/16" UNF)			1	254811%%	
89	Locknuts			2	254851%%	

% Cars numbered up to 77500123, 776000051, 77800002
%% Cars numbered from 77500124, 776000052, 77800003 onwards

*Asterisk indicates a new part which has not been used on any previous model.

SELECTOR LEVER AND RODS, AUTOMATIC TRANSMISSION, MK III MODELS

1270

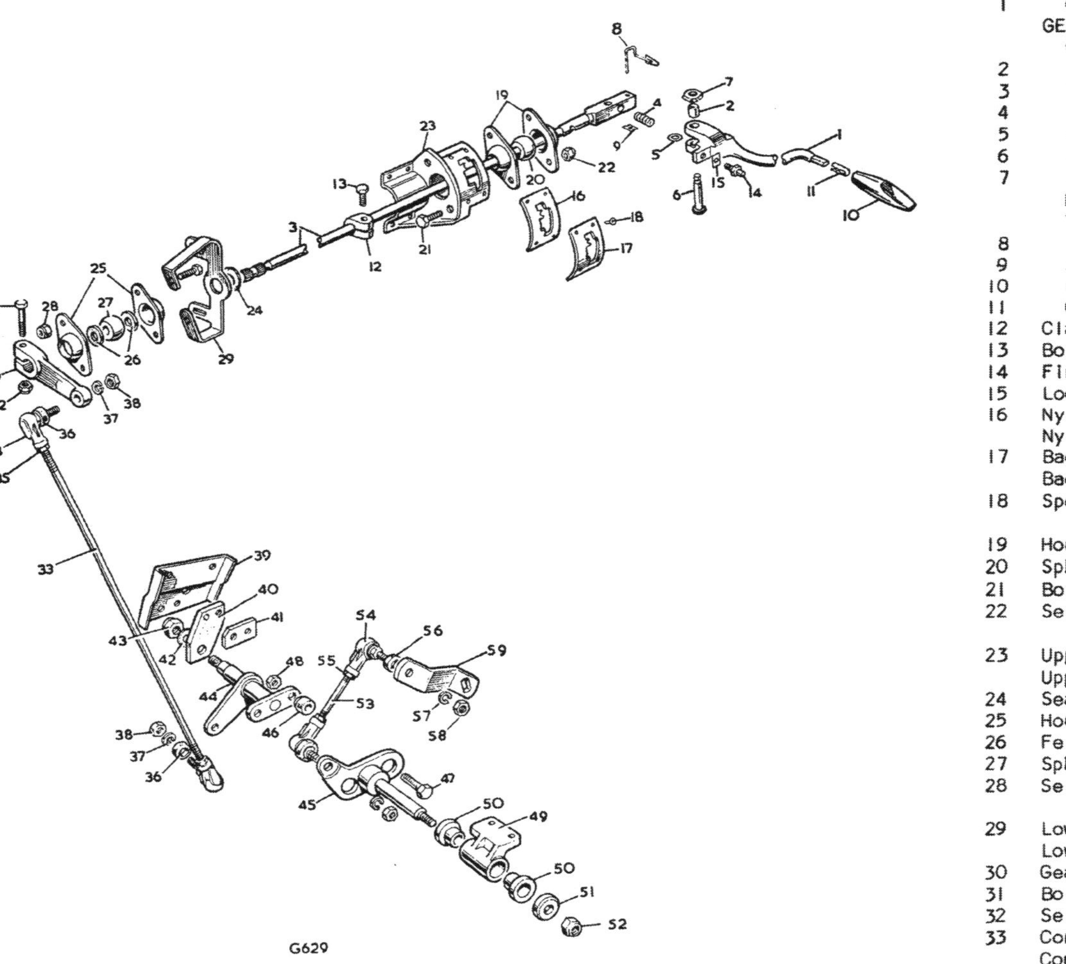

Plate Ref.	1 2 3 4 Description	Qty.	Part No.	Remarks
	GEAR SELECTOR LEVER COMPLETE ASSEMBLY	1	521682	Cars numbered up to:
1	GEAR LEVER ASSEMBLY	1	500872	79500536, 80001294
	GEAR SELECTOR LEVER COMPLETE ASSEMBLY	1	556457	Cars numbered from:
	GEAR LEVER ASSEMBLY	1	556458	79500537, 80001295
2	Bush for gear lever	2	243030	
3	Tube for gear lever	1	522023	
4	Spring for gear lever	1	243465	
5	Belleville washer	1	244027	
6	Joint pin } Fixing gear lever to tube end	1	244026	Cars numbered up to:
7	Special clip }	1	236961	79500536, 80001294
	Hinge pin for gear selection lever	1	556453	Cars numbered from:
	Tension pin for gear selection lever	1	556454	79500537, 80001295
8	Indicator arrow for gear selection	1	540286	
9	Speed nut fixing indicator to tube	1	556251	
10	Knob for gear lever	1	279918	
11	Clip for knob	1	279955	
12	Clamp for gear lever tube	1	279910	
13	Bolt (1/4" UNF x 3/4" long) fixing clamp to tube	1	255207	
14	Finger for gear selector	1	556567	
15	Locker for finger	1	2500	
16	Nylon liner for gear selector gate	1	556374	RH Stg
	Nylon liner for gear selector gate	1	556377	LH Stg
17	Backing plate for liner	1	556372	RH Stg
	Backing plate for liner	1	556376	LH Stg
18	Special screw fixing liner and backing plate to upper bracket and gate	4	500218	
19	Housing for spherical bearing } For gear change	2	217983	
20	Spherical bearing } at lever end	1	279903	
21	Bolt (5/16" UNF x 3/4" long) Fixing spherical bear-	2	255226	
22	Self-locking nut (5/16"UNF) ing to bracket and gate	2	252211	
23	Upper bracket and gate for gear change	1	556375	RH Stg
	Upper bracket and gate for gear change	1	556378	LH Stg
24	Sealing washer for gear lever tube at bottom	1	510802	
25	Housing for spherical bearing } For gear	1	217983	
26	Felt ring } change at	1	279902	
27	Spherical bearing } lower end	1	279903	
28	Self-locking nut (5/16" UNF) fixing spherical bearing to lower bracket	2	252211	
29	Lower bracket for gear change	1	539646	RH Stg
	Lower bracket for gear change	1	539647	LH Stg
30	Gear selection lever, lower	1	539644	
31	Bolt (5/16" UNF x 1 1/2" long) } Fixing lever	1	256222	
32	Self-locking nut (5/16"UNF) } to tube end	1	252161	
33	Control rod, compensator to gear selector lever	1	553189	RH Stg
	Control rod, compensator to gear selector lever	1	553229	LH Stg
34	Ball joint	2	532967	
35	Locknut (5/16" UNF) for ball joint	2	254851	
36	Rubber washer }	2	549202	
37	Spring washer } Fixing ball joints	2	3075	
38	Locknut (5/16" UNF) } to control rod	2	254851	

*Asterisk indicates a new part which has not been used on any previous model.

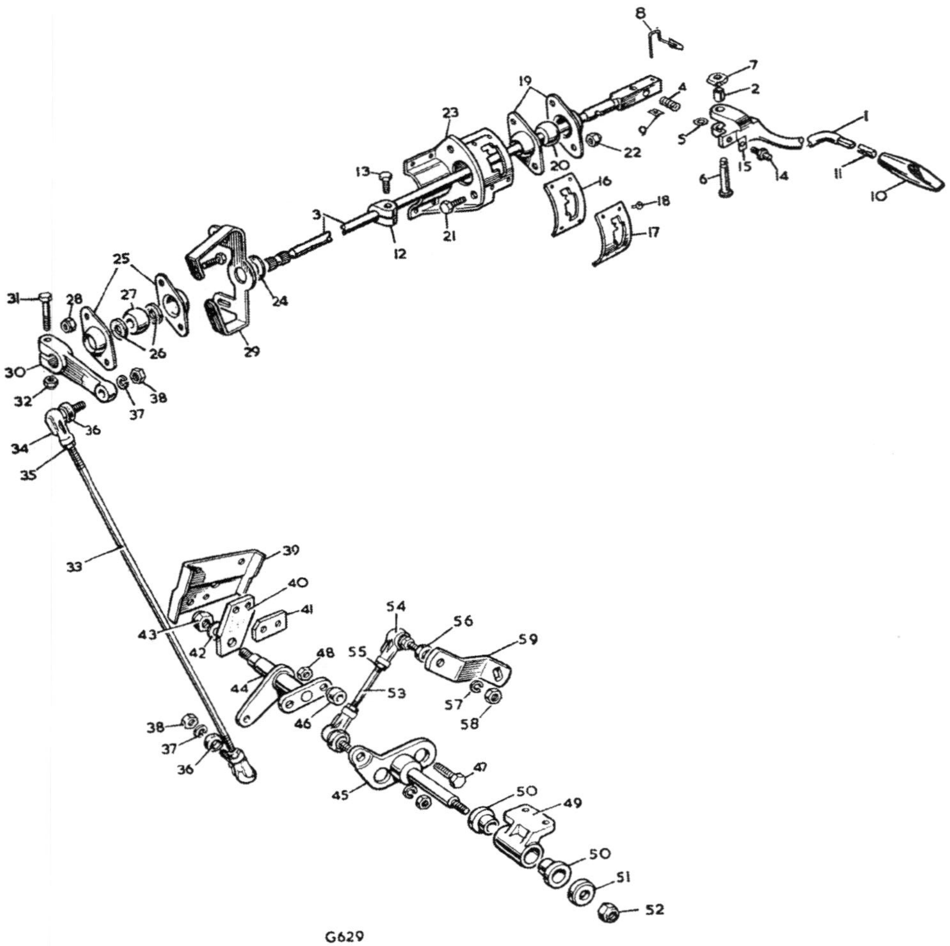

SELECTOR LEVER AND RODS, AUTOMATIC TRANSMISSION, MK III MODELS

G629

Plate Ref.	1 2 3 4 Description		Qty.	Part No.	Remarks
39	Support bracket for compensator shaft		1	520966	RH Stg
	Bolt (5/16" UNF x ¾" long)	Fixing bracket to gearbox cover	2	255226	
	Plain washer		2	4148	
	Self-locking nut (5/16" UNF)		2	252211	
	Support bracket for compensator shaft		1	556166	LH Stg
	Bolt (¼" UNF x ¾" long)	Fixing bracket to gearbox cover	2	255207	
	Plain washer		2	3467	
	Self-locking nut (¼" UNF)		2	252210	
40	Strap for compensator cross-shaft		1	553399	RH Stg
41	Squash plate		1	540906	
	Strap for compensator shaft		1	520972	LH Stg
	Strap support plate		1	520973	
	Bolt (¼" UNF x ⅞" long)	Fixing strap to bracket	2	255208	
	Plain washer		2	2213	
	Self-locking nut		2	252160	
42	Plain washer	Fixing strap to compensator	1	3840	
43	Self-locking nut (¼" UNF)		1	252160	
44	Cross-shaft, compensator, gear change		1	556184	RH Stg
	Cross-shaft, compensator, gear change		1	556187	LH Stg
45	Lever and shaft assembly		1	556183	RH Stg
	Crossover shaft		1	533047	LH Stg
46	Bush for shaft		2	242691	
	'O' ring for crossover shaft		2	267828	LH Stg
47	Bolt (⅜" UNF x 1⅜" long)	Fixing lever and shaft to compensator	2	256040	
48	Self-locking nut (⅜" UNF)		2	252162	
49	Bracket assembly, lower support		1	553505	RH Stg
	Bracket and tube crossover assembly		1	556188	LH Stg
50	Bush for cross-shaft		2	274185	
51	Washer	Fixing lever to support bracket	1	553488	RH Stg
52	Self-locking nut (5/16" UNF)		1	252161	
	Lever for gear change cross-shaft		1	556191	LH Stg
	Plain washer	Fixing lever to shaft	1	3830	
	Self-locking nut (5/16" UNF)		1	252161	
	Spring washer	Fixing bracket to bell housing	2	3075	RH Stg
	Set bolt(5/16" UNC x⅞"long)		2	253027	
	Spring washer	Fixing crossover bracket to bell housing	4	3075	LH Stg
	Set bolt(5/16" UNCx¾"long)		4	253025	
53	Control rod		1	553191	RH Stg
	Control rod		1	556154	LH Stg
54	Ball joint	Fixing control rod to gearbox selector lever	2	532967	
55	Locknut(5/16" UNF) fixing ball joints to rod		2	254851	
56	Rubber washer for ball joint		2	549202	
57	Spring washer		4	3075	
58	Locknut(5/16" UNF)		4	254851	
59	Control lever for gearbox		1	556167	

*Asterisk indicates a new part which has not been used on any previous model.

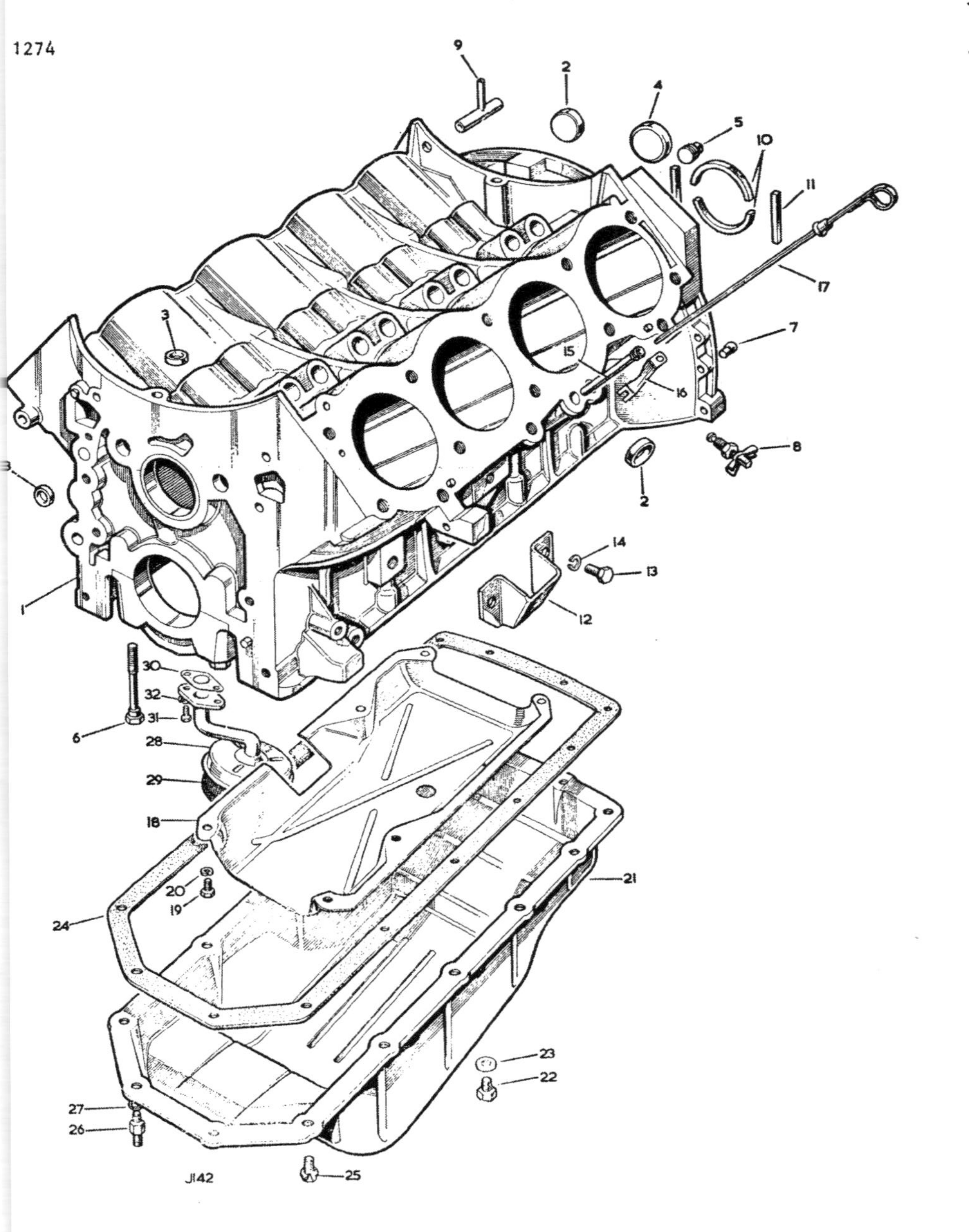

Plate Ref.	1 2 3 4 Description	Qty.	Part No.	Remarks
	ENGINE ASSEMBLY, 10.5:1 compression ratio	1	607740*	Engines numbered up to 84018583C
	ENGINE ASSEMBLY, 10.5:1 compression ratio	1	608285*	Engines numbered from 84018584D onwards
1	CYLINDER BLOCK ASSEMBLY	1	610010	Engines numbered up to 84018583C
	CYLINDER BLOCK ASSEMBLY	1	613991*	Engines numbered from 84018584D onwards
2	Cup plug, 1.5" dia	8	602152	
3	Cup plug, 0.609" dia	4	602147	
4	Cup plug, 1.812" dia	1	602146	
5	Threaded plug for tappet oil liner	2	602212	
6	Special bolt fixing main bearing caps	10	602130	
7	Dowel, locating flywheel housing	1	602141	
8	Drain tap	2	602915	
9	Inlet pipe for crankcase breather	1	603143	
10	Oil seal for crankshaft rear	2	602038	Engines numbered up to 84018583C
	Oil seal, lip type for crankshaft rear	1	611409*	Engines numbered from 84018584D onwards
11	Packing for oil seal, straight type	2	603726	Alternatives
	Packing for oil seal, cruciform type	2	611089	
12	Bracket, engine mounting, upper	2	560133	
13	Bolt (7/16" UNC x 1"long) } Fixing bracket to engine	4	253066	
14	Spring washer } Fixing bracket to engine	4	3077	
15	Oil level rod tube	1	611818*	Engines numbered up to 84016870C
	OIL LEVEL ROD TUBE ASSEMBLY	1	613533	Engines numbered from 84016871C
	'O' ring for tube	1	532319	
	Oil level rod tube	1	610198*	France only
16	Clamp bracket for tube	1	602449	
	Screw (10 UNC x 9/16" long) fixing bracket to rocker cover	1	78862	
	'O' ring for oil level rod tube	1	602545	
17	Oil level rod	1	603173	Engines numbered up to 84016870C
	Oil level rod	1	613532*	Engines numbered from 84016871C
18	Sump baffle plate	1	602290	
19	Bolt (¼" UNC x ⅜" long) } Fixing baffle plate to cylinder block	6	253206	
	Plain washer } Fixing baffle plate to cylinder block	6	3840	
20	Spring washer } Fixing baffle plate to cylinder block	6	3074	
21	Crankcase sump	1	603434	
	Reinforcing strip	1	603943	
	Special bolt and spring washer fixing reinforcing strip and sump	2	603944	
22	Drain plug	1	554164	
23	Joint washer for drain plug	1	213961	
24	Gasket, sump to cylinder block	1	GEG539	
25	Special bolt and spring washer fixing sump to cylinder block	11	602199	
26	Shouldered stud } Fixing sump and pipe brackets	3	603429	
27	Spring washer } Fixing sump and pipe brackets	3	3075	

*Asterisk indicates a new part which has not been used on any previous model.

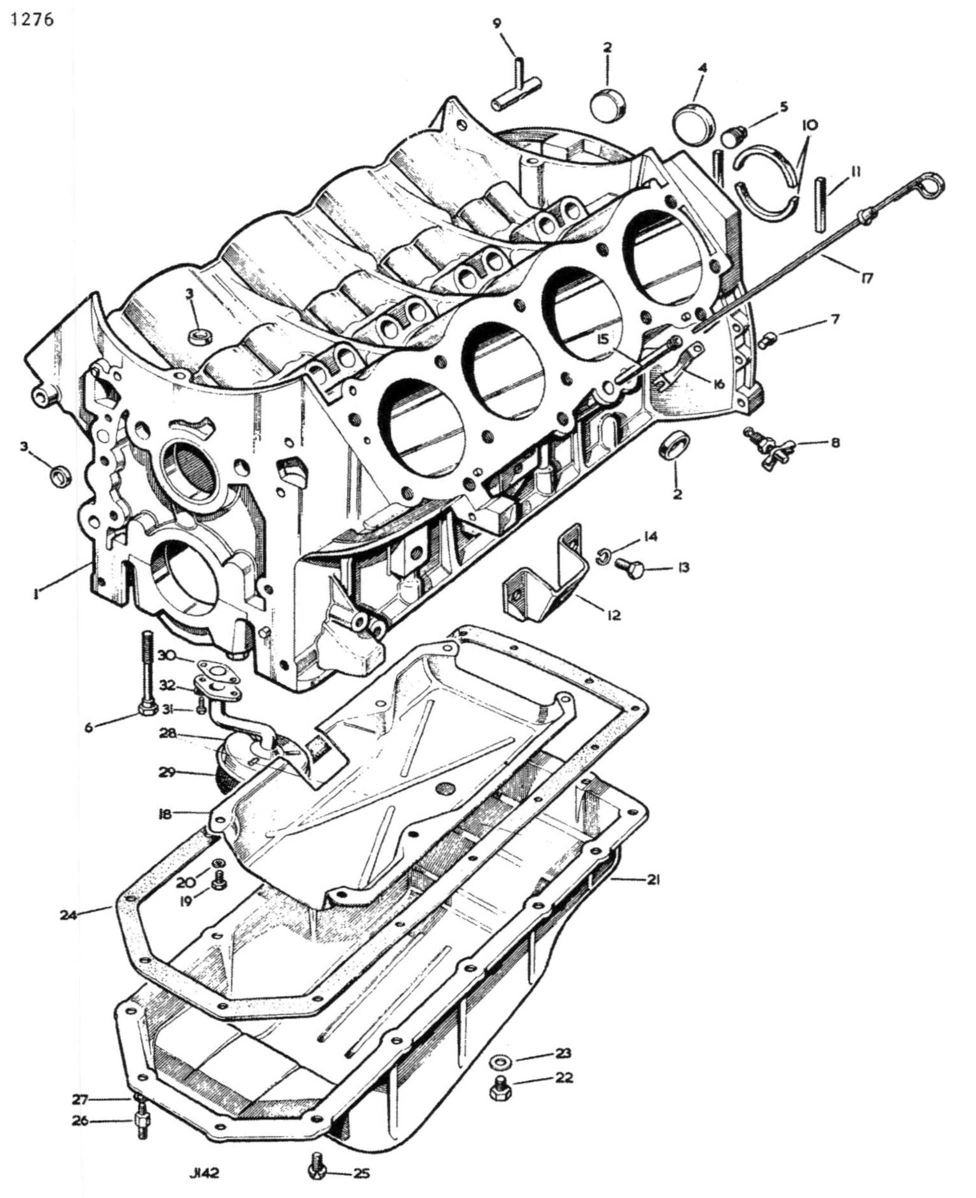

Plate Ref.	1 2 3 4	Description	Qty.	Part No.	Remarks
28		OIL SCREEN HOUSING ASSEMBLY	1	603625	
29		Oil screen	1	602070	
30		Gasket for oil screen housing	1	602068	
31		Bolt (¼" UNC x ⅜" Long) fixing oil screen housing assembly to block	2	253407	
32		Spring washer	2	3074	

*Asterisk indicates a new part which has not been used on any previous model.

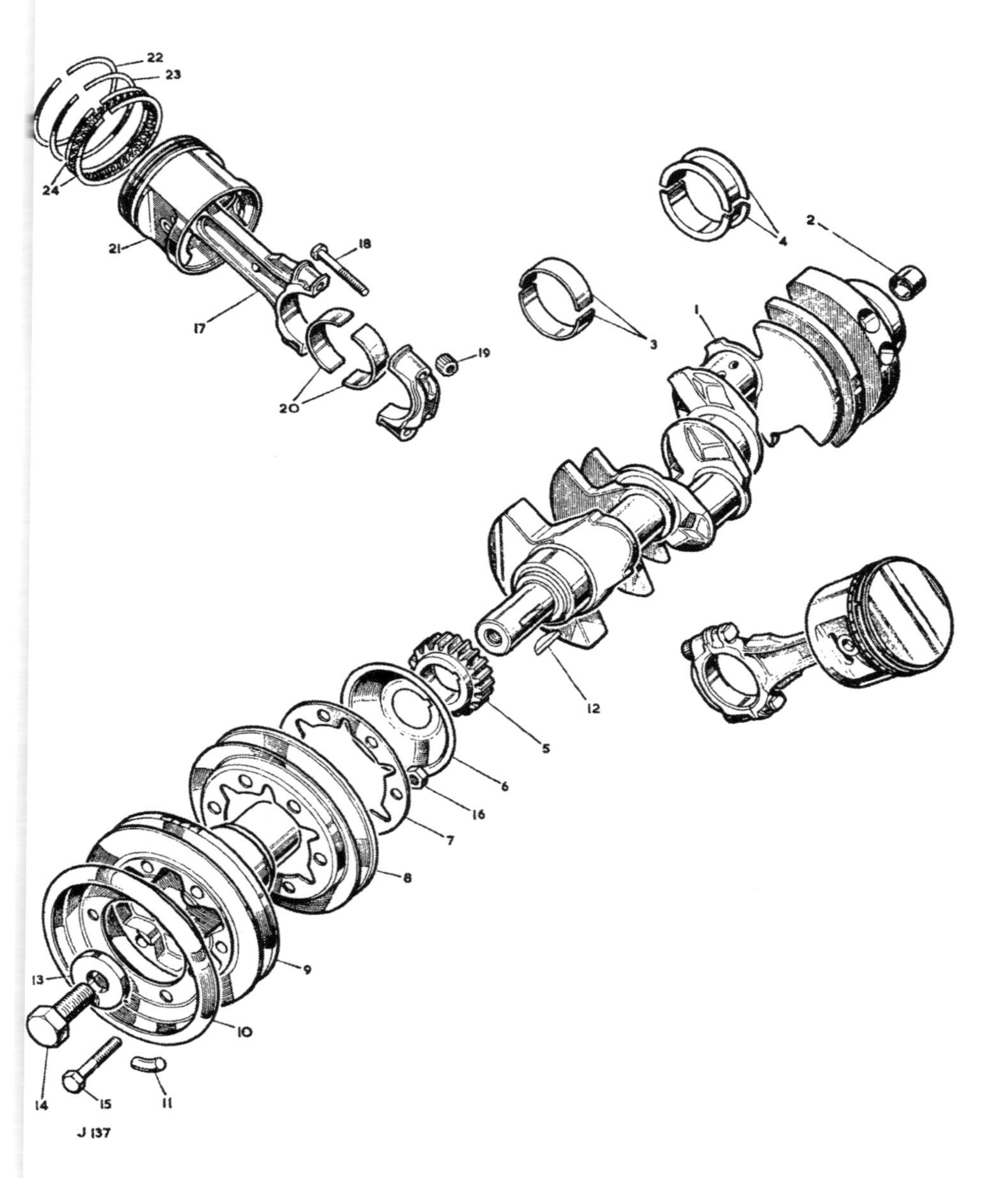

Plate Ref.	1 2 3 4 Description	Qty.	Part No.	Remarks
1	CRANKSHAFT ASSEMBLY, STD.	1	610193	Engines numbered up to 84014127C
	CRANKSHAFT ASSEMBLY, 0.010" U/S	1	606015	
	CRANKSHAFT ASSEMBLY, 0.020" U/S	1	606016	
	CRANKSHAFT ASSEMBLY, 0.030" U/S	1	606017	
	CRANKSHAFT ASSEMBLY, 0.040" U/S	1	606018	
2	Bush for crankshaft	1	610196	
	CRANKSHAFT ASSEMBLY, STD	1	611149	Engines numbered from 84014128C to 84018583C
	CRANKSHAFT ASSEMBLY, 0.010" U/S	1	606976	
	CRANKSHAFT ASSEMBLY, 0.020" U/S	1	606977	
	CRANKSHAFT ASSEMBLY, 0.030" U/S	1	606978	
	CRANKSHAFT ASSEMBLY, 0.40" U/S	1	606979	
	Bush for crankshaft	1	610196	
	CRANKSHAFT ASSEMBLY, STD	1	613275*	Engines numbered from 84018584D obwards
	CRANKSHAFT ASSEMBLY, .010" U/S	1	614299*	
	CRANKSHAFT ASSEMBLY, .020" U/S	1	614301*	
	CRANKSHAFT ASSEMBLY, .030" U/S	1	614303*	
	CRANKSHAFT ASSEMBLY, .040" U/S	1	614305*	
	Bush for crankshaft	1	610196	
3	Main bearing, STD } Pairs	4	606865	
	Main bearing, .010" U/S	4	606866	
	Main bearing, .020" U/S	4	606867	
	Main bearing, .030" U/S	4	606868	
	Main bearing, .040" U/S	4	606869	
4	Main bearing, flanged, STD journals and thrust flanges } Pairs	1	606870	
	Main bearing, flanged, 0.010" undersize journals and standard thrust flanges	1	606871	
	Main bearing, flanged, 0.020" undersize journals, 0.010" oversize thrust flange	1	606872	
	Main bearing, flanged, 0.030" undersize journals, 0.010" oversize thrust flange	1	606873	
	Main bearing, flanged, 0.040" undersize journals, 0.020" oversize thrust flange	1	606874	
5	Chainwheel on crankshaft	1	602372	
6	Oil thrower on crankshaft	1	602046	
7	Reinforcing plate	1	602587	Not part of engine assembly
8	Pulley driving power steering pump	1	603314	
9	Vibration damper and pulley	1	602578	
	Spring dowel pin for vibration damper	1	603535	
10	Balancing rim	1	603301	
11	Balance weight, .519" long } For rim	1	603523	As required
	Balance weight, 1.038" long	1	603524	
	Balance weight, 1.557" long	1	603525	
	Balance weight, 2.076" long	1	603526	
12	Key } Fixing vibration damper to crankshaft	1	602025	
13	Plain washer	1	602411	
14	Special bolt	1	602195	
15	Bolt (5/16" UNF x 1⅜" long) } Fixing pulley to vibration damper pulley	6	256221	
16	Nut (5/16" UNF)	6	254831	

*Asterisk indicates a new part which has not been used on any previous model.

Plate Ref.	1 2 3 4 Description	Qty.	Part No.	Remarks
17	CONNECTING ROD ASSEMBLY	8	602082	
18	Special bolt	16	602609	
19	Self-locking nut	16	602061	
20	Connecting rod bearing, std (Pairs)	8	606860	
	Connecting rod bearing, 0.010" U/S (Pairs)	8	606861	
	Connecting rod bearing, 0.020" U/S (Pairs)	8	606862	
	Connecting rod bearing, 0.030" U/S (Pairs)	8	606863	
	Connecting rod bearing, 0.040" U/S (Pairs)	8	606864	
21	PISTON ASSEMBLY STD GRADE Z	8	605152	Engines numbered up to 84014127C
	PISTON ASSEMBLY, STD GRADE A	8	605153	Engines numbered up to 84014127C
	PISTON ASSEMBLY, STD, GRADE B	8	605154	Engines numbered up to 84014127C
	PISTON ASSEMBLY, STD, GRADE C	8	605155	Engines numbered up to 84014127C
	PISTON ASSEMBLY, STD, GRADE D	8	605156	Engines numbered up to 84014127C
	PISTON ASSEMBLY, 0.010" O/S	8	605148	Engines numbered up to 84014127C
	PISTON ASSEMBLY, 0.020" O/S	8	605149	Engines numbered up to 84014127C
	PISTON ASSEMBLY, STD GRADE Z	8	606958	Engines numbered From 84014128C onwards
	PISTON ASSEMBLY, STD GRADE A	8	606959	Engines numbered From 84014128C onwards
	PISTON ASSEMBLY, STD GRADE B	8	606960	Engines numbered From 84014128C onwards
	PISTON ASSEMBLY, STD GRADE C	8	606961	Engines numbered From 84014128C onwards
	PISTON ASSEMBLY, STD GRADE D	8	606962	Engines numbered From 84014128C onwards
	PISTON ASSEMBLY, STD GRADE S	8	607269	Engines numbered From 84014128C onwards
	PISTON ASSEMBLY, 0.010" O/S	8	606963	Engines numbered From 84014128C onwards
	PISTON ASSEMBLY, 0.020" O/S	8	606964	Engines numbered From 84014128C onwards
22	Piston ring, compression, upper, std	8	610757	
	Piston ring, compression, upper, 0.010" O/S	8	611481	
	Piston ring, compression, upper, 0.020" O/S	8	611482	
23	Piston ring, compression, lower, std	8	602056	
	Piston ring, compression, lower, 0.010" O/S	8	603453	
	Piston ring, compression, lower, 0.020" O/S	8	603454	
24	Piston ring, scraper, std	8	603457	
	Piston ring, scraper, 0.010" O/S	8	603458	
	Piston ring, scraper, 0.020" O/S	8	603459	

*Asterisk indicates a new part which has not been used on any previous model.

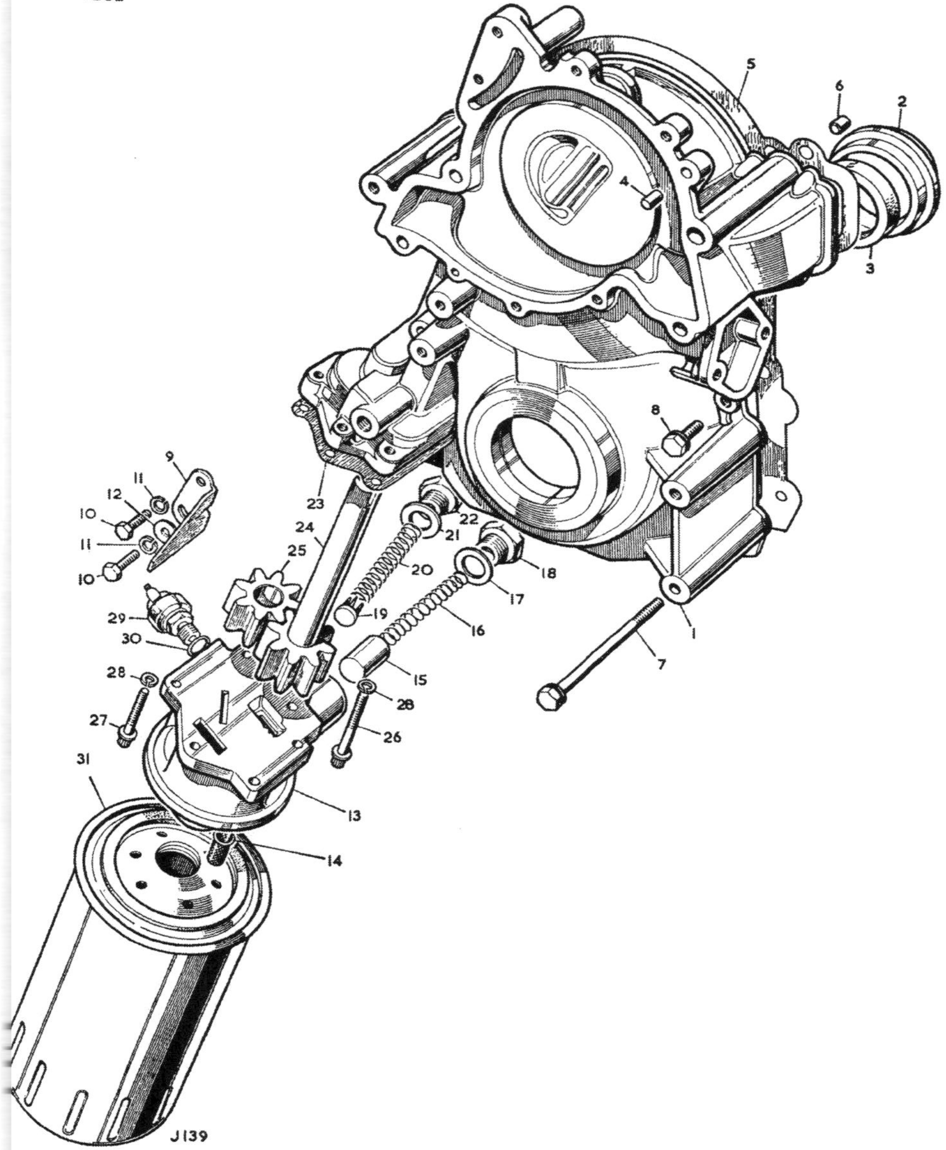

1282

J139

Plate Ref.	1 2 3 4 Description		Qty.	Part No.	Remarks
1	FRONT COVER ASSEMBLY		1	610391	
2	Oil thrower		1	602158	
3	Packing for crankshaft		1	602178	
4	Dowel locating water pump		2	602201	
5	Gasket for front cover		1	603775	
6	Dowel	Fixing front	2	602202	
7	Bolt (5/16" UNC x 3" long)	cover to	1	254030	
8	Bolt (5/16" UNCx1.12"long)	cylinder block	3	602234	
9	Timing pointer		1	602690	
10	Bolt (¼" UNC x ⅜" long)	Fixing timing	2	253206	
11	Spring washer	pointer to	2	3074	
12	Plain washer	front cover	1	3840	
	OIL PUMP COVER ASSEMBLY		1	603557	
13	OIL PUMP COVER		1	602324	
14	Oil strainer		1	602065	
15	Oil pressure relief valve		1	602064	
16	Spring for relief valve		1	602067	
17	Joint washer		1	603521	Not part of engine
18	Cap for relief valve		1	602071	assembly
19	Valve, nylon, oil filter by-pass		1	603558	
20	Spring, oil filter by-pass		1	602073	
21	Joint washer		1	603521	
22	Plug, oil filter by-pass		1	603208	
23	Gasket for oil pump cover		1	602072	
24	Oil pump shaft and gear		1	602018	
25	Oil pump idler gear		1	602017	
26	Bolt (¼" UNC x 1 9/16" long)	Fixing oil pump	1	602913	
27	Bolt (¼" UNC x 1" long)	cover to	5	602911	
28	Spring washer	front cover	6	3074	
29	Oil pressure switch		1	567920	
30	Joint washer for switch		1	243958	Not part of
	Oil pressure transmitter		1	555947	enginne assembly
	Shim washer for transmitter		1	512387	Coupe
	Adaptor for oil pressure transmitter		1	603144	
	Joint washer for adpator		1	603521	
31	Oil filter		1	GFE123	

*Asterisk indicates a new part which has not been used on any previous model.

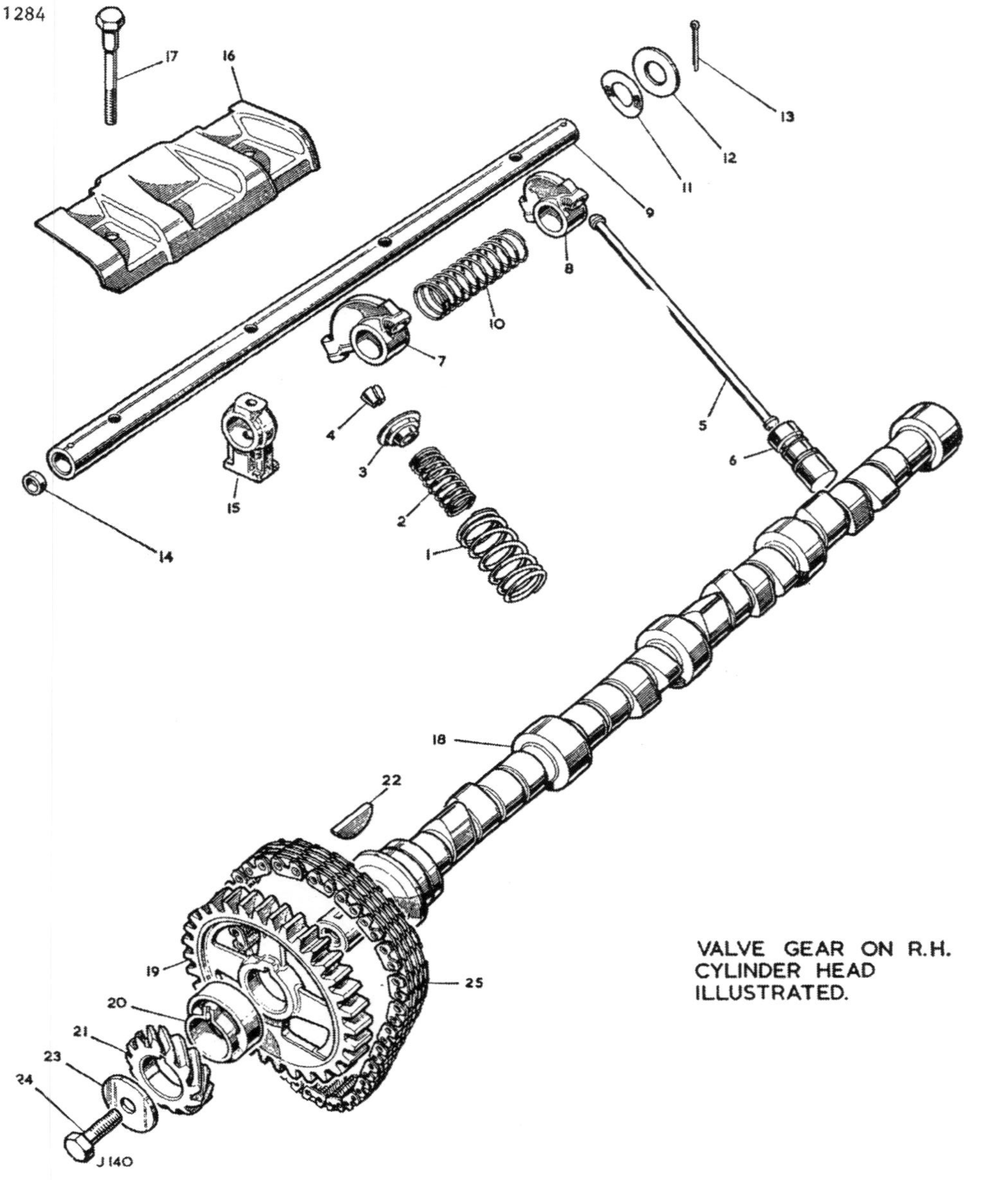

VALVE GEAR ON R.H.
CYLINDER HEAD
ILLUSTRATED.

VALVE GEAR, CAMSHAFT AND TIMING CHAIN

Plate Ref.	1 2 3 4 DESCRIPTION	Qty	Part No.	REMARKS
1	Valve spring, outer	16	602240	
2	Valve spring, inner	16	602241	
3	Cap for valve spring	16	602451	
4	Valve cotter, half	32	602303	
5	Push rod	16	603378	
6	Hydraulic tappet	16	602187	
7	Valve rocker, RH	8	602153	
8	Valve rocker, LH	8	602154	
9	Rocker shaft	2	602171	
10	Spring for valve rocker shaft	6	602142	
11	Waved washer, rocker shaft end	4	602148	
12	Washer } For valve	4	602186	
13	Split pin } rocker shafts	4	2981	
14	Plug for rocker shaft	4	603332	
15	Bracket for rocker shaft	8	603725	
16	Oil baffle for rocker arm	2	602172	
17	Special bolt fixing bracket to cylinder head	8	602097	
18	Camshaft	1	603724	
19	Chainwheel for camshaft	1	602373	
20	Cam, fuel pump drive	1	602149	
21	Gear, distributor drive	1	602159	
22	Key } Fixing chainwheel,	1	602025	
23	Washer } cam and gear	1	602510	
24	Special bolt } to camshaft	1	602227	
25	Timing chain	1	602371	

* Asterisk indicates a new part which has not been used on any previous Rover model

INLET MANIFOLD

1305

J152

INLET MANIFOLD 3½ LITRE MODELS

Plate Ref.	1 2 3 4 Description	Qty.	Part No.	Remarks
1	INLET MANIFOLD ASSEMBLY	1	603649	
2	Plug, .937" dia.	1	525497	
3	Plug, 1.5" dia.	1	602152	
4	Air bleed adaptor, .937" long	1	603430	Alternatives
	Air bleed adaptor, 1.5" long	1	603920	Alternatives
5	Stud (5/16" UNF/UNC) for carburetter	8	252514	
6	Bush for countershaft	2	522932	
	Blanking plate for thermostat switch hole	1	603664	Cars with automatic choke control
	Joint washer for blanking plate	1	236022	Cars with automatic choke control
	Set screw — Fixing blanking plate to inlet manifold	3	257064	Cars with automatic choke control
	Spring washer — Fixing blanking plate to inlet manifold	3	3073	Cars with automatic choke control
8	Gasket for inlet manifold	1	602216	
9	Seal for gasket	2	GEG645	
10	Clamp	2	602076	
11	Bolt and spring washer fixing clamp and gasket to cylinder block	2	602236	
12	Bolt (3/8" UNC x 1.50" long) — Fixing inlet manifold to cylinder head	10	254041	
13	Bolt (3/8" UNC x 1.88" long) — Fixing inlet manifold to cylinder head	2	602218	
14	Plug — Sealing redundant hole in inlet manifold	1	603224	Cars with manual choke control
15	Washer for plug — Sealing redundant hole in inlet manifold	1	232043	Cars with manual choke control
	Adaptor for delivery hose	1	603319	Cars with automatic choke control
16	Outlet pipe for heater	1	603440	Engine numbered up to Suffix letter 'C'
	Outlet pipe for heater	1	613377*	Engines numbered from Suffix letter 'D' onwards
17	Gasket for outlet pipe	1	603441	
18	Set bolt (1/4" UNC x 3/4" long) — Fixing outlet pipe to manifold	2	253007	
19	Spring washer — Fixing outlet pipe to manifold	2	3074	
20	Anchor bracket for countershaft spring	1	603561	
21	Anchor bracket for throttle return springs	2	602428	
22	Thermostat	1	GTS104	
23	Water outlet elbow	1	602645	Engines numbered up to 84006722
	Water outlet elbow	1	610388	Engines numbered from 84006723 onwards
24	Gasket for outlet elbow	1	GTG116	
25	Bolt (5/16" UNC x 7/8" long) — Fixing outlet elbow to manifold	2	253027	
26	Spring washer — Fixing outlet elbow to manifold	2	3075	
27	Hose for thermostat by-pass	1	602412	Engines numbered up to 84006722

*Asterisk indicates a new part which has not been used on any previous model.

INLET MANIFOLD

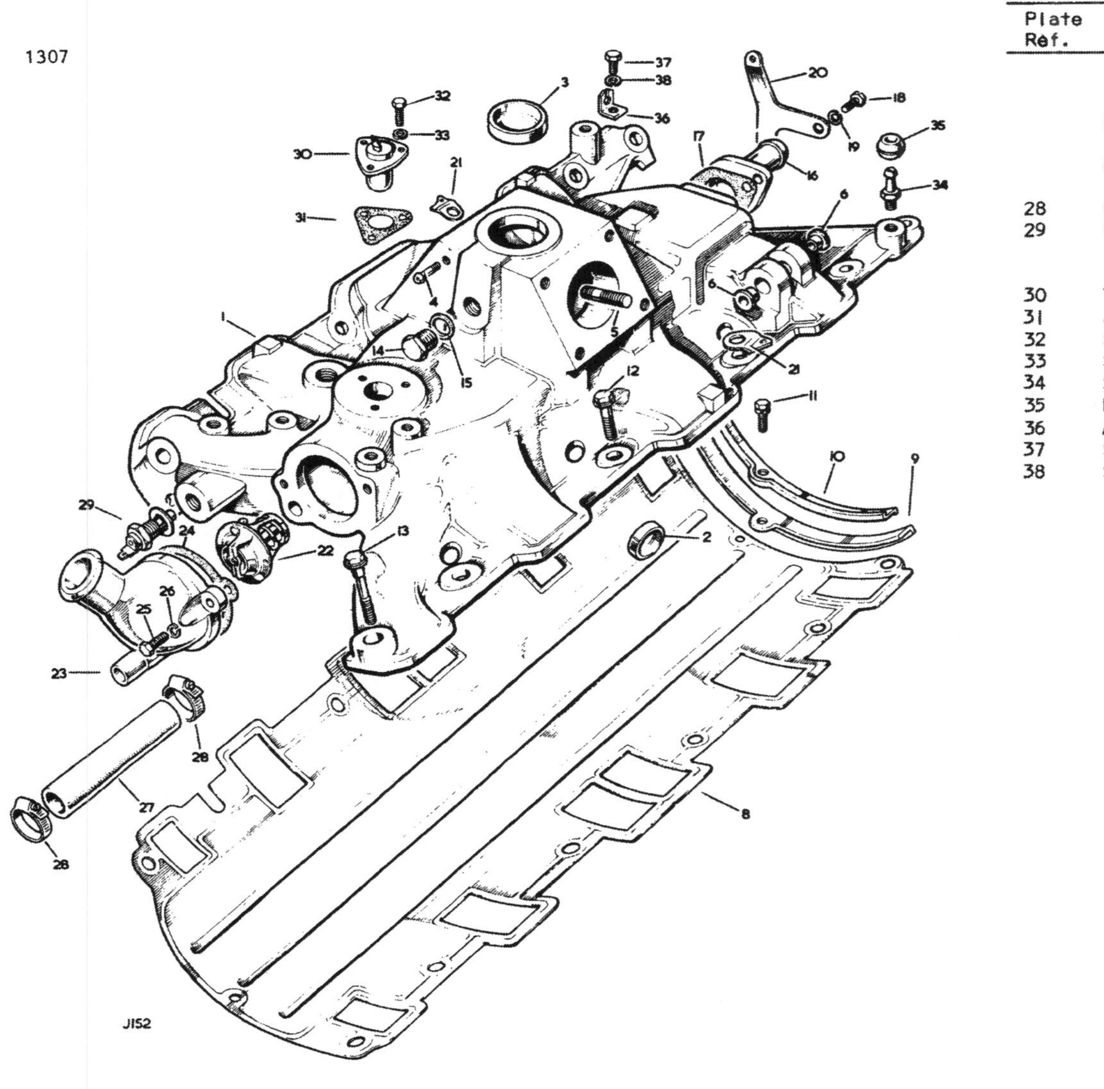

INLET MANIFOLD 3½ LITRE MODELS

Plate Ref.	1 2 3 4	Description	Qty.	Part No.	Remarks
		Hose for thermostat by-pass	1	610389	Engines numbered from 84006723 onwards
		Hose for therostat by-pass with unequal diameters	1	610392	
28		Hose clip fixing hose to front cover and elbow	2	GHC709	
29		WATER TEMPERATURE TRANSMITTER	1	568055	
		Joint washer for water temperature transmitter	1	568054	
30		Thermostat switch	1	545010	Cars with manual choke control
31		Joint washer for switch	1	236022	Cars with manual choke control
32		Set screw (10 UNC x 7/16" long) } Fixing switch to manifold	3	257064	Cars with manual choke control
33		Spring washer } Fixing switch to manifold	3	3073	Cars with manual choke control
34		Special stud for air cleaner fixing	1	603408	
35		Rubber grommet for stud	1	603851	
36		Anchor bracket for accelerator cable	1	603186	
37		Set bolt (¼" UNC x ⅜" long) } Fixing abutment bracket to manifold	1	253206	
38		Spring washer } Fixing abutment bracket to manifold	1	3074	

*Asterisk indicates a new part which has not been used on any previous model.

L.H. CYLINDER HEAD ILLUSTRATED

J138

Plate Ref.	1 2 3 4 Description	Qty.	Part No.	Remarks
1	CYLINDER HEAD ASSEMBLY	2	612572	
2	Valve guide	16	603554	
3	Valve seat insert, inlet	8	602052	
4	Valve seat insert, exhaust	8	602120	
5	Core plug, threaded	4	602123	
6	Cup plug	4	602289	
7	Gasket for cylinder head	2	GEG340	
8	Dowel	4	602040	
9	Special bolt, 2.28" long — Fixing cylinder heads to cylinder block	8	602191	
10	Special bolt, 2.74" long (ditto)	13	602192	
11	Special bolt, 3.94" long (ditto)	6	602193	
12	Special double-ended bolt (ditto)	1	602200	
13	Special washer (ditto)	28	602098	
	Special bolt, threaded head	1	602450	
14	Front lifting bracket, on LH head	1	603031	
15	Rear lifting bracket, on RH head	1	603032	
16	Bolt (3/8" UNC x 7/8" long) — Fixing lifting bracket to cylinder head	4	253046	
17	Spring washer (ditto)	4	3076	
18	Inlet valve	8	602166	
19	Exhaust valve	8	602165	
	Rocker cover assembly, RH	1	605896	
20	Rocker cover assembly, LH	1	605897	
21	Retainer for spark plug leads	2	603672	
22	Screw (10 UNC x .562" long) fixing retainer	2	602529	
23	Gasket for rocker cover	2	GEG436	
24	Screw (1/4" UNF x 1.312" long) — Fixing rocker cover to cylinder heads	4	603127	
25	Screw (1/4" UNC x 7/8" long) (ditto)	4	602530	
26	Spring washer (ditto)	8	3074	
27	Plain washer (ditto)	8	3911	
28	Oil filler cap for rocker cover, RH	1	574088	
29	'0' ring	1	564258	
	Gasket set, decarbonising	1	GEG165	

*Asterisk indicates a new part which has not been used or any previous model.

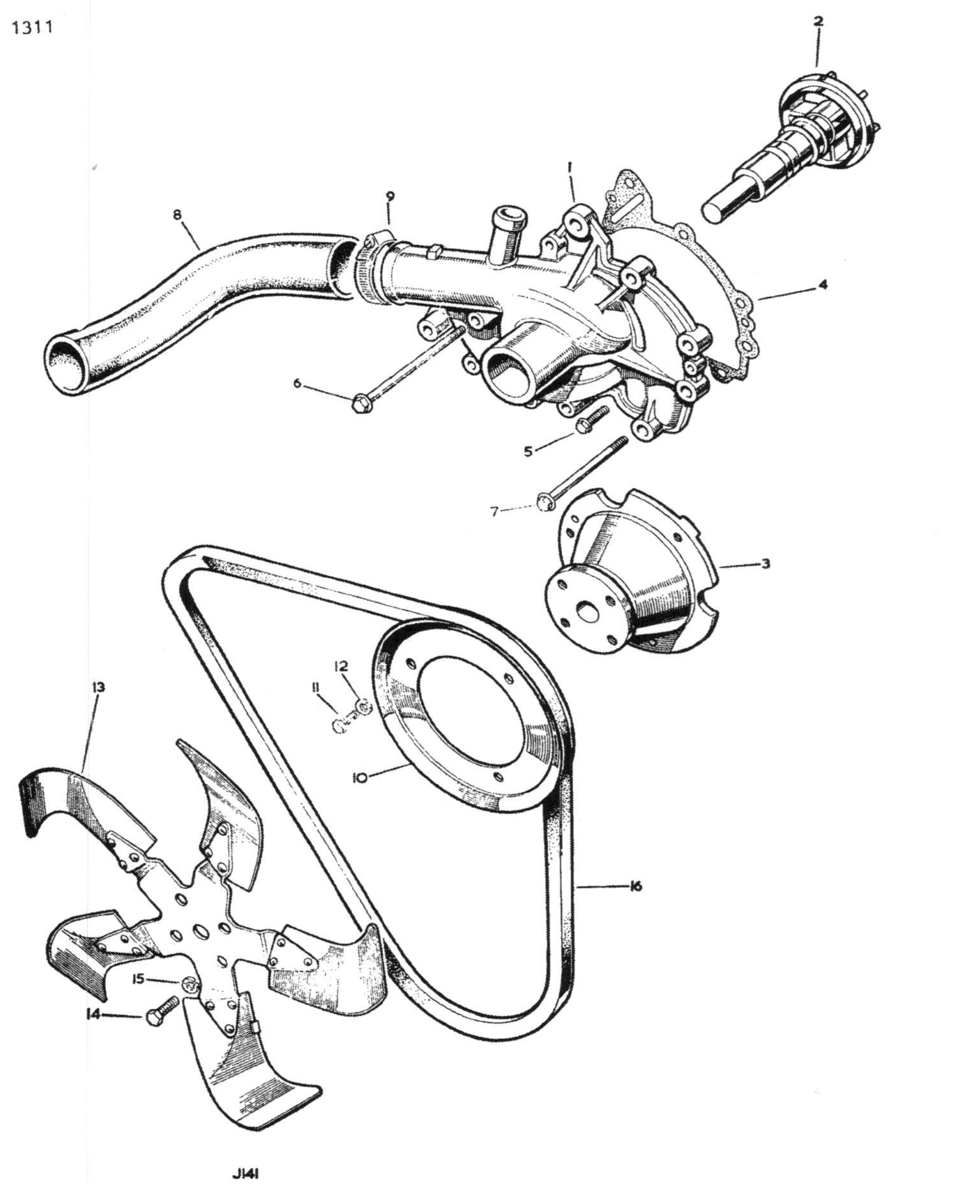

Plate Ref.	1 2 3 4 Description		Qty.	Part No.	Remarks
	WATER PUMP ASSEMBLY		1	GWP304	
3	Hub for fan and water pump pulley		1	603427	Engines numbered up to 84010541
	Hub for fan and water pump pulley		1	603743	Engines numbered from 84010542 onwards
4	Gasket for water pump		1	610756	
5	Bolt (¼" UNC x 1⅛" long)	Fixing water pump to front cover	5	253010	
6	Plain washer		5	2203	
7	Bolt (5/16" UNC x 5½" long)	Fixing water pump and front cover to block	1	253955	
8	Bolt (5/16" UNC x 4½" long)		3	254037	
9	Plain washer		4	2920	
10	Hose, pump to radiator		1	GRH451	Not part of engine assembly
11	Clip for hose		2	GHC913	
12	Pulley for water pump		1	602582	
13	Bolt (¼" UNF x ⅜" long)	Fixing pulley to hub	3	255206	Engines numbered up to 84010541
14	Spring washer		3	3074	
	Bolt (¼" UNC x ⅜" long)		3	253206	Engines numbered from 84010542 onwards
	Spring washer		3	3074	
15	Fan, 5-blade		1	612370	
16	Bolt (5/16" UNF x ¾" long)	Fixing fan to water pump pulley	4	255226	Engines numbered up to 84010541
17	Spring washer		4	3075	
	Bolt (5/16" UNC x ¾" long)		4	253026	Engines numbered from 84010542 onwards
	Spring washer		4	3075	
18	Driving belt for fan and alternator		1	GFB148	Not part of engine assembly

*Asterisk indicates a new part which has not been used on any previous model.

1313

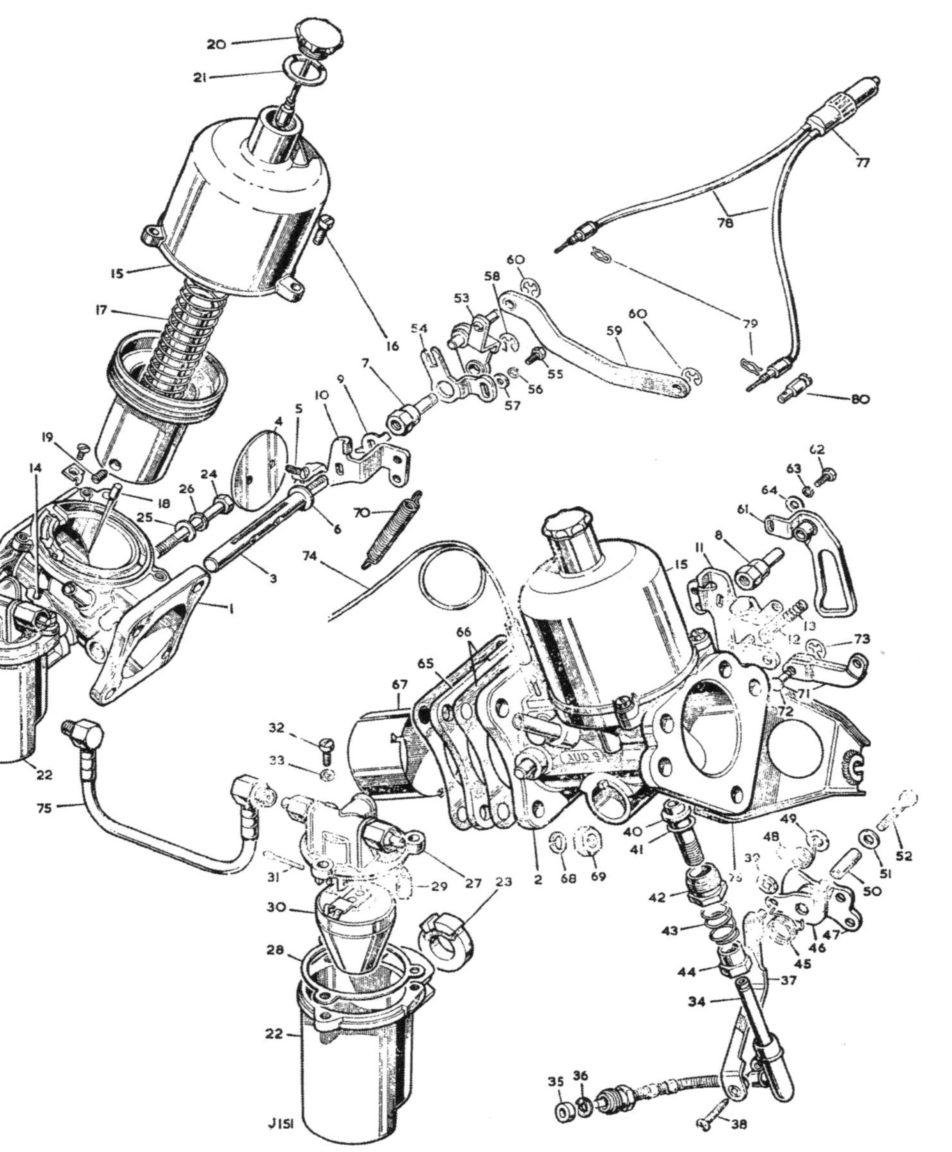

Plate Ref.	1 2 3 4 Description	Qty.	Part No.	Remarks
	CARBURETTER, RH	1	603700	Early models
	CARBURETTER, LH	1	603699	Early models
	CARBURETTER, RH	1	610516*	Late models
	CARBURETTER, LH	1	610515*	Late models
1	Carburetter body, RH	1	605441	
2	Carburetter body, LH	1	605442	
3	Throttle spindle	2	605443	
4	Throttle butterfly	2	245297	
5	Screw for throttle butterfly	4	262481	
6	Washer for throttle spindle	2	600103	
7	Special nut for spindle RH	1	603062	
8	Special nut for spindle LH	1	603241	
9	Tab washer for nut	1	600107	
10	Carburetter lever, RH	1	603055	
11	Carburetter lever, LH	1	603247	
12	Throttle adjustment screw	4	600111	2 off on cars with automatic choke
13	Spring for adjustment screw	4	262489	2 off on cars with automatic choke
14	Piston lift pin	2	601514	
	Spring (For piston lift pin)	2	262480	
	Circlip (For piston lift pin)	2	542318	
15	Suction chamber and piston complete	2	605495	
16	Special screw fixing suction chamber	6	600119	
17	Spring for piston (yellow)	2	262443	
18	Needle, KL	2	603701	
19	Special screw, .415" long, fixing needle	2	605865	Alternatives
	Special screw, .360" long, fixing needle	2	606108	Alternatives
20	Oil cap and damper complete	2	600120	
21	Washer for oil cap	2	600121	
22	Float chamber	2	605446	
	Adaptor for float chamber RH	1	605447	
23	Adaptor for float chamber LH	1	605448	
24	Special bolt (Fixing float chamber)	1	600124	
25	Plain washer (Fixing float chamber)	1	600126	
26	Spring washer (Fixing float chamber)	1	600125	
	Lid for float chamber RH	1	605459	
27	Lid for float chamber LH	1	605460	
28	Joint washer for float chamber	2	600128	
29	Needle valve and seat	2	601600	
30	Float and lever complete	2	600591	
31	Pin for float lever	2	267909	
32	Special screw (Fixing float chamber lid)	6	262493	
33	Spring washer (Fixing float chamber lid)	6	600131	
	JET ASSEMBLY, RH	1	605449	
34	JET ASSEMBLY, LH	1	605450	
35	Gland washer for pipe, float	2	600134	
36	Brass washer chamber end	2	600133	

*Asterisk indicates a new part which has not been used on any previous model.

1315

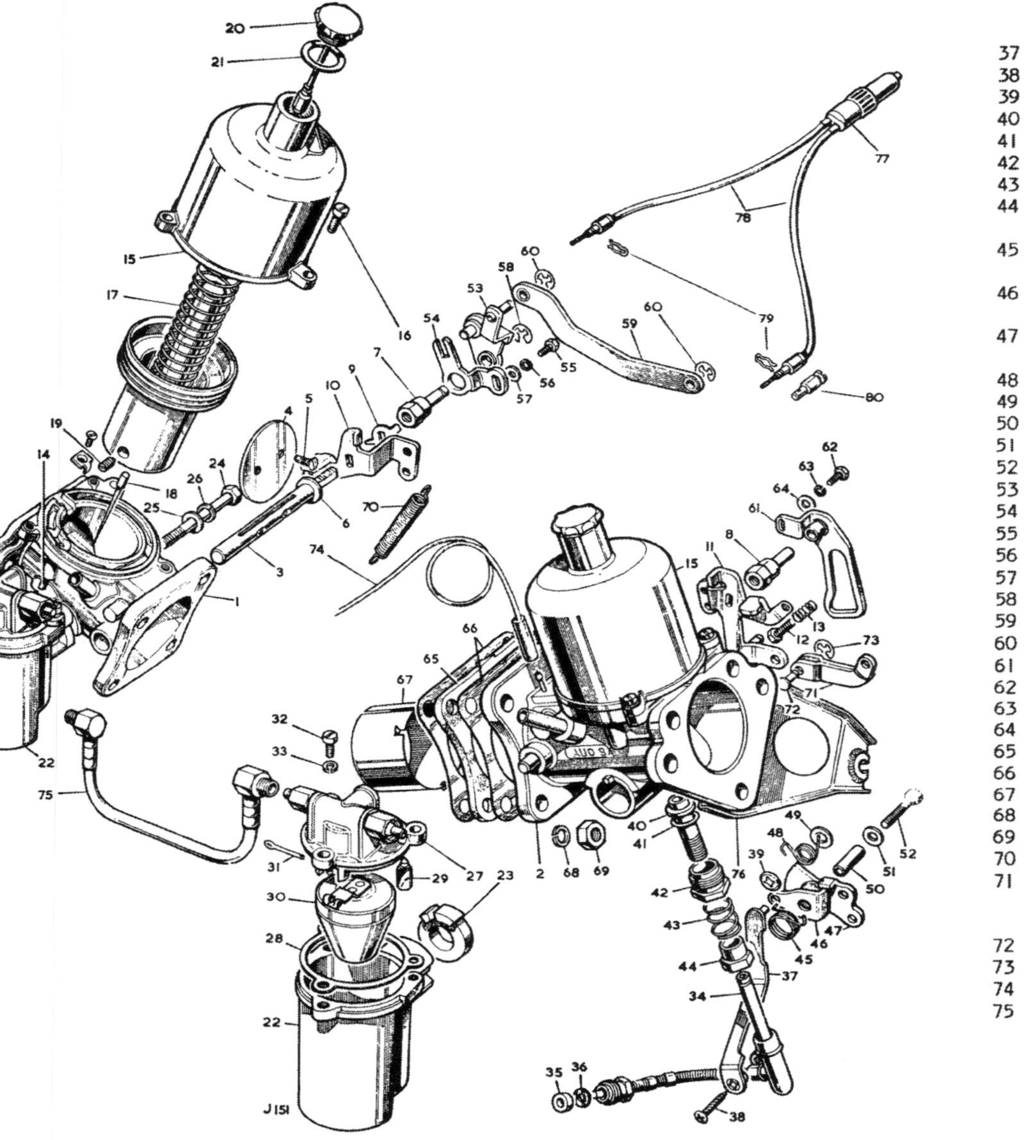

Plate Ref.	1 2 3 4 Description	Qty.	Part No.	Remarks
	Link bracket, RH,	1	605451	
37	Link bracket, LH	1	605452	
38	Self-tapping screw fixing bracket to jet	2	600140	
39	Starlock washer fixing link bracket	2	605298	
40	Jet bearing	2	600135	
41	Brass washer for jet bearing	2	600136	
42	Special nut fixing jet bearing	2	274958	
43	Lock spring } For jet	2	600138	
44	Adjusting nut } For jet	2	600137	
	Spring for pick-up lever, RH	1	605453	
45	Spring for pick-up lever, LH	1	600141	
	Pick-up lever and link, RH	1	605451	Cars with manual choke control
46	Pick-up lever and link, LH	1	605452	
	Cam lever, RH	1	605454	
47	Cam lever, LH	1	605455	
	Spring for cam lever, RH	1	605457	
48	Spring for cam lever, LH	1	605456	
49	Shim washer	4	600145	
50	Tube	2	600149	
51	Skid washer	2	600147	
52	Pivot bolt	2	600148	
53	Throttle lever, RH	1	603056	
54	Adjusting lever at RH throttle lever	1	603061	
55	Bolt (6 UNC x ⅜" long) } Fixing adjusting lever to RH throttle lever	1	257183	
56	Spring washer } Fixing adjusting lever to RH throttle lever	1	3072	
57	Plain washer } Fixing adjusting lever to RH throttle lever	1	3886	
58	Circlip for extension nut	1	603063	
59	Throttle link	1	610002	
60	Circlip fixing throttle link	2	602441	
61	Throttle lever complete, LH	1	603239	
62	Bolt (6 UNC x ⅜" long) } Fixing throttle lever	1	257183	
63	Spring washer } Fixing throttle lever	1	3072	
64	Plain washer } Fixing throttle lever	1	3886	
65	Insulator for carburetter	2	610849	
66	Joint washer for carburetter and insulator	4	242318	
67	Liner, carburetter to inlet manifold	2	602404	
68	Spring washer } Fixing carburetter to inlet manifold	8	3075	
69	Nut (5/16" UNF) } Fixing carburetter to inlet manifold	8	254831	
70	Return spring	2	610846	
71	Link for downshift cable	1	571340	
	Bolt (10 UNF x ¼" long) } Fixing downshift cable to link	1	257020	
	Self-locking nut (10 UNF) } Fixing downshift cable to link	1	251345	
72	Clevis pin } Fixing link LH carburetter lever	1	602648	Cars numbered up to suffix letter 'C'
73	Starlock washer } Fixing link LH carburetter lever	1	602441	
74	Vacuum pipe for distributor	1	603410	
75	Fuel pipe connecting carburetters	1	603030	

*Asterisk indicates a new part which has not been used on any previous model.

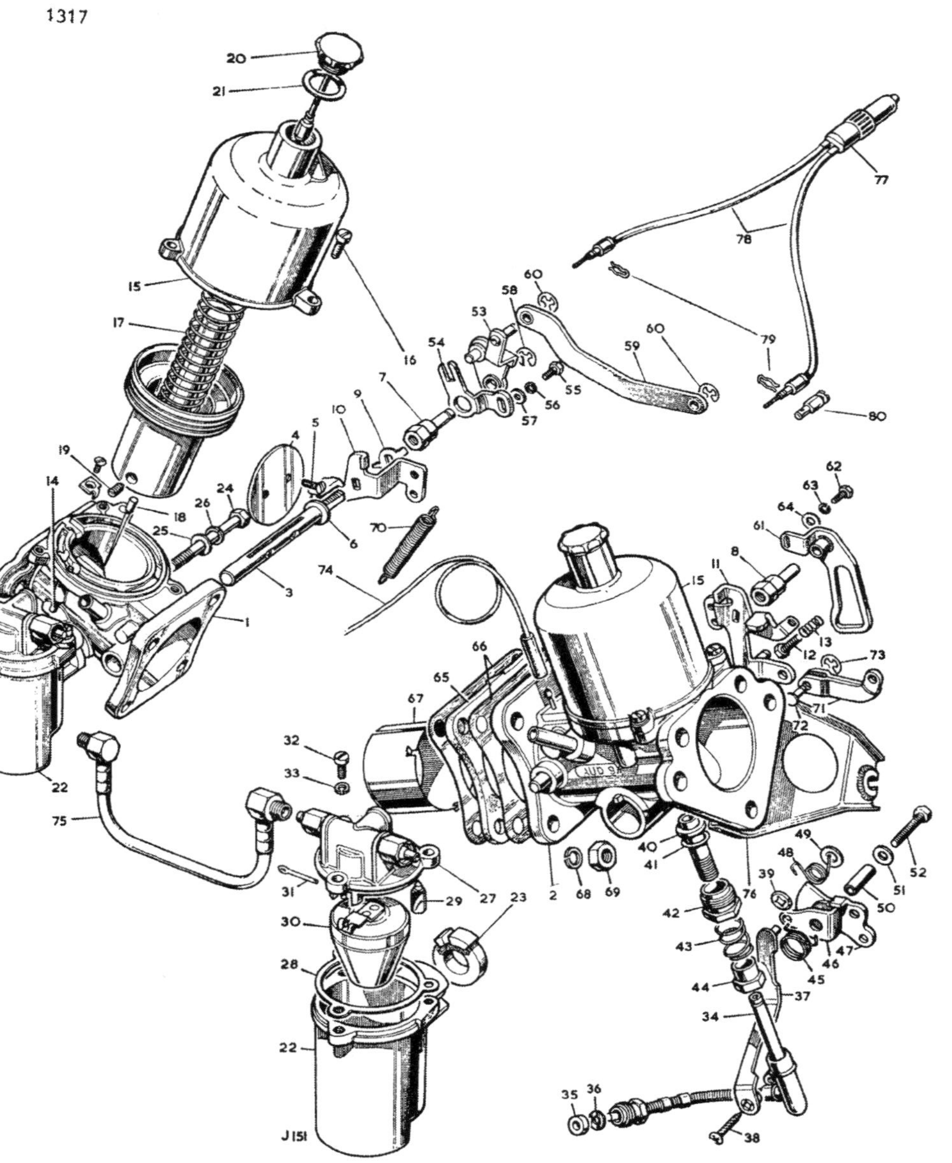

Plate Ref.	1 2 3 4 Description	Qty.	Part No.	Remarks
	Nylon vent pipe for carburetters	2	603679	
	Connecting hose for nylon vent pipe	2	603167	
	Clip] Fixing vent pipe to	2	603683	
	Grommet] inlet manifold	2	603689	
	Clip] Fixing vent pipes to	1	219676	
	Clip] cylinder block	1	603702	
76	Abutment bracket on LH carburetter for downshift cable	1	603786	Cars numbered up to suffix letter 'C'
77	Junction box for cold start cables	1	603331	
78	Cold start cable at carburetters	2	603400	
79	Spring clip fixing cables to bracket	2	538890	
80	Locking pin fixing cables to carburetter	2	553951	

*Asterisk indicates a new part which has not been used on any previous model.

Plate Ref.	1 2 3 4 Description		Qty.	Part No.	Remarks
1	AUTOMATIC ENRICHMENT DEVICE		1	606419	
	Gasket for float chamber lid and valve body		1	606064	
	Nylon filter		1	606061	
	Washer for plug		1	606060	
2	Bracket for A.E.D.		1	603719	
3	Bolt (¼" UNC x ⅜" long)	Fixing bracket to inlet manifold	2	253206	
4	Spring washer		2	3074	
5	Bolt (10 UNF x ½" long)	Fixing AED to bracket	2	257017	
6	Spring washer		2	3073	
7	Hot air pick-up at exhaust manifold	4 point fixing	1	612667	Engines numbered Up to 84016384
8	Gasket for hot air pick-up		1	612605	
9	Bolt (¼" UNC x ½" long)	Fixing hot air pick-up to exhaust manifold	4	253004	
10	Plain washer		4	2203	
11	Tab washer		4	603770	
	Hot air pick-up at exhaust manifold	6 point fixing	1	613466*	Engines numbered from 84016385 onwards
	Gasket for hot air pick-up		1	611429*	
	Bolt (¼" UNC x ½" long)	Fixing hot air pick-up to exhaust manifold	6	253004	
	Plain washer		6	2203	
	Tab washer		6	603770	
12	Hose, hot air pick-up to rear pipe		1	611039*	
13	Rear pipe		1	602644	
14	Hose, rear pipe to carburetter elbow		1	602646	
15	Insulation sleeve for upper pipe on hot air pick		1	602625	
16	Clip for upper pipe on hot air pick-up		1	603291	
17	Drive screw fixing clip		1	78604	
18	CONNECTING PIPE ASSEMBLY, hot air pick-up to A.E.D		1	610479	
19	Hose for connecting pipe		2	603290	
20	Delivery hose AED to inlet manifold		1	610214	Early models
	Cut-off valve for fuel to AED unit		1	610987	
	Hose for cut-off valve		1	611018	Late models
	Delivery hose for A.E.D unit		1	611017	
21	Clip for delivery hose		2	602198	4 off on late models
22	Balance pipe A.E.D to carburetter adaptor		1	603877	
23	Hose for balance pipe		2	603492	

*Asterisk indicates a new part which has not been used on any previous model.

1321

DISTRIBUTOR AND IGNITION HARNESS

1322

DISTRIBUTOR AND IGNITION HARNESS

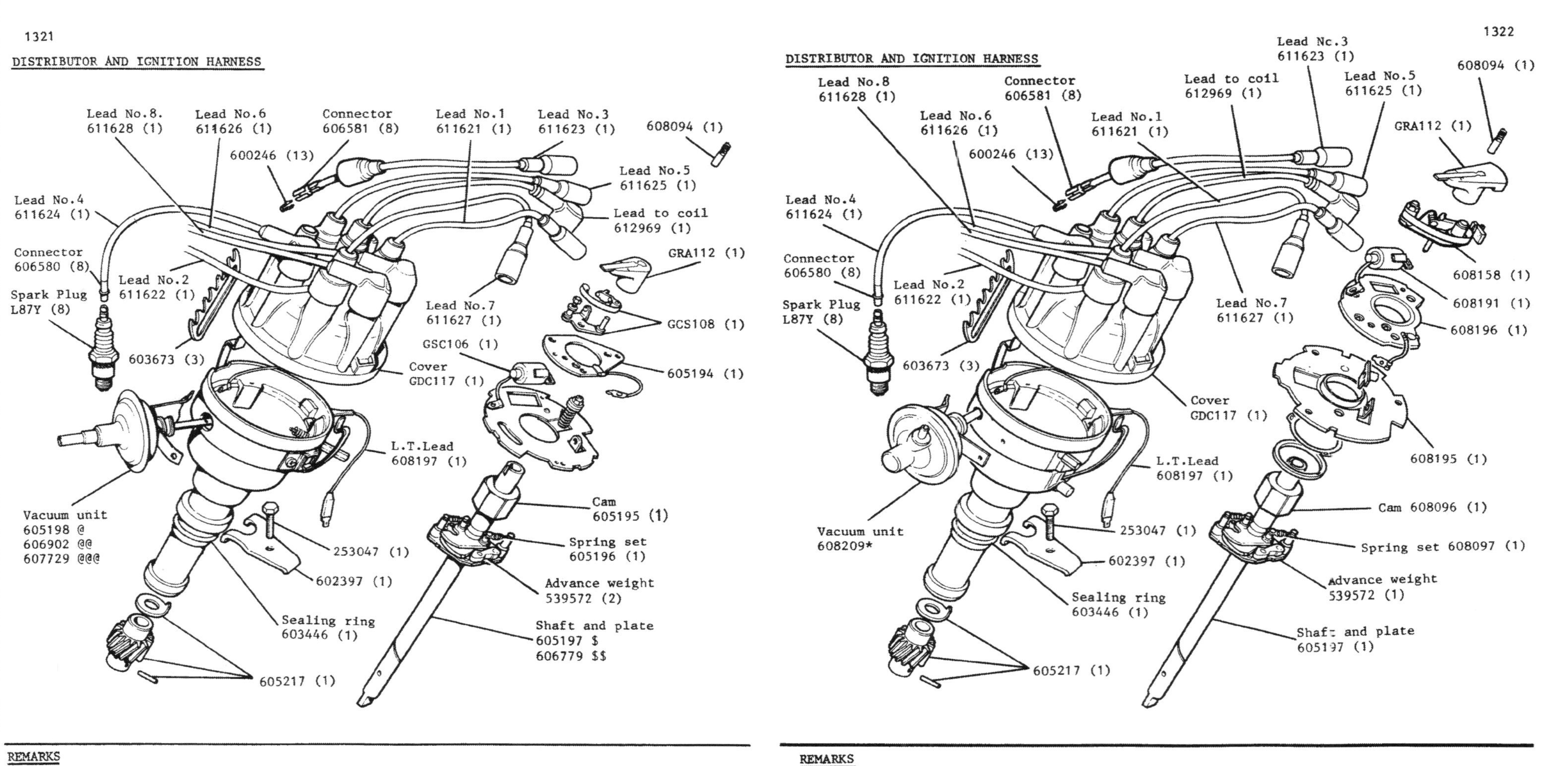

REMARKS

Distributor complete 611080. Applicable to 3½ Litre models with engines numbered up to 84017925.C

$ Applicable to all distributors except LU.41317. $$ Applicable to distributors LU.41317.

@ Part of Distributor LU.41176. @@ Part of Distributor LU.41278.

@@@ Part of Distributor LU.41317.

NOTE Early 3½ Litre models built to French specification differ from the above in the following details.

Cover screen for distributor 610186

Drive screw fixing cover 78581 (2)

H.T. lead coil to distributor 605982

Lead No.1 605983, Lead No.2 605984, Lead No.3 605985, Lead No.4 605986,

Lead No.5 605987, Lead No.6 605988, Lead No.7 605989, Lead No.8 605990

REMARKS

Distributor complete 614031*

Applicable to 3½ Litre models with engines numbered from 84017926.C onwards

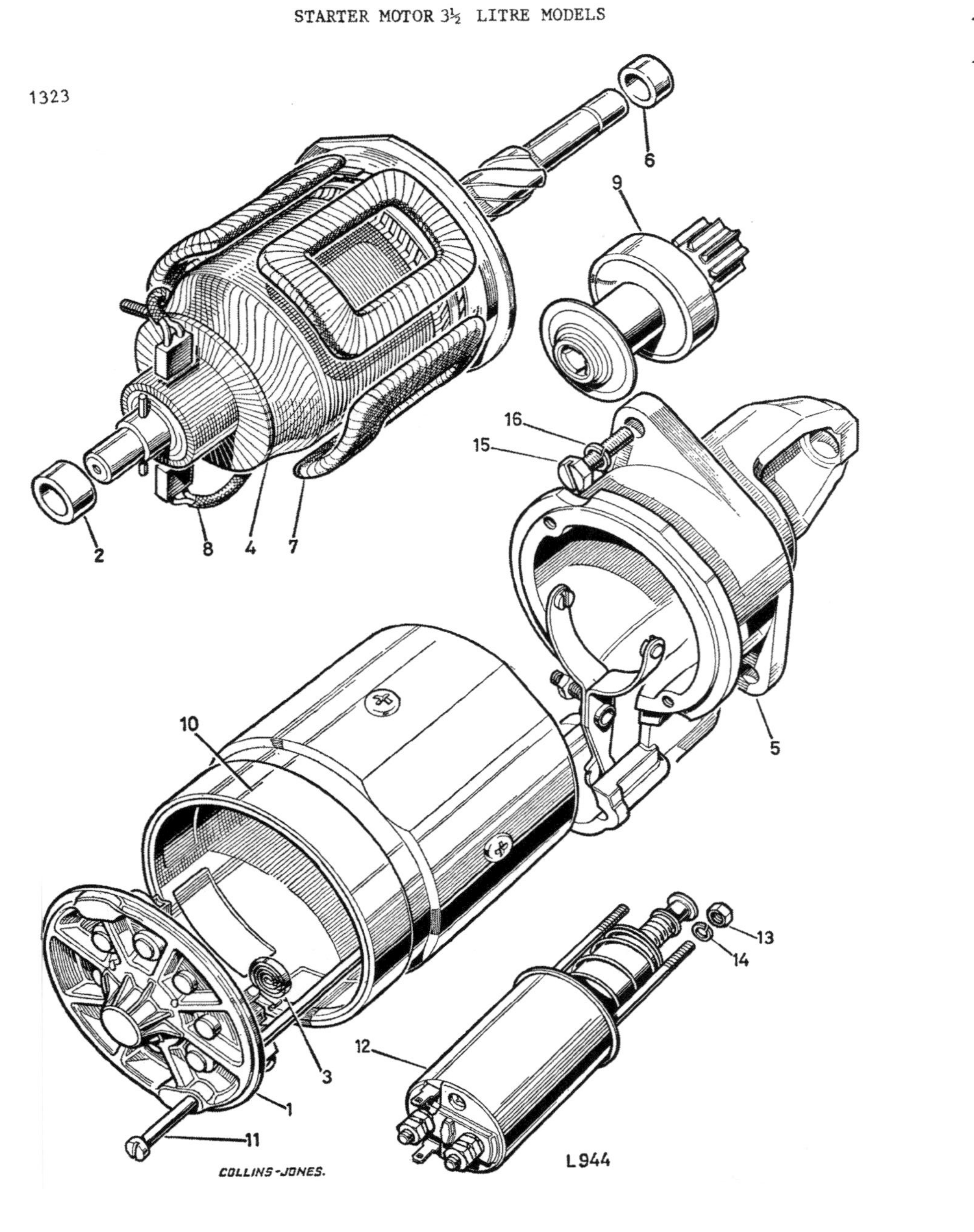

Plate Ref.	1 2 3 4 Description	Qty.	Part No.	Remarks
	STARTER MOTOR COMPLETE	1	559382	Not part of engine assembly
	BRACKET, COMMUTATOR END	1	605416	
	Bush, commutator end	1	605037	
	Spring set for brushes	1	601754	
	Armature	1	605421	
	BRACKET, DRIVE END	1	605418	
	Bush for bracket	1	605040	
	BRACKET, INTERMEDIATE	1	605419	
	Bush for intermediate bracket	1	605420	
	Field coil	1	605423	
	Brushes, set	1	GSB103	
	Drive assembly	1	605417	
	Cover band	1	605415	Engines numbered up to 84002574
	Bolt for starter	2	242675	
	Solenoid for starter motor	1	605630	
	Nut (¼" UNF) } Fixing solenoid to starter motor	2	254810	
	Spring washer } Fixing solenoid to starter motor	2	3074	
	Sundry parts kit	1	270251	
	STARTER MOTOR COMPLETE	1	568240	Not part of engine assembly
1	BRACKET, COMMUTATOR END	1	605917	
2	Bush, commutator end	1	605614	
3	Spring set for brushes	1	532569	
4	Armature	1	605920	
5	BRACKET, DRIVE END	1	605919	
6	Bush for bracket	1	605040	
7	Field coil	1	605916	Engines numbered from 84002575 to 84018390
8	Brushes, set	1	270225	
9	Drive assembly	1	605918	
10	Cover band	1	605915	
11	Bolt for starter	2	242675	
12	Solenoid for starter motor	1	605921	
13	Special nut } Fixing solenoid to starter motor	2	605705	
14	Spring washer } Fixing solenoid to starter motor	2	3074	
	Sundry parts kit	1	270251	
15	Special bolt } Fixing starter motor to cylinder block	2	602623	Not part of engine assembly
16	Spring washer } Fixing starter motor to cylinder block	2	3076	
	Cable strap } Fixing harness to starter motor	1	568680	
	Strap grip } Fixing harness to starter motor	1	568681	

*Asterisk indicates a new part which has not been used on any previous model.

STARTER MOTOR, 3M100 TYPE

1325

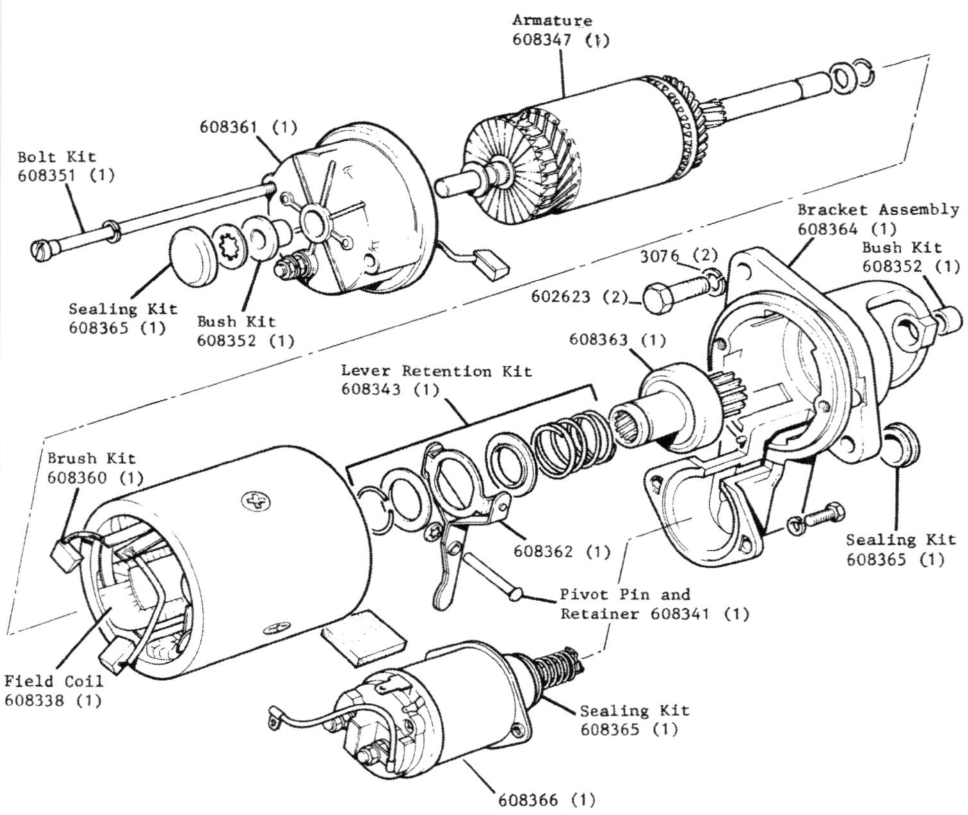

Armature 608347 (1)
608361 (1)
Bolt Kit 608351 (1)
Sealing Kit 608365 (1)
Bush Kit 608352 (1)
Bracket Assembly 608364 (1)
Bush Kit 608352 (1)
3076 (2)
602623 (2)
608363 (1)
Lever Retention Kit 608343 (1)
Brush Kit 608360 (1)
608362 (1)
Sealing Kit 608365 (1)
Pivot Pin and Retainer 608341 (1)
Field Coil 608338 (1)
Sealing Kit 608365 (1)
608366 (1)

REMARKS

Starter Motor Complete 586643

Sundry Parts Kit 608178

Applicable to 3½ Litre models with engines numbered from 84018391 onwards

This page is intentionally left bank

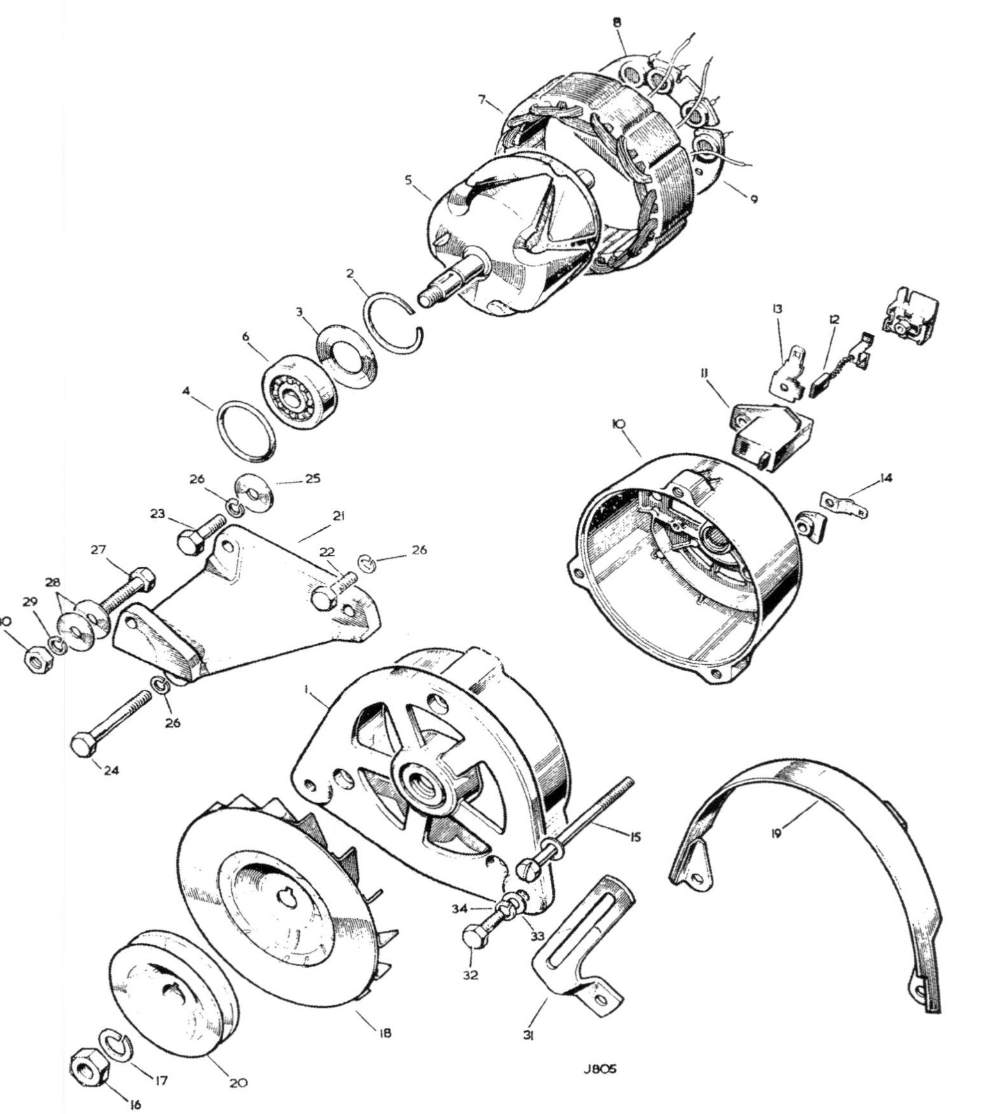

Plate Ref.	Description (1 2 3 4)		Qty.	Part No.	Remarks
	ALTERNATOR COMPLETE		1	559392	
1	BRACKET, DRIVE END		1	605411	
2	Circlip for bearing		1	601740	
3	Retaining plate for bearing		1	601741	
4	'O' ring		1	601742	
5	Rotor		1	601743	
6	Bearing for rotor, drive end		1	242672	
7	Stator		1	605404	
8	Heatsink with anode base diodes		1	605405	
9	Heatsink with cathode base diodes		1	605406	
10	Bracket slip ring end		1	605753	
11	Brushbox,		1	605407	
12	Brush, spring and Lucar		1	605408	
13	Main connector, Lucar		1	605409	
14	Connector, Lucar cut-out		1	563257	
15	Special bolt fixing brackets		3	605901	
16	Special nut for rotor		1	601738	
17	Spring washer		1	547604	Not part
	Sundry parts set		1	606822	of engine assembly
18	Fan		1	605410	
19	Guard for alternator fan		1	603781	
20	Pulley for alternator		1	602505	
21	Mounting bracket for alternator		1	602369	
22	Bolt (3/8" UNC x 1 3/8" long)	Fixing bracket to cylinder head	1	254040	
23	Bolt (5/16" UNC x 2" long)		1	254026	
24	Bolt (3/8" UNC x 2¼" long)		1	254046	
25	Plain washer		1	3036	
26	Spring washer		2	3076	
27	Bolt (5/16" UNF x 1½" long)	Fixing alternator to bracket	1	256222	
28	Plain washer		3	2220	
29	Spring washer		1	3075	
30	Nut (5/16" UNF)		1	254831	
31	Adjusting link for alternator		1	602398	
32	Bolt (5/16" UNC x 7/8" long)	Fixing adjusting link to alternator	1	253027	
33	Plain washer		1	4569	
34	Spring washer		1	3075	

*Asterisk indicates a new part which has not been used on any previous model.

1331

Plate Ref.	1 2 3 4 Description	Qty.	Part No.	Remarks
	AUTOMATIC TRANSMISSION AND TORQUE CONVERTER 1.FU series, Blue plate	1	605710	Cars numbered with suffix letter 'A'
	AUTOMATIC TRANSMISSION AND TORQUE CONVERTER 5.FU series, Red plate	1	605738	Cars numbered suffix letter 'B'
	AUTOMATIC TRANSMISSION AND TORQUE CONVERTER 267 series, French Blue plate	1	607081	Alternatives. Check before ordering
	AUTOMATIC TRANSMISSION AND TORQUE CONVERTER 303 series, Yellow plate	1	607290	Alternatives. Check before ordering
1	Torque converter, without cooling fins	1	607431	
2	Bolt (3/8" UNF x 3/8" long) } Fixing torque converter to flexible drive plate	4	255444	Part of Automatic Transmission
3	Plain washer } Fixing torque converter to flexible drive plate	4	2204	Part of Automatic Transmission
4	Housing with cooling apertures for torque converter	1	605360	Alternatives. Check before ordering. Part of Automatic Transmission
	Housing without cooling apertures for torque converter	1	607432	Alternatives. Check before ordering. Part of Automatic Transmission
5	Stone guard, plastic	2	605758	Engines numbered up to. Part of Automatic Transmission
	Bolt (3/8" UNC x 2¾" long) } Fixing gearbox to engine	2	254048	
6	Bolt (3/8" UNC x 2" long) } Fixing gearbox to engine	6	254045	
7	Spring washer } Fixing gearbox to engine	8	3076	
8	Reinforcing plate	1	602906	
9	Flexible drive plate	1	602903	
10	Bolt (7/16" UNF x 1.365" long) fixing flexible drive plate to crankshaft	6	602905	
11	STARTER RING GEAR ASSEMBLY	1	603340	
12	Dowel	2	529364	
13	Special bolt fixing starter ring to flexible drive plate	10	603223	Early models
	Special bolt } Fixing starter ring to flexible drive plate	10	610735	Late models
	Washer } Fixing starter ring to flexible drive plate	10	610736	Late models
14	Spacer	1	603295	
15	Front cover plate	1	571409	
16	Bolt (¼" UNC x ½" long) } Fixing front cover plate to gearbox and engine	4	253004	
17	Spring washer } Fixing front cover plate to gearbox and engine	4	3074	
18	Bolt (5/16" UNC x 3/8" long) } Fixing front cover plate to gearbox and engine	2	253025	
19	Spring washer } Fixing front cover plate to gearbox and engine	2	3075	
	Special bolt and washer fixing housing to casing	6	601003	

*Asterisk indicates a new part which has not been used on any previous model.

AUTOMATIC TRANSMISSION MAIN CASING 3½ LITRE MODELS

1333

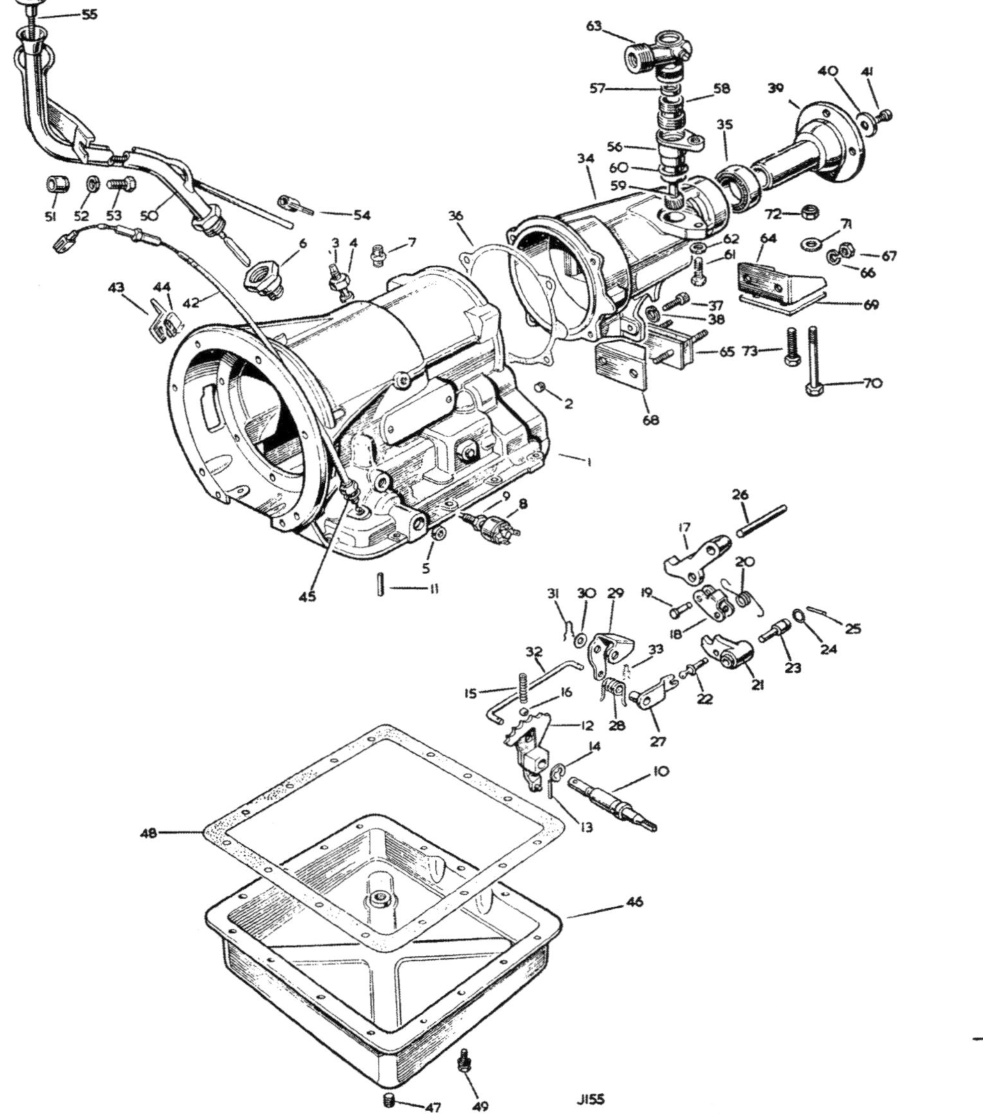

1334

AUTOMATIC TRANSMISSION MAIN CASING 3½ LITRE MODELS

Plate Ref.	1 2 3 4 Description	Qty.	Part No.	Remarks
1	MAIN CASE ASSEMBLY	1	605395	
2	Pipe plug	1	601006	
3	Screw, rear servo adjusting	1	601007	
4	Nut, locking adjusting screw	1	601008	
5	Oil seal for manual control shift	1	601009	
6	Adaptor for tube connector	1	601369	
7	Connector, tube outlet, oil cooler	2	601368	
8	INHIBITOR SWITCH	1	601010	
9	Locknut for switch	1	601925	
10	Shaft for manual valve lever	1	606258	
11	Pin retaining shaft	1	605384	
	Cup plug, 5/32" dia.	1	605385	
12	Lever for manual valve detent	1	605381	
13	Retaining pin for lever	1	601924	
14	Circlip	1	605383	
15	Spring for manual valve detent	1	601146	
16	Ball (¼" dia.)for manual valve detent	1	601148	
17	Pawl for parking brake	1	601085	
18	Link for parking brake	1	601086	
19	Pin for toggle link, pawl end	1	601087	
20	Spring, parking brake release	1	601088	
21	Toggle lever	1	601089	
22	Pin and spring for toggle lever, ball-ended	1	601695	
23	Pin for toggle lever, shouldered	1	601093	
24	'O' ring for toggle lever	1	601094	
25	Roll pin for toggle lever pin	1	601095	
26	Anchor pin for pawl	1	601096	
27	Toggle lift lever, forked	1	601097	
28	Spring for lift lever	1	601098	
29	Torsion lever for parking brake	1	605858	
30	Plain washer for torsion lever	1	601100	
31	Retaining spring for otrsion lever	1	601101	
32	Link for parking brake	1	605859	
33	Clip for linkage rod	2	601103	
34	EXTENSION HOUSING ASSEMBLY	1	605387	
35	Oil seal for housing	1	601160	
36	Gasket for extension casing	1	605428	
37	Special bolt fixing extension housing to case	4	605388	
38	Lock washer	4	605389	
39	Coupling flange	1	601381	
40	Plain washer for flange	1	601163	
41	Special screw fixing coupling flange	1	601164	
42	Downshift cable complete, long	1	605403	Cars numbered up to suffix letter 'C'
	Downshift cable complete, short	1	606514	Cars numbered from suffix letter 'D' onwards
	Downshift cable complete, short	1	607927	'303' transmission
43	Bracket for downshift cable	1	603545	Cars numbered up to suffix letter 'C' only
44	Clip fixing downshift cable to bracket	1	603546	
	Clip retaining downshift cable to adjuster	1	576307	
	Distance piece, bracket to cylinder head	1	603123	
45	Sealing washer for cable	1	601946	
46	Oil pan	1	601376	

*Asterisk indicates a new part which has not been used on any previous model.

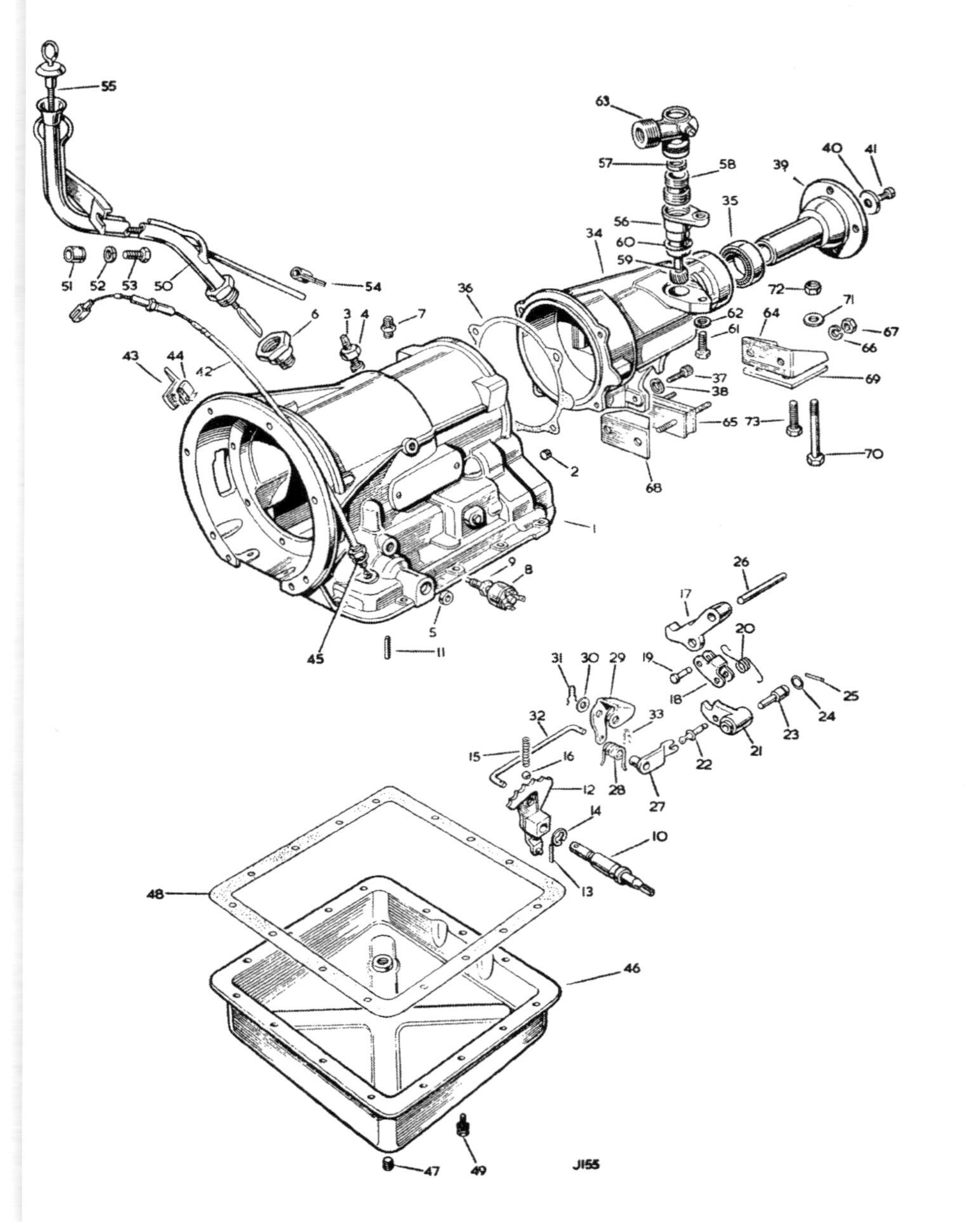

Plate Ref.	1 2 3 4	Description	Qty.	Part No.	Remarks
47		Drain plug, taper	1	601377	Alternatives
		Drain plug, pan head	1	606637	
48		Gasket for oil pan	1	601150	
49		Special bolt and washer fixing oil pan	13	601151	
		Special stud for oil pan fixing	2	605386	
50		Oil filler tube	1	561989	
51		Distance piece	1	561843	
52		Spring washer — Fixing oil filler tube	1	3074	
53		Set bolt (¼" UNC x 1⅜" long) to inlet manifold	1	253052	
54		Clip for breather tube	1	554423	
55		Oil level rod for gearbox	1	576954*	
56		Housing for speedometer drive	1	549065	
57		Oil seal, small, for housing	1	528747	
58		Adaptor for housing	1	549061	
59		Speedo driven gear and spindle	1	556609	
60		'O' ring for speedo housing	1	271013	
61		Bolt (¼" UNC x 1" long) — Fixing speedometer housing to extension casing	1	253009	
62		Spring washer — Fixing speedometer housing to extension casing	1	3074	
63		Right-angled drive for speedometer	1	277741	
64		Mounting bracket for gearbox	1	562195	
65		Rubber mounting	1	562194	
66		Spring washer — Fixing rubber mounting to bracket and sub-frame	4	3076	
67		Nut (⅜" UNF) — Fixing rubber mounting to bracket and sub-frame	4	254812	
68		Backplate for gearbox mounting	1	562314	
69		Shim plate, .128" thick — Fixing backplate to chassis sub-frame	1	562191	
		Shim plate, .064" thick — Fixing backplate to chassis sub-frame	1	562190	
		Shim plate, .032" thick — Fixing backplate to chassis sub-frame	1	562189	
70		Bolt (⅜" UNF x 4¾" long) — Fixing backplate to chassis sub-frame	1	256056	
71		Self-locking nut (⅜" UNF) — Fixing backplate to chassis sub-frame	3	252162	
72		Plain washer — Fixing backplate to chassis sub-frame	6	4094	
73		Bolt (⅜" UNF x 1" long) — Fixing backplate to chassis sub-frame	2	255247	
		Plain washer — Fixing automatic transmission to engine assembly	8	2210	
		Nut (⅜" UNF) — Fixing automatic transmission to engine assembly	12	254812	
		Plain washer — Fixing automatic transmission to engine assembly	1	2550	
		Nut (5/16" UNF) — Fixing automatic transmission to engine assembly	1	254831	

*Asterisk indicates a new part which has not been used on any previous model.

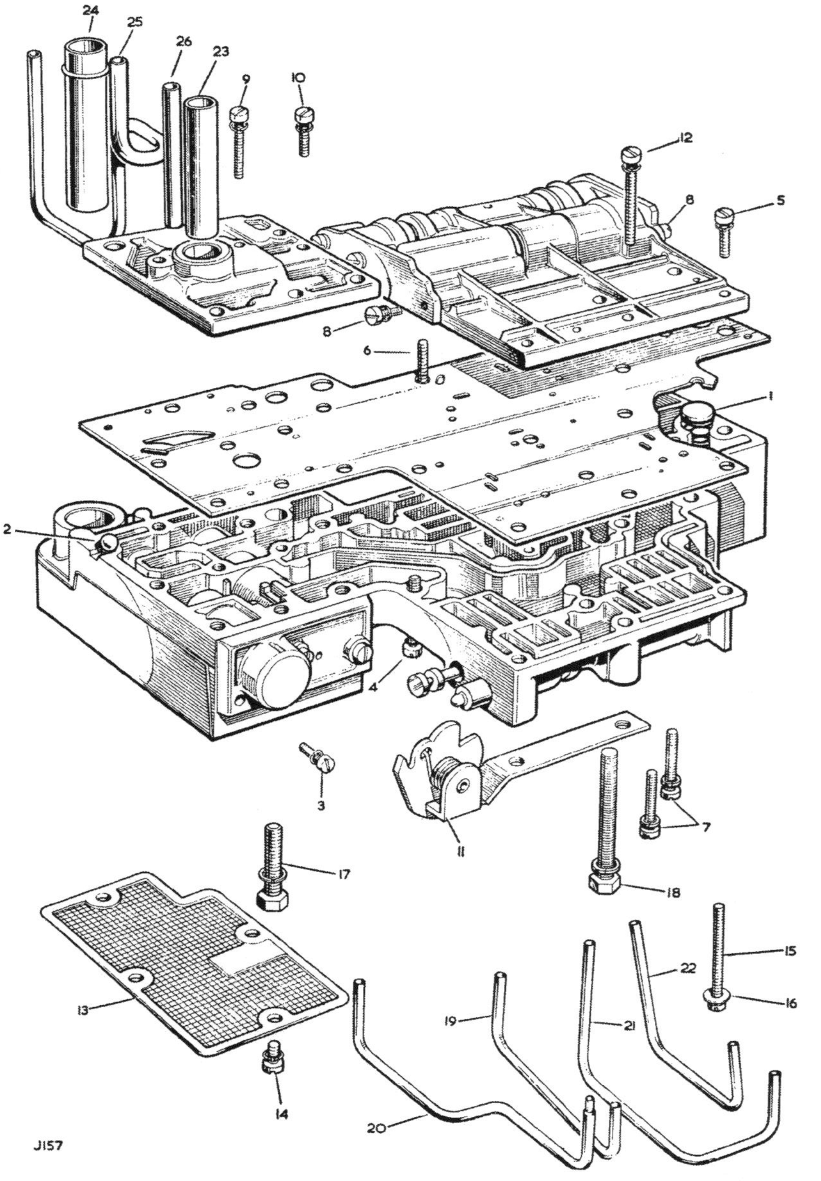

Plate Ref.	1 2 3 4 Description	Qty.	Part No.	Remarks
	VALVE BODIES ASSEMBLY	1	605463	Cars numbered suffix letter 'A'
	VALVE BODIES ASSEMBLY	1	606037	Cars numbered suffix letter 'B' onwards
1	Check valve	1	601114	
2	Ball (¼" diameter) steel	1	601157	Alternatives
	Ball (¼" diameter) nylon	1	606439	
3	Special screw and washer fixing end plate to valve body lower	3	605545	
4	Special screw and washer fixing valve body upper to valve body lower	5	605547	
5	Special screw and washer Fixing valve body upper to	2	605545	
6	Special screw and washer valve body lower	1	605548	
7	Special screw and washer fixing plate to valve body	2	605547	
8	Special screw and washer fixing end plates to upper valve bodies	6	605546	
9	Special screw and washer fixing oil tube Plate to valve body lower	2	605547	
10	Special screw and washer fixing oil tube plate to valve body lower	6	605545	
11	Cam assembly for downshift valve	1	605398	
12	Special screw and washer fixing downshift valve cam assembly	2	605549	
	Stop for throttle valve	1	605768	
	Stop plate for throttle valve	1	605769	
13	Oil strainer for front pump	1	601127	
14	Special screw and washer fixing front pump strainer	4	605546	
15	Special screw fixing valve body	2	605550	
16	Plain washer	2	605426	
17	Special bolt and washer, short, fixing valve boxy	1	601131	
18	Special bolt and washer, long, fixing valve body	2	601132	
19	Tube, front servo apply	1	601133	
20	Tube, front servo release	1	601134	Cars numbered suffix letter 'A'
	Tube, front servo release	1	601373	Cars numbered suffix letter 'B' onwards
21	Tube, rear clutch	1	601135	
22	Tube, rear servo	1	601139	
23	Tube, front pump outlet	1	601136	
24	Tube, front pump inlet	1	601137	
25	Tube, oil cooler inlet	1	605379	
26	Tube, converter out and in	1	601138	
	Tube 2nd gear hold	1	607727*	Cars numbered suffix letter 'B' onwards

*Asterisk indicates a new part which has not been used on any previous model.

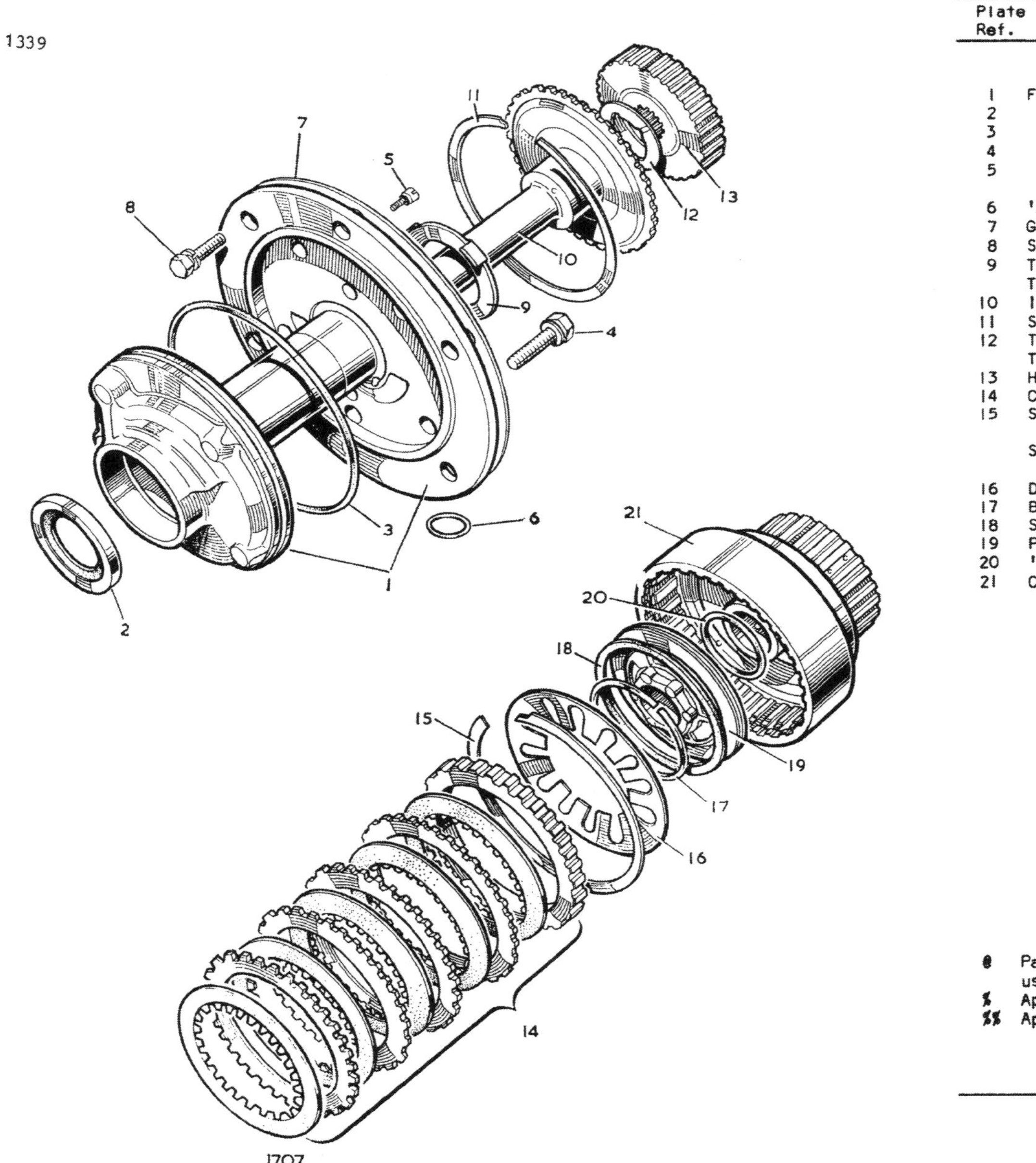

Plate Ref.	1 2 3 4 Description	Qty.	Part No.	Remarks
1	FRONT PUMP ASSEMBLY	1	601297	
2	Seal for pump	1	601012	
3	'O' ring for pump	1	606527	
4	Special bolt and washer (7/8" long)	5	601003	
5	Special screw and washer (3/8" long) fixing pump halves together	1	601014	
6	'O' ring for pump inlet tube	1	601015	
7	Gasket, pump to main case	1	601016	
8	Spécial bolt and washer fixing pump to case	6	601003	
9	Thrust washer	1	605392	@%
	Thrust washer	1	601386	@%%
10	Input shaft complete	1	605361	
11	Snap ring for input shaft	1	601025	
12	Thrust washer	1	605392	@%
	Thrust washer	1	601386	@%%
13	Hub for front clutch	1	601026	
14	Clutch plates (service kit) for front clutch	1	605393	
15	Snap ring for front clutch	1	601025	Transmission numbered up to 5 FU 2388
	Snap ring for front clutch	1	606449*	Transmission numbered from 5 FU 2389 and 7 FU 3003
16	Diaphragm spring and snap ring for front clutch	1	605973	
17	Bearing ring for clutch spring	1	601022	
18	Sealing ring, large, for piston	1	601023	
19	Piston and reed assembly for front clutch	1	601019	
20	'O' ring, small, for piston	1	601021	
21	Cylinder for front clutch	1	605981	

@ Part numbers 605392 and 601386 represent a Service kit of all thrust washers used in the transmission. Individual thrust washers are not serviced separately.
% Applicable to Automatic transmission part number 605158 only.
%% Applicable to Automatic transmissions except part number 605158.

*Asterisk indicates a new part which has not been used on any previous model.

Plate Ref.	1 2 3 4 Description	Qty.	Part No.	Remarks
1	Piston and reed assembly, rear clutch	1	601030	
2	'O' ring, small, for piston rear clutch	1	601029	
3	Sealing ring, large, for piston	1	601031	
4	Spring for rear clutch	1	601032	
5	Seat for spring	1	601033	
6	Snap ring	1	601034	
7	Clutch plates (service kit)	1	601427	Except '303' transmission
	Clutch plates (service kit) for use with 3 groove type front drum	1	608355*	'303' transmission
8	Snap ring, large	1	601025	
9	Thrust washer	1	605392	@%
	Thrust washer	1	601386	@%%
	Front drum assembly, single groove	1	605697	%
10	Front drum assembly, single groove	1	601429	%%
	Front drum assembly, 3 groove type	1	607673*	'303' type transmission
	Front drum assembly, 3 groove type includes spacer	1	606358	Late models
11	Oil seal ring	3	601035	
12	Forward sun gear assembly	1	605362	%
	Forward sun gear assembly	1	605597	%%
13	Oil seal ring, nylon	1	601039	
14	Oil seal ring for forward sun gear	2	601041	
15	Thrust needle bearing	1	605392	@%
16	Thrust needle bearing and plate	1	605392	@%
	Thrust needle bearing	1	601386	@%%
	Thrust needle bearing and plate	1	601386	@%%
17	Centre support assembly	1	601042	
18	Special bolt fixing centre support	2	601043	
19	Special lockwasher	2	601044	
20	One-way clutch assembly	1	601046	
21	PLANET GEARS AND REAR DRUM ASSEMBLY	1	605363	%
	PLANET GEARS AND REAR DRUM ASSEMBLY	1	607255	%%
22	Outer race assembly for one-way clutch	1	601045	
23	Snap ring for one-way clutch	1	601025	
24	Thrust needle bearing and plate	1	605392	@%
	Thrust needle bearing and plate	1	601386	@%%
25	Ring gear	1	601053	%
	Ring gear	1	605366	%%
26	Driven shaft assembly	1	605365	%
	Driven shaft assembly	1	605598	%%
27	Thrust washer for driven shaft	1	605392	@%
	Thrust washer for driven shaft	1	601386	@%%
28	Snap ring, .055" } Fixing ring gear to driven shaft	1	601054	} Alternatives
	Snap ring, .057"	1	601055	
	Snap ring, .059"	1	601056	
29	Oil sealing ring for driven shaft	3	601060	
30	Adaptor plate	1	605364	
31	Special screw and washer fixing adaptor plate to case	5	601050	
32	GOVERNOR ASSEMBLY	1	601378	

@ Part numbers 605392 and 601386 represent a Service kit of all thrust washers used in the transmission. Individual thrust washers are not serviced separately.

% Applicable to Automatic transmission part number 605158

%% Applicable to Automatic transmissions except part number 605158

*Asterisk indicates a new part which has not been used on any previous model.

Plate Ref.	1 2 3 4 Description	Qty.	Part No.	Remarks
33	Spring for governor	1	601154	
34	Retainer for governor spring	1	605972	Alternative
	Circlip for governor stem	1	606989*	Alternative
35	Special lockscrew fixing cover plate	2	601155	
36	Special screw and washer fixing governor body	2	601156	
37	Steel ball (¼" dia.)	1	601157	
38	Snap ring for governor	1	601034	
39	Gear for speedo drive	1	601158	
	SERVO ASSEMBLY, FRONT (non-self-adjusting)	1	605461	%
40	SERVO ASSEMBLY, FRONT (self-adjusting)	1	606458	%%
	SERVO ASSEMBLY, FRONT (self-adjusting)	1	606571	%%%
	Body for front servo	1	605462	For servo assemblies 605461 and 606571
41	Body for front servo	1	606459	For servo assembly 606458
42	Lever for front servo	1	605369	
43	Pivot pin for lever	1	601070	
	Adjusting screw for front servo	1	601071	For servo assembly 605461
44	Adjusting screw for front servo	1	606461	For servo assemblies 606458 and 606571
	Locknut for adjusting screw	1	601072	
45	Spring for adjusting screw	1	606462	For servo assemblies 606458 and 606571
46	Spring for front servo	1	601371	
	'O' ring, large, for front servo piston	1	605370	For servo assemblies 605461 and 606571
47	'O' ring, large, for front servo piston	1	605397	For servo assembly 606458
48	'O' ring, small, for front servo piston	1	605371	
49	Oil sealing ring for front servo piston sleeve (square section seal)	1	605397	Alternative. Check before ordering
	Oil sealing ring for front servo piston sleeve (round section seal)	1	605370	Alternative. Check before ordering
50	Snap ring for front servo piston	1	605373	
51	Cam plate	1	606460	For servo assembly 606458
	Cam plate	1	606572	For servo assembly 606571
52	Bolt fixing cam plate	2	606463	
53	Special bolt and washer (1⅜" long) fixing servo to case	1	601073	
	Special bolt and washer (1" long) fixing servo to case	1	601074	
54	Piston assembly (service kit)	1	605396	
55	Brake bank for front servo	1	601104	
56	Strut for brake band	1	601105	

% Automatic Transmissions numbered in the 3 FU and 5 FU series and up to 7 FU 10163
%% Automatic Transmission numbered from 7 FU 10164 to 7 FU 14967
%%% Automatic Transmissions numbered from 7 FU 14968 onwards

***Asterisk indicates a new part which has not been used on any previous model.**

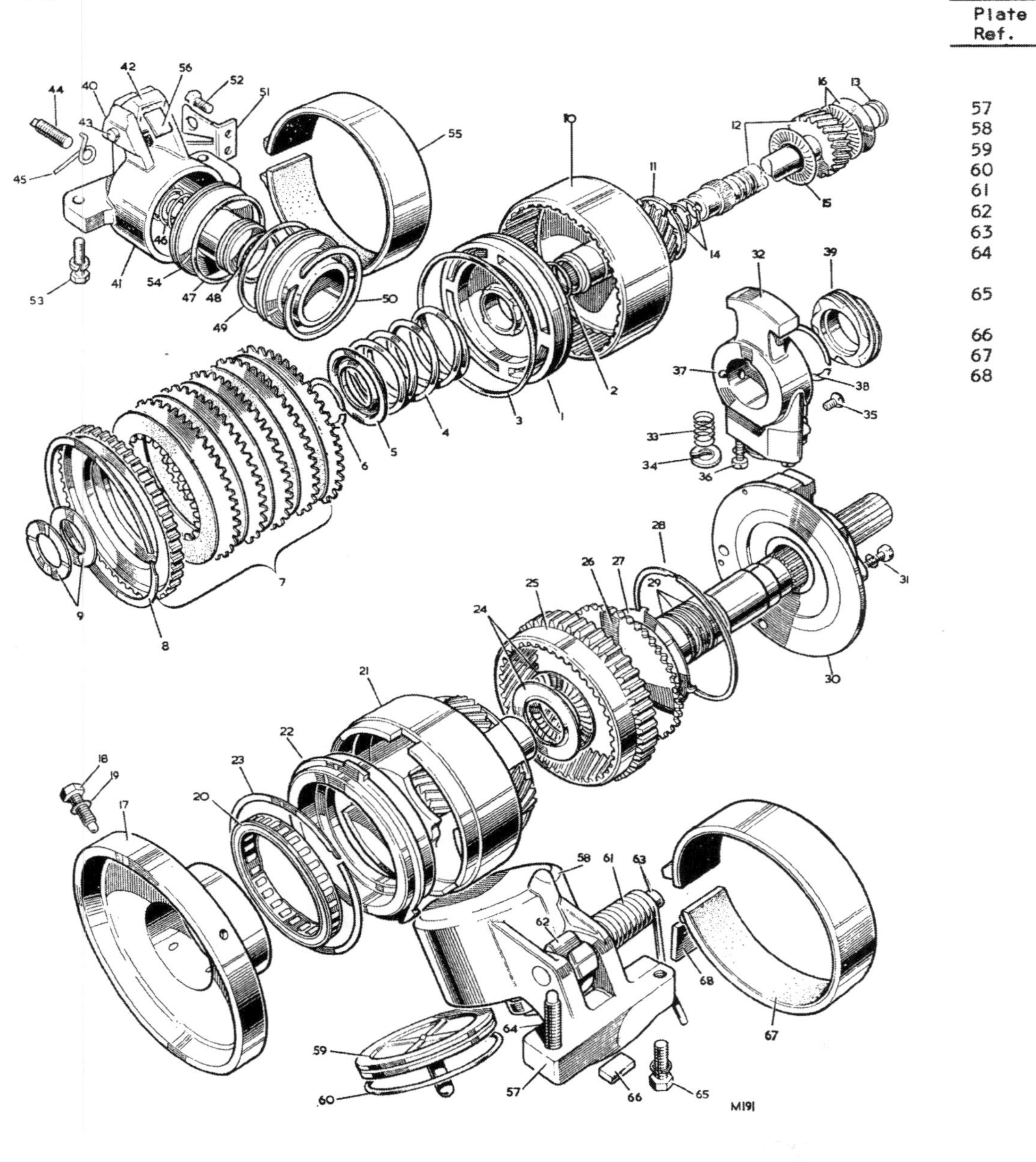

Plate Ref.	1 2 3 4 Description	Qty.	Part No.	Remarks
57	SERVO ASSEMBLY, REAR	1	601075	
58	Body for rear servo	1	601076	
59	Piston for rear servo	1	601077	
60	'O' ring for piston	1	601078	
61	Spring for rear servo	1	601079	
62	Lever for servo	1	601080	
63	Pivot pin for lever	1	601081	
64	Special bolt, rear, fixing rear servo to case	1	601082	
	Special spring washer	1	601083	
65	Special bolt and washer, rear, fixing rear servo to case	1	601084	
66	Magnet attached to rear bolt for rear servo	1	601152	
67	Brake band for rear servo	1	605377	
68	Strut for brake band	1	601105	

*Asterisk indicates a new part which has not been used on any previous model.

1347

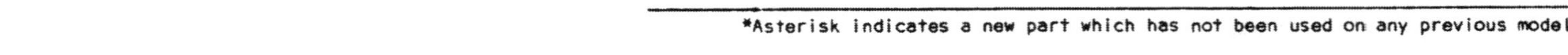

Plate Ref.	1 2 3 4 Description	Qty.	Part No.	Remarks
	GEAR SELECTOR LEVER ASSEMBLY	1	591351*	
1	Gear selector lever	1	591689*	
2	Plug for lever	1	561409	
3	Pin for selector gate	1	561475	
4	Spring for plug — For gear selector	1	556605	
5	Eye bolt — For gear selector	1	561561	Early models
	Eye bolt — For gear selector	1	591291*	Late models
6	Nut (⅜" UNF) for eye bolt lever	1	254852	
7	Spherical seat for lever	1	571933	
8	Locating pin for spherical seat	1	540959	
9	Retaining cap for spherical seat	1	556589	
10	Pivot housing	1	556592	
11	Self-locking nut (¼" UNF) fixing retaining cap to pivot housing	3	252160	
12	Control rod for gear selector lever	1	556594	
13	Knob for gear selector lever	1	576413	
14	Fibre washer for knob	1	267721	
15	'O' ring for knob	1	247186	
16	Button for knob	1	556593	Early models
	Button for knob	1	576699*	Late models
17	Rubber insert for knob	1	576643*	
18	HOUSING FOR GEAR SELECTOR LEVER	1	591388	Cars with engines numbered up to 84016745
	HOUSING FOR GEAR SELECTOR LEVER	1	593922	Cars with engines numbered from 84016746
19	Shield for gear selector lever	1	561499	Alternative to brush shield. Check before ordering.
20	Spring tensioner	1	561573	Alternative to brush shield. Check before ordering.
21	Side clip, short — Fixing tensioner to housing	1	561505	Alternative to brush shield. Check before ordering.
22	Side clip, long — Fixing tensioner to housing	1	561506	Alternative to brush shield. Check before ordering.
	Brush light shield for housing	2	591386	Alternative to stainless steel shield. Check before ordering.
23	Set bolt (5/16" UNC x ⅞" long) — Fixing gear selector lever housing to pivot housing	3	253027	
24	Plain washer — Fixing gear selector lever housing to pivot housing	3	2220	
25	ILLUMINATION LAMP FOR GEAR SELECTOR INDICATOR	1	567747	
26	Bulb, 12 volt, 3 watt festoon	1	67762	
27	Screw (10 UNF x ⅜" long) — Fixing lamps to housing	1	78555	
28	Shake-proof washer — Fixing lamps to housing	1	74236	
29	Indicator plate, P.R.N.D2.D1.L. for gear selector lever	1	563403	Cars numbered suffix letter 'A'
	Indicator plate, P.R.N.D.2.1. for gear selector lever	1	586422	Cars numbered suffix letter 'B'
30	Spring for indicator plate	1	570845	
31	Split pin	1	2389	
32	Compensator bracket for gear linkage	1	561269	
33	Set bolt (5/16" UNC x⅞" long) — Fixing bracket to torque converter housing	3	253026	
34	Spring washer — Fixing bracket to torque converter housing	3	3075	

*Asterisk indicates a new part which has not been used on any previous model.

Plate Ref.	1 2 3 4	Description	Qty.	Part No.	Remarks
35		Compensator block	2	561270	
36		Drive screw fixing compensator blocks together	2	78760	
37		Shaft and lever for compensator	1	576816	
38		Selector lever upper	1	576812	
39		Self-locking nut (¼" UNF) fixing upper lever to shaft	1	252160	
		Plain washer, upper lever to shaft	2	3840	Late models
40		Selector lever, lower	1	576814	
41		Self-locking nut (5/16" UNF) fixing lower lever to transmission unit	1	252161	
42		Lower control rod	1	540401	
43		Ball joint for rod	1	540402	
44		Locknut (¼" UNF) for ball joint	1	254850	
45		Self-locking nut (¼" UNF) fixing lower rod to levers	2	252160	
46		Tie-rod, compensator block to pivot housing	1	561016	
47		Ball joint for tie-rod, compensator end	1	540405	
48		Ball joint for tie-rod, pivot housing end	1	540402	
49		Locknut (¼" UNF) for ball joints	2	254850	
50		Plain washer } Fixing tie-rod	2	3840	
51		Self-locking nut (¼" UNF) } to compensator	1	252160	
52		Self-locking nut (¼" UNF) } fixing tie-rod to pivot housing	1	576000	
53		Control rod, gear selector lever to compensator	1	561013	
54		Ball joint for control rod	2	540402	
55		Locknut (¼" UNF) for ball joint	2	254850	
56		Self-locking nut (¼" UNF) fixing control rod to levers	2	252160	

L535

*Asterisk indicates a new part which has not been used on any previous model.

1352

D974

Plate Ref.	Description (1 2 3 4)	Qty.	Part No.	Remarks
	REAR AXLE COMPLETE ASSEMBLY, 3.9 ratio	1	279725	Mk I models without overdrive – For use with drum-type front brakes. Assembly includes brakes and differential
	REAR AXLE COMPLETE ASSEMBLY, 4.3 ratio	1	268670	Mk I models with overdrive – For use with drum-type front brakes. Assembly includes brakes and differential
	REAR AXLE COMPLETE ASSEMBLY, 4.3 ratio	1	512193	Mk I and MkIA models with overdrive – For use with disc-type brakes. Assembly includes brakes and differential
	REAR AXLE COMPLETE ASSEMBLY, 3.9 ratio	1	540217	Mk I and MkIA Mk II Automatic – For use with disc-type brakes. Assembly includes brakes and differential
	REAR AXLE COMPLETE ASSEMBLY, 4.3 ratio	1	540216	Mk II 4-speed – For use with disc-type brakes. Assembly includes brakes and differential
	REAR AXLE COMPLETE ASSEMBLY, 4.3 ratio	1	568000	Mk III 4-speed – For use with disc-type brakes. Assembly includes brakes and differential
	REAR AXLE COMPLETE ASSEMBLY, 3.54 ratio	1	553337	Mk III – For use with disc-type brakes. Assembly includes brakes and differential
	REAR AXLE COMPLETE ASSEMBLY, 3.54 ratio	1	561632	Early 3½ Litre
	REAR AXLE COMPLETE ASSEMBLY, 3.54 ratio	1	593630*	Late 3½ Litre
1	REAR AXLE CASING COMPLETE	1	561122	
2	Dowel for pinion housing	2	55705	When fitted
3	Special bolt, short – For differential housing	4	253805	Alternative to studs
4	Special bolt, long – For differential housing	6	253806	Alternative to studs
	Stud, 1¼" long – For differential housing	4	561195	Alternative to set bolts
	Stud, 1⅞" long – For differential housing	6	561196	Alternative to set bolts
5	Breather for axle casing	1	42431	
6	Drain plug for axle casing	1	241110	
7	Joint washer for plug	1	230509	
	AXLE SHAFT COMPLETE ASSEMBLY, LH	1	502907	These assemblies include brake anchor plates
	AXLE SHAFT COMPLETE ASSEMBLY, RH	1	502908	These assemblies include brake anchor plates
8	AXLE SHAFT ASSEMBLY, LH	1	540573	
9	AXLE SHAFT ASSEMBLY, RH	1	540572	
10	Stud for road wheel, threaded type	10	268574	@
	Stud for road wheel, serrated type	10	545511	@@
11	Extension piece for wheel changing	2	07860	
12	Bearing housing for rear wheel	2	515833	
13	'O' ring for bearing housing	2	GHS1006	
14	Distance washer for hub bearing	2	266584	
15	Bearing for rear hub	2	GHB161	
16	Sleeve for oil seal	2	515581	
	Felt seal for rear hub	2	518438	
17	Oil seal for hub	2	GHS166	
18	Hub retaining collar	2	07297	
19	Bolt (⅜" UNF x 1⅜" long) – Fixing bearing housing to axle case	12	256040	
20	Spring washer – Fixing bearing housing to axle case	12	3076	
21	Nut (⅜" UNF) – Fixing bearing housing to axle case	12	254812	
22	PROPELLER SHAFT, FRONT HALF	1	514481	Mk I and Mk IA Automatic and models without overdriv[e]

@ Mk I, Mk IA and Mk II axles numbered up to 77001615A, 77501812A

@@ Mk II axles and Mk III numbered from 77001616A, 77501813A onwards and 3½ Litre.

*Asterisk indicates a new part which has not been used on any previous model.

D974

Plate Ref.	1 2 3 4 Description	Qty.	Part No.	Remarks
	PROPELLER SHAFT, FRONT HALF	1	539812	Mk II Automatic
	PROPELLER SHAFT, FRONT HALF	1	553098	Mk III Automatic
	PROPELLER SHAFT, FRONT HALF	1	539811	Models with overdrive
	PROPELLER SHAFT, FRONT HALF	1	561022	3½ Litre
23	PROPELLER SHAFT, REAR	1	561968	
24	Splined end } For rear propeller shaft	1	600655	
25	Flange } For rear propeller shaft	2	504437	
26	Flange, front } For front propeller shaft	1	504438	3 Litre models without overdrive
	Flange, front } For front propeller shaft	1	504437	3 Litre models with overdrive
	Flange, front for front propeller shaft	1	600656	3½ Litre
27	JOURNAL COMPLETE FOR SHAFTS	3	GUJ108	
28	Circlip for journal	12	242522	
29	Grease nipple for journal	3	276201	
30	Grease nipple for propeller shaft sleeve	1	234532	
31	Dust shield } For rear of front propeller shaft	1	261774	
32	Bearing for shaft } For rear of front propeller shaft	1	261775	
33	Support tube and plate } For rear of front propeller shaft	1	504441	
34	Special support stud } For rear of front propeller shaft	2	235632	3 Litre
	Special support stud } For rear of front propeller shaft	2	561477	3½ Litre
35	Rubber bush } For rear of front propeller shaft	2	235919	
36	Plain washer } Fixing special stud	2	3933	
37	Spring washer } Fixing special stud	2	3076	
38	Nut (⅜" BSF) } Fixing special stud	2	2827	
39	Flange for front propeller shaft	1	261924	3 Litre
40	Woodruff key } Fixing flange	2	261925	3 Litre
41	Special castle nut } Fixing flange	1	243196	3 Litre
42	Split pin } Fixing flange	1	2428	3 Litre
	FLANGE, SPLINED FOR FRONT PROPELLER SHAFT	1	605477	3½ Litre
	Special bolt in flange } Fixing flange	4	539841	3½ Litre
	Special castle nut } Fixing flange	1	605478	3½ Litre
	Split pin } Fixing flange	1	2428	3½ Litre
43	Bolt (⅜" UNF x 1⅛" long) } Fixing propeller shaft together	4	237341	
44	Self-locking nut (⅜" UNF) } Fixing propeller shaft together	4	251322	
45	Bracket for propeller shaft centre bearing	1	534106	
46	Bolt (5/16" UNF x 1¼" long) } Fixing bracket to propeller shaft tunnel	4	256024	
47	Distance piece } Fixing bracket to propeller shaft tunnel	4	512230	
48	Grommet, male half } Fixing bracket to propeller shaft tunnel	4	76649	
49	Grommet, female half } Fixing bracket to propeller shaft tunnel	4	76650	
50	Plain washer, small } Fixing bracket to propeller shaft tunnel	4	3830	
51	Plain washer, large } Fixing bracket to propeller shaft tunnel	4	3638	
52	Spring washer } Fixing bracket to propeller shaft tunnel	4	3075	
53	Nut (5/16" UNF) } Fixing bracket to propeller shaft tunnel	4	254811	
54	Spring for propeller shaft support	1	248411	
55	Stud } Fixing spring to propeller shaft bearing and bracket	1	277131	
56	Bolt (¼" UNF x ¾" long) } Fixing spring to propeller shaft bearing and bracket	1	255207	
57	Plain washer } Fixing spring to propeller shaft bearing and bracket	1	4071	
58	Nut (¼" UNF) } Fixing spring to propeller shaft bearing and bracket	2	254810	
	Self-locking nut (¼" UNF)	1	252160	
59	Self-locking nut (⅜" BSF) fixing propeller shaft	8	251322	

*Asterisk indicates a new part which has not been used on any previous model.

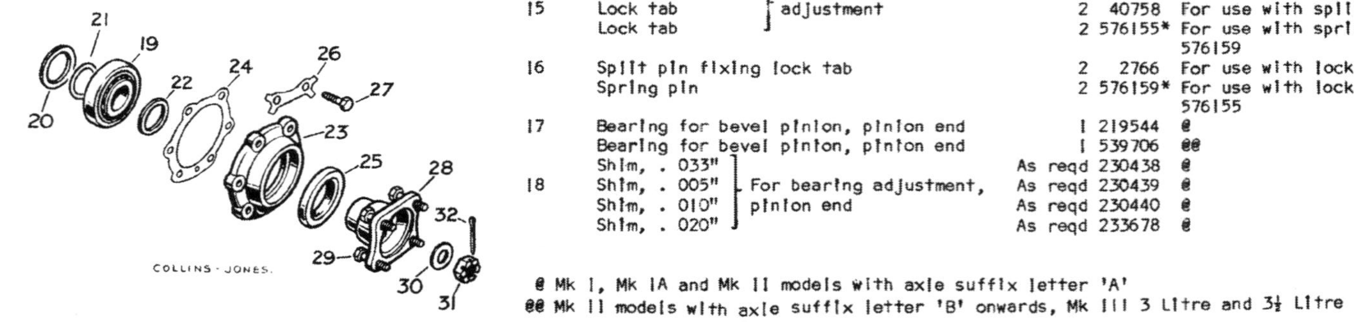

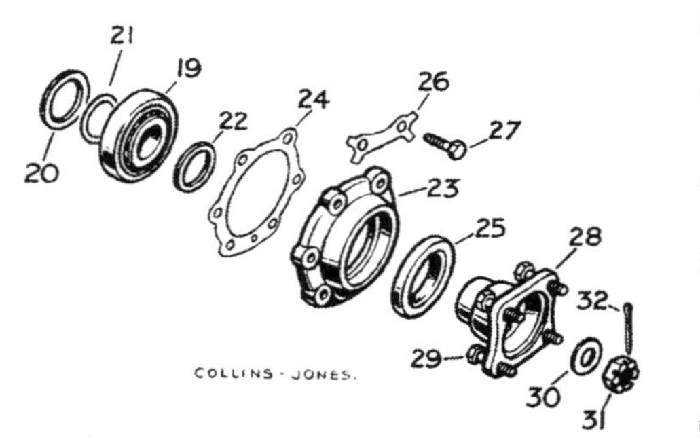

Plate Ref.	1 2 3 4 Description	Qty.	Part No.	Remarks
	DIFFERENTIAL ASSEMBLY, 3.9 ratio	1	540489	Mk I, Mk IA and Mk II 3 litre Automatic and 3 litre models without overdrive
	DIFFERENTIAL ASSEMBLY, 3.54 ratio	1	561174	Mk III 3 litre Automatic
	DIFFERENTIAL ASSEMBLY, 4.3 ratio	1	540490	3 Litre models with overdrive
	DIFFERENTIAL ASSEMBLY, 3.54 ratio	1	561755	Early 3½ Litre
	DIFFERENTIAL ASSEMBLY, 3.54 RATIO	1	593629*	Late 3½ Litre
	Crown wheel and bevel pinion, 3.9 ratio	1	504867	Mk I, Mk IA and Mk II 3 Litre Automatic and 3 Litre models without overdrive
	Crown wheel and bevel pinion, 3.54 ratio	1	553379	Mk III 3 Litre Automatic and 3½ Litre
1	Crown wheel and bevel pinion, 4.3 ratio	1	504868	3 Litre models with overdrive
2	Differential case	1	273441	Except Mk III 3 Litre and 3½ Litre Automatic
	Differential case	1	605229	Mk III 3 litre Automatic 3½ litre
3	Locker, double type } Fixing crown wheel to differential case	5	272922	
4	Set bolt (⅜" x 1⅛" long) } Fixing crown wheel to differential case	8	237339	
	Special fitting bolt, hexagon head } Fixing crown wheel to differential case	2	272934	
5	Differential wheel	2	533794	
6	Differential pinion	2	533777	
7	Spindle for pinion	1	11378	@
	Spindle for pinion	1	539703	@@
8	Plain pin } For spindle	1	11379	
9	Split pin } For spindle	1	2396	
	Thrust washer, .040" } For differential wheels	2	533786	
10	Thrust washer, .045" } For differential wheels	2	533787	As required
	Thrust washer, .050" } For differential wheels	2	533788	
11	BEVEL PINION HOUSING COMPLETE	1	528257	@
	BEVEL PINION HOUSING COMPLETE	1	576135	@@
12	Special set bolt fixing bearing cap	4	4072	
13	Bearing for differential	2	41045	
14	Serrated nut } For bearing adjustment	2	40757	
15	Lock tab } For bearing adjustment	2	40758	For use with split pin 2766
	Lock tab } For bearing adjustment	2	576155*	For use with spring pin 576159
16	Split pin fixing lock tab	2	2766	For use with lock tab 40758
	Spring pin	2	576159*	For use with lock tab 576155
17	Bearing for bevel pinion, pinion end	1	219544	@
	Bearing for bevel pinion, pinion end	1	539706	@@
	Shim, . 033" } For bearing adjustment, pinion end	As reqd	230438	@
18	Shim, . 005" } For bearing adjustment, pinion end	As reqd	230439	@
	Shim, . 010" } For bearing adjustment, pinion end	As reqd	230440	@
	Shim, . 020" } For bearing adjustment, pinion end	As reqd	233678	@

@ Mk I, Mk IA and Mk II models with axle suffix letter 'A'
@@ Mk II models with axle suffix letter 'B' onwards, Mk III 3 Litre and 3½ Litre

*Asterisk indicates a new part which has not been used on any previous model.

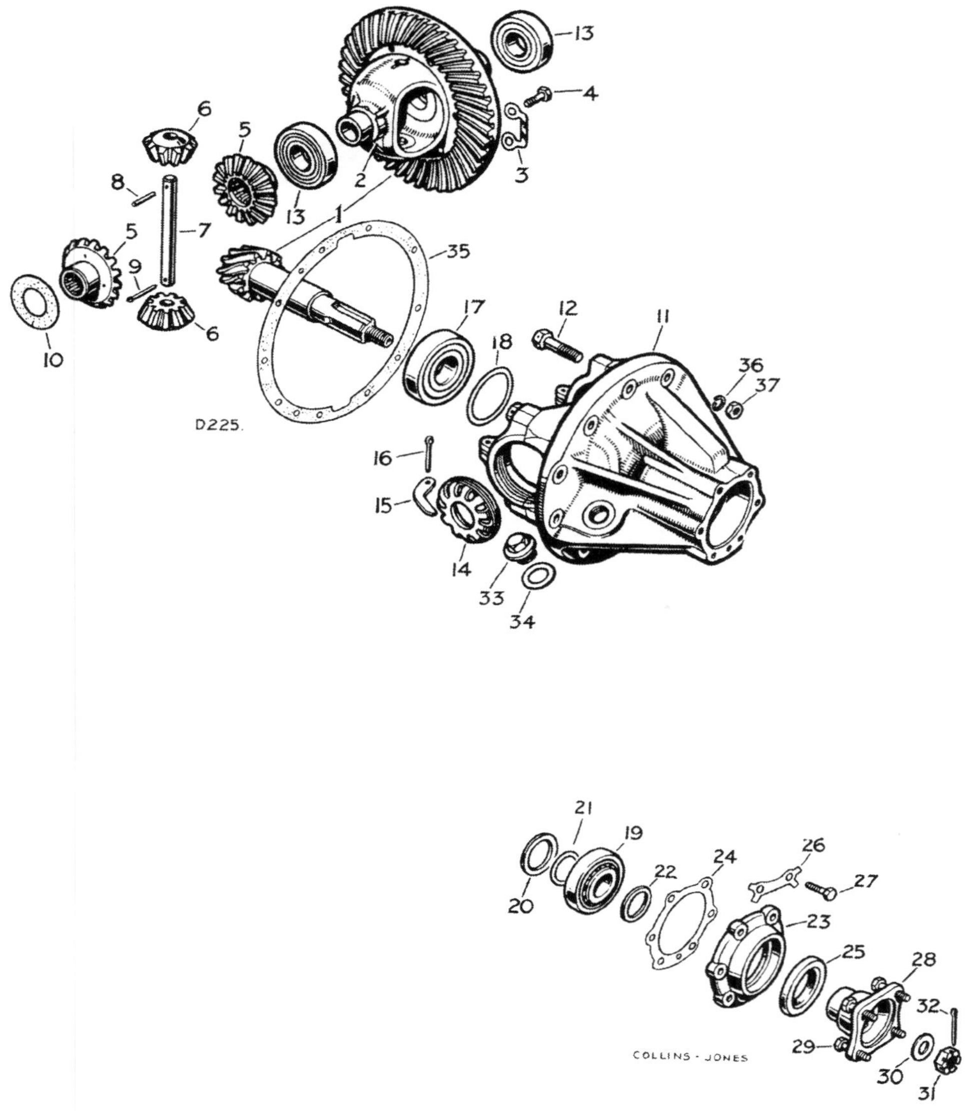

D225

COLLINS - JONES

Plate Ref.	Description		Qty.	Part No.	Remarks
	Shim, .020"	For bearing adjustment pinion end	As reqd	539711	@@
	Shim, .022"		As reqd	539713	@@
	Shim, .024"		As reqd	539715	@@
	Shim, .030"		As reqd	539717	@@
19	Bearing for bevel pinion, flange end		1	219550	@
	Bearing for bevel pinion, flange end		1	539707	@@
20	End washer for bearing shim		2	502248	@
	Shim, .003"	For bearing adjustment flange end	As reqd	219547	@
21	Shim, .005"		As reqd	219548	@
	Shim, .010"		As reqd	219549	@
	Shim, .020"		As reqd	233677	@
	Shim, .072"	For bearing adjustment, flange end	2	539718	@@
	Shim, .074"		2	539720	@@
	Shim, .076"		2	539722	@@
	Shim, .081"		2	539724	@@
22	Washer for front bearing		1	231242	@
	Washer for front bearing		1	539745	@@
23	Retainer for oil seal		1	236633	
24	Joint washer for oil seal retainer		1	553412	
25	Oil seal for pinion		1	217507	
26	Locker	Fixing oil seal retainer	3	41049	Alternatives
27	Set bolt (5/16" x ⅞" long)		6	237544	
	Set screw (5/16" BSF special	Fixing oil seal retainer	6	571916	*
	Spring washer		6	3075	
	Mudshield for oil seal		1	236072	Late 3½ Litre
28	Driving flange		1	236632	
29	Special bolt in flange (BSF thread)		4	237341	Alternatives
	Special bolt in flange (UNF thread)		4	539841	
30	Plain washer		1	513454	
31	Special castle nut	Fixing flange to bevel pinion	1	3259	
32	Split pin		1	2428	
33	Oil filler plug, large hexagon		1	05366	Alternatives
	Oil filler plug, small hexagon		1	533358	
34	Joint washer for plug		1	06009	
35	Joint washer for differential casing		1	07316	
36	Spring washer	Fixing differential to axle casing	10	3076	
37	Nut (⅜" UNF)		10	254162	

@ Mk I, Mk IA and Mk II models with axle suffix letter 'A'
@@ Mk II models with axle suffix letter 'B' onwards, Mk III 3 Litre and 3½ Litre

*Asterisk indicates a new part which has not been used on any previous model.

STUB AXLES AND FRONT HUBS

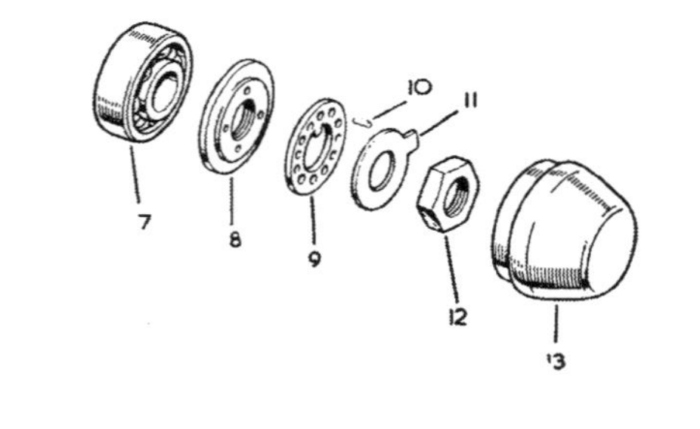

Plate Ref.	1 2 3 4 Description	Qty.	Part No.	Remarks
	Stub axle, RH	1	275791	For drum-type brakes
	Stub axle, LH	1	275792	For drum-type brakes
	Stub axle, RH — ⅜" dia. fixing holes — Alternative fixings. Check before ordering	1	512033	For disc type brakes
	Stub axle, LH — ⅜" dia. fixing holes — Alternative fixings. Check before ordering	1	512032	For disc type brakes
	Stub axle, RH — 7/16" dia. fixing holes — Alternative fixings. Check before ordering	1	527335	Mk I, Mk IA and 3 Litre with suffix letters 'A' and 'B' — For disc type brakes
	Stub axle, LH — 7/16" dia. fixing holes — Alternative fixings. Check before ordering	1	527336	Mk I, Mk IA and 3 Litre with suffix letters 'A' and 'B' — For disc type brakes
1	Stub axle, RH — Fully sealed lubrication	1	542517	Mk II with suffix letter 'C' Mk III 3 Litre and 3½ Litre — For disc type brakes
	Stub axle, LH — Fully sealed lubrication	1	542516	Mk II with suffix letter 'C' Mk III 3 Litre and 3½ Litre
	Plug for stub axle oil ways	4	255001	Mk I, Mk IA and Mk II 3 Litre with suffix letters 'A' and 'B' only
	Joint washer for plug	4	232037	Mk I, Mk IA and Mk II 3 Litre with suffix letters 'A' and 'B' only
	FRONT HUB ASSEMBLY	2	272998	For drum-type brakes
2	FRONT HUB ASSEMBLY	2	538289	For disc-type brakes
	Stud for road wheel	10	272997	For drum-type brakes
	Stud, threaded type, for road wheel	10	500789	For disc-type brakes
3	Stud, serrated type,for road wheel	10	545510	
4	Oil thrower ring for front hub	2	230202	Drum brakes
	Oil seal for inner bearing, 0.312" thick	2	235986	Alternatives Check before ordering
5	Oil seal for inner bearing, 0.375" thick	2	GHS150	Alternatives Check before ordering
6	Bearing for hub, inner	2	GHB159	
7	Bearing for hub, outer	2	GHB104	
8	Inner nut, LH thread	1	507648	
	Inner nut, RH thread	1	507647	
9	Key washer	2	10176	
10	Peg for key washer	2	03574	
11	Lock washer	2	10175	
12	Outer nut, LH thread	1	244477	
13	Hub cap, front	2	219809	

*Asterisk indicates a new part which has not been used on any previous model.

RADIUS ARMS AND FRONT TORSION BARS

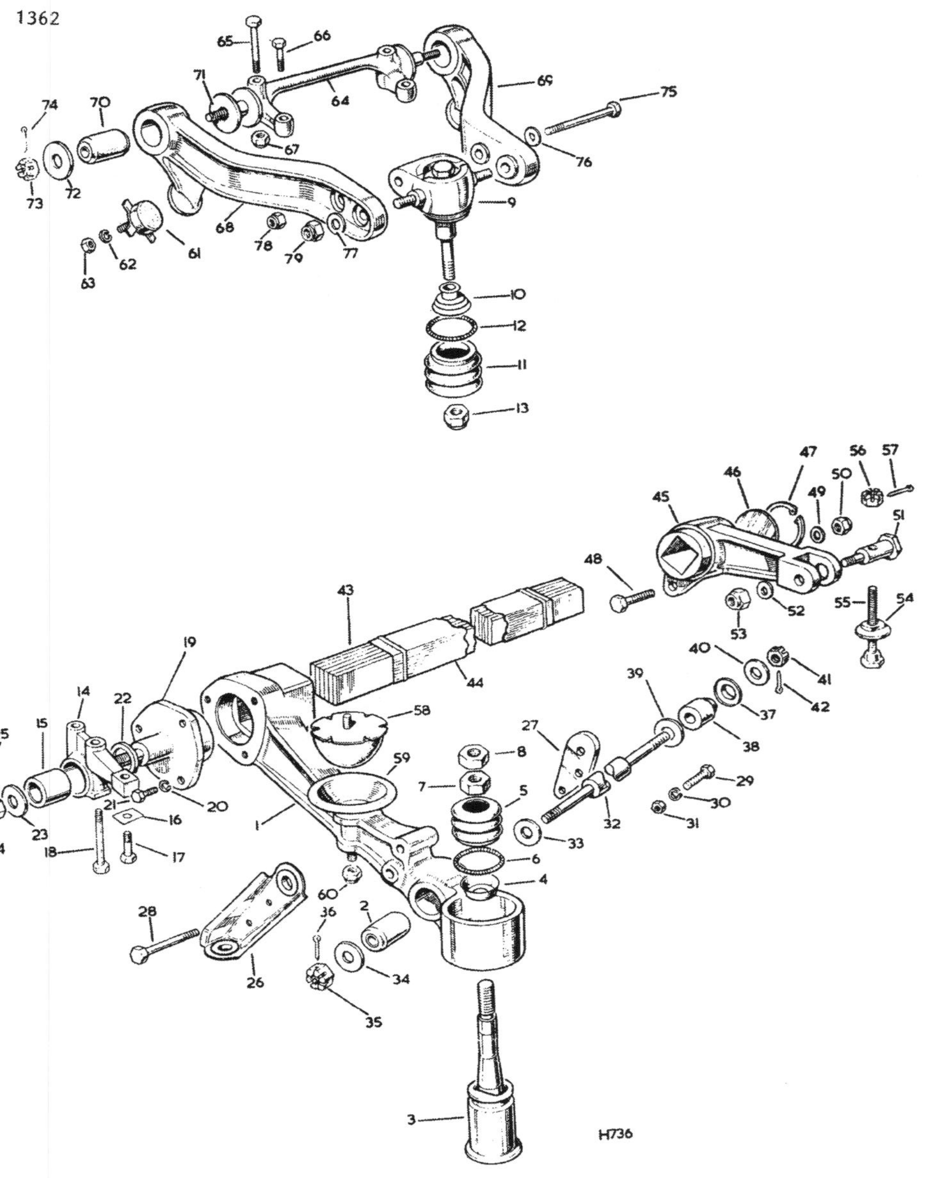

Plate Ref.	1 2 3 4 Description	Qty.	Part No.	Remarks
	BOTTOM LINK ASSEMBLY, RH	1	542262	
1	BOTTOM LINK ASSEMBLY, LH	1	542263	
2	Flexible bush for radius arm	2	548208	
	Bottom ball swivel complete	2	271601	%
3	Bottom ball swivel complete	2	542264	%%
4	Support collar for dust cover	2	600326	%%
5	Dust cover	2	600327	%%
6	Retaining ring for dust cover	2	600325	%%
	Oil seal for bottom ball swivel	2	521619	%
	Rubber boot for ball swivel	2	502009	%
	Retaining spring, small, for boot	2	502008	%
	Retaining spring, large, for boot	2	502007	%
7	Nut (¾" UNF) } Fixing bottom	2	254877	
8	Locknut (¾" UNF) } swivel to stub axle	2	512078	
	TOP BALL SWIVEL COMPLETE	2	268034	For drum-type brakes
	TOP BALL SWIVEL COMPLETE	2	516440	%For disc-type brakes
	Oil seal for top ball swivel	2	212669	%
	Rubber boot for ball swivel	2	507039	%
	Retaining spring, small, for boot	2	502008	%
	Retaining spring, large, for boot	2	502007	%
9	TOP BALL SWIVEL COMPLETE	2	542261	%%
10	Support collar for dust cover	2	600324	%%
11	Dust cover	2	600323	%%
12	Retaining ring for dust cover	2	600325	%%
13	Self-locking nut (7/16" UNF) fixing top swivel to stub axle	2	252163	
14	Link mounting pin housing complete	2	503289	
15	Flexible bush for link mounting pin	2	271311	
16	Tab washer } Fixing link mounting housing to chassis sub-frame	2	271464	
	Tab washer	4	512236	
17	Set bolt (⅜" UNF x 1¼" long)	2	255249	
18	Set bolt (⅜" UNF x 2¾" long)	4	256048	
19	Bottom link mounting pin	2	271308	
20	Spring washer } Fixing pin to bottom link	6	3076	
21	Set bolt (⅜" UNF x ⅞" long)	6	255046	
22	Plain washer, chamfered } Fixing link mounting pin housing to mounting pin	2	503885	
23	Plain washer	2	503883	
24	Castle nut	2	254956	
25	Split pin	2	2428	
	Shock absorber and anti-roll bar bracket, RH	1	503301	@
	Shock absorber and anti-roll bar bracket, RH	1	517197	@@
	Shock absorber and anti-roll bar bracket, LH	1	503300	@
26	Shock absorber and anti-roll bar bracket, LH	1	517198	@@
27	Shock absorber mounting plate	2	503287	

@ 3 Litre Cars numbered up to 625100983,626100244,628100107,630100699, 631100326, 633100055

@@ 3 Litre Cars numbered from 625100984, 626100245, 628100108, 630100700, 631100327, 633100056 onwards and 3½ Litre

% Mk I, Mk IA and Mk II 3 Litre cars with suffix letter 'A' and 'B'

%% Mk II cars with suffix letter 'C', Mk III and 3½ Litre

*Asterisk indicates a new part which has not been used on any previous model.

RADIUS ARMS AND FRONT TORSION BARS

H736

Plate Ref.	1 2 3 4 Description		Qty.	Part No.	Remarks
28	Bolt (3/8" UNF x 2½" long)	Fixing brackets to bottom links	2	256047	
29	Set bolt (3/8" UNF x 5/8" long)		4	255044	
30	Spring washer		6	3076	
31	Nut (3/8" UNF)		2	254812	
32	Radius arm complete		2	271828	
33	Plain washer, chamfered	Fixing radius arm to bottom link	2	503886	
34	Plain washer		2	503884	
35	Castle nut (½" UNF)		2	254974	
36	Split pin		2	2393	
37	Cup washer	Fixing radius arm to chassis sub-frame	4	271826	
38	Rubber pad		4	545923	
39	Plain washer,chamfered		2	503667	
40	Plain washer		2	512394	
41	Castle nut (7/16" UNF)		2	254953	
42	Split pin		2	2393	
	Torsion bar road spring, front RH		1	507290	
43	Torsion bar road spring, front LH		1	507289	
44	Grease sleeve for torsion bar		2	504913	
	Torsion bar adjuster lever, RH		1	500245	All Mk I and Mk IA, 3 Litre also Mk II 3 Litre with high suspension
	Torsion bar adjuster lever, LH		1	500244	
	Torsion bar adjuster lever, RH		1	514181	Mk II 3 Litre standard suspension
	Torsion bar adjuster lever, LH		1	514180	
45	Torsion bar adjuster lever, RH		1	500245	Mk III 3 Litre standard and high suspension
	Torsion bar adjuster lever, LH		1	500244	
	Torsion bar adjuster lever, RH		1	514181	3½ Litre standard and high suspension
	Torsion bar adjuster lever, LH		1	514180	
46	Retaining plate for torsion bar		2	500305	
47	Circlip fixing plate to lever		2	242431	
48	Bolt (½" UNF x 1½" long)	Clamping adjusting levers	2	255089	
49	Plain washer		2	2253	
50	Self-locking nut (½" UNF)		2	252164	
51	Trunnion for adjuster levers		2	500323	
52	Plain washer	Fixing trunnion to chassis sub-frame	2	2253	
53	Self-locking nut (½" UNF)		2	252164	
54	Adjuster lever pin seat, 2¼" dia.		2	541213	
55	Adjuster pin for levers		2	503965	
56	Castle nut (½" UNF)	Retaining adjuster pin	2	254974	
57	Split pin		2	2980	
	Bump rubber, front		2	276263	All Mk I and Mk IA, 3 Litre also Mk II and Mk III 3 Litre and 3½ Litre high suspension
58	Bump rubber, front		2	531884	Mk II, Mk III 3 Litre and 3½ Litre standard suspension
59	Bump stop for front spring		2	271024	Except high suspension
	Bump stop for front spring		2	538064	High suspension

NOTE: Cars with 'high' suspension may be identified by the letter 'H' stamped on the car serial number plate

*Asterisk indicates a new part which has not been used on any previous model.

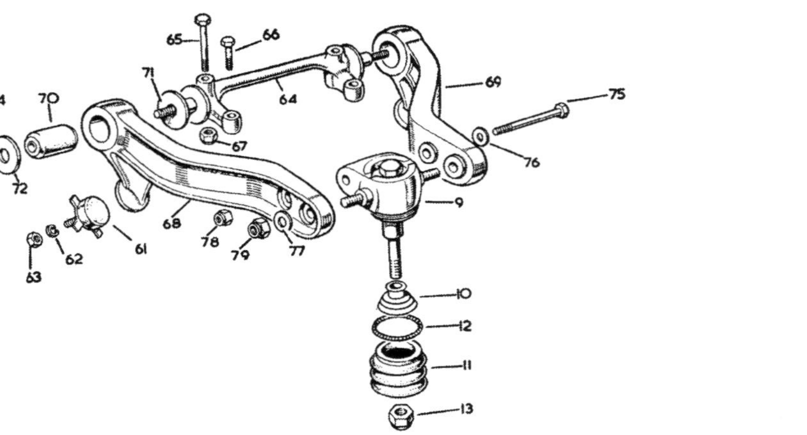

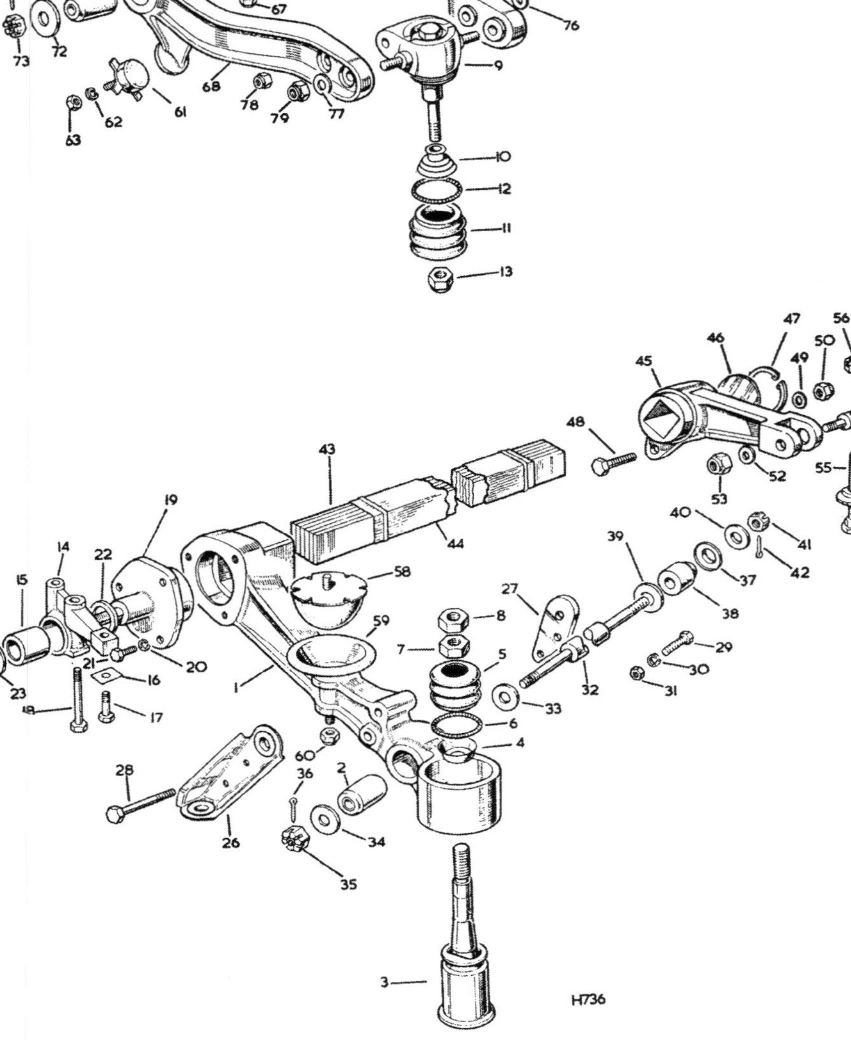

Plate Ref.	1 2 3 4 Description		Qty.	Part No.	Remarks
60	Self-locking nut fixing bum stop to bottom link		2	252163	
61	Rebound stop rubber		4	500901	
62	Spring washer	Fixing rubber to chassis sub-frame	4	3075	
63	Nut (5/16" UNF)		4	254831	
	Top link mounting arm, RH		1	518654	
64	Top link mounting arm, LH	Fixing mounting arm to chassis sub-frame	1	518655	
65	Bolt (5/16" UNF x 2¾" long)		4	245429	
66	Bolt (5/16" UNC x 1¾" long)		4	256424	
	Plain washer		8	3830	
67	Self-locking nut (5/16" UNF)		8	252161	
68	TOP LINK COMPLETE, FRONT LH		1	502310	
	TOP LINK COMPLETE, FRONT RH		1	502311	
69	TOP LINK COMPLETE, REAR LH		1	502306	
	TOP LINK COMPLETE, REAR RH		1	502307	
70	Bush for top link		4	548211	
71	Plain washer, thin	Fixing top links to mounting arm	4	2854	All 3 Litre and 3½ Litre Cars numbered up to:-
72	Plain washer, thick		4	243022	
73	Castle nut (7/16" UNF)		4	254953	
74	Split pin		4	2426	
75	Bolt (⅜" UNF x 3½" long)	Fixing top links to top swivel housing	2	256251	
76	Plain washer, small		4	2251	
77	Plain washer, large		4	2257	
78	Self-locking nut (⅜" UNF)		2	252162	
79	Self-locking nut (7/16"UNF)		4	252163	
	Plain washer, thin	Fixing front top links to mounting arm	2	2854	3½ Litre Cars numbered from
	Plain washer, thick		2	243022	
	Castle nut (7/16" UNF)		2	254953	
	Split pin		2	2426	
	Spacer washer	Fixing rear top links to mounting arm	2	4908*	
	Plain washer, large		2	2854	
	Plain washer, small		2	2599	
	Castle nut (7/16" UNF) thin		2	254903*	
	Split pin		2	2426	
	Packing strip	Fixing top links to top swivel housing	2	595665*	
	Location tube		2	595709*	
	Bolt (⅜" UNF x 3⅜in Long)		2	256052	
	Self-locking nut (⅜" UNF)		2	257405	
	Plain washer		2	2251	
	Spring washer		2	3077	
	Nut (7/16" UNF)		2	254803	
	Self-locking nut (7/16" UNF)		2	257407	
	Plain washer		2	3843	

*Asterisk indicates a new part which has not been used on any previous model.

Plate Ref.	1 2 3 4 Description		Qty.	Part No.	Remarks
1	Shock absorber, front		2	542858	3 Litre standard suspension
	Shock absorber, front		2	562371	3½Litre standard suspension
	Shock absorber, front		2	536595	3 Litre and 3½ Litre High suspension
2	Shock absorber, rear		2	542859	Standard suspension
	Shock absorber, rear		2	536596	High suspension
3	Washer, top	Fixing front shock absorbers at top	2	503281	
4	Washer, bottom		2	503282	
5	Rubber pad		4	505252	
6	Locknut (⅜" UNF)		2	254852	
7	Nut (⅜" UNF)		2	254812	
8	Rubber bush	Fixing front shock absorbers to brackets at bottom	4	243057	
9	Distance tube		2	503288	
10	Bolt (7/16" UNF x 2¾" long)		2	256067	
11	Self-locking nut (7/16" UNF)		2	252163	
12	Rubber bush	Fixing rear shock absorbers to rear axle case	4	243057	
13	Bolt (7/16" UNF x 6¾" long)		2	253830	
14	Plain washer		2	503282	
15	Self-locking nut (7/16" UNF)		2	252163	
16	Guide washer	Fixing rear shock absorber to body	4	240800	
17	Retainer for rubber, large		2	503230	
	Retainer for rubber, small		2	511378	
18	Rubber pad		4	578035*	
19	Locknut (⅜" UNF)		2	254852	
20	Nut (⅜" UNF)		2	254812	
21	Torsion rod, front		1	272087	3 Litre
	Torsion rod, front		1	557338	3½ Litre
22	Support bracket for rod	Fixing rod to chassis sub-frame	2	213612	Alternatives. Check before ordering
23	Support bracket for rod		1	538035	
	Support bracket for rod		1	213612	
24	Rubber bush for rod		2	266722	
25	Set bolt (⅜" UNF x 1"long)		4	255245	
26	Spring washer		4	3076	
27	Link pin	Fixing torsion rod to bracket on bottom link	2	272958@	
	Link pin		2	512183@@	
28	Nut (5/16" UNF) for link pin		2	254831	
29	Distance tube		2	272959@	
	Distance tube		2	512182@@	
30	Rubber buffer		8	07014	
31	Cup washer		8	07015	
32	Nut (5/16" UNF)		2	254831	

@ 3 Litre Cars numbered up to 625100983,626100244, 628100107, 630100699, 631100326, 633100055

@@ 3 Litre Cars numbered from 625100984, 626100245, 628100108, 630100700, 631100327, 633100056 onwards and 3½ Litre models

*Asterisk indicates a new part which has not been used on any previous model.

BRAKES, FRONT, DRUM TYPE

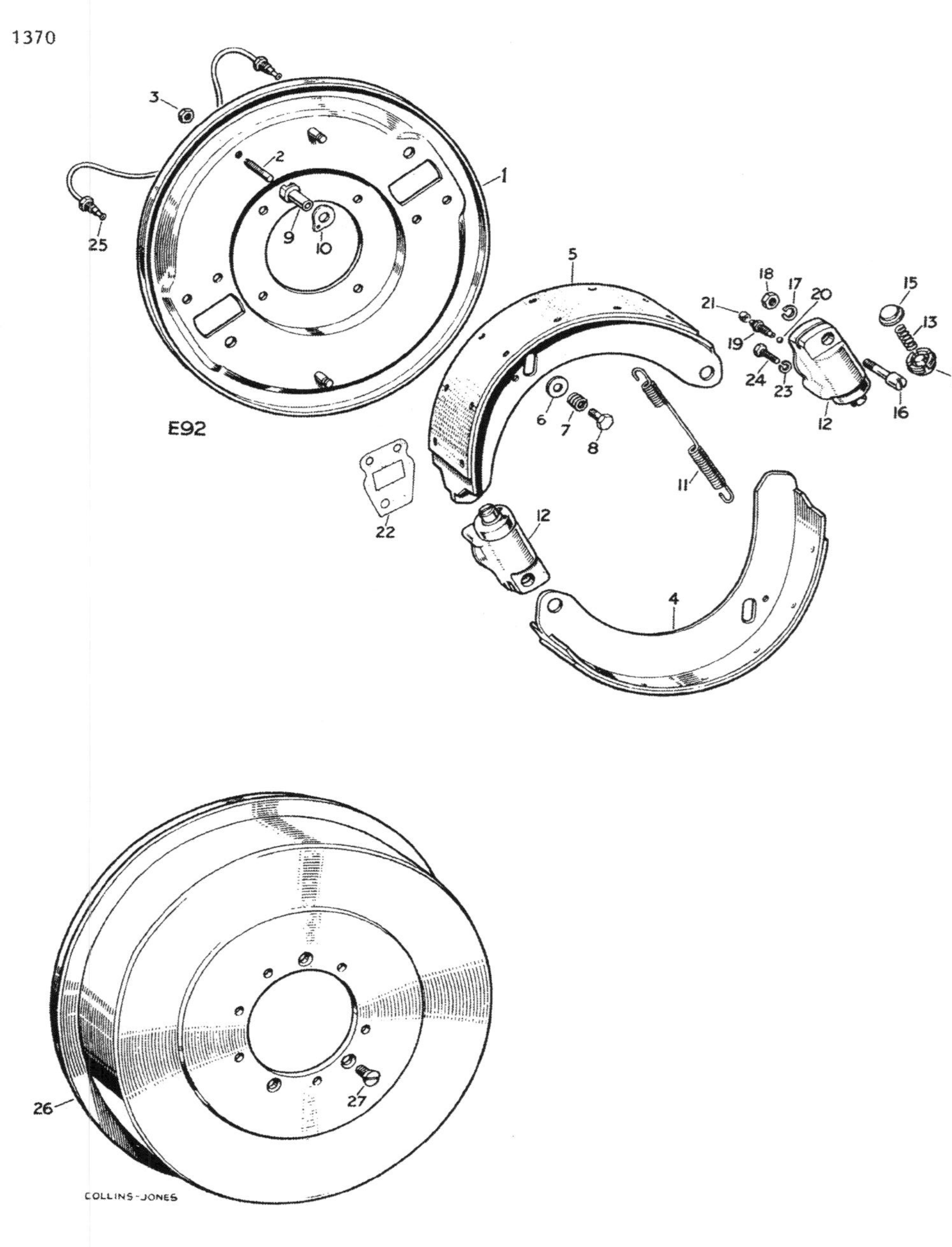

Plate Ref.	1 2 3 4 Description	Qty.	Part No.	Remarks
1	Brake anchor plate, front LH	1	504799	
	Brake anchor plate, front RH	1	504800	
	Bolt ($\frac{3}{8}$" x $2\frac{1}{4}$" long)	2	256446	Fixing brake anchor plate
	Bolt ($\frac{3}{8}$" x 2" long)	2	256445	Fixing brake anchor plate
	Bolt ($\frac{3}{8}$" x $1\frac{1}{8}$" long)	4	255048	Fixing brake anchor plate
	Self-locking nut ($\frac{3}{8}$")	8	252162	Fixing brake anchor plate
2	Steady post for brake shoe	4	510306	
3	Special nut fixing steady post	4	270681	
4	BRAKE SHOE ASSEMBLY, BOXED PAIR, LH	1	506936	
	BRAKE SHOE ASSEMBLY, BOXED PAIR, RH	1	506937	
5	Lining for brake shoe	4	268457	
	Rivet fixing lining	40	242467	
6	Friction washer	4	505015	For brake shoe automatic adjustment
7	Spring	4	505014	For brake shoe automatic adjustment
8	Special screw	4	505016	For brake shoe automatic adjustment
9	Link pin	4	510307	For brake shoe automatic adjustment
10	Link washer	4	504956	For brake shoe automatic adjustment
11	Pull-off spring for brake shoe	4	504998	
12	WHEEL CYLINDER ASSEMBLY, LH	2	279789	
	WHEEL CYLINDER ASSEMBLY, RH	2	279788	
13	Spring	4	268758	
14	Air excluder, seal	4	268756	
15	Air excluder, cylinder	4	268757	
16	Pivot pin	4	268460	For brake shoe
17	Spring washer	4	218970	For brake shoe
18	Nut ($\frac{1}{2}$")	4	3828	For brake shoe
19	Bleed screw	2	248909	
20	Steel ball for bleed screw	2	3746	
21	Dust cap for bleed screw	2	234957	
22	Gasket for wheel cylinder	4	504559	
23	Spring washer	8	3075	Fixing wheel cylinder
24	Special set bolt	8	255024	Fixing wheel cylinder
25	Connecting pipe for wheel cylinder	2	268759	
26	Brake drum	2	269370	
27	Set screw fixing brake drum	6	269738	
	Anti-squeal ring	2	512409	
	Asbestos tape for ring	2	512410	
	Bolt($\frac{1}{4}$" UNF x $1\frac{1}{8}$" long)	2	255010	Fixing tape and ring to brake anchor plate
	Nut ($\frac{1}{4}$" UNF)	4	254810	Fixing tape and ring to brake anchor plate

*Asterisk indicates a new part which has not been used on any previous model.

BRAKES, FRONT, DISC TYPE

1372

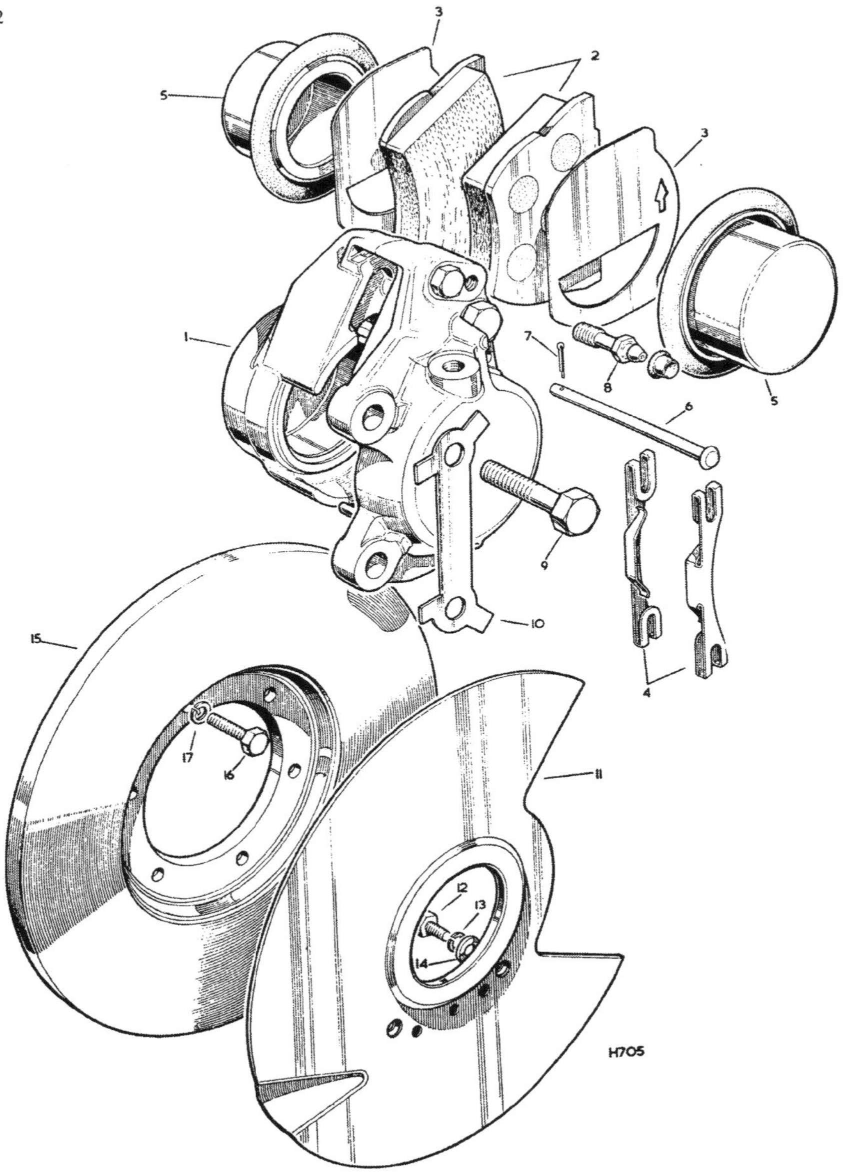

Plate Ref.	1 2 3 4 Description		Qty.	Part No.	Remarks
	FRONT BRAKE CALIPER, RH		1	605959	3 Litre
	FRONT BRAKE CALIPER, LH		1	605960	
	FRONT BRAKE CALIPER, RH		1	605961	3½ Litre
	FRONT BRAKE CALIPER, LH		1	605962	
	Pad for disc brake, Ferodo DS 5S, (set of 4)		1	513672	Alternatives for 3 Litre
2	Pad for disc brake, Don 105/3 (set of 4)		1	534466	
	Pad for disc brake, Ferodo 2424F (set of 4)		1	605208	Early 3½ Litre
	Pad for disc brake, Don 227 (set of 4)		1	GBP200*	Late 3½ Litre
	Pad for disc brake, Ferodo 2430F (set of 4)		1	606097	Alternative to standard pads on 3½ Litre
3	Shim kit for brake caliper, set of 4		1	607273*	
4	Anti-rattle spring for brake pads		4	536309	
	Piston for brake caliper		4	536559@	
	Piston for brake caliper		4	562024@@	
6	Pad retaining pin		4	513666	
7	Special split pin for retaining pin		4	513667	
8	Bleed screw for caliper		2	513668	
9	Set bolt (½" UNF x ½" long)	Fixing brake caliper to axle	4	253864	
10	Lock plate		2	512742	
	Shield for brake disc, RH	⅜" dia fixing holes	1	510903	1959-60
	Shield for brake disc, LH		1	510904	
11	Shield for brake disc, RH	7/16" dia fixing holes	1	527339	1961 onwards
	Shield for brake disc, LH		1	527340	
12	Set bolt (10 UNF x 9/16" long)	Additional fixing, disc shield to stub axle	2	257006	Mk IA cars numbered from 72500607, 72600097, 72800107, 73001022, 73100176, 73300170, Mk II and Mk III and 3½ Litre
13	Spring washer		2	3073	
14	Spacer		2	527343	
15	Disc for brake		2	548076	
16	Special set bolt	Fixing disc to hub	10	572973	
17	Special washer		10	572974	
	Overhaul kit for caliper		1	513673@	
	Overhaul kit for caliper		1	531939@@	

@ 3 Litre cars numbered up to 72500084A, 72600022A, 72800012A, 73000124A, 73100020A, 73300023A

@@ 3 Litre cars numbered from 72500085A, 72600023A, 72800013A, 7300125A, 73100021A, 83300024A onwards and 3½ litre

*Asterisk indicates a new part which has not been used on any previous model.

1374

E94

Plate Ref.	1 2 3 4	Description	Qty.	Part No.	Remarks
1		Brake anchor plate complete, rear LH	1	237710	
		Brake anchor plate complete, rear RH	1	237711	
		Bolt (3/8" x 1 3/8" long) } Fixing brake anchor plate	12	250542	
		Spring washer } Fixing brake anchor plate	12	3076	
		Nut (3/8") } Fixing brake anchor plate	12	2827	
2		Steady post for brake shoe	4	234192	
3		Tufnol bush for steady post	4	234888	
4		Special nut for steady post	4	254861	
5		BRAKE SHOE ASSEMBLY, BOXED PAIR	2	512532	
6		Lining for brake shoe	4	512171	
7		Rivet for lining	40	242034	
8		Pull-off spring for brake shoes, adjuster end	2	42318	
9		Pull-off spring for brake shoes, wheel cylinder end	2	242905	
10		Adjuster housing	2	234901	
11		Spring washer } Fixing adjuster housing	4	3076	
12		Special set bolt, } Fixing adjuster housing	4	255245	
13		Plunger, LH	2	234902	
14		Plunger, RH	2	234903	
15		Cone for adjuster	2	04965	
16		WHEEL CYLINDER ASSEMBLY, LH	1	268464	3 Litre with drum-type front brakes
		WHEEL CYLINDER ASSEMBLY, RH	1	268465	3 Litre with drum-type front brakes
		WHEEL CYLINDER ASSEMBLY, LH	1	501256	3 Litre with disc-type front brakes
		WHEEL CYLINDER ASSEMBLY, RH	1	501257	3 Litre with disc-type front brakes
		WHEEL CYLINDER ASSEMBLY, RH	1	561630	3½ Litre
		WHEEL CYLINDER ASSEMBLY, LH	1	561629	3½ Litre
17		Spring	2	268769	3 Litre drum-type front brakes
		Spring	2	246320	3 Litre disc-type front brakes
		Spring	2	605138	3½ Litre
18		Tappet	4	502291	
19		Roller,	4	234959	
20		Draw link for brake rod	2	234961	Mk I, Mk IA and Mk II 3 Litre
		Draw link for brake rod	2	542965	Mk III 3 Litre and 3½ Litre
21		Cover plate	2	234960	
22		Special screw } Fixing cover plate	8	234962	
23		Shakeproof washer } Fixing cover plate	8	234963	
25		Bleed screw	2	248909	
26		Steel ball for bleed screw	2	3746	
27		Dust cap for bleed screw	2	234957	
28		Dust cover plate	2	268930	Mk I, Mk IA and Mk II 3 Litre

*Asterisk indicates a new part which has not been used on any previous model.

Plate Ref.	1 2 3 4	Description	Qty.	Part No.	Remarks
		Dust cover plate	2	545536	Mk III 3 Litre
		Rubber dust cover, LH	1	542961	Mk III 3 Litre
		Rubber dust cover, RH	1	542962	Mk III 3 Litre
		Retaining plate for dust cover, LH	1	542963	Mk III 3 Litre
		Retaining plate for dust cover, RH	1	542964	Mk III 3 Litre
		Dust cover plate	2	605225	3½ Litre
		Rubber dust cover, LH	1	605141	3½ Litre
		Rubber dust cover, RH	1	605140	3½ Litre
		Retaining plate for dust cover	2	605224	3½ Litre
29		Spring washer } Fixing wheel cylinder	6	2284	
30		Self-locking nut } Fixing wheel cylinder	6	273081	
31		Brake drum	2	540504	
32		Set screw fixing brake drum	6	269738	

*Asterisk indicates a new part which has not been used on any previous model.

D975

Plate Ref.	Description	Qty.	Part No.	Remarks
	Brake pipe, four-way junction to RH front hose	1	504512	For drum-type brakes
	Brake pipe, four-way junction to LH front hose	1	504513	For drum-type brakes
	Brake pipe, four-way junction to RH front hose	1	522784@	For disc-type brakes
	Brake pipe, four-way junction to LH front hose	1	535045@	For disc-type brakes
1	Brake pipe, four-way junction to RH front hose	1	522784@@	For disc-type brakes
2	Brake pipe, four-way junction to LH front hose	1	522785@@	For disc-type brakes
	Brake pipe, servo to four-way junction	1	504511@	
3	Brake pipe, servo to four-way junction	1	514138@@	
	Brake pipe, four-way junction to rear connector	1	537211	
4	Brake pipe, four-way junction to rear connector	1	514139@@	
5	Union connector at rear	1	504765	
6	Brake pipe, union to rear hose	1	504515	
7	Brake pipe, three-way junction to servo	1	504510%	
	Brake pipe, three-way junction to servo	1	542873%%	
8	Brake pipe, three-way junction to master cylinder	1	243194	RH Stg
	Brake pipe, three-way junction to master cylinder	1	542872%%	RH Stg
	Brake pipe, three-way junction to master cylinder	1	504832%	LH Stg 4- speed
	Brake pipe, three-way junction to master cylinder	1	557103%%	LH Stg 4- speed
	Brake pipe, three-way junction to master cylinder	1	531413%	LH Stg Automatic
	Brake pipe, three-way junction to master cylinder	1	557102%%	LH Stg Automatic except Mk III
	Brake pipe, three-way junction to master cylinder	1	557102	LH Stg Mk III Automatic
9	Three-way junction on dash	1	268964%	
	Three-way junction on dash	1	234928%%	
10	Bolt (5/16" UNF x ⅜" long) Fixing three-way junction	1	255225%	
	Bolt (5/16" UNF x ¾" long) Fixing three-way junction	1	255026%%	
11	Spring washer Fixing three-way junction	1	3075	
12	Nut (5/16" UNF) Fixing three-way junction	1	254810	

@ Cars numbered up to 625000378, 626000119, 628000046, 630000162, 631000040, 633000094

@@ Cars numbered from 625000379, 626000120, 628000047, 630000163, 631000041, 633000095 onwards

% Mk I, Mk IA and Mk II cars with suffix letter 'A' and 'B'

%% Mk II cars with suffix letter 'C' onwards and Mk III

*Asterisk indicates a new part which has not been used on any previous model.

1380

D975.

Plate Ref.	1 2 3 4 Description		Qty.	Part No.	Remarks
13	Four-way junction on chassis sub-frame,		1	241690	
14	Bolt (¼" UNF x 1¼" long)	Fixing four-way junction to sub-frame	1	256001	
15	Spring washer		1	3074	
16	Nut (¼" UNF)		1	254810	
17	Stop lamp switch		1	545280	Mk I, Mk IA and Mk II cars with suffix letters 'A' and 'B'
	Stop lamp switch		1	560793	Mk II cars with suffix letter 'C' and Mk III
	Washer for switch		1	233220	
	Brake hose for front wheel cylinders		2	268341	For drum-type brakes
18	Brake hose for front caliper		2	512247	For disc-type brakes
19	Shakeproof washer	Fixing hose to bracket	2	233305	
20	Special nut		2	254852	
21	Gasket for hose		4	233220	
23	Banjo	Fixing brake hose to front calipers	2	233299	For disc-type brakes
24	Banjo bolt		2	512235	
25	Joint washer, small		2	216914	
26	Joint washer, large		2	512387	
27	Brake pipe, tee piece to wheel cylinder, RH rear		1	504516	
28	Brake pipe, tee piece to wheel cylinder, LH rear		1	504517	
29	Tee piece for rear pipes		1	234928	
30	Bolt (5/16" UNF x ⅞" long)	Fixing tee piece to support on rear axle	1	255227	
31	Plain washer		1	2550	
32	Spring washer		1	3075	
33	Nut (5/16" UNF)		1	254811	
34	Brake hose complete to rear axle		1	506673	
35	Shakeproof washer	For rear hose	1	233305	
36	Special nut		1	254852	
37	Gasket		1	233220	
38	Plate for hose to rear axle		1	505043	
39	Bolt (⅜" UNF x ⅝" long)	Fixing plate to body	1	255044	
40	Spring washer		1	3076	
41	Nut (⅜" UNF)		1	254812	
42	Pipe clip	Fixing brake pipes to dash	As reqd	233274	LH Stg
43	Pipe clip		As reqd	41379	
	Drive screw fixing clips		As reqd	78322	
	Bolt (2 BA x ⅝" long)	Fixing clips	As reqd	237121	LH Stg
	Spring washer		As reqd	3073	
	Nut (2 BA)		As reqd	2247	
44	Viblock clip for brake pipes on rear axle		3	11820	

*Asterisk indicates a new part which has not been used on any previous model.

1382

BRAKE PIPES, FRONT, 3½ LITRE

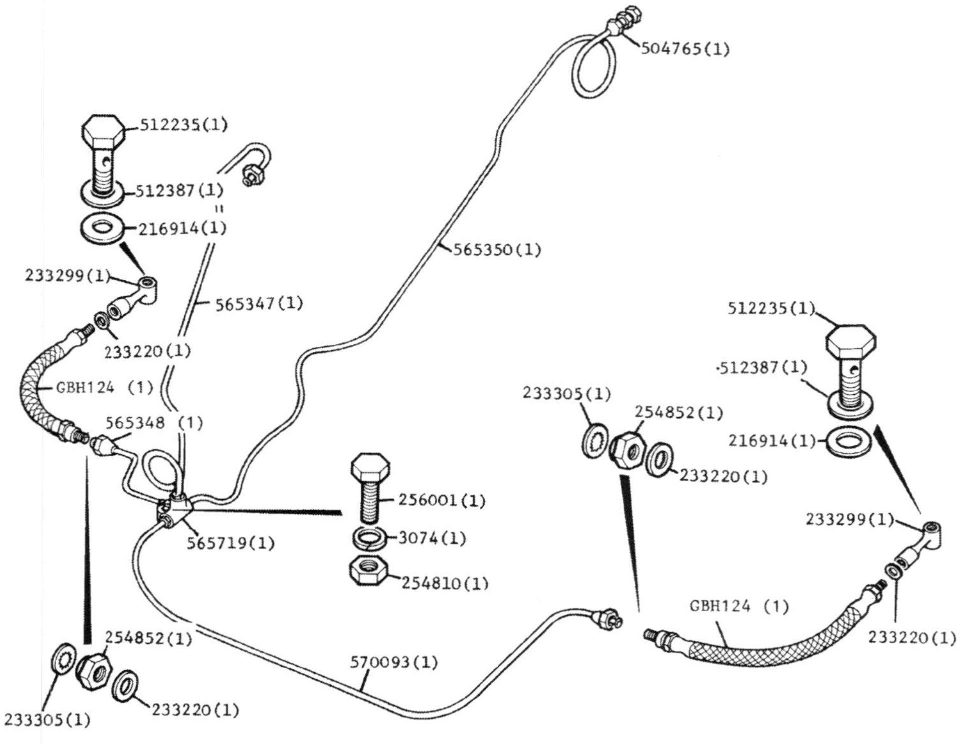

512235(1)
512387(1)
216914(1)
233299(1)
233220(1)
GBH124 (1)
565348 (1)
565347(1)
565350(1)
504765(1)
565719(1)
256001(1)
3074(1)
254810(1)
254852(1)
233305(1)
233220(1)
570093(1)
233305(1)
254852(1)
233220(1)
512235(1)
512387(1)
216914(1)
233299(1)
GBH124 (1)
233220(1)

REMARKS

13 83

BRAKE PIPES, REAR, 3½ LITRE

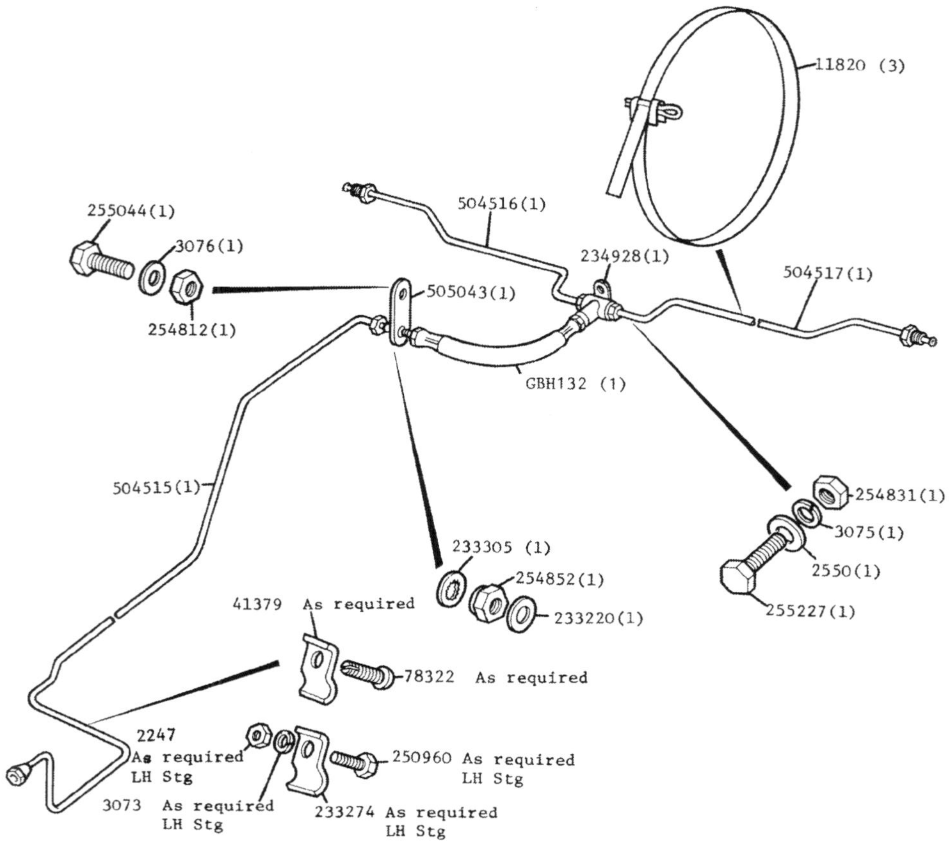

11820 (3)
504516(1)
255044(1)
3076(1)
254812(1)
505043(1)
234928(1)
504517(1)
GBH132 (1)
504515(1)
254831(1)
3075(1)
2550(1)
255227(1)
233305 (1)
254852(1)
233220(1)
41379 As required
78322 As required
2247 As required LH Stg
3073 As required LH Stg
250960 As required LH Stg
233274 As required LH Stg

REMARKS

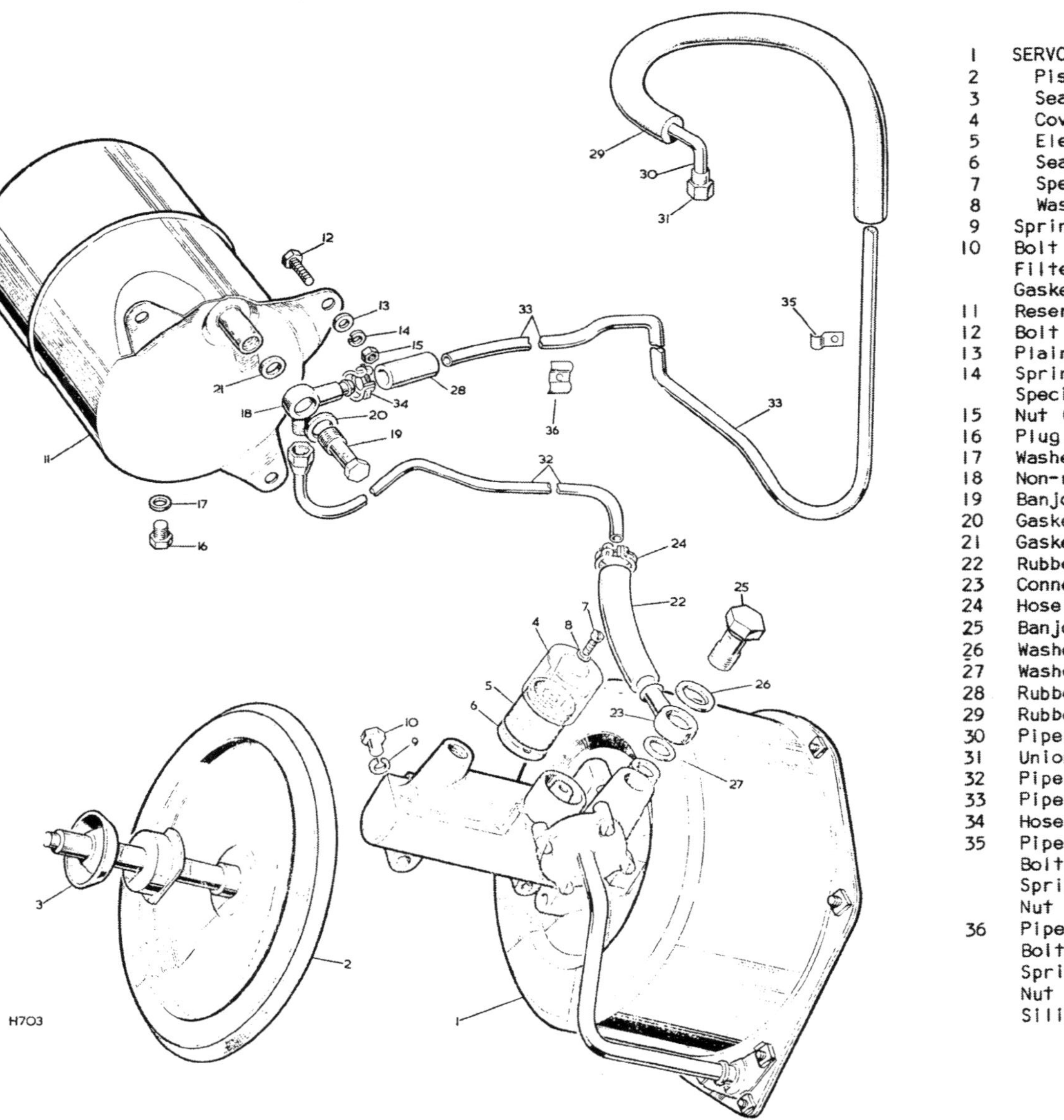

Plate Ref.	1 2 3 4 Description	Qty.	Part No.	Remarks
1	SERVO UNIT COMPLETE	1	545900	
2	Piston and guide assembly	1	535766	
3	Seal for piston	1	535767	
4	Cover for filter	1	600560	
5	Element forfilter	1	600561	
6	Seal for filter	1	600562	
7	Special screw	1	601698	
8	Washer for screw	1	601699	
9	Spring washer } Fixing	3	3076	
10	Bolt ($\frac{3}{8}$" UNFx$\frac{3}{8}$"long) } servo unit	3	255245	
	Filter unit for servo	1	275070	
	Gasket, filter to servo unit	1	267604	
11	Reservoir tank for servo unit	1	501531	
12	Bolt (5/16" UNF x $\frac{3}{4}$" long) }	3	255226	
13	Plain washer } Fixing reservoir	3	2223	
14	Spring washer } tank to RH	2	3075	
	Special earthing washer } wing valance	2	510912	
15	Nut (5/16" UNF) }	3	254831	
16	Plug for test hole in tank	1	250976	
17	Washer for plug	1	243957	
18	Non-return valve for reservoir tank	1	501273	
19	Banjo bolt } Fixing	1	233218	
20	Gasket, large } non-return valve	1	233303	Cars numbered up to
21	Gasket, small to tank	1	275189	625000378, 626000119,
22	Rubber hose, tank to servo	1	504835	628000046, 630000162,
23	Connection on servo	1	267487	631000040, 633000094
24	Hose clip, rubber hose to pipe and connection	2	50302	
25	Banjo bolt }	1	267490	
26	Washer, large } Fixing hose to servo inlet	1	267604	
27	Washer, small }	1	267605	
28	Rubber hose, tank end	1	275072	
29	Rubber hose, manifold end	1	277148	
30	Pipe, manifold end	1	504191	
31	Union nut for pipe, manifold end	1	277332	
32	Pipe, reservoir to servo	1	504186	
33	Pipe, tank hose to manifold hose	1	504187	
34	Hose clip for manifold to tank hoses	4	50302	
35	Pipe clip fixing manifold to tank pipe to dash	1	216708	
	Bolt (2 BA x $\frac{3}{8}$" long) } Fixing	1	250960	
	Spring washer } manifold to tank	1	3073	
	Nut (2 BA) } pipe to dash	1	2247	
36	Pipe clip for reservoir tank pipes	1	504193	
	Bolt (2 BA x $\frac{3}{8}$" long) } Fixing	1	250960	
	Spring washer } pipes to	1	3073	
	Nut (2 BA) } wing valance	1	2247	
	Silicone MS4 grease for servo unit, $\frac{1}{2}$ oz tube	1	270656	

*Asterisk indicates a new part which has not been used on any previous model.

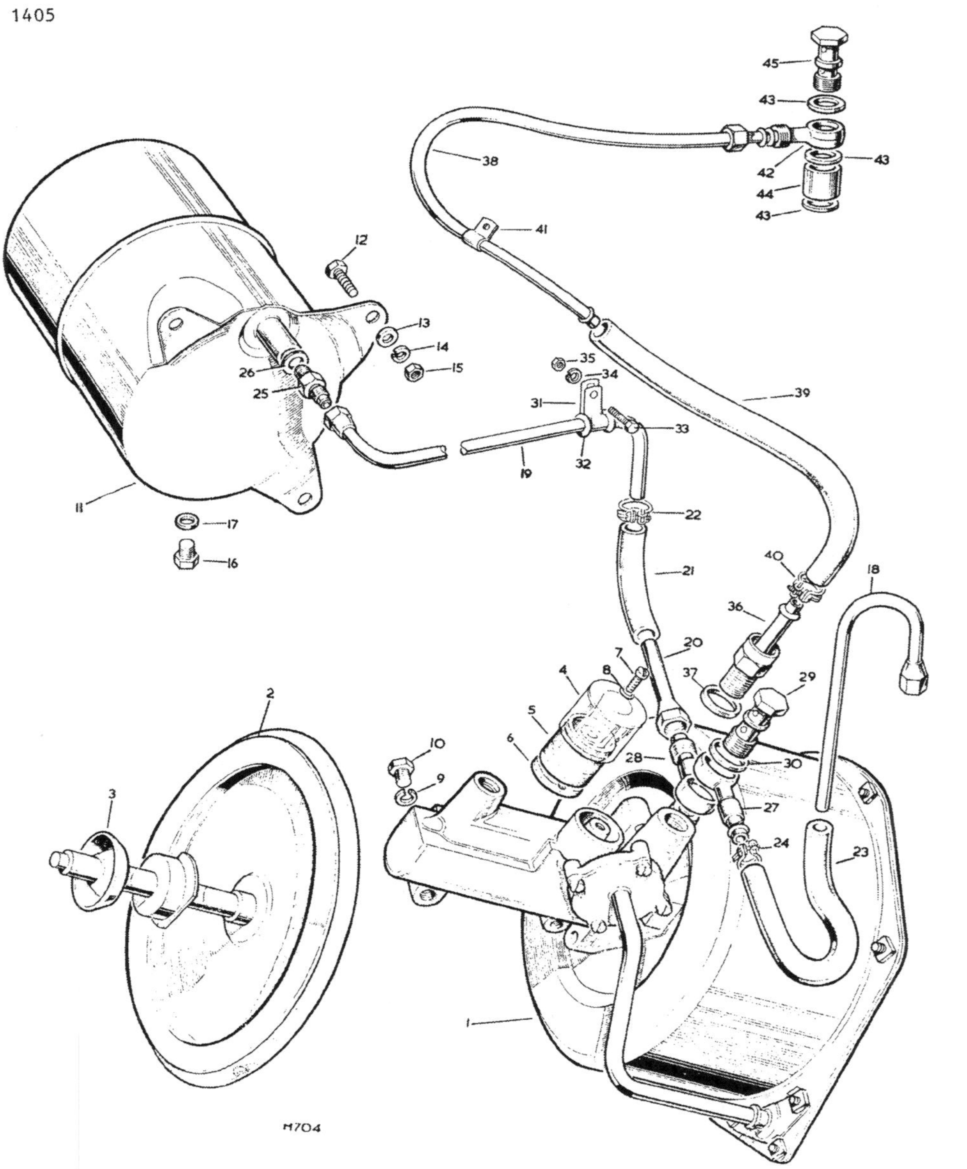

Plate Ref.	1 2 3 4 Description	Qty.	Part No.	Remarks
1	SERVO UNIT COMPLETE	1	545900	
2	Piston and guide assembly	1	535766	
3	Seal for piston	1	535767	
	FILTER ASSEMBLY FOR SERVO UNIT	1	520072	Clip-type fixing
	Filter element for servo	1	514091	Clip-type fixing
4	Cover for filter	1	600560	Screw-type fixing
5	Element for filter	1	600561	Screw-type fixing
6	Seal for filter	1	600562	Screw-type fixing
7	Special screw } Fixing filter cover	1	601698	Screw-type fixing
8	Washer for screw } Fixing filter cover	1	601699	Screw-type fixing
9	Spring washer } Fixing servo unit	3	3076	
10	Bolt (3/8" UNF x 5/8" long) } Fixing servo unit	3	255045	
11	Reservoir tank for servo unit	1	501531	Cars numbered from 625000379, 626000120, 628000047, 63000163, 631000041, 63300095 onwards and Mk IA only
12	Bolt (5/16" UNF x 3/4" long) } Fixing reservoir tank to RH wing valance	3	255226	ditto
13	Plain washer } Fixing reservoir tank to RH wing valance	3	2223	ditto
14	Spring washer } Fixing reservoir tank to RH wing valance	2	3075	ditto
	Special earthing washer } Fixing reservoir tank to RH wing valance	2	510912	ditto
15	Nut (5/16" UNF) } Fixing reservoir tank to RH wing valance	3	254831	ditto
16	Plug for test hole in tank	1	250976	ditto
17	Washer for plug	1	243957	ditto
18	Pipe, manifold to servo	1	513347	ditto
19	Pipe, reservoir tank to hose	1	516272	ditto
20	Pipe, hose to servo	1	516273	ditto
21	Hose connecting servo pipes	1	516274	ditto
22	Clip for hose	2	50302	ditto
23	Hose for pipe, manifold to servo	1	277147	ditto
24	Clip for hose	2	50302	ditto
25	Adaptor for brake reservoir tank	1	514141	ditto
26	Joint washer for adaptor	1	243967	
27	Non-return valve for servo unit	1	513342	
28	Connection on servo for tank to servo pipe	1	513341	
29	Banjo bolt } Fixing connection and non-return valve to servo unit	1	513343	
30	Washer } Fixing connection and non-return valve to servo unit	3	267604	
31	Clip } Fixing pipe, tank to servo to wing valance	1	56667	
32	Rubber grommet } Fixing pipe, tank to servo to wing valance	1	06860	
33	Bolt (2 BA x 3/8" long) } Fixing pipe, tank to servo to wing valance	1	250960	
34	Spring washer } Fixing pipe, tank to servo to wing valance	1	3073	
35	Nut (2 BA) } Fixing pipe, tank to servo to wing valance	1	2247	
36	Non-return valve, servo to manifold	1	560163	Mk II and Mk III
37	Sealing washer, valve to servo unit	1	267604	Mk II and Mk III
38	Pipe complete, manifold to servo	1	536257	Mk II
	Pipe complete, manifold to servo	1	557135	Mk III
39	Hose connecting manifold pipe to servo	1	565476	Mk II and Mk III
40	Clip for hose	2	50302	Mk II and Mk III
41	Clip, vacuum pipe to manifold	1	50642	Mk II and Mk III
42	Banjo manifold to servo pipe	1	513341	Mk II and Mk III
43	Sealing washer } Fixing banjo to manifold	3	267604	Mk II and Mk III
44	Spacer } Fixing banjo to manifold	1	548081	Mk II and Mk III
45	Banjo bolt } Fixing banjo to manifold	1	513343	Mk II and Mk III
	Piston conversion kit	1	535673	
	Repair kit for servo unit	1	600554	
	Silicone MS4 grease for servo unit, 1/2 oz tube	1	270656	

*Asterisk indicates a new part which has not been used on any previous model.

1407

Plate Ref.	1 2 3 4 Description		Qty.	Part No.	Remarks
1	BRAKE SERVO UNIT		1	578667	
	Major repair kit		1	607133	
	Air control valve kit		1	601907	
2	Plain washer	Fixing servo unit to mounting bracket	3	2550	
3	Spring washer		3	3075	
4	Nut (5/16" UNF)		3	254831	
5	Mounting bracket for servo		1	562192	
6	Bolt (¼" UNF x ⅝" long)		1	255206	
7	Plain washer	Fixing servo mounting bracket to valance	2	4071	
8	Spring washer		1	3074	
9	Nut (¼" UNF)		1	254810	
10	Mounting bracket for hose		1	562296	
11	Clamp		1	562448	
12	Bolt (¼" UNF x ¾" long)	Fixing clamp to mounting bracket	1	255207	
13	Spring washer		1	3074	
14	Nut (¼" UNF)		1	254810	
15	Reservoir tank for servo		1	562500	
	Drain plug	For reservoir tank	1	250976	
	Washer		1	243957	
16	Bolt (5/16" UNF x 1½" long)	Fixing vacuum reservoir to wing valance	2	256022	
17	Bolt (5/16" UNF x ¾" long)		1	255026	
18	Plain washer		3	2550	
19	Spring washer		3	3075	
20	Nut (5/16" UNF)		3	254831	
21	Non-return valve	For vacuum reservoir tank	1	565376	
22	Washer for banjo bolt		2	267604	
23	Banjo bolt		1	534156	
24	Pipe complete, inlet manifold to reservoir tank		1	565352	
25	Vacuum hose, reservoir tank to pipe		1	565354	
26	Vacuum hose, inlet manifold non-return valve to reservoir supply tank		1	565355	
27	Hose clip, hose to non-return valve		1	554260	
	Hose clip, vacuum hose to pipe		1	572548	
28	Non-return valve at inlet manifold		1	578620	
29	Washer for non-return valve		1	232043	
30	Pipe complete, reservoir tank to servo unit		1	565353	
31	Vacuum hose, servo unit to pipe		1	565354	
32	Clip fixing pipe to servo bracket		1	552848	
33	Clip fixing servo pipes to wing valance		1	548088	
34	Drain screw fixing clip to valance		1	78322	
	Pipe complete servo to master cylinder		1	565346	RH Stg
	Pipe complete servo to master cylinder		1	565363	LH Stg

*Asterisk indicates a new part which has not been used on any previous model.

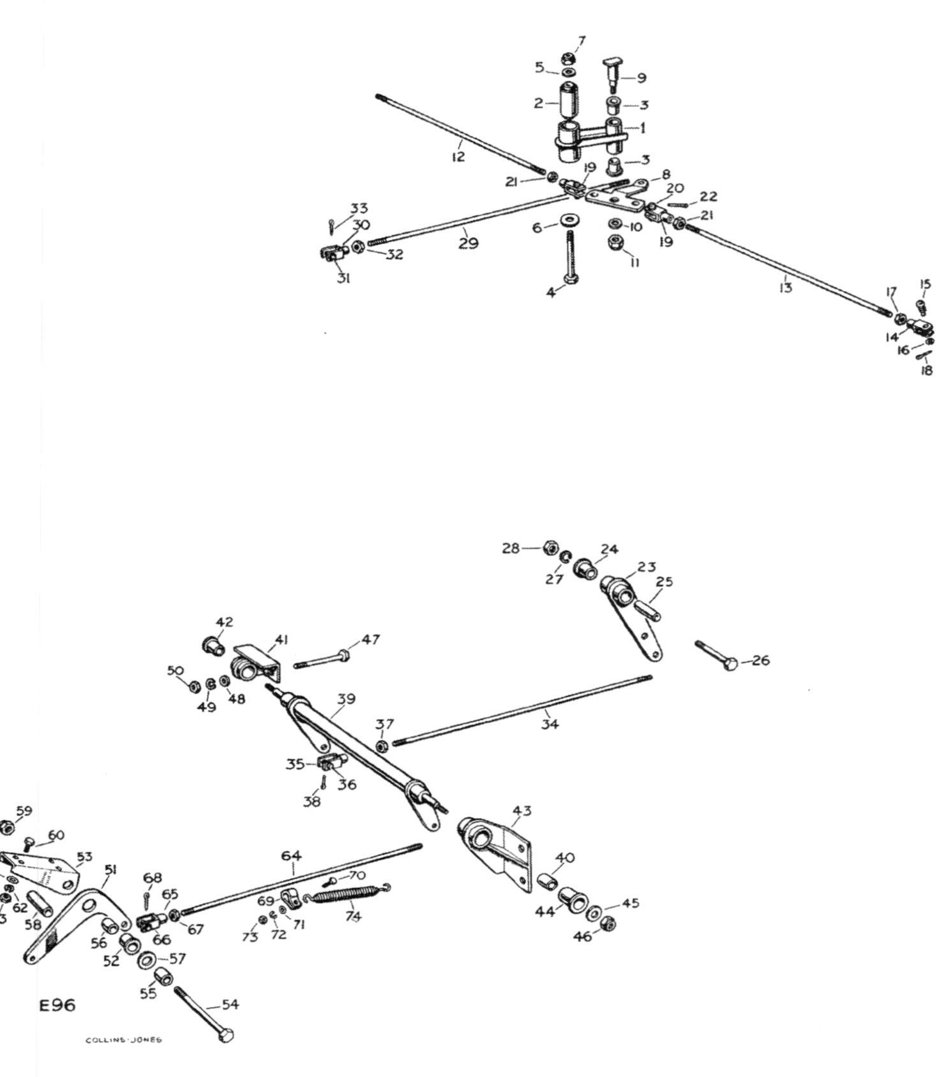

Plate Ref.	1 2 3 4 DESCRIPTION		Qty	Part No.	REMARKS
1	LINK PLATE ASSEMBLY FOR BRAKE		1	241850	
2	Rubber bush for link plate		1	242564	
3	Bush for link plate		2	238793	
4	Bolt ($\frac{7}{16}$" UNF x 3" long)		1	256468	
5	Plain washer, top, small	Fixing link	1	3261	
6	Plain washer, bottom, large	to axle	1	2441	
7	Self-locking nut ($\frac{7}{16}$" UNF)		1	252163	
8	Balance lever for brake		1	263373	
9	Pin for brake balance lever		1	278654	
10	Plain washer	Fixing balance lever	1	3830	
11	Self-locking nut ($\frac{7}{16}$" UNF)	to link plate	1	252161	
12	Brake cross rod, short		1	278577	Mk I, Mk IA and Mk II
	Brake cross rod, short		1	542967	Mk III
13	Brake cross rod, long		1	278576	Mk I, Mk IA and Mk II
	Brake cross rod, long		1	542966	Mk III
14	Clevis fork end		2	275199	
15	Clevis pin and spring	Fixing axle	2	216421	
16	Plain washer	cross rods to	2	2920	Mk I, Mk IA and Mk II
17	Locknut ($\frac{5}{16}$" UNF)	draw link	2	254851	
18	Split pin for clevis		2	2392	
	Clevis fork end		2	247926	
	Clevis pin and spring	Fixing axle cross	2	230279	Mk III
	Locknut ($\frac{1}{4}$" UNF)	rods to draw link	2	254850	
	Split pin for clevis		2	2389	
19	Clevis fork end		2	247926	
20	Clevis pin and spring	Fixing brake	2	230279	
21	Locknut ($\frac{1}{4}$" UNF)	rod to levers	2	254850	
22	Split pin for clevis		2	2389	
23	HANDBRAKE RELAY LEVER COMPLETE ..		1	501260	
24	Bush for relay lever		2	503194	
25	Bush		1	278884	
26	Bolt ($\frac{3}{8}$" UNF x $1\frac{7}{8}$" long)	Fixing relay lever to	1	256044	
27	Spring washer	bracket on body floor	1	3076	
28	Nut ($\frac{3}{8}$" UNF)		1	254812	
29	Brake rod, relay to compensator		1	501128	
30	Clevis fork end		2	247926	
31	Clevis pin and spring	Fixing rod to relay	2	230279	
32	Locknut ($\frac{1}{4}$" UNC)	and balance lever	2	254850	
33	Split pin for clevis		2	2389	
34	Brake rod, relay to cross-shaft		1	501127	
35	Clevis fork end		2	247926	
36	Clevis pin and spring	Fixing rod to relay lever	2	230279	
37	Locknut ($\frac{1}{4}$" UNC)	and cross-shaft	2	254850	
38	Split pin for clevis		2	2389	

* Asterisk indicates a new part which has not been used on any previous Rover model

1411

E96

Plate Ref.	1 2 3 4 DESCRIPTION	Qty	Part No.	REMARKS
39	Cross-shaft complete for handbrake	1	501115	
40	Sleeve for bearing	2	501108	
41	BRACKET AND BUSH COMPLETE FOR CROSS-SHAFT, OUTER	1	501102	
42	Bush for cross-shaft bracket	1	501107	
43	BRACKET AND BUSH COMPLETE FOR CROSS-SHAFT, INNER	1	501110	
44	Bush for cross-shaft bracket	1	501107	
45	Plain washer } Fixing cross-shaft to bracket and bush	2	4094	
46	Self-locking nut ($\frac{3}{8}$" UNF) } Fixing cross-shaft to bracket and bush	2	252162	
47	Bolt ($\frac{3}{8}$" UNF x $3\frac{1}{4}$" long) } Fixing outer bracket to body cross-member	2	256250	
48	Plain washer } Fixing outer bracket to body cross-member	2	4094	
49	Spring washer } Fixing outer bracket to body cross-member	2	3076	
50	Nut ($\frac{3}{8}$" UNF) } Fixing outer bracket to body cross-member	2	254812	
51	LEVER AND BUSH COMPLETE FOR HANDBRAKE AT DASH	1	501118	
52	Bush for handbrake lever	1	511127	
	Thrust washer for handbrake lever	1	511128	
53	Mounting bracket for handbrake lever	1	501125	
54	Bolt ($\frac{3}{8}$" UNF x $4\frac{1}{4}$" long) } Fixing lever and bush to mounting bracket	1	256254	
55	Spacing collar } Fixing lever and bush to mounting bracket	1	504129	
56	Sleeve for bearing } Fixing lever and bush to mounting bracket	1	501121	
57	Plain washer } Fixing lever and bush to mounting bracket	2	2251	
58	Distance tube in bracket } Fixing lever and bush to mounting bracket	1	278481	
59	Self-locking nut ($\frac{3}{8}$" UNF) } Fixing lever and bush to mounting bracket	1	252162	
60	Bolt ($\frac{5}{16}$" UNF x $\frac{5}{8}$" long) } Fixing mounting bracket and dip switch bracket to dash, lower	2	255225	
	Bolt ($\frac{5}{16}$" UNF x $\frac{7}{8}$" long) } Fixing mounting bracket and dip switch bracket to dash, lower	2	255227	
61	Plain washer } Fixing mounting bracket and dip switch bracket to dash, lower	4	2550	
62	Spring washer } Fixing mounting bracket and dip switch bracket to dash, lower	4	3075	
63	Nut ($\frac{5}{16}$" UNF) } Fixing mounting bracket and dip switch bracket to dash, lower	4	254811	
64	Brake rcd, cross-shaft to lever and bush at dash	1	501124	
65	Clevis fork end } Fixing brake rod to cross-shaft lever and lever and bush at dash	2	247926	
66	Clevis pin and spring } Fixing brake rod to cross-shaft lever and lever and bush at dash	2	230279	
67	Locknut ($\frac{5}{16}$" UNF) } Fixing brake rod to cross-shaft lever and lever and bush at dash	2	254850	
68	Split pin } Fixing brake rod to cross-shaft lever and lever and bush at dash	2	2389	
69	Anchor clip for return spring	1	238858	
70	Bolt (2 BA x $\frac{1}{2}$" long) } Fixing clip to rod	1	234603	
71	Plain washer } Fixing clip to rod	1	3867	
72	Spring washer } Fixing clip to rod	1	3073	
73	Nut (2 BA) } Fixing clip to rod	1	2247	
74	Return spring for brake	1	06295	

* Asterisk indicates a new part which has not been used on any previous Rover model

J337

Plate Ref.	1 2 3 4	Description		Qty.	Part No.	Remarks
1		LINK PLATE ASSEMBLY FOR BRAKE		1	562073	
2		Rubber bush for link plate		1	548207	
3		Bush for link plate		2	238793	
4		Bolt (7/16" UNF x 3" long)	Fixing link to axle	1	256468	
5		Plain washer, top, small		1	3261	
6		Plain washer, bottom, large		1	2441	
7		Self-locking nut (7/16" UNF)		1	252163	
8		Balance lever for brake		1	263373	
9		Pin for brake balance lever		1	278654	
10		Plain washer	Fixing balancer lever to link plate	1	3830	
11		Self-locking nut(5/16"UNF)		1	252161	
12		Brake cross rod, short		1	562076	
13		Brake cross rod, long		1	562077	
14		Clevis fork end	Fixing axle cross rods to draw link	2	247926	
15		Clevis pin and spring		2	230279	
16		Locknut (¼" UNF)		2	254850	
17		Split pin for clevis		2	2389	
18		Clevis fork end	Fixing brake rod to levers	2	247926	
19		Clevis pin and spring		2	230279	
20		Locknut (¼" UNF)		2	254850	
21		Split pin for clevis		2	2389	
22		Brake rod, relay to compensator		1	562066	
23		Clevis fork end	Fixing rod to relay and balance lever	2	247926	
24		Clevis pin and spring		2	230279	
25		Locknut (¼" UNC)		2	254850	
26		Split pin for clevis		2	2389	
27		Mounting bracket for relay lever		1	560317	
28		Relay lever assembly		1	562343	
29		Split pin	Fixing pivot pin to lever and bracket	1	3498	
30		Plain washer		2	3833	
31		Bolt (5/16" UNF x ¾" long)	Fixing mounting bracket to heel board	2	255226	
32		Spring washer		2	3075	
33		Nut (5/16" UNF)		2	254831	
34		Brace, relay lever to mounting bracket		1	562306	
35		Brake rod, relay to cross-shaft		1	562067	
36		Clevis fork end	Fixing rod to relay lever and cross-shaft	2	247926	
37		Clevis pin and spring		2	230279	
38		Locknut (¼" UNC)		2	254850	
39		Split pin for clevis		2	2389	
40		MOUNTING BRACKET FOR HANDBRAKE ROD		1	560316	
41		Bearing for bracket		1	553851	

*Asterisk indicates a new part which has not been used on any previous model.

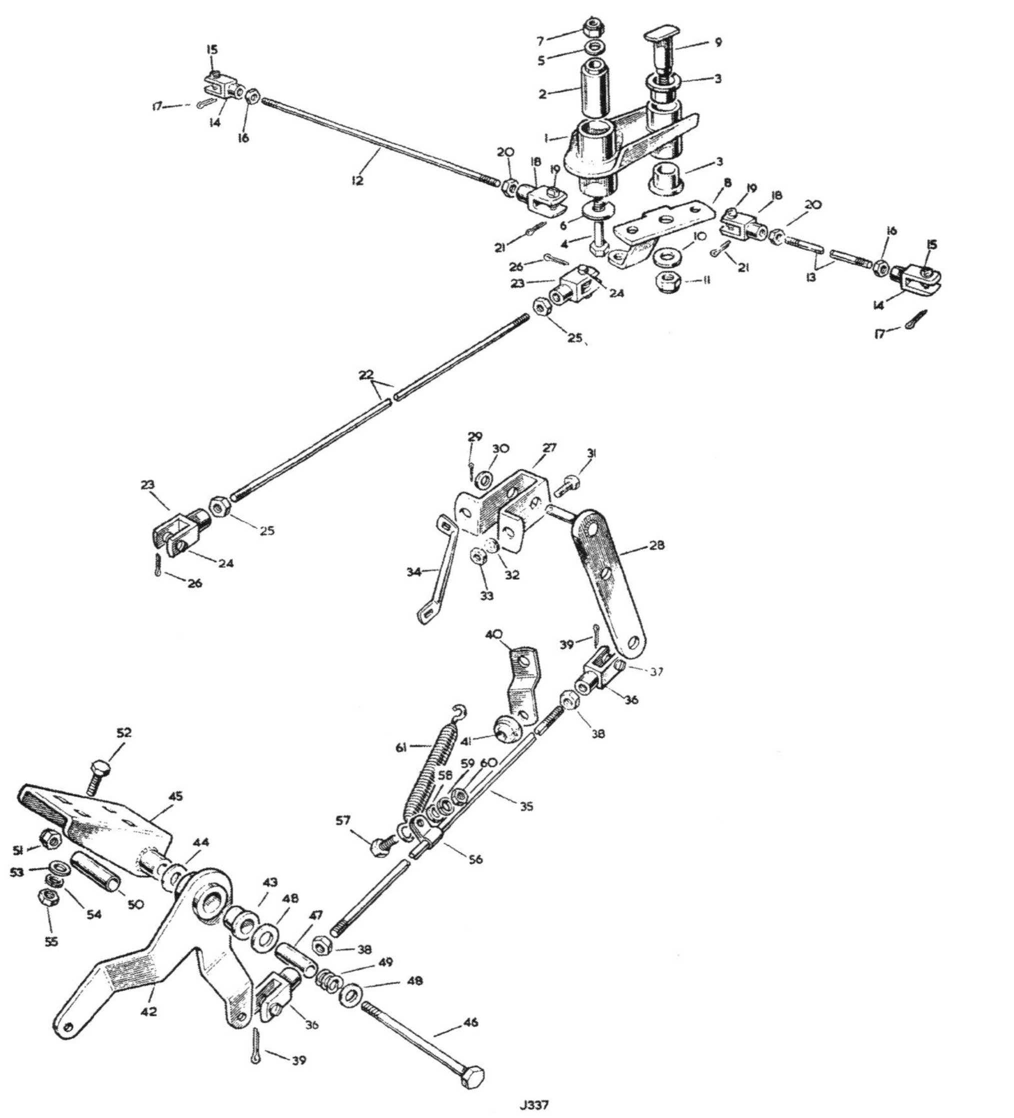

Plate Ref.	1 2 3 4 DESCRIPTION		Qty	Part No.	REMARKS
42	LEVER AND BUSH COMPLETE FOR HANDBRAKE AT DASH		1	560232	
43	Bush for handbrake lever		1	511127	
44	Thrust washer for handbrake lever		1	511128	
45	Mounting bracket for handbrake lever		1	560235	
46	Bolt (⅜" UNF x 4¼" long)	Fixing lever and bush to mounting bracket	1	256056	
47	Sleeve for bearing		1	565402	
48	Plain washer		2	2251	
49	Compression spring		1	565403	
50	Distance tube in bracket		1	278481	
51	Self-locking nut (⅜" UNF)		1	252162	
52	Bolt (5/16" UNF x ⅝" long)	Fixing mounting bracket and dip switch bracket to dash, lower	2	255225	
	Bolt (5/16" UNF x ⅞" long)		2	255227	
53	Plain washer		4	2550	
54	Spring washer		4	3075	
55	Nut (5/16" UNF)		4	254831	
56	Anchor clip for return spring		1	238858	
57	Bolt (2 BA x ½" long)	Fixing clip to rod	1	234603	
58	Plain washer		1	3867	
59	Spring washer		1	3073	
60	Nut (2 BA)		1	2247	
61	Return spring for brake		1	06295	

* Asterisk Indicates a new part which has not been used on any previous Rover model

HANDBRAKE

1417

F583

COLLINS-JONES

Plate Ref. 1 2 3 4	Description	Qty.	Part No.	Remarks
	HANDBRAKE ASSEMBLY	1	557133	
1	Trigger for handbrake handle	1	505505	
2	Special screw fixing trigger	1	505506	
3	Spring for trigger push rod	1	505507	
4	Push rod	1	505508	
5	Adjusting nut for push rod	1	505510	
6	Spring for operating wedge	1	505509	
7	Operating wedge	1	505511	
8	Pawl	1	505512	
9	Spring for pawl	1	505513	
10	Rivet fixing pawl	1	213147	
11	Bolt (¼" UNF x 1" long) ⎤ Fixing handbrake to mounting bracket at front end	1	255209	
12	Spring washer	1	3074	
13	Nut (¼" UNF)	1	254810	
14	Switch for warning light	1	267175	Except 1959-60 models and 1961 4-speed
15	Nut (¼" UNF) fixing switch to handbrake	1	254810	
	Mounting bracket for handbrake at handle end	1	521341@	RH STG
	Mounting bracket for handbrake at handle end	1	521342@	LH STG
16	Mounting bracket for handbrake at handle end	1	530123@@	RH Stg Mk I, Mk IA and Mk II 3 Litre
	Mounting bracket for handbrake at handle end	1	530124@@	LH Stg
	Mounting bracket for handbrake at handle end	1	521341	RH Stg Mk III 3 Litre and 3½ Litre
	Mounting bracket for handbrake at handle end	1	521342	LH Stg
17	Distance tube ⎤ Mounting bracket to dash	1	277020	
	Distance tube	1	506786@	
	Distance tube	1	506787@	
18	Bolt (5/16" UNF x 1¼" long) ⎤ Fixing mounting bracket to dash rail and handbrake to mounting bracket	1	256220	
	Bolt (5/16" UNF x 1½" long)	2	256222@	
19	Spring washer	3	3075	1 off from@@
20	Plain washer	8	2550@	
21	Nut (5/16" UNF)	3	254831@	
	Set bolt (¼" UNF x ¾" long) ⎤ Fixing mounting bracket to dash rail	2	244207@@	
	Plain washer	4	3840@@	
	Spring washer	2	3074@@	
	Fan disc washer	2	510170@@	
22	Mounting bracket for handbrake at front end	1	507016	RH Stg
	Mounting bracket for handbrake at front end	1	507017	LH Stg
23	Plain washer ⎤ Fixing mounting bracket to dash	1	3840	
24	Spring washer	1	3074	
25	Nut (¼" UNF)	1	254810	
26	Grommet for handbrake cable in dash	1	263524	
27	Handbrake cable complete	1	557236	RH Stg 3 Litre
	Handbrake cable complete	1	562289	RH Stg 3½ Litre
	Handbrake cable complete	1	557237	LH Stg 3 Litre
	Handbrake cable complete	1	562301	LH Stg 3½ Litre
28	Pulley for handbrake cable ⎤ For pulley	2	530878	
29	Bush	4	278292	
30	Cover plate	2	278470	
	Cover plate on dash, outer	1	502458	LH Stg
31	Mounting bracket for pulley	1	501695	RH Stg
	Mounting bracket for pulley on dash, outer	1	505981	LH Stg

@ 3 Litre Cars numbered up to 625901065,626900182,628900123,630900512,631900086,633900091

@@3 Litre Cars numbered from 625901066, 626900183, 628900124, 630900513, 631900087, 633900092 onwards and 3½ Litre

*Asterisk indicates a new part which has not been used on any previous model.

HANDBRAKE

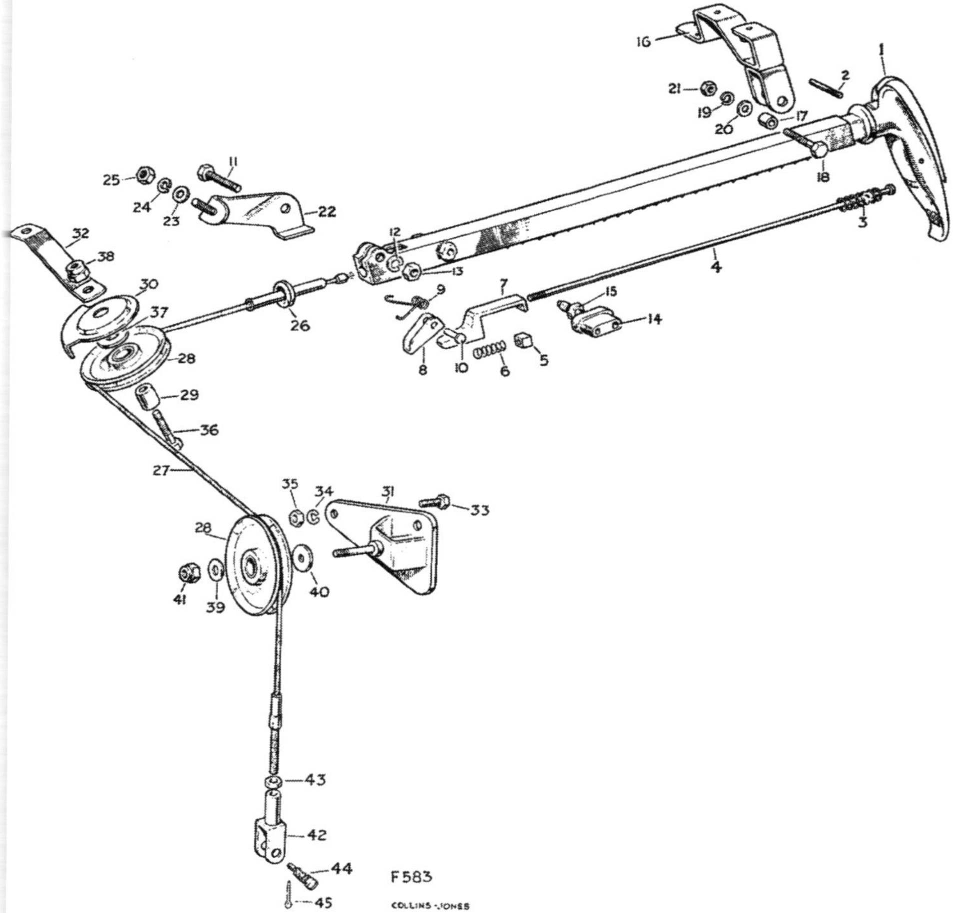

1419

F583

COLLINS-JONES

Plate Ref.	1 2 3 4 Description	Qty.	Part No.	Remarks
32	Support bracket for pulley mounting bracket on wing valance	1	506655	RH Stg
	Support bracket for pulley mounting bracket on wing valance	1	572614	LH Stg
33	Bolt (5/16" UNF x 5/8" long) } Fixing pulley mounting bracket and clutch pedal bracket to dash	3	255225	
34	Spring washer } Fixing pulley mounting bracket and clutch pedal bracket to dash	3	3075	
35	Nut (5/16" UNF) } Fixing pulley mounting bracket and clutch pedal bracket to dash	3	254831	
36	Bolt (5/16" UNF x 1½" long) } Fixing pulley to bracket on wing valance	1	256222	
37	Plain washer, large } Fixing pulley to bracket on wing valance	2	3868	
38	Self-locking nut (5/16" UNF) } Fixing pulley to bracket on wing valance	1	252161	
39	Plain washer, small } Fixing pulley to bracket on dash	1	2550	
40	Plain washer, large } Fixing pulley to bracket on dash	2	3868	
41	Self-locking nut (5/16" UNF) } Fixing pulley to bracket on dash	1	252161	
	Pulley for handbrake cable at dash, inner	1	530878	LH Stg
	Bush } For pulley	1	278292	LH Stg
	Cover plate } For pulley	1	278470	LH Stg
	Mounting bracket for pulley	1	502459	LH Stg
	Bolt (5/16" UNF x 3/8" long) } Fixing bracket to dash, inner	2	255225	LH Stg
	Spring washer } Fixing bracket to dash, inner	3	3075	LH Stg
	Nut (5/16" UNF) } Fixing bracket to dash, inner	3	254831	LH Stg
	Plain washer, small } Fixing pulley to bracket on dash, inner	1	2550	LH Stg
	Plain washer, large } Fixing pulley to bracket on dash, inner	2	3868	LH Stg
	Self-locking nut (5/16" UNF) } Fixing pulley to bracket on dash, inner	1	252161	LH Stg
42	Clevis fork end } Fixing handbrake cable to lever at dash bottom	1	278514	
43	Locknut for clevis } Fixing handbrake cable to lever at dash bottom	1	254850	
44	Clevis pin complete } Fixing handbrake cable to lever at dash bottom	1	230279	
45	Split pin } Fixing handbrake cable to lever at dash bottom	1	2389	

*Asterisk indicates a new part which has not been used on any previous model.

E688

Plate Ref.	1 2 3 4 Description	Qty.	Part No.	Remarks
	BRAKE PEDAL AND BRACKET ASSEMBLY	1	500334	RH Stg 3 Litre 4-speed
	BRAKE PEDAL AND BRACKET ASSEMBLY	1	500715	LH Stg
	BRAKE PEDAL AND BRACKET ASSEMBLY	1	500722	3 Litre Automatic
	BRAKE PEDAL AND BRACKET ASSEMBLY	1	560219	3½ Litre
1	PEDAL BRACKET FOR BRAKE	1	500332	
2	Bush for bracket	2	272439	
3	Brake pedal	1	500328	RH Stg 3 Litre
	Brake pedal	1	500713	LH Stg 4-speed
	Brake pedal	1	500719	Automatic
4	Pivot pin for pedal	1	278676	
5	Pin locating pedal pivot pin	1	250345	
	SWITCH ASSEMBLY ASSEMBLY FOR STOP LIGHT	1	567880	
	Buffer	1	537109	3½ Litre
	Nut	1	537110	
	Spacer for switch	1	563316	Early 3½ Litre only
	CLUTCH PEDAL AND BRACKET ASSEMBLY	1	500335	RH Stg
	CLUTCH PEDAL AND BRACKET ASSEMBLY	1	500716	LH Stg
6	PEDAL BRACKET FOR CLUTCH	1	500333	3 Litre 4-speed
7	Bush for bracket	2	272439	
8	Clutch pedal	1	500329	RH Stg
	Clutch pedal	1	500692	LH Stg
9	Pivot pin for pedal	1	278676	
10	Pin locating pedal pivot pin	1	250345	
11	Rubber pad for brake pedal	1	500336	RH Stg 3 Litre 4-speed
	Rubber pad for brake pedal	1	500339	Automatic
	Rubber pad for clutch pedal	1	502046	RH Stg 3 Litre
	Rubber pad for clutch pedal	1	500336	LH Stg 4-speed
12	Bolt (¼" UNF x 1¼" long) For	2	255211	3 Litre
13	Locknut (¼" UNF) pedal stop	2	254850	1 off on AUtomatic
14	Rubber pad for pedal stop bolt	2	275678	
	Bolt (5/16" UNF x 1⅜" long) For pedal	1	560223	3½ Litre
	Locknut (5/16" UNF) stop	1	254831	
15	Distance piece for brake master cylinder	1	500341	3 litre
	Distance piece for brake master cylinder	1	511977	3½ litre
16	Distance piece for clutch master cylinder	1	511977	3 Litre 4-speed
17	BRAKE MASTER CYLINDER	1	520851	
18	Return spring for piston	1	217698	3 Litre
19	Seal spacer and end cap	1	516410	
20	Push rod and retaining washer for piston	1	523083	
	BRAKE MASTER CYLINDER	1	562239	
	Pivot pin Fixing end eye	1	562346	
	Plain washer to pedal lever	1	2550	3 Litre
	Split pin	1	2422	
21	CLUTCH MASTER CYLINDER	1	514132	
22	Return spring for piston	1	545931	
23	Push rod for piston	1	274715	3 Litre
24	Filler cap	1	500201	4-speed
25	Filter	1	518682	
26	Sealing washer for filter	1	264767	
27	Adaptor for pipe	1	276061@	
28	Joint washer for adaptor	1	233220@	

@ 3 Litre Cars numbered up to 625901036, 626900175, 628900120

*Asterisk indicates a new part which has not been used on any previous model.

1423

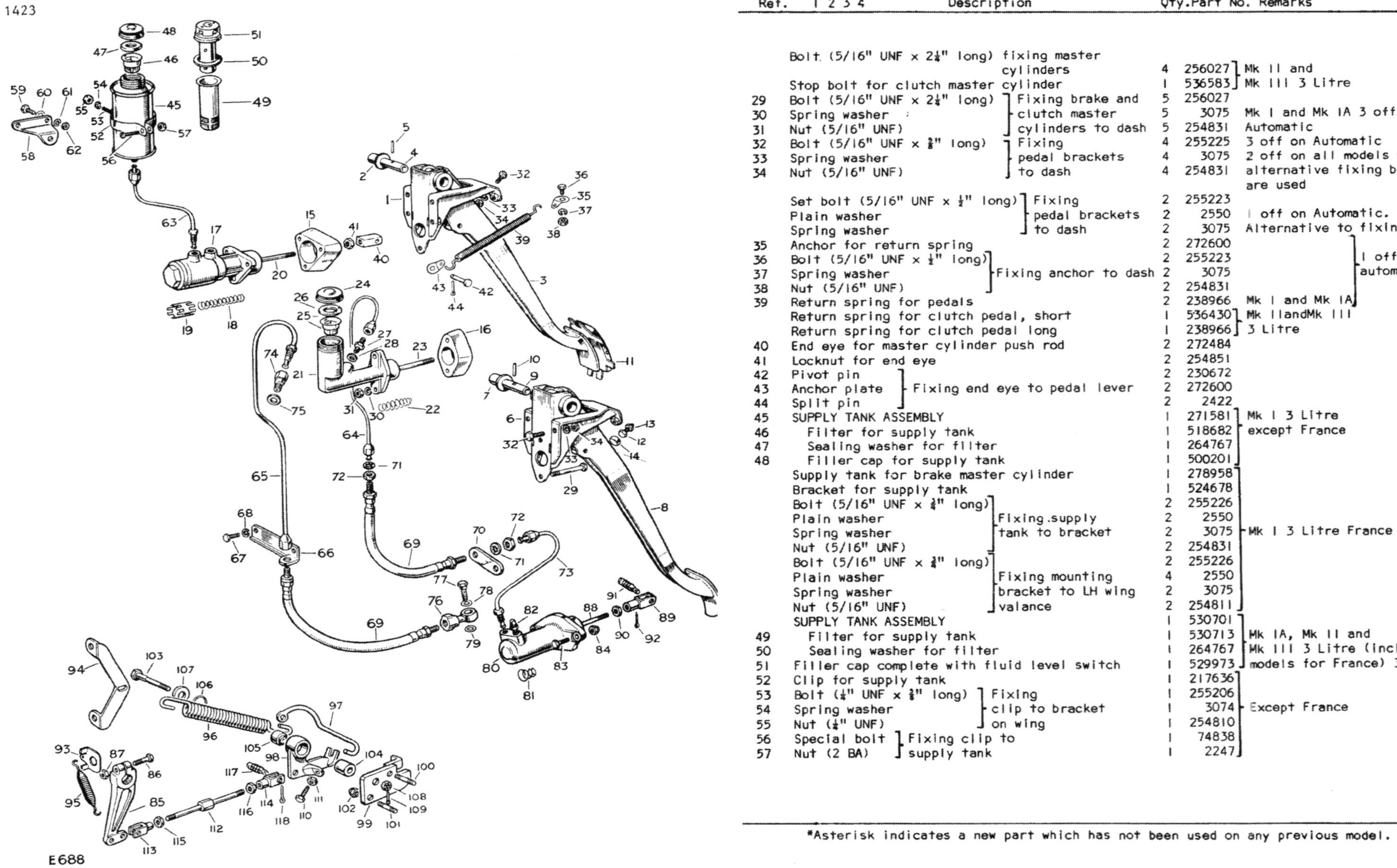

Plate Ref.	1 2 3 4 Description	Qty.	Part No.	Remarks
	Bolt (5/16" UNF x 2¼" long) fixing master cylinders	4	256027	Mk II and
	Stop bolt for clutch master cylinder	1	536583	Mk III 3 Litre
29	Bolt (5/16" UNF x 2¼" long) } Fixing brake and clutch master cylinders to dash	5	256027	
30	Spring washer } Fixing brake and clutch master cylinders to dash	5	3075	Mk I and Mk IA 3 off on
31	Nut (5/16" UNF) } Fixing brake and clutch master cylinders to dash	5	254831	Automatic
32	Bolt (5/16" UNF x ⅞" long) } Fixing pedal brackets to dash	4	255225	3 off on Automatic
33	Spring washer } Fixing pedal brackets to dash	4	3075	2 off on all models when
34	Nut (5/16" UNF) } Fixing pedal brackets to dash	4	254831	alternative fixing below are used
	Set bolt (5/16" UNF x ½" long) } Fixing pedal brackets to dash	2	255223	1 off on Automatic.
	Plain washer } Fixing pedal brackets to dash	2	2550	Alternative to fixings above
	Spring washer } Fixing pedal brackets to dash	2	3075	
35	Anchor for return spring	2	272600	
36	Bolt (5/16" UNF x ½" long) } Fixing anchor to dash	2	255223	1 off on automatic
37	Spring washer } Fixing anchor to dash	2	3075	
38	Nut (5/16" UNF) } Fixing anchor to dash	2	254831	
39	Return spring for pedals	2	238966	Mk I and Mk IA
	Return spring for clutch pedal, short	1	536430	Mk IIandMk III
	Return spring for clutch pedal long	1	238966	3 Litre
40	End eye for master cylinder push rod	2	272484	
41	Locknut for end eye	2	254851	
42	Pivot pin } Fixing end eye to pedal lever	2	230672	
43	Anchor plate } Fixing end eye to pedal lever	2	272600	
44	Split pin } Fixing end eye to pedal lever	2	2422	
45	SUPPLY TANK ASSEMBLY	1	271581	Mk I 3 Litre
46	Filter for supply tank	1	518682	except France
47	Sealing washer for filter	1	264767	
48	Filler cap for supply tank	1	500201	
	Supply tank for brake master cylinder	1	278958	
	Bracket for supply tank	1	524678	
	Bolt (5/16" UNF x ¾" long) } Fixing supply tank to bracket	2	255226	
	Plain washer } Fixing supply tank to bracket	2	2550	
	Spring washer } Fixing supply tank to bracket	2	3075	Mk I 3 Litre France only
	Nut (5/16" UNF) } Fixing supply tank to bracket	2	254831	
	Bolt (5/16" UNF x ¾" long) } Fixing mounting bracket to LH wing valance	2	255226	
	Plain washer } Fixing mounting bracket to LH wing valance	4	2550	
	Spring washer } Fixing mounting bracket to LH wing valance	2	3075	
	Nut (5/16" UNF) } Fixing mounting bracket to LH wing valance	2	254811	
	SUPPLY TANK ASSEMBLY	1	530701	
49	Filter for supply tank	1	530713	Mk IA, Mk II and
50	Sealing washer for filter	1	264767	Mk III 3 Litre (including
51	Filler cap complete with fluid level switch	1	529973	models for France) 3½ Litre
52	Clip for supply tank	1	217636	
53	Bolt (¼" UNF x ⅜" long) } Fixing clip to bracket on wing	1	255206	
54	Spring washer } Fixing clip to bracket on wing	1	3074	Except France
55	Nut (¼" UNF) } Fixing clip to bracket on wing	1	254810	
56	Special bolt } Fixing clip to supply tank	1	74838	
57	Nut (2 BA) } Fixing clip to supply tank	1	2247	

*Asterisk indicates a new part which has not been used on any previous model.

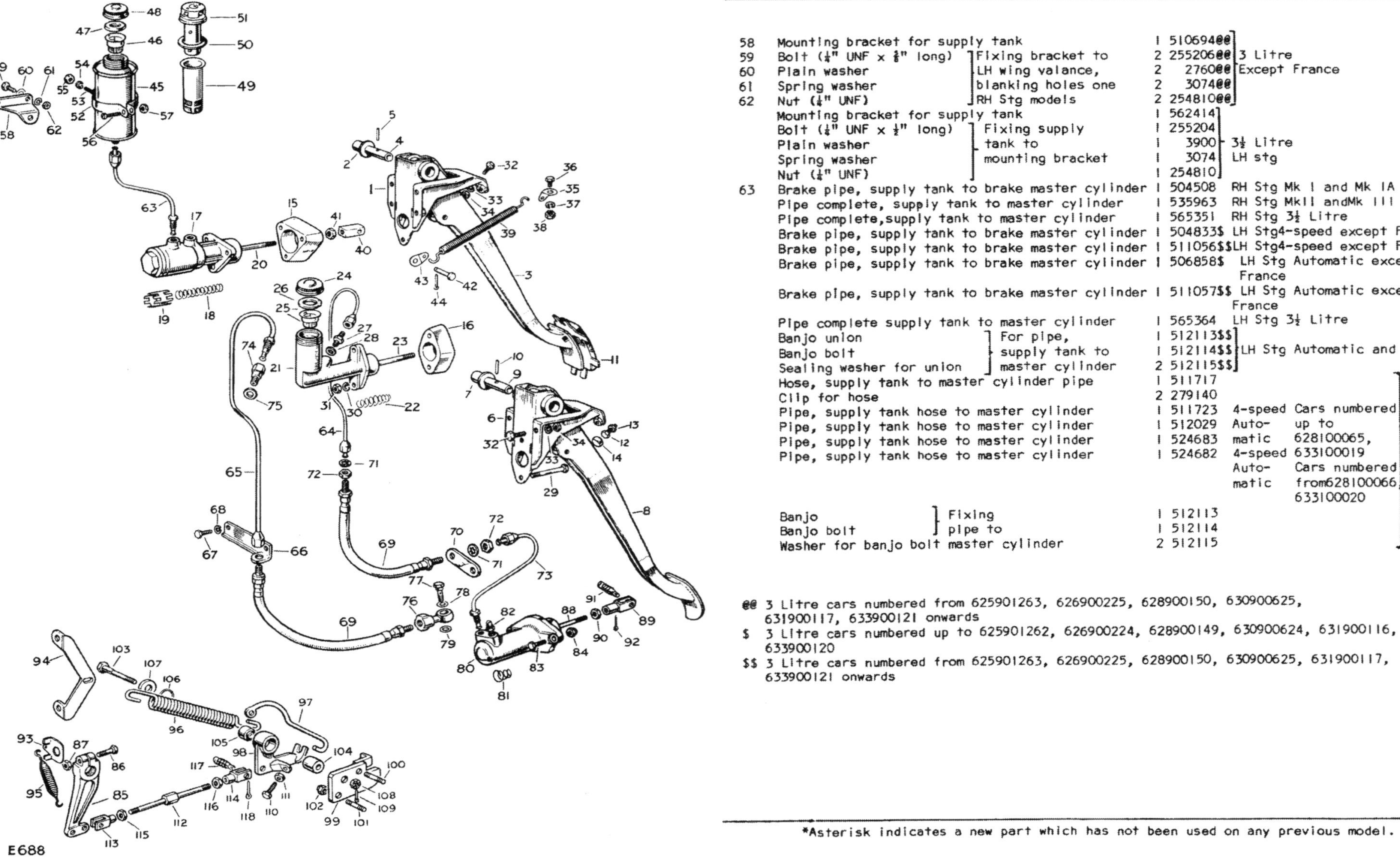

Plate Ref.	1 2 3 4 Description	Qty.	Part No.	Remarks
58	Mounting bracket for supply tank	1	510694@@	3 Litre Except France
59	Bolt (¼" UNF x ⅜" long) — Fixing bracket to LH wing valance, blanking holes one RH Stg models	2	255206@@	3 Litre Except France
60	Plain washer	2	2760@@	3 Litre Except France
61	Spring washer	2	3074@@	3 Litre Except France
62	Nut (¼" UNF)	2	254810@@	3 Litre Except France
	Mounting bracket for supply tank	1	562414	3½ Litre LH stg
	Bolt (¼" UNF x ½" long) — Fixing supply tank to mounting bracket	1	255204	3½ Litre LH stg
	Plain washer	1	3900	3½ Litre LH stg
	Spring washer	1	3074	3½ Litre LH stg
	Nut (¼" UNF)	1	254810	3½ Litre LH stg
63	Brake pipe, supply tank to brake master cylinder	1	504508	RH Stg Mk I and Mk IA 3Litre
	Pipe complete, supply tank to master cylinder	1	535963	RH Stg MkII andMk III 3Litre
	Pipe complete,supply tank to master cylinder	1	565351	RH Stg 3½ Litre
	Brake pipe, supply tank to brake master cylinder	1	504833$	LH Stg4-speed except France
	Brake pipe, supply tank to brake master cylinder	1	511056$$	LH Stg4-speed except France
	Brake pipe, supply tank to brake master cylinder	1	506858$	LH Stg Automatic except France
	Brake pipe, supply tank to brake master cylinder	1	511057$$	LH Stg Automatic except France
	Pipe complete supply tank to master cylinder	1	565364	LH Stg 3½ Litre
	Banjo union — For pipe, supply tank to master cylinder	1	512113$$	LH Stg Automatic and France
	Banjo bolt	1	512114$$	LH Stg Automatic and France
	Sealing washer for union	2	512115$$	LH Stg Automatic and France
	Hose, supply tank to master cylinder pipe	1	511717	Mk I models France only
	Clip for hose	2	279140	Mk I models France only
	Pipe, supply tank hose to master cylinder	1	511723	4-speed Cars numbered up to 628100065, 633100019 — Mk I models France only
	Pipe, supply tank hose to master cylinder	1	512029	Automatic Cars numbered up to 628100065, 633100019 — Mk I models France only
	Pipe, supply tank hose to master cylinder	1	524683	4-speed Cars numbered from628100066, 633100020 — Mk I models France only
	Pipe, supply tank hose to master cylinder	1	524682	Automatic Cars numbered from628100066, 633100020 — Mk I models France only
	Banjo — Fixing pipe to master cylinder	1	512113	Mk I models France only
	Banjo bolt	1	512114	Mk I models France only
	Washer for banjo bolt	2	512115	Mk I models France only

@@ 3 Litre cars numbered from 625901263, 626900225, 628900150, 630900625, 631900117, 633900121 onwards

$ 3 Litre cars numbered up to 625901262, 626900224, 628900149, 630900624, 631900116, 633900120

$$ 3 Litre cars numbered from 625901263, 626900225, 628900150, 630900625, 631900117, 633900121 onwards

*Asterisk indicates a new part which has not been used on any previous model.

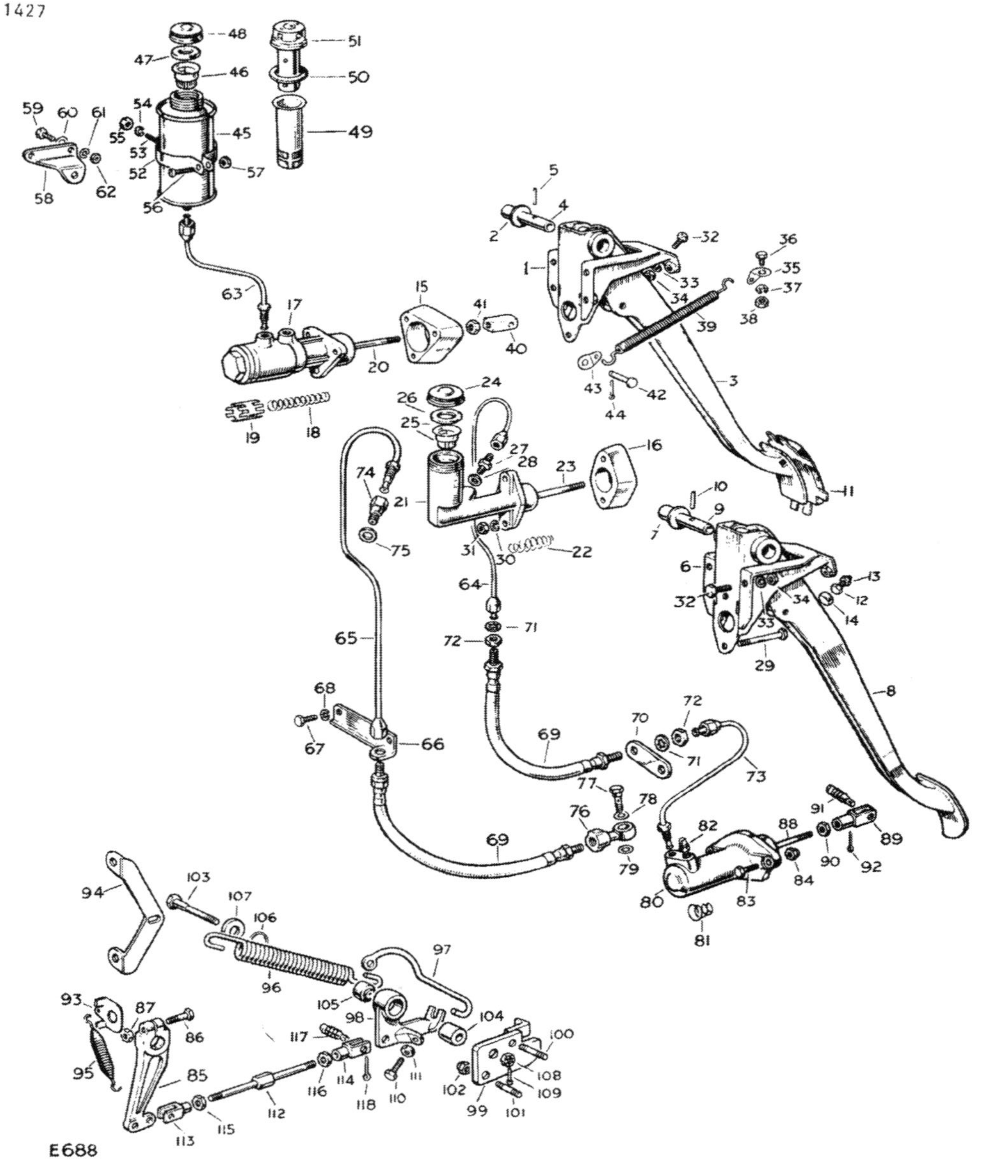

E688

Plate Ref.	1 2 3 4 Description	Qty.	Part No.	Remarks
	Clip fixing pipe to wing valance	1	50183	Mk I models.
	Drive screw fixing clip	1	78322	France only
	Pipe complete, clutch master cylinder to hose	1	504830@	RH Stg
	Pipe complete, clutch master cylinder to hose	1	504831@	LH Stg
64	Pipe complete, clutch master cylinder to hose	1	512648%	RH Stg
	Pipe complete, clutch master cylinder to hose	1	512974%	LH Stg
65	Pipe complete, clutch master cylinder to hose	1	522723	RH Stg ⎤ Cars numbered from 625101046 626100263 628100110 onwards
	Pipe complete, clutch master cylinder to hose	1	522724	LH Stg ⎥
	Pipe complete, clutch master cylinder to hose	1	557104	LH Stg Mk III ⎦
66	Abutment bracket for clutch hose	1	522725	
67	Bolt (¼" UNF x ½" long) ⎤ Fixing abutment bracket	1	255204	
68	Spring washer ⎦ to bulkhead	1	3074	
	Hose, pipe to clutch slave cylinder	1	504834@	
69	Hose, pipe to clutch slave cylinder	1	512649@@	
70	Bracket for hose at bell housing	1	505043@	
	Shakeproof washer ⎤ Fixing pipe and hose	2	233305@	
	Special nut ⎦ to bracket and dash	2	254852@	
71	Shakeproof washer ⎤ Fixing pipe and hose	1	512651@	
72	Special nut ⎦ to bracket and dash	1	216912@@	
73	Pipe complete, hose to clutch slave cylinder	1	506674@	
	Adaptor for clutch slave cylinder	1	512650%	
74	Adaptor for clutch master cylinder	1	512650@@	RH Stg
75	Gasket for adaptor	2	233220%	1 off on LH Stg
76	Banjo for slave cylinder	1	545537	Mk I and Mk IA
	Banjo for clutch slave cylinder	1	538068	Mk II and Mk III
77	Banjo bolt	1	512235	Cars numbered from
78	Gasket ⎤ For banjo on	1	512387	625101046, 626100263
79	Gasket ⎦ slave cylinder	1	216914	628100110 onwards
	Gasket for hose to master cylinder adaptor	1	216914@@	
	Banjo bolt ⎤ For clutch	1	512235@@	⎤
	Banjo union ⎦ master cylinder	1	512113@@	⎥ LH Stg
	Gasket ⎤ For	1	216914@@	⎥
	Gasket ⎦ banjo	1	512387@@	⎦
80	CLUTCH SLAVE CYLINDER	1	512119	Mk I and Mk IA
	CLUTCH SLAVE CYLINDER	1	533562	Mk II and Mk III
81	Return spring for piston	1	277073	
82	Bleed screw	1	248909	Mk I and Mk IA
	Bleed screw	1	556508	Mk II and
	Bracket for slave cylinder	1	533561	Mk III
83	Bolt (5/16" UNF x 1⅛" long) ⎤ Fixing slave	2	255231	Mk I and Mk IA
	Bolt (5/16" UNF x 1¼" long) ⎥ cylinder to	2	255030	Mk II and Mk III
84	Self-locking nut (5/16" UNF) ⎦ bell housing	2	252211	

(Items from "Pipe complete, clutch master cylinder to hose" (504830@) through 84: 3 Litre 4-speed models)

@ 3 Litre cars numbered up to 625901036, 626900175, 628900120

@@ 3 Litre cars numbered from 625901037, 626900176, 628900121 onwards

% Cars numbered from 625901037 to 625101045, to 626100262, 625100261 to 628100109

*Asterisk indicates a new part which has not been used on any previous model.

E688

Plate Ref.	1 2 3 4 Description	Qty.	Part No.	Remarks
	Operating lever for clutch	1	501171@	3 Litre 4-speed models
85	Operating lever for clutch	1	507863@@	
86	Bolt (5/16" UNF x 1⅜" long) } Fixing operating lever to withdrawal shaft	1	256221	
87	Nut (5/16" UNF) } Fixing operating lever to withdrawal shaft	1	254831	
88	Push rod for slave cylinder	1	501170	
89	Clevis for push rod	1	504036	Alternatives. Check before ordering
	Insulator assembly for clutch slave cylinder piston rod	1	528286	Alternatives. Check before ordering
	Plain washer for piston rod	1	2550	Alternatives. Check before ordering
90	Locknut (5/16" UNF) for push rod	1	254831	
91	Clevis pin } Fixing clevis to operating lever	1	216421	
92	Split pin } Fixing clevis to operating lever	1	2392	
93	Anchor for clutch return spring	1	504061@	
94	Anchor for clutch return spring	1	510893$	
	Anchor for clutch return spring	1	514245	Cars numbered from 625101046,626100263 628100110 onwards
95	Clutch return spring	1	070488	
96	Clutch balance spring	1	507857@@	
97	Hook for balance spring at over-centre lever	1	512158@@	
98	Over-centre lever for clutch	1	512152@@	
99	Support bracket for over-centre lever	1	512150@@	
100	Stud, long } Fixing support bracket	1	513950@@	
101	Stud, short } Fixing support bracket	1	512950@@	
102	Self-locking nut } Fixing support bracket	2	251321@@	
103	Pivot bolt } Fixing over-centre lever to support bracket	1	512472@@	
104	Distance piece } Fixing over-centre lever to support bracket	1	512471@@	
105	Needle bearing } Fixing over-centre lever to support bracket	1	512161@@	
106	'O' ring } Fixing over-centre lever to support bracket	2	271013@@	
107	Special washer } Fixing over-centre lever to support bracket	1	512157@@	
108	Castle nut } Fixing over-centre lever to support bracket	1	254972@@	
109	Split pin } Fixing over-centre lever to support bracket	1	2423@@	
110	Stop bolt } For over-centre lever	1	255011@@	
111	Locknut } For over-centre lever	1	254810@@	
112	Adjuster rod, over-centre lever to operating lever	1	510896@@	
113	Clevis, front, for adjuster rod	1	510937@@	
114	Clevis, rear, for adjuster rod	1	275199@@	
115	Locknut, front, for adjuster rod	1	510938@@	
116	Locknut, rear, for adjuster rod	1	254851@@	
117	Clevis pin } Fixing clevis to over-centre and operating levers	2	216421@@	
118	Split pin } Fixing clevis to over-centre and operating levers	2	2392@@	
	Clip for brake pipes on dash	1	501830@@	
	Clip for brake pipes on dash	1	243395@@	LH Stg

@ 3 Litre cars numbered up to 625901036, 626900175, 628900120
@@3 Litre cars numbered from 625901037, 626900176, 628900121 onwards
$ 3 Litre cars numbered from 625901037 to 625101045, 626900176 to 626100262, 628900121 to 628100109

*Asterisk indicates a new part which has not been used on any previous model.

STEERING BOX, MK I, MK IA AND MK II MODELS WITH SUFFIX LETTER 'A'
MANUAL STEERING

1431

F641

Plate Ref.	1 2 3 4 Description	Qty.	Part No.	Remarks
	STEERING BOX ASSEMBLY, 17.6:1 ratio	1	529367	RH Stg
	STEERING BOX ASSEMBLY, 17.6:1 ratio	1	529368	LH Stg
1	End cover plate, upper	1	507777	
2	End cover plate, lower	1	271378	
3	Joint washer steel, .003"	As reqd	261858	
	Joint washer, steel, .010"	As reqd	521522	
4	Joint washer, thick	As reqd	261857	
	Joint washer, thin	As reqd	271379	
5	Stud for end cover plate	8	507778	
6	Spring washer } Fixing end cover plate	8	3075	
7	Nut (5/16" Whit) } Fixing end cover plate	8	3800	
8	Rocker shaft	1	507775@	
	Rocker shaft	1	538052@@	
	Adjuster cap for rocker shaft	1	538059@@	
9	Bush for rocker shaft	1	261850	
10	Oil seal for rocker shaft	1	507776	
11	Adjusting screw for rocker shaft	1	261873@	
12	Nut for adjusting screw	1	534299@	
	Spring for rocker shaft	1	538042@@	
	Distance piece for spring cover plate	1	538040@@	
	Cover plate for rocker shaft spring	1	538041@@	
	Joint washer for cover plate	2	538060@@	
	Set bolt (5/16" UNF x ¾" long) } Fixing cover plate to top cover	2	255226@@	
	Spring washer } Fixing cover plate to top cover	2	3075@@	
13	MAIN NUT ASSEMBLY	1	507770	RH Stg
	MAIN NUT ASSEMBLY	1	507771	LH Stg
14	Set bolt (2BA x ½" long) } Fixing retainer	2	234603	
15	Lock washer for bolt } Fixing retainer	2	261868	
16	Steel ball (⅜") for main nut	12	1643	
17	Roller for main nut	1	261869	
18	Inner column	1	507772	RH Stg
	Inner column	1	507773	LH Stg
19	Oil seal for inner column	1	507774	
20	Adjustable ball race	2	271384	
21	Steel balls (.280") for adjustable race	20	260823	
22	Stud for top cover plate	4	507779	
23	Top cover plate	1	261878@	RH Stg
	Top cover plate	1	261879@	LH Stg
24	Joint washer for top cover plate	1	261880@	
	Top cover plate	1	538054@@	RH Stg
	Top cover plate	1	538061@@	LH Stg
	Joint washer, paper, thick	As reqd	538055@@	
	Joint washer, paper, thin	As reqd	538056@@	
	Joint washer, steel, thin	As reqd	538057@@	
	Joint washer, steel, thick	As reqd	538058@@	
25	Spring washer } Fixing top cover plate	4	3075	
26	Nut (5/16" Whit) } Fixing top cover plate	4	3800	
27	Oil filler plug	1	260817	

@ Cars numbered up to 72500976, 72600139, 72800207, 73001449, 73100248, 73300327

@@ Cars numbered from 72500977, 72600140, 72800208, 73001450, 73100249, 73300328 onwards

*Asterisk indicates a new part which has not been used on any previous model.

STEERING BOX, MK I, MK IA AND MK II MODELS WITH SUFFIX LETTER 'A' MANUAL STEERING

1433

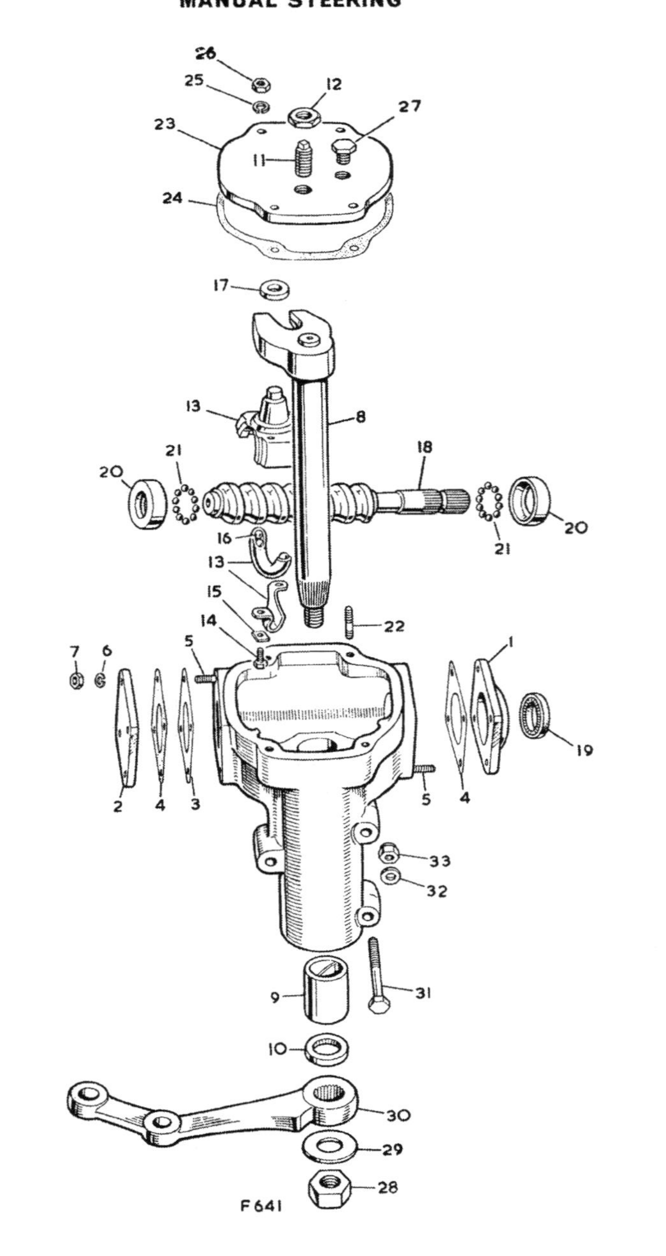

STEERING BOX, MK I, MK IA AND MK II 3 LITRE MODELS WITH SUFFIX LETTER 'A' MANUAL STEERING

1434

Plate Ref.	1 2 3 4	Description		Qty.	Part No.	Remarks
28	Special nut	Fixing drop arm to steering box		1	254877	
29	Lock washer			1	4115	
30	Drop arm			1	517015	RH Stg
	Drop arm			1	517016	LH Stg
31	Bolt (5/16" UNF x 6" long)	Fixing steering box to chassis sub-frame		3	253874	
	Packing washer			As reqd	3843	
32	Plain washer			3	3261	
33	Self-locking nut (7/16" UNF)			3	252163	

*Asterisk indicates a new part which has not been used on any previous model.

STEERING BOX, MK II MODELS WITH SUFFIX LETTER 'B' ONWARDS
MANUAL STEERING

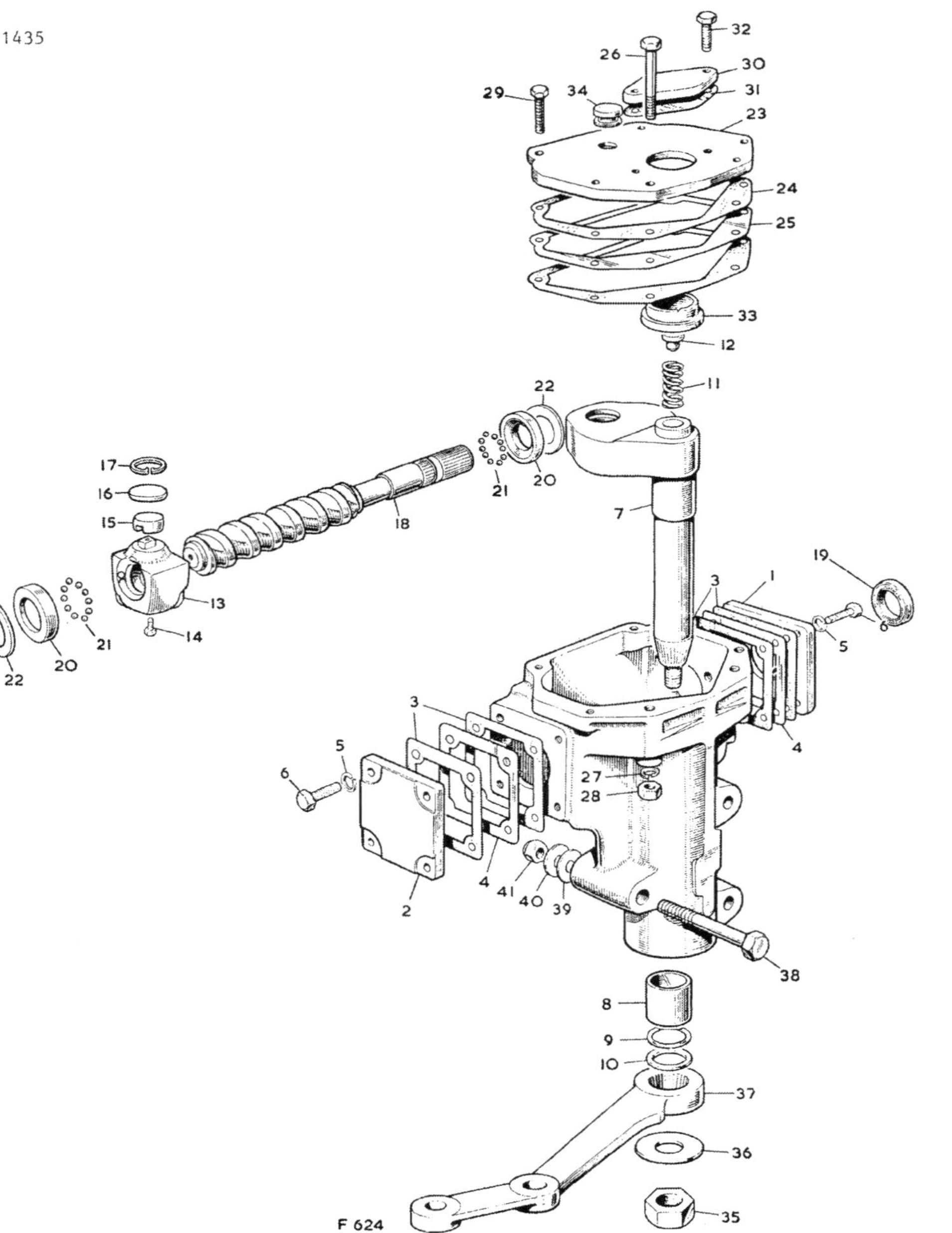

Plate Ref.	1 2 3 4 Description	Qty.	Part No.	Remarks
	STEERING BOX ASSEMBLY, 20.3:1 ratio	1	536524	RH Stg
	STEERING BOX ASSEMBLY, 20.3:1 ratio	1	536525	LH Stg
1	End cover plate, upper	1	507777	
2	End cover plate, lower	1	271378	
3	Joint washer, paper, .010" } For end cover plate	As reqd	261857	
	Joint washer, paper, .003" } For end cover plate	As reqd	271379	
4	Joint washer, steel, .003" } For end cover plate	As reqd	261858	
	Joint washer, steel .010" } For end cover plate	As reqd	538053	
5	Spring washer	8	3075	
6	Special set bolt ($\frac{3}{4}$" long) fixing end cover plate	8	541958	
7	Rocker shaft	1	541933	
8	Bush for rocker shaft	1	541944	
9	Oil seal for rocker shaft	1	541948	
10	Washer for oil seal	1	541942	
11	Spring for rocker shaft	1	541949	
12	Button for rocker shaft	1	541941	
13	MAIN NUT ASSEMBLY	1	541956	RH Stg
	MAIN NUT ASSEMBLY	1	541957	LH Stg
14	Bolt (2BA x 5/16" long) fixing retainer	2	251017	
	Steel ball (9/32") for main nut	31	260823	
15	Roller for main nut	1	541947	
16	Thrust washer for rocker shaft head	1	541950	
17	Circlip for rocker shaft	1	541951	
18	Inner column	1	541932	RH Stg
	Inner column	1	541955	LH Stg
19	Oil seal for inner column	1	507774	
20	Adjustable ball race	2	271384	
21	Steel ball (9/32") for adjustable race	20	260823	
22	Spacer washer for ball race	2	541943	
23	Top cover plate	1	541939	
	Joint washer, steel, .010" } For top cover plate	As reqd	541935	
24	Joint washer, paper, .005" } For top cover plate	As reqd	541936	
25	Joint washer, steel, .002/3" } For top cover plate	As reqd	541937	
	Joint washer, steel, .004/5" } For top cover plate	As reqd	541938	
26	Dowel bolt ($2\frac{3}{8}$" long) for cover plate	2	541960	
27	Spring washer	2	3075	
28	Nut (5/16" UNF) for dowel bolt	2	254811	
29	Special set bolt (1" long) fixing top cover plate	4	271382	
30	Cover plate for rocker shaft spring	1	541940	
31	Joint washer for cover plate	1	541934	
32	Bolt (5/16" UNF x $\frac{7}{8}$" long) fixing cover plate to top cover	2	255225	
33	Bush for top cover plate	1	541945	
34	Oil filler plug	1	541946	
35	Special nut } Fixing drop arm to rocker shaft	1	254877	
36	Lock washer } Fixing drop arm to rocker shaft	1	4115	
37	Drop arm	1	536857	RH Stg
	Drop arm	1	536858	LH Stg
38	Bolt (7/16" UNF x 6" long) } Fixing steering box to chassis sub-frame	3	253874	
39	Packing washer } Fixing steering box to chassis sub-frame	As reqd	3843	
40	Plain washer } Fixing steering box to chassis sub-frame	3	3261	
41	Self-locking nut (7/16" UNF) } Fixing steering box to chassis sub-frame	3	252163	

*Asterisk indicates a new part which has not been used on any previous model.

F613

Plate Ref.	1 2 3 4 Description	Qty.	Part No.	Remarks
1	OUTER COLUMN ASSEMBLY	1	510681@	RH Stg] Cars without column grommet
	OUTER COLUMN ASSEMBLY	1	514618@	LH Stg] Cars without column grommet
	OUTER COLUMN ASSEMBLY	1	542927@@	RH Stg] Cars with column grommet
	OUTER COLUMN ASSEMBLY	1	542928@@	LH Stg] Cars with column grommet
2	Dowel for shrouds and bracket	2	500993	
3	Bearing for outer column	1	279072	2 off from @@
4	Dowel for steering column support bracket	1	510682	
5	Steering column shaft	1	512090@	Cars without column grommet
	Steering column shaft	1	529194@@	Cars with column grommet
6	Spring washer for steering column shaft	1	541235	2 off from @@
7	Clamping ring for shaft	1	529193@@	] Cars with column grommet
8	Shakeproof washer] Fixing clamping ring to shaft	2	78249@@	] Cars with column grommet
9	Set screw (4BA x ⅜" long)] Fixing clamping ring to shaft	2	78496@@	] Cars with column grommet
10	Steering coupling flange	2	523077	
11	Flexible disc for coupling	1	520833	
12	Locating plate for coupling	2	522951	
13	Bolt (5/16" UNF x 1¾" long)] Fixing flanges, locating plates and flexible disc together	4	256421	
14	Self-locking nut (5/16" UNF)] Fixing flanges, locating plates and flexible disc together	4	252161	
15	Bolt (5/16" UNF x 1⅜" long)] Fixing coupling to steering box and steering column shaft	2	256223	
16	Self-locking nut (5/16" UNF)] Fixing coupling to steering box and steering column shaft	2	252161	
17	Steering wheel	1	536498	
	Steering wheel assembly	1	525887	For use when steering wheel extension assembly is fitted
18	Plain washer] Fixing steering wheel	1	278461	
19	Special nut] Fixing steering wheel	1	278773	
20	Steering column mounting bracket, lower angle	1	359527	RH Stg
21	Steering column mounting bracket, tie bracket	1	356213	RH Stg
	Steering column mounting bracket, lower angle	1	359528	LH Stg
	Steering column mounting bracket, tie bracket	1	356214	LH Stg
22	Support bracket] Steering column mounting bracket to main instrument mounting bracket	1	359474	RH Stg] Mk II
	Support bracket] Steering column mounting bracket to main instrument mounting bracket	1	359475	LH Stg] Mk II
23	Tapped plate, steering column mounting bracket to parcel shelf	1	359483	] Mk II
	Shim, steering column mounting bracket to parcel shelf	1	359884	] Mk II
	Bolt (¼" UNF x 2" long)] Fixing support bracket to steering mounting bracket and main instrument bracket	4	255204	] Mk II
	Spring washer] Fixing support bracket to steering mounting bracket and main instrument bracket	4	3074	] Mk II
	Plain washer] Fixing support bracket to steering mounting bracket and main instrument bracket	4	3840	] Mk II
	Nut (¼" UNF)] Fixing support bracket to steering mounting bracket and main instrument bracket	4	254810	] Mk II
24	Set bolt (¼" UNF x ⅜" long)] Fixing bracket to angle, dash and parcel shelf	6	255406	
25	Plain washer, small] Fixing bracket to angle, dash and parcel shelf	4	3946	
26	Plain washer, large] Fixing bracket to angle, dash and parcel shelf	4	3900	
27	Spring washer] Fixing bracket to angle, dash and parcel shelf	6	3074	

@ Cars numbered up to 72500551, 72600090, 72800081, 73000766, 73100150, 73300152

@@ Cars numbered from 72500552, 72600091, 72800082, 73000767, 73100151, 73300153 onwards

*Asterisk indicates a new part which has not been used on any previous model.

1439

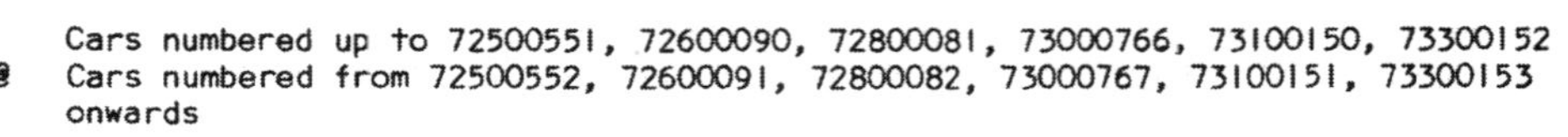
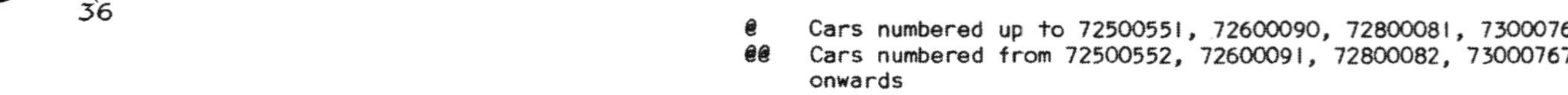

F613.

Plate Ref.	1 2 3 4 Description		Qty.	Part No.	Remarks
28	Support bracket for steering column		1	510678	
29	Set bolt (1/4" UNF x 3/4" long)	Fixing support bracket to mounting bracket	4	255207	
30	Plain washer		4	3840	
31	Spring washer		4	3074	
32	Anchor bracket for steering column		1	279058@	Cars without column grommet
33	Set bolt (5/16" UNF x 3/4" long)	Fixing anchor bracket to dash	2	255225@	
34	Plain washer		2	2550@	
35	Spring washer		2	3075@	
36	Grommet for steering column		1	532057@@	
37	Felt washer for grommet		1	529192@@	
38	Hose clip, grommet to outer column		1	503146@@	
39	Set bolt (5/16" UNF x 3/4" long)	Fixing grommet to dash	2	255225@@	
40	Plain washer		2	2550@@	
41	Spring washer		2	3075@@	
42	Nut (5/16" UNF)		2	254811@@	
43	Clamp fixing outer column to bracket		1	279061@	Cars without column grommet
44	Bolt (5/16" UNF x 7/8" long)	In clamp	1	255227@	
45	Nut (5/16" UNF)		1	254811@	
46	Clip for steering column		1	500420	
47	Tape for steering column clip		1	510685	
48	Bolt (5/16"UNF x 3/4" long)	Fixing clip to support	2	255226	
49	Spring washer		2	3075	
50	Nut (5/16" UNF)		2	254811	
51	Clamp plate for steering column		1	503594	Mk I and Mk IA 4-speed
	Clamp plate		1	536529	Mk II
	Switch mounting plate		1	536528	
52	Strengthening plate for steering column		1	503595	Mk I and Mk IA
53	Bolt (2BA x 3/8" long)	Fixing plates to outer column	4	250960	
54	Self-locking nut (2BA)		4	251335	

@ Cars numbered up to 72500551, 72600090, 72800081, 73000766, 73100150, 73300152
@@ Cars numbered from 72500552, 72600091, 72800082, 73000767, 73100151, 73300153 onwards

*Asterisk indicates a new part which has not been used on any previous model.

1441

Plate Ref.	1 2 3 4 Description	Qty.	Part No.	Remarks
	POWER STEERING UNIT ASSEMBLY	1	578547	RH Stg
	POWER STEERING UNIT ASSEMBLY	1	578548	LH Stg
	STEERING BOX HOUSING	1	524001@	
1	STEERING BOX HOUSING	1	605240@@	RH Stg
	STEERING BOX HOUSING	1	524002@	
	STEERING BOX HOUSING	1	605241@@	LH Stg
2	Tube locating pin	1	600705	
3	Oil seal for rocker shaft	1	524013	
4	Retaining washer for oil seal	1	524014	
5	Circlip for oil seal	1	524015	
	BEARING HOUSING COMPLETE	1	524027@	
6	BEARING HOUSING COMPLETE	1	605242@@	
7	Sealing ring for bearing housing	1	524034	
	Seal, inner, rubber and plastic	1	524031@	
8	'U'-type gland seal	1	605243@@	
	Ball bearing (5/32") for bearing housing	30	3750	
	Tube for piston	1	524035@	
	Tube for piston	1	524036@	
9	Tube and side plate complete	1	605250@@	RH Stg
	Tube and side plate complete	1	605251@@	LH Stg
10	Piston	1	524037	
11	Ring for piston	1	524038	
12	Seal, rubber and plastic, for piston	1	524039	
13	Seal, rubber and plastic, for valve side plate, inner	1	524039	
14	Ball bearing (5/32") for side plate, inner	30	3750	
	VALVE AND SPOOL COMPLETE, less seals	1	532274@	
15	VALVE AND SPOOL COMPLETE, less seals	1	605400@@	RH Stg
	VALVE AND SPOOL COMPLETE, less seals	1	605903@@	LH Stg
16	Valve seal 'O' ring	2	524050	Not part of 605400,605903
17	Seal plate for valve spring	2	524049	
18	Valve plate spring, pair	1	524047	
19	Ball cups for spool assembly	2	524022	
20	Ball bearing (5/16")	10	3050	
21	Shim, .002" (For spool assembly)	As reqd	524023	
	Shim, .003" (For spool assembly)	As reqd	524024	
	Shim, .005" (For spool assembly)	As reqd	524025	
	Shim, .010" (For spool assembly)	As reqd	530287	
	Shim, .015" (For spool assembly)	As reqd	530290	
22	Sealing ring for valve, outer	1	605247@@	Not part of 605400,605903
	Sealing ring for valve outer	1	536532@	Not part of 532274
	'O' ring for valve spool	1	601612	
23	Spool locking pin	4	524026	
	Sealing ring for valve centre and inner	2	536532@	
24	Sealing ring, round section, for valve, centre and inner	2	605244@@	

@ Power steering units numbered up to 46320

@@ Power steering units numbered from 46321 onwards

*Asterisk indicates a new part which has not been used on any previous model.

J390

1443

J390

Plate Ref.	1 2 3 4 Description		Qty.	Part No.	Remarks
25	Cam for steering box		1	541872	RH Stg
	Cam for steering box		1	541873	LH Stg
26	Bush, cam end seal		1		Supplied with end cover 606994 only
27	'O' ring for end seal bush		1	524053	
28	Washer for end seal bush		1	524052	
29	Circlip for end seal bush		1	524054	
30	Oil seal for rocker shaft		1	524068	
31	Rocker shaft and cam roller complete		1	524061	
	Seal, rubber and plastic for end cap		1	524031@	Note on steering units numbered from@@ the seal is supplied complete with end cover 606994only
33	Washer for sealing ring		1	524030	
34	Circlip fixing washer end seal		1	524057	
	End cover for steering box		1	524055@	
35	End cover, seal and bush for steering box		1	606994*@@	
36	Bolt (5/16" UNF x 1" long)	Fixing end cover	6	255228	
	Bolt (5/16" UNF x 1 3/8" long)	Fixing end cover	2	256223	
37	Spring washer	Fixing end cover	8	3075	
38	Rubber dust cover		1	605245@@	
39	Top cover complete		1	524078	RH Stg
	Top cover complete		1	524079	LH Stg
40	Alloy gasket for top rocker cover		1	605248	
41	Seal for top cover		1	605249	
42	Locknut for rocker shaft adjusting screw		1	524067	
43	Special screw	Fixing top cover	1	524085	
44	Bolt (5/16" UNF x 1 3/8" long)	Fixing top cover	6	256421	
	Bolt (5/16" UNF x 1 3/8" long)	Fixing top cover	2	256223	
45	Spring washer	Fixing top cover	7	3075	
46	Bleed screw for top cover		1	524056	
47	Dust cover for bleed screw		1	234957	
48	Drop arm	For steering unit	1	516264	RH Stg
	Drop arm	For steering unit	1	516265	LH Stg
49	Sealing ring for drop arm		1	524086	
50	Lock washer for drop arm		1	524087	
51	Nut fixing drop arm		1	524858	
52	Bolt (7/16" UNF x 3" long)	Fixing steering box to chassis sub-frame	1	256068	
53	Bolt (7/16" UNF x 6¼" long)	Fixing steering box to chassis sub-frame	2	253819	
54	Spring washer	Fixing steering box to chassis sub-frame	1	3077	
55	Self-locking nut (7/16" UNF)	Fixing steering box to chassis sub-frame	2	252163	
	Oil seal kit, power steering unit		1	542276@	
	Oil seal kit, power steering unit		1	605399@@	
	Oil seal kit and end cover assembly		1	607151*@@	

@ Power steering units numbered up to 46320
@@ Power steering units numbered from 46321 onwards

*Asterisk indicates a new part which has not been used on any previous model.

OIL RESERVOIR, POWER STEERING UNIT

H645

Plate Ref.	Description (1 2 3 4)	Qty.	Part No.	Remarks
	OIL RESERVOIR COMPLETE	1	510961	'Holbourn Eaton' type reservoir 1961, Mk IA and Mk II
	Filler cap and dipstick complete	1	524091	
	Rubber gasket for top cover	1	524096	
	Rubber gasket, large, for inlet and outlet bracket	2	524104	
	Rubber gasket, small, for inlet and outlet bracket	2	524105	
	Element for filter	1	524106	
	Sealing washer, small } For pipe adaptor	1	524110	
	Sealing washer, large } For pipe adaptor	1	524112	
	OIL RESERVOIR COMPLETE	1	557192@	Late Mk II and Mk III
1	OIL RESERVOIR COMPLETE	1	560142*@@	
	OIL RESERVOIR COMPLETE	1	562279	3½ Litre
2	Filter element	1	601444@@	'Hydrosteer' type reservoir
3	Spring for filter element	1	601445@@	
4	Special nut fixing filter	1	601448@@	
5	Washer for nut	1	601447@@	
6	Sealing washer for cover	1	601446@@	
7	Filler cap and dipstick assembly	1	601450@@	
8	Sealing washer for filler cap	1	601449@@	
	Mounting bracket for oil reservoir and screen washer	1	519263	Except Mk III @@
9	Mounting plate for oil reservoir	1	562280*@@	
	Bolt (¼" UNF x 1⅛" long) } Fixing oil reservoir to bracket	3	256200	'Holbourn Eaton' type reservoir } Check before ordering
	Spring washer } Fixing oil reservoir to bracket	2	3074	
	Nut (¼" UNF) } Fixing oil reservoir to bracket	3	254810	
10	Set bolt (¼" UNF x ½" long) } Fixing oil reservoir to bracket	3	255204	'Hydrosteer' type reservoir } Check before ordering
11	Spring washer } Fixing oil reservoir to bracket	3	3074	
	Set bolt (¼" UNF x ⅝" long) } Fixing mounting bracket to LH wing valance	1	255207	
	Spring washer } Fixing mounting bracket to LH wing valance	1	3074	
	Plain washer } Fixing mounting bracket to LH wing valance	2	3840	
	Nut (¼" UNF) } Fixing mounting bracket to LH wing valance	1	254810	
	Delivery hose, pump to steering box	1	518941	RH Stg Mk I and Mk IA
12	Delivery hose, pump to steering box	1	536260	RH Stg Mk II and MkIII 3Litre
	Delivery hose, pump to steering box	1	518962	LH Stg Mk I and Mk IA
	Delivery hose, pump to steering box	1	536263	LH Stg Mk II and MkIII
	Delivery hose, pump to steering box	1	562401	RH Stg } 3½ Litre
	Delivery hose, pump to steering box	1	565716	LH Stg } 3½ Litre

@ Mk III cars numbered up to 79500576, 79600037, 79800099, 80001558, 80100101, 80300137, 80500425, 80600007, 80800065, 81001117, 81100048, 81300044

@@ Mk III cars numbered from 79500577, 79600038, 79800100, 80001559, 80100102, 80300138, 80500426, 80600008, 80800066, 81001118, 81100049, 81300045 onwards

*Asterisk indicates a new part which has not been used on any previous model.

OIL RESERVOIR, POWER STEERING UNIT

1447

Plate Ref.	1 2 3 4 Description	Qty.	Part No.	Remarks
13	Union — At steering box for	1	518944	
14	Sealing washer — delivery hose	1	518953	
15	Sealing washer for delivery hose, pump end	1	518954	
16	Inlet hose, reservoir to pump	1	518938	3 Litre
	Inlet hose, reservoir to pump	1	562140	3½ Litre
17	Hose clip for inlet hose	2	517821	
18	Return hose, steering box to reservoir	1	518945	RH Stg — 3 Litre
	Return hose, steering box to reservoir	1	518963	LH Stg
	Return hose steering box to reservoir	1	518945	RH Stg — 3½ Litre
	Return hose steering box to reservoir	1	565735	LH Stg
19	Hose clip, return hose to reservoir	2	517821	
20	Banjo connection — Fixing return	1	518946	3 Litre
	Banjo connection — hose to	1	565423	3½ Litre
21	Sealing washer — steering	2	518954	
22	Banjo bolt — box end cover	1	518947	
23	Clip for hose, front cross-member	2	4170	3 Litre
24	Drive screw fixing clip	2	78121	
	Clip for delivery hose	1	530886	1961 and Mk IA 3 Litre
25	Clip for hose, front cross-member	2	50690	RH Stg
	Clip for hose, front cross-member	2	56666	
	Clip for high-pressure delivery hose at air cleaner support	1	524935	
	Bolt (5/16" UNF x 4" long) fixing clip to air cleaner support	1	256034	LH Stg 1961 3 Litre
	Bolt (¼" UNF x ½" long)	1	255204	
	Spring washer — Clamping hose in clip	1	3074	
	Nut (¼" UNF)	1	254810	
	Hose mounting bracket on engine foot	1	530891	
	Delivery hose clip	1	523183	
	Bolt (5/16" UNF x ¾" long)	3	255226	LH Stg Mk IA 3 Litre
	Plain washer — Fixing bracket end	3	2220	
	Spring washer — clip to engine foot	3	3075	
	Nut (5/16" UNF)	3	254831	
26	Hose mounting bracket on engine foot	1	277099	
27	Bolt (¼" UNF x ¾" long)	1	255207	
28	Plain washer — Fixing bracket to	2	3840	
29	Spring washer — engine foot	1	3074	
30	Nut (¼" UNF)	1	254810	Mk II and Mk III
31	Delivery hose clip	1	536262	3 Litre
32	Bolt (¼" UNF x ¾" long)	1	255207	
33	Plain washer — Fixing hose clip	2	3840	
34	Spring washer — to bracket	1	3074	
35	Nut (¼" UNF)	1	254810	
	Clip for hose at horn bracket	1	248122	
	Clip fixing delivery hose to sump	1	536262	3½ Litre
	Clip for hose at steering box	1	50641	

*Asterisk indicates a new part which has not been used on any previous model.

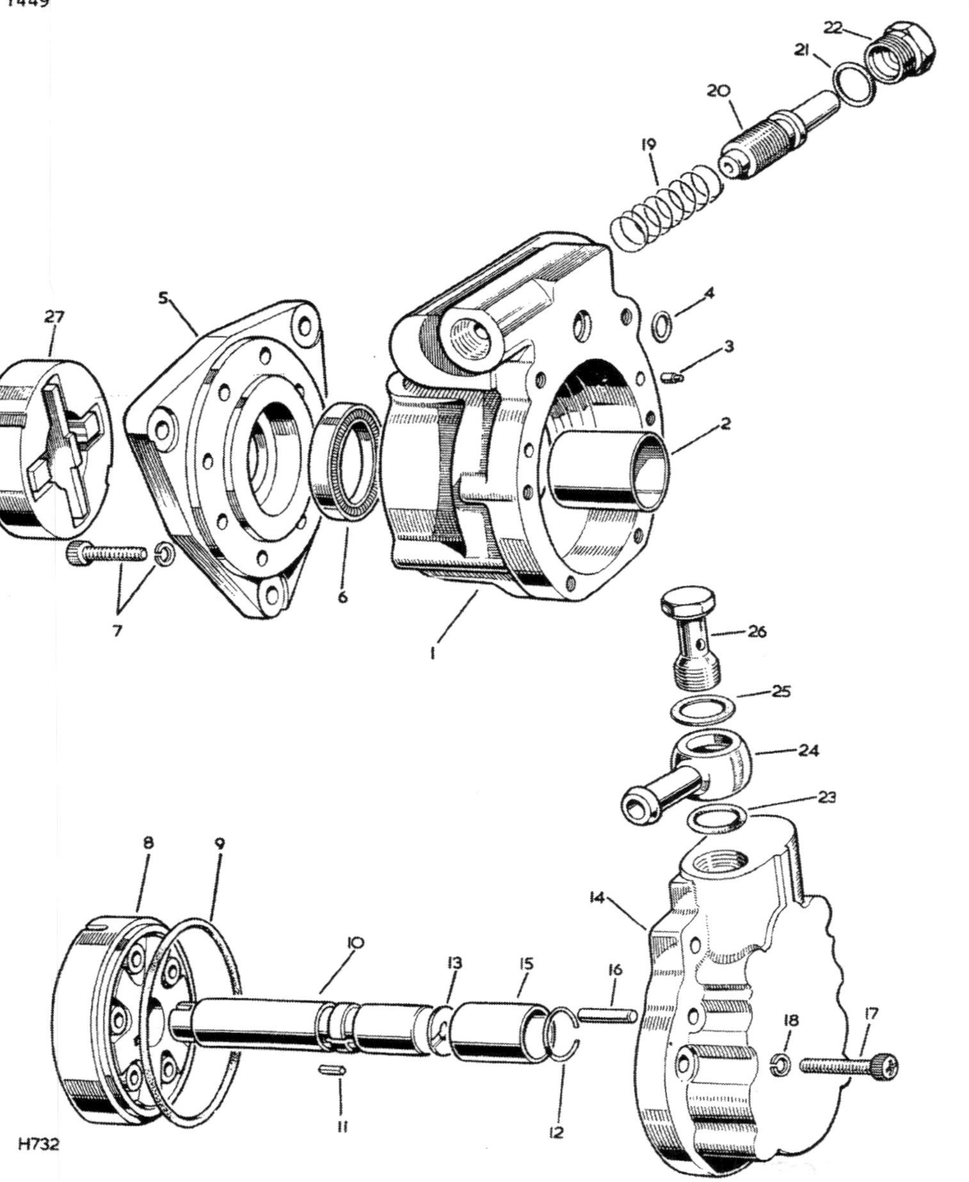

Plate Ref.	1 2 3 4 Description	Qty.	Part No.	Remarks
	PUMP FOR POWER STEERING	1	538349	
1	Body for pump	1	536357	
2	Bush for shaft in body	1	536358	
3	Cam lock peg in pump body	1	536376	
4	'O' ring for by-pass in pump body	1	536385	
5	End plate	1	536379	
6	Oil seal, end plate to body	1	536359	
7	Special screw and washer fixing end plate	4	536380	
8	Vane carrier and roller vanes	1	536364	
9	'O' ring for vane carrier to body	1	536366	
10	Shaft ⎤ For vane	1	536378	
11	Drive pin ⎦ carrier	1	536377	
12	Retaining clip on vane shaft	1	536375	
13	Thrust washer, shaft to end cover	1	536374	
14	Cover for pump	1	536360	
15	Bush for shaft in end cover	1	536361	
16	Dowel for fixing end cover to body	2	536362	
17	Special screw ⎤ Fixing end	6	536370	
18	Lock washer ⎦ cover to body	6	536371	
19	Control spring for valve assembly	1	536373	
20	Valve assembly in pump body	1	536363	
21	Valve cap 'O' ring	1	536367	
22	Valve cap in pump body	1	536372	
23	Seal for adaptor in cover	1	536368	
24	Adaptor for end cover	1	536381	
25	Fibre washer for adaptor	1	536369	
26	Adaptor screw fixing adaptor to end cover	1	536382	
27	Coupling, dynamo to pump	1	501009	
	Stud (5/16" UNF x 1⅜" long) ⎤ Fixing	3	500667	
	Spring washer ⎬ pump to	3	3075	
	Nut (5/16" UNF) ⎦ dynamo	3	254811	
	Oil seal kit	1	536365	

*Asterisk indicates a new part which has not been used on any previous model.

PUMP FOR POWER STEERING 3½ LITRE

Plate Ref.	1 2 3 4	Description	Qty.	Part No.	Remarks
		PUMP ASSEMBLY	1	602348	
1		BODY ASSEMBLY	1	605171	
2		Oil seal } For body	1	536359	
3		Bush } For body	1	536361	
4		'O' ring, body to cover	1	536366	
5		END COVER ASSEMBLY	1	605172	
6		Bush } For end cover	1	536361	
7		Dowel } For end cover	2	536362	
8		Valve assembly	1	605173	
9		Spring, flow control	1	536373	
10		'O' ring for valve cap	1	605177	
11		Valve cap	1	605178	
12		Seal for adaptor	1	536368	
13		Adaptor banjo	1	536381	
14		Fibre washer for adaptor	2	536369	
15		Banjo bolt	1	536382	
16		'O' ring for by-pass in pump body	1	605176	
17		Shaft	1	605180	
18		Drive pin fixing vane carrier	1	605179	
19		Bearing	1	605181	
20		Vane carrier and roller vane	1	605174	
21		Cam lock peg	1	536376	
22		Special screw } Fixing end cover to body	6	536370	
23		Lock washer } Fixing end cover to body	6	536371	
24		End plate	1	605182	
25		Special screw and washer fixing end plate	4	536380	
26		Key for pump shaft	1	554880	
27		Pulley	1	602349	
28		Bolt (5/16" UNC x 7/8" long) } Fixing pulley to steering pump	1	253027	
29		Spring washer } Fixing pulley to steering pump	1	3075	
30		Plain washer } Fixing pulley to steering pump	1	2217	
31		Mounting arm for pump rear	1	602470	
32		Bolt (1/4" UNF x 5/8" long) } Fixing mounting arm to pump	3	255206	
33		Spring washer } Fixing mounting arm to pump	3	3074	

*Asterisk indicates a new part which has not been used on any previous model.

Plate Ref.	1 2 3 4 Description		Qty.	Part No.	Remarks
34	Mounting bracket, rear		1	602471	
35	Bolt (5/16" UNC x 7/8" long)	Fixing mounting bracket, rear, to cylinder block	2	253027	
36	Plain washer		2	2550	
37	Spring washer		2	3075	
38	Bolt (3/8" UNF x 1 1/8" long)	Fixing pump to rear bracket	1	255248	
39	Spring washer		1	3076	
40	Nut (3/8" UNF)		1	254812	
41	Mou nting bracket, front		1	602472	
42	Bolt (3/8" UNC x 7/8" long)	Fixing mounting bracket, front, to cylinder block	1	253046	
43	Bolt (5/16" UNC x 3.82" long)		1	602388	
44	Spring washer		1	3075	
45	Spring washer		1	3076	
46	Bolt (3/8" UNF x 1 1/8" long)	Fixing steering pump to brackets	1	255248	
47	Spring washer		2	3076	
48	Nut (3/8" UNF)		2	254812	
49	Bolt (3/8" UNF x 1 1/8" long)	For steering pump driving belt adjustment	1	255248	
50	Plain washer		1	2251	
51	Spring washer		1	3076	
52	Nut (3/8" UNF)		1	254812	
53	Belt, driving steering pump		1	602473	

*Asterisk indicates a new part which has not been used on any previous model.

F613

Plate Ref.	1 2 3 4 Description	Qty.	Part No.	Remarks
	OUTER COLUMN ASSEMBLY	1	510681@	RH Stg Cars without column grommet
	OUTER COLUMN ASSEMBLY	1	514618@	LH Stg Cars without column grommet
1	OUTER COLUMN ASSEMBLY	1	542927@@	RH Stg Cars with column grommet
	OUTER COLUMN ASSEMBLY	1	542928@@	LH Stg Cars with column grommet
2	Dowel for shrouds and brackets	2	500993	
3	Bearing for outer column	1	279072	2 off from @@
4	Dowel for steering column support bracket	1	510682	
	Steering column shaft	1	513057@	Cars without column grommet
5	Steering column shaft	1	529209@@	Cars with column grommet
	Steering column shaft	1	553758	Late Mk III 3 Litre and 3½ Litre
6	Spring washer for steering column shaft	1	551984	2 off from @@
7	Clamping ring for shaft	1	529193@@	Cars with column grommet
8	Shakeproof washer } Fixing clamping ring to shaft	2	782498@@	Cars with column grommet
9	Set screw (4 BA x ⅜" long) } Fixing clamping ring to shaft	2	784968@@	Cars with column grommet
10	Steering coupling flange	2	523077	
11	Flexible disc for coupling	1	520833	
12	Locating plate for coupling	2	522951	
13	Bolt (5/16" UNF x 1⅜" long) } Fixing flanges, locating plates and flexible disc together	4	256421	
14	Self-locking nut (5/16" UNF) } Fixing flanges, locating plates and flexible disc together	4	252161	
15	Bolt (5/16" UNF x 1¾" long) } Fixing coupling to steering box and steering column shaft	2	256223	
16	Self-locking nut (5/16" UNF) } Fixing coupling to steering box and steering column shaft	2	252161	
17	Steering wheel	1	536498	
18	Plain washer } Fixing steering wheel	1	278461	
19	Special nut } Fixing steering wheel	1	278773	
20	Steering column mounting bracket, lower angle	1	359527	RH Stg
21	Steering column mounting bracket tie bracket	1	356213	RH Stg
	Steering column mounting bracket, lower angle	1	359528	LH Stg
	Steering column mounting bracket, tie bracket	1	356214	LH Stg
22	Support bracket } Steering column mounting bracket to main instrument mounting bracket	1	359474	RH Stg
	Support bracket } Steering column mounting bracket to main instrument mounting bracket	1	359475	LH Stg
23	Tapped plate, steering column mounting bracket to parcel shelf	1	359483	
	Shim	1	359884	
	Bolt (¼" UNF x 2" long) } Fixing support bracket to steering mounting bracket and main instrument bracket	4	255204	
	Spring washer } Fixing support bracket to steering mounting bracket and main instrument bracket	4	3074	
	Plain washer } Fixing support bracket to steering mounting bracket and main instrument bracket	4	3840	
	Nut (¼" UNF) } Fixing support bracket to steering mounting bracket and main instrument bracket	4	254810	
24	Set bolt (¼" UNF x ⅝" long) } Fixing bracket to angle, dash and parcel shelf	6	255406	
25	Plain washer, small } Fixing bracket to angle, dash and parcel shelf	4	3946	
26	Plain washer, large } Fixing bracket to angle, dash and parcel shelf	4	3900	
27	Spring washer } Fixing bracket to angle, dash and parcel shelf	6	3074	
28	Support bracket for steering column	1	510678	
29	Set bolt (¼" UNF x ¾" long) } Fixing support bracket to mounting bracket	4	255207	
30	Plain washer } Fixing support bracket to mounting bracket	4	3840	
31	Spring washer } Fixing support bracket to mounting bracket	4	3074	

@ 3 Litre cars numbered up to 72500509, 72600073, 72800060, 73000712, 73100136, 73300138

@@ 3 Litre Cars numbered from 72500510, 72600074, 72800061, 73000713, 73100137, 73300139 onwards also 3½ Litre

*Asterisk indicates a new part which has not been used on any previous model.

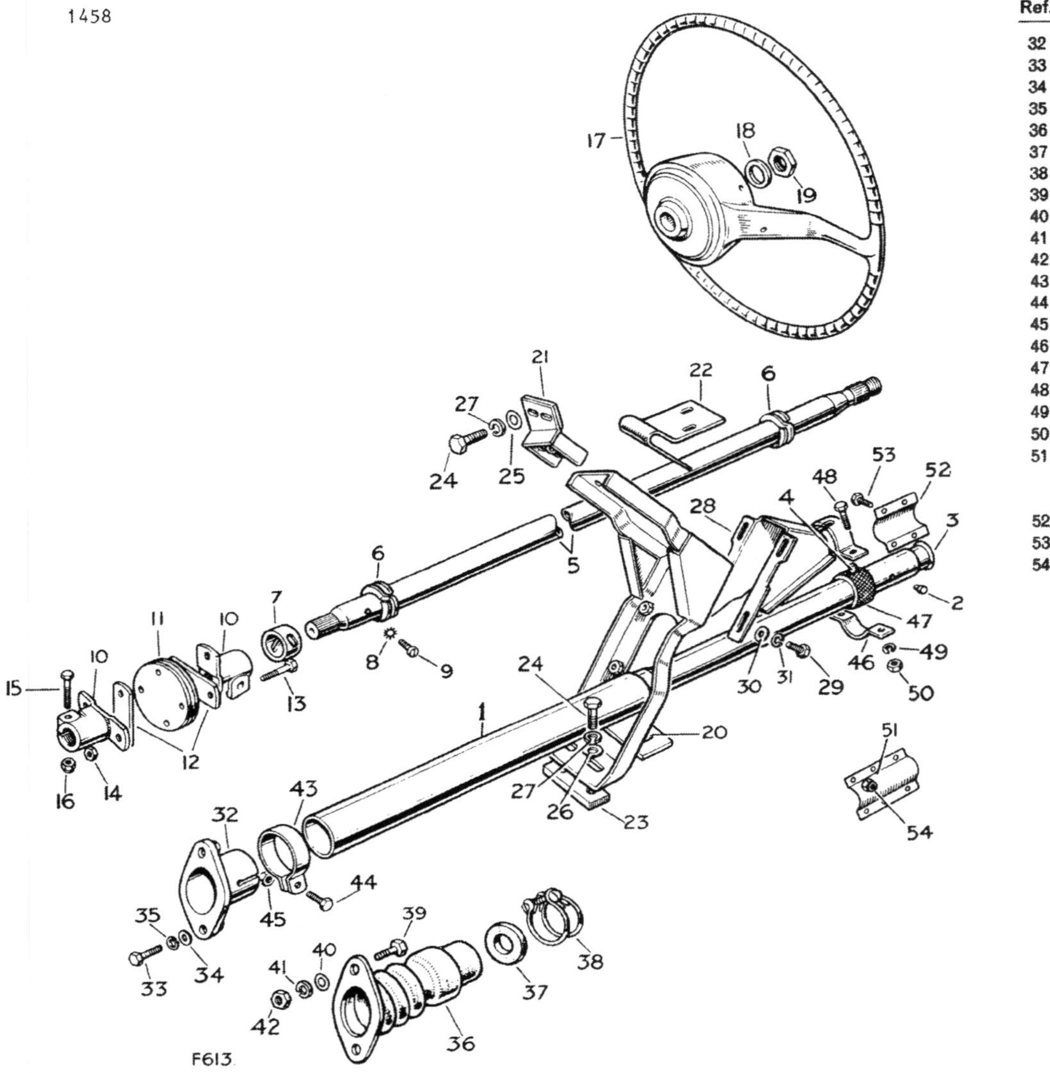

STEERING OUTER COLUMN AND SHAFT, POWER STEERING

Plate Ref.	1 2 3 4 DESCRIPTION		Qty	Part No.	REMARKS
32	Anchor bracket for steering column		1	279058†	
33	Set bolt ($\frac{5}{16}$" UNF x $\frac{5}{8}$" long)	Fixing	2	255025†	Cars without column grommet
34	Plain washer	anchor bracket	2	2550†	
35	Spring washer	to dash	2	3075†	
36	Grommet for steering column		1	532057††	
37	Felt washer for grommet		1	529192††	
38	Hose clip, grommet to outer column		1	50314††	
39	Set bolt ($\frac{5}{16}$" UNF x $\frac{5}{8}$" long)		2	255025††	Cars with column grommet
40	Plain washer	Fixing grommet	2	2550††	
41	Spring washer	to dash	2	3075††	
42	Nut ($\frac{5}{16}$" UNF)		2	254811††	
43	Clamp fixing outer column to bracket		1	279061†	
44	Bolt ($\frac{5}{16}$" UNF x $\frac{7}{8}$" long)	Fixing	1	255227†	Cars without column grommet
45	Nut ($\frac{5}{16}$" UNF)	clamp	1	254811†	
46	Clip for steering column		1	500420	
47	Tape for steering column clip		1	535823	
48	Bolt ($\frac{5}{16}$" UNF x $\frac{3}{4}$" long)	Fixing clip	2	255226	
49	Spring washer	to support	2	3075	
50	Nut ($\frac{5}{16}$" UNF)	bracket	2	254811	
51	Clamp plate for steering column		1	503594	Mk I and Mk IA 4-speed
	Clamp plate for steering column		1	536529	Mk II and
	Switch mounting plate		1	536528	Mk III
52	Strengthening plate for steering column		1	503595	Mk I and Mk IA
53	Bolt (2 BA x $\frac{5}{8}$" long)	Fixing plates to	4	250960	
54	Self-locking nut (2 BA)	outer column	4	251335	

* Asterisk indicates a new part which has not been used on any previous Rover model
† Cars numbered up to 72500509, 72600073, 72800060, 73000712, 73100136, 73300138
†† Cars numbered from 72500510, 72600074, 72800061, 73000713, 73100137, 73300139 onwards

STEERING LINKAGE, MANUAL STEERING

1460

F 397

Plate Ref.	1 2 3 4 Description	Qty.	Part No.	Remarks
	STEERING TRACK ROD ASSEMBLY	1	542324	
1	Steering track rod only	1	279352	
2	Clamp bracket for steering damper	1	542323	
3	BALL JOINT ASSEMBLY, RH thread, straight	1	276567	
4	BALL JOINT ASSEMBLY, LH thread, straight	1	276568	
5	Rubber cover for ball joint	2	214649	
6	Spring ring, cover to body	2	214685	
7	Spring ring, cover to ball	2	214684	
8	Locknut(9/16" UNF), RH thread } Fixing ball joints to track rod	1	254855	
9	Locknut(9/16" UNF), LH thread } Fixing ball joints to track rod	1	275781	
10	Plain washer } Fixing ball joints to drop arms	2	3843	
11	Special castle nut } Fixing ball joints to drop arms	2	276482	
12	Split pin } Fixing ball joints to drop arms	2	2393	
13	Bolt(5/16" UNF x 1" long) } Fixing clamp bracket to track rod	2	255228	Mk II cars with suffix letter 'B' onwards (cars with steering damper)
14	Plain washer } Fixing clamp bracket to track rod	2	2220	Mk II cars with suffix letter 'B' onwards (cars with steering damper)
15	Self-locking nut (5/16" UNF) } Fixing clamp bracket to track rod	2	252161	Mk II cars with suffix letter 'B' onwards (cars with steering damper)
16	Steering damper	1	535824	Mk II cars with suffix letter 'B' onwards (cars with steering damper)
17	Plain washer, small } Fixing steering damper to clamp bracket and support bracket	2	2208	Mk II cars with suffix letter 'B' onwards (cars with steering damper)
18	Plain washer, large } Fixing steering damper to clamp bracket and support bracket	2	2251	Mk II cars with suffix letter 'B' onwards (cars with steering damper)
19	Self-locking nut(⅜" UNF) } Fixing steering damper to clamp bracket and support bracket	2	252162	Mk II cars with suffix letter 'B' onwards (cars with steering damper)
	STEERING LINK ASSEMBLY	2	542731	
20	Steering link only	2	279350	
21	BALL JOINT ASSEMBLY, RH thread, cranked	4	542726	
22	Rubber cover for ball joint	4	276351	
23	Spring ring, cover to body	4	275397	
24	Spring ring, cover to ball	4	214684	
25	Locknut fixing ball joint to steering link	4	254856	
26	Plain washer } Fixing ball joints to drop arms and track rod levers	4	3843	
27	Special castle nut } Fixing ball joints to drop arms and track rod levers	4	276482	
28	Split pin } Fixing ball joints to drop arms and track rod levers	4	2393	
	STEERING RELAY ASSEMBLY	1	530772@@	RH Stg; Mk I, Mk IA and Mk II cars with suffix letter 'A' (cars without steering damper)
	STEERING RELAY ASSEMBLY	1	530773@@	LH Stg; Mk I, Mk IA and Mk II cars with suffix letter 'A' (cars without steering damper)
	HOUSING ASSEMBLY FOR STEERING RELAY	1	507597@	RH Stg; Mk I, Mk IA and Mk II cars with suffix letter 'A' (cars without steering damper)
	HOUSING ASSEMBLY FOR STEERING RELAY	1	537069@@	RH Stg; Mk I, Mk IA and Mk II cars with suffix letter 'A' (cars without steering damper)
	HOUSING ASSEMBLY FOR STEERING RELAY	1	507598@	LH Stg; Mk I, Mk IA and Mk II cars with suffix letter 'A' (cars without steering damper)
	HOUSING ASSEMBLY FOR STEERING RELAY	1	530770@@	LH Stg; Mk I, Mk IA and Mk II cars with suffix letter 'A' (cars without steering damper)
29	Bush for housing, top	1	276732@	Mk I, Mk IA and Mk II cars with suffix letter 'A' (cars without steering damper)
	Bush for housing, top	1	530766@@	Mk I, Mk IA and Mk II cars with suffix letter 'A' (cars without steering damper)
	Bush for housing, bottom	1	276731@	Mk I, Mk IA and Mk II cars with suffix letter 'A' (cars without steering damper)
	Bush for housing, bottom	1	530767@@	Mk I, Mk IA and Mk II cars with suffix letter 'A' (cars without steering damper)
	Covershell plug for relay housing	1	530771@@	Mk I, Mk IA and Mk II cars with suffix letter 'A' (cars without steering damper)
30	Shaft for steering relay lever	1	507238@	Mk I, Mk IA and Mk II cars with suffix letter 'A' (cars without steering damper)
	Shaft for steering relay lever	1	530768@@	Mk I, Mk IA and Mk II cars with suffix letter 'A' (cars without steering damper)
	Oil seal for shaft	1	212669	Mk I, Mk IA and Mk II cars with suffix letter 'A' (cars without steering damper)
31	Thrust washer for relay shaft	1	507319	Mk I, Mk IA and Mk II cars with suffix letter 'A' (cars without steering damper)

@ Mk I and Mk IA cars numbered up to 72500815, 72600118, 72800155, 73001172, 73100219, 73300253

@@ Mk IA cars numbered from 72500816, 72600119, 72800156, 73001173, 73100220, 73300254 onwards

*Asterisk indicates a new part which has not been used on any previous model.

STEERING LINKAGE, MANUAL STEERING

1462

Plate Ref.	1 2 3 4 Description	Qty.	Part No.	Remarks
	Shim for relay shaft	As reqd	507599	Mk I, Mk IA & Mk II cars with suffix letter 'A' (cars without steering damper)
32	Rubber sealing ring for relay shaft	1	515573	
33	Mounting plate for relay damper	1	507591	
34	Wooden friction disc for damper, .080" thick	As reqd	507579@	
	Ferodo friction disc, .080" thick	2	530765@@	
35	Wooden friction disc for damper, .093" thick	As reqd	507578@	
	Ferodo friction disc, .093" thick	2	530764@@	
36	Rotor plate for damper	2	507492	
37	Fixed plate, intermediate for damper	1	507533	
38	Fixed plate, upper, for damper	1	507534	
39	Pressure plate for damper	1	507495	
40	Adjuster plate for damper	1	507493	
41	Damper housing	1	507494	
42	Bolt (¼" Whit x ¾" long) } For damper adjustment	1	507797	Alternatives
43	Nut (¼" Whit)	1	2531	Check
	Bolt(¼" UNC x ¾" long)	1	253007	before
	Nut(¼" UNC)	1	256850	ordering
44	Rubber cover for damper	1	507236	
45	Bolt(¼" UNF x 2⅜" long) } Fixing damper to relay	4	253862	
	Self-locking nut (¼" UNF)	4	252160	
	STEERING RELAY ASSEMBLY	1	522844	RH Stg — Mk II cars with suffix letter 'B' onwards (cars with steering damper)
	STEERING RELAY ASSEMBLY	1	522843	LH Stg
46	HOUSING ASSEMBLY FOR STEERING RELAY	1	530769	RH Stg
	HOUSING ASSEMBLY FOR STEERING RELAY	1	530770	LH Stg
47	Bush for housing, top	1	530766	
48	Bush for housing, bottom	1	530767	
49	Shaft for steering relay lever	1	276733	
50	Oil seal for shaft	1	212669	
51	Thrust washer for relay shaft	1	507319	
52	Shim for relay shaft	As reqd	507599	
53	Set bolt (¼" UNC x ⅜" long) } For relay top cap	1	253002	
54	Washer for set bolt	1	232037	
55	Bolt(¼" UNF x 1⅛" long) } Fixing cap to relay	4	256200	
56	Self-locking nut (¼" UNF)	4	252160	
57	Drop arm for relay shaft	1	517016	RH Stg
	Drop arm for relay shaft	1	517015	LH Stg
58	Nut(⅜" UNF) } Fixing arm to shaft	1	254857	
59	Lock washer	1	4115	
60	Bolt (7/16" UNF x 4½" long) } Fixing relay to chassis sub-frame	2	256075	
61	Bolt(7/16" UNF x 6" long)	1	253874	
	Plain washer	2	217245	
62	Plain washer	3	3261	
63	Self-locking nut (7/16" UNF)	3	252163	
64	Special bolt } In chassis sub-frame for steering lock adjustment	2	274476	
65	Locknut	2	254811	
	Track rod lever, LH	1	276783	For drum-type brakes
	Track rod lever, RH	1	276782	

@ Mk I and Mk IA cars numbered up to 72500815, 72600118, 72800155, 73001172, 73100219, 73300253

@@ Mk IA cars numbered from 72500816, 72600119, 72800156, 73001173, 73100220, 73300254 onwards

*Asterisk indicates a new part which has not been used on any previous model.

STEERING LINKAGE, MANUAL STEERING

1464

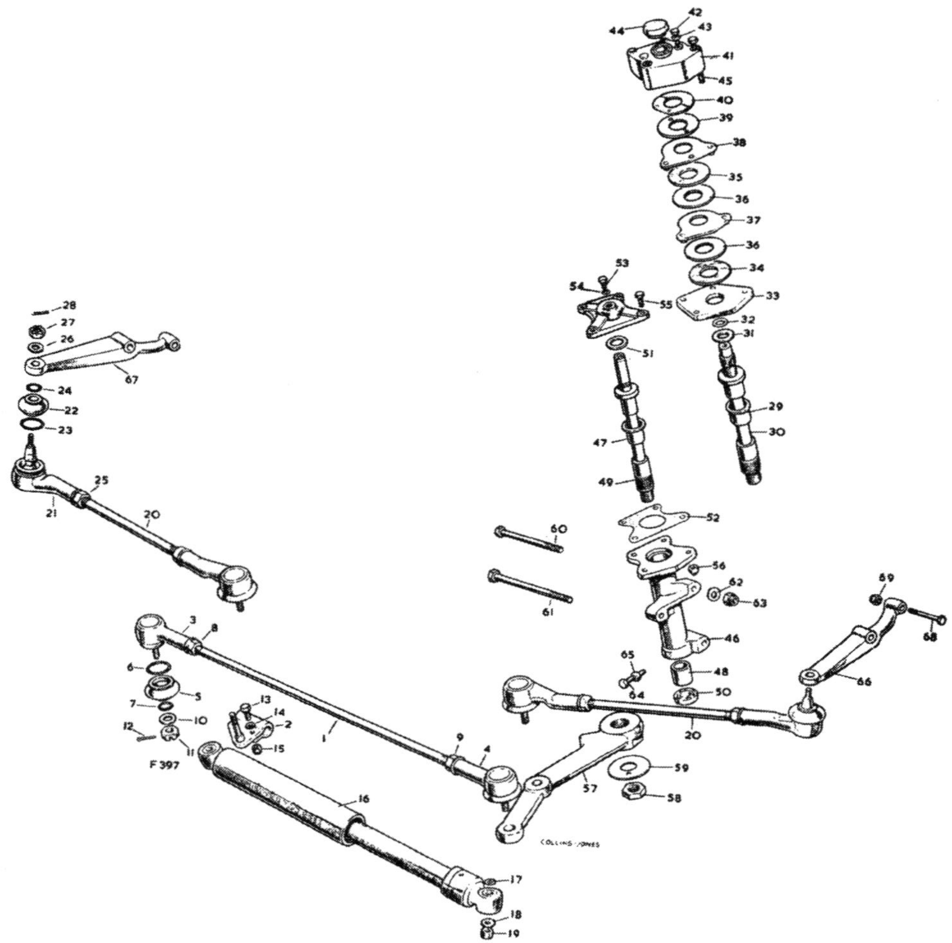

Plate Ref.	1 2 3 4 Description		Qty.	Part No.	Remarks	
	Bolt(3/8" UNF x 1 1/8" long)	Fixing track rod levers and brake anchor plate to stub axle	4	255248	For drum-type brakes	
	Bolt(3/8" UNF x 2" long)		2	256445		
	Bolt(3/8" UNF x 2 1/4" long)		2	256446		
	Self-locking nut (3/8" UNF)		8	252162		
	Track rod lever, LH	With 3/8" dia. fixing holes	1	505550	1959-60	For disc-type brakes
	Track rod lever, RH		1	505549		
66	Track rod lever, LH	With 7/16" dia. fixing holes	1	515426	1961 onwards	
67	Track rod lever, RH		1	515425		
	Bolt (3/8" UNF x 2 1/4" long)	Fixing track rod lever	4	256446		
	Self-locking nut(3/8" UNF)		4	252162		
68	Bolt(7/16" UNF x 2 1/4" long)	Fixing track rod lever with 7/16" dia. holes	4	253907		
69	Self-locking nut (7/16" UNF)		4	252163		

*Asterisk indicates a new part which has not been used on any previous model.

F 397

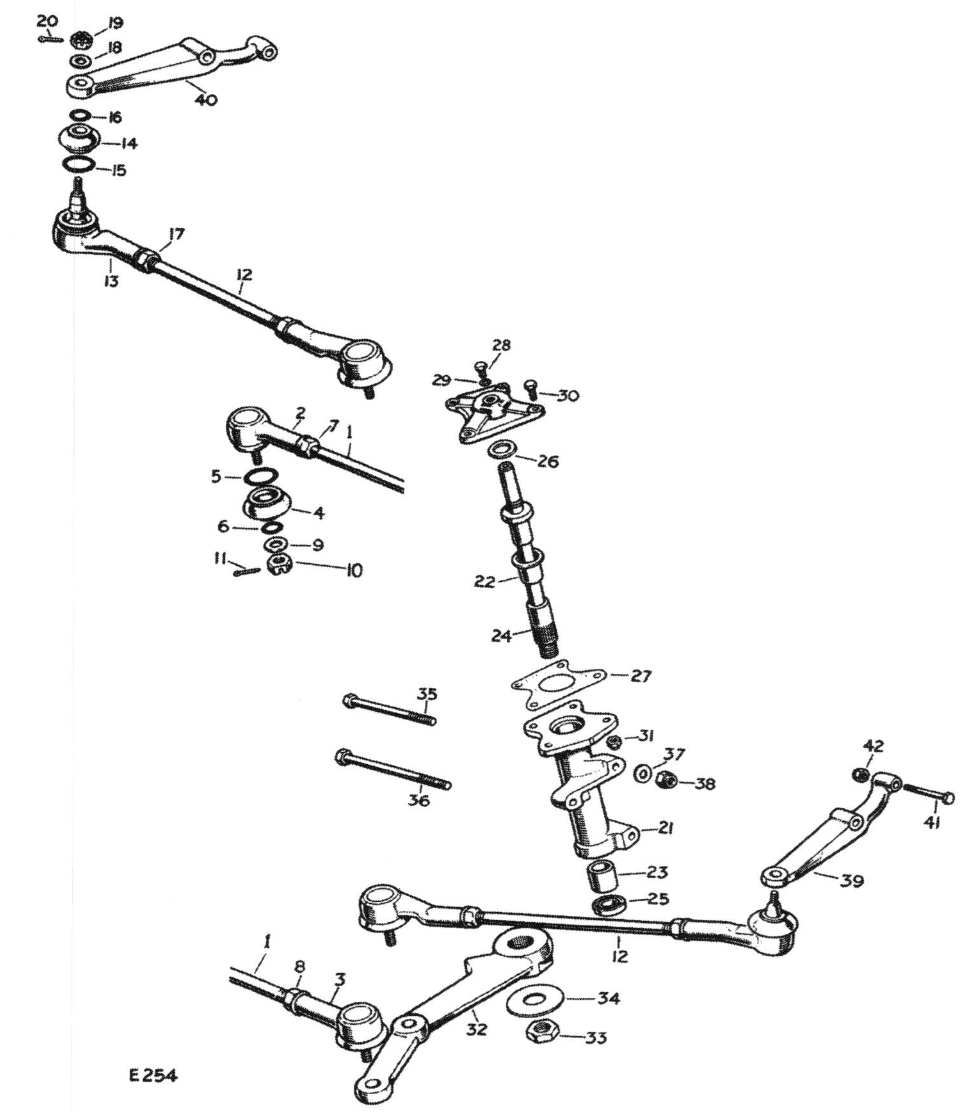

Plate Ref.	1 2 3 4 Description		Qty.	Part No.	Remarks
	STEERING TRACK ROD ASSEMBLY		1	518996	
1	Steering track rod only		1	279352	
2	BALL JOINT ASSEMBLY, RH thread, straight		1	276567	
3	BALL JOINT ASSEMBLY, LH thread, straight		1	276568	
4	Rubber cover for ball joint		2	214649	
5	Spring ring, cover to body		2	214685	
6	Spring ring, cover to ball		2	214684	
7	Locknut (7/16" UNF),RH thread	Fixing ball joints to track rod	1	254855	
8	Locknut (7/16" UNF),LH thread		1	275781	
9	Plain washer	Fixing ball joints to drop arms	2	3843	
10	Special castle nut		2	276482	
11	Split pin		2	2393	
	STEERING LINK ASSEMBLY		2	542727	
12	Steering link only		2	279350	
13	BALL JOINT ASSEMBLY, RH thread, cranked		4	542726	
14	Rubber cover for ball joint		4	276351	
15	Spring ring, cover to body		4	275397	
16	Spring ring, cover to ball		4	214684	
17	Locknut fixing ball joint to steering link		4	254856	
18	Plain washer	Fixing ball joints to drop arms and track rod levers	4	3843	
19	Special castle nut		4	276482	
20	Split pin		4	2393	
	STEERING RELAY ASSEMBLY		1	522844	RH Stg
	STEERING RELAY ASSEMBLY		1	522843	LH Stg
21	HOUSING ASSEMBLY FOR STEERING RELAY		1	530769	RH Stg
	HOUSING ASSEMBLY FOR STEERING RELAY		1	530770	LH Stg
22	Bush for housing, top		1	530766	
23	Bush for housing, bottom		1	530767	
24	Shaft for steering relay lever		1	276733	
25	Oil seal for shaft		1	212669	
26	Thrust washer for relay shaft		1	507319	
27	Shim for relay shaft		As reqd	507599	
28	Set bolt (¼" UNC x ⅜" long)	For relay top cap	1	253002	
29	Washer for set bolt		1	232037	
30	Bolt (¼" UNF x 1⅛" long)	Fixing cap to relay	4	256200	
31	Self-locking nut (¼" UNF)		4	252160	
32	Drop arm for relay shaft		1	517016	RH Stg
	Drop arm for relay shaft		1	517015	LH Stg
33	Nut (¾" UNF)	Fixing arm to shaft	1	254857	
34	Lock washer		1	4115	
35	Bolt (7/16" UNF x 5¼" long)	Fixing relay to chassis sub-frame	2	256077	
36	Bolt (7/16" UNF x 6¼" long)		1	253819	
37	Spacer washer		3	519001	
38	Self-locking nut (7/16" UNF)		3	252163	
39	Track rod lever, LH		1	515426	
40	Track rod lever, RH		1	515425	
41	Bolt (7/16" UNF x 2¼" long)	Fixing track rod lever	4	253907	
42	Self-locking nut (7/16" UNF)		4	252163	

*Asterisk indicates a new part which has not been used on any previous model.

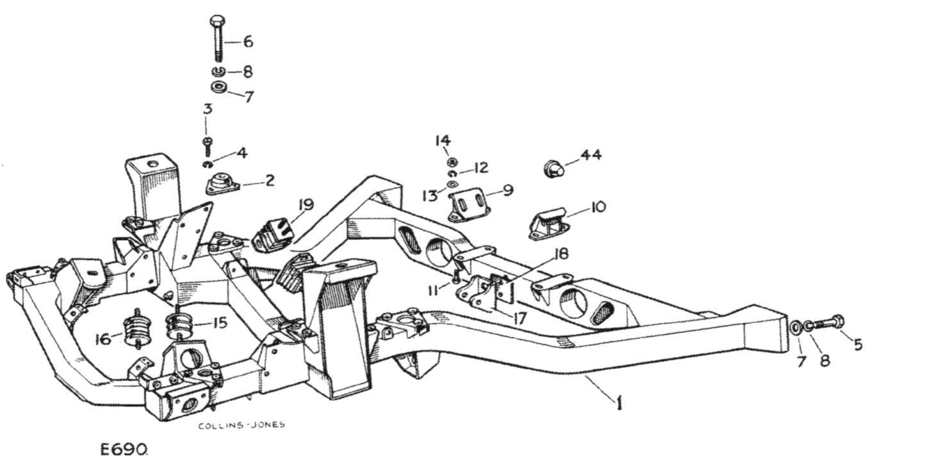

E690

Plate Ref.	1 2 3 4 Description		Qty.	Part No.	Remarks
1	Chassis sub-frame complete		1	530459	Mk I, Mk IA and Mk II
	Chassis sub-frame complete		1	600982	Mk III
	Conversion kit for chassis sub-frame		1	530149@	Use in conjunction with chassis sub-frame 530459
2	Mounting for chassis sub-frame		6	500256	4 off on 4-speed non-overdrive models
	Mounting for chassis sub-frame		2	518700	4-speed models without overdrive numbered from 625001974, 626000610, 628000289 onwards
3	Set bolt (5/16" UNF x ¾" long)	Fixing mounting to chassis sub-frame	18	253870	
4	Spring washer	Fixing mounting to chassis sub-frame	17	3075	
	Special earthing washer	Fixing mounting to chassis sub-frame	2	510912	
	Adaptor bracket for horn mounting		1	555970	Mk III
5	Set bolt (½" UNF x 3½" long)	Fixing chassis sub-frame to body	4	256289	Alternatives. Check before ordering
	Set bolt (½" UNF x 3¼" long)	Fixing chassis sub-frame to body	4	256088	Alternatives. Check before ordering
6	Set bolt (½" UNF x 4" long)	Fixing chassis sub-frame to body	2	256091	
7	Plain washer	Fixing chassis sub-frame to body	8	2222	
8	Spring washer	Fixing chassis sub-frame to body	6	3078	
	Special earthing washer for sub-frame		6	510912	
9	Support bracket for gearbox, RH		1	503126@	
10	Support bracket for gearbox, LH		1	503127@	
11	Bolt (5/16" UNF x ⅝" long)	Fixing support brackets to chassis sub-frame and mounting rubbers to support brackets	8	255225@	
12	Spring washer	Fixing support brackets to chassis sub-frame and mounting rubbers to support brackets	8	3075@	
13	Plain washer	Fixing support brackets to chassis sub-frame and mounting rubbers to support brackets	8	3868@	
14	Nut (5/16" UNF)	Fixing support brackets to chassis sub-frame and mounting rubbers to support brackets	8	254811@	
15	Engine mounting rubber, front		2	240140@	
16	Engine mounting rubber, front		2	522695@@	
	Plain washer	Fixing engine mounting	4	2251	
	Spring washer	Fixing engine mounting	4	3076	
	Nut (⅜" BSF)	Fixing engine mounting	4	2827@	
	Nut (⅜" UNF) fixing engine mounting		4	254812@@	
	Special washer, rubber mounting to engine foot		2	553976	Mk III Automatic
17	Stabiliser support bracket		1	520916@@	
18	Shim, support bracket to rear cross-member		As reqd	519018@@	
	Shim (10 SWG) support bracket to rear cross-member		As reqd	528197	
	Bolt (⅜" UNF x 1¾" long)	Fixing support bracket at rear cross-member	2	256043@@	
	Spring washer	Fixing support bracket at rear cross-member	2	3076@@	
	Plain washer	Fixing support bracket at rear cross-member	2	2251@@	
19	Engine mounting rubber, rear		2	527181@@	Except Mk III Automatic
	Engine mounting rubber, rear		2	562351	Mk III Automatic

@ Cars numbered up to 625101045, 626100262, 628100109, 630100840, 631100353, 633100078

@@Cars numbered from 625101046, 626100263, 628100110, 630100841, 631100354, 633100079 onwards

*Asterisk indicates a new part which has not been used on any previous model.

Plate Ref.	1 2 3 4 Description	Qty.	Part No.	Remarks
	Plain washer] Fixing mounting rubbers to engine mounting bracket, RH rear	2	2550@@	
	Spring washer	2	3075@@	
	Nut (5/16" UNF)	2	254831@@	
	Bolt (5/16" UNF x ⅞" long)] Fixing mounting rubber to sub-frame, RH rear	2	255227@@	
	Plain washer	2	2550@@	
	Spring washer	2	3075@@	
	Nut (5/16" UNF)	2	254831@@	
	Set bolt (5/16" UNF x 1" long)] Fixing mounting rubber to fly-wheel housing LH	2	255228@@	
	Plain washer	2	2550@@	
	Spring washer	2	3075@@	
	Plain washer] Fixing mounting rubber to sub-frame LH rear	2	2550@@	
	Spring washer	2	3075@@	
	Nut (5/16" UNF)	2	254831@@	
20	ROAD SPRING, REAR	2	542825	Mk I and Mk IA
	ROAD SPRING, REAR	2	536714	Mk II standard suspension
	ROAD SPRING, REAR	2	553899	Mk III standard suspension
	ROAD SPRING, REAR	2	542825	Mk II and Mk III high suspension
21	Main leaf for spring	2	507037	Mk I and Mk IA
	Main leaf for spring	2	600611	Mk II standard suspension
	Main leaf for spring	2	600972	Mk III standard suspension
	Main leaf for spring	2	507037	] Mk II and Mk III high suspension
	Auxiliary leaf for spring	2	538407	
22	Bush for spring	2	502709	
23	Centre dowel for spring	2	256049	
24	Washer for centre dowel	4	510940	
25	Grease sleeve for rear springs	4	525153	Mk I and Mk IA
	Grease sleeve for rear springs	4	530806	Mk II and Mk III
26	Lace for grease sleeve	4	504436	
27	Taper wedge for rear spring	2	510941	Mk I and Mk IA
28	Bottom plate for spring	2	500770	
29	'U' bolt] Fixing spring to axle	4	510978	
30	Nut (7/16" UNF)	8	254823	
31	Locknut (7/16" UNF)	8	254873	
	Rubber mounting for rear end of spring, colour purple	2	504543$	] NOTE: Cars fitted with heavy-duty shock absorbers use 504543
32	Rubber mounting for rear end of spring, colour pink	2	516850$$	
33	Packing washer for rubber mounting	4	507085	Only required with mounting 516850

@@ Cars numbered from 625101046, 626100263, 628100110, 630100841, 631100354, 633100079

$ Cars numbered up to 625000743, 626000251, 628000086, 630000309, 631000092, 633000225

$$ Cars numbered from 625000744, 626000252, 628000087, 630000310, 631000093, 633000226 onwards

NOTE: Cars with 'high' suspension may be identified by the letter 'H' stamped on the car serial number plate

*Asterisk indicates a new part which has not been used on any previous model.

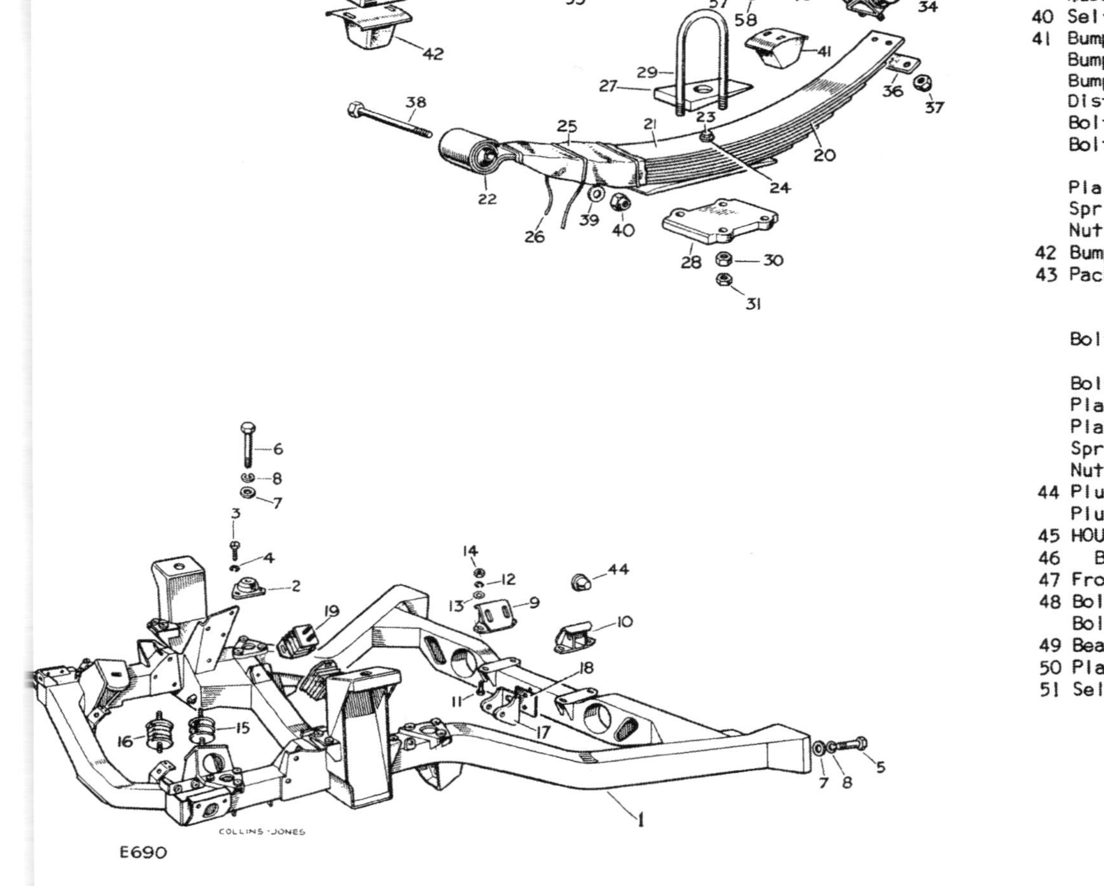

Plate Ref.	1 2 3 4 Description		Qty.	Part No.	Remarks
34	Plain washer	Fixing mounting to bracket on body	4	2251	
35	Self-locking nut (⅜" UNF)		4	252212	
36	Clamp plate	Fixing rear end of spring to rubber mounting	2	502841	
37	Self-locking nut (⅜" UNF)		4	252212	
38	Shackle pin (5½" long), spring to front brackets		2	522991	
39	Packing washer	For shackle pins	4	3895	Use with 5¼" shackle pin
	Distance piece		4	522992	Use with 5½" shackle pin
	Rubber washer		4	525405	
40	Self-locking nut (9/16" UNF)		2	252215	
41	Bump rubber for rear axle		2	503140	Mk I and Mk IA
	Bump rubber for rear axle		2	531885	Mk II and Mk III
	Bump rubber for rear axle		2	503140	Mk II and Mk III high suspension
	Distance piece for rear bump rubber		2	538063	
	Bolt (5/16" UNF × ¾" long)	Fixing bump rubber to body	4	255225	Mk I and Mk IA
	Bolt (5/16" × 1⅛" long)		4	255029	Mk II and Mk III high suspension
	Plain washer		4	2220	
	Spring washer		4	3075	
	Nut (5/16" UNF)		4	254831	
42	Bump rubber on body floor for rear axle		1	538267	
43	Packing block for use with early-type bump rubber		1	511039	Mk I and Mk IA Not required with 538267
	Bolt (¼" UNF × 1⅜" long) fixing packing block and rubber		2	256004	Alternatives
	Bolt (¼" UNF × ¾" long)	Fixing bump rubber to body floor	2	255207	
	Plain washer, small		2	3946	
	Plain washer, large		2	3821	
	Spring washer		2	3074	
	Nut (¼" UNF)		2	254810	
44	Plug for 1.3/16" dia. jacking tube		4	272537	Alternatives Check before ordering
	Plug for 1¼" dia. jacking tube		4	548386	
45	HOUSING FOR SPARE WHEEL		1	541711	
46	Bush for spare wheel housing		2	500800	
47	Front mounting bracket for spare wheel housing		2	351715	
48	Bolt (5/16" UNF × 1¼" long)	Fixing mounting bracket to housing at pivot point	2	256220	Mk I and Mk IA
	Bolt (5/16" UNF × ¾" long)		2	255226	Mk II and Mk III
49	Bearing sleeve		2	500799	
50	Plain washer		4	3830	
51	Self-locking nut (5/16" UNF)		2	252161	

*Asterisk indicates a new part which has not been used on any previous model.

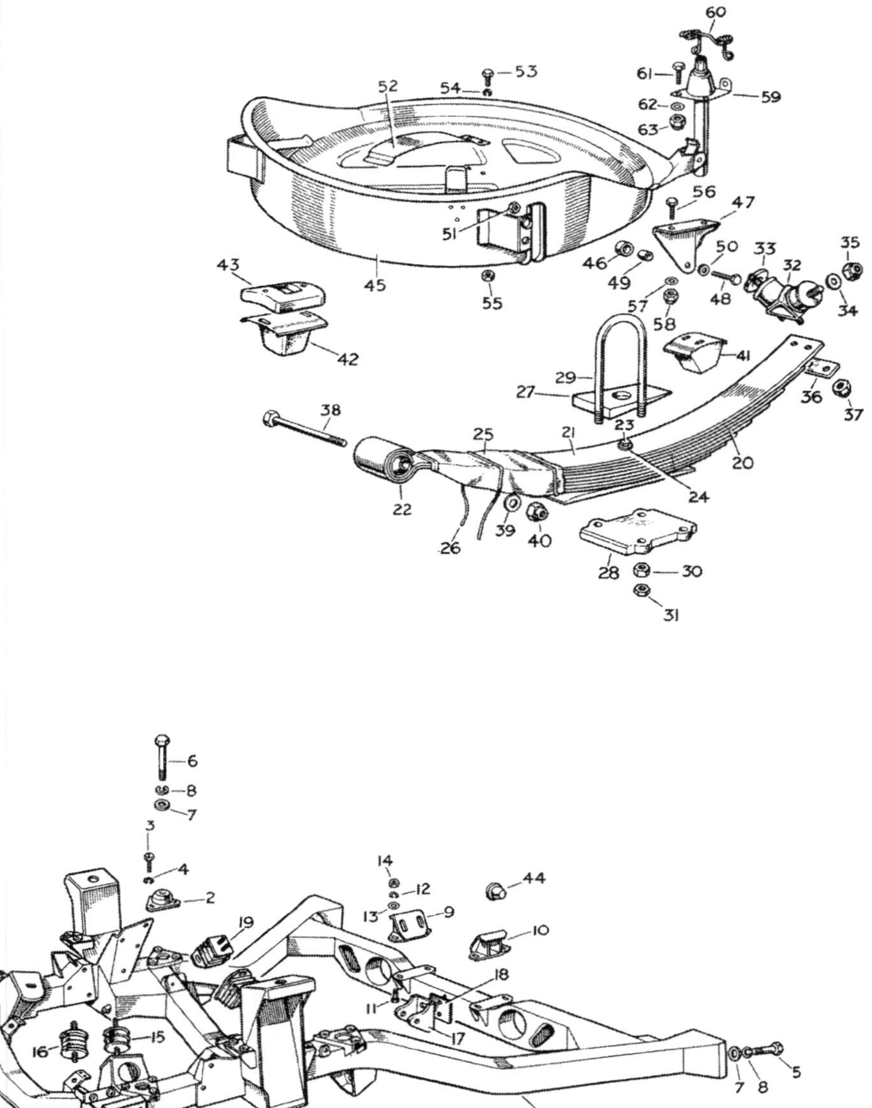

Plate Ref.	1 2 3 4 Description	Qty.	Part No.	Remarks
52	Ejection spring for spare wheel	1	501734	
53	Bolt (¼" UNF × ½" long) } Fixing spring to spare wheel carrier	2	255204	
54	Spring washer	2	3074	
55	Nut (¼" UNF)	2	254810	
56	Bolt (5/16" UNF × ¾" long) } Fixing mounting brackets to body	4	255226	Mk I and Mk IA
	Bolt (5/16" UNF × 1¼" long)	4	255030	Mk II
	Packing washer	8	2216	Mk II
57	Plain washer	8	3868	
58	Self-locking nut (5/16" UNF)	4	252161	
59	Spare wheel housing operating screw	1	541715	
60	Locking spring for spare wheel housing operating screw	1	524635	
61	Bolt (¼" UNF × ½" long) } Fixing screw and spring to boot floor	4	255204	
62	Plain washer	4	2213	
63	Self-locking nut (¼" UNF)	2	252160	
	Protection tube (PVC) for spare wheel retaining bracket	1	383283	
	Road wheel	5	537208	
	Tyre, 15" × 6.70", tubeless, Dunlop C41 'Gold Seal' or Avon HM, ribbed nylon, 4-ply	5	534349	Mk I and Mk IA
	Tyre, 15" × 6.70", Dunlop 'Roadspeed' or Avon 'Turbospeed', less tube	5	532555	Mk II and Mk III Alternatives
	Inner tube, 15" × 6.70"	5	535745	Mk II and Mk III Alternatives
	Tyre, 15" × 6.70", tubeless, Dunlop 'Roadspeed'	5	556842	Mk II and Mk III Alternatives
	Tyre, 15" × 6.70", tubeless, Avon Whitewall	5	530661	Optional Equipment
	Tyre, 15" × 6.70", tubeless, Dunlop C41 'Gold Seal' Whitewall	5	534350	Optional Equipment
	Tyre, 15" × 6.70", tubeless, Avon 6-ply, heavy-duty	5	505233	Optional Equipment
	Tyre, 15" × 6.70", tubeless, Dunlop C41 'Gold Seal' heavy-duty, 6-ply	5	534351	Optional Equipment
	Tyre, 15" × 6.70", tubeless, Dunlop C41 'Gold Seal' heavy-duty, 6-ply, Whitewall	5	534352	Optional Equipment
	Tyre, 15" × 7.10", tubeless, Dunlop or Avon	5	504030	Optional Equipment
	Tyre, 15" × 6.70", Dunlop 'Roadspeed' Whitewall, 4-ply	5	535744	Optional Equipment
	Inner tube for Dunlop 'Roadspeed' tyre	5	535745	Optional Equipment
	Tyre, 15" × 6.70", tubeless, Dunlop 'Roadspeed' Whitewall	5	556843	Optional Equipment
	Tyre, 15" × 6.70", tubeless, Dunlop C41	5	548163	New Zealand
	Valve complete for tubeless tyre, self-sealing type	5	267665	

*Asterisk indicates a new part which has not been used on any previous model.

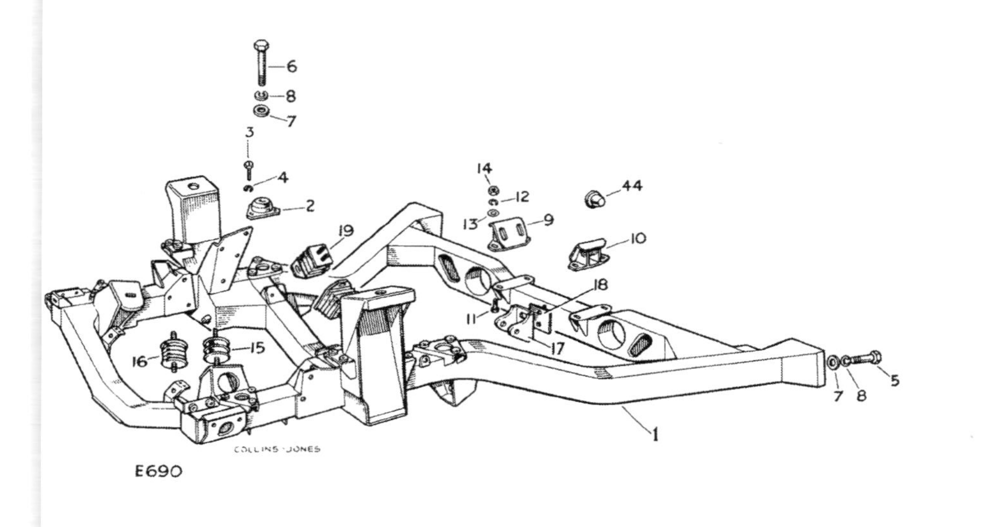

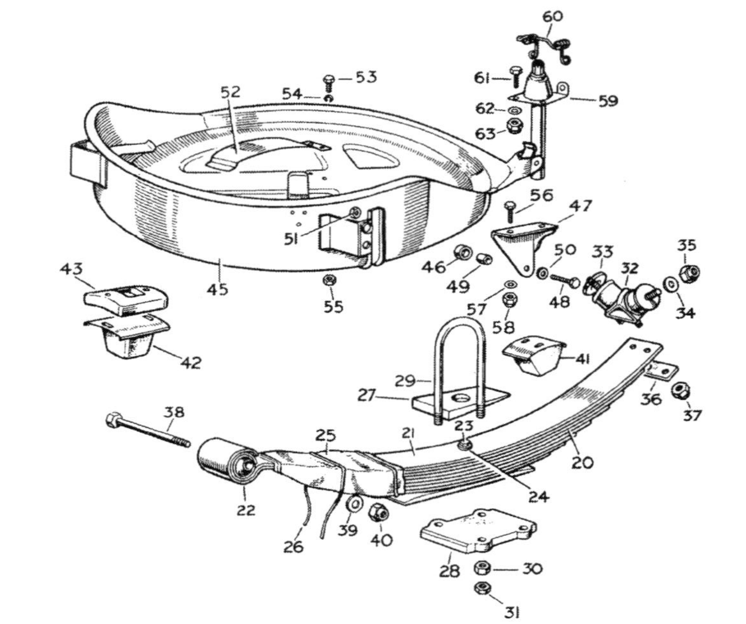

Plate Ref.	1 2 3 4	Description	Qty.	Part No.	Remarks
		Balance weight for road wheel, ½ oz	As reqd	503771	
		Balance weight for road wheel, 1 oz	As reqd	503772	
		Balance weight for road wheel, 1½ oz	As reqd	503773	
		Balance weight for road wheel, 2 oz	As reqd	503774	
		Balance weight for road wheel, 2½ oz	As reqd	503775	
		Balance weight for road wheel, 3 oz	As reqd	503776	
		Hub cover plate	4	277467	Mk I only
		Hub cover plate and wheel trim complete	4	513022	Mk IA, Mk II and Mk III
		Badge for hub cover plate	4	279324	
		Rubber washer } Fixing badge to hub cover plate	4	279064	
		Retaining ring } Fixing badge to hub cover plate	4	277468	
		Wheel trim	5	501708	Mk I only
		Clip for wheel trim	20	529074	Mk I only
		Nut for road wheel	20	542214	

*Asterisk indicates a new part which has not been used on any previous model.

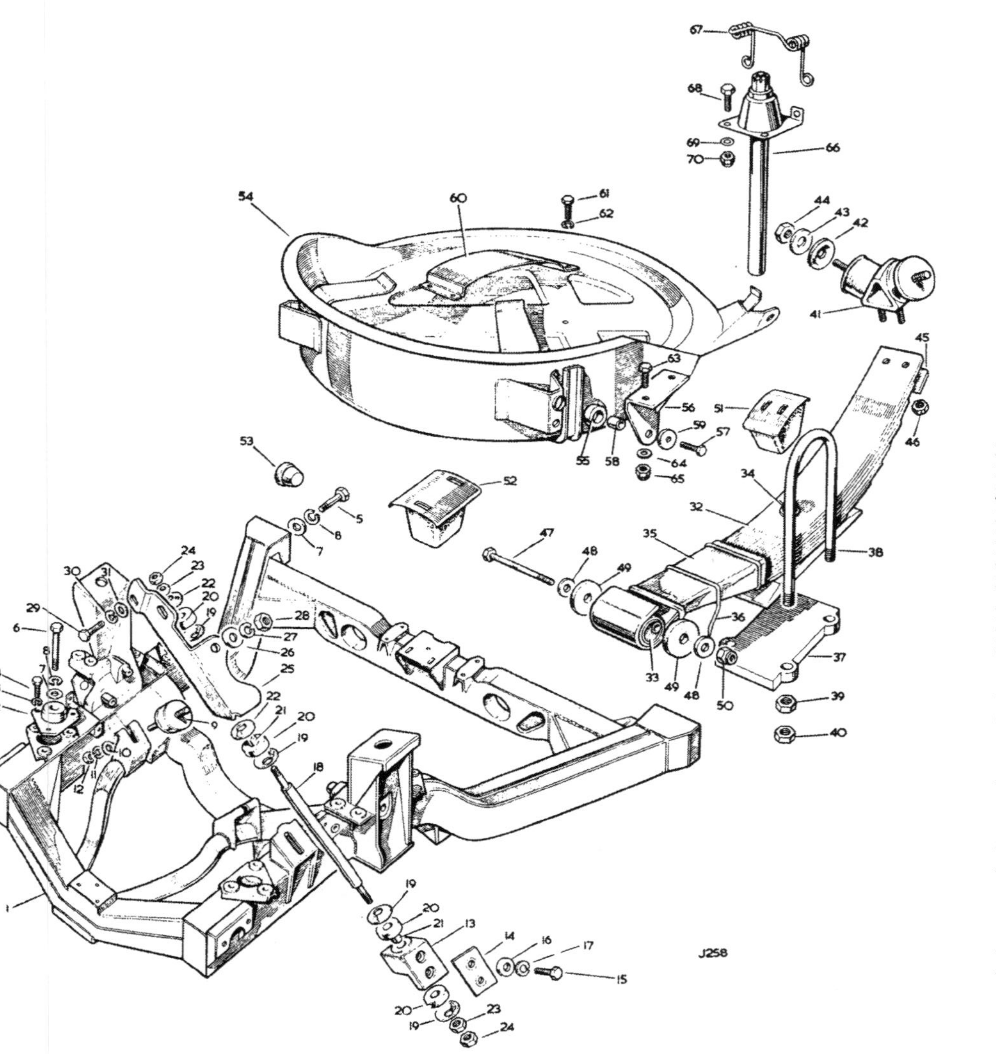

Plate Ref.	1 2 3 4 Description		Qty.	Part No.	Remarks
1	Chassis sub-frame complete		1	560296	
2	Mounting for chassis sub-frame		6	562349	
3	Spring washer	Fixing mounting to chassis sub-frame	17	3075	
4	Set bolt (5/16" UNF x¾" long)		17	253870	
	Set bolt (½" UNF x 3¼" long)	Fixing chassis sub-frame to body	2	256088	
5	Set bolt (½" UNF x 3½" long)		2	256289	
6	Set bolt (½" UNF x 4" long)		2	256091	
7	Plain washer		6	2222	
8	Spring washer		6	3078	
9	Engine mounting rubber, front		2	555494	
10	Plain washer	Fixing engine mounting rubbers to bracket and sub-frame	2	560322	
11	Spring washer		4	3076	
12	Nut (⅜" UNF)		4	254812	
	Packer		2	565600	
13	Bracket for engine tie bar		1	565412	
14	Clamp bracket		1	565410	
15	Bolt(⅜" UNF x 1" long)	Fixing tie bar bracket to chassis sub-frame	2	255247	
16	Plain washer		2	2211	
17	Spring washer		2	3076	
18	Engine tie bar		1	565365	
19	Cup washer	Fixing engine tie rod to bracket	4	240800	
20	Rubber bush		4	514280	
21	Distance tube		2	545544	
22	Guide washer		2	07015	
23	Nut (5/16" UNF)		2	254831	
24	Locknut (5/16" UNF)		2	254861	
25	Bracket at cylinder head for tie bar		1	603297	
26	Plain washer	Fixing tie bar bracket to cylinder head	1	3036	
27	Spring washer		1	3076	
28	Nut (⅜" UNF)		1	254812	
29	Bolt (⅜" UNC x 1⅛" long)	Fixing tie bar and lifting brackets to cylinder head	1	253048	
	Bolt (⅜" UNC x 1¼" long)		1	253049	
30	Plain washer		1	3036	
31	Spring washer		1	3076	
32	ROAD SPRING, REAR		2	553899	
	Main leaf for spring		2	600972	
33	Bush for spring		2	502709	
34	Centre dowel		2	256049	
	Washer for centre dowel		4	510940	
35	GREASE SLEEVE FOR REAR SPRING		4	530806	
36	Lace for grease sleeve		4	504436	
37	Bottom plate for spring		2	500770	
38	'U' bolt	Fixing rear spring to axle	4	510978	
39	Nut (7/16" UNF)		8	254823	
40	Locknut (7/16" UNF)		8	254873	
41	Rubber mounting for rear end of spring		2	516850	

*Asterisk indicates a new part which has not been used on any previous model.

J258

Plate Ref.	1 2 3 4 DESCRIPTION		Qty	Part No.	REMARKS
42	Packing washer	Fixing mounting to bracket on body	4	507085	
43	Plain washer		4	2251	
44	Self-locking nut (3/8" UNF)		4	252212	
45	Clamp plate	Fixing rear end of spring to rubber mounting	2	502841	
46	Self-locking nut (3/8" UNF)		2	252212	
47	Shackle pin	Fixing front end of spring to body	2	522991	
48	Distance piece		4	522992	
49	Rubber washer		4	525405	
50	Self-locking nut (5/16" UNF)		2	252215	
51	Bump rubber for rear axle		2	531885	
	Bolt (5/16" UNF x 3/4" long)	Fixing bump rubber to body	4	255225	
	Spring washer		4	3075	
	Plain washer		4	2220	
	Nut (5/16" UNF)		4	254831	
52	Bump rubber on body floor for rear axle		1	538267	
	Bolt (1/4" UNF x 3/4" long)	Fixing bump rubber to body floor	2	255207	
	Plain washer, small		2	3946	
	Plain washer, large		2	3821	
	Spring washer		2	3074	
	Nut (1/4" UNF)		2	254810	
53	Plug for jacking tube		4	548386	
54	HOUSING FOR SPARE WHEEL		1	541711	
55	Bush for spare wheel housing		2	500300	
56	Front mounting bracket for spare wheel housing ..		2	351715	
57	Bolt (5/16" UNF x 3/4" long)	Fixing mounting bracket to housing at pivot point	2	255226	
58	Bearing sleeve		2	500799	
59	Plain washer		4	3830	
	Self-locking nut		2	252161	
60	Ejection spring for spare wheel		1	501734	
61	Bolt (1/4" UNF x 1/2" long)	Fixing spring to spare wheel carrier	2	255204	
62	Spring washer		2	3074	
	Nut (1/4" UNF)		2	254810	
63	Bolt (5/16" UNF x 1¼" long)	Fixing mounting brackets to body	4	255030	
	Packing washer		8	2216	
64	Plain washer		8	3868	
65	Self-locking nut (5/16" UNF)		4	252161	
66	Spare wheel housing operating screw		1	541715	
67	Locking spring for spare wheel housing operating screw		1	524635	
68	Bolt (1/4" UNF x 1/2" long)	Fixing screw and spring to boot floor	4	255004	
69	Plain washer		4	3840	
70	Self-locking nut (1/4" UNF)		2	252160	
	Protection tube (PVC) for spare wheel retaining bracket..		1	383283	

*Asterisk indicates a new part which has not been used on any previous model.

Plate Ref.	1 2 3 4	Description	Qty.	Part No.	Remarks
		Road wheel, Rostyle-type	5	565544	
		Tyre, 15" x 6.70" Dunlop 'Roadspeed' or Avon 'Turbospeed', less tube	5	532555	Alternatives
		Inner tube, 15" x 6.70"	5	535745	Alternatives
		Tyre, 15" x 6.70", tubeless, Dunlop 'Roadspeed'	5	556842	Alternatives
		Tyre, 15" x 6.70" Avon Turbospeed tubeless	5	556842	Alternatives
		Valve complete for tubeless tyre	5	548133	Alternatives
		Balance weight for road wheel, ½ oz	As reqd	503771	Early models
		Balance weight for road wheel, 1 oz	As reqd	503772	Early models
		Balance weight for road wheel, 1½ oz	As reqd	503773	Early models
		Balance weight for road wheel, 2 oz	As reqd	503774	Early models
		Balance weight for road wheel, 2½ oz	As reqd	503775	Early models
		Balance weight for road wheel, 3 oz	As reqd	503776	Early models
		Balance weight for road wheel 15 grams	As reqd	578337	Self-adhesive type late models
		Balance weight for road wheel 20 grams	As reqd	578338	Self-adhesive type late models
		Balance weight for road wheel 30 grams	As reqd	578339	Self-adhesive type late models
		Balance weight for road wheel 40 grams	As reqd	578340	Self-adhesive type late models
		Balance weight for road wheel 50 grams	As reqd	578341	Self-adhesive type late models
		WHEEL TRIM ASSEMBLY	4	565520	
		Motif for wheel trim centre	4	562127	
		Nut for Rostyle-type road wheel	20	570461	

*Asterisk indicates a new part which has not been used on any previous model.

ACCELERATOR AND FIXINGS, MK1 MK1A AND MK 11 3 LITRE MODELS WITH SUFFIX LETTERS 'A' and 'B'

F597

ACCELERATOR AND FIXINGS, MK I, MK IA AND MK II 3 LITRE MODELS WITH SUFFIX LETTERS 'A' AND 'B'

Plate Ref.	Description 1 2 3 4	Qty.	Part No.	Remarks
1	Shaft and lever for accelerator	1	500583	RH Stg Mk I and Mk IA
	Shaft and lever for accelerator	1	534018	RH Stg } Mk II
	Shaft and lever for accelerator	1	534090	LH Stg } 4-speed
	Accelerator shaft	1	501731	LH Stg Mk I and Mk IA
	Accelerator shaft	1	534088	LH Stg Mk II 4-speed
	Shaft and lever for accelerator	1	502785	LH Stg Automatic, 1959-60
	Shaft and lever for accelerator	1	520150	LH Stg 1961 and Mk IA Automatic
	Shaft and lever for accelerator	1	534163	LH Stg Mk II Automatic
	Accelerator pedal lever complete	1	502005	LH Stg Mk I and Mk IA
	Accelerator pedal lever complete	1	534089	LH Stg Mk II
2	Bearing bracket for accelerator shaft	2	500580	1 off on LH Stg
	Bearing bracket for accelerator shaft, pedal end	1	504040	LH Stg
3	Grommet for bearing	2	500987	
4	Bearing, nylon, for accelerator shaft	2	500581	
5	Felt washer for accelerator shaft	2	231293	1 off on Automatic LH Stg
6	Stop clip for accelerator shaft	1	231294	} RH Stg
7	Bolt (1/4" UNF x 1 1/8" long) } Fixing stop clip to accelerator shaft	1	256200	} RH Stg
8	Nut (1/4" UNF) }	1	254810	} RH Stg
	Spring dowels fixing accelerator lever and pedal lever to shaft	2	534021	Mk II
9	Grommet for accelerator pedal shaft	1	502033	
10	Bolt (1/4" UNF x 9/16" long) } Fixing accelerator pedal shaft to dash	4	255005	} 2 off on Automatic LH Stg
11	Spring washer }	4	3074	} 2 off on Automatic LH Stg
12	Nut (1/4" UNF) }	4	254810	} 2 off on Automatic LH Stg
	Bolt (5/16" UNFx9/16" long) }	1	255024	} LH Stg
	Spring washer }	1	3075	} LH Stg
	Nut (5/16" UNF) }	1	254831	} LH Stg
13	Pedal for accelerator	1	548064	
14	Screw (1/4" UNF x 1 1/4" long) } Fixing pedal to floor	2	252115	
15	Countersunk washer }	2	76898	
16	Tapped plate }	1	352567	
17	Screw (1/4" UNF x 1/2" long) fixing tapped plate	1	78230	
18	Link, accelerator pedal to lever	1	517378	
19	Spring washer } Fixing link to pedal	2	3073	
20	Nut (2 BA) }	2	2247	
21	Plain washer } Fixing link to accelerator lever	2	3946	
22	Split pin }	1	2388	
	Bracket for accelerator pedal stop	1	502949	LH Stg
23	Bracket for accelerator pedal spring loaded stop	1	502754	Automatic RH Stg
24	Bolt (1/4" UNF x 3/8" long) } Fixing bracket to floor	3	255206	} Except RH Stg with manual gear change
25	Plain washer }	6	3840	} Except RH Stg with manual gear change
26	Spring washer }	3	3074	} Except RH Stg with manual gear change
27	Nut (1/4" UNF) }	3	254810	} Except RH Stg with manual gear change
	Accelerator pedal stop	1	502967	} LH Stg models with manual gear change except with overdrive
	Adjusting screw for stop	1	274476	} LH Stg models with manual gear change except with overdrive
	Locknut (5/16" UNF) for adjusting screw	1	254851	} LH Stg models with manual gear change except with overdrive
	Set bolt (10 UNF x 1/2" long) } Fixing pedal stop to bracket	2	257017	} LH Stg models with manual gear change except with overdrive
	Plain washer }	2	3902	} LH Stg models with manual gear change except with overdrive
	Spring washer }	2	3073	} LH Stg models with manual gear change except with overdrive

*Asterisk indicates a new part which has not been used on any previous model.

ACCELERATOR AND FIXINGS, MK1 MK1A AND MK 11 3 LITRE MODELS
WITH SUFFIX LETTERS 'A' and 'B'

1507

F597

Plate Ref.	1 2 3 4 Description	Qty.	Part No.	Remarks
28	Housing for accelerator pedal spring-loaded stop	1	507928	Mk I and Mk IA Automatic
	Housing for accelerator pedal spring-loaded stop	1	532276	Mk II Automatic
29	Plunger to pedal stop housing	1	507927	Mk I and Mk IA Automatic
	Plunger for pedal stop housing	1	532273	Mk II Automatic
30	Spring for pedal stop	1	510467	1959-60 Automatic
	Spring for pedal stop	1	520277	1961 Mk I and Mk IA Automatic
	Spring, short, for pedal stop	1	523145	Mk II Automatic
	Ball for pedal stop	2	3739	Mk II Automatic
	Spring, long, for pedal stop	1	523144	Mk II Automatic
31	Screw(10 UNF x ¾" long) fixing housing	2	78371	
	LEVER ASSEMBLY ON ACCELERATOR SHAFT	1	502338	Mk I and Mk IA 4-speed and LH Stg Automatic
	LEVER ASSEMBLY ON ACCELERATOR SHAFT	1	534019	Mk II 4-speed and LH Stg Automatic
32	LEVER ASSEMBLY ON ACCELERATOR SHAFT	1	502486	RH Stg Mk I and Mk IA Automatic
	LEVER ASSEMBLY ON ACCELERATOR SHAFT	1	534114	Mk II Automatic RH Stg
33	Ball end for lever	1	1481	
34	Cotter — Fixing lever to accelerator shaft	2	511946	1 off on RH Stg 4-speed models
35	Plain washer — Fixing lever to accelerator shaft	2	2203	1 off on RH Stg 4-speed models
36	Nut (¼" UNF) — Fixing lever to accelerator shaft	2	254810	1 off on RH Stg 4-speed models
	Countershaft for accelerator	1	500575	Mk I and Mk IA
37	Countershaft for accelerator	1	525191	Mk II
38	Felt washer for countershaft	2	231293	
	Stop clip for accelerator countershaft	1	231294	
	Bolt (¼" UNF x 1⅛" long) — Fixing stop clip to countershaft	1	256200	
	Nut (¼" UNF) — Fixing stop clip to countershaft	1	254810	
	Bracket for countershaft on wing valance	1	502109	Mk I and Mk IA
39	Bracket for countershaft on wing valance	1	535959	Mk II
40	Rubber grommet — For counterhsaft bracket	1	500987	
41	Bearing, nylon — For counterhsaft bracket	1	500581	
	Backing plate — For counterhsaft bracket	1	502110	Mk I and Mk IA
42	Bolt (¼" UNF x ¾" long) — Fixing bracket and backing plate to RH wing valance	2	255207	
43	Plain washer — Fixing bracket and backing plate to RH wing valance	4	3840	
44	Spring washer — Fixing bracket and backing plate to RH wing valance	2	3074	
45	Nut (¼" UNF) — Fixing bracket and backing plate to RH wing valance	2	254810	
46	LEVER ASSEMBLY TO PEDAL SHAFT	1	242315	
47	Ball end for lever	1	1481	
	LEVER ASSEMBLY FOR THROTTLE SPRING	1	237610	Mk I and Mk IA 2 off on models with overdrive
	Ball end for lever	1	1481	
48	LEVER ASSEMBLY, CARBURETTER END	1	230937@	
	LEVER ASSEMBLY, CARBURETTER END	1	524962@@	
49	Ball end for lever	1	1481	

@ Cars numbered up to 625101045, 626100262, 628100109, 630100840, 631100353, 633100078

@@ Cars numbered from 625101046, 626100263, 628100110, 630100841, 631100354, 633100079 onwards

*Asterisk indicates a new part which has not been used on any previous model.

F597

Plate Ref.	1 2 3 4 Description	Qty.	Part No.	Remarks
50	Bolt (¼" UNF x 1⅜" long) } Fixing levers to countershaft	3	256200	4 off on models with overdrive
51	Nut (¼" UNF)	3	254810	
	Bracket for accelerator counterhsaft on cylinder head	1	500573	Mk I and Mk IA without overdrive
	Bracket for countershaft and switch	1	500574	Mk I and Mk IA with overdrive
	Bracket for accelerator countershaft on cylinder head	1	525185	Mk II Automatic
52	Bracket for countershaft and overdrive switch	1	536470	Mk II 4-speed
53	Grommet for bearing	1	500987	
54	Bearing, nylon, for countershaft	1	522839	
	LEVER ASSEMBLY FOR COUNTERSHAFT	1	536472	} Mk II 4-speed
55	Lever	1	536473	
56	Ball end	1	1481	
57	Cam for shaft	1	536475	
58	Shaft	1	536474	
59	Bolt (¼" UNF x 1⅜" long) } Fixing lever to shaft	1	256000	Mk II 4-speed
60	Nut (¼" UNF)	1	254810	
	Pin fixing cam to shaft	1	250343	
61	Bearing for shaft	2	506450	
62	Washer between bearing and cam	1	4075	
	Washer between bearing and lever	1	2550	
63	Set bolt (¼" UNF x ¾" long) fixing bracket to cylinder head	1	253007	
64	Spring washer } Fixing b racket to cylinder head or inlet manifold	3	3074	
	Nut (¼" UNF)	3	254810	
	LEVER ASSEMBLY FOR ROTARY SWITCH	1	273965	} Models with overdrive
	Ball end for rotary switch lever	1	273964	
	Plain washer } Fixing ball end to rotary switch lever	2	3902	Mk I and Mk IA
	Spring washer	1	3073	
	Nut (2 BA)	1	2247	
65	LEVER ASSEMBLY ON COUNTERSHAFT FOR OVERDRIVE SWITCH CONTROL ROD	1	237610	
66	Ball end for lever	1	1481	Mk II
67	Bolt (¼" UNF x 1⅜" long) } Fixing lever	1	255010	
68	Nut (¼" UNF)	1	254810	
	Control rod, switch to lever	1	273092	Mk I and Mk IA
69	Control rod, switch to lever	1	241938	Mk II
70	Ball socket } For control rod	2	531324	
71	Nut (2 BA)	2	2247	
72	Anchor for throttle return spring	1	237612	} Mk I and Mk IA
73	Bolt (2 BA x ½" long) } Fixing anchor to dash	2	234603	
74	Spring washer	2	3073	Mk I and Mk IA
75	Nut (2 BA)	2	2247	
	Return spring for throttle	1	234688	
76	Return spring for throttle	1	535941	Mk II
	Link for spring	1	502477	Mk I and Mk IA
	Control rod, accelerator lever to countershaft	1	241938	
77	Control rod, accelerator lever to countershaft	1	535999	Mk II

*Asterisk indicates a new part which has not been used on any previous model.

F597

Plate Ref.	1 2 3 4 Description	Qty.	Part No.	Remarks
	Control rod, countershaft to carburetter	1	273090@	Mk I and Mk IA 4-speed
78	Control rod, countershaft to carburetter	1	524972@@	non- overdrive and Mk II 4-speed
	Ball joint for link and rod	5	531324	3 off on Automatic and
	Locknut (2 BA) for ball joint	5	2247	overdrive models
	CONTROL ROD COUNTERSHAFT TO CARBURETTER	1	524968	All Automatic and
79	Ball joint } For	2	531324	Mk I and Mk IA
80	Locknut (2 BA) } control rod	2	2247	overdrive models
81	ACCELERATOR SHAFT RELAY EXTENSION COMPLETE	1	504447	1959-60 — RH Stg Auto-matic
	ACCELERATOR SHAFT RELAY ASSEMBLY	1	520058	1961 Mk I and Mk IA — RH Stg Auto-matic
	ACCELERATOR SHAFT RELAY ASSEMBLY	1	534113	Mk II — RH Stg Auto-matic
82	Ball end	1	1481	RH Stg Auto-matic
	Accelerator shaft and relay lever	1	502785	1959-60 — LH Stg Automatic
	Accelerator shaft relay and assembly	1	520150	1961 Mk I and Mk IA — LH Stg Automatic
	Accelerator shaft relay assembly	1	534163	Mk II — LH Stg Automatic
83	Bearing bracket for extension shaft	1	504040	RH Stg
84	Bearing, nylon, for bracket	1	500581	RH Stg
85	Rubber grommet for bearing	1	500987	RH Stg
86	Bolt (5/16" UNF x 9/16" long) } Fixing bearing bracket to dash	3	255024	RH Stg
87	Spring washer } Fixing bearing bracket to dash	3	3075	RH Stg
88	Nut (5/16" UNF) } Fixing bearing bracket to dash	3	254831	RH Stg
89	Felt washer for extension shaft	1	231293	RH Stg
90	Stop clip for extension shaft	1	231294	RH Stg
91	Bolt (¼" UNF x 1⅛" long) } Fixing stop clip to shaft	1	256200	RH Stg
92	Nut (¼" UNF) } Fixing stop clip to shaft	1	254810	RH Stg
	GEAR CONTROL RELAY BOX ASSEMBLY	1	511344	1959-60 — Auto-matic
	GEAR CONTROL RELAY BOX ASSEMBLY	1	520059	1961 Mk I only — Auto-matic
	GEAR CONTROL RELAY BOX ASSEMBLY	1	526334	Mk IA and Mk II — Auto-matic
93	ACCELERATOR RELAY BOX COMPLETE	1	502771	Auto-matic
94	Bush for relay box	2	238793	Mk I — Auto-matic
95	MANUAL CONTROL LEVER COMPLETE	1	505705	Auto-matic
96	Bearing for control lever	1	238793	Auto-matic
97	Gearbox control lever, large	1	501688%	Auto-matic
	Gearbox dog complete	1	511343	LH Stg and %% RH Stg — Auto-matic
98	Ball end	1	1481	Auto-matic
99	ACCELERATOR DOG COMPLETE	1	511877	1959-60 — Auto-matic
	ACCELERATOR DOG COMPLETE	1	520055	1961 onwards — Auto-matic
100	Bearing for lever	1	501652	Auto-matic
101	Shaft for relay box	1	501655	Auto-matic
102	Pin fixing shaft to gearbox control lever	1	50439	Auto-matic
	Spring for relay shaft	1	527366	Mk IA and Mk II — Auto-matic
103	Bolt (¼" UNF x 9/16" long) } Fixing relay box to dash	4	255205	Auto-matic
104	Plain washer } Fixing relay box to dash	4	3840	Auto-matic
105	Spring washer } Fixing relay box to dash	4	3074	Auto-matic
106	Nut (¼" UNF) } Fixing relay box to dash	4	254810	Auto-matic
107	Control rod, accelerator relay box	1	512190	1959-60 — Auto-matic
	Control rod, accelerator to relay box	1	520054	1961 onwards — Auto-matic
	Adjuster for rod	1	277778	Auto-matic

@ Cars numbered up to 625101045, 626100262, 628100109, 630100840, 631100353, 63310078
@@Cars numbered from 625101046, 626100263, 628100110, 630100841, 631100354, 63310079 onwards
% Cars numbered up to 630000622, 631000150
%%Cars numbered from 630000623, 631000151 onwards

*Asterisk indicates a new part which has not been used on any previous model.

ACCELERATOR AND FIXINGS, MK1 MK1A AND MK 11 3 LITRE MODELS WITH SUFFIX LETTERS 'A' and 'B'

1513

F597

ACCELERATOR AND FIXINGS, MK I, MK IA AND MK II 3 LITRE MODELS WITH SUFFIX LETTERS 'A' AND 'B'

1514

Plate Ref.	Description (1 2 3 4)	Qty.	Part No.	Remarks
	Ball end (For control rod)	1	233109	Automatic
	Locknut (For control rod)	1	233108	Automatic
108	Ball end for rod	2	531324	1 off with 512190; Automatic
109	Locknut for ball end	2	2247	1 off with 512190; Automatic
110	Control cable for second speed lock	1	504145	1959-60; Automatic
	Manual control	1	520151	1961 Mk I only; Automatic
111	Clip fixing control cable dash	1	505789	Automatic
112	Bolt (¼" UNF x ½" long) (Fixing clip to dash)	1	255204	Automatic
	Drive screw (Fixing clip to dash)	1	77789	Automatic
113	Spring washer (Fixing clip to dash)	1	3074	Automatic
114	Nut (¼" UNF) (Fixing clip to dash)	1	254810	Automatic
115	Bolt (10 UNF x ½" long) (Fixing clip to cable)	1	257017	Automatic
116	Spring washer (Fixing clip to cable)	1	3073	Automatic
	Plain washer (Fixing clip to cable)	2	3885	Automatic
117	Nut (10 UNF) (Fixing clip to cable)	1	257023	Automatic
118	Grommet in dash for control cable	1	510779	Automatic
119	Overriding spring on cable	1	504417	Automatic
	Bolt (8 UNC) (on end of cable retaining overriding spring)	1	257307	Automatic
	Nipple (on end of cable retaining overriding spring)	1	356765	Automatic
120	Return spring for second speed lock control lever	1	504078	Automatic
121	Anchor plate for return spring	1	504076	Automatic
122	Anchor pin (Fixing anchor plate to second speed lock control lever)	1	504077	Mk I; Automatic
123	Plain washer (Fixing anchor plate to second speed lock control lever)	2	2203	Automatic
124	Control rod, relay box to gearbox	1	504453	1959-60; Automatic
	Control rod, relay box to gearbox	1	520061	1961 onwards; Automatic
125	Ball joint (For control rod)	2	531324	Automatic
126	Locknut (2 BA) (For control rod)	2	2247	Automatic
127	Ball end	1	273964	Automatic
128	Spring washer (Fixing ball end to lever on gearbox)	1	3073	Automatic
129	Nut (2 BA) (Fixing ball end to lever on gearbox)	1	2247	Automatic
130	Spring for governor control lever	1	514535	Automatic
	Split pin fixing spring to lever	1	4342	1961 onwards; Automatic
	Split pin fixing spring to gearbox mounting bracket	1	4286	Automatic

*Asterisk indicates a new part which has not been used on any previous model.

ACCELERATOR AND FIXINGS, MK II MODELS WITH SUFFIX LETTER 'C' AND MK III MODELS

1515

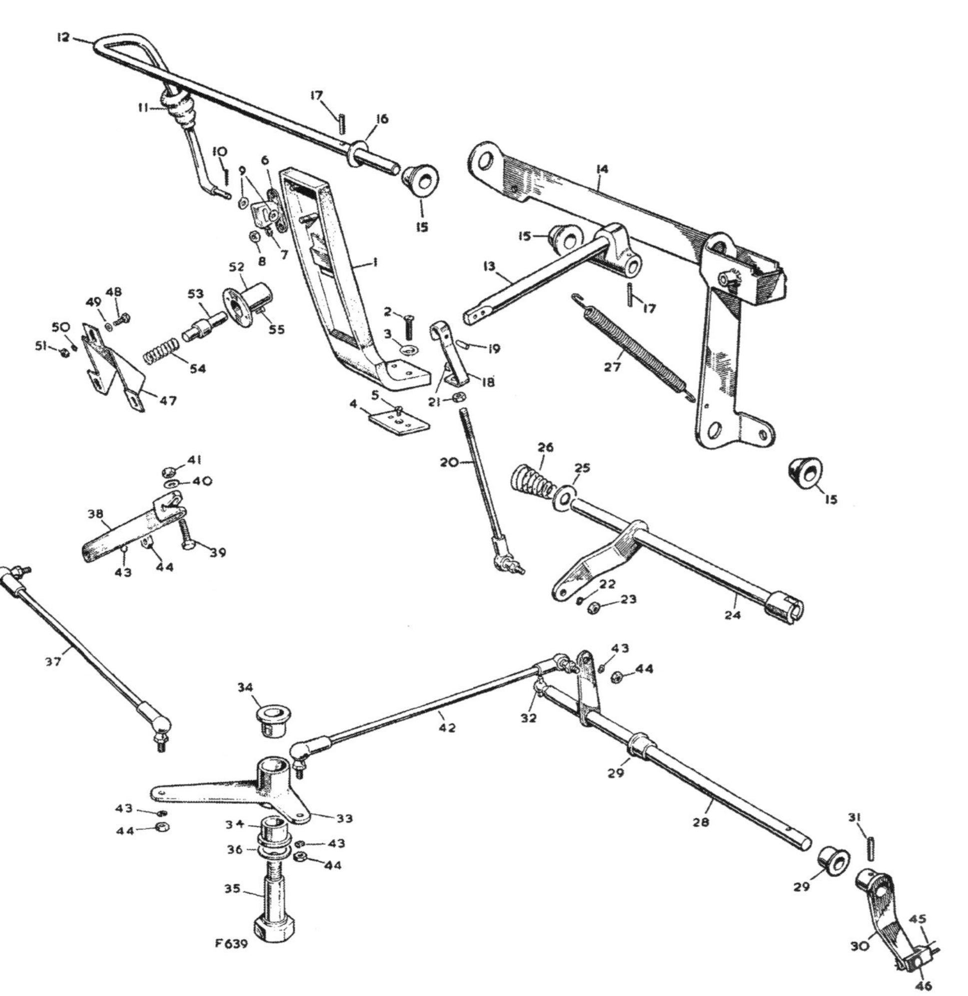

Plate Ref.	1 2 3 4 Description	Qty.	Part No.	Remarks
1	Accelerator pedal	1	273428	
2	Screw ($\frac{1}{4}$" UNF x 1$\frac{1}{8}$" long)	2	252114	
3	Countersunk washer	2	76898	
4	Tapped plate	1	352567	
5	Screw ($\frac{1}{4}$" UNF x $\frac{1}{2}$" long) fixing tapped plate	1	78230	
6	Link, accelerator pedal to lever	1	517378	
7	Spring washer } Fixing link to pedal	2	3073	
8	Nut (2 BA) } Fixing link to pedal	2	2247	
9	Plain washer } Fixing link to accelerator lever	2	3946	
10	Split pin } Fixing link to accelerator lever	1	2388	
11	Grommet for accelerator lever	1	502033	
12	Accelerator shaft and pedal lever	1	542789	RH Stg
	Accelerator pedal lever	1	534089	LH Stg
13	Accelerator lever on pedal lever	1	542748	RH Stg
	Accelerator lever	1	542750	LH Stg
14	Mounting bracket for accelerator shaft	1	542742	RH Stg
	Mounting bracket for accelerator shaft	1	542743	LH Stg
15	Bearing for mounting bracket	3	545571	
16	Plain washer for shaft	1	4426	
17	Spring dowel fixing shaft	2	534021	
	Bolt (5/16" UNF x $\frac{3}{4}$" long) } Fixing mounting bracket and clutch pedal bracket to dash	2	255225	
	Bolt (5/16" UNF x 9/16" long) } Fixing mounting bracket and clutch pedal bracket to dash	2	255024	3 off on LH Stg
	Spring washer } Fixing mounting bracket and clutch pedal bracket to dash	3	3075	
	Nut (5/16" UNF) } Fixing mounting bracket and clutch pedal bracket to dash	3	254831	
18	Stirrup for control lever	1	542781	
19	Spiral pin fixing stirrup to lever	1	542827	
20	Adjustable control rod	1	542782	
21	Locknut for adjustable control rod	2	257023	
22	Spring washer } Fixing control rod to extension shaft lever	1	3073	
23	Nut (10 UNF) } Fixing control rod to extension shaft lever	1	257023	
24	Extension shaft	1	545983	RH Stg
	Extension shaft	1	545982	LH Stg
25	Plain washer } For extension shaft	1	4514	
26	Conical spring } For extension shaft	1	542861	
27	Return spring for pedal shaft lever	1	545572	
	Bracket for overdrive switch	1	545957	4-speed
	Bolt ($\frac{1}{4}$" UNC x $\frac{5}{8}$" long) } Fixing bracket to flywheel housing	2	253006	4-speed
	Spring washer } Fixing bracket to flywheel housing	2	3074	4-speed
	Cam and lever assembly for switch	1	542763	4-speed
	Accelerator cross-shaft in flywheel housing	1	542758	RH Stg 4-speed
	Accelerator cross-shaft in flywheel housing	1	542759	LH Stg 4-speed
28	Accelerator cross-shaft in flywheel housing	1	542756	RH Stg Automatic
	Accelerator cross-shaft in flywheel housing	1	542757	LH Stg Automatic
29	Bush for cross-shaft	2	511127	
30	Lever for cross-shaft	1	542760	Except Mk III Automatic
	Lever for cross-shaft	1	553813	Mk III Automatic
31	Spring dowel fixing lever	1	534021	
32	Spiral pin for cross-shaft	1	542783	

*Asterisk indicates a new part which has not been used on any previous model.

ACCELERATOR AND FIXINGS, MK II MODELS WITH SUFFIX LETTER 'C' AND MK III MODELS

1517

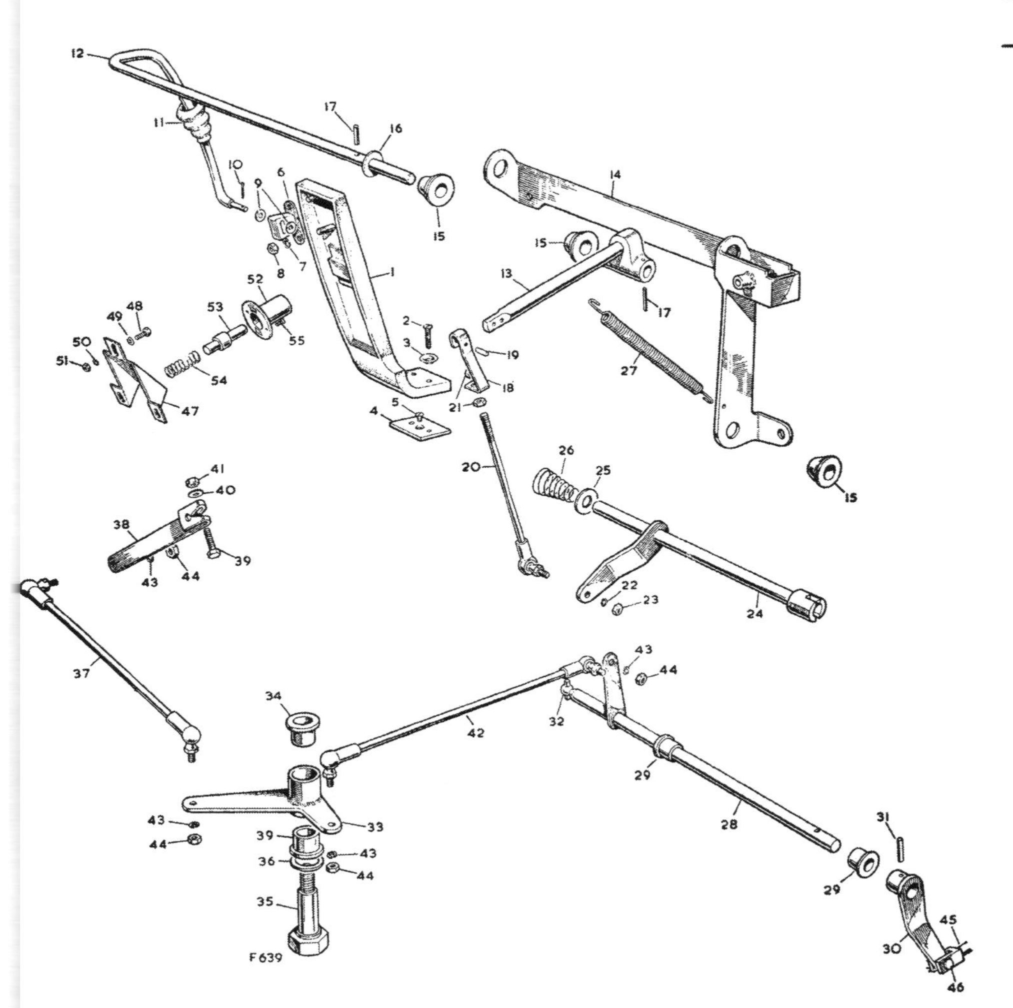

Plate Ref.	1 2 3 4 Description		Qty.	Part No.	Remarks
33	BELL CRANK LEVER AND BEARINGS ASSEMBLY		1	542777	
34	Bearing for bell crank		2	238793	
35	Centre pin for bell crank		1	542776	
36	Plain washer		1	3899	
37	Control rod assembly, carburetter to bell crank		1	542778	
38	Throttle lever for carburetter		1	542786	Except Mk III Automatic
	Throttle lever for carburetter		1	557343	Mk III Automatic
39	Bolt (10 UNF x 7/8" long)	Fixing lever to carburetter	1	257021	
40	Plain washer	Fixing lever to carburetter	1	2874	
41	Nut (10 UNF)	Fixing lever to carburetter	1	257023	
42	CONTROL ROD ASSEMBLY, BELL CRANK TO CROSS SHAFT		1	542779	Except Mk III Automatic
	CONTROL ROD ASSEMBLY, BELL CRANK TO CROSS-SHAFT		1	557109	Mk III Automatic
	Rod	For control rod assembly	1	557110	Mk III Automatic
	Stirrup	For control rod assembly	1	557113	Mk III Automatic
	Spring	For control rod assembly	1	557111	Mk III Automatic
	Plain washer	For control rod assembly	1	2874	Mk III Automatic
	Spring dowel	For control rod assembly	1	557114	Mk III Automatic
	Clevis pin	Fixing control rod assembly	1	557112	Mk III Automatic
	Split pin	Fixing control rod assembly	1	2389	Mk III Automatic
43	Spring washer	Fixing control rods to levers	4	3073	
44	Nut (10 UNF)	Fixing control rods to levers	4	257023	
45	Control cable, cross-shaft to gearbox		1	542785	Mk II Automatic
46	Clevis pin fixing cable to levers on cross-shaft and gearbox		2	542784	Mk II Automatic
	Split pin		2	2388	Mk II Automatic
47	Bracket for accelerator pedal stop		1	502754	RH Stg
48	Bolt (1/4" UNF x 5/8" long)	Fixing bracket to body floor	3	255206	
49	Plain washer	Fixing bracket to body floor	3	3840	
50	Spring washer	Fixing bracket to body floor	3	3074	
51	Nut (1/4" UNF)	Fixing bracket to body floor	3	254810	
	Accelerator pedal stop		1	502967	4-speed
	Adjusting screw		1	274476	4-speed
	Locknut (5/16" UNF)		1	254851	4-speed
	Set bolt (10 UNF x 1/2" long)	Fixing pedal stop to bracket	2	257017	4-speed
	Spring washer	Fixing pedal stop to bracket	2	3073	4-speed
	Plain washer	Fixing pedal stop to bracket	2	3902	4-speed
52	Housing for accelerator pedal spring-loaded stop		1	507928	Automatic
53	Plunger for pedal stop housing		1	507927	Automatic
54	Spring for pedal stop		1	520277	Automatic
55	Screw (10 UNF x 3/4" long) fixing housing		2	78371	Automatic

*Asterisk indicates a new part which has not been used on any previous model.

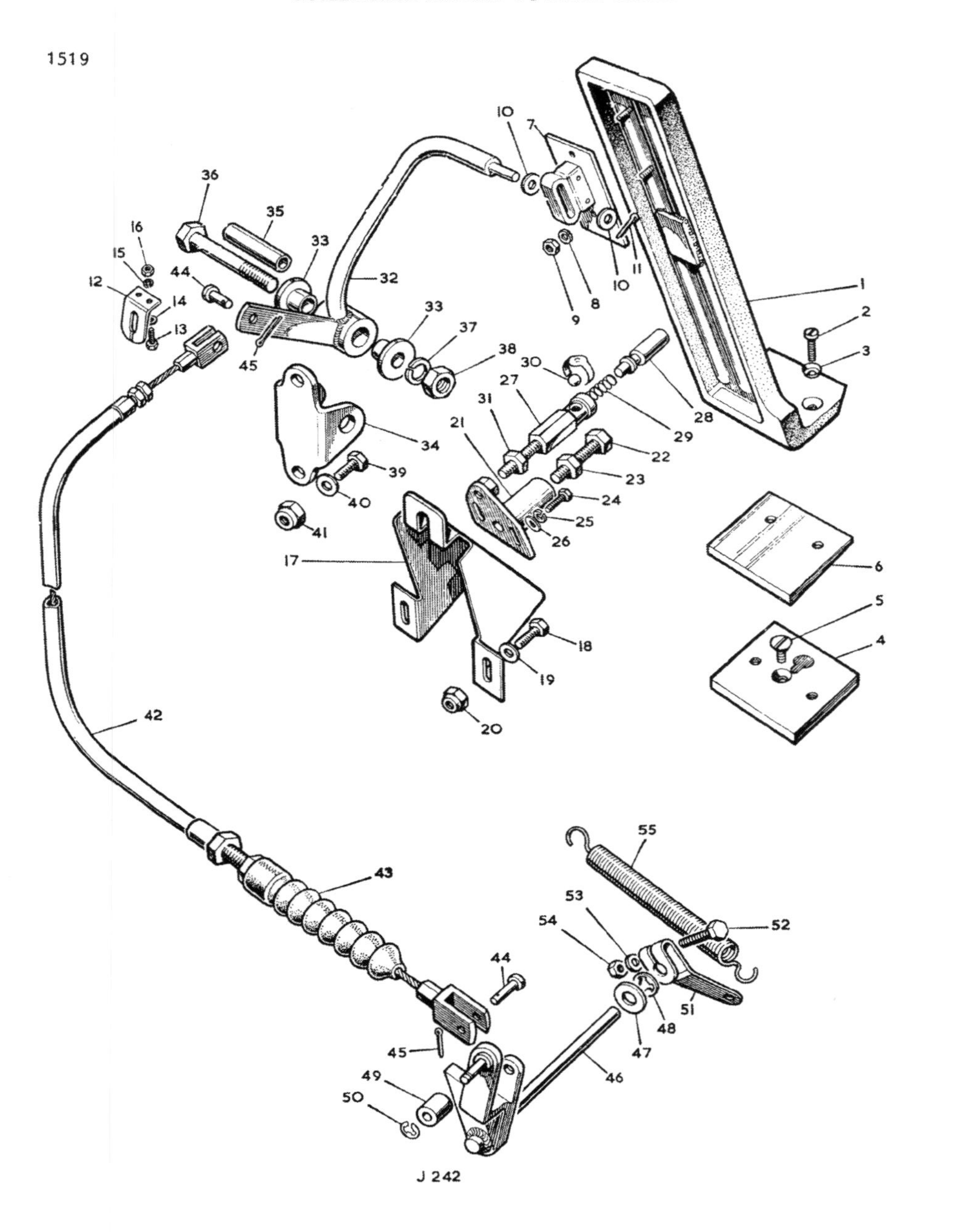

Plate Ref.	1 2 3 4 Description		Qty.	Part No.	Remarks
1	Accelerator pedal		1	548064	
2	Screw (¼" UNF x 1⅛" long)	Fixing pedal to floor	2	252114	
3	Countersunk washer		2	76898	
4	Tapped plate		1	352567	
5	Screw (¼" UNF x ½" long) fixing tapped plate		1	78230	
6	Packing for accelerator pedal		1	562463	
7	Link, accelerator pedal to lever		1	562464	
8	Spring washer	Fixing link to pedal	2	3073	
9	Nut (2 BA)		2	2247	
10	Plain washer	Fixing link to accelerator lever	2	3946	
11	Split pin		1	2388	
12	Abutment bracket		1	560354	
13	Bolt (¼" UNF x ½" long)	Fixing abutment bracket to parcel shelf	2	255204	
14	Plain washer		2	3467	
15	Spring washer		2	3074	
16	Nut (¼" UNF)		2	254810	
17	Bracket for accelerator pedal stop		1	562462	
18	Bolt (¼" UNF x ⅝" long)	Fixing bracket to body floor	3	255206	
19	Plain washer		3	3840	
20	Self-locking nut (¼" UNF)		3	252160	
21	Accelerator pedal stop		1	562469	
22	Adjusting screw		1	274476	
23	Locknut (5/16" UNF)		1	254861	
24	Set bolt (10 UNF x ½" long)	Fixing pedal stop to bracket	2	257017	
25	Spring washer		2	3073	
26	Plain washer		2	3851	
27	PEDAL DETENT ASSEMBLY		1	562472	
28	Plunger		1	562474	
29	Return spring		1	562475	
30	Ball bearing		1	3750	
31	Locknut (5/16" UNF) pedal detent to bracket		1	254861	
32	ACCELERATOR LEVER ASSEMBLY		1	562459	RH Stg
	ACCELERATOR LEVER ASSEMBLY		1	565429	LH Stg
33	Bush for accelerator lever		2	503194	
34	Pivot bracket		1	560253	
35	Distance piece		1	278884	
36	Bolt (⅜" UNF x 2" long)	Pivot fixing through lever and pivot bracket	1	256045	
37	Spring washer		1	3076	
38	Nut (⅜" UNF)		1	254812	
39	Bolt (5/16" UNF x ¾" long)	Fixing pivot bracket to dash	2	255226	
40	Plain washer		2	2223	
41	Self-locking nut (5/16" UNF)		2	252161	
42	Accelerator cable complete		1	560256	RH Stg
	Accelerator cable complete		1	562048	LH Stg
	Grommet for accelerator cable		1	562106	

*Asterisk indicates a new part which has not been used on any previous model.

521

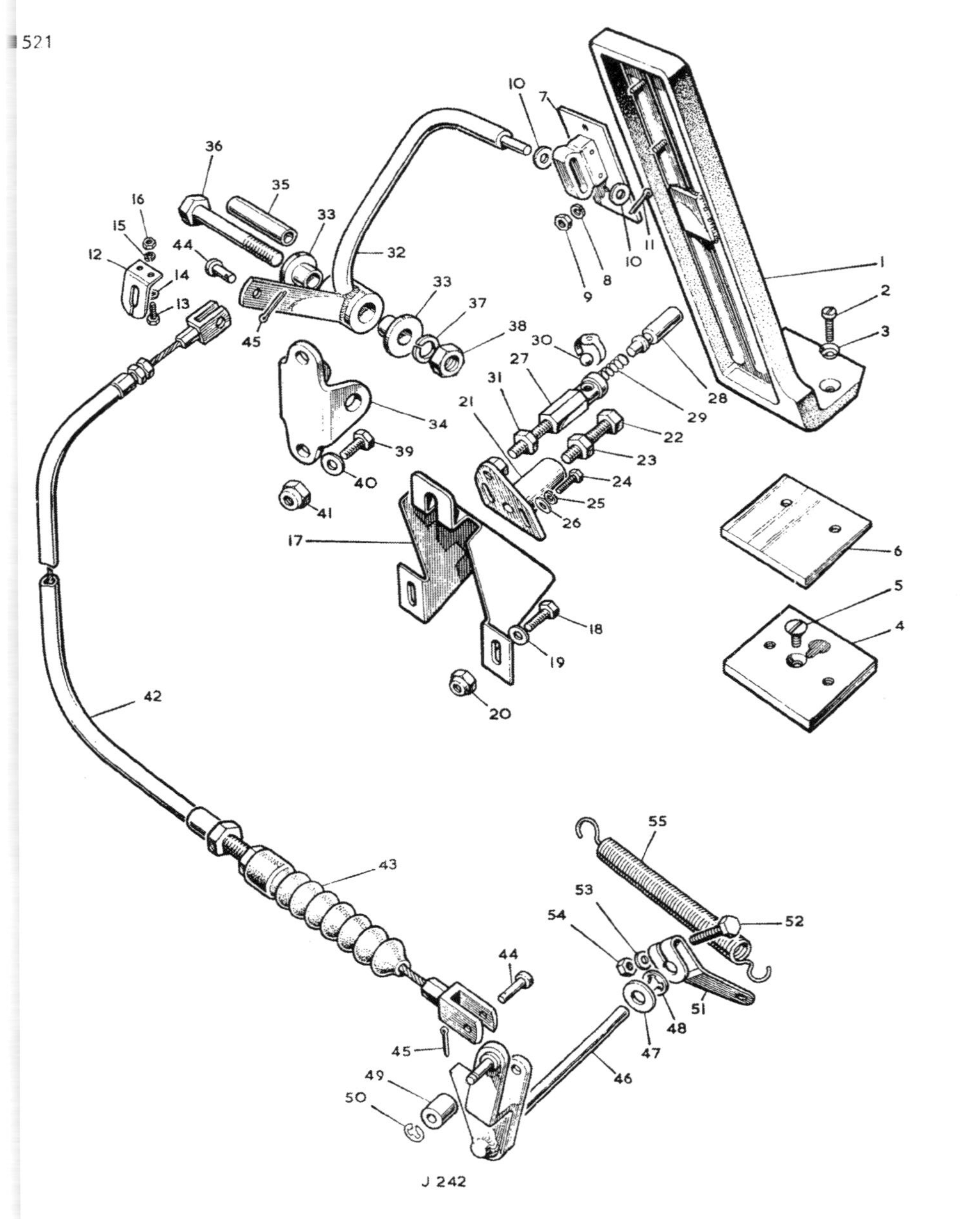

Plate Ref.	1 2 3 4 Description		Qty.	Part No.	Remarks
44	Clevis pin	For accelerator cable	2	562481	
45	Split pin	For accelerator cable	2	3359	
	Accelerator coupling shaft		1	610997	Cars numbered suffix letter 'D' onwards
	Bracket for coupling shaft and down shift cable abutment		1	610993	Cars numbered suffix letter 'D' onwards
	Bolt (3/8" UNC x 1 3/8" long)	Fixing bracket to cylinder head	1	253048	Cars numbered suffix letter 'D' onwards
	Plain washer thin	Fixing bracket to cylinder head	1	4094	Cars numbered suffix letter 'D' onwards
	Plain washer thick	Fixing bracket to cylinder head	1	4266	Cars numbered suffix letter 'D' onwards
	Spring washer	Fixing bracket to cylinder head	1	3076	Cars numbered suffix letter 'D' onwards
	Plastic bearing for coupling shaft		1	553851	Cars numbered suffix letter 'D' onwards
	Tubular pin fixing coupling shaft to countershaft		1	611021	Cars numbered suffix letter 'D' onwards
	Spring for downshift lever		1	558145	Cars numbered suffix letter 'D' onwards
	Clip for down shift cable		1	576788	Cars numbered suffix letter 'D' onwards
	Mounting bracket for accelerator lever		1	562385	LH Stg
	Bolt (5/16" UNF x 3/8" long)	Fixing mounting bracket to dash	4	255225	LH Stg
	Plain washer	Fixing mounting bracket to dash	4	2223	LH Stg
	Self-locking nut(5/16" UNF)	Fixing mounting bracket to dash	4	252161	LH Stg
46	Countershaft and lever		1	603782	
47	Plain washer	Fixing countershaft and lever to manifold	1	2220	
48	Circlip	Fixing countershaft and lever to manifold	1	521453	
49	Roller for counterhsaft		1	603237	
50	Circlip fixing roller		1	521452	
51	Lever for throttle spring		1	541343	
52	Bolt (10 UNF x 3/4" long)	Fixing lever to countershaft	1	257020	
53	Plain washer	Fixing lever to countershaft	1	3902	
54	Nut(10 UNF)	Fixing lever to countershaft	1	257023	
55	Return spring for countershaft		1	603622	

*Asterisk indicates a new part which has not been used on any previous model.

EXHAUST MANIFOLD AND FRONT EXHAUST PIPE 3 LITRE MODELS

1523

F584

EXHAUST MANIFOLD AND FRONT EXHAUST PIPE 3 LITRE MODELS

Plate Ref.	1 2 3 4 Description	Qty.	Part No.	Remarks
1	EXHAUST MANIFOLD ASSEMBLY	1	504183	Except Mk II and Mk III NADA } Not part of engine assembly
	EXHAUST MANIFOLD ASSEMBLY (Vitreous enamelled)	1	550416	Mk II and Mk III NADA } Not part of engine assembly
2	Stud for exhaust pipe	3	252623	
3	Joint washer, front	1	GEG640	
4	Joint washer, centre	1	GEG640	
5	Joint washer, rear	1	GEG640	
6	Clamp for manifold	5	09161	Not part of engine assembly
7	Spring washer } Fixing clamps	5	3076	
8	Nut (⅜" UNF) } Fixing clamps	5	254812	
9	Plain washer } Fixing ends of manifold	2	2220	
10	Spring washer } Fixing ends of manifold	2	3075	
11	NUT (5/16" UNF) } Fixing ends of manifold	2	254831	
12	Front exhaust pipe complete	1	GEX1195	
13	Spring washer } Fixing pipe to manifold	3	3075	
14	Nut (5/16" UNF) } Fixing pipe to manifold	2	253809	
	MOUNTING BRACKET ASSEMBLY FOR FRONT EXHAUST PIPE	1	525119@	5/16" UNF fixings. Alternative to 10 UNF. Cars numbered up to 72500804, 72600116, 72800156, 73001157, 73100218, 73300256
	Exhaust mounting bracket	1	525524@	
	Exhaust mounting plate	1	522868@	
	Bolt (5/16" UNF x ⅞" long) } Fixing bracket to mounting plate	3	255227@	
	Plain washer } Fixing bracket to mounting plate	6	2550@	
	Spring washer } Fixing bracket to mounting plate	3	3075@	
	Nut (5/16" UNF) } Fixing bracket to mounting plate	3	254831@	
	Mounting bracket for front exhaust pipe	1	557132	10 UNF mounting fixings. Alternative to 5/16". Cars numbered from 72500805, 72600117, 72800157, 7300158, 73100219, 73300257 onwards
16	Mounting plate for front exhaust pipe	1	533928	
17	Packing washer between bracket and plate	2	533920	
18	Bolt (10 UNF x 1" long) } Fixing mounting bracket to mounting plate	3	257022	
19	Plain washer } Fixing mounting bracket to mounting plate	3	3885	
20	Distance piece } Fixing mounting bracket to mounting plate	3	533921	
21	Plain washer } Fixing mounting bracket to mounting plate	3	2760	
22	Spring washer } Fixing mounting bracket to mounting plate	3	533957	
23	Plain washer } Fixing mounting bracket to mounting plate	3	2983	
24	Self-locking nut (10 UNF)	3	250742	
25	Clamp plate for mounting bracket	1	517414@	
	Plain washer, bracket to bell housing	2	2251@	
26	Bolt (¼" UNF x ⅞" long) } Fixing clamp to exhaust pipe and mounting bracket	3	255208@	
27	Spring washer } Fixing clamp to exhaust pipe and mounting bracket	3	3075@	
28	Nut (¼" UNF) } Fixing clamp to exhaust pipe and mounting bracket	3	254810@	

@ Cars numbered from 625001885, 626000578, 628000271, 630001091, 631000273, 633000534 onwards

*Asterisk indicates a new part which has not been used on any previous model.

INTERMEDIATE EXHAUST PIPE AND FRONT SILENCER, MK I AND MK IA MODELS

1525

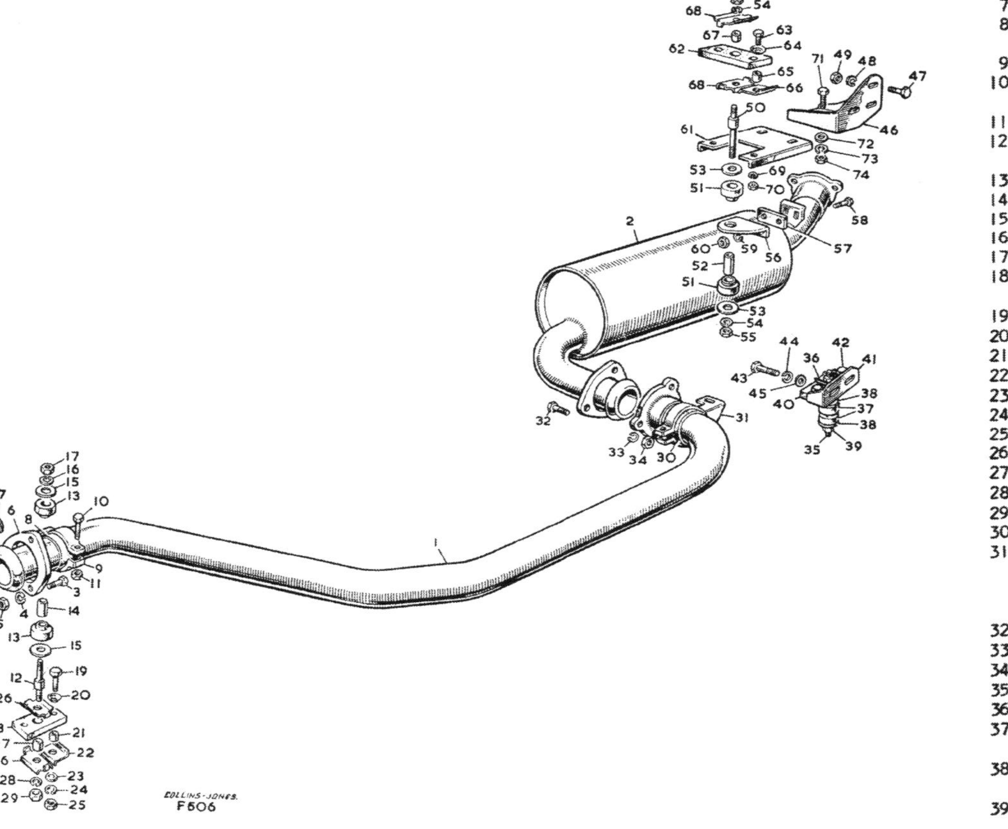

Plate Ref.	1 2 3 4 Description		Qty.	Part No.	Remarks
1	Intermediate exhaust pipe and clamping bracket		1	GEX 1206	
	Intermediate exhaust pipe and clamping bracket		1	GEX 1196	
2	Exhaust silencer, front		1	GEX 3165	
3	Bolt (5/16" UNF x 1¼" long)	Fixing front pipe to intermediate pipe	3	256220	
4	Spring washer		3	3075	
5	Nut (5/16" UNF)		3	254831	
6	Flange		1	273468	
7	Split ring, halves		2	541405	
8	Asbestos packing for intermediate exhaust pipe mounting clamp		1	506049@	
9	Mounting clamp front		1	510731@	
10	Bolt(5/16" UNF x 1⅜" long)	Fixing mounting clamp and packing to intermediate exhaust pipe	1	256223@	
11	Nut (5/16" UNF)		1	254831@	
12	Special stud for intermediate exhaust pipe mounting		1	506051@	
13	Rubber bush	Fixing mounting stud to exhaust pipe	2	506050@	
14	Distance tube		1	512816@	
15	Plain washer		2	3841@	
16	Spring washer		1	3075@	
17	Nut (5/16" UNF)		1	254831@	
18	Flexible mounting strap for intermediate exhaust pipe		1	502808@	
19	Bolt(¼" UNF x 1" long)	Fixing flexible mounting strap to chassis sub-frame bracket	2	255209@	
20	Cup washer		2	502811@	
21	Distance tube		2	502809@	
22	Support plate		2	502810@	
23	Plain washer		2	3840@	
24	Spring washer		2	3074@	
25	Nut (¼" UNF)		2	254810@	
26	Clamp plate	Fixing mounting stud to flexible mounting	2	505391@	
27	Distance tube		1	505912@	
28	Spring washer		1	3075@	
29	Nut (5/16" UNF)		1	254831@	
30	Asbestos packing for mounting clamp		1	506049	
31	Mounting clamp, rear intermediate pipe		1	506860	
	Bolt (5/16" UNFx 1⅜" long)	Fixing mounting clamp and packing to exhaust pipe	1	256223	
	Nut (5/16" UNF)		1	254831	
32	Bolt(5/16" UNF x 1¼" long)	Fixing intermediate pipe to front silencer	3	256220	
33	Spring washer		3	3075	
34	Nut (5/16" UNF)		3	254831	
35	Special stud	Fixing rear mounting clamp to intermediate exhaust pipe	1	506051	
36	Clamp plate		2	505391	
37	Rubber bush		2	506050	
	Distance tube		1	512816	
38	Plain washer		2	3841	
	Spring washer		2	3075	
39	Nut (5/16" UNF)		2	254831	

@ Cars numbered up to 625001884, 626000577, 628000270, 630001090, 631000272, 633000533

@@ Cars numbered from 625001885, 626000578, 628000271, 630001091, 631000273, 633000534 onwards

*Asterisk indicates a new part which has not been used on any previous model.

INTERMEDIATE EXHAUST PIPE AND FRONT SILENCER, MK I AND MK IA MODELS

1527

INTERMEDIATE EXHAUST PIPE AND FRONT SILENCER, MK I AND MK IA 3 LITRE MODELS 1528

Plate Ref.	1 2 3 4 Description		Qty.	Part No.	Remarks
40	Flexible mounting strap		1	502808	
41	Support bracket		1	559611	
42	Bolt (¼" UNF x 1" long)	Fixing flexible mounting strap to support bracket	2	255209	
	Cup washer		2	502811	
	Distance tube		2	502809	
	Support plate		2	502810	
	Spring washer		2	3074	
	Plain washer		2	3840	
	Nut (¼" UNF)		2	254810	
43	Bolt (⅜" UNF x ⅞" long)	Fixing support bracket to chassis sub-frame	2	255246	
44	Spring washer		2	3076	
45	Plain washer		2	4094	
46	Support bracket on body for front exhaust silencer flexible mounting strap		1	538263	
47	Bolt (5/16" UNF x ¾" long)	Fixing support bracket for front silencer to body	3	255226	
48	Spring washer		3	3075	
49	Nut (5/16" UNF)		3	254831	
50	Special stud for front exhaust silencer mounting		1	506051	
51	Rubber bush	Fixing mounting stud to mounting bracket on exhaust pipe	2	506050	
52	Distance tube		1	512816	
53	Plain washer		2	3841	
54	Spring washer		2	3075	
55	Nut (5/16" UNF)		2	254811	
56	Mounting bracket for front exhaust silencer mounting		1	506805	
57	Asbestos packing strip for mounting		1	506843	
58	Bolt (¼" UNF x ¾" long)	Fixing mounting bracket and packing to exhaust pipe	2	255207	
59	Spring washer		2	3074	
60	Nut (¼" UNF)		2	254810	
61	Mounting bracket for front exhaust silencer flexible mounting strap		1	506874	
62	Flexible mounting strap for front exhaust silencer		1	502808	
63	Bolt (¼" UNF x 1" long)	Fixing flexible mounting strap to mounting bracket	2	255209	
64	Cup washer		2	502811	
65	Distance tube		2	502809	
66	Support plate		2	502810	
67	Distance tube		1	505912	
68	Clamp plate		2	505391	
69	Spring washer		2	3074	
70	Nut (¼" UNF)		2	254810	
71	Bolt (¼" UNF x ¾" long)	Fixing mounting bracket to support bracket on body	2	255207	
72	Plain washer		2	3840	
73	Spring washer		2	3074	
74	Nut (¼" UNF)		2	254831	

*Asterisk indicates a new part which has not been used on any previous model.

INTERMEDIATE EXHAUST PIPE AND FRONT SILENCER, MK II AND MK III MODELS

1529

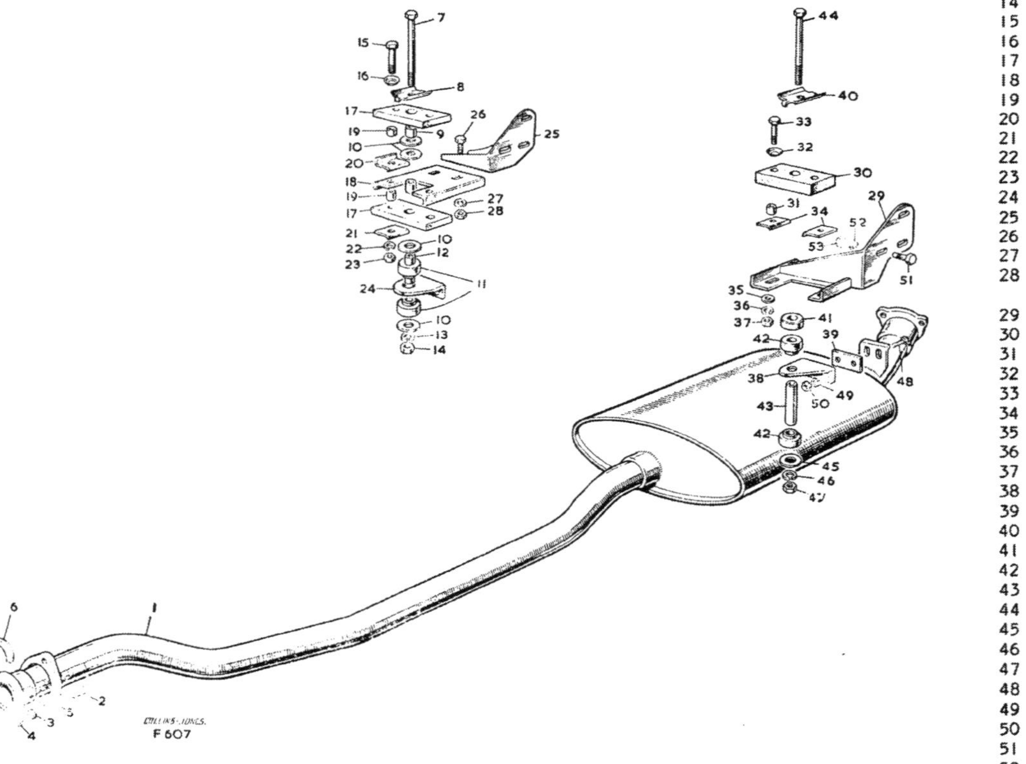

INTERMEDIATE EXHAUST PIPE AND FRONT SILENCER, MK II AND MK III 3 LITRE MODELS 1530

Plate Ref.	1 2 3 4 Description		Qty.	Part No.	Remarks
1	In ermediate exhaust pipe and silencer		1	536304	
2	Bolt (5/16" UNF x 1¼" long)	Fixing intermediate	3	256222	
3	Spring washer	exhaust pipe	3	3075	
4	Nut (5/16" UNF)	and silencer	3	254831	
5	Flange	to front	1	273468	
6	Split ring, halves	exhaust pipe	2	273469	
7	Bolt (5/16" UNF x 2½" long)		1	256228@	
8	Clamp plate		1	505391@	
9	Distance tube		1	505912@	
10	Plain washer	Fixing flexible	4	3841@	
11	Rubber bush	mounting straps to	2	506050@	
12	Distance tube	mounting bracket	1	512816@	
13	Spring washer		1	3075@	
14	Nut (5/16" UNF)		1	254831@	
15	Bolt (¼" UNF x 1⅜" long)		2	256202@	
16	Cup washer		2	502811@	
17	Flexible mountin g strap	Fixing flexible	2	502808@	
18	Mounting bracket	mounting straps	1	506874@	
19	Distance tube	to mounting	2	502809@	
20	Support plate, upper	bracket and	2	502810@	
21	Support plate, lower	support bracket	1	538262@	
22	Spring washer		2	3074@	
23	Nut (¼" UNF)		2	254810@	
24	Mounting bracket		1	506805	
25	Suspension bracket assembly		1	538263@	
26	Bolt (5/16" UNF x ¾" long)	Fixing support	2	255226@	
27	Spring washer	bracket to	2	3075@	
28	Nut (5/16" UNF)	mounting bracket	2	254831@	
	MOUNTING BRACKET ASSEMBLY		1	542296@@	
29	Suspension bracket		1	542292@@	
30	Mounting strap		1	541770@@	
31	Distance tube		2	541773@@	
32	Cup washer		2	502811@@	
33	Bolt(¼" UNF x 1)" long)	Fixing mounting	2	256201@@	
34	Support plate	strap to	2	542654@@	
35	Plain washer	suspension bracket	2	3840@@	
36	Spring washer		2	3074@@	
37	Nut (¼" UNF)		2	254810@@	
38	Mounting bracket		1	506805@@	
39	Asbestos packing strip		1	506843@@	
40	Clamp plate		1	505391@@	
41	Spacer		1	541774@@	
42	Rubber bush	Fixing mounting	2	506050@@	
43	Distance tube	bracket and	1	541772@@	
44	Bolt(5/16" UNFx2½"long)	packing to	1	256228@@	
45	Plain washer	mounting strap	1	3841@@	
46	Spring washer		1	3075@@	
47	Nut (5/16" UNF)		1	254811@@	
48	Bolt(¼" UNF x ¾" long)	Fixing mounting	2	255207	
49	Spring washer	bracket to exhaust	2	3074	
50	Nut(¼" UNF)	pipe bracket	2	254810	
51	Bolt(5/16" UNF x ¾" long)	Fixing	2	255226	
52	Spring washer	suspension bracket	2	3075	
53	Nut (5/16" UNF)	to body	2	254831	

@ Mk II cars with suffix letters'A' and'B' @@ Mk II cars with suffix letter 'C' onwards

*Asterisk indicates a new part which has not been used on any previous model.

F608

Plate Ref.	Description (1 2 3 4)	Qty.	Part No.	Remarks
	Exhaust silencer, rear (4½"dia) and tail pipe	1	523406@	
1	Exhaust silencer, rear (5½"dia) and tail pipe	1	529898@@	MkII 3 litre cars with suffix letters 'A' and 'B'
	Exhaust silencer, rear, and tail pipe	1	GEX3168%%	
2	Bolt (5/16"UNF x 1½"long) – Fixing rear silencer to intermediate exhaust pipe	3	256222	
3	Spring washer – Fixing rear silencer to intermediate exhaust pipe	3	3075	
4	Nut (5/16" UNF) – Fixing rear silencer to intermediate exhaust pipe	3	254831	
5	Clamp bracket on tail pipe	1	538532	
6	Bolt (5/16"UNF x 1¼"long) – Fixing clamp bracket to tail pipe	1	256220	
7	Nut (5/16" UNF) – Fixing clamp bracket to tail pipe	1	254831	
	Special stud for tail pipe mounting	1	506051@	
8	Special stud for tail pipe mounting	1	530404@@	MkII cars with suffix letters 'A' and 'B'
	Bolt (5/16" UNF x 2½" long) for pipe mounting	1	256228%%	
9	Rubber bush	2	506050	
	Distance piece – Fixing special stud or mounting bolt to flexible strap and to mounting bracket on exhaust pipe	1	541774%%	
10	Distance tube – Fixing special stud or mounting bolt to flexible strap and to mounting bracket on exhaust pipe	1	512816	
	Distance tube – Fixing special stud or mounting bolt to flexible strap and to mounting bracket on exhaust pipe	1	541772%%	
11	Plain washer – Fixing special stud or mounting bolt to flexible strap and to mounting bracket on exhaust pipe	2	3841	1 off from %%
12	Spring washer – Fixing special stud or mounting bolt to flexible strap and to mounting bracket on exhaust pipe	2	3075	1 off from %%
13	Nut (5/16" UNF) – Fixing special stud or mounting bolt to flexible strap and to mounting bracket on exhaust pipe	2	254831	1 off from %%
14	Mounting bracket for tail pipe mounting	1	506805	
15	Asbestos packing strip for mounting	1	506843	
16	Bolt (¼"UNF x ¾"long) – Fixing mounting bracket and packing to clamp bracket on tail pipe	2	255207	
17	Spring washer – Fixing mounting bracket and packing to clamp bracket on tail pipe	2	3074	
18	Nut (¼" UNF) – Fixing mounting bracket and packing to clamp bracket on tail pipe	2	254810	
19	Mounting bracket for tail pipe flexible mounting strap	1	559611*	
20	Bolt (5/16"UNF x ¾"long) – Fixing mounting bracket to support bracket on body	2	255226	
21	Plain washer – Fixing mounting bracket to support bracket on body	2	2550	
22	Spring washer – Fixing mounting bracket to support bracket on body	2	3075	
23	Nut (5/16" UNF) – Fixing mounting bracket to support bracket on body	2	254831	
	Flexible mounting strap for tail pipe	1	502808%	
	Flexible mounting strap for tail pipe	1	541770%%	

@Cars numbered up to 72500342A, 72600055A, 72800031A, 73000513A, 73100061A, 73311191A

@@Cars numbered from 72500343A, 72600056A, 72800032A, 73000514A, 73100062A, 73300092A onwards and Mk II models

%Mk I, Mk IA and Mk II cars with suffix letters 'A' and 'B'
%%Mk II cars with suffix letter 'C' and Mk III

*Asterisk indicates a new part which has not been used on any previous model.

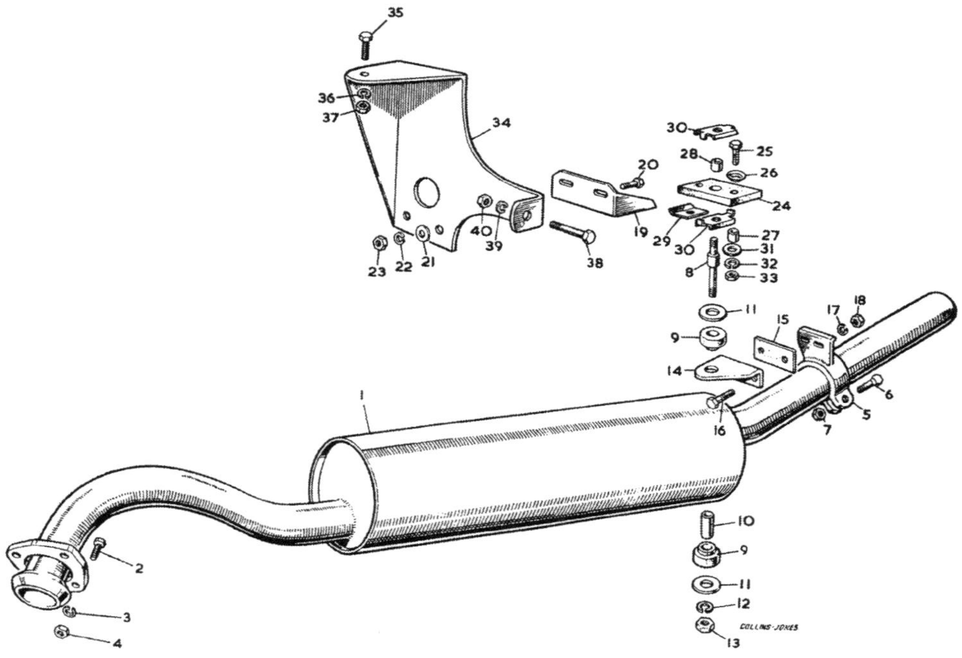

REAR EXHAUST SILENCER AND TAIL PIPE 3 LITRE MODELS

1533

F608

REAR EXHAUST SILENCER AND TAIL PIPE 3 LITRE MODELS

Plate Ref.	1 2 3 4	Description		Qty.	Part No.	Remarks
25		Bolt (¼" UNF x 1" long)	Fixing flexible mounting strap to mounting bracket	2	255209$	
		Bolt (¼" UNF x 1¼" long)		2	256201$$	
26		Cup washer		2	502811	
27		Distance tube		2	502809$	
		Distance tube		2	541773$$	
28		Distance tube		1	505912$	
29		Support plate		2	502810$	
		Support plate		2	542654$$	
30		Clamp plate		2	505391	1 off from $$
31		Plain washer		2	3840	
32		Spring washer		2	3074	
33		Nut (¼" UNF)		2	254810	
34		Support bracket for rear exhaust silencer		1	502237@	
		Support plate, tapped		1	530777@@	
35		Bolt (5/16" UNF x ⅞" long)	Fixing support bracket to body floor	2	255226	
36		Spring washer		2	3075	
37		Nut (5/16" UNF)		2	254831	
38		Bolt (5/16" UNF x 3¼" long)	Fixing support bracket to body sill	1	256031	
39		Spring washer		1	3075	
40		Nut (5/16" UNF)		1	254831	

@Cars numbered up to 72500342A, 72600055A, 72800031A, 73000513A, 73100061A, 73300091A
@@Cars numbered from 72500343A, 72600056A, 72800032A, 73000514A, 73100062A, 73300092A onwards and Mk II models
$Mk I, Mk IA and Mk II cars with suffix letters 'A' and 'D'
$$Mk II cars with suffix letter 'C' and Mk III

*Asterisk indicates a new part which has not been used on any previous model.

1535

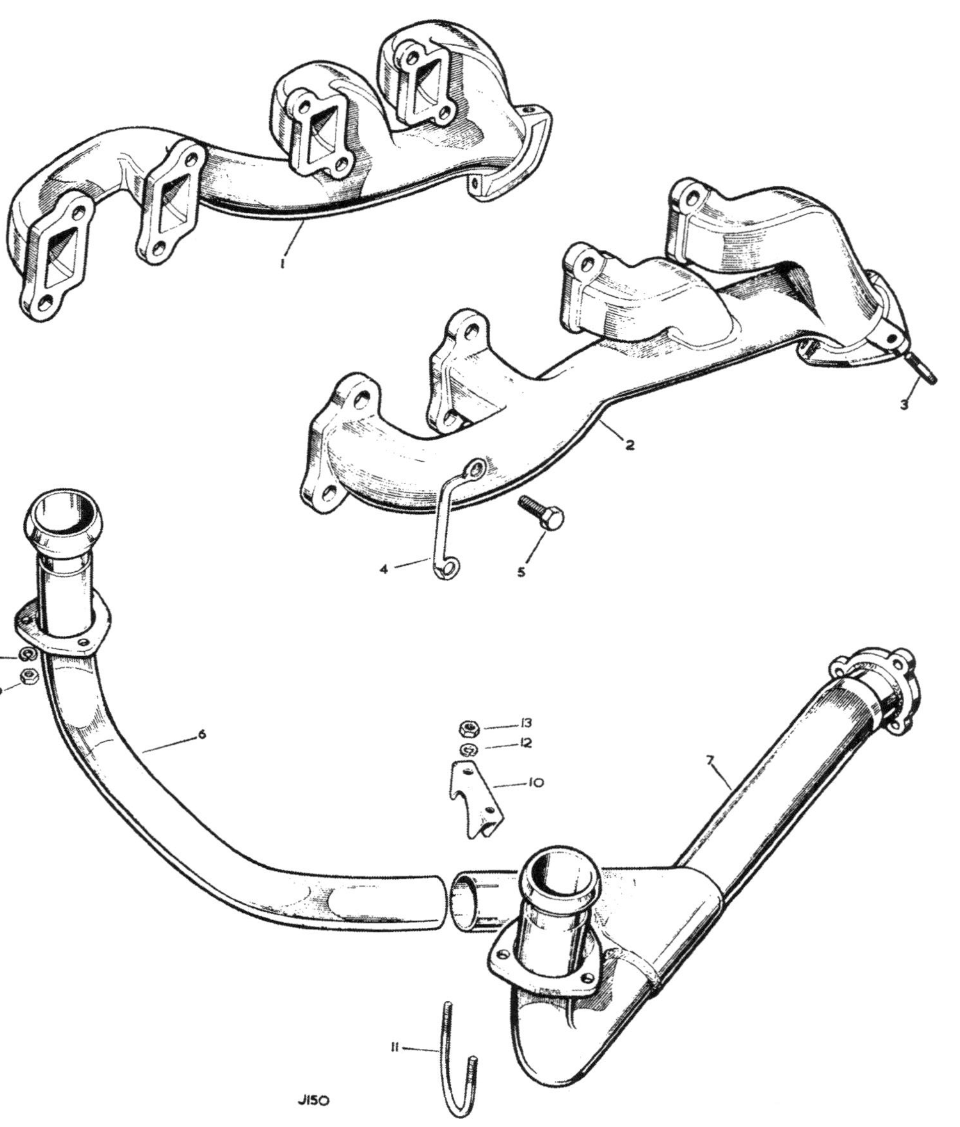

Plate Ref.	1 2 3 4 Description		Qty.	Part No.	Remarks
1	EXHAUST MANIFOLD ASSEMBLY, RH		1	602908	
2	EXHAUST MANIFOLD ASSEMBLY, LH		1	610413	Cars with manual choke
	EXHAUST MANIFOLD ASSEMBLY, LH		1	602907*	Cars with automatic enrichment device engines numbered up to 84016384
	EXHAUST MANIFOLD ASSEMBLY, LH		1	611561*	Engines numbered from 84016385
	Special screw (¼" UNC x ⅜" long) for redundant holes in exhaust manifold, LH		4	610411*	Cars with manual choke
3	Stud for exhaust down pipe		6	252623	
4	Locking plate	Fixing exhaust manifolds to cylinder heads	2	602951	
	Locking plate		6	613659	
5	Bolt (⅜" UNCx1¼"long)		16	253048	
6	Front exhaust pipe, RH		1	GEX1202	
7	Front exhaust pipe, LH		1	GEX1201	
8	Spring washer	Fixing pipes to exhaust manifolds	6	3075	
9	Nut (5/16" UNF)		6	253809	
10	Clamp bracket		1	562321	
11	'U' clamp	Fixing front exhaust pipes together	1	562322	
12	Spring washer		2	3075	
13	Nut (5/16" UNF)		2	254831	

*Asterisk indicates a new part which has not been used on any previous model.

CENTRE EXHAUST PIPE AND SILENCER

1537

REAR EXHAUST SILENCER AND TAIL PIPES

1538

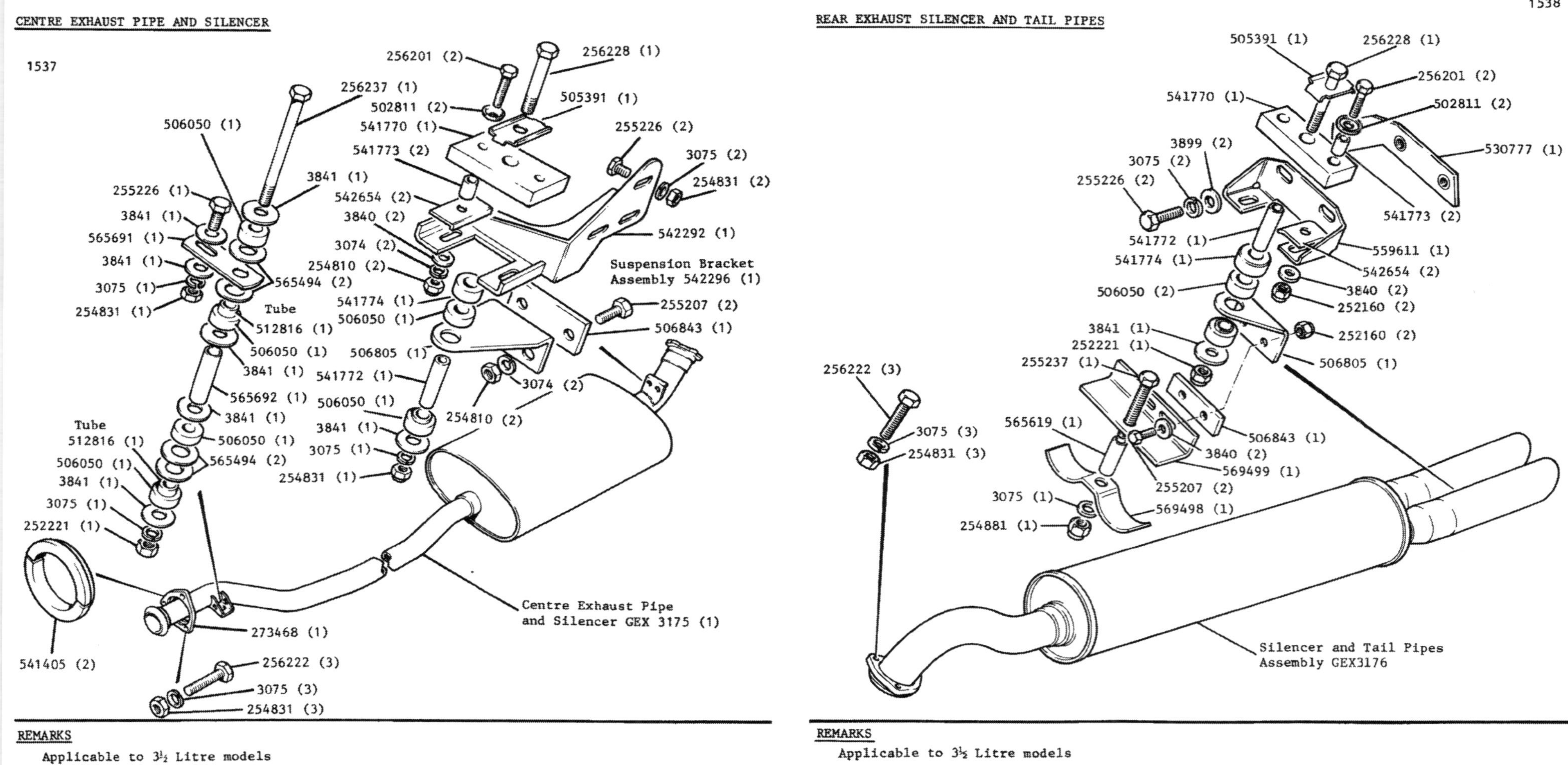

REMARKS

Applicable to 3½ Litre models

REMARKS

Applicable to 3½ Litre models

1539

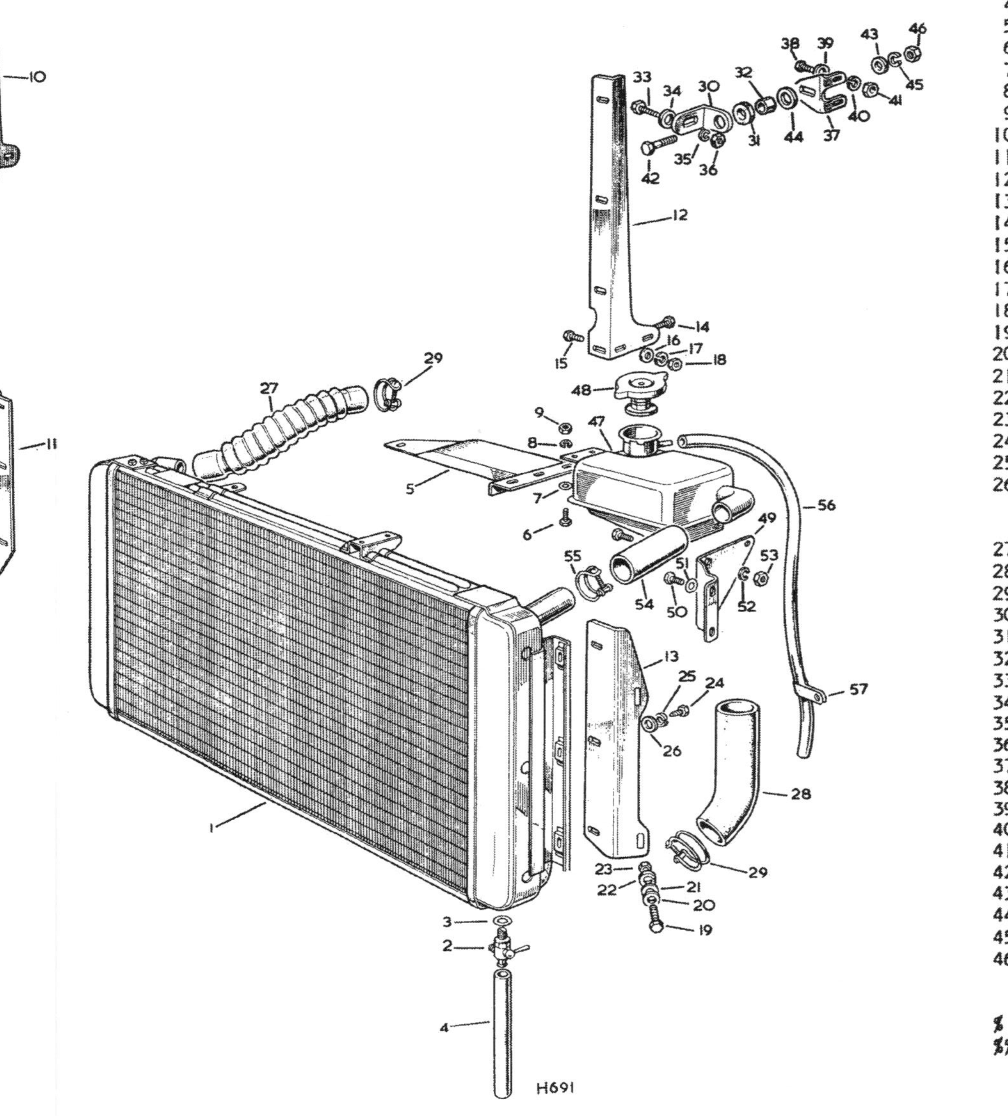

Plate Ref.	1 2 3 4 Description		Qty.	Part No.	Remarks
1	Block complete for radiator		1	545531	
2	Drain tap for radiator		1	500853	
3	Joint washer for drain tap		1	213959	
4	Extension tube for drain tap		1	572506	
5	Fan cowl and heater tank support		1	524755	
6	Bolt (5/16" UNF x ⅝" long)	Fixing cowl to radiator block and header tank	5	255225	
7	Plain washer		5	3830	
8	Spring washer		5	3075	
9	Nut (5/16" UNF)		3	254831	
10	Mounting bracket for radiator, RH		1	274479%	
11	Mounting bracket for radiator, RH		1	542929%%	
12	Mounting bracket for radiator, LH		1	274480%	
13	Mounting bracket for radiator, LH		1	542930%%	
14	Bolt (5/16" UNF x ¾" long)	Fixing mounting brackets to chassis sub-frame	4	255226%	
15	Bolt (5/16" UNF x 1" long)		2	255228%	
16	Plain washer		6	2550%	
17	Spring washer		6	3075%	
18	Nut (5/16" UNF)		6	254831%	
19	Bolt (⅜" UNF x ¾" long)	Fixing mounting brackets to body valance	4	255245%%	
20	Plain washer		4	3036%%	
21	Plain washer		4	2251%%	
22	Spring washer		4	3076%%	
23	Nut (⅜" UNF)		4	254812%%	
24	Set bolt (5/16" UNF x ⅝" long)	Fixing block to mounting brackets	6	253825	
25	Spring washer		6	3075	
26	Plain washer		6	2266%%	
	Plain washer		6	3830%	
	Top water hose		1	GRH376	Mk I and Mk IA
27	Top water hose		1	GRH377	Mk II and Mk III
28	Bottom water hose		1	GRH382	
29	Clip for hose		4	GHC1217	
30	Stay for radiator at side		2	501195%	
31	Rubber grommet	In stay for connecting bolt	2	250996%	
32	Distance piece		2	501197%	
33	Bolt (5/16" UNF x ¾" long)	Fixing stay to radiator mounting	2	255226%	
34	Plain washer		2	3830%	
35	Spring washer		2	3075%	
36	Nut (5/16" UNF)		2	254831%	
37	Stay for radiator at valance side		2	532674%	
38	Bolt (5/16" UNF x ⅝" long)	Fixing stay to valance	4	255225%	
39	Plain washer		4	2250%	
40	Spring washer		4	3075%	
41	Nut (5/16" UNF)		4	254831%	
42	Bolt (5/16" UNF x 1" long)	Connecting radiator stays together	2	255228%	
43	Plain washer, small		2	3830%	
44	Plain washer, large		2	2266%	
45	Spring washer		2	3075%	
46	Nut (5/16" UNF)		2	254831	

% Mk I, Mk IA and Mk II cars with suffix letters 'A' and 'B'
%% Mk II cars with suffix letter 'C' and Mk III

*Asterisk indicates a new part which has not been used on any previous model.

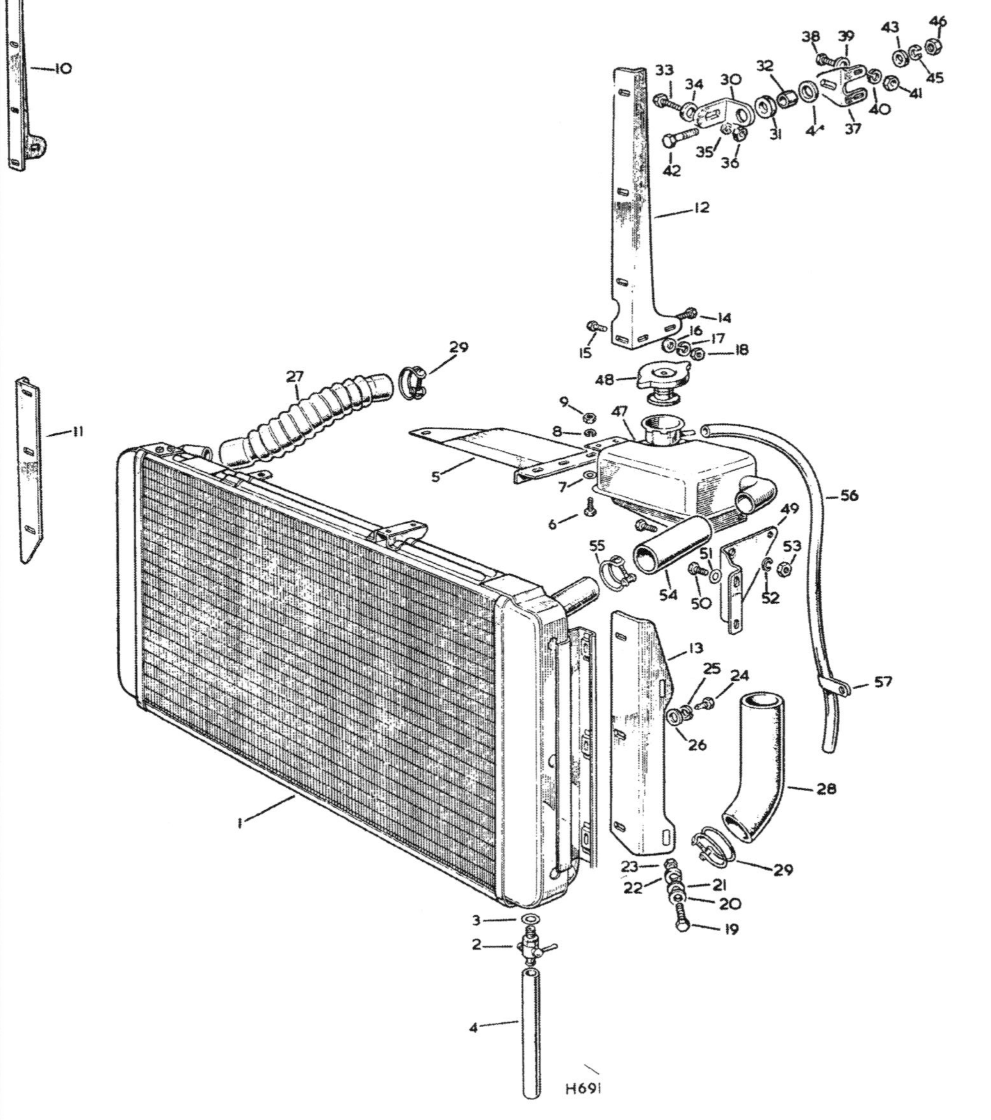

Plate Ref.	1 2 3 4 Description		Qty.	Part No.	Remarks
47	HEADER TANK FOR RADIATOR		1	527186	
	Filler cap for radiator header tank		1	GRC120@	
48	Filler cap for radiator header tank		1	GRC108@@	
	Joint washer for cap		1	504442	
49	Bottom support bracket for header tank		1	273279	
50	Bolt (5/16" UNF x $\frac{7}{8}$" long)	Fixing support bracket to header tank and radiator mounting bracket	4	255225	
51	Plain washer		2	3830	
52	Spring washer		4	3075	
53	Nut (5/16" UNF)		2	254831	
54	Hose connecting header tank to radiator		1	273310	
55	Clip for hose		2	50314	
56	Hose for overflow pipe		1	352539	
57	Clip for hose		2	41375	

@ Cars numbered up to 625101045, 626100262, 628100109, 630100840, 631100353, 633100078

@@ Cars numbered from 625101046, 626100263, 628100110, 630100841, 631100354, 633100079 onwards

*Asterisk indicates a new part which has not been used on any previous model.

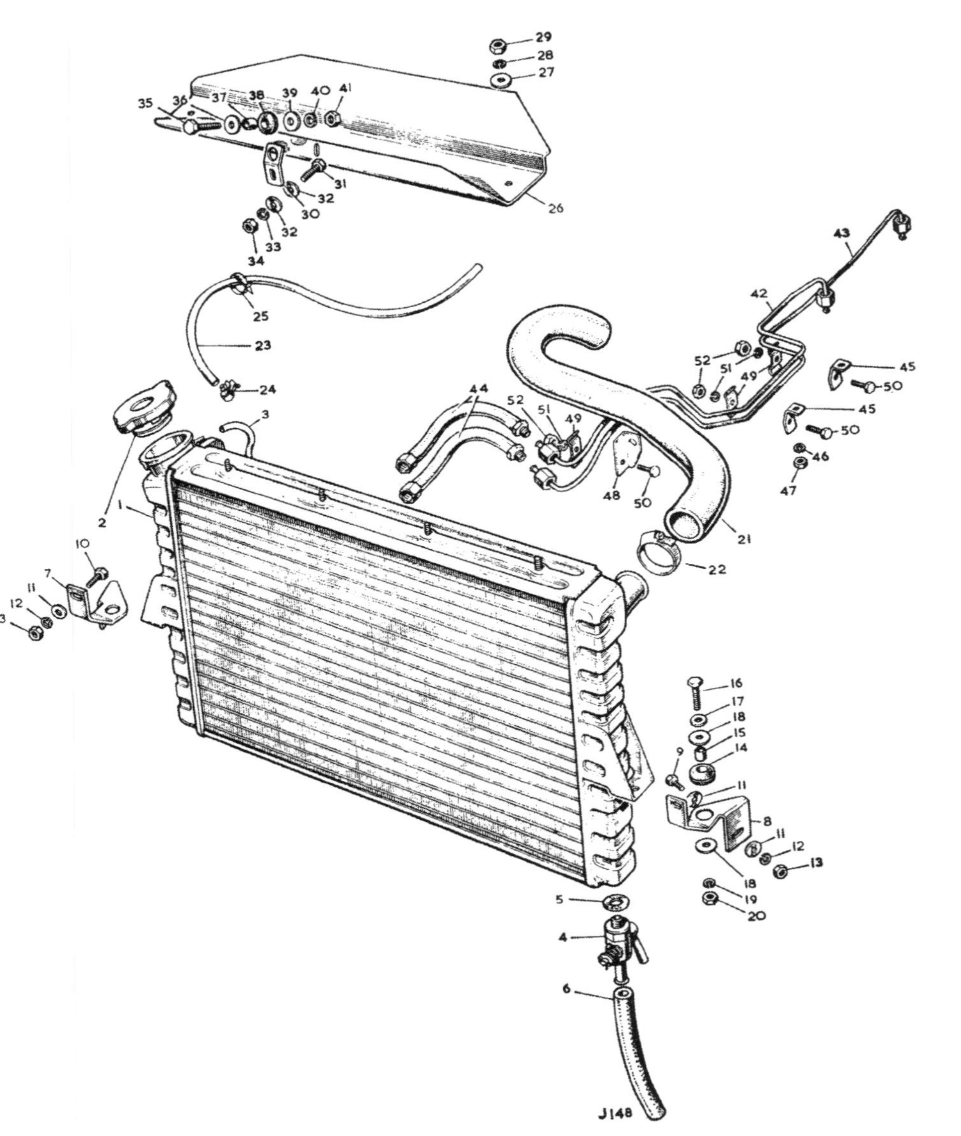

Plate Ref.	Description		Qty.	Part No.	Remarks
1	BLOCK COMPLETE FOR RADIATOR AND OIL COOLER		1	578510	
2	Radiator filler cap		1	GRC110	
3	Overflow hose		1	570033	
4	Drain tap for radiator		1	500853	Early models
5	Joint washer for drain tap		1	2339	Early models
6	Extension tube for drain tap		1	572506	Early models
	Drain tap for radiator		1	538608	Late models
	Extension tube for drain tap		1	572506	Late models
7	Mounting bracket for radiator, RH		1	560240	
8	Mounting bracket for radiator, LH		1	560239	
9	Bolt (5/16" UNF x ⅝" long)	Fixing mounting brackets to body	2	255225	
10	Bolt (5/16" UNF x ¾" long)	Fixing mounting brackets to body	2	255226	
11	Plain washer	Fixing mounting brackets to body	6	3830	
12	Spring washer	Fixing mounting brackets to body	4	3075	
13	Nut (5/16" UNF)	Fixing mounting brackets to body	4	254831	
14	Grommet	Fixing radiator mounting brackets to brackets on radiator block	2	250995	
15	Distance piece	Fixing radiator mounting brackets to brackets on radiator block	2	555437	
16	Bolt(5/16" UNF x 1" long)	Fixing radiator mounting brackets to brackets on radiator block	2	255228	
17	Plain washer	Fixing radiator mounting brackets to brackets on radiator block	2	3830	
18	Plain washer	Fixing radiator mounting brackets to brackets on radiator block	4	2223	
19	Spring washer	Fixing radiator mounting brackets to brackets on radiator block	2	3075	
20	Nut (5/16" UNF)	Fixing radiator mounting brackets to brackets on radiator block	2	254831	
21	Top water hose		1	GRH444	
22	Clip for hose		2	GHC913	
23	Hose, radiator to inlet manifold		1	565686	
24	Clip fixing hose to radiator and manifold		2	572546*	
25	Cleat		1	240430	
26	Fan guard		1	562287	
27	Plain washer	Fixing fan guard to radiator	4	3830	
28	Spring washer	Fixing fan guard to radiator	4	3074	
29	Nut (¼" UNF)	Fixing fan guard to radiator	4	254810	
30	Mounting bracket for radiator top		1	562290	
31	Bolt (5/16" UNF x ¾" long)	Fixing mounting bracket to bonnet platform	1	255226	
32	Plain washer	Fixing mounting bracket to bonnet platform	2	2550	
33	Spring washer	Fixing mounting bracket to bonnet platform	1	3075	
34	Nut (5/16" UNF)	Fixing mounting bracket to bonnet platform	1	254831	
35	Bolt (5/16" UNF x 1" long)	Fixing mounting bracket to fan guard	1	255228	
36	Plain washer	Fixing mounting bracket to fan guard	1	3830	
37	Distance piece	Fixing mounting bracket to fan guard	2	555437	
38	Grommet	Fixing mounting bracket to fan guard	1	250997	
39	Plain washer	Fixing mounting bracket to fan guard	2	2223	
40	Spring washer	Fixing mounting bracket to fan guard	1	3075	
41	Nut (5/16" UNF)	Fixing mounting bracket to fan guard	1	254831	

*Asterisk indicates a new part which has not been used on any previous model.

RADIATOR, OIL COOLER AND FIXINGS

1545

Plate Ref.	1 2 3 4	Description	Qty.	Part No.	Remarks
42		Feed pipe complete for oil cooler	1	562328	
43		Return pipe complete for oil cooler	1	562329	
44		Flexible pipe	2	562413	
45		Mounting bracket for oil cooler pipe clip at sump	2	562331	
46		Spring washer ⎤ Fixing mounting	2	3074	
47		Nut (¼" UNF) ⎦ bracket to sump	2	254810	
48		Mounting bracket for oil cooler pipe clip at cylinder block	1	562332	
49		Clip for oil cooler pipes	3	509412	
50		Bolt (10 UNF x ½" long) ⎤ Fixing pipe	3	257017	
51		Spring washer ⎬ clips to	3	3073	
52		Nut (10 UNF) ⎦ mounting brackets	3	257023	

*Asterisk indicates a new part which has not been used on any previous model.

OIL COOLER, AUTOMATIC TRANSMISSION, MK III MODELS

1547

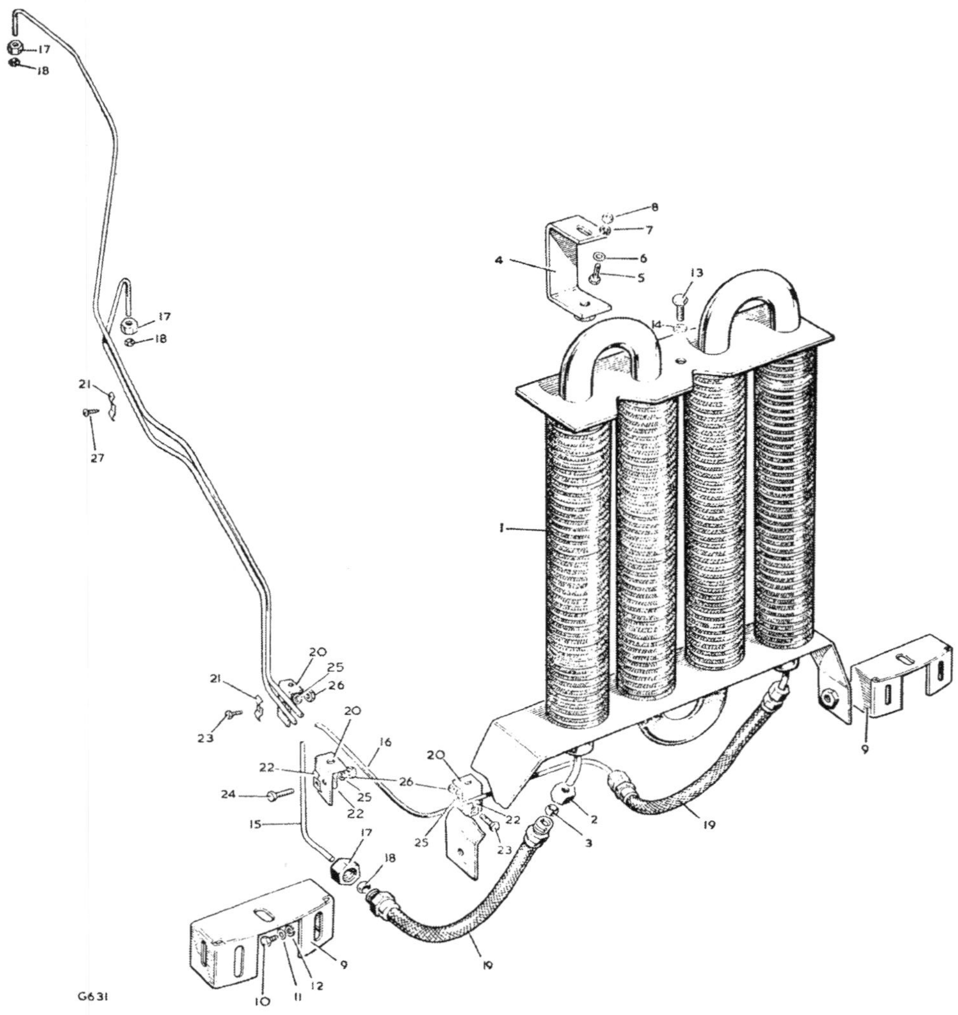

G631

Plate Ref.	1 2 3 4 Description		Qty.	Part No.	Remarks
1	OIL COOLER		1	553868	
2	Union nut	Oil cooler connections	2	553794	
3	Olive for union nut		2	553815	
4	Hanger bracket for oil cooler		1	553869	
5	Bolt (¼" UNF x ⅝" long)	Fixing hanger bracket to body at bonnet locking platform	1	255206	
6	Plain washer		1	3946	
7	Spring washer		1	3074	
8	Locknut		1	254850	
9	Mounting bracket for oil cooler and radiator grille		2	553886	
10	Bolt(¼" UNF x ⅝" long)	Fixing oil cooler to mounting brackets	2	255206	
11	Plain washer		2	3946	
12	Spring washer		2	3074	
13	Bolt (¼" UNF x ⅝" long)	Fixing oil cooler to hanger bracket	1	255206	
14	Spring washer		1	3074	
15	FEED PIPE COMPLETE		1	553745	
16	RETURN PIPE COMPLETE		1	548499	
17	Union nut	For feed and return pipes	4	553794	
18	Olive		4	553815	
19	Flexible pipe, oil connection		2	557393	
20	Support bracket for pipe clip		3	548500	
21	Pipe clip (double)		2	509412	
22	Pipe clip (single)		3	216708	
23	Bolt (2 BA x ⅝" long)	Fixing clips to support brackets at sump fixings	2	250960	
24	Bolt (2 BA x ¾" long)		1	250961	
25	Spring washer		3	3073	
26	Nut (2 BA)		3	2247	
27	Drive screw fixing clip to flywheel housing		1	78152	

*Asterisk indicates a new part which has not been used on any previous model.

PETROL TANK AND PIPES, MK I AND MK IA MODELS

1549

F621.

COLLINS-JONES.

Plate Ref.	1 2 3 4 Description		Qty.	Part No.	Remarks
	FUEL TANK COMPLETE		1	274307●	
1	FUEL TANK COMPLETE		1	515070●●	
2	Drain plug for tank		1	2885	
3	Joint washer for drain plug		1	2347	
4	FEED ELBOW AND RESERVE TAP		1	503669	
5	Joint washer for elbow		1	267837	
6	Spring washer	Fixing elbow	2	3101	
7	Screw (3 BA x 9/16" long)	to tank	2	3972	
8	GAUGE UNIT FOR FUEL TANK		1	503599	
9	Joint washer for gauge unit		1	56398	
	Rubber ring for gauge unit float		1	551320	
10	Spring washer	Fixing gauge	6	3101	
11	Set screw (3 BA x⅜" long)	unit to tank	1	513238	
12	Bolt (5/16" UNF x 1¼" long)		3	256220	
13	Rubber packing	Fixing	3	272853	
14	Plain washer	fuel tank	6	2266	
15	Nut (5/16" UNF)	to body	4	254831	
16	Self-locking nut (5/16"UNF)		1	252161	
17	Expansion tank for fuel tank		1	515071●●	
18	Bolt (¼" UNF x 9/16" long)	Fixing	4	255005●●	
19	Plain washer	expansion tank	4	3840●●	
20	Spring washer	to body	4	3074●●	
21	Breather hose for tank		2	542403●●	
22	Overflow hose for tank, plastic, 25" long		1	515123●●	
	Reinforcement sleeve for overflow hose, 16" long		1	522706●●	
	Overflow hose, plastic, 4½" long		1	542404	
23	Hose clip for breather and overflow hoses		6	50300●●	
24	Overflow pipe and bracket, copper, long		1	562265	
25	Clip for overflow pipe		1	515502●●	
26	Grommet for overflow pipe		1	515770●●	
	Drive screw fixing overflow pipe to parcel shelf frame		1	78322	
27	Bolt (¼" UNF x ½" long)	Fixing	1	255204●●	
28	Spring washer	overflow pipe clip	1	3074●●	
29	Nut (¼" UNF)	to body	1	254810●●	
30	Bolt (2 BA x ½" long)	Fixing	1	234603●●	
31	Spring washer	overflow pipe	1	3073●●	
32	Nut (2 BA)	to clip	1	2247●●	
	FUEL FILLER CAP AND TUBE, less barrel lock		1	352436●	
	FUEL FILLER CAP AND TUBE, less barrel lock		1	357447●●	
33	FILLER CAP COMPLETE, less barrel lock		1	358200	
34	Rubber seal for cap		1	358177	
	Polyether foam insert for filler cap seal		1	357446	
35	Return spring for catch		1	352641	
36	Filler tube and fixings		1	352642●	
	Filler tube and fixings		1	356511●●	
37	Barrel lock for fuel filler, 'FP' series		1	352910	Alternatives. State key
	Barrel lock fuel filler, 'FR' and 'FS' series		1	606024*	number when ordering
38	Drainage tube for fuel filler		1	532539	
39	Pipe, fuel tank to filler		1	513966	

● Cars numbered up to 625001035, 626000370, 628000121, 630000523, 631000128, 633000350

●● Cars numbered from 625001036, 626000371, 628000122, 630000524, 631000129, 633000351 onwards

*Asterisk indicates a new part which has not been used on any previous model.

PETROL TANK AND PIPES, MK I AND MK IA MODELS

1551

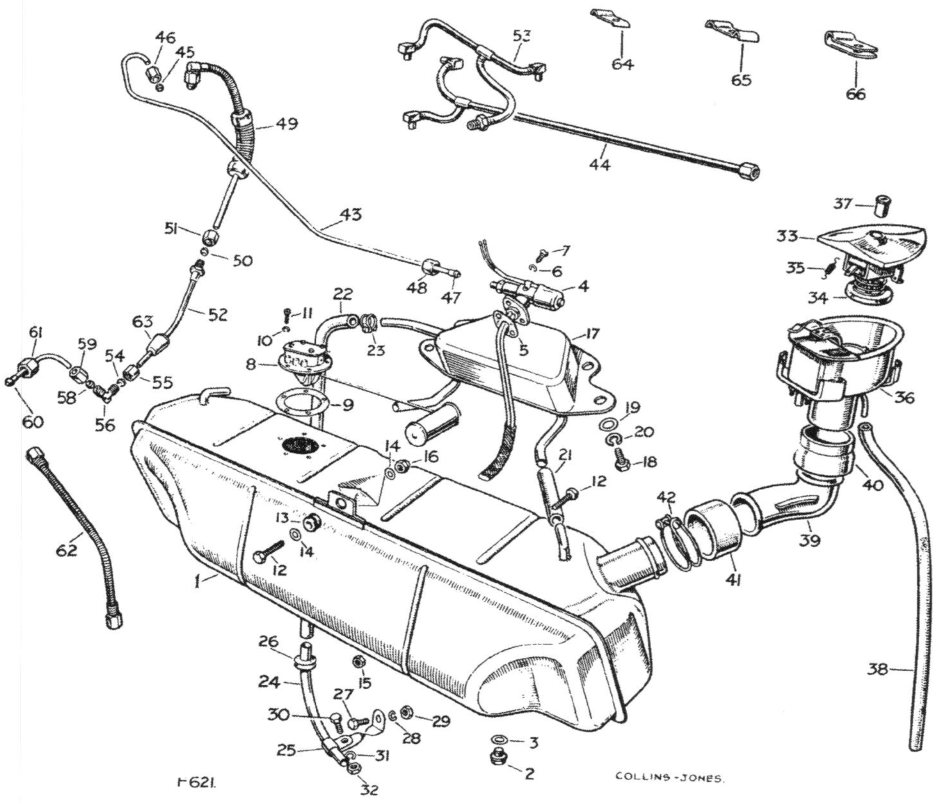

FUEL TANK AND PIPES, MK I AND MK IA 3 LITRE MODELS

Plate Ref.	1 2 3 4 Description	Qty.	Part No.	Remarks
40	Hose, filler end	1	513870	
41	Hose, tank end	1	513555	
42	Clip for hose, tank and filler end	4	271870	
	Clip, large, for rubber connection	1	50334	Cars with rubber connection tank to filler
	Clip, small for rubber connection	1	50331	
	Tube for air vent pipe	1	513951@	
	Clip for air vent hose	2	503028@	
43	FUEL PIPE COMPLETE, tank to pump	1	504188	Mk I
44	FUEL PIPE COMPLETE, tank to pump	1	529137	Mk IA
45	Olive — For pipe	1	270105	
46	Union nut — For pipe	1	270115	
47	Nipple — For pipe	1	2728	
48	Union nut — For pipe	1	270141	
49	Flexible fuel pipe with air cushion	1	502767	Mk I
	Flexible fuel pipe with air cushion	1	529153	Mk IA
50	Olive — For pipe	1	270105	
51	Union nut — For pipe	1	270115	
52	Fuel pipe complete, elbow to flexible pipe	1	504189	Mk I
53	Fuel pipe complete, elbow to flexible pipe	1	529138	Mk IA
54	Olive — For pipe	1	270105	
55	Union nut — For pipe	1	270115	
56	Elbow for fuel pipes	1	268661	
57	FUEL PIPE COMPLETE, elbow to filter	1	504190	
58	Olive — For pipe	1	270105	
59	Union nut — For pipe	1	270115	
60	Nipple — For pipe	1	2728	
61	Union nut — For pipe	1	270141	
62	Flexible fuel pipe, carburetter to filter	1	276266	
63	Grommet in boot floor (¾" dia.)	1	312937	Alternatives Check before ordering
	Grommet in boot floor (1" dia.)	1	522848	
64	Clip — Fuel pipe to body	3	216708	
65	Clip — Fuel pipe to body	4	243395	
66	Clip — Fuel pipe to body	1	215791	
	Drive screw — Fixing clips	6	78322	
	Bolt (2 BA x ½" long) — Fixing clips	1	250959	
	Spring washer — Fixing clips	1	3073	
	Nut — Fixing clips	1	2247	
	Set bolt (¼" UNF x ¾" long) — Fixing clips	1	255206	
	Plain washer — Fixing clips	1	3900	
	Spring washer — Fixing clips	1	3074	
	Nut (¼" UNF) — Fixing clips	1	254810	

@ Cars numbered up to 625001035, 626000370, 628000121, 630000523, 631000128, 633000350

*Asterisk indicates a new part which has not been used on any previous model.

PETROL TANK AND PIPES, MK II AND MK III MODELS

1553

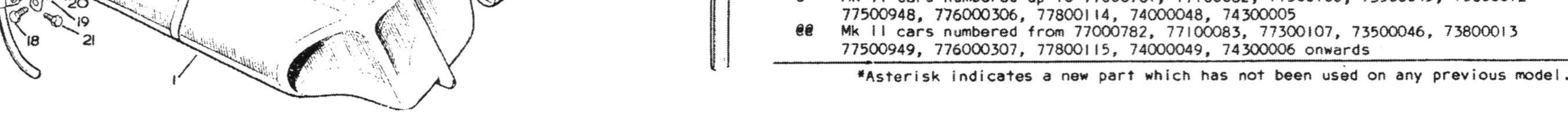

FUEL TANK AND PIPES, MK II and MK III 3 LITRE MODELS

Plate Ref.	1 2 3 4 Description	Qty.	Part No.	Remarks
1	FUEL TANK COMPLETE	1	542508	
	Drain plug for tank	1	2885	
	Joint washer for drain plug	1	2347	
2	Main feed elbow and tube	1	536451	
3	Reserve fuel feed pipe, nylon	1	536455	
4	Fuel filter	1	536457	
5	Union for reserve fuel feed	1	536456	
6	Main connector union	1	267529%	
	Main connector union	1	525530%%	
7	Washer for union	2	243967	
8	Joint washer } Fixing fuel feed and reserve assembly to fuel tank	1	267837	
	Screw }	2	3972	
	Spring washer }	2	3101	
9	GAUGE UNIT FOR FUEL TANK	1	563321	
10	Locking ring for gauge unit	1	519964	
11	Sealing ring for gauge unit	1	519965	
	Joint washer for gauge unit	1	56398	
	Rubber ring for gauge unit float	1	551320	
	Bolt (5/16" UNF x 1¼" long) } Fixing fuel tank to body	3	256020	
	Rubber packing }	3	272853	
	Plain washer }	6	2266	
	Nut (5/16" UNF) }	4	254831	
	Self-locking nut(5/16"UNF)	1	252161	
12	'Y' piece for breather and overflow hoses	1	535634	
	Drive screw } Fixing 'Y' piece	2	78155	
	Shakeproof washer }	2	74236	
13	Breather hose for tank, 14" long	2	542403	
14	Overflow hose for tank, 4½" long	1	542404	
15	Overflow pipe	1	562265	
16	Clip for overflow pipe	1	515502	
17	Grommet for overflow pipe	1	515770	
	Drive screw fixing overflow pipe to parcel shelf frame	1	78322	
18	Bolt (¼" UNF x ½" long) } Fixing overflow pipe clip to body	1	255204	
19	Spring washer }	1	3074	
20	Nut (¼" UNF) }	1	254810	
21	Bolt (2 BA x ½" long) } Fixing overflow pipe to clip	1	234603	
22	Spring washer }	1	3073	
23	Nut (2 BA) }	1	2247	
	FUEL FILLER CAP AND TUBE, less barrel lock	1	357447@	} Mk II cars with chrome filler cap
	FILLER CAP COMPLETE, less barrel lock	1	358200@	}
	Rubber seal for cap	1	358177@	}
	Polyether foam insert for filler cap seal	1	357446@	}
	Return spring for catch	1	352641@	}
	Filler tube and fixings	1	356511@	}
	Barrel lock for fuel filler	1	600874	}
24	Fuel filler lid	1	381223@@	} Mk II cars with concealed filler cap
25	Fuel filler box	1	351696@@	}
26	Locking bolt for lid	1	381222@@	}
	'O' ring for fuel filler locking bolt	1	359244	}

% Mk II cars with suffix letters 'A' and 'B'
%% Mk II cars with suffix letter 'C' and Mk III
@ Mk II cars numbered up to 77000781, 77100082, 77300106, 73500045, 73800012 77500948, 776000306, 77800114, 74000048, 74300005
@@ Mk II cars numbered from 77000782, 77100083, 77300107, 73500046, 73800013 77500949, 776000307, 77800115, 74000049, 74300006 onwards

*Asterisk indicates a new part which has not been used on any previous model.

PETROL TANK AND PIPES, MK II AND MK III MODELS

1555

H694

FUEL TANK AND PIPES, MK II AND MK III 3 LITRE MODELS

Plate Ref.	1 2 3 4 Description	Qty.	Part No.	Remarks
27	Circlip fixing locking bolt	2	381352@@	
28	Striker bracket	1	381344@@	
	Screw (10 UNF x ½" long) } Fixing striker bracket to filler lid	2	542390@@	
	Spring washer } Fixing striker bracket to filler lid	2	3073@@	
	Plain washer } Fixing striker bracket to filler lid	2	3685@@	
29	FUEL FILLER CAP	1	381399@@	
30	Rubber sealing washer for cap	1	384757@@	
31	Hinge complete for lid	1	381337@@	
	Shim for hinge } Fixing hinge to filler lid	2	381220@@	Mk II cars with concealed filler cap
	Screw (10 UNF x ½" long) } Fixing hinge to filler lid	2	542390@@	
	Plain washer } Fixing hinge to filler lid	2	3685@@	
	Spring washer } Fixing hinge to filler lid	2	3073@@	
	Sealing washer } Fixing hinge to filler box	2	381353@@	
	Plain washer } Fixing hinge to filler box	2	3685@@	
	Nylock nut (10 UNF) } Fixing hinge to filler box	2	251345@@	
32	Torsion spring for lid	1	381221@@	
33	Anti-rattle pad, fuel filler, hinge to bowl	1	359246@@	
34	Anti-rattle pad, fuel filler, lid to bowl	1	359245@@	
35	FUEL FILLER UNIT AND CHROME CAP	1	384385	
36	Seal for filler cap	1	380886	
37	LOCKING BOLT FOR FUEL FILLER	1	384722	
	'O' ring for locking bolt	1	384717	
	Retaining clip	1	384719	Mk III
38	Seal, fuel filler unit to body	1	384387	
39	Drive screw } Fixing fuel filler unit to body	4	534907	
40	Plain washer } Fixing fuel filler unit to body	4	4563	
41	Clinch nut } Fixing fuel filler unit to body	4	364435	
	Drainage tube for fuel filler	1	352539@	Cap with chrome filler cap
42	Drainage tube for fuel filler	1	359247@@	Cars with concealed fille cap
	Pipe, fuel tank to filler	1	513966@	
43	Pipe, fuel tank to filler	1	541866@@	
	Hose, filler end	1	513870@	Mk II cars with chrome filler cap
44	Hose, filler end	1	541871@@	Mk II cars with concealed filler cap
	Hose, filler end	1	513555	
45	Hose, tank end	1	513555	
46	Internal filler hose, plastic	1	536521	
47	Clip for hose, tank and filler end	4	271870	
48	Fuel pipe complete, outlet to filter bowl	1	536436	
49	Olive } Fixing pipe to pump	1	270105	
50	Nut (¼" BSP) } Fixing pipe to pump	1	50421	
51	Fuel pipe complete, tank to pump (main)	1	542848	
52	Fuel pipe complete, tank to pump (reserve)	1	542847	

@ Mk II cars numbered up to 77000781, 77100082, 77300106, 73500045, 73800012 77500948, 77600306, 77800114, 74000048, 74300005

@@ Mk II cars numbered from 77000782, 77100083, 77300107, 73500046, 73800013 77500949, 77600307, 77800115, 74000049, 74300006 onwards

*Asterisk indicates a new part which has not been used on any previous model.

PETROL TANK AND PIPES, MK II AND MK III MODELS

1557

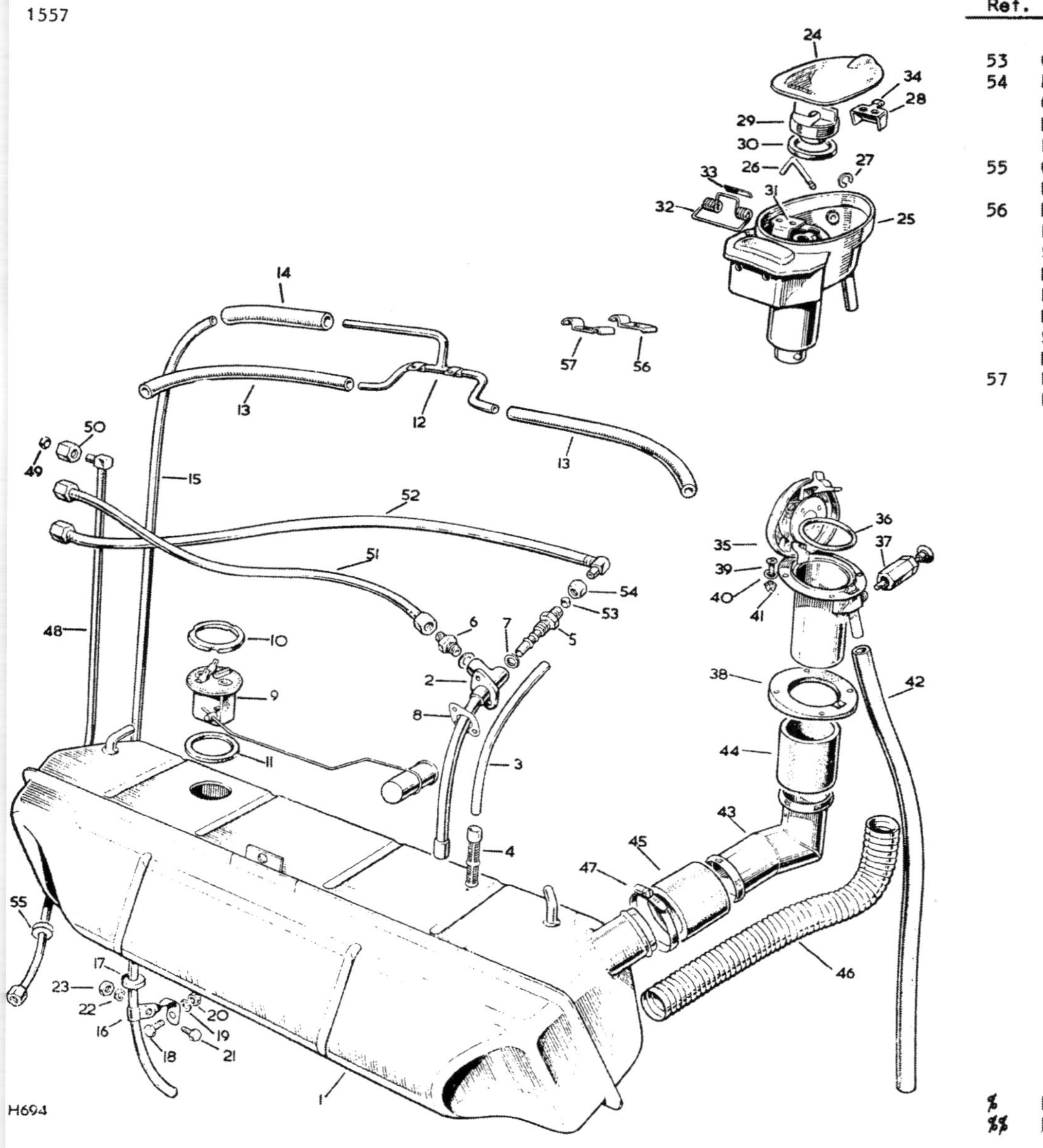

Plate Ref.	1 2 3 4 Description		Qty.	Part No.	Remarks
53	Olive	Fixing pipe to pump and tank	2	270105%	
54	Nut (¼" BSP)		2	50421%	
	Olive	Fixing pipes to pump and tank	5	542846%%	
	Nut		5	542845%%	
	Fuel pipe, nylon, carburetter to filter		1	541491	
55	Grommet in boot floor(1" dia.)		1	522848	
	Pipe clip, single,		4	50642	
56	Pipe clip, single		1	530716	
	Bolt (¼" UNF x ½" long)	Fixing front single clips to body	2	255204	
	Spring washer		2	3074	
	Nut (¼" UNF)		2	254810	
	Bolt (¼" UNF x ⅝" long)	Fixing rear single clip to body	1	255206	
	Plain washer		1	3900	
	Spring washer		1	3074	
	Nut (¼" UNF)		1	254810	
57	Pipe clip, double		3	530717	
	Drive screws for clips		4	78322	

% Mk II cars with suffix letters 'A' and 'B'
%% Mk II cars with suffix letter 'C' onwards

*Asterisk indicates a new part which has not been used on any previous model.

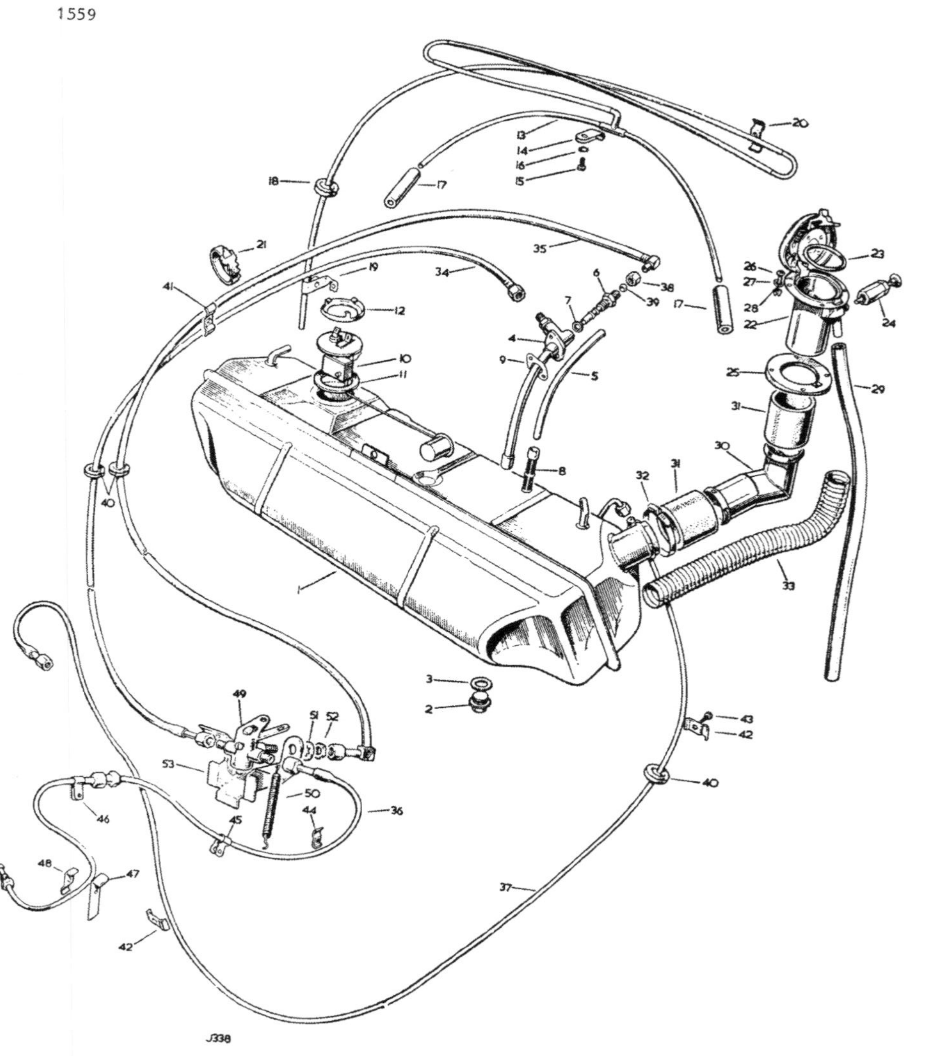

Plate Ref.	1 2 3 4	Description	Qty.	Part No.	Remarks
1		FUEL TANK COMPLETE	1	562173	
2		Drain plug for tank	1	2885	
3		Joint washer for drain plug	1	2347	
4		Main feed tube complete	1	562150	
5		Reserve feed pipe	1	536455	
6		Union for reserve feed	1	562148	
7		Washer for union	1	213959	
8		Fuel filter	1	536457	
9		Joint washer } Fixing fuel feed and reserve assembly to tank	1	267837	
		Screw(3 BA x 9/16"long) } Fixing fuel feed and reserve assembly to tank	2	535822	
		Spring washer } Fixing fuel feed and reserve assembly to tank	2	3101	
10		GAUGE UNIT FOR FUEL TANK	1	563321	
11		Joint washer for gauge unit	1	519965	
12		Locking ring fixing gauge unit	1	559247	
		Bolt(5/16"UNF x 1¼"long) } Fixing fuel tank to body	3	256020	
		Rubber packing } Fixing fuel tank to body	3	272853	
		Plain washer } Fixing fuel tank to body	6	2266	
		Nut (5/16" UNF) } Fixing fuel tank to body	4	254831	
		Self-locking nut(5/16"UNF) } Fixing fuel tank to body	1	252161	
		Plain washer } Fixing fuel tank to body	1	3898	
13		Vent pipe complete	1	562171	
14		Clip for vent pipe	2	557148	
15		Drive screw } Fixing 'T' piece to body	2	78155	
16		Shakeproof washer } Fixing 'T' piece to body	2	74236	
17		Breather hose	2	557187	
18		Grommet for overflow pipe	1	232916	
19		Clip for overflow pipe	1	562172	
		Bolt(¼"UNF x ½"long) } Fixing clip to body	1	255204	
		Spring washer } Fixing clip to body	1	3074	
		Nut (¼"UNF) } Fixing clip to body	1	254810	
		Bolt(10 UNF x ½"long) } Fixing overflow pipe to clip	1	257017	
		Spring washer } Fixing overflow pipe to clip	1	3073	
		Nut (10 UNF) } Fixing overflow pipe to clip	1	257023	
20		Clip fixing vent pipe to body	2	509412	
		Drive screw fixing clip	2	78640	
21		Cleat	2	240430	
22		FUEL FILLER UNIT AND CHROME CAP	1	384385	
23		Seal for filler cap	1	380886	
24		LOCKING BOLT FOR PETROL FILLER UNIT	1	384722	
		'O' ring for locking bolt	1	384717	
		Retaining clip	1	384719	
25		Seal, fuel filler unit to body	1	384387	
26		Drive screw } Fixing fuel filler unit to body	4	534907	
27		Plain washer } Fixing fuel filler unit to body	4	4563	
28		Clinch nut } Fixing fuel filler unit to body	4	364435	
29		Drainage tube for fuel filler unit	1	359247	

*Asterisk indicates a new part which has not been used on any previous model.

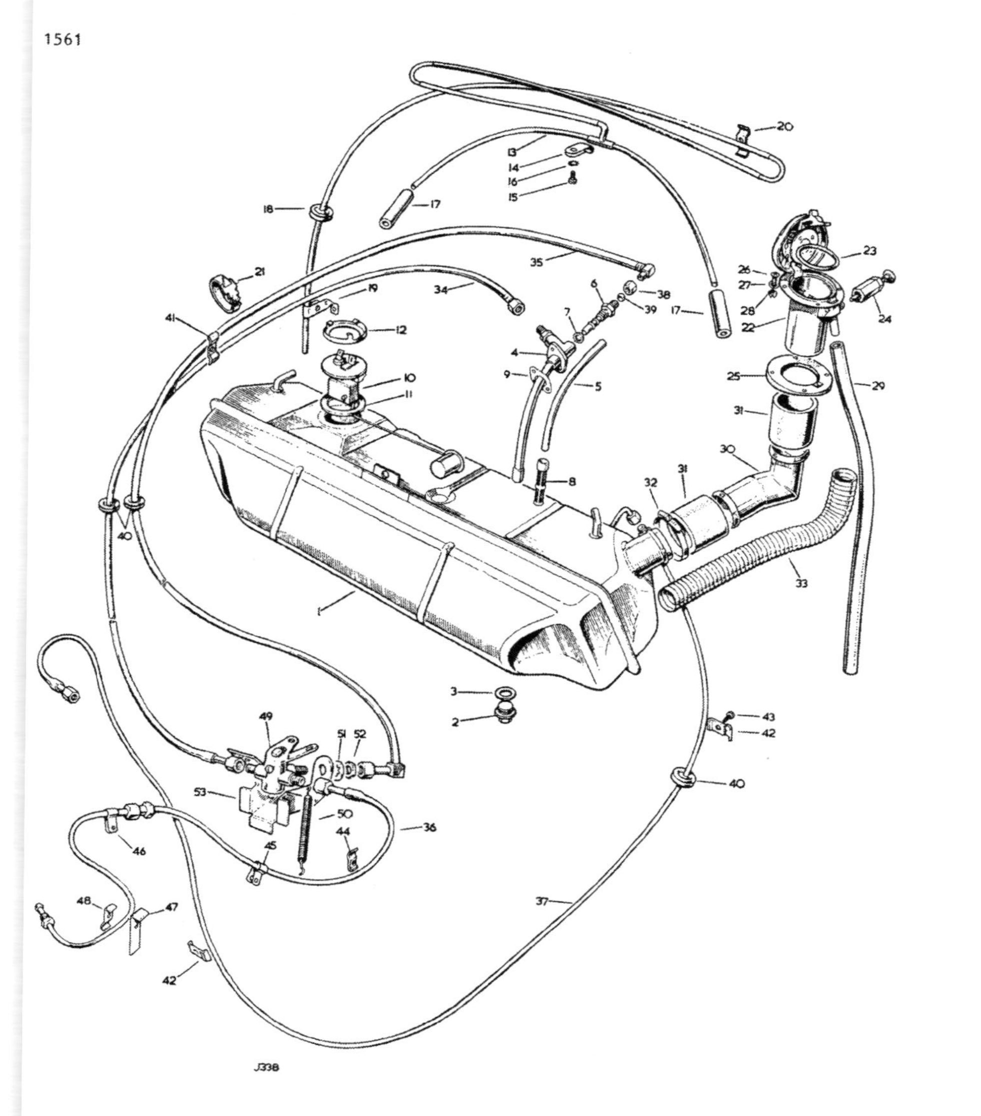

Plate Ref.	1 2 3 4	Description		Qty.	Part No.	Remarks
30		Pipe, fuel tank to filler unit		1	541866	
31		Hose, pipe to tank and pipe to filler unit		2	513555	
32		Clip for hose		4	271870	
33		Internal filler hose, plastic		1	536521	
34		Main fuel feed pipe, tank to reserve tap		1	562152	
35		Reserve fuel feed pipe, tank to reserve tap		1	562153	
36		Main fuel feed pipe, reserve tap to bundy tube		1	562154	
		Fuel pipe, bundy, main feed pipe to fuel pump		1	562372	
37		Fuel return pipe		1	562156	
38		Union nut	Fixing reserve feed pipe to tank	1	603431	
39		Olive		1	542846	
		Union nut	Fixing return pipe to engine and tank	2	534790	
		Olive		2	534797	
40		Grommet for boot floor		3	522848	
41		Pipe clip, double		6	548088	
		Drive screw fixing double clips		6	78322	
		Screw (10 UNF x ½" long)	Fixing double clip to rear shock absorber bracket	1	257017	
		Nut (10 UNF)		1	257023	
		Spring washer		1	3073	
		Plain washer		1	3903	
42		Pipe clip, single	Fixing fuel return pipe to body	13	50183	
43		Drive screw		13	78322	
44		Pipe clip		2	530716	
		Drive screw fixing clip		2	78322	
45		Pipe clip at sub-frame		1	3621	
		Pipe clip	Fixing fuel return pipe to fuel tank	1	50183	
		Screw (10 UNF x 7/16" long)		1	257004	
		Nut (10 UNF)		1	257023	
		Spring washer		1	3073	
46		Pipe clip		2	215791	
		Bolt (¼" UNF x ½" long)	Fixing clip to body	2	255204	
		Spring washer		2	3074	
47		Bracket for clip		1	562174	
48		Clip		1	216708	
		Bolt (10 UNF x ½" long)	Fixing clip to bracket	1	257017	
		Spring washer		1	3073	
		Nut (10 UNF)		1	257023	
49		Fuel reserve tap		1	565421	
50		Return spring, tap to cross-member		1	546004	
51		Shakeproof washer	Fixing reserve tap to bracket	1	72001	
52		Nut (7/16" UNF)		1	534978	
53		Bracket for fuel reserve tap		1	386109	
		Nylon clip at air cleaner, fixing fuel return pipe		2	603329	

*Asterisk indicates a new part which has not been used on any previous model.

PETROL SEDIMENT BOWL

Plate Ref.	1 2 3 4	Description	Qty.	Part No.	Remarks
		PETROL SEDIMENT BOWL COMPLETE	1	267494	AC Delco type
1		Body only	1	268793	AC Delco type
2		Bowl only	1	236891	AC Delco type
3		Joint washer for bowl	1	241225	AC Delco type
4		Gauze for bowl	1	241223	AC Delco type
5		Retainer for bowl	1	268797	AC Delco type
		Body only	1	514269	Wipac type
		Bowl only	1	236896	Wipac type
		Joint washer for bowl	1	236895	Wipac type
		Gauze for bowl	1	514270	Wipac type
		Retainer for bowl	1	240324	Wipac type
		Screw cap for retainer	1	240325	Wipac type
6		Bolt (¼" UNF x ¾" long)	2	255207	
7		Plain waasher — Fixing filter to wing valance	4	3840	
8		Spring washer — Fixing filter to wing valance	2	3074	
9		Nut (¼" UNF) — Fixing filter to wing valance	2	254810	
10		Double-ended union for petrol filter	2	267529	Mk I and Mk IA
		Double-ended union, filter bowl outlet	1	525530	Mk II with 'Sureflex' — Alternative
		Double-ended union, filter bowl outlet	1	267529	Mk II with nylon pipe — Alternative
		Double-ended union, filter bowl inlet	1	267529	Mk II and Mk III
		Elbow — Outlet union to pipe	1	50499	Mk II and Mk III
		Nut (¼" BSP) — Outlet union to pipe	1	50421	Mk II and Mk III
11		Joint washer for union	2	243967	

*Asterisk indicates a new part which has not been used on any previous model.

PETROL PUMP, MK1 AND MK 1A 3 LITRE MODELS

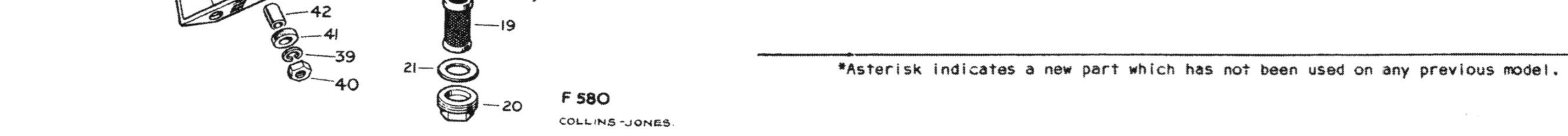

PETROL PUMP, Mk I AND Mk IA 3 LITRE

Plate Ref.	1 2 3 4 Description	Qty.	Part No.	Remarks
	PETROL PUMP COMPLETE, type HP, twin contacts	1	537979	2 off on Mk IA
1	Coil complete	1	524125	SU pump AUA 82
2	Spring for armature (short spring)	1	524503	SU pump AUA 82
	Coil complete	1	530884	SU pumps AUA 182 and AUB 182
	Spring for armature	1	530885	SU pumps AUA 182 and AUB 182
3	Diaphragm complete	1	262142	SU pumps AUA 50 and AUA 82
	Diaphragm complete	1	530883	SU pumps AUA182 and AUB182
4	Roller for diaphragm	11	260583	
5	Plate body	1	262299	
6	Joint washer for plate body	1	262300	
7	Body	1	262287	
8	Screw fixing coil housing to body	6	262301	
9	Special screw — For earth terminal	1	262829	
10	Special spring washer — For earth terminal	1	262303	
11	Valve cage	1	262288	
12	Disc for valve	2	262297	
13	Spring clip retaining valve disc	1	262289	
14	Washer for valve cage	1	262290	
15	Outlet union	1	262294	
16	Washer for outlet union	1	262293	
17	:nlet union	1	262296	
18	Washer for inlet union	1	262298	
19	Filter	1	262292	
20	Plug for filter	1	262295	
21	Washer for filter plug	1	262293	
22	Contact set complete	1	268674	
23	Special screw for contact blade	1	538506	
24	Moulding for end plate	1	260585	
25	Screw fixing moulding	2	262308	
26	Terminal screw	1	262307	
27	Cover for end plate, black	1	269848	
28	Condenser for petrol pump	1	245382	
29	Clip for condenser	1	245384	
30	Terminal nut	1	262306	
31	Tag for terminal	1	262305	
32	Rubber mounting plate	1	212557	
33	Rubber bush — Fixing pump to mounting bracket	2	212558	
34	Plain washer — Fixing pump to mounting bracket	2	2228	
35	Special set bolt — Fixing pump to mounting bracket	2	212556	
36	Bracket for petrol pump	1	279369	
37	Bolt (5/16" UNF x 1⅛" long) — Fixing bracket to body	2	255029	
38	Plain washer — Fixing bracket to body	2	2223	
39	Spring washer — Fixing bracket to body	2	3075	
40	Nut (5/16" UNF) — Fixing bracket to body	4	254831	
41	Rubber washer — Fixing bracket to body	4	504432	
42	Distance tube — Fixing bracket to body	2	504431	
43	Mounting bracket for petrol pump	1	529366	Mk IA

*Asterisk indicates a new part which has not been used on any previous model.

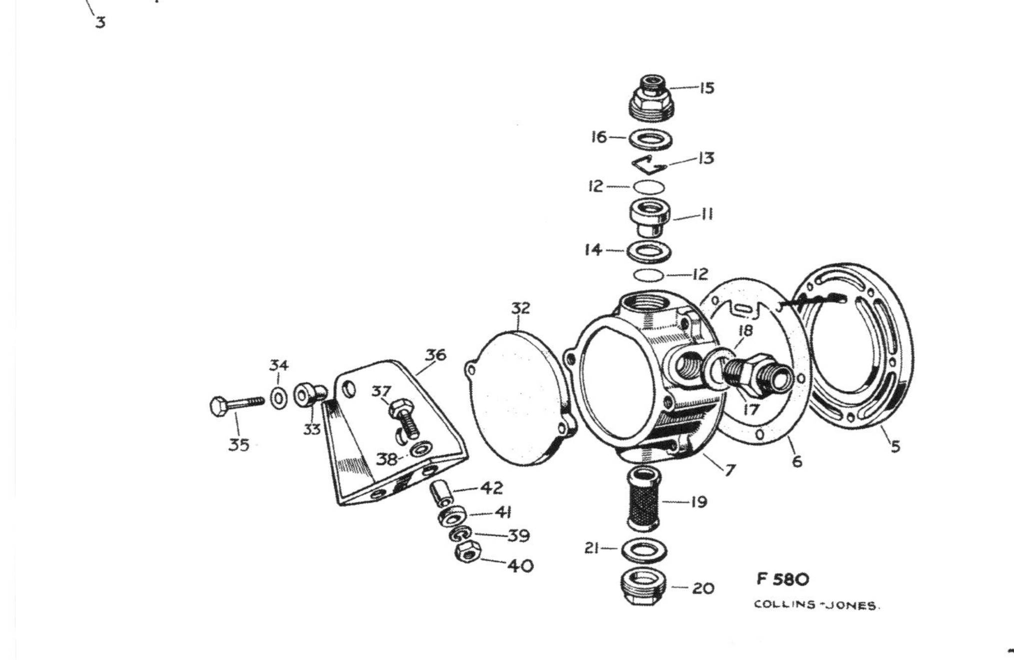

Plate Ref.	1 2 3 4 DESCRIPTION		Qty	Part No.	REMARKS
44	Clamp bracket for petrol pump		1	525249	Mk IA
45	Rubber strip	Mounting petrol pump to clamp bracket	1	534200	
46	Bolt (¼" UNF x ⅞" long)		1	255208	
47	Self-locking nut (¼" UNF)		1	252160	
48	Bolt (¼" UNF x ¾" long)	Fixing mounting bracket and clamp bracket together and fixing mounting bracket to body	2	255207	
49	Bolt (¼" UNF x ⅝" long)		2	255206	
50	Spring washer		4	3074	
51	Plain washer		4	3840	
52	Nut (¼" UNF)		4	254811	

* Asterisk indicates a new part which has not been used on any previous Rover model

1569

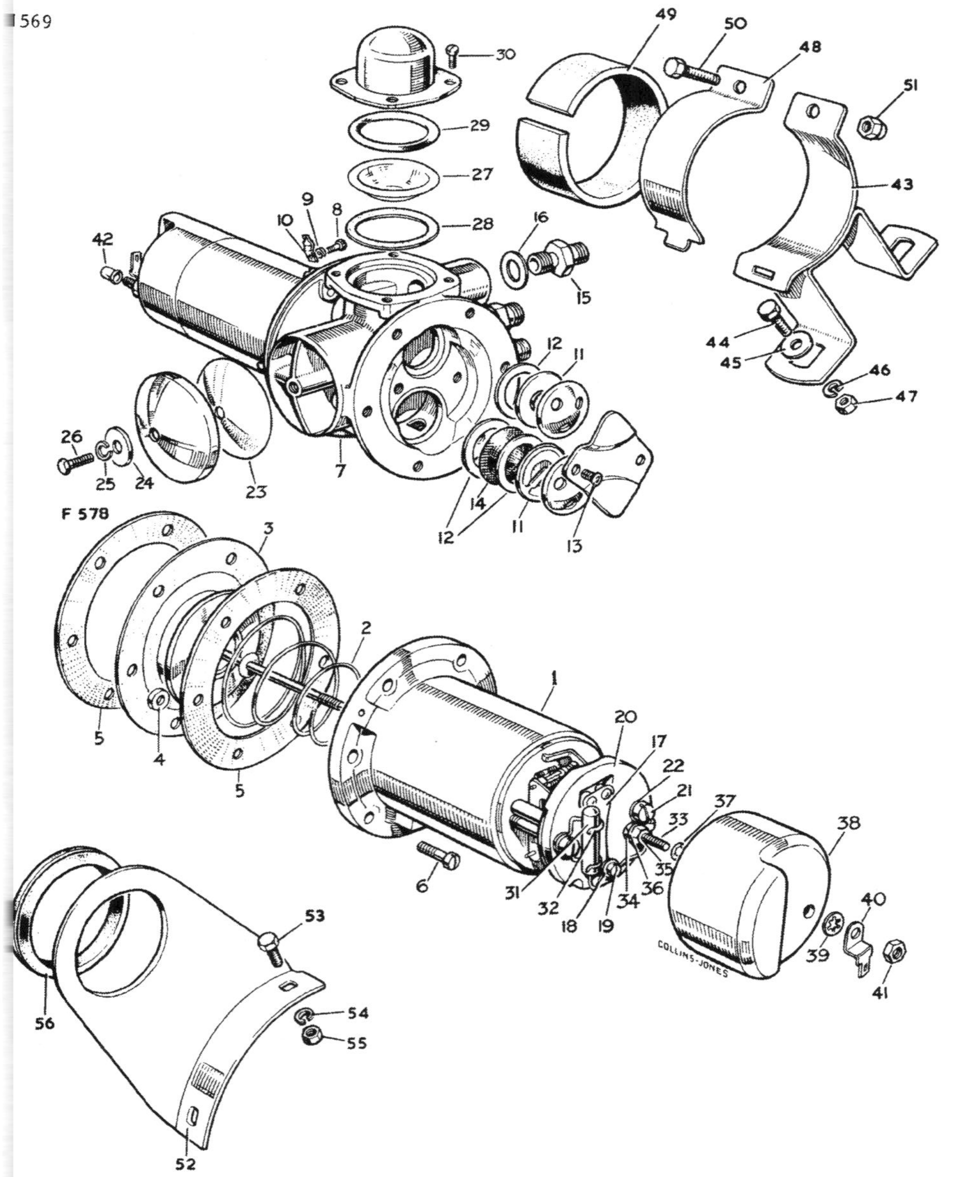

Plate Ref.	1 2 3 4	Description	Qty.	Part No.	Remarks
		PETROL PUMP, DOUBLE ENTRY TYPE	1	545306	
1		Coil complete	2	530884	
2		Spring for armature	2	530885	
3		Diaphragm complete	2	542320	
4		Roller for diaphragm	22	260583	
5		Joint washer, diaphragm to body	2	538499	
6		Screw, fixing coil housing to body	12	538501	
7		Body	1	538485	
8		Special screw	1	538502	For earth terminal, SU AUA 699
9		Spring washer	1	538503	For earth terminal, SU AUA 699
10		Lucar connector	1	538510	For earth terminal, SU AUA 699
11		Valve assembly	4	538486	
12		Sealing washer for valve assembly	6	538487	
13		Screw for valve assembly	4	538489	
14		Filter	2	538488	
15		Unions, inlet and outlet	3	538498	
16		Sealing ring for unions	3	538497	
17		Contact set complete	2	538504	
18		Dished washer	2	538507	
19		Screw	2	538506	
20		Pedestal assembly	2	260585	
21		Screw	4	262308	Fixing pedestal assembly
22		Spring washer	2	262303	Fixing pedestal assembly
23		Joint washer for dished cover	1	538490	
24		Dished washer	1	538491	Fixing dished cover to body
25		Spring washer	1	538492	Fixing dished cover to body
26		Bolt	1	538493	Fixing dished cover to body
27		Diaphragm for air bottle	1	538494	
28		Joint washer, diaphragm to housing	1	542319	
29		Sealing ring, for air bottle cover	1	538495	
30		Screw, fixing air bottle cover	4	538496	
31		Condenser	1	245382	
32		Clip for condenser	1	245384	
33		Terminal screw	2	262307	
34		Spring washer for terminal	2	262303	
35		Terminal nut	2	262306	
36		Lead washer	2	262305	
37		Washer for terminal screw	2	538508	
38		Cover, black	2	538509	
39		Shakeproof washer	2	274951	
40		Lucar connector	2	538510	
41		Terminal nut Lucar connector	2	262309	
42		Insulating sleeve, for terminal	2	538511	
43		Petrol pump mounting bracket	1	536431	
44		Bolt (5/16" UNF x ¾" long)	2	255226	Fixing bracket to body
45		Plain washer	2	2249	Fixing bracket to body
46		Spring washer	2	3075	Fixing bracket to body
47		Nut (5/16" UNF)	2	254831	Fixing bracket to body

*Asterisk indicates a new part which has not been used on any previous model.

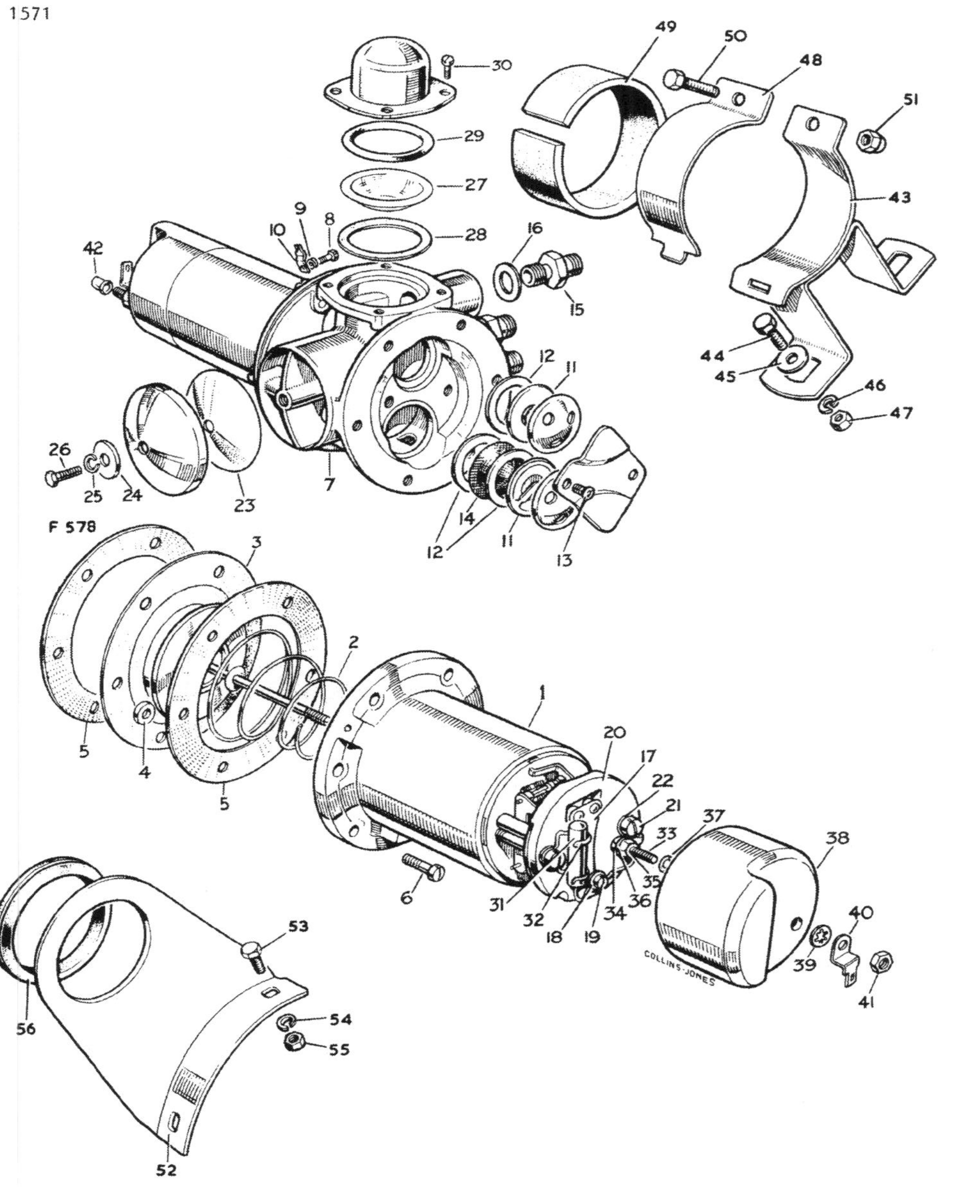

F 578

Plate Ref.	1 2 3 4	DESCRIPTION	Qty	Part No.	REMARKS
48	Clamp plate for pump		1	529141	
49	Rubber strip	} Fixing clamp plate and petrol pump to mounting plate	1	534200	
50	Bolt (¼" UNF x ⅞" long)		1	255208	
51	Self-locking nut (¼" UNF)		1	252160	
52	Petrol pump mounting bracket		1	536434	
53	Bolt (¼" UNF x $\frac{9}{16}$" long)	} Fixing mounting bracket to body	2	255205	
54	Spring washer		2	3074	
55	Nut (¼" UNF)		2	254810	
56	Grommet for petrol pump mounting		1	536562	

* Asterisk indicates a new part which has not been used on any previous Rover model

AIR SILENCER AND OIL BATH CLEANER MK1 3 LITRE MODELS

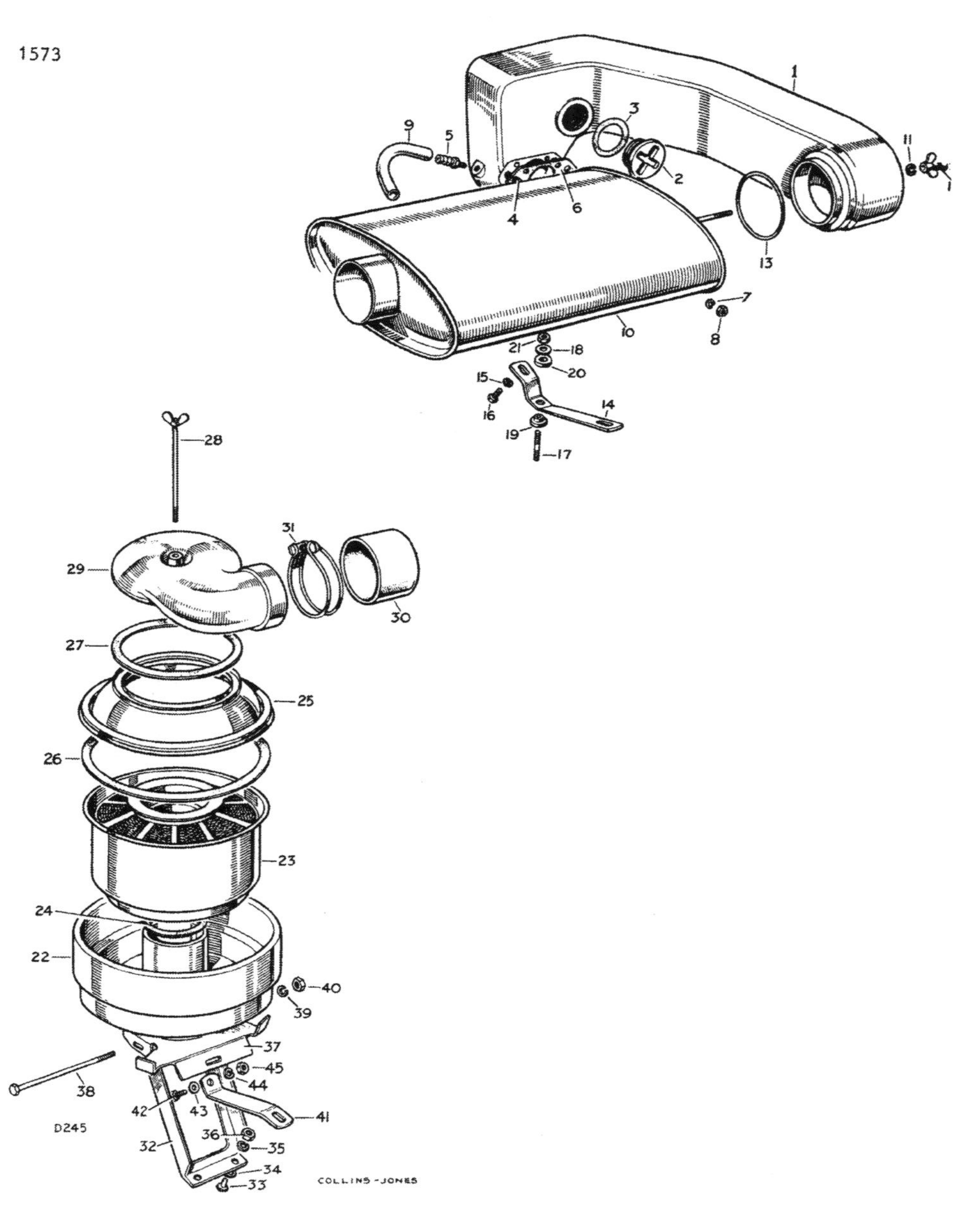

AIR SILENCER AND OIL BATH CLEANER, MK I 3 LITRE MODELS

Plate Ref.	1 2 3 4 Description	Qty.	Part No.	Remarks
1	ELBOW ASSEMBLY FOR AIR SILENCER	1	500666	
2	Plug for elbow	3	279241	
3	Joint washer for plug	3	279858	
4	Stud for carburetter	2	500667	
5	Hose adaptor	1	242319	Not required on RHStg cars with power steering
	Hose adaptor	1	523184	RHStg cars with power steering
	Clip fixing delivery hose	1	523183	RHStg cars with power steering
6	Joint washer, elbow to carburetter	1	277518	
7	Spring washer } Fixing elbow to carburetter	2	3075	
8	Nut (5/16" UNF) } Fixing elbow to carburetter	2	254831	
9	Air hose, adaptor to carburetter	1	505033	
10	AIR SILENCER	1	277094	
11	Spring washer } Fixing elbow to air silencer	1	3074	
12	Nut (¼" UNC) } Fixing elbow to air silencer	1	256850	
13	Rubber ring } Fixing elbow to air silencer	1	500686	
14	Air silencer support bracket	1	503662	
15	Spring washer } Fixing bracket to silencer	2	3075	
16	Set bolt (5/16" UNF x ⅜"long) } Fixing bracket to silencer	2	255021	
17	Stud } Fixing air silencer to carburetter distance piece	1	512700	
18	Plain washer } Fixing air silencer to carburetter distance piece	2	2223	
19	Rubber bush } Fixing air silencer to carburetter distance piece	1	212558	
20	Rubber washer } Fixing air silencer to carburetter distance piece	1	503663	
21	Nut (5/16"UNF) } Fixing air silencer to carburetter distance piece	1	254831	
22	OIL BATH AIR CLEANER COMPLETE	1	269222	
23	Filter element	1	269785	
24	Gasket for element	1	262638	
25	Top cover for filter	1	261628	
26	Gasket for cover	1	261629	
27	Gasket for outlet pipe	1	261630	
28	Centre bolt for air cleaner	1	269786	
29	Outlet pipe for air cleaner	1	277096	
30	Rubber connecting tube for air silencer	1	231789	
31	Clip for connecting tube	2	50347	
32	Support bracket for air cleaner bracket	1	500550	
33	Bolt (5/16"UNF x 1" long) } Fixing support bracket to engine foot	2	255228	
34	Plain washer } Fixing support bracket to engine foot	2	2550	
	Distance washer } Fixing support bracket to engine foot	2	4075	
35	Spring washer } Fixing support bracket to engine foot	1	3075	
	Special earthing washer } Fixing support bracket to engine foot	2	510912	
36	Nut (5/16" UNF) } Fixing support bracket to engine foot	2	254831	
37	Bracket for oil bath air cleaner	1	274415	
38	Bolt (5/16"UNF x 3¾"long) } Fixing bracket to support bracket	1	256233	
39	Spring washer } Fixing bracket to support bracket	1	3075	
40	Nut (5/16" UNF) } Fixing bracket to support bracket	1	254831	
41	Connecting bracket, dynamo to oil bath air cleaner	1	277099	
42	Bolt (¼"UNF x ⅜"long) } Fixing connecting bracket to dynamo and air cleaner bracket	2	255206	
43	Plain washer } Fixing connecting bracket to dynamo and air cleaner bracket	1	3840	
44	Spring washer } Fixing connecting bracket to dynamo and air cleaner bracket	2	3074	
45	Nut (¼" UNF) } Fixing connecting bracket to dynamo and air cleaner bracket	2	254810	

*Asterisk indicates a new part which has not been used on any previous model.

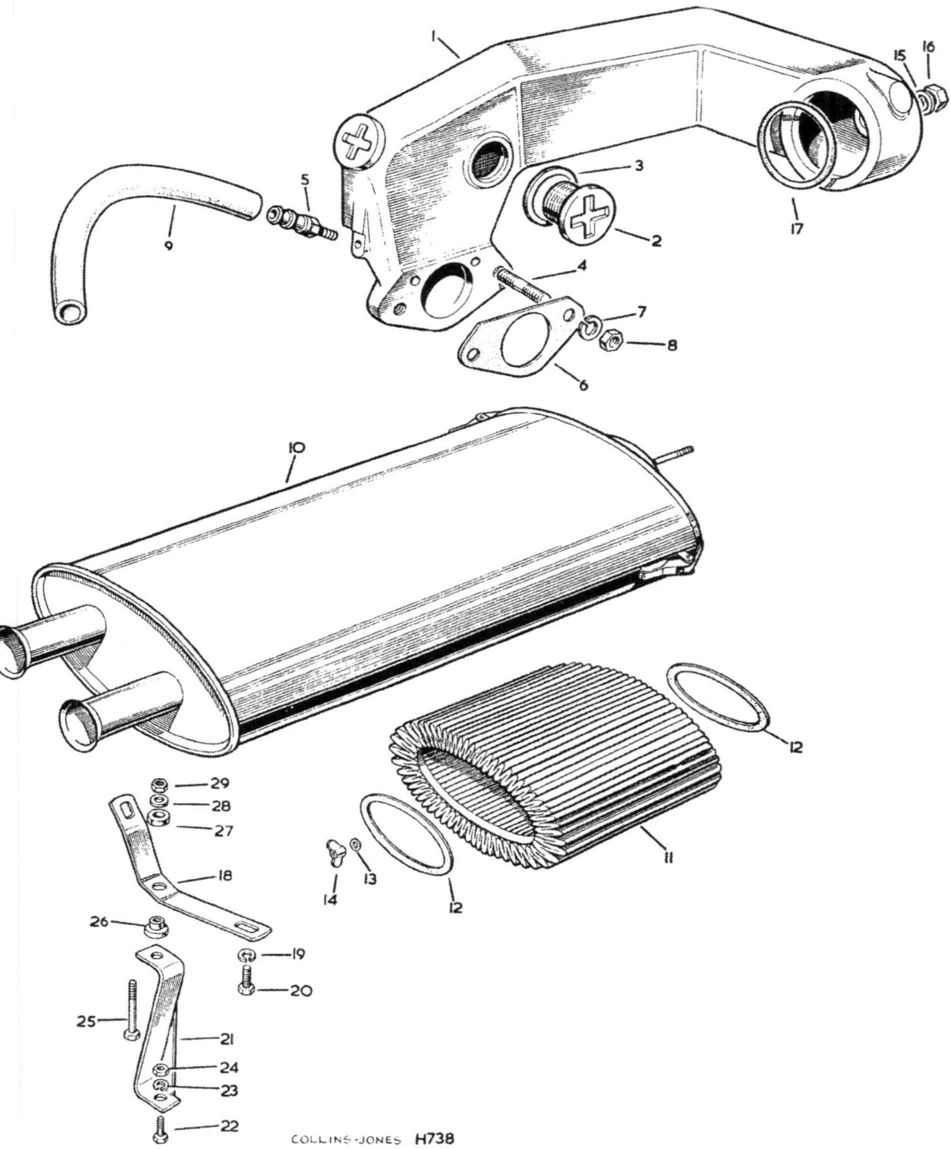

Plate Ref.	1 2 3 4 DESCRIPTION		Qty	Part No.	REMARKS
	ELBOW ASSEMBLY FOR AIR SILENCER		1	530703†	Mk IA
	ELBOW ASSEMBLY FOR AIR SILENCER		1	532664††	Mk IA
1	ELBOW ASSEMBLY FOR AIR SILENCER		1	534015	Mk II and Mk III except NADA
	ELBOW ASSEMBLY FOR AIR SILENCER		1	542416	Mk II and Mk III NADA
2	Plug for elbow		3	279241	
3	Joint washer for plug		3	279858	
4	Stud for carburetter		2	500667	
5	Hose adaptor		1	242319	
6	Joint washer, elbow to carburetter		1	277518	
7	Spring washer	Fixing elbow to carburetter	2	3075	
8	Nut (5/16" UNF)	Fixing elbow to carburetter	2	254811	
9	Air hose, adaptor to carburetter		1	505033	
10	AIR CLEANER AND SILENCER		1	532661	
11	Element for air cleaner		1	532138	
12	Rubber washer for element		2	532139	
13	Plain washer		1	3840	
14	Wing nut (1/4" UNC)		1	250044	
15	Plain washer	Fixing elbow to air silencer	1	3840	
16	Nut (1/4" UNC)	Fixing elbow to air silencer	1	256850	
17	Rubber ring for air silencer elbow		1	500686	
18	Air cleaner support bracket, upper		1	523417	
19	Spring washer	Fixing bracket to air silencer	2	3075	
20	Set bolt (5/16" UNF x 5/8" long)	Fixing bracket to air silencer	2	255021	
21	Bracket, lower, air cleaner support to dynamo bracket		1	530569	Mk IA
22	Bolt (5/16" UNF x 7/8" long)	Fixing bracket to dynamo bracket	1	255225	Mk IA
23	Spring washer	Fixing bracket to dynamo bracket	1	3075	Mk IA
24	Nut (5/16" UNF)	Fixing bracket to dynamo bracket	1	254811	Mk IA
25	Bolt (5/16" UNF x 1 1/8" long)	Fixing air silencer to bracket	1	255029	Mk IA
26	Rubber bush	Fixing air silencer to bracket	1	212558	Mk IA
27	Rubber washer	Fixing air silencer to bracket	1	503663	Mk IA
28	Plain washer	Fixing air silencer to bracket	2	2223	Mk IA
29	Nut (5/16" UNF)	Fixing air silencer to bracket	1	254811	Mk IA
	Plain washer	Fixing silencer to inlet manifold	1	2223	Mk II and Mk III
	Rubber bush	Fixing silencer to inlet manifold	1	212558	Mk II and Mk III
	Rubber washer	Fixing silencer to inlet manifold	1	503663	Mk II and Mk III
	Spring washer	Fixing silencer to inlet manifold	1	3075	Mk II and Mk III
	Nut (5/16" UNF)	Fixing silencer to inlet manifold	1	254811	Mk II and Mk III

* Asterisk indicates a new part which has not been used on any previous Rover model
† Cars numbered up to 72500322A, 72600045A, 72800024A, 73000431A, 73100050A, 73300063A
†† Cars numbered from 72500323A, 72600046A, 72800025A, 73000432A, 73100051A, 73300064A
NADA indicates parts peculiar to cars exported to the North American dollar area

AIR CLEANER AND SILENCER AND FUEL PUMP

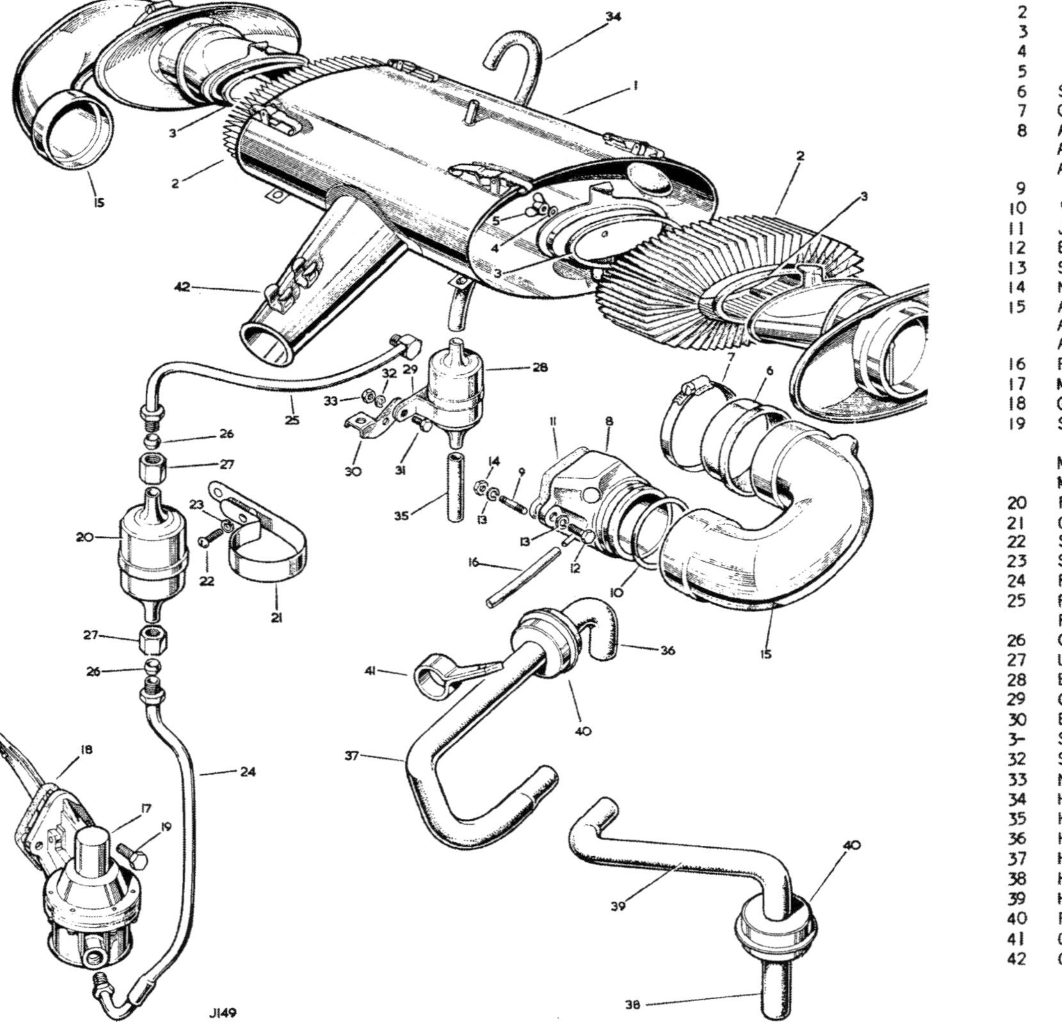

Plate Ref.	1 2 3 4 Description	Qty.	Part No.	Remarks
1	AIR CLEANER AND SILENCER	1	603618	
	Grommet for air cleaner body	1	603574	
2	Element for air cleaner	2	605191	
3	Rubber washer for element	4	532139	
4	Plain washer	2	3840	
5	Wing nut (¼" UNC)	2	250044	
6	Seal, air intake elbow to air silencer	2	603069	
7	Clip fixing seal to air cleaner	2	50344	
8	AIR INTAKE ADAPTOR ASSEMBLY	2	603685	Cars with manual choke
	AIR INTAKE ADAPTOR ASSEMBLY RH	1	603686	Cars with automatic choke
	AIR INTAKE ADAPTOR ASSEMBLY LH	1	603875	
9	Stud	1	610512	
10	'O' ring, adaptors to elbow	2	602634	
11	Joint washer	2	602633	
12	Bolt (5/16" UNC x 1" long) — Fixing adaptors to carburetters	4	253028	
13	Spring washer	6	3075	
14	Nut (5/16" UNF)	4	254831	
15	Air intake elbow	2	602619	Cars with manual choke
	Air intake elbow RH	1	602619	Cars with automatic choke
	Air intake elbow LH	1	603907	
16	Pressure balance pipe	1	603167	
17	Mechanical fuel pump	1	602410	
18	Gasket for fuel pump	1	602180	
19	Special bolt(5/16" UNC x 1.12"long)fixing fuel pump to front cover	2	602234	
	Minor repair kit — For fuel pump	1	605167	
	Major overhaul kit	1	605168	
20	Fuel filter	1	606168	
21	Clip fixing fuel filter	1	603184	
22	Screw(10 UNF x ⅜" long) — Fixing clip to front lifting bracket	2	78555	
23	Spring washer	2	3073	
24	Fuel pipe, fuel pump to filter	1	603564	
25	Fuel pipe, filter to carburetter	1	603562	Cars with manual choke
	Fuel pipe, filter to carburetters	1	603563	Cars with automatic choke
26	Olive — For fuel pipes	2	542846	
27	Union nut	2	603431	
28	Breather filter for crankcase	1	606168	
29	Clip fixing filter	1	603185	
30	Bracket for clip on inlet manifold	1	603162	
3-	Set bolt (¼" UNF x ½" long) — Fixing clip to	1	255204	
32	Spring washer	1	3074	
33	Nut (¼" UNF)	1	254810	
34	Hose, long, breather inlet	1	611107	
35	Hose, short, breather inlet	1	611108	
36	Hose breather outlet, RH	1	611092	
37	Hose carburetter breather outlet, RH	1	611093	
38	Hose, breather outlet, LH	1	611095	
39	Hose, carburetter breather outlet, LH	1	611094	
40	Flame trap	2	603330	
41	Clip for breather hose	1	603160	
42	Clip for cold start cable junction box	2	603376	

*Asterisk indicates a new part which has not been used on any previous model.

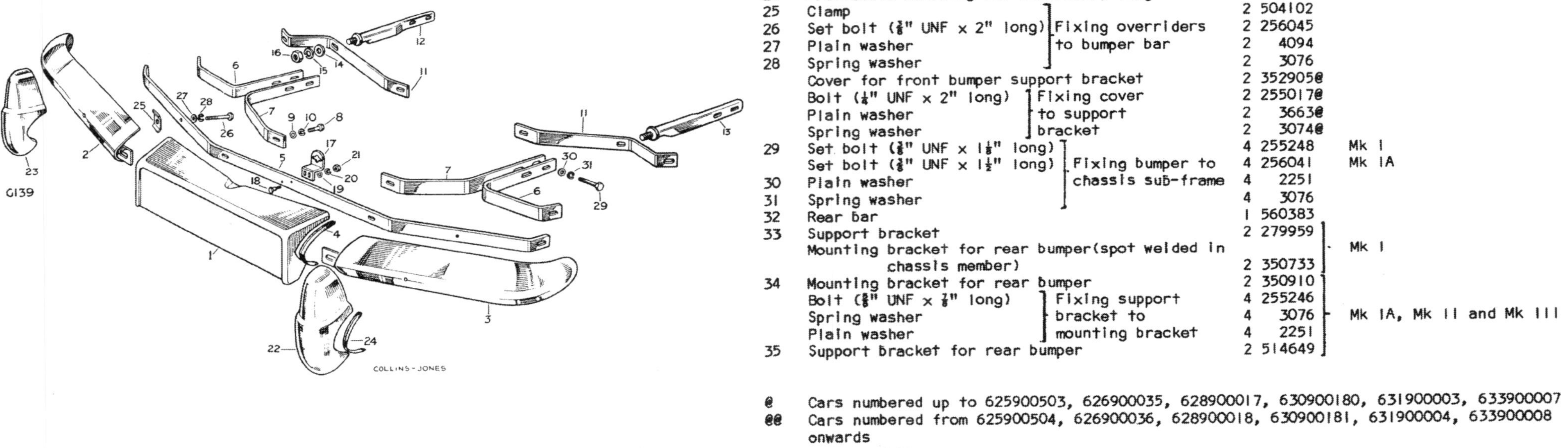

Plate Ref.	1 2 3 4 Description		Qty.	Part No.	Remarks
1	Centre bar	For front bumper	1	279946	
2	Outer bar, RH	For front bumper	1	507033	
3	Outer bar, LH	For front bumper	1	507032	
4	Protective moulding for centre bar		2	503604	
	Support bracket, centre		1	279947@	
5	Support bracket, centre		1	510049@@	
	Support bracket, outer		2	279949@	
6	Support bracket, outer		2	510050@@	Mk I
7	Support bracket, inner		2	279948	Mk I
8	Set bolt (⅜" UNF x 1¼" long)	Fixing bars to brackets	4	255249	Mk I
9	Plain washer	Fixing bars to brackets	4	2251	Mk I
10	Spring washer	Fixing bars to brackets	4	3076	Mk I
11	Support bracket for front bumper		2	514648	Mk IA, Mk II and Mk III
12	Mounting bracket, RH, for front bumper		1	518956	Mk IA, Mk II and Mk III
13	Mounting bracket, LH, for front bumper		1	518957	Mk IA, Mk II and Mk III
14	Plain washer	Fixing support bracket to mounting brackets	2	3300	Mk IA, Mk II and Mk III
15	Spring washer	Fixing support bracket to mounting brackets	2	3080	Mk IA, Mk II and Mk III
16	Nut (⅜" UNF)	Fixing support bracket to mounting brackets	2	254806	Mk IA, Mk II and Mk III
17	Bracket for starting handle		1	500656	
18	Bolt (5/16" UNF x 1" long)	Fixing starting handle bracket	2	255228	
19	Plain washer	Fixing starting handle bracket	2	2550	
20	Spring washer	Fixing starting handle bracket	2	3075	
21	Nut (5/16" UNF)	Fixing starting handle bracket	2	254831	
22	Overrider complete, front LH		1	500157	
23	Overrider complete, front RH		1	500156	
	Protective moulding for overrider, short		2	503602	
24	Protective moulding for overrider, long		2	503603	
25	Clamp	Fixing overriders to bumper bar	2	504102	
26	Set bolt (⅜" UNF x 2" long)	Fixing overriders to bumper bar	2	256045	
27	Plain washer	Fixing overriders to bumper bar	2	4094	
28	Spring washer	Fixing overriders to bumper bar	2	3076	
	Cover for front bumper support bracket		2	352905@	
	Bolt (¼" UNF x 2" long)	Fixing cover to support bracket	2	255017@	
	Plain washer	Fixing cover to support bracket	2	3663@	
	Spring washer	Fixing cover to support bracket	2	3074@	
29	Set bolt (⅜" UNF x 1⅛" long)	Fixing bumper to chassis sub-frame	4	255248	Mk I
	Set bolt (⅜" UNF x 1½" long)	Fixing bumper to chassis sub-frame	4	256041	Mk IA
30	Plain washer	Fixing bumper to chassis sub-frame	4	2251	
31	Spring washer	Fixing bumper to chassis sub-frame	4	3076	
32	Rear bar		1	560383	
33	Support bracket		2	279959	Mk I
	Mounting bracket for rear bumper(spot welded in chassis member)		2	350733	Mk I
34	Mounting bracket for rear bumper		2	350910	Mk IA, Mk II and Mk III
	Bolt (⅜" UNF x ⅞" long)	Fixing support bracket to mounting bracket	4	255246	Mk IA, Mk II and Mk III
	Spring washer	Fixing support bracket to mounting bracket	4	3076	Mk IA, Mk II and Mk III
	Plain washer	Fixing support bracket to mounting bracket	4	2251	Mk IA, Mk II and Mk III
35	Support bracket for rear bumper		2	514649	Mk IA, Mk II and Mk III

@ Cars numbered up to 625900503, 626900035, 628900017, 630900180, 631900003, 633900007

@@ Cars numbered from 625900504, 626900036, 628900018, 630900181, 631900004, 633900008 onwards

*Asterisk indicates a new part which has not been used on any previous model.

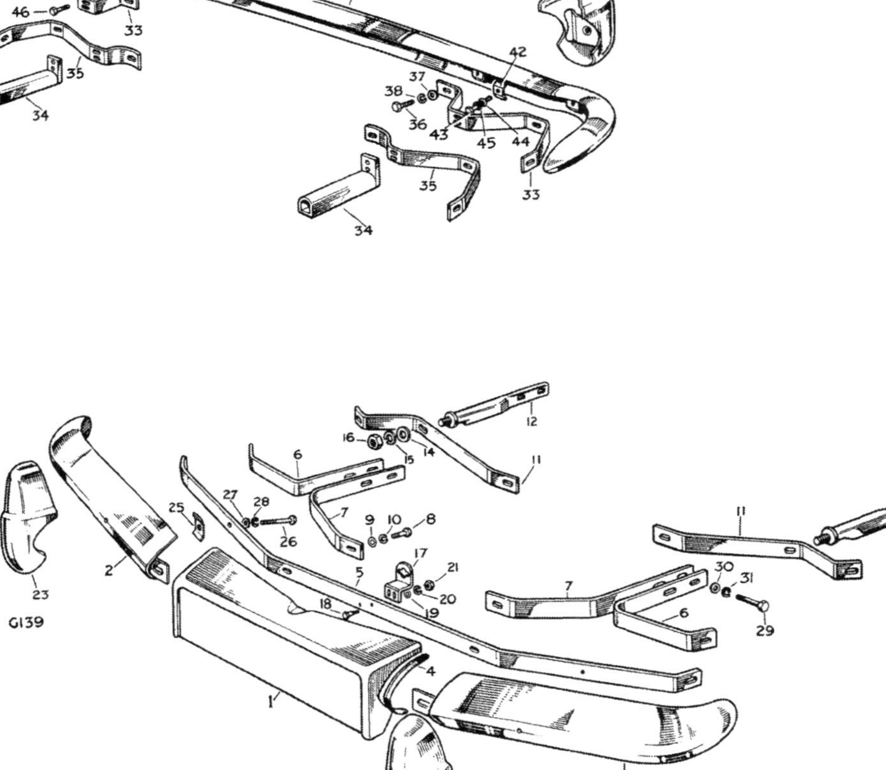

Plate Ref.	1 2 3 4 Description		Qty.	Part No.	Remarks
36	Set bolt ($\frac{3}{8}$" UNF x $\frac{7}{8}$" long)	Fixing bar to brackets	6	255246	
37	Plain washer	Fixing bar to brackets	6	2251	
38	Spring washer	Fixing bar to brackets	6	3076	
39	Overrider complete, rear LH		1	500156	
40	Overrider complete, rear RH		1	500157	
	Blanking plate for rear overrider, RH		1	541267	Mk II and Mk III Export
	Blanking plate for rear overrider, LH		1	541268	Mk II and Mk III Export
	Protective moulding for overrider, short		2	503602	
41	Protective moulding for overrider, long		2	503603	
42	Clamp	Fixing overrider to bumper bar	2	504102	
43	Set bolt ($\frac{3}{8}$" UNF x $1\frac{1}{4}$" long)	Fixing overrider to bumper bar	2	256043	
44	Plain washer	Fixing overrider to bumper bar	2	4049	
45	Spring washer	Fixing overrider to bumper bar	2	3076	
	Cover for rear bumper support bracket, RH		1	352906	1959 models only
	Cover for rear bumper support bracket, RH		1	352907	1959 models only
46	Bolt($\frac{3}{8}$" UNF x $\frac{7}{8}$" long)	Fixing rear bumper and support bracket cover to body	4	255246	
47	Plain washer	Fixing rear bumper and support bracket cover to body	4	2251	
48	Spring washer	Fixing rear bumper and support bracket cover to body	4	3076	
49	Nut ($\frac{3}{8}$" UNF)	Fixing rear bumper and support bracket cover to body	4	254812	

*Asterisk indicates a new part which has not been used on any previous model.

Plate Ref.	1 2 3 4 Description		Qty.	Part No.	Remarks
1	Centre bar	For front bumper	1	562224	
2	Outer bar, RH		1	560381	
3	Outer bar, LH		1	560380	
4	Protective moulding for centre bar		2	503604	
5	Support bracket centre		1	560382	
6	Support bracket for front bumper		2	514648	
7	Mounting bracket, RH, for front bumper		1	518956	
8	Mounting bracket, LH, for front bumper		1	518957	
9	Bolt(3/8" UNF x 1¼" long)	Fixing outer support brackets to centre support bracket	4	255249	
10	Spring washer		4	3076	
11	Plain washer		4	2251	
12	Plain washer	Fixing support bracket to mounting brackets	2	3300	
13	Spring washer		2	3080	
14	Nut (3/8" UNF)		2	254806	
15	Set bolt (3/8" UNF x 1½" long)	Fixing bumper to chassis sub-frame	4	256041	
16	Spring washer		4	3076	
17	Plain washer, large		4	2219	
18	Plain washer, small		6	2251	
19	OVERRIDER ASSEMBLY, RH FRONT		1	562226	
20	OVERRIDER ASSEMBLY, LH FRONT		1	562227	
21	Rubber for overrider		2	565366	
22	Spring washer	Fixing rubber to overrider	2	3074	
23	Nut (¼" UNF)		2	254810	
24	Clamp		4	504102	
25	Set bolt(5/16"UNFx2¼"long)	Fixing front overrider to bumper bar	4	256027	
26	Spring washer		4	3075	
27	Plain washer		4	2266	
28	Moulding for overrider		4	503603	
29	Rear bumper bar		1	560383	
30	Support bracket for rear bumper		2	514649	
31	Set bolt (3/8" UNF x 7/8" long)	Fixing bar to brackets	6	255046	
32	Spring washer		6	3076	
33	Plain washer		6	2251	
34	Mounting bracket for rear bumper		2	350910	
35	Bolt (3/8" UNF x 7/8" long)	Fixing support bracket to mounting bracket	4	255046	
36	Spring washer		4	3076	
37	Plain washer		4	2251	
38	OVERRIDER ASSEMBLY, RH REAR		1	562227	
39	OVERRIDER ASSEMBLY, LH REAR		2	562226	
40	Rubber for overrider		2	565366	
41	Spring washer	Fixing rubber to overrider	2	3074	
42	Nut(¼" UNF)		2	254810	
43	Moulding for overrider		4	503603	
	Clamp		4	504102	
	Set bolt (5/16" UNF x 2" long)	Fixing rear overrider to rear bumper	4	256226	
	Spring washer		4	3075	
	Plain washer		4	2266	

J334

*Asterisk indicates a new part which has not been used on any previous model.

H384

Plate Ref.	1 2 3 4 Description	Qty.	Part No.	Remarks
	HEADLAMP COMPLETE	2	502095	RH Stg Mk I and Mk IA
1	HEADLAMP COMPLETE, 60/45 watt	2	605069*	RH Stg Mk II and Mk III — Alternatives. Check before ordering
	HEADLAMP COMPLETE, 75/50 watt	2	559254	RH Stg Mk III — Alternatives. Check before ordering
	HEADLAMP COMPLETE	2	605071	LH Stg except France
	HEADLAMP COMPLETE	2	605072*	LH Stg France
	HEADLAMP SHELL AND CONNECTOR	2	605073*	America. Dollar area
2	Special screw for light unit adjustment	2	601977	Mk II and Mk III
	Adaptor for bulb	2	507766	RH Stg Mk I and Mk IA
3	Adaptor for bulb	2	600226	RH Stg Mk II and Mk III, also Export
	Sleeve for terminal	6	507807	
	Bulb for headlamp	2	505198	France
	Bulb for headlamp	2	GLB416	RH Stg Mk I and Mk IA
	Bulb for headlamp	2	503354	LH Stg except Europe
	Bulb for headlamp	2	GLB410	LH Stg Europe except France
4	Light unit, 60/45 watt	2	GLU101	RH Stg Mk II and Mk III — Alternative. Check before ordering
	Light unit, 75/45 watt sealed beam	2	GLU106	RH Stg Mk III — Alternative. Check before ordering
	Light unit — 'Block-pattern' lens	2	GLU507	RH Stg Mk I and Mk IA
	Light unit — 'Block-pattern' lens	2	507808	LH Stg except Europe and France
	Light unit — 'Block-pattern' lens	2	507804	LH Stg Europe and France
	Light unit, 50/40 watt sealed beam	2	541537	America. Dollar area. Mk II and Mk III
	Spring retaining bulb	2	507805	
	Rim retaining light unit	2	274379	Mk I and Mk IA
5	Rim retaining light unit	2	545150	Mk II and Mk III
6	Special screw for light unit rim	6	262100	
	Special screw — For light unit adjustment	6	261940	Mk I and Mk IA
	Cup washer for screw — For light unit adjustment	6	261943	Mk I and Mk IA
	Spring for special screw — For light unit adjustment	6	261941	Mk I and Mk IA
7	Rim for headlamp	2	507764	
	Rubber gasket for headlamp rim	2	262101	Mk I and Mk IA
	Rubber gasket for body sealing rim	2	507767	Mk I and Mk IA
8	Rubber gasket for body sealing rim	2	529001	Mk II and Mk III
9	Adaptor plate	2	601976	Mk II and Mk III
10	Screw (10 UNF x ½" long) — Fixing headlamp	10	78348	
11	Spring washer — Fixing headlamp	10	3073	
12	Nut (10 UNF) — Fixing headlamp	10	257023	
	SIDELAMP COMPLETE	2	352349	
13	GLASS AND RIM FOR SIDELAMP	2	507606	
14	Glass for sidelamp	2	507607	
15	Plastic ring for glass	2	507608	

*Asterisk indicates a new part which has not been used on any previous model.

Plate Ref.	1 2 3 4 Description	Qty.	Part No.	Remarks
16	Rubber seal for glass	2	507609	
17	Bulb holder, interior, complete	2	244700	
18	Bulb for sidelamp	2	GLB989	
19	Boot cover for sidelamp	2	507610	
20	Sealing washer for lamp body	2	507613	
21	Bezel for sidelamp	2	507612	
22	Sealing rubber for bezel	2	507611	
23	Red indicator light for sidelamp	2	510801	
24	Screw, self-tapping, fixing indicator to bezel	2	356578	
25	Screw (10 UNF x ⅜" long) } Fixing sidelamp	4	78291	
26	Spring washer } Fixing sidelamp	4	3073	
27	Nut (10 UNF) } Fixing sidelamp	4	257023	
	FLASHER LAMP, FRONT RH } Amber lens	1	541680	Except America Dollar area and Italy
28	FLASHER LAMP, FRONT LH } Amber lens	1	541682	Except America Dollar area and Italy
	FLASHER LAMP, FRONT RH } Clear lens	1	541681	America. Dollar area and Italy
	FLASHER LAMP, FRONT LH } Clear lens	1	541683	America. Dollar area and Italy
29	Bulb holder, interior, complete	2	514144	
30	Bulb for flasher lamp	2	GLB382	
	Lens for flasher lamp, amber, RH	1	507615	Except America Dollar area
31	Lens for flasher lamp, amber, LH	1	507620	Except America Dollar area
	Lens for flasher lamp, clear, RH	1	511985	America. Dollar area and Italy
	Lens for flasher lamp, clear, LH	1	511986	America. Dollar area and Italy
32	Sealing rubber for glass	2	507616	
33	Special screw fixing glass	4	507614	
	Terminal sleeve	2	271932	
	Boot cover for flasher lamp	2	514147	Mk I and Mk IA
	Sealing rubber for flasher lamp	2	507619	Mk I and Mk IA
	Rubber boot sealing RH lamp	1	541703	MK II and Mk III
34	Rubber boot sealing LH lamp	1	541704	MK II and Mk III
35	Drive screw fixing flasher lamp	4	77704	

*Asterisk indicates a new part which has not been used on any previous model.

1609

HEAD, SIDE, FRONT FLASHER AND FOG LAMPS 3½ LITRE MODELS

Plate Ref.	1 2 3 4 Description	Qty.	Part No.	Remarks
1	HEADLAMP COMPLETE	2	559254	RH Stg
	HEADLAMP COMPLETE	2	605071*	LH Stg except France
	HEADLAMP COMPLETE	2	605072*	LH Stg France only
	HEADLAMP COMPLETE	2	605073*	America. Dollar area
2	Special screw for light unit adjustment	2	601977	
3	Adaptor for bulb	2	600226	RH Stg
	Sleeve for terminal	6	507807	
	Bulb for headlamp	2	GLB411	France
	Bulb for headlamp	2	GLB410	LH Stg Europe except France
4	Light unit, 75/45 watt sealed beam	2	GLU106	RH Stg
	Light unit, 'Block-pattern' lens	2	507808	LH Stg except Europe and France
	Light unit 'Block-pattern' lens	2	507804	LH Stg Europe and France
	Light unit, 50/40 watt sealed beam	2	541537	America. Dollar area
	Spring retaining bulb	2	507805	
5	Rim retaining light unit	2	545150	
6	Special screw for light unit rim	6	262100	
7	Rim for headlamp	2	507764	
8	Rubber gasket for body sealing rim	2	529001	
9	Adaptor plate	2	601976	
10	Screw (10 UNF x ½" long) } Fixing headlamp	10	78348	
11	Spring washer } Fixing headlamp	10	3073	
12	Nut (10 UNF) } Fixing headlamp	10	257023	
	SIDELAMP COMPLETE	2	352349	
13	GLASS AND RIM FOR SIDELAMP	2	507606	
14	Glass for sidelamp	2	507607	
15	Plastic ring for glass	2	507608	
16	Rubber seal for glass	2	507609	
17	Bulb holder, interior, complete	2	244700	
18	Bulb for sidelamp	2	GLB989	
19	Boot cover for sidelamp	2	507610	
20	Sealing washer for lamp body	2	507613	
21	Bezel for sidelamp	2	507612	
22	Sealing rubber for bezel	2	507611	
23	Red indicator light for sidelamp	2	545083	
24	Screw, self-tapping, fixing indicator to bezel	2	356578	
25	Screw (10 UNF x ⅜" long) } Fixing sidelamp	4	78291	
26	Spring washer } Fixing sidelamp	4	3073	
27	Nut (10 UNF) } Fixing sidelamp	4	257023	

*Asterisk indicates a new part which has not been used on any previous model.

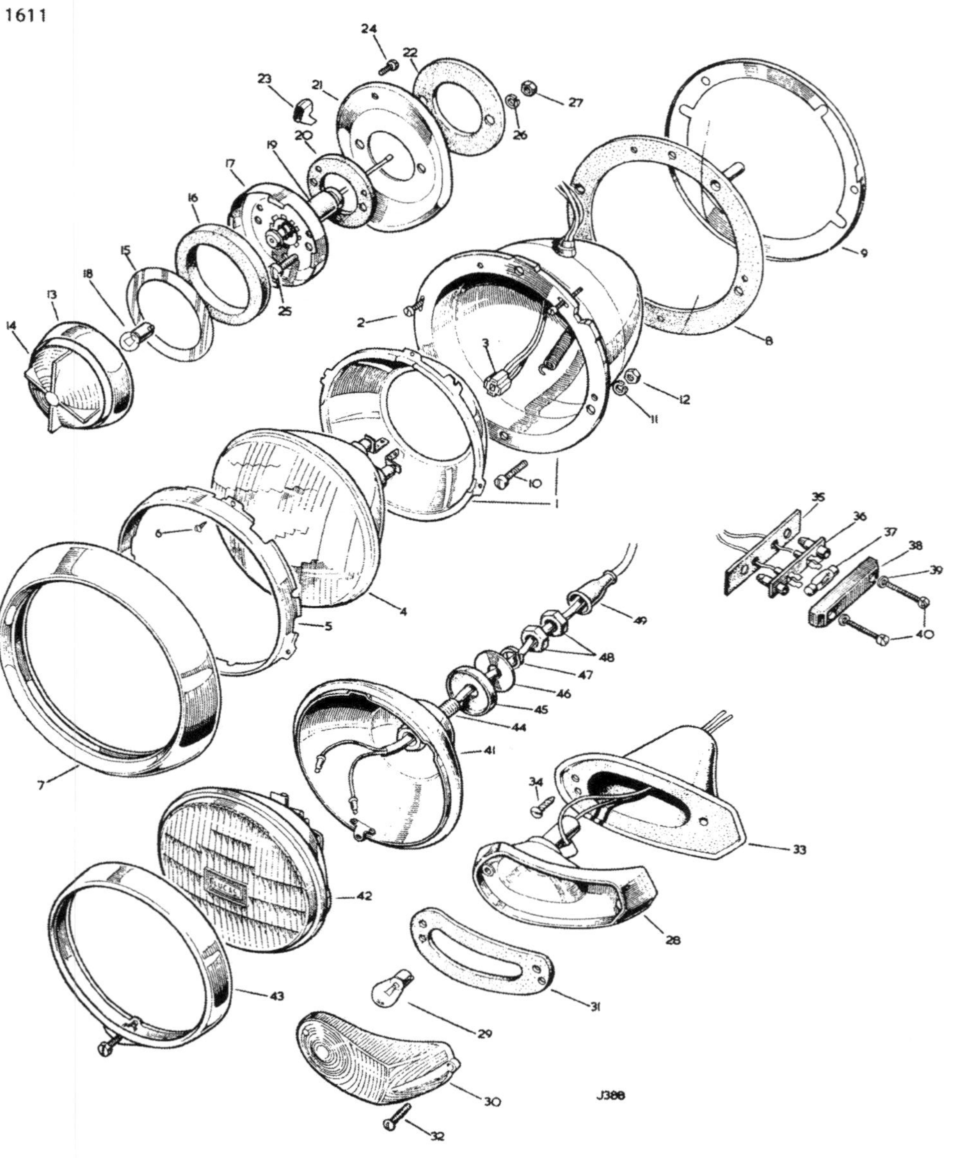

Plate Ref.	1 2 3 4 Description		Qty.	Part No.	Remarks
	FLASHER LAMP, FRONT RH	Amber lens	1	541680	Except America
28	FLASHER LAMP, FRONT LH		1	541682	Dollar area and Italy
	FLASHER LAMP, FRONT RH	Clear lens	1	541681	America. Dollar
	FLASHER LAMP, FRONT LH		1	541683	area and Italy
	Bulb holder, interior, complete		2	514144	
29	Bulb for flasher lamp		2	GLB382	
	Lens for flasher lamp, amber, RH		1	507615	Except America
30	Lens for flasher lamp, amber, LH		1	507620	Dollar area
	Lens for flasher lamp, clear, RH		1	511985	America Dollar
	Lens for flasher lamp, clear, LH		1	511986	area and Italy
31	Sealing rubber for glass		2	507616	
32	Special screw fixing glass		4	507614	
	Terminal sleeve		2	271932	
	Rubber boot sealing RH lamp		1	541703	
33	Rubber boot sealing, LH lamp		1	541704	
34	Drive screw fixing flasher lamp		4	77704	
	FLASHER REPEATER LAMP ASSEMBLY		4	566005	
35	Rubber seal for base		4	566008	
36	BASE COMPLETE FOR REPEATER LAMP		4	566007	
37	Bulb, festoon, 3 watt		4	GLB256	
38	Lens for repeater lamp		4	566000	
39	Rubber washer	Fixing lamp to body	8	551430	
40	Drive screw		8	78785	
41	FOG LAMP ASSEMBLY		2	563388	Except France
	FOG LAMP ASSEMBLY Amber lens		2	570543	France
42	Light unit		2	605233	Except France
	Light unit, Amber lens		2	606127	France
43	Rim		2	605232	
44	Special bolt fixing lamp		2	605234	
45	Rubber seating washer		2	605235	
46	Plate	Fixing lamp to body	2	605236	
47	Spring washer		2	3077	
48	Nut		4	605237	
49	Grommet for cable entry		2	240408	
	Blade adaptor	For fog lamp connections	4	568004	
	Female		4	568005	

*Asterisk indicates a new part which has not been used on any previous model.

STOP, TAIL, REAR FLASHER AND REVERSE LAMPS

1613

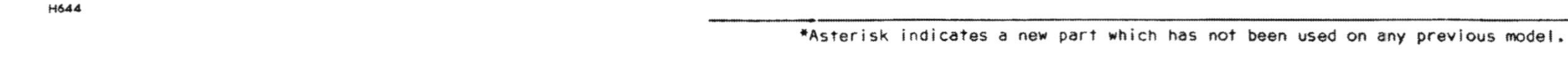

Plate Ref.	1 2 3 4 Description	Qty.	Part No.	Remarks
	STOP/TAIL AND FLASHER LAMP COMPLETE	2	352330	Except America. Dollar area
	STOP/TAIL AND FLASHER LAMP COMPLETE	2	356218	America. Dollar area
1	Bulb, top, single filament	2	GLB382	
2	Bulb, bottom, double filament	2	GLB380	
3	RIM FOR LAM	2	507628	
4	Special screw fixing rim	2	507629	
5	Washer for screw	2	551430	
6	Reflector for lamp	2	507630	
7	Flasher glass, top	2	507631	Except America. Dollar area
	Flasher glass, top, red	2	511984	America. Dollar area
8	Sealing rubber for flasher glass	2	507632	
9	Stop/tail, glass, bottom	2	507633	
10	Sealing rubber for stop/tail glass	2	507634	
	Bulb holder, interior, for flasher	2	514144	
	Bulb holder, interior, for stop/tail	2	264782	
11	Terminal sleeve	2	271932	
12	Boot cover for lamp	4	507618	
13	Sealing rubber for stop/tail lamp	2	507635	
14	Back plate for tail lamp aperture in body	2	356223	
15	Sealing rubber for back plate	2	356224	
	Drive screw fixing back plate and rubber to body	8	77707	
16	Set screw (10 UNF x ⅜" long)] Fixing stop/tail lamp to body	8	78272	
17	Shakeproof washer] Fixing stop/tail lamp to body	8	74236	
18	Plain washer] Fixing stop/tail lamp to body	8	3852	
	REAR NUMBER PLATE AND REVERSING LAMP	1	519741	
19	Bulb for reverse	1	GLB382	
20	Bulb for number plate	2	GLB989	
21	Cover for lamp	1	507622	
22	Special screw fixing cover	2	507621	
23	Shakeproof washer	2	605031	
24	Glass for lamp	1	507623	
25	Sealing rubber for glass	1	507624	
26	Bracket and bulb holders	1	510910	
27	Bulb holder, interior, for reverse	1	507626	
	Terminal sleeve	1	271932	
28	Body for lamp	1	510911	
29	Sealing rubber for lamp	1	507627	
30	Plain washer] Fixing rear number plate lamp to boot lid	4	3946	
31	Spring washer] Fixing rear number plate lamp to boot lid	4	3074	
32	Nut (¼" UNF)] Fixing rear number plate lamp to boot lid	4	254810	
33	BOOT LIGHT COMPLETE	1	236278	
34	Bulb for boot light	1	GLB989	
35	Drive screw fixing boot light	2	78001	

*Asterisk indicates a new part which has not been used on any previous model.

WINDSCREEN WIPER

Plate Ref.	1 2 3 4 Description	Qty.	Part No.	Remarks
	SCREEN WIPER MOTOR	1	352020	Mk I and Mk IA
	SCREEN WIPER MOTOR	1	380024	Mk II cars with suffix letters 'A' and 'B'
1	SCREEN WIPER MOTOR	1	601696	Mk II cars with suffix letter 'C', Mk III 3 Litre and 3½ Litre
	Special stud, grommet and fixings for wiper motor	3	356198	Mk I and Mk IA
2	Sepcial stud, grommet and fixings for wiper motor	3	600487	Mk II and Mk III 3 Litre and 3½ Litre
	Flexible drive cable (49 13/32" long)	1	356415	Mk I and Mk IA
3	Flexible drive cable (47 29/32" long)	1	380025	Mk II and Mk III 3 Litre and 3½ Litre
	Mounting bracket for wiper motor	1	350491	Mk I and Mk IA
4	Mounting bracket for wiper motor	1	531658	Mk II and Mk III 3 Litre and 3½ Litre
	Bolt (¼" UNF x ⅝" long) } Fixing bracket to wing and valance	4	255206	2 off from @@
	Plain washer } Fixing bracket to wing and valance	6	3900	2 off from @@
	Spring washer } Fixing bracket to wing and valance	3	3074	2 off from @@ Mk I and Mk IA
	Shakeproof washer } Fixing bracket to wing and valance	1	78114@	
	Nut (¼" UNF) } Fixing bracket to wing and valance	2	254810	
5	Bolt (¼" UNF x ⅝" long) } Fixing mounting bracket to valance and wing	2	255206	Mk II and Mk III 3 Litre and 3½ Litre
6	Plain washer } Fixing mounting bracket to valance and wing	2	3900	Mk II and Mk III 3 Litre and 3½ Litre
7	Spring washer } Fixing mounting bracket to valance and wing	2	3074	Mk II and Mk III 3 Litre and 3½ Litre
8	Rubber grommet for wiper in dash	1	276054	
	WHEELBOX FOR WIPER	2	340452	Mk I and Mk IA
9	WHEELBOX FOR WIPER	2	536970	Mk II and Mk III
	Spindle and gear for wheelbox	2	507636	Mk I and Mk IA
10	Spindle and gear for wheelbox	2	538411	Mk II and Mk III 3 litre and 3½ Litre
11	Steel bush for wheelbox	2	352923	
	Union for outer casing	1	352608@	
	Outer casing, union to wheelbox	1	352532@	
	Outer casing, motor to wheelbox	1	352440@	
	Outer casing, motor to wheelbox	1	352925@@	Mk I and Mk IA
12	Outer casing, motor to wheelbox	1	537085	Mk II and Mk III 3 Litre and 3½ Litre
13	Outer casing, wheelbox to wheelbox	1	352441	
14	Outer casing, wheelbox end	1	575047	
	Wiper arm	2	352990	Mk I and Mk IA
15	Wiper arm	2	531653	Mk II and Mk III 3 Litre and 3½ Litre
	Wiper blade	2	GWB123	Mk I and Mk IA
16	Wiper blade	2	GWB142	Mk II and Mk III 3 Litre and 3½ Litre

@ 3 Litre cars numbered up to 625901262,626900224,628900149, 630900624, 631900116,633900120
@@ 3 Litre cars numbered from 625901263,626900225, 628900150,630900625,631900117,633900121 onwards
NADA indicates parts peculiar to cars exported to the North American dollar area

*Asterisk indicates a new part which has not been used on any previous model.

H382

WINDSCREEN WIPER

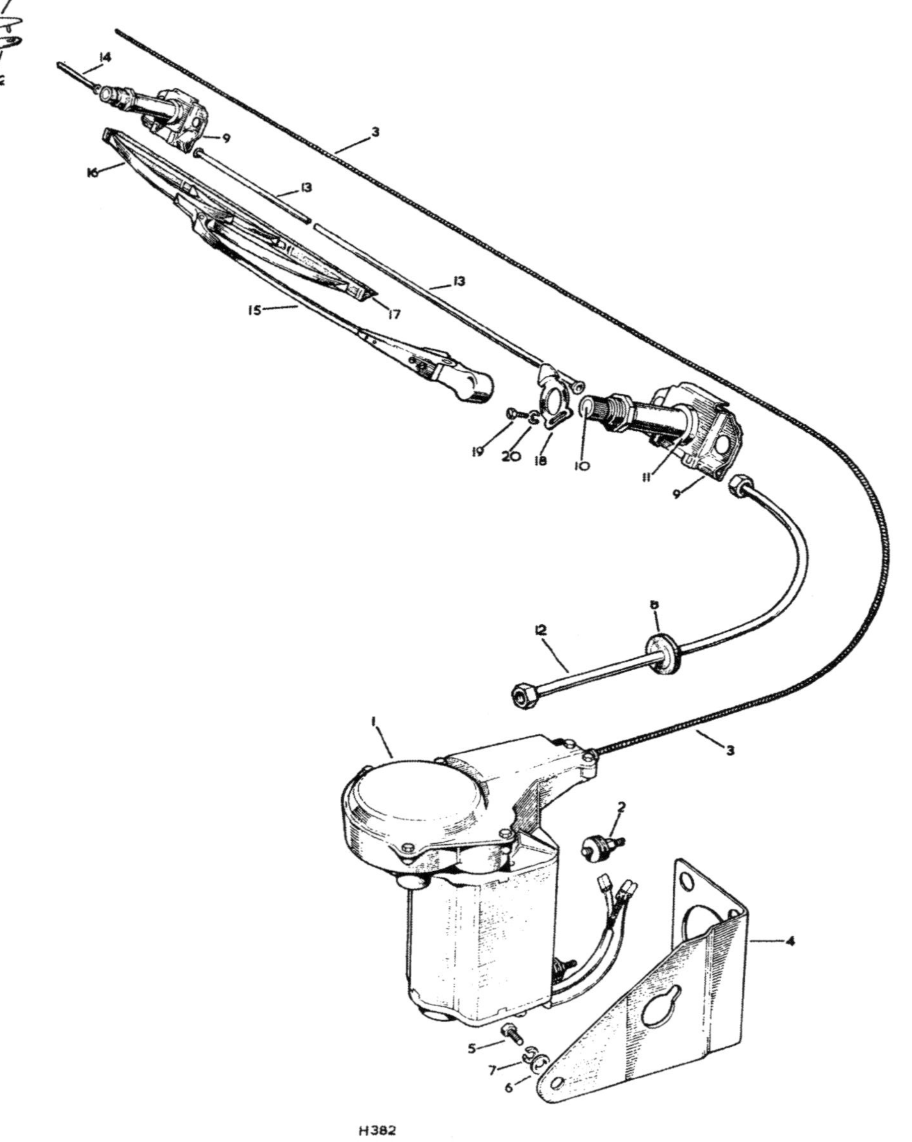

H382

Plate Ref.	1 2 3 4 Description	Qty.	Part No.	Remarks
	Rubber for wiper blade (11" long)	1	524676	Mk I and Mk IA
17	Rubber for wiper blade (12" long)	2	600441	Mk II and Mk III 3 Litre and 3½ Litre
18	Ramp for wiper blade	2	352578	
19	Special screw (4 UNC x ½" long) } Fixing ramp to	2	78341	
20	Spring washer } air intake grille	2	3071	
21	Parking ramp for RH wiper blade	1	352975	
22	Rubber seal for parking ramp	1	352976	
	Spigot fixing parking ramp to body	2	352985	
	Cable for two-speed wiper conversion	1	541705	} Mk II and Mk III NADA
	Switch for screen wiper (fitted in the foglamp switch position on the underside of instrument housing	1	512253	}

*Asterisk indicates a new part which has not been used on any previous model.

WINDSCREEN WASHER, MANUAL

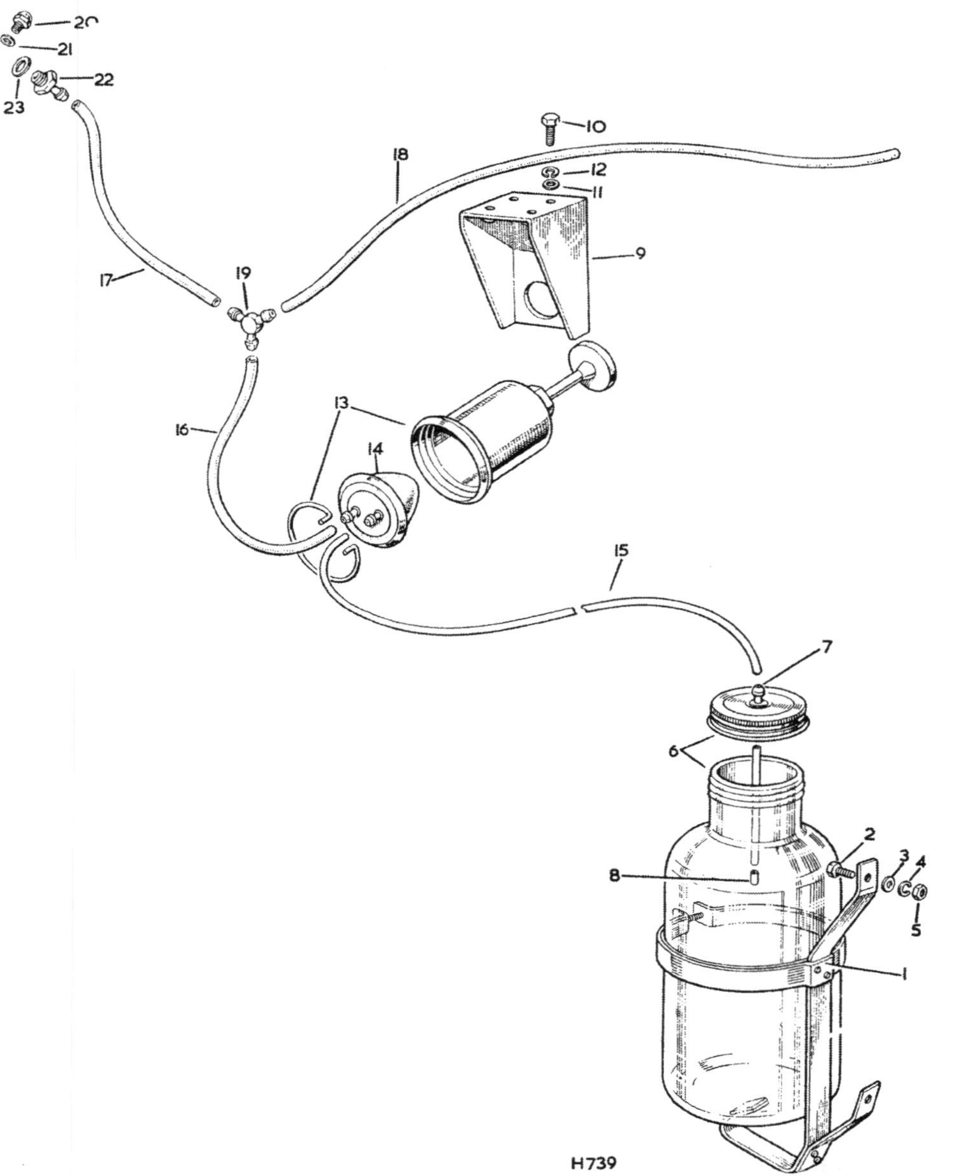

Plate Ref.	1 2 3 4 Description		Qty.	Part No.	Remarks
1	Bracket for windscreen washer reservoir		1	502970	
2	Bolt (¼" UNF x ⅝" long)	Fixing bracket to LH wing valance	2	255206	
3	Plain washer		2	3840	
4	Spring washer		2	3074	
5	Nut (¼" UNF)		2	254810	
6	WINDSCREEN WASHER RESERVOIR COMPLETE		1	265835	
7	Suction valve, top		1	524121	
8	Suction valve, bottom		1	243475	
9	Bracket for windscreen washer pump		1	502973	RH Stg
	Bracket for windscreen washer pump		1	504024	LH Stg
10	Bolt (¼" UNF x ⅝" long)	Fixing bracket to parcel shelf	4	255206	
11	Plain washer		4	3840	
12	Spring washer		4	3074	
	Nut (¼" UNF)		4	254810	LH Stg
13	WINDSCREEN WASHER PUMP UNIT		1	502971	
14	Pump for unit		1	507993	
15	Pipe, reservoir to pump unit		1	502974	
16	Pipe, pump unit to 'Y' piece		1	512466	
17	Pipe, 'Y' piece to RH nozzle		1	502976	
18	Pipe, 'Y' piece to LH nozzle		1	502977	
19	'Y' piece connector		1	536879	
20	WINDSCREEN WASHER NOZZLE		2	352191	
21	Joint washer for nozzle		2	4191	
22	DELIVERY VALVE FOR WINDSCREEN WASHER		2	352192	
23	Joint washer for delivery valve		2	52111	

*Asterisk indicates a new part which has not been used on any previous model.

WINDSCREEN WASHER, ELECTRIC

1621

F593

Plate Ref.	1 2 3 4	Description	Qty.	Part No.	Remarks
1		ELECTRIC SCREEN WASHER AND FIXING BRACKET	1	514709	MkI and Mk IA
		Glass bowl for screen washer	1	534311	
		Motor for screen washer	1	605633	
		Bracket for screen washer	1	605634	
2		ELECTRIC SCREEN WASHER AND FIXING BRACKET	1	547100	Mk II and Mk III 3 Litre and 3½ Litre cars numbered suffix letter 'A'
		Motor for screen washer	1	600663	
		Cover and pump for screen washer	1	600664	
		Screw (10 UNF x ½" long) — Fixing screen washer to LH wing valance	3	78275	Mk I
		Spring washer — Fixing screen washer to LH wing valance	3	3073	
		Nut (10 UNF) — Fixing screen washer to LH wing valance	2	257023	
3		Mounting bracket for screen washer	1	519768	Mk IA and Mk II
		Mounting bracket for screen washer	1	551597	Mk III 3 litre and 3½ litre cars numbered suffix letter 'A'
4		Screw (10 UNF x ½" long) — Fixing mounting bracket to LH wing valance	3	257017	Mk IA and Mk II
5		Spring washer — Fixing mounting bracket to LH wing valance	3	3073	
6		Plain washer — Fixing mounting bracket to LH wing valance	3	2874	
7		Screw (10 UNF x ½" long) — Fixing reservoir to mounting bracket	3	257017	
8		Spring washer — Fixing reservoir to mounting bracket	3	3073	
9		Nut (10 UNF) — Fixing reservoir to mounting bracket	3	257023	
10		Plastic tubing for screen washer	As reqd	83950	State length required
11		'T' piece connector	1	536879	Washer jets mounted in air intake grille
		'T' piece connector	1	538969	Cars with washer jets in bonnet panel
12		Grommet for tubing in dash	1	514740	
		Jet for windscreen washer, fixed type	2	514710	Washer jets mounted in air intake grille
13		Jet for windscreen washer, adjustable	2	601568	
		Jet for windscreen washer	2	545203	Cars with washer jets in bonnet panel
14		Joint washer for jet	2	4191	Washer jets mounted in air intake grille
		Delivery valve nipple for fixed-type jet	2	514716	
15		Delivery valve nipple for adjustable-type jet	2	528982	
16		Joint washer for delivery valve	2	4352	
17		Switch for windscreen washer	1	514706	
18		Escutcheon plate for switch	1	514723	
19		Clip fixing screen washer tubing and harness to LH wing valance	3	514708	
20		Cleat fixing screen washer tubing and harness adjacent to LH cold air duct	1	240429	Mk I, Mk IA and Mk II
		Cleat fixing screen washer tube to main harness at RH bonnet hinge	1	240430	Mk III 3 litre and 3½ litre cars numbered suffix letter 'A'
		Cleat fixing screen washer tube to fuel pipe at filter bowl	1	240429	
		Grommet for tubing in bonnet	3	531689	

*Asterisk indicates a new part which has not been used on any previous model.

Impellor
606998 (1)

Motor Pump and Cap
573222 (1)

240430 (1)

84220
As required

570510 (1)

531689 (3)

78660 (3)

78136 (3)

568028 (1)

84221
As required

573260 (1)

568027 (2)

Reservoir only
567972 (1)

570758 (1)

Reservoir and pump
complete 573120 (1)

514723 (1)

514706 (1)

REMARKS

Applicable to 3½ Litre models

This page is intentionally left bank

GENERAL ELECTRICAL EQUIPMENT, MK I MODELS

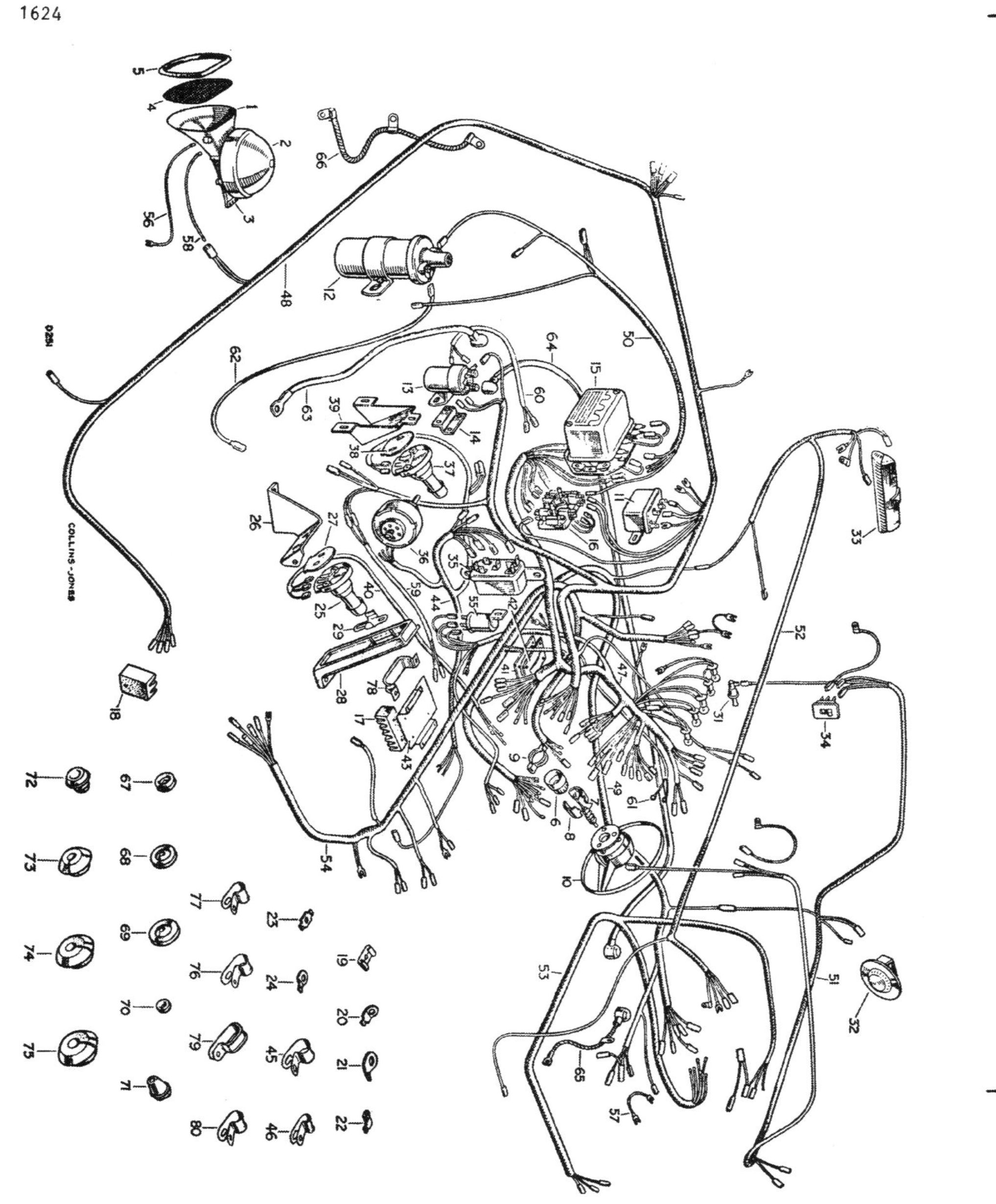

1624

Plate Ref.	1 2 3 4 Description	Qty.	Part No.	Remarks
1	HORN, HIGH NOTE, RH	1	601972	
	HORN, LOW NOTE, LH	1	601973	
	Contact set for horn	2	271500	
2	Cover for horn	2	262205	
3	Bracket for horn	2	262110	
4	Gauze for horn	2	266162	
5	Ring for gauze on high note horn	1	271845	
	Ring for gauze on low note horn	1	271844	
	Bolt (¼" UNF x ⅜" long) } Fixing horns to mounting bracket on chassis sub-frame	4	255206	
	Fan disc washer } Fixing horns to mounting bracket on chassis sub-frame	4	510170	
	Nut (¼" UNF) } Fixing horns to mounting bracket on chassis sub-frame	4	254810	
	HORN CONTACT AND SLIP RING COMPLETE	1	502964	
6	Horn contact, slip ring	1	277912	
7	Horn contact, cable and rotor, top half	1	279104	
8	Horn slip ring rotor, bottom half	1	606249	
9	Flasher trip ring	1	548069	
	Shakeproof washer } Fixing ring to steering column shaft	2	78249	
	Set screw (4BAx⅜" long) } Fixing ring to steering column shaft	2	75804	
10	Horn ring complete	1	275999	
	Special screw fixing horn ring	3	542695	
11	Relay for horn	1	502094	
	Shakeproof washer } Fixing relay	2	311373	
	Drive screw } Fixing relay	2	78417	
12	Ignition coil	1	503276	
	Boot for HT terminal nut	1	506679	
	Cable nut for coil	1	510237	
	Split washer for coil	1	214279	
	Bolt (¼" UNF x ⅞" long) } Fixing coil to engine	2	255207	
	Plain washer } Fixing coil to engine	2	3840	
	Spring washer } Fixing coil to engine	2	3074	
	Nut (¼" UNF) } Fixing coil to engine	2	254810	
13	Starter solenoid	1	567969	
14	Bracket for solenoid	1	513678	
	Bolt (10 UNF x 5/16" long) } Fixing solenoid and bracket to dash	4	257014	
	Shakeproof washer } Fixing solenoid and bracket to dash	4	311396	
	Nut (10 UNF) } Fixing solenoid and bracket to dash	4	257023	
15	VOLTAGE CONTROL BOX	1	502092	For C45 PV5 dynamo, Part No. 503278
	VOLTAGE CONTROL BOX	1	512250	For C45 PV6 dynamo, Part No. 512248
	Cover for box (drive screw fixing)	1	504444	
	Cover for regulator box (2 BA screw fixing)	1	513937	
	Rubber seal for cover	1	507810	
	Set bolt (¼" UNF x ⅞" long) } Fixing voltage control box	3	255207	
	Spring washer } Fixing voltage control box	3	3074	
16	FUSE BOX	1	505156	Models without overdrive
	Fuse, 35-amp	As reqd	12738	Models without overdrive
	Fuse, 50-amp	As reqd	515083	Models without overdrive
	Cover for fuse box	1	505158	Models without overdrive
	FUSE BOX	1	606254	Models with overdrive
	Fuse, 35-amp	As reqd	12738	Models with overdrive
	Cover for fuse box	1	505159	Models with overdrive
	drive screw locating fuse box	2	77704	

*Asterisk indicates a new part which has not been used on any previous model.

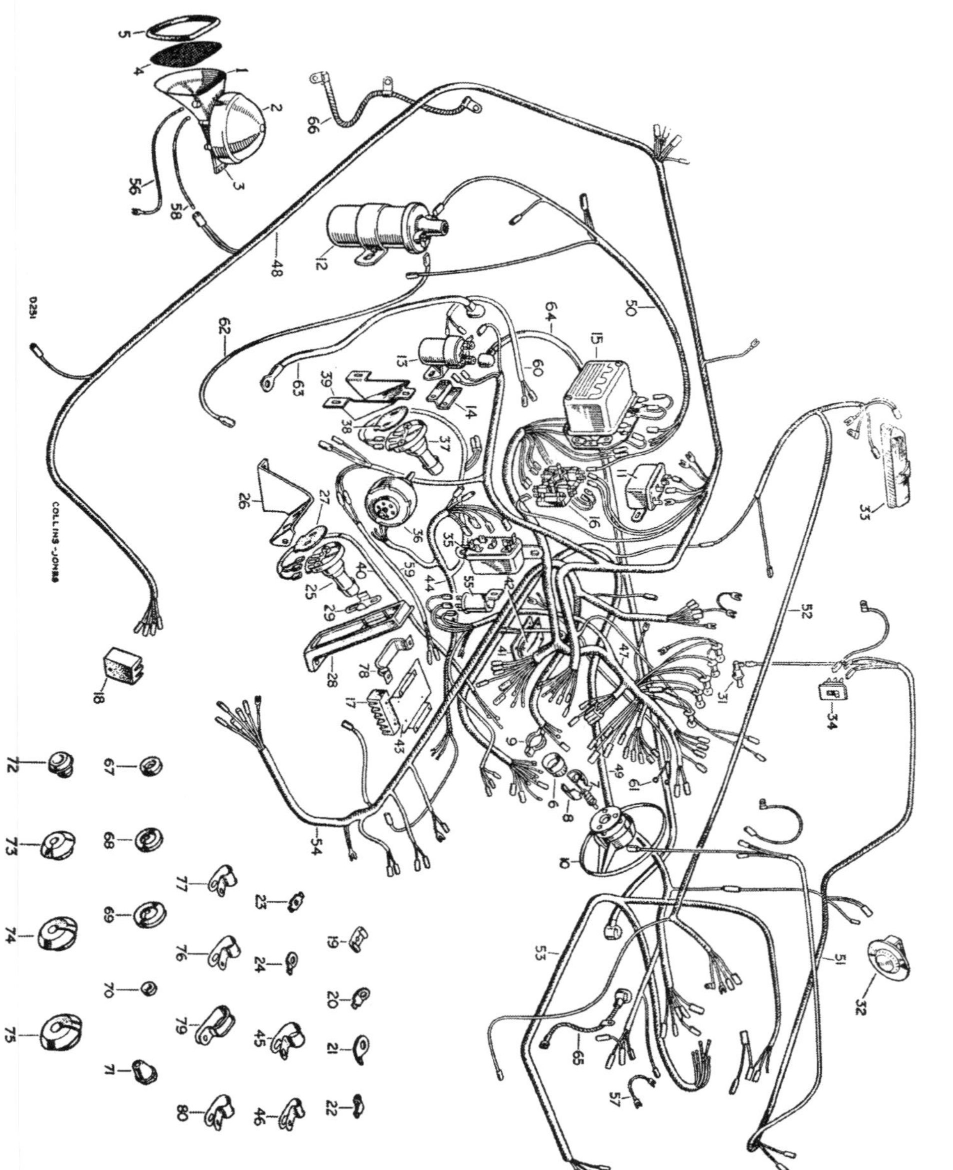

Plate Ref.	1 2 3 4 Description	Qty.	Part No.	Remarks
	Drive screw fixing fuse box	2	78153	
17	Terminal block, six-way	2	513250	
18	Terminal block, two-way	9	513249	
	Cover plate for terminal block under parcel shelf	1	510182	
	Drive screw fixing terminal block	18	77704	
19	Earthing clip on front and rear wing valance	7	236366	
	Drive screw } Fixing earthing clip	4	78417	
	Shakeproof washer } Fixing earthing clip	4	74236	
20	Special earthing blade on front wing valance and dash	2	505204	
	Drive screw } Fixing earthing blade to valance and dash	2	78417	
	Shakeproof washer } Fixing earthing blade to valance and dash	2	74236	
21	Special earthing blade at voltage regulator, flasher unit and horn fixing	4	505205	
22	Special earthing blade at fuel gauge unit fixing screw and ignition coil	2	505203	
23	Special earthign blade at instrument unit	3	505195	
24	Special earthing blade at petrol pump fixing	1	507110	
25	Dip switch	1	502087	
	Screw (10 UNF x 1" long) } Fixing dip switch	2	78325	
	Plain washer } Fixing dip switch	2	2874	RH Stg
	Spring washer } Fixing dip switch	2	3073	RH Stg
	Nut (10 UNF) } Fixing dip switch	2	257023	RH Stg
26	BRACKET FOR DIP SWITCH	1	502155	RH Stg
	BRACKET FOR DIP SWITCH	1	504848	LH Stg
27	Rubber seal for bracket	1	502896	
	Bolt ($\frac{1}{4}$" UNF x $\frac{3}{8}$" long) } Fixing bracket and seal to floor	1	255206	4 off on LH Stg
	Plain washer } Fixing bracket and seal to floor	2	3900	4 off on LH Stg
	Spring washer } Fixing bracket and seal to floor	1	3074	4 off on LH Stg
	Nut ($\frac{1}{4}$" UNF) } Fixing bracket and seal to floor	1	254810	4 off on LH Stg
28	Pedal for dip switch	1	503514	
29	Clip, pedal to dip switch	1	503515	
	Plain washer } Fixing clip to pedal	2	3902	
	Spring washer } Fixing clip to pedal	2	3073	
	Nut (2 BA) } Fixing clip to pedal	2	2247	
	Screw ($\frac{1}{4}$" UNF x $1\frac{1}{4}$" long) } Fixing pedal to floor	2	252115	
	Countersunk washer } Fixing pedal to floor	2	66316	LH Stg
	Spring washer } Fixing pedal to floor	2	3074	LH Stg
	Nut ($\frac{1}{4}$" UNF) } Fixing pedal to floor	2	254810	LH Stg
31	Pillar switch 'A' and 'D' posts, 9/16" dia. barrel	4	352403	
	Pillar switch 'A' and 'D' posts, $\frac{1}{2}$" dia. barrel	4	541670	
32	ROOF LAMP, REAR	2	352013@	
33	ROOF LAMP COMPLETE WITH FESTOON-TYPE BULB	4	567752	2 off up to @@
	Lens for roof lamp, rear	2	356110	
	Lens for roof lamp, front, with bayonet-type bulb	2	514185	4 off from @@
	Lens for roof lamp with festoon-type bulb	4	600458	
	Bulb, bayonet-type, for roof lamp	4	242491	
	Bulb 6 watt festoon-type, for roof lamp	4	GLB254	Alternative Check before ordering
	Bulb, 7 watt festoon-type for roof lamp	4	567804	Alternative Check before ordering

@ Cars numbered up to 625000979, 626000332, 628000110, 630000459, 631000115, 633000310

@@ Cars numbered from 625000980, 626000333, 628000111, 630000460, 631000116, 633000311 onwards

*Asterisk indicates a new part which has not been used on any previous model.

GENERAL ELECTRICAL EQUIPMENT, MK I MODELS

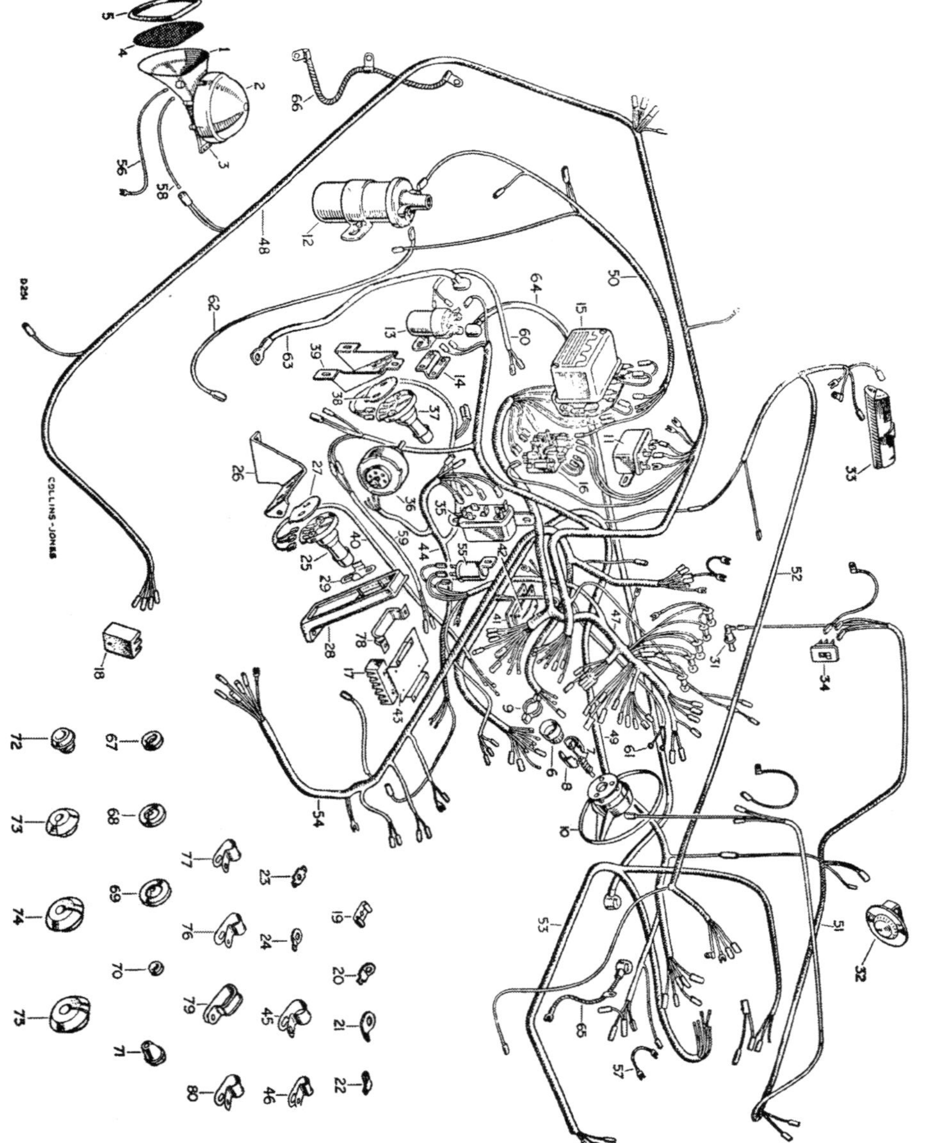

Plate Ref.	1 2 3 4 Description	Qty.	Part No.	Remarks
	Screw fixing rear lamp	4	78140@	
	Screw fixing front lamp	4	78126	8 off from @@
	Seal for roof lamp, front	2	355235@	
34	Door pillar switch	2	352585@	
	Battery	1	537036	Except America. Dollar area
	Battery	1	514720	America. Dollar area
35	Relay unit for overdrive	1	502910	Models with overdrive
36	Rotary switch for overdrive	1	273334	Models with overdrive
37	Kickdown switch for overdrive	1	502895	Models with overdrive
48	Rubber sealing pad for kickdown switch	1	502896	Models with overdrive
	Screw (10 UNF x 1" long)fixing switch to bracket	2	78325	Models with overdrive
39	Bracket for kickdown switch	1	502754	RH Stg models with overdrive
	Bolt (¼" UNF x ⅜" long) } Fixing kickdown bracket	3	255206	RH Stg models with overdrive
	Plain washer } Fixing kickdown bracket	6	3840	RH Stg models with overdrive
	Spring washer } Fixing kickdown bracket	3	3074	RH Stg models with overdrive
	Nut (¼" UNF) } Fixing kickdown bracket	3	254810	RH Stg models with overdrive
40	Lead for dip switch	1	503363	RH Stg
	Lead for dip switch	1	510122	LH Stg
41	Support bracket for harness at parcel shelf	1	352589	
42	Rubber pad for harness at parcel shelf	1	352590	
43	Support bracket for electrical connections	1	352588	
	Screw(10 UNF x ⅜" long) } Fixing brackets and pad to parcel shelf	8	78301	
	Plain washer } Fixing brackets and pad to parcel shelf	8	3902	
	Spring washer } Fixing brackets and pad to parcel shelf	8	3073	
	Nut (10 UNF) } Fixing brackets and pad to parcel shelf	8	257023	
44	Harness for overdrive	1	503397	RH Stg
	Harness for overdrive	1	503398	LH Stg
45	Clip } Fixing overdrive harness to gearbox and extension casing	1	3621	Models with overdrive
46	Clip } Fixing overdrive harness to gearbox and extension casing	1	219676	Models with overdrive
47	Dash harness	1	503360	RH Stg
	Dash harness	1	503361	LH Stg
48	Headlamp and sidelamp harness	1	503362	Rh Stg
	Headlamp and sidelamp harness	1	510125	LH Stg
	Lead, headlamp switch to tail lamp	1	510142	America. Dollar area
49	Body harness	1	503368@	RH Stg
	Body harness	1	510124@	LH Stg
	Body harness	1	514901@@	RH Stg
	Body harness	1	514902@@	LH Stg
50	Engine harness	1	503370	
51	Body harness, rear, for interior lights	1	503365@	
52	Body harness, front, for interior lights	1	503366@	
	Harness for interior light having bayonet-type bulb	1	513199@@	
	Harness for interior light having festoon-type bulb	1	528932	Use in conjunction with pillar switch with ½" dia. barrel
	Harness for interior light having festoon-type bulb	1	526417	Use in conjunction with pillar switch with 9/16" dia. barrel

@ Cars numbered up to 625000979,626000332,628000110,630000459,631000115,633000310
@@Cars numbered from 625000980,626000333,628000111,630000460,631000116,633000311 onwards

*Asterisk indicates a new part which has not been used on any previous model.

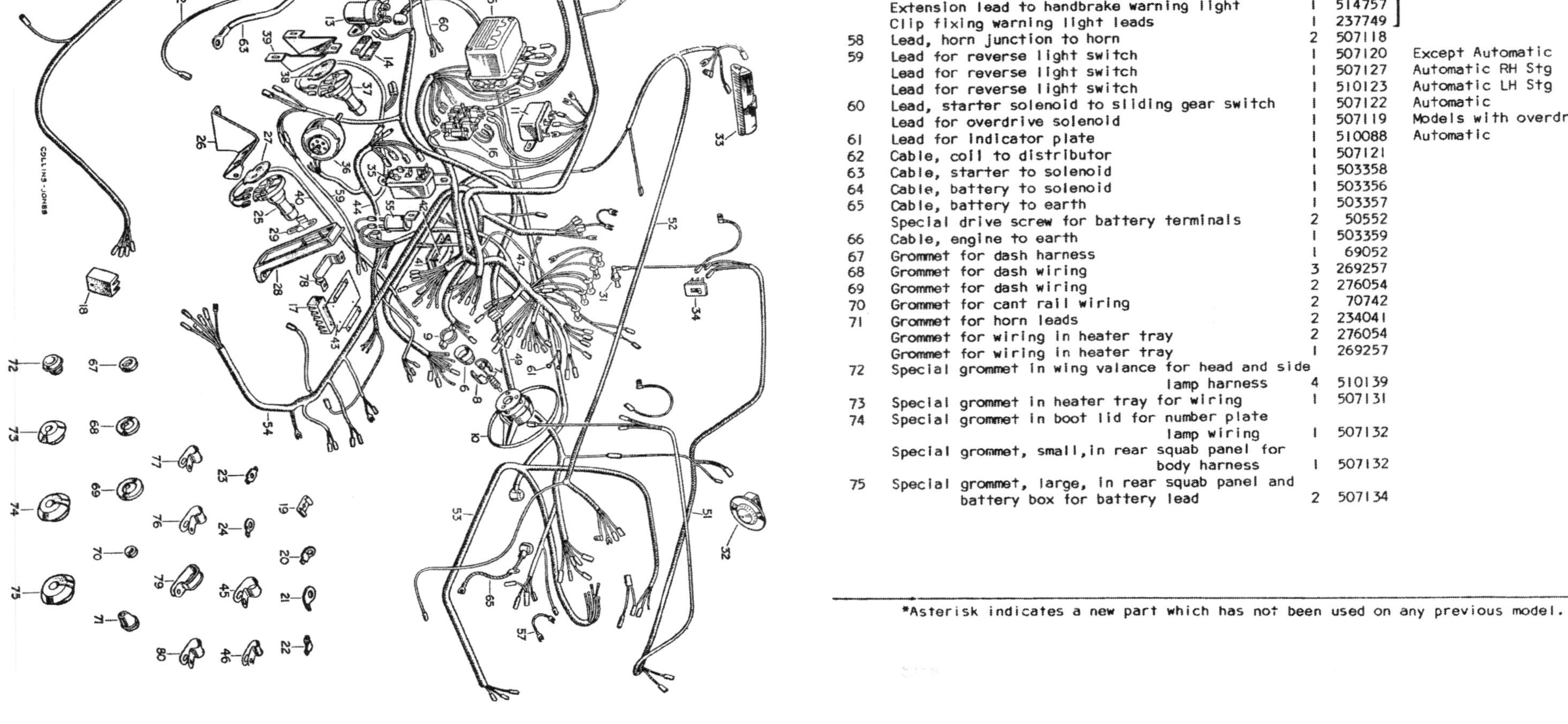

Plate Ref.	1 2 3 4 Description	Qty.	Part No.	Remarks
53	Harness for boot	1	519826	
54	Harness for wiper motor and heater	1	503367	RH Stg } For manual-type windscreen washer
	Harness for wiper motor and heater	1	510126	LH Stg } For manual-type windscreen washer
	Harness for wiper motor,heater and screen washer	1	514705	RH Stg } For electric-type windscreen washer
	Harness for wiper motor, heater and screen washer	1	514745	LH Stg } For electric-type windscreen washer
	Harness for windscreen washer	1	514711	} For electric-type windscreen washer
	Clip for wiper harness at wiper motor mounting bracket	1	3619	
55	Flasher unit	1	502096	
	Drive screw } Fixing flasher unit	1	78140	
	Shakeproof washer } Fixing flasher unit	1	78249	
	Plain washer } Fixing flasher unit	1	4203	
56	Earth lead for horn	2	507116	
	Earth lead for horn at steering column coupling	1	505170	
57	Earth lead for petrol pump	1	507117	
	Earth lead to handbrake warning light switch	1	514772	} 1961 Automatic only
	Lead, switch to handbrake warning light	1	514753	} 1961 Automatic only
	Lead, feed to handbrake warning light	1	514750	} 1961 Automatic only
	Extension lead to handbrake warning light	1	514757	} 1961 Automatic only
	Clip fixing warning light leads	1	237749	} 1961 Automatic only
58	Lead, horn junction to horn	2	507118	
59	Lead for reverse light switch	1	507120	Except Automatic
	Lead for reverse light switch	1	507127	Automatic RH Stg
	Lead for reverse light switch	1	510123	Automatic LH Stg
60	Lead, starter solenoid to sliding gear switch	1	507122	Automatic
	Lead for overdrive solenoid	1	507119	Models with overdrive
61	Lead for indicator plate	1	510088	Automatic
62	Cable, coil to distributor	1	507121	
63	Cable, starter to solenoid	1	503358	
64	Cable, battery to solenoid	1	503356	
65	Cable, battery to earth	1	503357	
	Special drive screw for battery terminals	2	50552	
66	Cable, engine to earth	1	503359	
67	Grommet for dash harness	1	69052	
68	Grommet for dash wiring	3	269257	
69	Grommet for dash wiring	2	276054	
70	Grommet for cant rail wiring	2	70742	
71	Grommet for horn leads	2	234041	
	Grommet for wiring in heater tray	2	276054	
	Grommet for wiring in heater tray	1	269257	
72	Special grommet in wing valance for head and side lamp harness	4	510139	
73	Special grommet in heater tray for wiring	1	507131	
74	Special grommet in boot lid for number plate lamp wiring	1	507132	
	Special grommet, small,in rear squab panel for body harness	1	507132	
75	Special grommet, large, in rear squab panel and battery box for battery lead	2	507134	

*Asterisk indicates a new part which has not been used on any previous model.

GENERAL ELECTRICAL EQUIPMENT, MK I MODELS

632

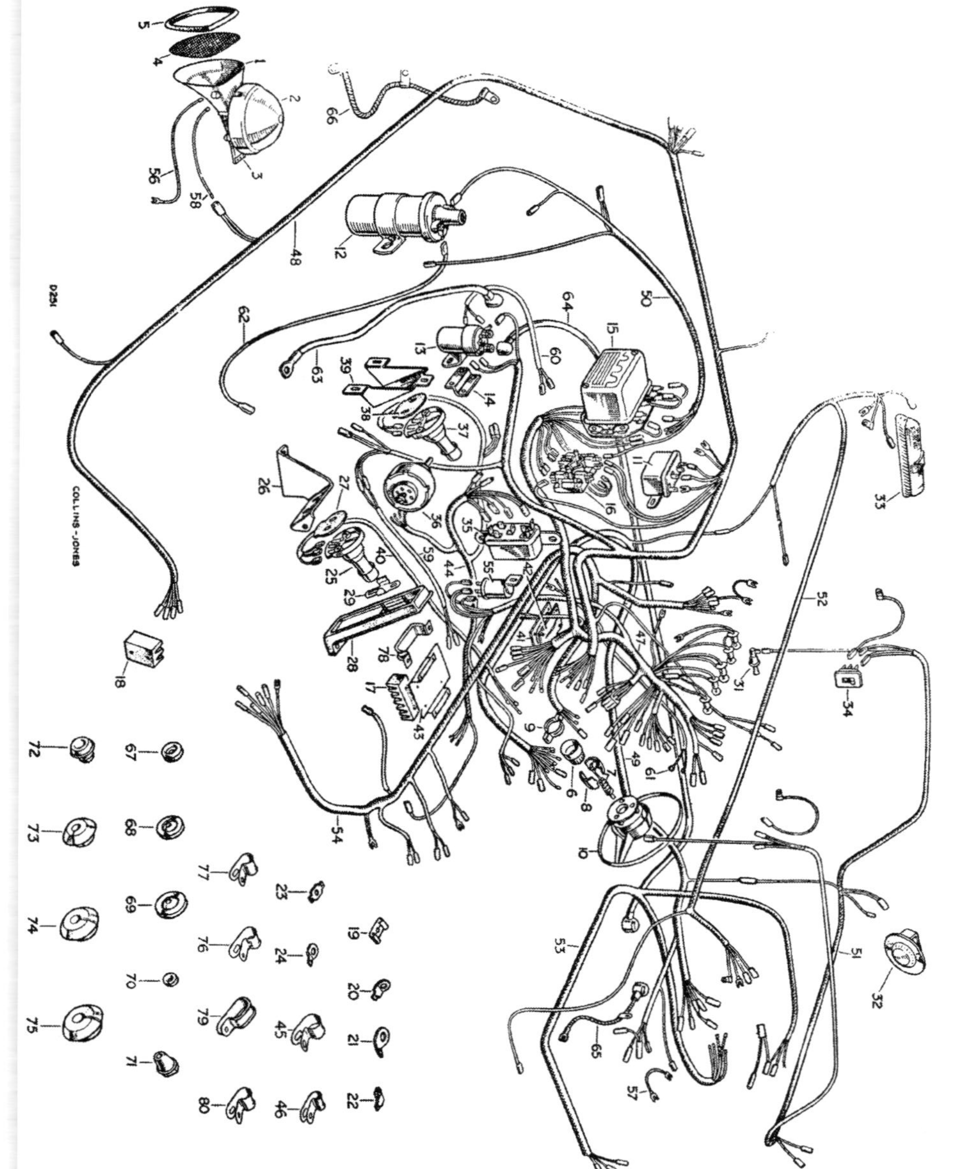

Plate Ref.	1 2 3 4	Description	Qty.	Part No.	Remarks
		Cable clip at voltage control box fixing bolt	1	3621	
76		Cable clip at cold air duct for engine harness	3	56666	
		Drive screw fixing clip	3	78417	
77		Cable clip on boot lid hinge	2	8885	
		Drive screw fixing clip	2	78006	
78		Cable clip under parcel shelf	2	507124	1 off on LH Stg
		Drive screw } Fixing clip to parcel shelf	4	78348	} 2 off on LH Stg
		Spring washer } Fixing clip to parcel shelf	4	3073	} 2 off on LH Stg
		Nut (10 UNF) } Fixing clip to parcel shelf	4	257023	} 2 off on LH Stg
		Cable clip for windscreen wiper leads	3	56666	} LH Stg
		Drive screw fixing clip	3	78417	} LH Stg
		Cable clip for dip switch leads at parcel shelf mounting bolt	1	50642	} LH Stg
		Special grommet for stop/tail lamp leads	2	510139	
		Cable clip for starter cable at flywheel housing	1	3551	
		Cable clip under parcel shelf	1	8885	
		Drive screw fixing clip to parcel shelf	1	78155	
79		Cable clip at cold air duct	1	3409	
		Drive screw fixing clip	1	78417	
80		Cable clip at support bracket on parcel shelf	1	50640	

*Asterisk indicates a new part which has not been used on any previous model.

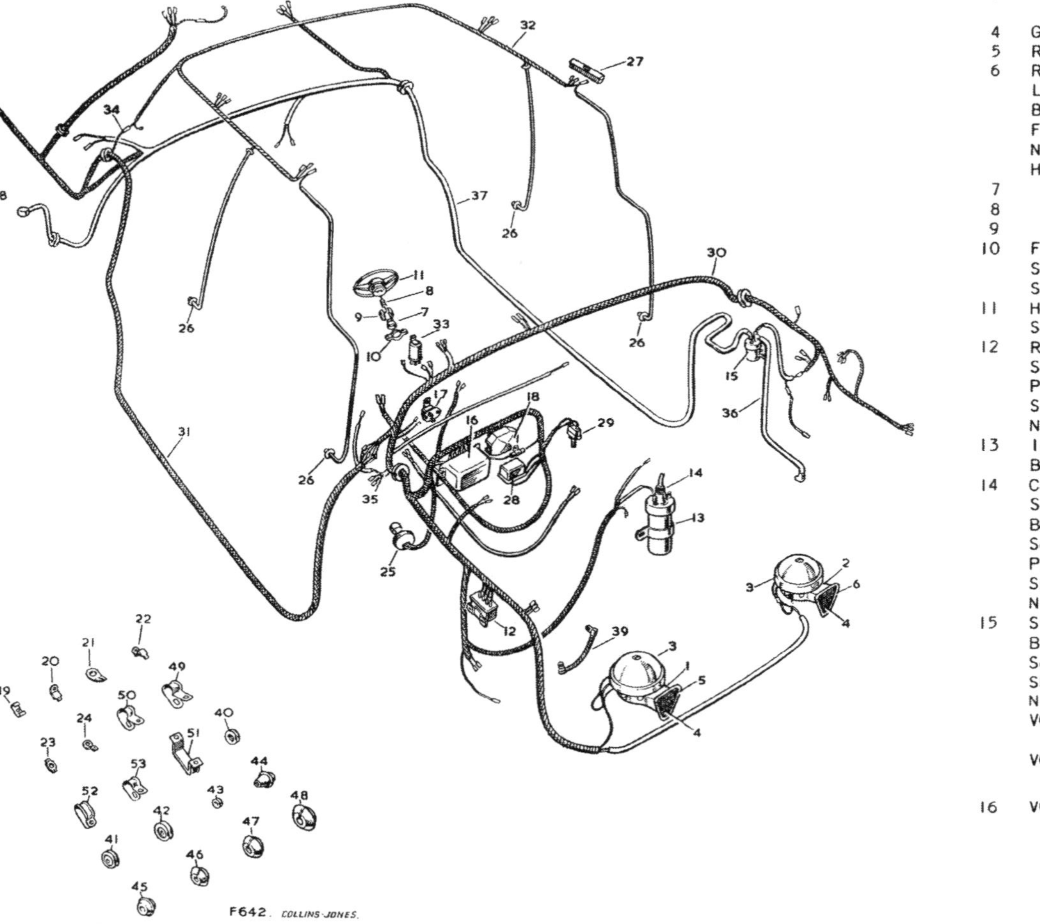

F642. COLLINS-JONES.

Plate Ref.	1 2 3 4 Description		Qty.	Part No.	Remarks
1	HORN, HIGH NOTE, RH		1	601972	
2	HORN, LOW NOTE, LH		1	601973	
	Contact set for horn		2	271500	
	Cover for horn		2	262205	Mk IA
3	Cover for horn		1	600018	Mk II
	Bracket for horn		2	262110	
4	Gauze for horn		2	266162	
5	Ring for gauze on high note horn		1	271845	
6	Ring for gauze on low note horn		1	271844	
	Lucar blade, earth to horn fixings		2	505205	
	Bolt(¼" UNF x ⅝" long)	Fixing horns to	4	255206	
	Fan disc washer	mounting bracket on	4	510170	
	Nut (¼" UNF)	chassis sub-frame	4	254810	
	HORN CONTACT AND SLIP RING COMPLETE		1	502964	
7	Horn contact, slip ring		1	277912	
8	Horn contact, cable and rotor, top half		1	279104	
9	Horn slip ring rotor, bottom half		1	277916	
10	Flasher trip ring		1	548069	
	Shakeproof washer	Fixing ring to steering	2	78249	
	Set screw (4BA x ⅜"long)	column shaft	2	75804	
11	Horn ring complete		1	275999	
	Special screw fixing horn ring		3	542695	
12	Relay for horn		1	502094	
	Screw(10 UNF x ½" long)	Fixing horn	2	78348	
	Plain washer	relay to	2	3816	
	Spring washer	RH inner	2	3073	
	Nut (10 UNF)		2	257023	
13	Ignition coil		1	503276	
	Boot for HT terminal nut		1	506679	
14	Cable nut for coil		1	510237	
	Split washer for coil		1	214279	
	Bolt (¼" UNF x ¾" long)		2	255207	Mk IA
	Set bolt (¼" UNC x ½" long)		2	253004	Mk II
	Plain washer	Fixing coil to	2	3840	
	Spring washer	engine	2	3074	
	Nut (¼" UNF)		2	254810	Mk IA
15	Starter solenoid		1	519952	
	Bracket for solenoid		1	513678	
	Screw(10 UNF x ½" long)	Fixing solenoid	4	257017	
	Shakeproof washer	and bracket	4	311396	
	Nut (10 UNF)	to dash	4	257023	
	VOLTAGE CONTROL BOX		1	502092	For C45 PV5 dynamo, 503278 — Mk IA
	VOLTAGE CONTROL BOX		1	512250	For C45 PV6 dynamo, 512248 — Mk IA
16	VOLTAGE CONTROL BOX		1	559265	Mk II
	Cover for box(drive screw fixing)		1	504444	
	Cover for regulator box (2 BA screw fixing)		1	513937	Mk IA
	Cover for regulator box		1	600442	Mk II
	Rubber seal for cover		1	507810	

*Asterisk indicates a new part which has not been used on any previous model.

GENERAL ELECTRICAL EQUIPMENT, MK IA AND MK II MODELS

1636

Plate Ref.	1 2 3 4 Description	Qty.	Part No.	Remarks
	Set bolt (¼" UNF x ¾" long) – Fixing voltage control box	3	255207	
	Spring washer – Fixing voltage control box	3	3074	
	Screw (10 UNF x 1⅜" long) – Fixing voltage control to cold box air duct	3	78477	Mk II
	Plain washer – Fixing voltage control to cold box air duct	3	3685	
	Riv-nut – Fixing voltage control to cold box air duct	3	532848	
17	Voltage regulator for instruments	1	555758	
	Bracket for regulator	1	536975	
	Drive screw – Fixing regulator to bracket	1	78140	
	Shakeproof washer – Fixing regulator to bracket	1	78249	
	Plain washer – Fixing regulator to bracket	1	4203	
	FUSE BOX	1	505156	Mk IA without overdrive
	Fuse, 35-amp	As reqd	12738	
	Fuse, 50-amp	As reqd	515083	
	Cover for fuse box	1	505158	
	FUSE BOX	1	505157	Mk IA with overdrive
	Fuse, 35-amp	As reqd	12738	
	Cover for fuse box	1	505159	
18	FUSE BOX	1	530047	Mk II with suffix letters 'A' and 'B'
	Fuse, 35-amp	3	12738	
	Cover for fuse box	1	505158	
	Drive screw locating fuse box	2	77704	Mk IA and Mk II with suffix letters 'A' and 'B'
	Drive screw fixing fuse box	2	78153	
	FUSE BOX	1	545293	Mk II with suffix letter 'C' onwards
	Cover for fuse box	1	600472	
	Fuse, 35-amp	1	12738	
	Fuse, 5-amp	1	541566	
	Drive screw fixing fuse box	2	78691	
19	Earthing clip on front and rear wing valance	7	236366	
	Drive screw – Fixing earthing clip	4	78417	
	Shakeproof washer – Fixing earthing clip	4	74236	
20	Special earthing blade on front wing valance and dash	2	505204	
	Drive screw – Fixing earthing blade to valance and dash	2	78137	
	Shakeproof washer – Fixing earthing blade to valance and dash	2	74236	
21	Special earthing blade at voltage regulator, flasher unit and horn fixing	4	505205	
	Special earth blade at speedometer	2	505195	Mk II 4-speed(3 off on Automatic)
	Insulating sleeve for spare terminal on voltage regulator	1	531509	
22	Special earthing blade at fuel gauge unit fixing screw and ignition coil	2	505203	
23	Special earthign blade at instrument unit	3	505195	
24	Special earthign blade at petrol pump fixing	1	507110	
25	Dip switch	1	502087	
	Screw(10 UNF x 1" long) – Fixing dip switch	2	78325	
	Plain washer – Fixing dip switch	2	2874	RH Stg
	Spring washer – Fixing dip switch	2	3073	
	Nut (10 UNF) – Fixing dip switch	2	257023	

*Asterisk indicates a new part which has not been used on any previous model.

F642. COLLINS-JONES.

Plate Ref.	1 2 3 4 Description		Qty.	Part No.	Remarks
	Bracket for dip switch		1	502155	RH Stg
	Bracket for dip switch		1	504848	LH Stg
	Rubber seal for bracket		1	502896	
	Bolt (¼" UNF x ⅜" long)	Fixing bracket and seal to floor	1	255206	4 off on LH Stg
	Plain washer		2	3900	
	Spring washer		1	3074	
	Nut (¼" UNF)		1	254810	
	Pedal for dip switch		1	503514	
	Clip, pedal to dip switch		1	503515	
	Plain washer	Fixing clip to pedal	2	3902	
	Spring washer		2	3073	
	Nut (2 BA)		2	2247	
	Screw(¼" UNF x 1¼" long)	Fixing pedal to floor	2	252115	
	Countersunk washer		2	66316	
	Spring washer		2	3074	LH Stg
	Nut(¼" UNF)		2	254810	
26	Pillar switch 'A' and 'D' posts, ½" dia. barrel		4	519847	Mk IA and Mk II
	Cap for switch		4	551260	
27	ROOF LAMP COMPLETE WITH FESTOON-TYPE BULB		4	567752	
	Lens for roof lamp		4	600458	
	Bulb 6 watt festoon-type, for roof lamp		4	311240	Alternative. Check before ordering
	Bulb,7 watt festoon-type, for roof lamp		4	567804	
	Screw fixing lamp		8	77704	
	Plain washer		8	3886	
	Battery		1	278283	Except America. Dollar area
	Battery		1	514720	America. Dollar area
28	Relay unit for overdrive		1	502910	Mk IA and Mk II with overdrive
	Rotary throttle switch for overdrive		1	273334	Mk IA with overdrive
29	Throttle switch		1	510267	Mk II 4-speed
	Kickdown switch for overdrive		1	502895	Mk IA with overdrive
	Rubber sealing pad for kickdown switch		1	502896	
	Screw(10 UNF x 1" long)fixing switch to bracket		2	78325	
	Plain washer	Fixing switch to bracket	2	3911	Mk II 4-speed
	Nut (¼" UNF)		2	254810	
	Bracket for kickdown switch		1	502754	RH Stg with overdrive
	Bolt (¼" UNF x ⅜" long)	Fixing kickdown bracket	3	255206	
	Plain washer		6	3840	
	Spring washer		3	3074	
	Nut (¼" UNF)		3	254810	
	Support bracket for harness at parcel shelf		1	352589	
	Rubber pad for harness at parcel shelf		1	352590	
	Support bracket for electrical connections		1	352588	
	Screw(10 UNF x ⅜" long)	Fixing brackets and pad to parcel shelf	8	78301	
	Plain washer		8	3902	
	Spring washer		8	3073	
	Nut (10 UNF)		8	257023	
	Harness for overdrive		1	530052	RH Stg
	Harness for overdrive		1	530053	LH Stg Mk IA

*Asterisk indicates a new part which has not been used on any previous model.

GENERAL ELECTRICAL EQUIPMENT, MK IA AND MK II MODELS

1640

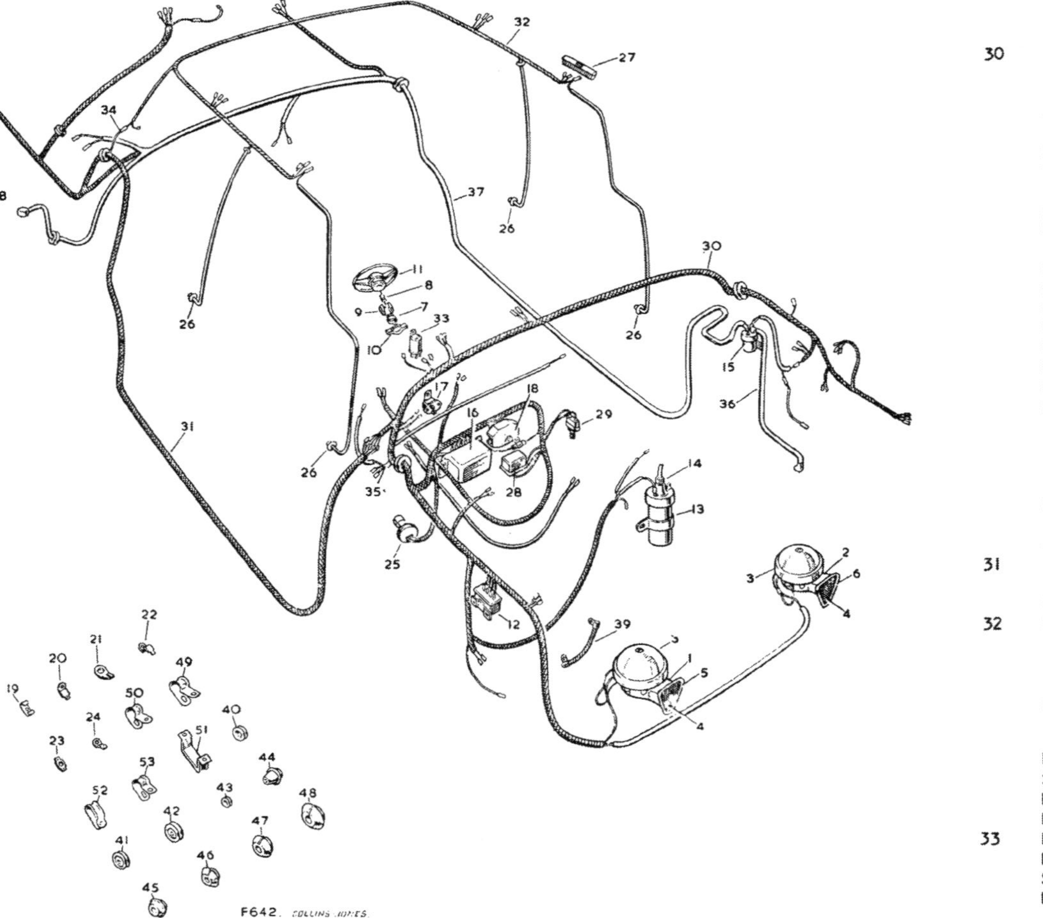

F642.

Plate Ref.	1 2 3 4 Description	Qty.	Part No.	Remarks
	Clip } Fixing overdrive harness to gearbox and extension casing	1	3621	With overdrive
	Clip	1	219676	
	Main harness	1	531510	RH Stg } Mk IA cars with red handbrake warning light
	Main harness	1	531511	LH Stg
	Main harness	1	519770	RH Stg } Mk IA cars with amber handbrake warning light
	Main harness	1	529991	LH Stg
30	Main harness	1	532862	RH Stg } Mk II saloon 4-speed } Mk II cars with suffix letter 'A' and 'B'
	Main harness	1	536001	LH Stg
	Main harness	1	532849	RH Stg Mk II saloon Automatic
	Main harness	1	532864	LH Stg
	Main harness	1	536166	RH Stg Mk II coupe Automatic
	Main harness	1	536164	LH Stg
	Main harness	1	536165	RH Stg Mk II coupe 4-speed
	Main harness	1	536161	LH Stg
	Main harness	1	547404	RH Stg Mk II saloon 4-speed } Mk II cars with suffix letter 'C' onwards
	Main harness	1	547405	LH Stg
	Main harness	1	547406	RH Stg Mk II saloon Automatic
	Main harness	1	547407	LH Stg
	Main harness	1	547408	RH Stg Mk II coupe 4-speed
	Main harness	1	547409	LH Stg
	Main harness	1	547410	RH Stg Mk II coupe Automatic
	Main harness	1	547411	LH Stg
	Lead, headlamp switch to tail lamp	1	531540	Mk IA } NADA
	Lead, headlamp switch to tail lamp	1	536067	Mk II with suffix letters 'A' and 'B'
	Lucar blade in-line connection	1	532819	
	Harness, side lamp switch to body connection	1	545068	Mk II with suffix letter 'C' onwards
	Body and boot harness	1	531512	Mk IA
31	Body harness	1	532865	Mk II saloon
	Body harness	1	536162	Mk II coupe
	Harness for interior light	1	528932	Mk IA
32	Harness for interior light	1	551246	Mk II
	Panel connector harness	1	519772	Mk IA
	Panel connector harness	1	532846	Mk II saloon
	Panel connector harness	1	536163	Mk II coupe
	Harness for bonnet light	1	555558	Coupe
	Harness for electrical intermediate gear lock, switch to solenoid	1	519973	} Automatic
	Earth lead for intermediate gear lock solenoid	1	526370	
	Set bolt (¼" UNC x 7/16" long) } Fixing earth lead to gearbox	1	253241	
	Plain washer	1	4049	
	Fan disc washer	1	512305	
33	Flasher unit	1	502096	
	Drive screw } Fixing flasher unit	1	78140	
	Shakeproof washer	1	78249	
	Plain washer	1	4203	

*Asterisk indicates a new part which has not been used on any previous model.

F642

Plate Ref.	1 2 3 4 Description	Qty.	Part No.	Remarks
	Earth lead for horn	2	507116	Mk IA
	Earth lead for horn at steering column coupling	1	505170	
34	Earth lead for petrol pump	1	507117	
	Earth lead between twin petrol pumps	1	528921	Mk IA
35	Lead, handbrake switch to warning light junction extension	1	531639	
	Lead, handbrake switch to A3 junction extension	1	531638	
	Lead, second gear lock, switch to A3 fuse	1	531640	RH Stg
	Lead, second gear lock, switch to A3 fuse	1	531641	LH Stg
	Lead, second gear lock switch to solenoid extension	1	531544	LH Stg
	Lead, horn junction to horn	2	507118	Mk IA
	Lead for reverse light switch	1	507120	Mk IA 4-speed
	Lead for reverse light switch	1	555658	4-speed
	Lead for reverse light switch	1	507127	Automatic RH Stg
	Lead for reverse light switch	1	510123	Automatic LH Stg
	Lead, starter solenoid to sliding gear switch	1	507122	Automatic
	Lead for overdrive solenoid	1	507119	Mk IA with overdrive
	Lead for indicator plate	1	510088	Mk IA Automatic
	Lead for indicator plate	1	536135	Mk II Automatic
	Lead, main harness to ignition switch	1	551609	RH Stg
	Lead, main harness to ignition switch	1	551610	LH Stg
	Cable, coil to distributor	1	507121	
	Cable, starter to solenoid	1	530016	RH Stg } Mk IA
	Cable, starter to solenoid	1	531603	LH Stg } Mk IA
36	Cable, starter to solenoid	1	532877	Mk II
	Cable, battery to solenoid	1	530017	Mk IA
37	Cable, battery to solenoid	1	532876	Mk II
38	Cable, battery to earth	1	503357	Mk IA and Mk II
	Special drive screw for battery terminals	2	50552	
39	Cable, engine to earth	1	503359	
40	Grommet for dash harness	1	69052	
41	Grommet for dash wiring	3	269257	
42	Grommet for dash wiring	2	276054	
43	Grommet for cant rail wiring	2	70742	
44	Grommet for horn leads	2	234041	Mk IA
	Grommet for wiring in heater tray	2	276054	
	Grommet for wiring in heater tray	1	269257	
	Grommet for heater box base	1	269257	Mk II
	Grommet for front scuttle	1	531505	Mk II Automatic
	Grommet for front scuttle	1	269257	Mk II 4-speed
	Grommet for cigar lighter in rear seat	1	219680	Mk II coupe
45	Special grommet in wing valance for head and side lamp harness	4	510139	

NADA indicates parts peculiar to cars exported to the North American dollar area

*Asterisk indicates a new part which has not been used on any previous model.

GENERAL ELECTRICAL EQUIPMENT, MK IA AND MK II MODELS

1644

Plate Ref.	1 2 3 4 Description	Qty.	Part No.	Remarks
46	Special grommet in heater tray for wiring	1	507131	
47	Special grommet in boot lid for number plate lamp wiring	1	507132	
	Special grommet, small, in rear squab panel for body harness	1	507132	
48	Grommet, split, for battery box	1	507134	
	Grommet, rear LH shelf	1	276054	
	Cable clip at voltage control box fixing bolt	1	3621	
49	Cable clip at cold air duct for engine harness	3	56666	
	Drive screw fixing clip	3	78417	
50	Cable clip on boot lid hinge	2	8885	
	Drive screw fixing clip	2	78006	
51	Cable clip under parcel shelf	2	507124	1 off on LH Stg
	Drive screw } Fixing clip to parcel shelf	4	78348	} 2 off on LH Stg
	Spring washer } Fixing clip to parcel shelf	4	3073	} 2 off on LH Stg
	Nut (10 UNF) } Fixing clip to parcel shelf	4	257023	} 2 off on LH Stg
	Cable clip for windscreen wiper leads	3	56666	} LH Stg
	Drive screw fixing clip	3	78417	} LH Stg
	Cable clip for dip switch leads at parcel shelf mounting bolt	1	50642	} LH Stg
	Special grommet for stop/tail lamp leads	2	510139	
	Cable clip for starter cable at flywheel housing	1	3551	
	Cable clip under parcel shelf	1	8885	
	Drive screw fixing clip to parcel shelf	1	78155	
52	Cable clip at cold air duct	1	3409	
	Drive screw fixing clip	1	78417	
53	Cable clip at support bracket on parcel shelf	1	50640	
	Clip at rear end of automatic gearbox securing solenoid feed lead	1	50638	} Mk IA
	Clip for main harness at lower screen washer bracket fixing bolt	1	50639	} Mk IA
	Cable clip at cold air duct	2	50639	} Mk II
	Drive screw } Fixing clip	1	70747	} Mk II
	Drive screw } Fixing clip	1	78155	} Mk II
	Cable clip for starter cable at front scuttle	1	50640	} Mk II
	Screw (10 UNF x ½" long) } Fixing starter cable clips	4	78348	} Mk II
	Plain washer } Fixing starter cable clips	4	3685	} Mk II
	Spring washer } Fixing starter cable clips	4	3073	} Mk II
	Nut (10 UNF) } Fixing starter cable clips	4	257023	} Mk II
	Cable clip at cold air duct for bonnet light harness	2	237749	Mk II
	Cable clip under rear seat for cigar lighter leads	2	237749	coupe

*Asterisk indicates a new part which has not been used on any previous model.

GENERAL ELECTRICAL EQUIPMENT, MK III MODELS

1646

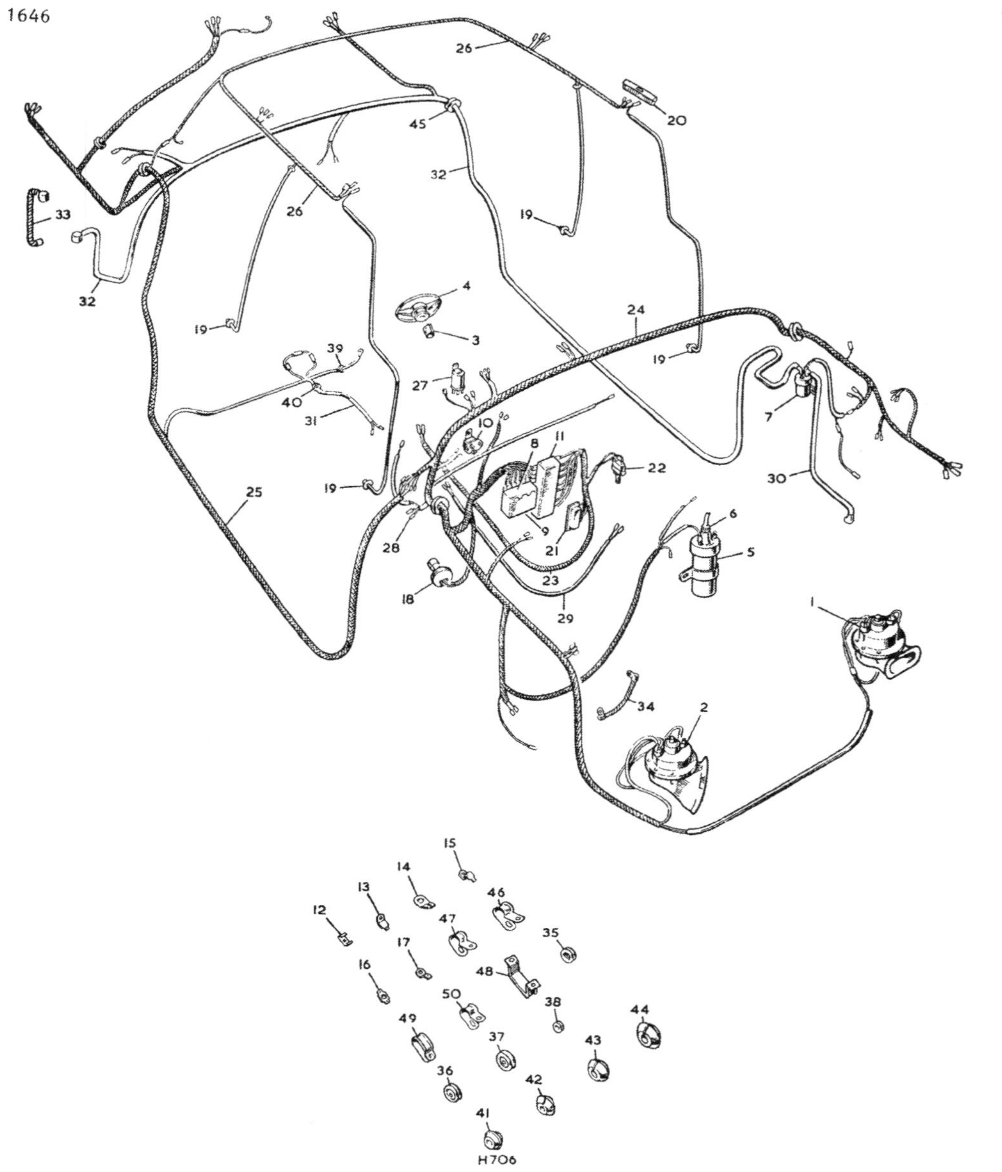

Plate Ref.	1 2 3 4 Description		Qty.	Part No.	Remarks
1	HORN, LOW NOTE		1	555679	
2	HORN, EXTRA LOW NOTE		1	555972	
	Lucar blade, earth to horn fixings		2	505205	
	Bolt (¼"UNF x ⅝"long)	Fixing horns to mounting bracket on chassis sub-frame	4	255206	
	Fan disc washer		4	510170	
	Nut (¼"UNF)		4	254810	
	HORN CONTACT AND SLIP RING COMPLETE		1	502964	Early models
	Horn contact, slip ring		1	277912	
	Horn contact, cable and rotor, top half		1	279104	
	Horn slip ring rotor, bottom half		1	277916	
	Flasher trip ring		1	548069	
3	Flasher trip ring		1	553775	Late models
	Shakeproof washer	Fixing ring to steering column shaft	2	78249	
	Set screw(4BA x ⅜"long)		2	75804	
4	Horn ring complete		1	275999	
	Special screw fixing horn ring		3	542695	
5	Ignition coil		1	GCL 101	
	Boot for HT terminal nut		1	506679	
6	Cable nut for coil		1	510237	
	Split washer for coil		1	214279	
	Set bolt(¼"UNC x ½"long)	Fixing coil to engine	2	253004	
	Plain washer		2	3840	
	Spring washer		2	3074	
7	Starter solenoid		1	567969	
	Bracket for solenoid		1	513678	
	Screw(10UNF x ½"long)	Fixing solenoid and bracket to dash	4	257017	
	Shakeproof washer		4	311396	
	Nut (10 UNF)		4	257023	
8	VOLTAGE CONTROL BOX		1	559265	
9	Cover for control box		1	600442	
	Rubber seal for cover		1	507810	
	Screw(10UNF x 1⅜"long)	Fixing voltage control box to cold air duct	3	78477	
	Plain washer		3	3685	
	Riv-nut		3	532848	
10	Voltage regulator for instruments		1	555758	
	Bracket for regulator		1	536975	
	Drive screw	Fixing regulator to bracket	1	78140	
	Shakeproof washer		1	78249	
	Plain washer		1	4203	
11	FUSE BOX COMPLETE		1	551574	
	Fuse, 5-amp		3	541566	
	Fuse, 10-amp		2	541567	
	Fuse, 15-amp		3	541570	
	Fuse, 25-amp		2	40167	
	Fuse, 35-amp		1	12738	
	Cover for fuse box		1	601477	
	Drive screw fixing fuse box		2	78489	

*Asterisk indicates a new part which has not been used on any previous model.

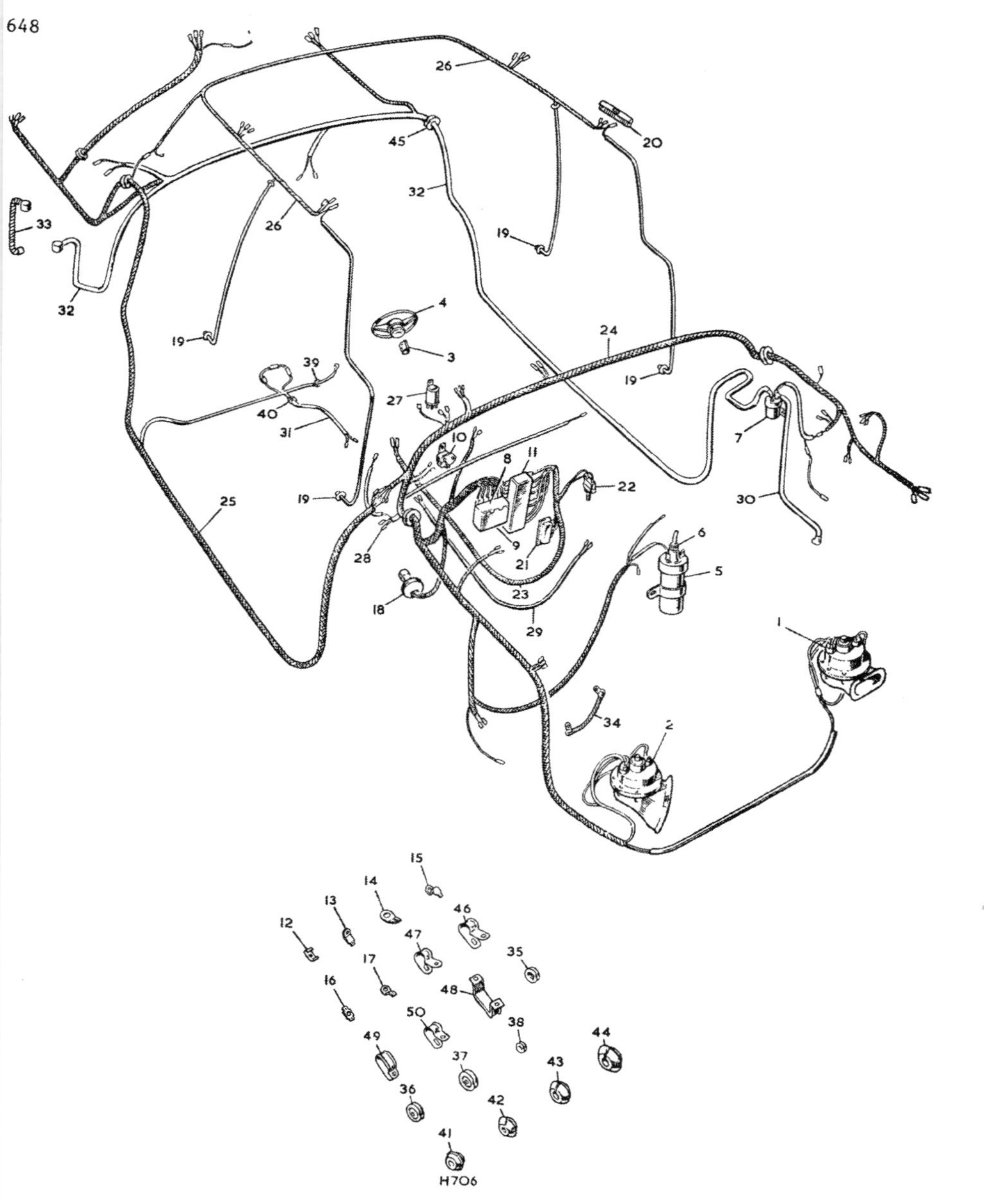

Plate Ref.	1 2 3 4 Description	Qty.	Part No.	Remarks
12	Earthing clip on front and rear wing valance	7	236366	
	Drive screw } Fixing earthing clip	4	78417	
	Shakeproof washer } Fixing earthing clip	4	74236	
13	Special earthing blade on front wing valance and dash	2	505204	
	Drive screw } Fixing earthing blade to valance and dash	2	78137	
	Shakeproof washer } Fixing earthing blade to valance and dash	2	74236	
14	Special earthing blade at voltage regulator, flasher unit and horn fixing	4	505205	
	Special earth blade at speedometer	2	505195	4-speed (3 off on Automatic)
	Insulating sleeve for spare terminal on voltage regulator	1	531509	
15	Special earthing blade at fuel gauge unit fixing screw and ignition coil	2	505203	
16	Special earthing blade at instrument unit	3	505195	
17	Special earthing blade at petrol pump fixing	1	507110	
18	Dip switch	1	502087	
	Screw (10 UNF x 1"long) } Fixing dip switch	2	78325	
	Plain washer } Fixing dip switch	2	2874	RH Stg
	Spring washer } Fixing dip switch	2	3073	RH Stg
	Nut (10 UNF) } Fixing dip switch	2	257023	RH Stg
	Bracket for dip switch	1	502155	RH Stg
	Bracket for dip switch	1	504848	LH Stg
	Rubber seal for bracket	1	502896	
	Bolt(¼"UNF x ⅜"long) } Fixing bracket and seal to floor	1	255206	4 off on LH Stg
	Plain washer } Fixing bracket and seal to floor	2	3900	4 off on LH Stg
	Spring washer } Fixing bracket and seal to floor	1	3074	4 off on LH Stg
	Nut (¼"UNF) } Fixing bracket and seal to floor	2	254810	4 off on LH Stg
	Pedal for dip switch	1	503514	
	Clip, pedal to dip switch	1	503515	
	Plain washer } Fixing clip to pedal	2	3902	
	Spring washer } Fixing clip to pedal	2	3073	
	Nut (2 BA) } Fixing clip to pedal	2	2247	
	Screw (¼"UNF x 1¼"long) } Fixing pedal to floor	2	252114	
	Countersunk washer } Fixing pedal to floor	2	66316	
	Spring washer } Fixing pedal to floor	2	3074	LH Stg
	Nut (¼" UNF) } Fixing pedal to floor	2	254810	LH Stg
19	Pillar switch	4	559061	
	Cap for switch	4	551260	
20	ROOF LAMP COMPLETE	4	567752	
	Lens for roof lamp	4	600458	
	Bulb, 6 watt festoon	4	GLB 254	Alternatives, check before ordering
	Bulb, 7 watt festoon	4	567804	Alternatives, check before ordering
	Screw fixing lamp	8	77704	
	Plain washer	8	3886	

*Asterisk indicates a new part which has not been used on any previous model.

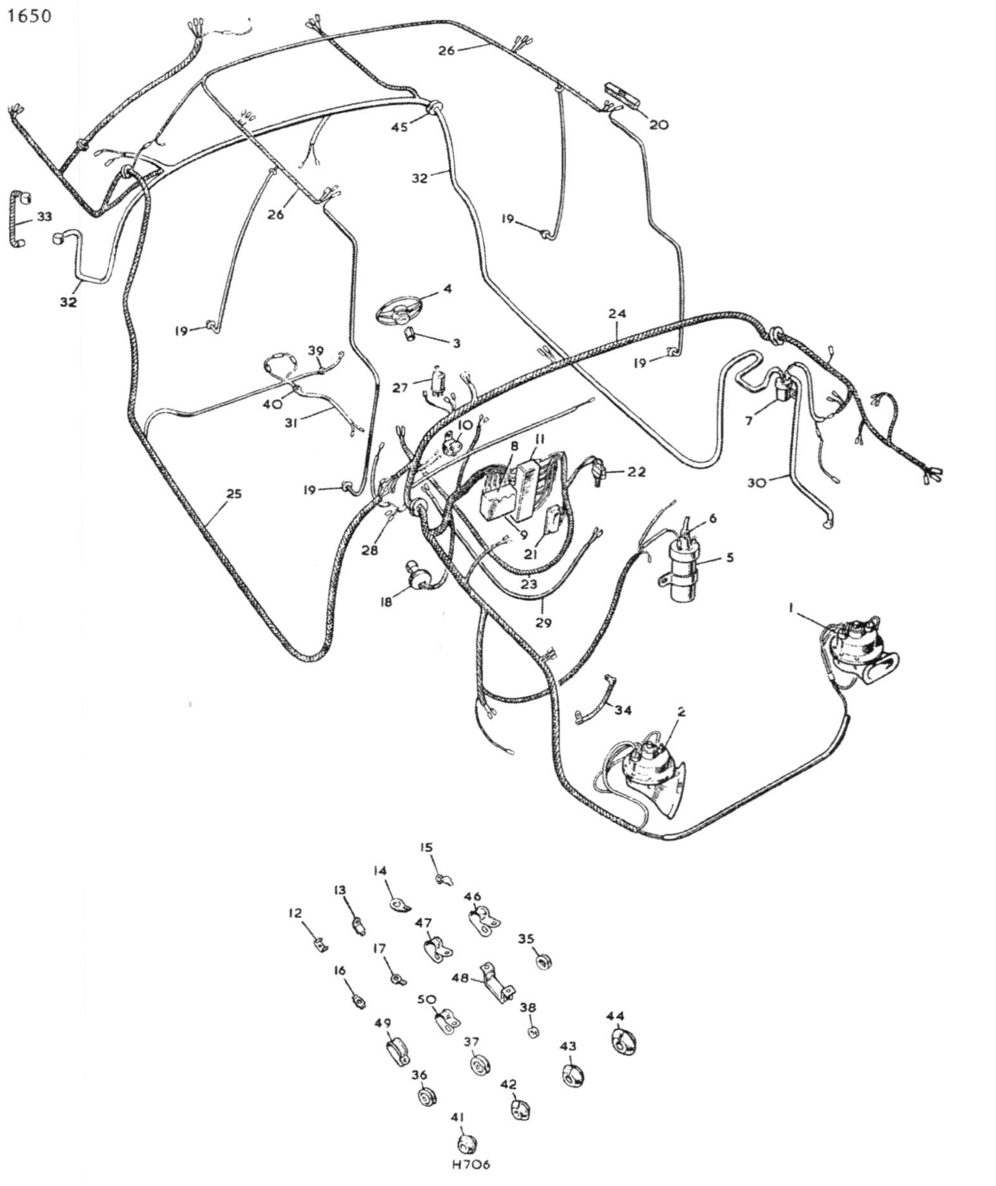

Plate Ref.	1 2 3 4 Description	Qty.	Part No.	Remarks
	BATTERY	1	567884	Except NADA
	BATTERY	1	514720	NADA
	Special screw for battery lugs	2	50552	4-speed
21	Relay unit for overdrive	1	555681	4-speed
	Drive screw } Fixing relay unit	2	78494	4-speed
	Shakeproof washer } Fixing relay unit	2	74236	4-speed
22	Throttle switch	1	510267	4-speed
	Bracket for throttle switch	1	545957	4-speed
	Bolt (¼" UNC x ⅜" long) } Fixing bracket to flywheel housing	2	253006	4-speed
	Spring washer } Fixing bracket to flywheel housing	2	3074	4-speed
23	Harness for overdrive	1	551612	RH Stg 4-speed
	Harness for overdrive	1	551613	LH Stg 4-speed
	Clip } Fixing overdrive harness to gearbox and extension casing	1	3621	4-speed
	Clip } Fixing overdrive harness to gearbox and extension casing	1	219676	4-speed
24	Main harness	1	559282	RH Stg } Saloon, 4-speed and Automatic
	Main harness	1	559284	LH Stg } Saloon, 4-speed and Automatic
	Main harness	1	559283	RH Stg } Coupe. 4-speed and Automatic
	Main harness	1	559285	LH Stg } Coupe. 4-speed and Automatic
	Lead, sidelamp, headlamp switch to tail light	1	551646	
	Lucrimp connector for lead	1	541684	NADA
25	Body harness	1	547478	Saloon
	Body harness	1	547479	Coupe
26	Harness for interior lights	1	551246	
	Panel connector harness	1	541700	
	Harness for bonnet light	1	555558	Coupe
	Lead, overdrive feed	1	551627	4-speed
	Harness for inhibitor switch to main harness	1	537113	Automatic
	Lead, main harness to ignition switch	1	551609	RH Stg Automatic
	Lead, main harness to ignition switch	1	551610	LH Stg Automatic
	Lead for indicator plate	1	547484	Automatic
27	Flasher unit	1	502096	
	Drive screw } Fixing flasher unit	1	78140	
	Shakeproof washer } Fixing flasher unit	1	78249	
	Plain washer } Fixing flasher unit	1	4203	
	Earth lead for horn at steering column coupling	1	505170	
28	Lead, handbrake switch to warning light junction extension	1	531639	
	Lead, handbrake switch to junction	1	551180	
29	Lead for reverse light switch	1	555658	4-speed
	Cable coil to distributor	1	507121	
30	Cable, starter to solenoid	1	532877	
	Lucar earth blade at flasher unit fixing	1	505205	
31	Harness for rear heater	1	555774	
32	Cable, battery positive to solenoid	1	551490	
33	Cable, battery negative to earth	1	551492	
34	Cable, engine to earth	1	503359	
35	Grommet for dash harness	1	69052	
36	Grommet for dash wiring	3	269257	
37	Grommet for dash wiring	2	276054	
38	Grommet for cant rail wiring	2	70742	

*Asterisk indicates a new part which has not been used on any previous model.

GENERAL ELECTRICAL EQUIPMENT, MK III MODELS

1652

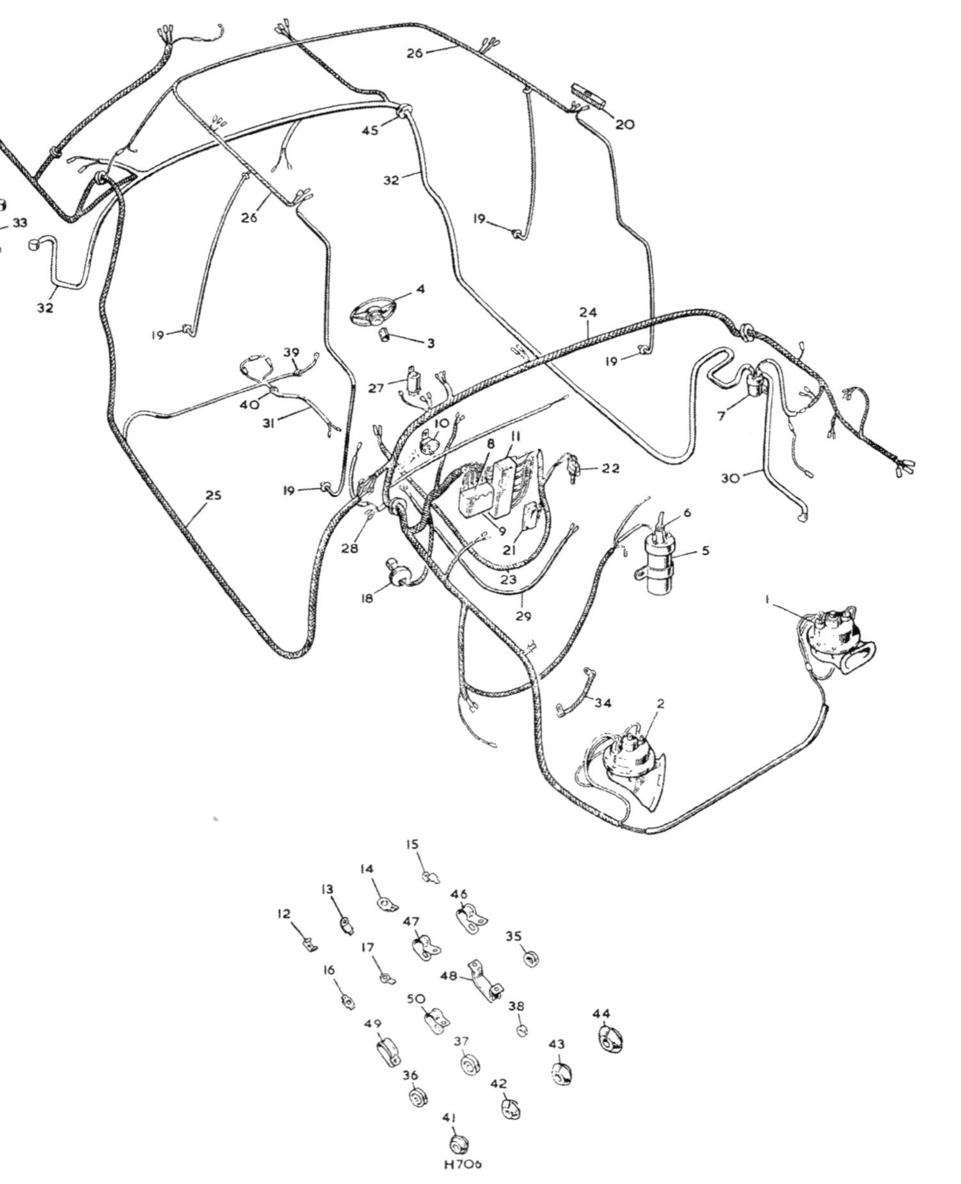

Plate Ref.	1 2 3 4	Description	Qty.	Part No.	Remarks
		Grommet for wiring in heater tray.	2	276054	
		Grommet for wiring in heater tray	1	269257	
		Grommet for heater box base	1	269257	
		Grommet for front scuttle	1	531505	Automatic
		Grommet for front scuttle	1	269257	4-speed
39		Grommet for cigar lighter in rear seat	1	219680	Coupe
40		Grommet for rear seat heelboard	1	22066	
41		Special grommet in wing valance for head and side lamp harness	4	510139	
42		Special grommet in heater tray for wiring,	1	507131	
43		Special grommet in boot lid for number plate lamp wiring,	1	507132	
		Special grommet, small, in rear squab panel for body harness	1	507132	
44		Grommet, split, for battery box	1	507134	
45		Grommet, rear LH shelf	1	276054	
		Cable clip at voltage control box fixing bolt	1	3621	
46		Cable clip at cold air duct for engine harness,	3	56666	
		Drive screw fixing clip	3	78417	
47		Cable clip on boot lid hinge	2	8885	
		Drive screw fixing clip	2	78006	
48		Cable clip under parcel shelf	2	507124	1 off on LH Stg
		Drive screw } Fixing clip to parcel shelf	4	78348	} 2 off on LH Stg
		Spring washer } Fixing clip to parcel shelf	4	3073	} 2 off on LH Stg
		NUT (10 UNF) } Fixing clip to parcel shelf	4	257023	} 2 off on LH Stg
		Cable clip for windscreen wiper leads	3	56666	} LH Stg
		Drive screw fixing clip	3	78417	} LH Stg
		Cable clip for dip switch leads at parcel shelf mounting bolt	1	50642	} LH Stg
		Special grommet for stop/tail lamp leads	2	510139	
		Cable clip for starter cable at flywheel housing	1	3551	
		Cable clip under parcel shelf	1	8885	
		Drive screw fixing clip to parcel shelf	1	78155	
49		Cable clip at cold air duct	1	3409	
		Drive screw fixing clip	1	78417	
50		Cable clip at support bracket on parcel shelf	1	50640	
		Cable clip at cold air duct	2	50639	
		Drive screw } Fixing clip	1	70747	
		Drive screw } Fixing clip	1	78155	
		Cable clip for starter cable at front scuttle	1	50640	
		Screw (10 UNF x $\frac{1}{2}$" long) } Fixing starter cable clips	4	78348	
		Plain washer } Fixing starter cable clips	4	3685	
		Spring washer } Fixing starter cable clips	4	3073	
		Nut (10 UNF) } Fixing starter cable clips	4	257023	
		Cable clip at cold air duct for bonnet light harness	2	237749	
		Cable clip under rear seat for cigar lighter leads	2	237749	Coupe
		PVC Sheathing (100 foot roll) for electrical wiring protection	1	601694*	State length required

*Asterisk indicates a new part which has not been used on any previous model.

Plate Ref.	1 2 3 4 Description		Qty.	Part No.	Remarks
1	Horn, low note		1	559443	
2	Horn, extra low note		1	559444	
	Mounting bracket for horns		1	562136	
	Bolt (¼" UNF x ⅜" long)	Fixing	3	255206	
	Spring washer	mounting bracket	3	3074	
	Plain washer	to chassis	3	3912	
	Bolt (¼" UNF x ⅜" long)	Fixing horns	4	255206	
	Spring washer	to mounting	4	3074	
	Nut (¼" UNF)	bracket	4	254810	
3	Flasher trip ring		1	553775	
	Shakeproof washer	Fixing trip ring	2	78249	
	Set screw (4 BA x ⅜" long)	to shaft	2	77701	
4	Horn ring complete		1	275999	
	Special screw fixing horn ring		3	542695	
5	Ignition coil		1	570763	
	Bolt (5/16" UNF x 1" long) top		1	255228	
	Bolt (5/16" UNF x ⅜" long) lower		1	255025	Except cars with
	Spring washer	Fixing coil	2	3075	thiefproof coil
	Spacer, lower		1	559445	
	Plain washer		4	2550	
	Nut (5/16" UNF)		4	254831	
	THIEFPROOF IGNITION COIL AND SWITCH		1	586196*	
	Special key for ignition switch		1	607435*	
	Connector, 17½ amp	Fixing coil	1	510119	Late models
	Sleeve for connector	terminal to harness	1	545496	
	Mounting bracket for thiefproof coil		1	586181*	
	Bolt (5/16" UNF x ⅜" long)		2	255205	RH Stg only
	Plain washer	Fixing coil to	2	3899	Cars numbered
	Spring washer	mounting bracket	2	3075	from:-
	Nut (5/16" UNF)		2	254831	
	Clip for armoured cable		1	586174*	
	Screw (10 UNF x ½" long)	Fixing cable clip	1	257017	
	Spring washer		1	3073	
	Nut (10 UNF)		1	257023	
	Alternator control box		1	551179	Cars numbered up to suffix letter 'B'
	Alternator control box		1	586659*	Cars numbered from suffix letter 'C' onwards
	Drive screw fixing voltage control box to cold air duct		2	78494	
7	Voltage regulator for instruments		1	555758	
	Bracket for regulator at facia bracket		1	536975	
	Fan disc washer at bracket fixing		1	513282	
	Set screw (4 BA x ⅜" long)	Fixing regulator	1	77701	
8	Spring washer	to bracket	1	3072	
	FUSE BOX COMPLETE		1	563248	
	Fuse, 35 amp		1	12738	
	Fuse, 25 amp		2	40167	
	Fuse, 15 amp		4	541570	
	Fuse, 10 amp		2	541567	
	Fuse, 5 amp		3	541566	
	Cover for fuse box		1	563249	
	Drive screw fixing fuse box		2	78494	

*Asterisk indicates a new part which has not been used on any previous model.

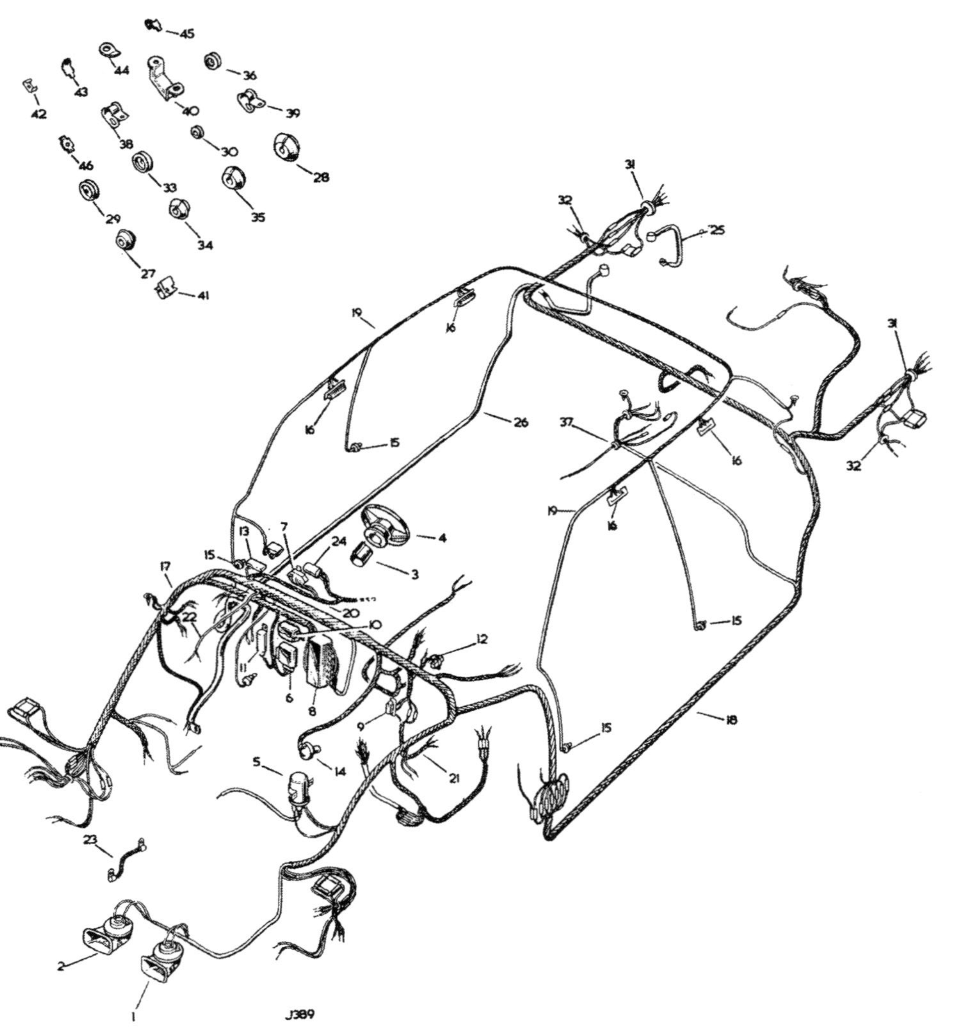

Plate Ref.	1 2 3 4 Description		Qty.	Part No.	Remarks
9	Relay for starter		1	563417	
	Bolt(10 UNF x 5/16" long)	Fixing relay to cold	2	257014	
	Shakeproof washer	air duct, LH side	2	311373	
10	Relay for alternator		1	555817	
	Drive screw	Fixing relay to cold	2	78494	
	Shakeproof washer	air duct, RH side	2	311373	
11	Relay for alternator warning light		1	555777	
	Drive screw	Fixing relay	1	78494	
	Plain washer	to cold air duct,	1	3816	
	Shakeproof washer	RH side	1	311373	
12	Ballast resistor for ignition coil		1	555863	Cars numbered up to suffix 'B' only
13	Shunt resistance for ammeter		1	559331	
	Insulating sleeve		2	559308	
	Bolt(10 UNF x ½" long)		2	257005	
	Plain washer	Fixing shunt to	2	3902	
	Spring washer	bracket adjacent	2	3073	
	Nut (10 UNF)	to steering column	2	257023	
14	Dip switch		1	502087	
	Screw(10 UNF x 1½" long)		2	78613	
	Plain washer	Fixing dip switch	2	3885	
	Spring washer	to bracket	2	3073	
	Nut (10 UNF)		2	257023	
	MOUNTING BRACKET FOR DIP SWITCH		1	502155	RH Stg
	MOUNTING BRACKET FOR DIP SWITCH		1	504848	LH Stg
	Rubber seal for bracket		1	502896	
	Bolt (¼" UNF x ⅜" long)		1	255206	4 off on LH Stg
	Plain washer	Fixing bracket and	2	3900	
	Plain washer	seal to floor	1	4075	
	Spring washer		1	3074	
	Nut (¼" UNF)		1	254810	
	Pedal for dip switch		1	503514	
	Clip, pedal to dip switch		1	503515	
	Plain washer		2	3902	
	Spring washer	Fixing clip to pedal	2	3073	
	Nut (2 BA)		2	2247	
	Screw(¼" UNF x 1¼" long)		2	252115	
	Countersunk washer	Fixing pedal	2	66316	
	Spring washer	to floor	2	3074	LH Stg
	Nut (¼" UNF)		2	254810	
15	Pillar switch for courtesy lights		4	559061	
	Drive screw	Fixing switch to	4	78551	
	Fan disc washer	door pillar	4	536993	
16	ROOF LAMP COMPLETE		4	567752	
	Lens for roof lamp		4	600458	
	Bulb, festoon type		4	567804	
	Drive screw	Fixing	8	77704	
	Plain washer	roof lamp	8	3886	

*Asterisk indicates a new part which has not been used on any previous model.

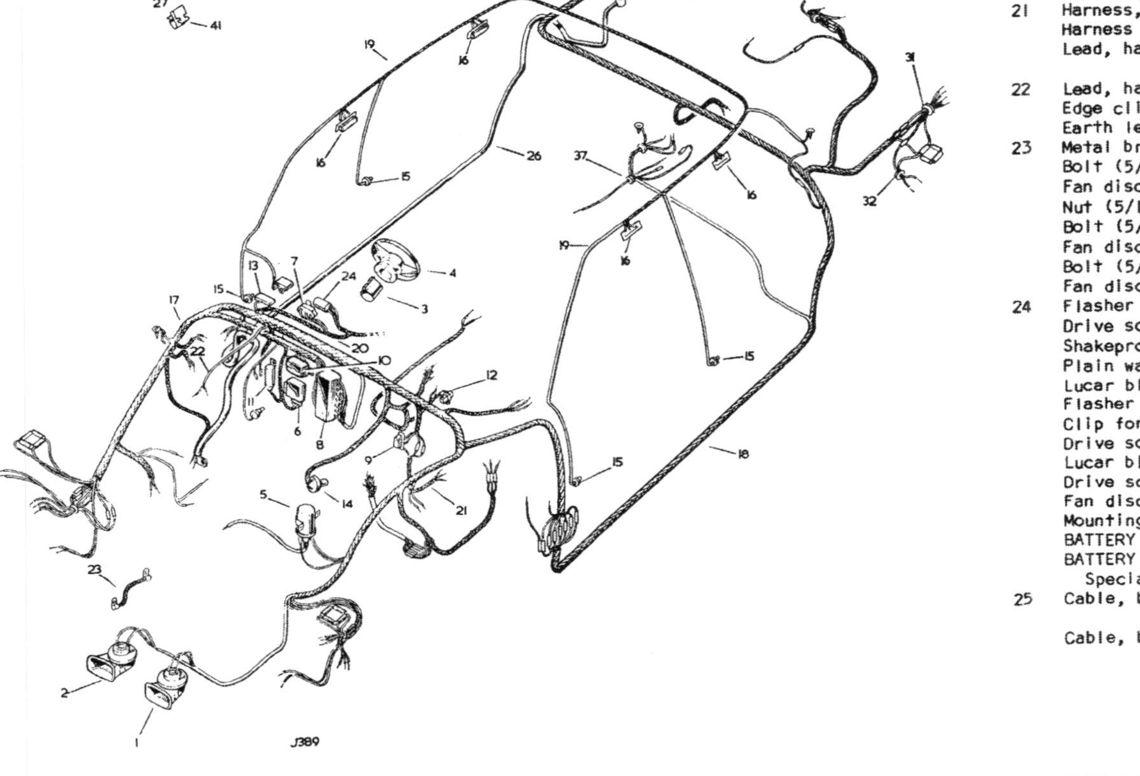

Plate Ref.	1 2 3 4 Description	Qty.	Part No.	Remarks
17	Main harness	1	565986	RH Stg Saloon — Cars numbered up to suffix letter 'B'
	Main harness	1	565988	LH Stg Saloon — Cars numbered up to suffix letter 'B'
	Main harness	1	565987	RH Stg Coupe — Cars numbered up to suffix letter 'B'
	Main harness	1	565989	LH Stg Coupe — Cars numbered up to suffix letter 'B'
	Main harness	1	568280*	RH Stg Saloon — Cars numbered from suffix letter 'C'
	Main harness	1	568281*	LH Stg Saloon — Cars numbered from suffix letter 'C'
	Main harness	1	568282*	RH Stg Coupe — Cars numbered from suffix letter 'C'
	Main harness	1	568283*	LH Stg Coupe — Cars numbered from suffix letter 'C'
	Main harness	1	586069*	RH Stg Saloon — Cars with thiefproof ignition coil
	Main harness	1	586071*	RH Stg Coupe — Cars with thiefproof ignition coil
18	Body harness	1	559340	Saloon
	Body harness	1	559339	Coupe
19	Harness for interior lights	1	551246	
20	Harness, panel connector	1	559286	
21	Harness, inhibitor switch to main harness	1	559342	
	Harness for bonnet lights	1	555558	Coupe
	Lead, handbrake switch to warning light junction extension	1	531639	
22	Lead, handbrake switch to junction	1	551180	
	Edge clip fixing leads	1	237749	
	Earth lead for horn at steering column coupling	1	505170	
23	Metal braid, engine to body frame	1	559449	
	Bolt (5/16" UNF x 3/8" long) — Fixing metal braid to wing	1	255225	
	Fan disc washer — Fixing metal braid to wing	2	510912	
	Nut (5/16" UNF) — Fixing metal braid to wing	1	254831	
	Bolt (5/16" UNF x 3/8" long) Fixing metal braid to upper mounting bracket	1	255225	
	Fan disc washer — Fixing metal braid to upper mounting bracket	2	510912	
	Bolt (5/16" UNFx 3/4"long) — Fixing metal braid to sub-frame	1	253870	
	Fan disc washer — Fixing metal braid to sub-frame	2	510912	
24	Flasher unit	1	GFU103	Cars numbered up to:- 84002707, 84501895, 84100129, 84600101, 84300135, 84800100
	Drive screw — Fixing flasher unit	1	78140	Cars numbered up to:- (as above)
	Shakeproof washer — Fixing flasher unit	1	78249	Cars numbered up to:- (as above)
	Plain washer — Fixing flasher unit	1	4203	Cars numbered up to:- (as above)
	Lucar blade at flasher unit fixing	1	505205	Cars numbered up to:- (as above)
	Flasher unit	1	GFU108	Cars numbered from:- 84002708, 84501896, 84100130, 84600102, 84300136, 84800101
	Clip for flasher unit	1	567959	Cars numbered from:- (as above)
	Drive screw fixing flasher unit	1	78522	Cars numbered from:- (as above)
	Lucar blade — Adjacent to flasher unit	2	505204	Cars numbered from:- (as above)
	Drive screw — Adjacent to flasher unit	1	78522	Cars numbered from:- (as above)
	Fan disc washer — Adjacent to flasher unit	1	513282	Cars numbered from:- (as above)
	Mounting bracket for flasher unit	1	568187	Cars numbered from:- (as above)
	BATTERY	1	567884	Except Canada
	BATTERY	1	567895*	Canada
	Special screw for battery lugs	2	50552	
25	Cable, battery negative to earth, drive screw fixing	1	551492	Alternatives
	Cable, battery negative to earth, clamp type connector	1	573203*	Alternatives

*Asterisk indicates a new part which has not been used on any previous model.

Plate Ref.	1 2 3 4 Description	Qty.	Part No.	Remarks
	Bolt (5/16" UNC x ⅜" long) } Fixing earth cable eyelet to boot floor tag	1	253225	
	Fan disc washer } Fixing earth cable eyelet to boot floor tag	2	510912	
	Nut (5/16" UNF) } Fixing earth cable eyelet to boot floor tag	1	254831	
26	Cable, battery positive to solenoid, drive screw fixing	1	559409	Alternative except Canada
	Cable, battery positive to solenoid, clamp type connector	1	573202*	Alternative except Canada
	Cable, battery positive to solenoid, drive screw fixing	1	568245*	Alternatives Canada
	Cable, battery positive to solenoid, clamp type type connector	1	586004*	Alternatives Canada
27	Grommet in wing valance for lamp harness	4	510139	
28	Grommet, split, for battery box	1	507134	
29	Grommet for wiring in heater tray	1	269257	
30	Grommet for cant rail wiring	2	70742	
31	Grommet for stop/tail flasher leads	2	510139	
32	Grommet for repeater flasher lamp leads	4	233243	
33	Grommet for LH parcel shelf	1	276054	
34	Special grommet in heater tray	1	507131	
35	Special grommet in boot lid for number plate lamp wiring	1	507132	
36	Grommet for dash harness	1	69052	
37	Grommet in heelboard for rear heater harness	1	22066	
38	Cable clip } On boot lid hinge for rear lamp harness	2	8885	
	Drive screw } On boot lid hinge for rear lamp harness	2	78006	
	Clip for horn leads at mounting bracket	1	523203	
	Bolt (¼" UNF x ½" long) } Fixing clip to bracket	1	255204	
	Spring washer } Fixing clip to bracket	1	3074	
	Nut (¼" UNF) } Fixing clip to bracket	1	254810	
39	Cable clip at control box fixing	1	513220	
40	Cable clip for starter cable at cold air duct	1	50640	
	Cable clip for starter lead and harness at RH wing valance	1	563289	
	Screw (6 UNC x ⅜" long) } Fixing cable clip to valance	2	78620	
	Shakeproof washer } Fixing cable clip to valance	2	78660	
	Nut (6 UNC) } Fixing cable clip to valance	2	257191	
41	Cable clip under rear seat for heater harness	2	237749	
	Drive screw } Earth connection for rear heater harness	1	77704	
	Shakeproof washer } Earth connection for rear heater harness	1	78249	
	Cable cleat for front lamp leads	2	240430	
42	Earthing clip on front and rear wing valance	7	236366	
43	Special earthing blade on wing valance and dash	2	505204	
44	Special earthing blade at flasher unit and horn fixing	3	505205	
45	Special earthing blade at fuel gauge unit and ignition coil	2	505203	
46	Special earthing blade at instrument unit	3	505195	

J389

*Asterisk indicates a new part which has not been used on any previous model.

H692

Plate Ref.	1 2 3 4 Description		Qty.	Part No.	Remarks
1	Facia frame		1	356778%	
	Facia frame		1	357518%%	
	Bracket, RH	For facia frame	1	352271	
2	Bracket, LH	For facia frame	1	352272	
	Bracket, RH	Facia frame to dash top	1	356770@@	
	Bracket, LH	Facia frame to dash top	1	356771@@	
	Bolt(1/4" UNF x 7/8" long)	Fixing brackets to facia frame	4	255206	
	Plain washer	Fixing brackets to facia frame	8	4071	
	Packing washer	Fixing brackets to facia frame	2	3840	
	Spring washer	Fixing brackets to facia frame	4	3074	
	Nut(1/4" UNF)	Fixing brackets to facia frame	4	254810	
	Drive screw fixing brackets to dash		6	78167	
	Bracket, facia frame to dash front		4	356772@@	
	Bolt(10 UNF x 3/8" long)	Fixing brackets and facia frame	16	257019@@	
	Plain washer	Fixing brackets and facia frame	16	3931@@	
	Spring washer	Fixing brackets and facia frame	16	3074@@	
3	FACIA BOARD COMPLETE		1	352339@	RH Stg
	FACIA BOARD COMPLETE		1	352547@	LH Stg
	FACIA BOARD COMPLETE		1	356706@@	RH Stg
	FACIA BOARD COMPLETE		1	356707@@	LH Stg
4	Beading for top edge, RH		1	352236@	
5	Beading for top edge, LH		1	352235@	
6	End finisher for LH side		1	352252@	RH Stg
	End finisher for RH side		1	352251@	LH Stg
	Side moulding for RH side		1	356716@@	RH Stg
	Side moulding for LH side		1	356717@@	LH Stg
	Woodscrew fixing side moulding		2	227818@@	
	Woodscrew fixing finisher and beading		7	65557	
7	Moulding for bottom of facia board		1	352255@	
	Moulding for bottom of facia board, RH		1	356718@@	
	Moulding for bottom of facia board, LH		1	356719@@	
	Woodscrew fixing moulding		5	78285@	
	Woodscrew fixing bottom moulding		4	227818@@	
8	Fixing bracket for facia board		2	352984	
	Woodscrew fixing bracket to board		4	78285	
	Set bolt(10 UNF x 1/2" long)	Fixing facia board to frame at bottom	2	257017	
	Spring washer	Fixing facia board to frame at bottom	2	3073	
	Plain washer	Fixing facia board to frame at bottom	2	3816	
	Nut(10 UNF)	Fixing facia board to frame at bottom	2	257023	
	Drive screw fixing facia board to frame at top		2	77923	
9	Main mounting bracket for instrument unit		1	352404	RH Stg
	Main mounting bracket for instrument unit		1	352918	LH Stg

@ Cars numbered up to 625001352, 626000431, 628000175, 630000748, 631000192, 633000443

@@ Cars numbered from 625001353, 626000432, 628000176, 630000749, 631000193, 633000444 onwards

% Cars numbered up to 625101045, 626100262, 628100109, 630100840, 631100353, 633100078

%% Cars numbered from 625101046, 626100263, 628100110, 630100841, 631100354, 633100079 and Mk IA models

*Asterisk indicates a new part which has not been used on any previous model.

FACIA FRAME AND GLOVE BOXES, MK I AND MK IA MODELS

1664

H692

FACIA FRAME AND GLOVE BOXES, MK I AND Mk IA 3 LITRE MODELS

Plate Ref.	1 2 3 4 Description	Qty.	Part No.	Remarks
10	Bracket for instrument unit mounting bracket	2	385501	
	Insulator for instrument unit mounting bracket	2	512372	
	Set bolt (¼" UNF x ⅝" long) } Fixing brackets to main bracket	4	255206	
	Plain washer } Fixing brackets to main bracket	4	3946	
	Spring washer } Fixing brackets to main bracket	4	3074	
	Bolt(10 UNF x ⅜"long) } Fixing mounting bracket and facia frame	10	275019	
	Plain washer, small } Fixing mounting bracket and facia frame	12	3851	
	Plain washer, large } Fixing mounting bracket and facia frame	2	3852	
	Spring washer } Fixing mounting bracket and facia frame	10	3073	
	Nut (10 UNF) } Fixing mounting bracket and facia frame	4	257023	
	Set bolt (¼" UNF x ⅝" long) } Fixing facia frame to dash front	4	255206	
	Plain washer } Fixing facia frame to dash front	4	3665	
	Spring washer } Fixing facia frame to dash front	4	3074	
11	Carburetter cold start control	1	502088	RH Stg
	Carburetter cold start control	1	504025	LH Stg
12	Rubber sleeve for control	1	237108	
13	Switch for cold start control	1	279625	
14	Shakeproof washer for switch	1	233017	
15	Bracket for cold start control	1	352141	
	Set bolt (10 UNF x ½" long) } Fixing bracket to parcel shelf channel	2	257017	
	Plain washer } Fixing bracket to parcel shelf channel	2	3816	
	Spring washer } Fixing bracket to parcel shelf channel	2	3073	
16	Moulding for cold start control	1	352140	
	Set screw (10 UNF x ¼" long) } Fixing moulding to bracket	2	78278	
17	Distance piece } Fixing moulding to bracket	1	352142	
	Shakeproof washer } Fixing moulding to bracket	2	74236	
18	End plate, RH, for moulding	1	352138	RH Stg
	End plate, LH, for moulding	1	352139	LH Stg
	Shakeproof washer } Fixing end plate to moulding	2	78249	
	Nut (6 UNC) } Fixing end plate to moulding	2	257203	
19	Grommet for cold start control in dash	1	233244	
20	CLOCK	1	545119	
	Bulb for clock	1	503352	
21	Retaining plate for clock	1	352394	
22	PLASTIC MOUNTING BODY COMPLETE FOR CLOCK	1	352391	
	Grub screw fixing clock and retaining plate	1	352415	
23	Distance collar for mounting body	2	352908	
24	Bracket retaining clock body	1	352393	
	Screw(10 UNF x ⅜" long) } Fixing retaining bracket to facia rail	2	78272	
	Spring washer } Fixing retaining bracket to facia rail	2	3073	
25	Grommet for clock harness	1	69052	
26	GLOVE BOX LID CENTRE ASSEMBLY	1	352335@	
	GLOVE BOD LID CENTRE ASSEMBLY	1	356699$	
	GLOVE BOX LID CENTRE ASSEMBLY	1	357303$$	

@ Cars numbered up to 625001352, 626000431, 628000175, 630000748, 631000192, 633000443

@@ Cars numbered from 625001353, 626000432, 628000176, 630000749, 631000193, 633000444 onwards

$ Cars numbered from 625001353, 630000749 up to 625101046, 630100840
626000432, 631000193 up to 626100262, 631100353
628000176, 633000444 up to 628100109, 633100078

$ Cars numbered up to 625101045, 626100262, 628100109, 630100840, 631100353, 633100078

$$ Cars numbered from 625101046, 626100263, 628100110, 630100841, 631100354, 633100079 and Mk IA models

*Asterisk indicates a new part which has not been used on any previous model.

FACIA FRAME AND GLOVE BOXES, MK I AND MK IA MODELS

1666

H692

Plate Ref.	1 2 3 4 Description	Qty.	Part No.	Remarks
	Glove box lid, centre	1	352216@	
	Glove box lid, centre	1	356700$	
	Glove box lid, centre	1	357304%%	
27	Handle for glove box	1	352234@	
	Handle for glove box	1	356730@@	
	Woodscrew fixing handle	2	22780@@	
	Hinge leaf, RH	1	352232%	
	Hinge leaf, LH	1	352233%	
	Hinge leaf, RH	1	356988%%	
	Hinge leaf, LH	1	356989%%	
	Woodscrew } Fixing hinge leaf	6	78315@	
	Woodscrew } Fixing hinge leaf	6	200160@@	
	Woodscrew } Fixing hinge leaf	2	78314	
28	Beading for lid top	2	352237@	
29	End finisher for lid, RH	1	352251@	
	End finisher for lid, LH	1	352252@	
	Side moulding for lid, RH	1	356701@@	
	Side moulding for lid, LH	1	356702@@	
	Woodscrew, short } Fixing beading, side moulding and finisher	6	65557@	
	Woodscrew, long } Fixing beading, side moulding and finisher	6	22781@	
30	Moulding for lid	1	352254@	
	Moulding for lid, bottom	1	356703@@	
	Woodscrew fixing moulding	6	78283@	
	Woodscrew fixing bottom moulding	3	22781@@	
	Striking plate for glove box lid, centre	1	352288	
	Cover for striking plate	1	352289	
	Woodscrew fixing cover to lid	2	22780	
31	GLOVE BOX LID SIDE ASSEMBLY	1	352336@	RH Stg } Less barrel lock
	GLOVE BOX LID SIDE ASSEMBLY	1	352552@	LH Stg } Less barrel lock
	GLOVE BOX LID SIDE ASSEMBLY	1	356686$	RH Stg } Less barrel lock
	GLOVE BOX LID SIDE ASSEMBLY	1	356687$	LH Stg } Less barrel lock
	GLOVE BOX LID SIDE ASSEMBLY	1	357204%%	RH Stg } Less barrel lock
	GLOVE BOX LID SIDE ASSEMBLY	1	357203%%	LH Stg } Less barrel lock
	Glove box lid, side	1	352219@	RH Stg
	Glove box lid, side	1	352549@	LH Stg
	Glove box lid, side	1	356688$	RH Stg
	Glove box lid, side	1	356689$	LH Stg
	Glove box lid, side	1	357309$$	RH Stg
	Glove box lid, side	1	357308$$	LH Stg
32	Handle for glove box	1	352234@	
	Handle for glove box	1	356730@@	
	Woodscrew fixing handle	2	22780@@	
33	Beading for lid top	2	352239@	
34	End finisher for lid, RH	1	352251@	RH Stg
	End finisher for lid, LH	1	352252@	LH Stg

@ Cars numbered up to 625001352, 626000431, 628000175, 630000748, 631000192, 633000443

@@ Cars numbered from 625001353, 626000432, 628000176, 630000749, 631000193, 633000444 onwards

% Cars numbered up to 625101045, 626100262, 628100109, 630100840, 631100353, 633100078

%% Cars numbered from 625101046, 626100263, 628100110, 630100841, 631100354, 633100079 and Mk IA models

$ Cars numbered from 625001353, 630000749, up to 625101046, 630100840
626000432, 631000193, up to 626100262, 631100353
628000176, 633000444, up to 628100109, 633100078

*Asterisk indicates a new part which has not been used on any previous model.

FACIA FRAME AND GLOVE BOXES, MK I AND MK IA MODELS

1668

H692

Plate Ref.	1 2 3 4 Description	Qty.	Part No.	Remarks
	Side moulding for lid, RH	1	35669000@@	RH Stg
	Side moulding for lid, LH	1	35669300@@	RH Stg
	Side moulding for lid, RH	1	35669200@@	LH Stg
	Side moulding for lid, LH	1	35669100@@	LH Stg
	Woodscrew, short — Fixing beading, side moulding and finisher	3	65557	
	Woodscrew, long — Fixing beading, side moulding and finisher	7	22781	
35	Moulding for lid	1	352253@	
	Moulding for lid, bottom	1	356694@@	
	Woodscrew fixing moulding	7	78283@	
	Woodscrew fixing bottom moulding	4	227810@@	
	Hinge leaf, RH	1	352232%	
36	Hinge leaf, LH	1	352233%	
	Hinge leaf, RH	1	356988%%	
	Hinge leaf, LH	1	356989%%	
	Woodscrew — Fixing hinge leaf	6	78315@	
	Woodscrew — Fixing hinge leaf	6	200160@@	
	Woodscrew — Fixing hinge leaf	2	78314	
37	Lock housing for glove box lid, side	1	359819	
	Woodscrew fixing housing	3	22808	
	Striking plate for glove box lid, side	1	352288	
	Cover for striking plate	1	352289	
	Woodscrew fixing cover to lid	2	22780	
	Barrel lock for glove box lid, 'FP' series	1	311186	Alternatives. State key number when ordering
	Barrel lock for glove box lid, 'FR' series	1	358252	Alternatives. State key number when ordering
38	Rubber buffer for glove box lid	4	310877	
39	Hinge pin for glove box lid	4	352909	
	Plain washer — Fixing hinge pin to facia frame	8	4071	
	Spring washer — Fixing hinge pin to facia frame	4	3074	
	Nut (¼" UNF) — Fixing hinge pin to facia frame	4	254810	
	Locknut (¼" UNF) — Fixing hinge pin to facia frame	4	254860	
40	Spring catch for glove boxes	2	352287	
	Bolt (8 UNC x ⅜" long) — Fixing spring catch to facia frame	4	257135	
	Plain washer — Fixing spring catch to facia frame	8	3902	
	Shakeproof washer — Fixing spring catch to facia frame	4	74236	
	Tapped plate — Fixing spring catch to facia frame	2	356236	
	Retainer plate for glove box lid lock	1	352648	
	Screw (8 UNC x ½" long) — Fixing retainer plate to facia frame	1	78408	
	Plain washer — Fixing retainer plate to facia frame	1	3902	
	Spring washer — Fixing retainer plate to facia frame	1	3073	
	Nut (8 UNC) — Fixing retainer plate to facia frame	1	257143	
41	Finisher for heater control, 'demist'	1	352294@	
42	Finisher for heater control, 'body heat'	1	352295@	
	Finisher for heater control, 'demist'	1	356728@@	
	Finisher for heater control, 'body heat'	1	356729@@	
	Spring washer — Fixing finishers to facia frame	2	3073	
	Nut (10 UNF) — Fixing finishers to facia frame	2	257023	
	Set screw (6UNF x¼" long) — Fixing finishers to facia frame	2	78284	
	Spring washer — Fixing finishers to facia frame	2	3072	

% Cars numbered up to 625101045, 626100262, 628100109, 630100840, 631100353, 633100078

%% Cars numbered from 625101046, 626100263, 628100110, 630100841, 631100354, 633100079 and Mk IA models

@ Cars numbered up to 625001352, 626000431, 628000175, 630000748, 631000192, 633000443

@@ Cars numbered from 625001353, 626000432, 628000176, 630000749, 631000193, 633000444 onwards

*Asterisk indicates a new part which has not been used on any previous model.

FACIA FRAME AND GLOVE BOXES, MK I AND MK IA MODELS

Plate Ref.	1 2 3 4 Description	Qty.	Part No.	Remarks
43	CORNER FILLET ASSEMBLY, RH, FOR FACIA	1	352337@	RH Stg
	CORNER FILLET ASSEMBLY, RH, FOR FACIA	1	352982@	LH Stg
	CORNER FILLET ASSEMBLY, LH, FOR FACIA	1	352338@	RH Stg
	CORNER FILLET ASSEMBLY, LH, FOR FACIA	1	352981@	LH Stg
	CORNER FILLET ASSEMBLY, RH, FOR FACIA	1	356708@@	RH Stg
	CORNER FILLET ASSEMBLY, LH, FOR FACIA	1	356709@@	LH Stg
	CORNER FILLET ASSEMBLY, LH, FOR FACIA	1	356711@@	RH Stg
	CORNER FILLET ASSEMBLY, RH, FOR FACIA	1	356710@@	LH Stg
	End finisher for corner fillet, LH	1	356727@@	LH Stg
	End finisher for corner fillet, RH	1	356726@@	RH Stg
44	Front moulding for RH fillet	1	352256@	
	Front moulding for LH fillet	1	352257@	
	Front moulding for RH fillet	1	356722@@	
	Front moulding for LH fillet	1	356723@@	
45	Rear moulding for RH fillet	1	352258@	
	Rear moulding for LH fillet	1	352259@	
	Rear moulding for RH fillet	1	356724@@	
	Rear moulding for LH fillet	1	356725@@	
	Woodscrew fixing mouldings	6	78285	
	End moulding for RH fillet	1	356726@@	RH Stg
	End moulding for LH fillet	1	356727@@	LH Stg
	Woodscrew fixing end moulding	1	227810@@	
46	Beading for RH fillet	1	352241@	
	Beading for LH fillet	1	352242@	
	Woodscrew fixing beading	4	65557	
47	Front fixing plate for fillet, passenger's side	1	352230	
	Woodscrew fixing plate	2	78286	
	Locating plate, front, for corner fillet, driver's side	1	356672@@	
	Stud plate for corner fillet, driver's side	1	356752@@	
	Woodscrew fixing stud and locating plates	4	200030@@	
	Fixing plate, front, for corner fillet	1	3522300@@	LH Stg
	Locating plate for corner fillet, passenger's side	1	356620@@	
	Woodscrew fixing locating plate	2	202130@@	
48	Fixing bracket for RH fillet	1	352228	
	Fixing bracket for LH fillet	1	352229	
49	Rear fixing bracket for corner fillet	1	356222	
	Woodscrew fixing brackets	4	78285	
	Plate fixing facia board to corner fillet	1	352979	
	Woodscrew fixing plate	6	20003	
	Drive screw } Fixing fillets to facia	6	63325	
	Drive screw } frame and 'A' post	2	71422	
	Plain washer }	4	3867	
	Spring washer } Fixing RH corner fillet	1	3073	
	Nut (10 UNF) } to facia frame	1	257023	
	Drive screw }	4	63325	

@ Cars numbered up to 625000863, 626000300, 628000105, 63000105, 630000402, 631000105, 633000287

@@ Cars numbered from 625000864, 626000301, 628000106, 630000403, 631000106, 633000288 onwards

% Cars numbered up to 625101045, 626100262, 628100109, 630100840, 631100353, 633100078

%% Cars numbered from 625101046, 626100263, 628100110, 630100841, 631100354, 633100079 and Mk IA models

*Asterisk indicates a new part which has not been used on any previous model.

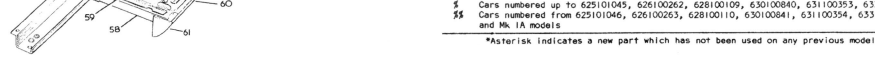

FACIA FRAME AND GLOVE BOXES, MK I AND MK IA MODELS

1672

H692

Plate Ref.	1 2 3 4 Description		Qty.	Part No.	Remarks
50	Mounting channel for parcel shelf		1	352155@	RH Stg
	Mounting channel for parcel shelf		1	356652%	RH Stg
	Mounting channel for parcel shelf		1	352643	LH Stg
	Mounting channel for parcel shelf		1	356653	LH Stg
	Mounting channel for parcel shelf		1	356962%%	
	Bolt(¼" UNF x ½" long)	Fixing channel to parcel shelf	14	255204	
	Plain washer		28	3946	
	Spring washer		14	3074	
	Nut (¼" UNF)		14	254810	
51	Frame for tool tray		1	352147@	
	Frame for tool tray		1	357506@@*	
	Bolt(¼" UNF x ⅜" long)	Fixing tool tray frame to parcel shelf	4	255206	
	Plain washer		6	3467	
	Spring washer		4	3074	
	Nut (¼" UNF)		2	254810	
52	Bracket for cigar lighter		1	352137%	
	Bracket for cigar lighter		1	356964%%	
	Set bolt(10 UNF x ½" long)	Fixing bracket to parcel shelf channel	2	257017	
	Plain washer		2	3816	
	Spring washer		2	3073	
53	Moulding for cigar lighter holder		1	352136	
	Set screw(10 UNF x ½" long)	Fixing holder to bracket	2	78276	
	Shakeproof washer		2	74236	
54	CIGAR LIGHTER ILLUMINATED-TYPE		1	565850	
	Bulb for cigar lighter		1	530054	
	Element for cigar lighter		1	600280	
	Knob for cigar lighter(black and silver)		1	600593	
	Flange for knob		1	600602	
55	End plate, RH, for moulding		1	352138%	RH Stg
	End plate, LH, for moulding		1	352139%	LH Stg
	Shakeproof washer	Fixing end plate to moulding	2	78249%	
	Nut (6 UNC)		2	257203%	
56	Ashtray body		2	359440	
57	Slide for ashtray		3	356811	
	Set screw (6 UNC x ⅜"long)	Fixing ashtray to parcel shelf channel	3	78317%	
	Drive screw		3	73644%	6 off when additional ashtray is fitted
	Set screw (6 UNF x ⅜" long)	Fixing ashtray slides to parcel shelf channel	6	78317%%	
	Drive screw		6	73644%%	
58	Tool tray		1	356649	
59	Pad for tool tray		1	356666	
60	Moulding for tool tray		1	352146@	
	Front moulding for tool tray		1	356657@@	
	Set screw (10 UNF x ½" long)	Fixing moulding to tool tray	3	78276@	
	Shakeproof washer		3	74236@	
	Clip fixing moulding to tool tray		4	311212@@	

@ Cars numbered up to 625000863, 626000300, 628000105, 630000402, 631000105, 633000287

@@ Cars numbered from 625000864, 626000301, 628000106, 630000403, 631000106, 633000288 onwards

% Cars numbered up to 625101045, 626100262, 628100109, 630100840. 631100353, 633100078

%% Cars numbered from 625101046, 626100263, 628100110, 630100841, 631100354, 633100079 onwards

*Asterisk indicates a new part which has not been used on any previous model.

FACIA FRAME AND GLOVE BOXES, MK I AND MK IA MODELS

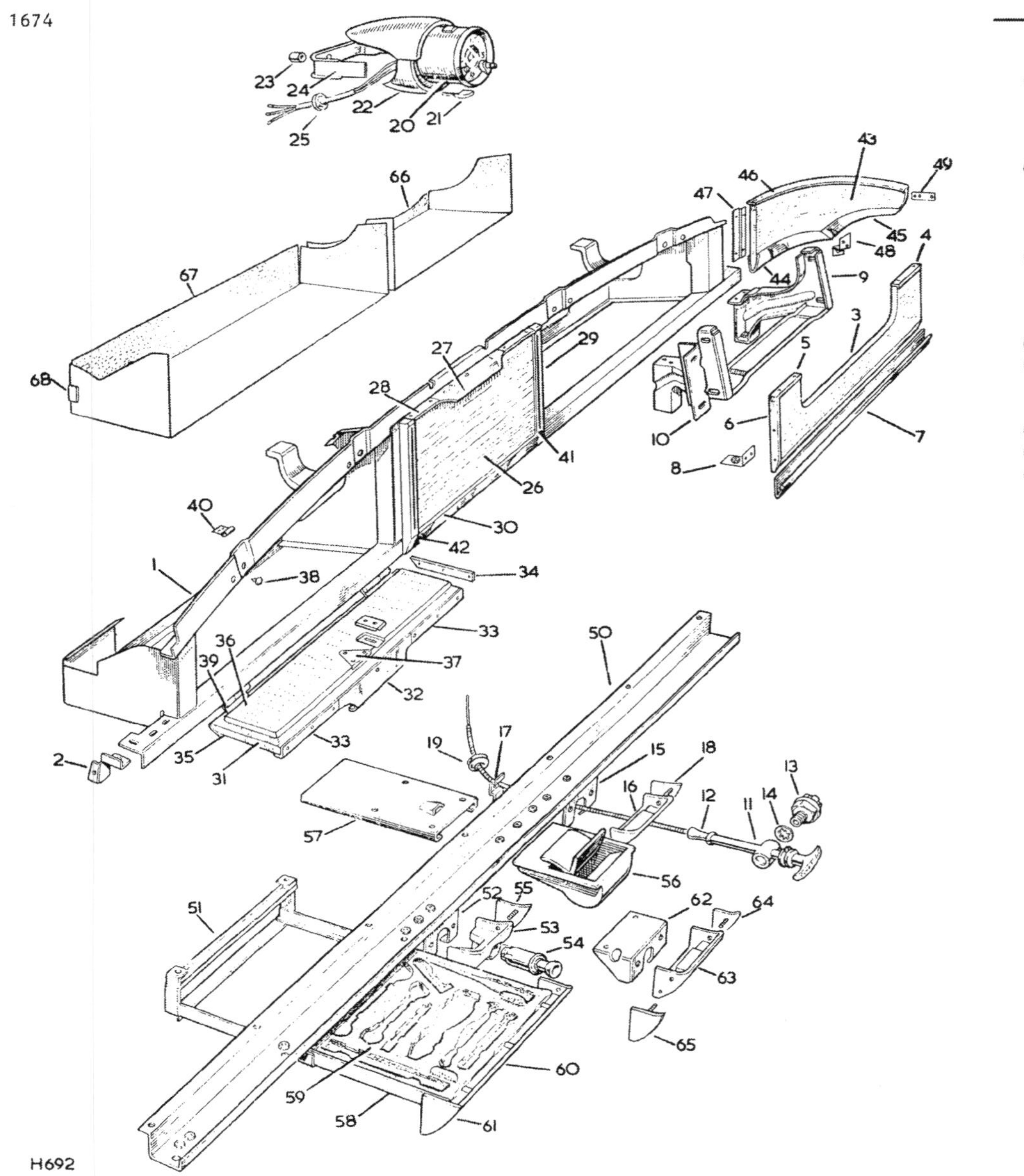

H692

Plate Ref.	1 2 3 4 Description		Qty.	Part No.	Remarks
61	End plate for moulding		1	352139@	RH Stg
	End plate for moulding		1	352138@	LH Stg
	Shakeproof washer	Fixing end plate to moulding	2	78249@	
	Nut (6 UNC)		2	257203@	
62	Bracket for intermediate gear lock		1	352141	Mk I Automatic
	Set bolt (10 UNF x ½" long)	Fixing bracket to parcel shelf channel	2	257017	Mk I Automatic
	Plain washer		2	3816	Mk I Automatic
	Spring washer		2	3073	Mk I Automatic
63	Moulding for intermediate gear lock		1	352140	Mk I Automatic
	Set screw (10 UNF x ¼" long)	Fixing moulding to bracket	2	78278	Mk I Automatic
	Distance piece		1	352142	Mk I Automatic
	Shakeproof washer		2	74236	Mk I Automatic
64	End plate, RH	For moulding	1	352138	Mk I Automatic
65	End plate, LH		1	352139	Mk I Automatic
	Shakeproof washer	Fixing end plate to moulding	4	78249	Mk I Automatic
	Nut (6 UNC)		4	257203	Mk I Automatic
66	Glove box trim complete, centre		1	354972	
67	Glove box trim complete, side		1	354969	
68	Fixing bracket for glove box trim		4	354970	
	Rivet fixing bracket		4	78118	

@ Cars numbered up to 625000863, 626000300, 628000105, 630000402, 631000105, 633000287

*Asterisk indicates a new part which has not been used on any previous model.

FACIA FRAME AND GLOVE BOXES, MK II AND MK III MODELS

1676

H693

Plate Ref.	1 2 3 4 Description		Qty.	Part No.	Remarks
1	Facia frame		1	381146	Mk II
	Facia frame		1	384073	Mk III
	Bracket, RH	Facia frame to dash top	1	356770	
2	Bracket, LH	Facia frame to dash top	1	356771	
	Bolt (¼" UNF x ⅝" long)	Fixing brackets to facia frame	4	255206	
	Plain washer	Fixing brackets to facia frame	8	4071	
	Packing washer	Fixing brackets to facia frame	2	3840	
	Spring washer	Fixing brackets to facia frame	4	3074	
	Nut (¼" UNF)	Fixing brackets to facia frame	4	254810	
	Drive screw fixing brackets to dash		6	78167	
	Bracket, facia frame to dash front		4	356772	
	Bolt(10 UNF x ⅝" long)	Fixing brackets and facia frame	16	257019	
	Plain washer	Fixing brackets and facia frame	16	3665	
	Spring washer	Fixing brackets and facia frame	16	3074	
3	Catch bracket for glove box lid lock		1	359509	
4	Shim for catch bracekt	Fixing catch bracket to facia frame	As reqd	359585	
	Bolt(10 UNF x ½" long)	Fixing catch bracket to facia frame	2	257017	
	Plain washer	Fixing catch bracket to facia frame	2	3816	
	Spring washer	Fixing catch bracket to facia frame	2	3073	
5	FACIA BOARD COMPLETE		1	385880	RH Stg
	FACIA BOARD COMPLETE		1	385881	LH Stg
6	Moulding, top inner		1	358964	
	Drive screw fixing moulding		1	69994	
	Fixing bracket facia board to facia frame		2	359570	
	Woodscrew fixing bracket to facia frame		6	20213	
	Set bolt(10 UNF x ½" long)	Fixing facia board to frame at bottom	2	257017	
	Spring washer	Fixing facia board to frame at bottom	2	3073	
	Plain washer	Fixing facia board to frame at bottom	2	3816	
	Nut (10 UNF)	Fixing facia board to frame at bottom	2	257023	
	Drive screw fixing facia board to frame at top		2	77892	
7	Main mounting bracket for instrument unit		1	359481@	RH Stg
	Main mounting bracket for instrument unit		1	385980@@	RH Stg
	Main mounting bracket for instrument unit		1	359482@	LH Stg
	Main mounting bracket for instrument unit		1	385981@@	LH Stg
8	Bracket for instrument unit mounting bracket		2	352411	Mk II
	Bracket for instrument unit mounting bracket		2	385501@	Mk III
	Adjustment bracket, RH	For main mounting bracket	1	385972@@	
	Adjustment bracket, LH	For main mounting bracket	1	385973@@	
	Insulator for instrument unit mounting bracket		2	512372	
	Set bolt (¼" UNF x ⅝" long)	Fixing brackets to main bracket	4	255206	
	Plain washer	Fixing brackets to main bracket	4	3946	
	Spring washer	Fixing brackets to main bracket	4	3074	

@ Mk II and Mk III cars numbered up to 79500576, 79600038, 79800103, 80001558, 80100103, 80300151, 80500518, 80800068, 81001067, 61100036, 81300049

@@ Mk III cars numbered from 79500577, 79600039, 79800104, 80001559, 80100104, 80300152, 80500519, 80800069, 81001068, 81100037,81300050 onwards

*Asterisk indicates a new part which has not been used on any previous model.

FACIA FRAME AND GLOVE BOXES, MK II AND MK III MODELS

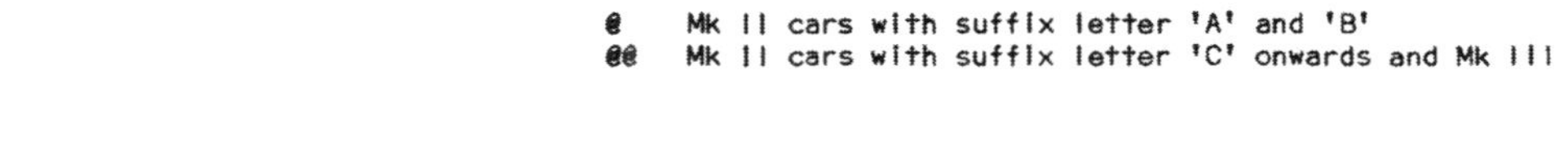

Plate Ref.	1 2 3 4 Description		Qty.	Part No.	Remarks
	Bolt(10 UNF x $\frac{3}{8}$" long)	Fixing mounting bracket and facia frame	10	257019	
	Plain washer, small	Fixing mounting bracket and facia frame	12	3851	
	Plain washer, large	Fixing mounting bracket and facia frame	2	3852	
	Spring washer	Fixing mounting bracket and facia frame	10	3073	
	Nut (10 UNF)	Fixing mounting bracket and facia frame	4	257023	
	Set bolt ($\frac{1}{4}$" UNF x $\frac{3}{8}$" long)	Fixing facia frame to dash front	4	255206	
	Plain washer	Fixing facia frame to dash front	4	3665	
	Spring washer	Fixing facia frame to dash front	4	3074	
9	CLOCK, chrome bezel		1	545119	Mk II
	Bulb for clock		1	503352	Mk II
10	Retaining plate for clock		1	352394	Mk II
11	PLASTIC MOUNTING BODY COMPLETE FOR CLOCK		1	381145	Mk II
	Grub screw fixing clock and retaining plate		1	352415	Mk II
12	Distance collar for mounting body		2	352908	Mk II
13	Bracket retaining clock body		1	352393	Mk II
	Screw(10 UNF x $\frac{3}{4}$" long)	Fixing retaining bracket to facia rail	2	78272	Mk II
	Spring washer	Fixing retaining bracket to facia rail	2	3073	Mk II
14	CLOCK, black bezel		1	551386	Mk III
	Bulb for clock		1	503352	Mk III
15	Retaining plate		1	352394	Mk III
16	Bracket for clock		1	384076	Mk III
	Woodscrew fixing bracket		2	20114	Mk III
17	Grub screw fixing clock to corner fillet		1	250015	Mk III
18	Grommet for clock harness		1	69052	
	Carburetter cold start control		1	536390@	RH Stg
	Carburetter cold start control		1	504025@	LH Stg
19	Carburetter cold start control		1	557141@@	RH Stg
	Carburetter cold start control		1	557142@@	LH Stg
	Rubber sleeve for control		1	237108	
	Switch for cold start control		1	279625@	
20	Switch for cold start control		1	563318@@	
	Shakeproof washer for switch		1	233017	
	Bracket for cold start control		1	352141@	
21	Bracket for cold start control		1	383331@@	
	Set bolt (10 UNF x $\frac{1}{2}$" long)	Fixing bracket to parcel shelf channel	2	257017	
	Plain washer	Fixing bracket to parcel shelf channel	2	3816	
	Spring washer	Fixing bracket to parcel shelf channel	2	3073	
22	Moulding for cold start control		1	359444	
	Set screw(10 UNF x $\frac{1}{4}$" long)	Fixing moulding to bracket	2	78278	
	Distance piece	Fixing moulding to bracket	1	352142	
	Shakeproof washer	Fixing moulding to bracket	2	74236	
23	End plate, RH, for moulding		1	352138	RH Stg
24	End plate, LH, for moulding		1	352139	LH Stg
	Shakeproof washer	Fixing end plate to moulding	4	78249	
	Nut(6 UNC)	Fixing end plate to moulding	4	257203	

@ Mk II cars with suffix letter 'A' and 'B'

@@ Mk II cars with suffix letter 'C' onwards and Mk III

*Asterisk indicates a new part which has not been used on any previous model.

FACIA FRAME AND GLOVE BOXES, MK II AND MK III MODELS

1680

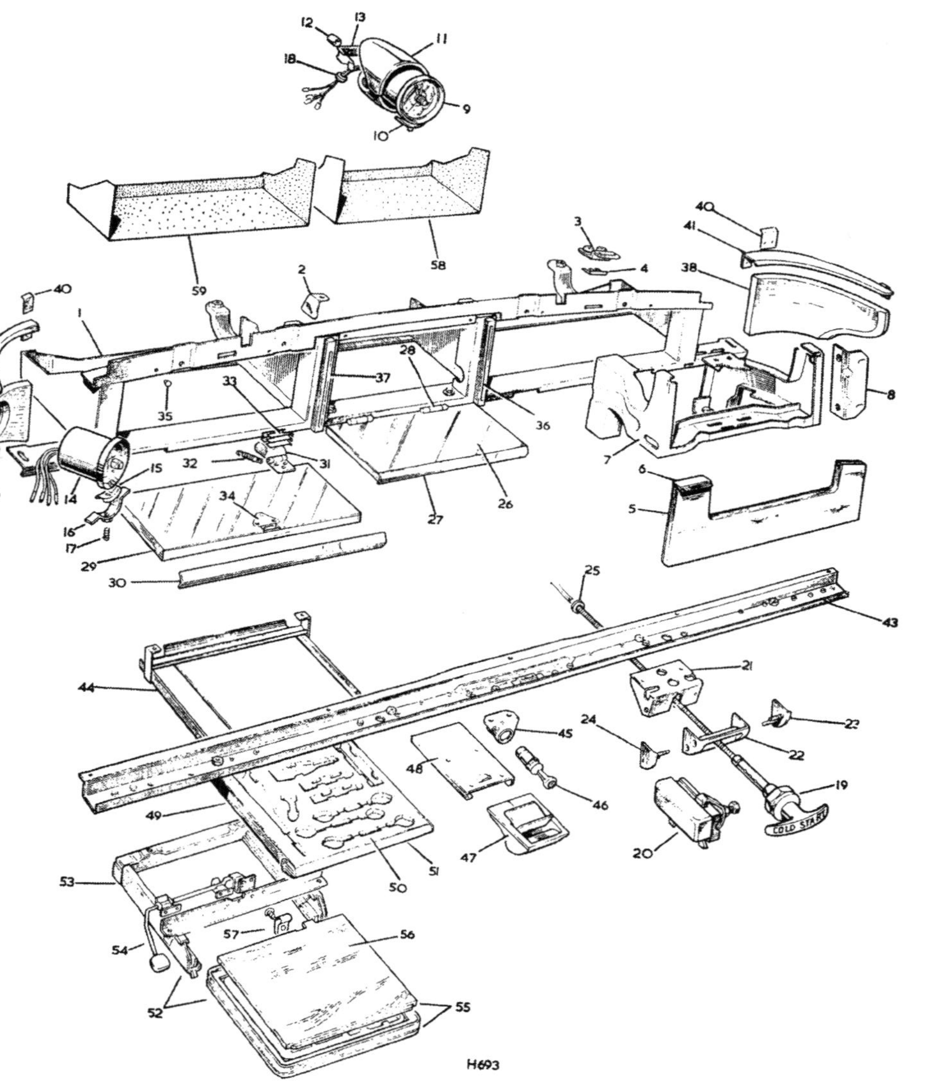

Plate Ref.	1 2 3 4 Description	Qty.	Part No.	Remarks
25	Grommet for cold start control in dash	1	233244	
26	GLOVE BOX LID ASSEMBLY, CENTRE	1	385878	
27	Moulding, top	1	357990	
28	Hinge for glove box lid	2	358969	
	Drive screw fixing hinge and moulding	10	64358	
29	GLOVE BOX LID ASSEMBLY, SIDE	1	385875	RH Stg
	GLOVE BOX LID ASSEMBLY, SIDE	1	359155	LH Stg
30	Moulding, top	1	358948	RH Stg
	Moulding, top	1	358947	LH Stg
31	Hinge for glove box lid	2	358969	
32	Spring for hinge	2	359218	
	Drive screws fixing hinge and moulding	12	64358	
33	Shim } Fixing glove box lid hinges to facia frame	8	359584	
	Bolt (10 UNF x ⅜" long) } Fixing glove box lid hinges to facia frame	4	257019	
	Spring washer } Fixing glove box lid hinges to facia frame	4	3073	
	Plain washer } Fixing glove box lid hinges to facia frame	8	3816	
	Nut (10 UNF) } Fixing glove box lid hinges to facia frame	4	257023	
34	Lock housing for glove box lid	1	359819	
	Drive screw fixing housing	3	69268	
	Barrel lock for glove box lid, 'FR' series	1	358252	Alternatives. State key number when ordering
	Barrel lock for glove box lid, 'FS' series	1	600874	Alternatives. State key number when ordering
35	Rubber buffer for glove box lid	4	310877	
36	Finisher for heater control, 'demist'	1	359151	
37	Finisher for heater control, 'body heat'	1	359152	
	Spring washer } Fixing finishers to facia frame	2	3073	
	Nut (10 UNF) } Fixing finishers to facia frame	2	257023	
	Nut (6 UNF) } Fixing finishers to facia frame	2	257191	
	Spring washer } Fixing finishers to facia frame	2	3072	
	CORNER FILLET ASSEMBLY, RH	1	359894	RH Stg
	CORNER FILLET ASSEMBLY, RH	1	359896	LH Stg Mk II
	CORNER FILLET ASSEMBLY, RH	1	385870	LH Stg Mk III
	CORNER FILLET ASSEMBLY, LH	1	359897	RH Stg Mk II
	CORNER FILLET ASSEMBLY, LH	1	385871	RH Stg Mk III
	CORNER FILLET ASSEMBLY, LH	1	359895	LH Stg
38	Corner fillet, RH	1	385869	RH Stg
	Corner fillet, RH	1	359870	LH Stg Mk II
	Corner fillet, RH	1	384078	LH Stg Mk III
	Corner fillet, LH	1	359871	Mk II
39	Corner fillet, LH	1	385873	RH Stg Mk III
	Fixing bracket lower, RH fillet	1	352228	
	Fixing bracket lower, LH fillet	1	352229	
	Woodscrew fixing lower bracket	2	20213	
40	Locating plate, top	2	356620	
	Woodscrew fixing location plate	4	20213	
	Locating plate, front	2	356672	
	Fixing plate for fillets	2	352230	
	Stud plate (10 UNF) for RH fillet	1	356752	RH Stg
	Woodscrew fixing plates	6	20003	
41	Moulding for corner fillet, top RH	1	358862	
42	Moulding for corner fillet, top LH	1	358863	
	Drive screw fixing mouldings	4	77707	

*Asterisk indicates a new part which has not been used on any previous model.

FACIA FRAME AND GLOVE BOXES, MK II AND MK III MODELS

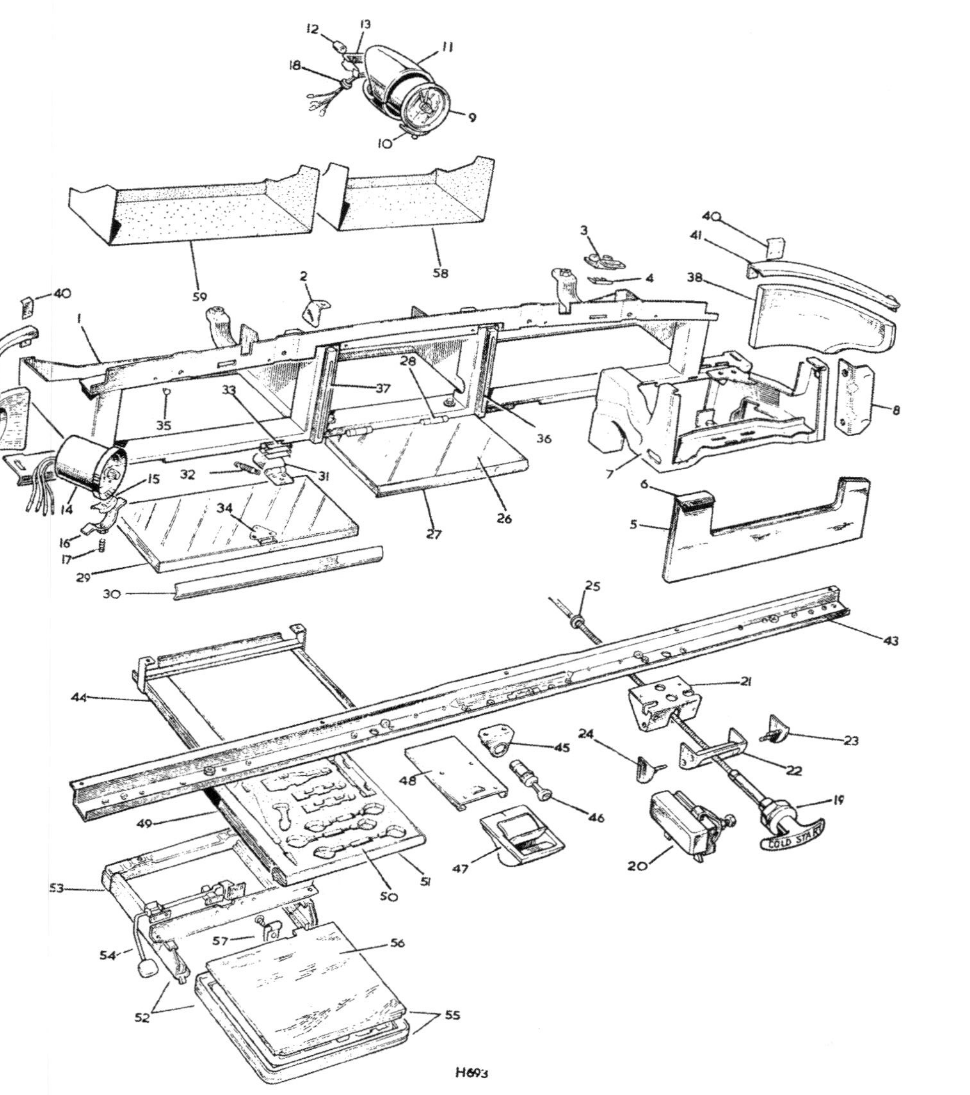

Plate Ref.	1 2 3 4 Description	Qty.	Part No.	Remarks
	Plain washer — Fixing RH corner fillet to facia frame	4	3867	
	Spring washer — Fixing RH corner fillet to facia frame	1	3073	
	Nut (10 UNF) — Fixing RH corner fillet to facia frame	1	257023	
	Drive screw — Fixing RH corner fillet to facia frame	1	63325	
	Drive screw — Fixing LH corner fillet to facia frame	3	63325	
	Woodscrew — Fixing LH corner fillet to facia frame	2	20003	
43	Mounting channel for parcel shelf	1	356962	Mk II
	Mounting channel for parcel shelf	1	383821	Mk III
	Bolt (¼"UNF x ½"long) — Fixing channel to parcel shelf	10	255204	
	Plain washer — Fixing channel to parcel shelf	20	4071	
	Spring washer — Fixing channel to parcel shelf	10	3074	
	Nut (¼"UNF) — Fixing channel to parcel shelf	10	254810	
44	Frame for tool tray	1	357506	Mk II
	Bolt(¼"UNF x ⅜"long) — Fixing tool tray frame to parcel shelf	4	255206	Mk II
	Plain washer — Fixing tool tray frame to parcel shelf	6	3467	Mk II
	Spring washer — Fixing tool tray frame to parcel shelf	4	3074	Mk II
	Nut (¼"UNF) — Fixing tool tray frame to parcel shelf	2	254810	Mk II
45	Bracket for cigar lighter	1	356964	
	Set bolt(10 UNF x ½"long) — Fixing bracket to parcel shelf channel	2	257017	
	Plain washer — Fixing bracket to parcel shelf channel	2	3816	
	Spring washer — Fixing bracket to parcel shelf channel	2	3073	
46	CIGAR LIGHTER	1	565850	
	Bulb for cigar lighter	1	530054	
	Element for cigar lighter	1	600280	
	Knob for cigar lighter(black and silver)	1	600593	
	Flange for knob	1	600602	
47	Ashtray body	2	359440	
48	Slide for ashtray	3	356811	2 off on Mk III
	Set screw(6UNF x ⅜"long) — Fixing ashtray slides to parcel shelf channel	6	78317	
	Drive screw — Fixing ashtray slides to parcel shelf channel	6	73644	
49	Tool tray	1	356649	Mk II
50	Pad for tool tray	1	356666	Mk II
51	Moulding for tool tray	1	352146	Mk II
	Drive screw fixing moulding	4	78174	Mk II
52	PICNIC AND TOOL TRAY COMPLETE ASSEMBLY	1	385926	Mk III
53	SUPPORT FRAME FOR PICNIC AND TOOL TRAY	1	385761	Mk III
54	Release lever for picnic and tool tray	1	384394	Mk III
55	PICNIC AND TOOL TRAY ASSEMBLY	1	385925	Mk III
56	Top only for picnic tray	1	385759	Mk III
57	Catch for picnic and tool tray	1	385746	Mk III
	Shim, support frame to picnic tray	4	312908	Mk III
58	Glove box trim complete, centre	1	359948	
59	Glove box trim complete, side	1	359951	
	Fixing bracket for glove box trim	4	354970	
	Rivet fixing bracket	4	78118	

*Asterisk indicates a new part which has not been used on any previous model.

1705

FACIA FRAME ASSEMBLY

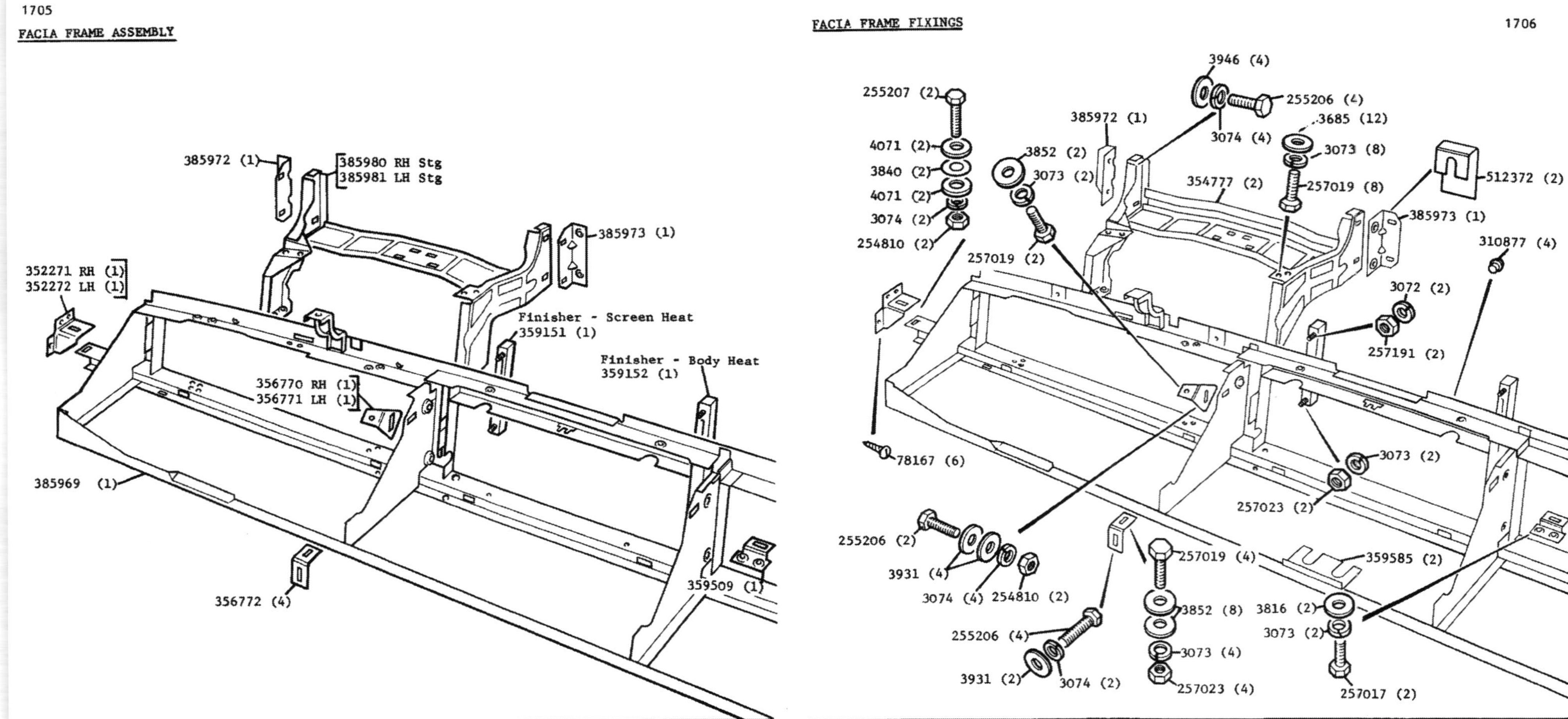

REMARKS
Applicable to 3½ Litre Saloon and Coupe models

FACIA FRAME FIXINGS

1706

REMARKS
Applicable to 3½ Litre Saloon and Coupe models

1707

FACIA BOARD AND CORNER FILLETS

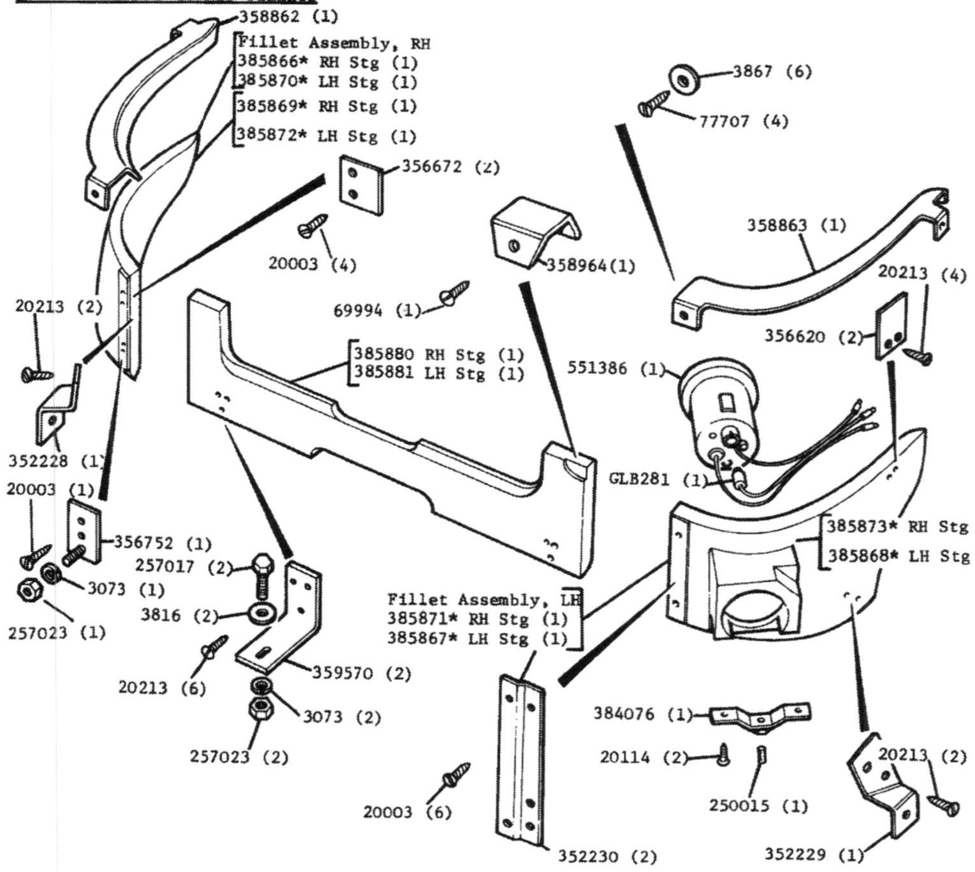

358862 (1)
Fillet Assembly, RH
385866* RH Stg (1)
385870* LH Stg (1)
385869* RH Stg (1)
385872* LH Stg (1)
356672 (2)
3867 (6)
77707 (4)
20003 (4)
358964(1)
358863 (1)
69994 (1)
20213 (4)
20213 (2)
385880 RH Stg (1)
385881 LH Stg (1)
551386 (1)
356620 (2)
GLB281 (1)
352228 (1)
20003 (1)
356752 (1)
257017 (2)
3073 (1)
3816 (2)
257023 (1)
385873* RH Stg
385868* LH Stg
Fillet Assembly, LH
385871* RH Stg (1)
385867* LH Stg (1)
20213 (6)
359570 (2)
3073 (2)
257023 (2)
384076 (1)
20114 (2)
250015 (1)
20213 (2)
20003 (6)
352230 (2)
352229 (1)

REMARKS

Applicable to 3½ Litre Saloon and Coupe models

NOTE Bracket 352228 is part of Fillet Assembly, RH 385866 and 385870

1708

GLOVE BOX LIDS AND TRIM

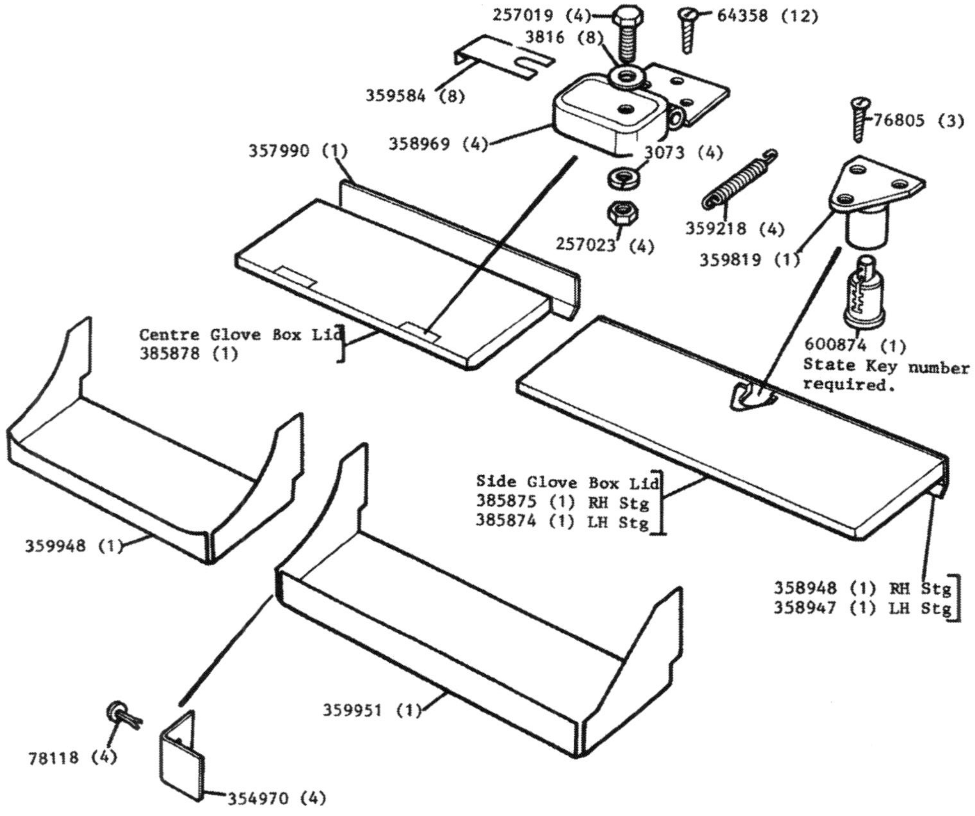

257019 (4)
64358 (12)
3816 (8)
359584 (8)
76805 (3)
358969 (4)
357990 (1)
3073 (4)
359218 (4)
257023 (4)
359819 (1)
Centre Glove Box Lid
385878 (1)
600874 (1)
State Key number required.
Side Glove Box Lid
385875 (1) RH Stg
385874 (1) LH Stg
359948 (1)
358948 (1) RH Stg
358947 (1) LH Stg
359951 (1)
78118 (4)
354970 (4)

REMARKS

Applicable to 3½ litre Saloon and Coupe models.

FUEL RESERVE AND COLD START CONTROLS, BRAKE WARNING LIGHT.

1709

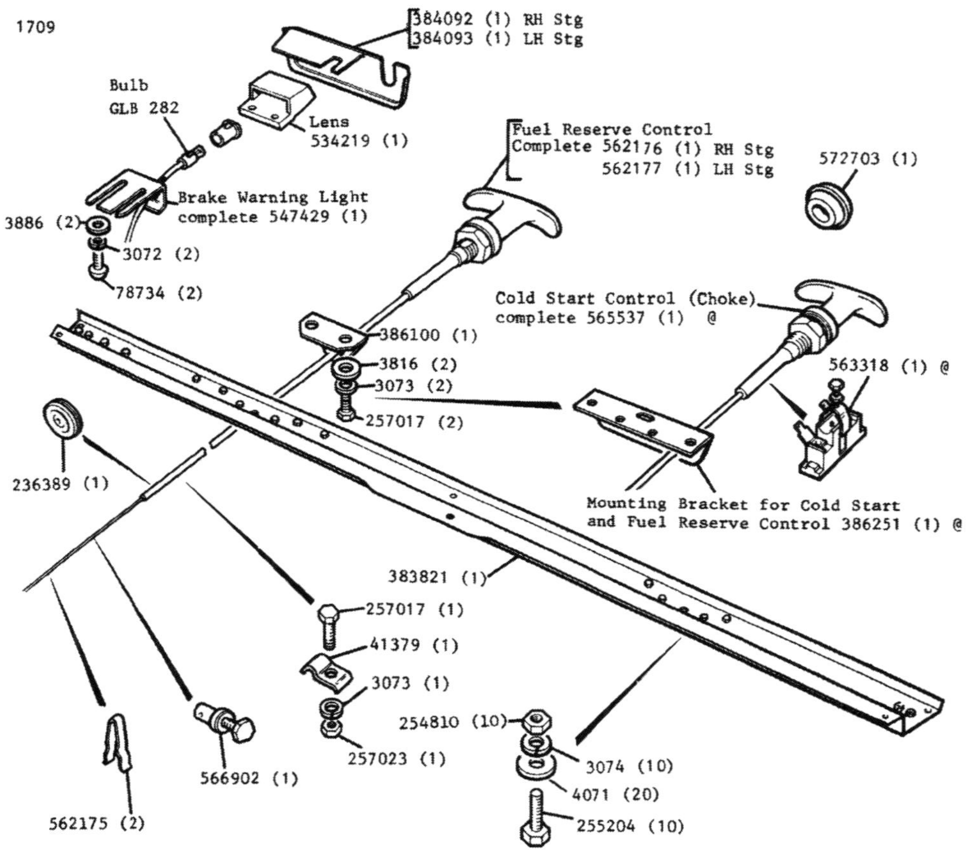

384092 (1) RH Stg
384093 (1) LH Stg
Bulb GLB 282
Lens 534219 (1)
Fuel Reserve Control Complete 562176 (1) RH Stg
562177 (1) LH Stg
572703 (1)
Brake Warning Light complete 547429 (1)
3886 (2)
3072 (2)
78734 (2)
Cold Start Control (Choke) complete 565537 (1) @
386100 (1)
3816 (2)
3073 (2)
257017 (2)
563318 (1) @
236389 (1)
Mounting Bracket for Cold Start and Fuel Reserve Control 386251 (1) @
383821 (1)
257017 (1)
41379 (1)
3073 (1)
254810 (10)
257023 (1)
3074 (10)
4071 (20)
255204 (10)
566902 (1)
562175 (2)

REMARKS
Applicable to 3½ litre Saloon and Coupe models.
@ Cars fitted with Manual Choke control only.

PICNIC AND TOOL TRAY

1710

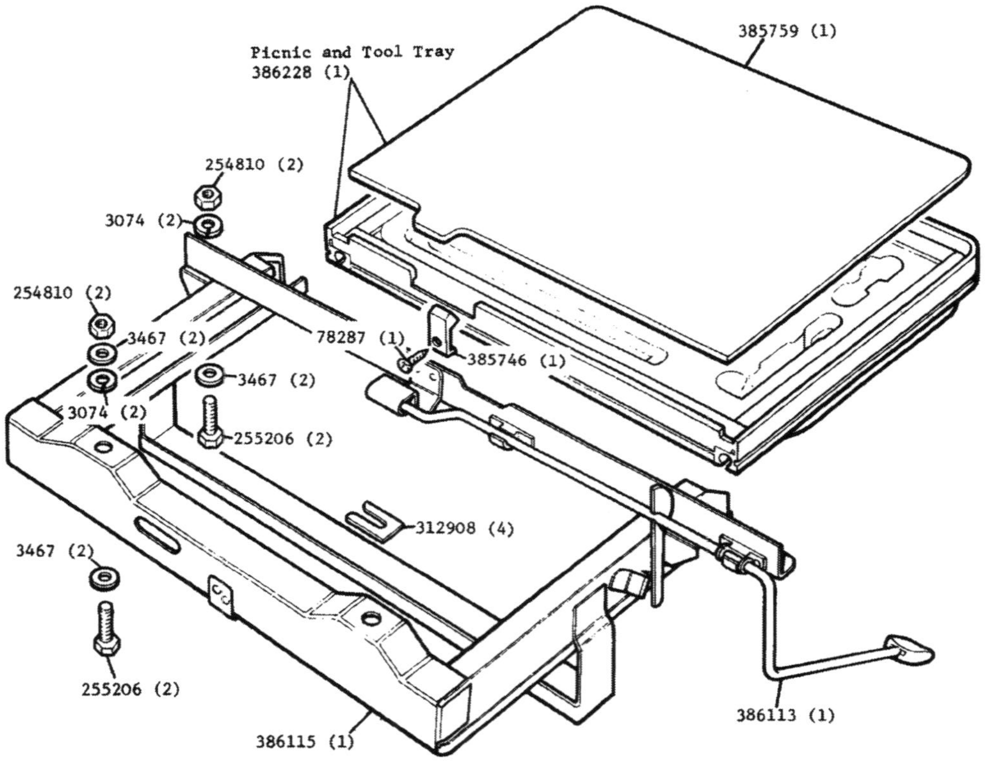

385759 (1)
Picnic and Tool Tray 386228 (1)
254810 (2)
3074 (2)
254810 (2)
3467 (2)
78287 (1)
385746 (1)
3467 (2)
3074 (2)
255206 (2)
312908 (4)
3467 (2)
255206 (2)
386113 (1)
386115 (1)

REMARKS
Applicable to 3½ Litre Saloon and Coupe models

Picnic, Tool Tray and Support Frame Complete Assembly 386116 (1)

INSTRUMENT PANEL AND FITTINGS, MK I AND MK IA MODELS

1711

MK II AND MK III

MK I AND MK IA

H702

Plate Ref.	1 2 3 4 Description	Qty.	Part No.	Remarks
1	INSTRUMENT UNIT COMPLETE	1	352371	RH Stg } 1959
	INSTRUMENT UNIT COMPLETE	1	352494	LH Stg } 1959
	INSTRUMENT UNIT COMPLETE	1	356748	RH Stg } 1960 onwards
	INSTRUMENT UNIT COMPLETE	1	356749	LH Stg } 1960 onwards
2	Face panel for instrument unit	1	380709	
	Loose facing for face panel	1	352370	
	Special set screw, face panel to body	2	78451	
	Set bolt($\frac{1}{4}$" UNF x $\frac{3}{8}$" long) } Upper fixings for instrument unit to dash	2	255206	
	Plain washer } Upper fixings for instrument unit to dash	2	4071	
	Spring washer } Upper fixings for instrument unit to dash	2	3074	
	Set bolt ($\frac{1}{4}$" UNF x $\frac{3}{8}$" long) } Lower fixings for instrument unit to dash	2	78208	
	Shakeproof washer } Lower fixings for instrument unit to dash	2	78114	
3	Bezel for speedometer and grouped instrument panel	2	277960	
	GROUPED INSTRUMENT PANEL COMPLETE	1	503192	Mk I
	GROUPED INSTRUMENT PANEL COMPLETE	1	530076	Mk IA
4	Ammeter	1	503393	Mk I
	Ammeter	1	530077	Mk IA
5	Fuel gauge	1	503341	
6	Thermometer	1	276917	
	Bulb for grouped instrument panel	1	232590	
7	Plastic panel plate, RH	1	513321	
	Plastic panel plate, LH	1	513322	
	Bulb for panel plate	4	GLB281	
8	Switch for lamps	1	503329	
9	Switch for oil level	1	503330	
10	Switch for wiper	1	515706	
11	Switch for panel light	1	502082	
12	Switch for fuel reserve	1	503329	
13	Switch for ignition and starter, less barrel lock	1	551508	
	Insulating sleeve for panel switch Lucar blades	4	519870	
14	Barrel lock for ignition and starter switch, 'FP' series	1	59895	} Alternatives. State key number when ordering
	Barrel lock for ignition and starter switch, 'FR' series	1	358176	} Alternatives. State key number when ordering
15	Bezel for panel switches	6	537136	
	Protection disc for panel switch bezel	6	512281	
16	Sealing washer for bezel, RH	1	510227	
	Sealing washer for bezel, LH	1	510226	
17	WARNING LIGHT, IGNITION	1	503331	
18	WARNING LIGHT, OIL	1	503332	
19	WARNING LIGHT, CHOKE	1	503334	
	Glass for warning light, ignition red	1	507798	
	Glass for warning light, oil, green	1	507800	
	Glass for warning light, choke, amber	1	507801	

*Asterisk indicates a new part which has not been used on any previous model.

INSTRUMENT PANEL AND FITTINGS, MK I AND MK IA MODELS

1713

MKII AND MKIII

MK I AND MK IA

H702

Plate Ref.	1 2 3 4 Description		Qty.	Part No.	Remarks
	Bezel for warning light		3	276574	
	Bulb for warning light		3	GLB987	
	Bulbholder for warning light		3	507799	
20	WARNING LIGHT, FLASHER		1	503333	
21	WARNING LIGHT, HEADLAMP		1	503335	
	Glass and rim for warning light, flasher green		1	507802	
	Glass and rim for warning light, headlamp, red		1	507803	
	Bulb for warning light		2	GLB987	
	Bulbholder for warning light		2	507799	
22	Spacing bracket for warning light, with three holes		1	502530	
23	Spacing bracket for warning light, with two holes		1	502531	
	HANDBRAKE WARNING LIGHT, RED		1	238015	1961 Automatic only
24	HANDBRAKE WARNING LITHT, RED		1	528960	Mk IA
	Bulb for warning light		1	232590	1961 Automatic and all Mk IA models with red handbrake warning light
	Bezel and glass, red		1	262919	1961 Automatic and all Mk IA models with red handbrake warning light
25	Mounting bracket for warning light		1	514748	RH Stg; 1961 Automatic and all Mk IA models with red handbrake warning light
	Mounting bracket for warning light		1	514760	LH Stg; 1961 Automatic and all Mk IA models with red handbrake warning light
	Set bolt (¼" UNF x ⅞" long)	Fixing mounting bracket and warning light bracket to dash rail	1	255207	1961 Automatic and all Mk IA models with red handbrake warning light
	Set bolt (¼" UNF x ⅜" long)	Fixing mounting bracket and warning light bracket to dash rail	1	255208	1961 Automatic and all Mk IA models with red handbrake warning light
	Spring washer	Fixing mounting bracket and warning light bracket to dash rail	1	3074	1961 Automatic and all Mk IA models with red handbrake warning light
	Nut (¼" UNF)	Fixing mounting bracket and warning light bracket to dash rail	1	254810	1961 Automatic and all Mk IA models with red handbrake warning light
	HANDBRAKE AND BRAKE FLUID LEVEL WARNING LIGHT		1	526393	Mk IA models with amber handbrake warning light
	Bulb for warning light		1	503352	Mk IA models with amber handbrake warning light
	Bezel and glass assembly, amber		1	534219	Mk IA models with amber handbrake warning light
	Screw (4 BA x ⅜" long)	Fixing warning light to handbrake mounting bracket	2	77701	Mk IA models with amber handbrake warning light
	Plain washer	Fixing warning light to handbrake mounting bracket	2	3886	Mk IA models with amber handbrake warning light
	Spring washer	Fixing warning light to handbrake mounting bracket	2	3072	Mk IA models with amber handbrake warning light
26	Bracket for headlamp switch		1	502316	RH Stg
	Bracket for headlamp switch		1	502315	LH Stg
	Spring washer	Fixing bracket	1	3073	
	Set screw (10 UNFx5/16" long)	Fixing bracket	1	78307	
27	Switch for headlamp		1	502086	RH Stg
	Switch for headlamp		1	511311	LH Stg
28	SWITCH FOR FLASHING INDICATORS		1	512263	
	Contact set for indicator switch		1	507864	
	Contact, slip ring for switch		1	518961	
	Set screw (2 BA x ½" long) fixing switch		2	78318	
	Switch for overdrive		1	511311	With overdrive
	Bracket for overdrive switch		1	538297	RH Stg; With overdrive
	Bracket for overdrive switch		1	538296	LH Stg; With overdrive
	Special set screw	Fixing bracket to shroud	1	78307	With overdrive
	Spring washer	Fixing bracket to shroud	1	3073	With overdrive
	Switch for intermediate gear lock		1	514834	RH Stg; Mk IA Automatic
	Switch for intermediate gear lock		1	514835	LH Stg; Mk IA Automatic
	Connector harness for instrument panel		1	530050	Mk I
	Connector harness for instrument panel		1	519772	Mk IA
	SPEEDOMETER COMPLETE, MPH 1344 rpm	For use with 3.9ratio axle	1	504794	Except Automatic and overdrive models; 1959
	SPEEDOMETER COMPLETE, KPH 820 rpk	For use with 3.9ratio axle	1	504795	Except Automatic and overdrive models; 1959
	SPEEDOMETER COMPLETE, MPH 1152 rpm	For use with 3.9ratio axle	1	500757	Automatic; 1959
	SPEEDOMETER COMPLETE, KPH 700 rpk	For use with 3.9ratio axle	1	500758	Automatic; 1959

*Asterisk indicates a new part which has not been used on any previous model.

1715

INSTRUMENT PANEL AND FITTINGS, MK I AND MK IA MODELS

MK II AND MK III

MK I AND MK IA

Plate Ref.	1 2 3 4 Description	Qty.	Part No.	Remarks
	SPEEDOMETER COMPLETE, MPH 1248 rpm For use with 4.3 ratio axle	1	279379	Models with overdrive 1959
	SPEEDOMETER COMPLETE, KPH 780 rpk For use with 4.3 ratio axle	1	279380	Models with overdrive 1959
	Flexible trip control for speedometer	1	512865	1959
	SPEEDOMETER COMPLETE, MPH 1376 rpm For use with 3.9ratio axle	1	512692	Except Automatic and overdrive models 1960 onwards
	SPEEDOMETER COMPLETE, KPH 860 rpk For use with 3.9ratio axle	1	512693	Except Automatic and overdrive models 1960 onwards
29	SPEEDOMETER COMPLETE, MPH 1152 rpm For use with 3.9ratio axle	1	512696	Automatic 1960 onwards
	SPEEDOMETER COMPLETE, KPH 697 rpk For use with 3.9ratio axle	1	512697	Automatic 1960 onwards
	SPEEDOMETER COMPLETE, MPH 1280 rpm For use with 4.3ratio axle	1	512694	Models with overdrive 1960 onwards
	SPEEDOMETER COMPLETE, KPH 775 rpk For use with 4.3ratio axle	1	512695	Models with overdrive 1960 onwards
	Flexible trip control for speedometer	1	512698	
	Bulb for speedometer	1	232590	
	Felt washer for speedometer	1	241387	
	SPEEDOMETER CABLE	1	270716	Except Automatic and over-drive
	Cable, inner	1	270718	Except Automatic and over-drive
	Cable, outer	1	270717	Except Automatic and over-drive
	SPEEDOMETER CABLE	1	279268	Automatic
30	Cable, inner	1	501352	Automatic
31	Cable, outer	1	501353	Automatic
	SPEEDOMETER CABLE	1	278983	Models with overdrive
	Cable, inner	1	515665	Models with overdrive
	Cable, outer	1	505549	Models with overdrive
	Right-angled drive for speedometer at gearbox	1	277741	Models with overdrive
32	Retaining plate for speedometer cable	1	232566	Except Automatic and over-drive
	Spring washer — Fixing cable to gearbox	3	3073	Except Automatic and over-drive
	Set bolt (2 BA x ¼"long) — Fixing cable to gearbox	3	250957	Except Automatic and over-drive
33	Grommet for speedometer cable	1	69052	
34	Clip for speedometer cable at bell housing	1	512419	LH Stg Automatic
35	Rubber grommet for clip	1	06860	LH Stg Automatic
	Cable cleat fixing speedometer cable to wiper leads at dash	1	240431	LH Stg Automatic
	Clip for speedometer cable at gearbox	1	50637	Except Automatic
	Bolt (¼" UNF x ⅜" long) — Fixing clip to gearbox	1	255206	Except Automatic
	Spring washer — Fixing clip to gearbox	1	3074	Except Automatic
	Nut (¼" UNF) — Fixing clip to gearbox	1	254810	Except Automatic
36	Grommet for water temperature pipe	1	510779	
	Clip — For water temperature pipe at accelerator countershaft bracket	1	214228	
	Grommet — For water temperature pipe at accelerator countershaft bracket	1	214229	
37	Plastic cover for steering column, RH	1	384583	Except Automatic LH Stg
	Plastic cover for steering column, RH	1	384584	Automatic LH Stg
	Plastic cover for steering column, LH	1	384585	4-speed RH Stg and LH Stg
38	Plastic cover for steering column, LH	1	384511	Automatic RH Stg
	Plastic cover for steering column, LH	1	384510	Automatic LH Stg
	Spring washer — Fixing plastic covers together	4	3073	
	Set screw(10 UNF x 1⅞" long) — Fixing plastic covers together	4	78312	
39	Retaining plate for plastic covers	1	502326	Mk I
	Retaining plate and switch mounting bracket	1	516382	RH Stg Mk IA
	Retaining plate and switch mounting bracket	1	516383	LH Stg Mk IA
	Spring washer — Fixing covers to retainer	4	3073	
	Set screw(10 UNF x 5/16"long) — Fixing covers to retainer	4	78307	
40	Finishing bead for plastic covers	1	352503	

*Asterisk indicates a new part which has not been used on any previous model.

INSTRUMENT PANEL AND FITTINGS, MK I AND MK IA MODELS

Plate Ref.	1 2 3 4 Description		Qty.	Part No.	Remarks	
41	Cover plate with hole and slot		1	352386	Alternative Check before ordering	
	Cover plate, plain		1	352387		
	Cover plate with one hole		1	352496		
	Cover plate with two holes		1	352497		
	Spring washer	Fixing cover plate	4	3073	2 off on Automatic	
	Nut (10 UNF)		4	257023		
42	Gear change indicator plate		1	545258	RH Stg	Automatic
	Gear change indicator plate		1	545259	LH Stg	
43	Bracket for indicator plate		1	505192	RH Stg	
	Bracket for indicator plate		1	505193	LH Stg	
	Screw (4 BA x ⅜" long)	Fixing bracket to indicator plate	2	77633		
	Spring washer		2	3072		
	Nut (4 BA)		2	4023		
44	Bulbholder for indicator plate		1	505182	RHD	
	Bulbholder for indicator plate		1	505189	LHD	
	Bulb for gear change indicator		1	503358		

*Asterisk indicates a new part which has not been used on any previous model.

INSTRUMENT PANEL AND FITTINGS, MK II AND MK III MODELS

1719

MKII AND MKIII

MKI AND MKIA

H702

Plate Ref.	1 2 3 4 Description	Qty.	Part No.	Remarks
	Instrument unit complete	1	359185	RH Stg saloon
	Instrument unit complete	1	359395	LH Stg saloon
45	Instrument unit complete	1	359888	RH Stg coupe
	Instrument unit complete	1	359396	LH Stg coupe
	Face panel for instrument unit	1	380739	
	Face panel for instrument unit	1	387089*@@	
46	Face panel for instrument unit	1	387090*	Mk III
	Special set screw fixing face panel to body	2	78451	
	Set bolt (1/4" UNF x 3/8" long) } Upper fixings for instrument unit to dash	2	255206	
	Plain washer } Upper fixings for instrument unit to dash	2	4071	
	Spring washer } Upper fixings for instrument unit to dash	2	3074	
	Set bolt (1/4" UNF x 3/8" long) } Lower fixings for instrument unit to dash	2	78208	
	Shakeproof washer } Lower fixings for instrument unit to dash	2	78114	
	GROUPED INSTRUMENT PANEL COMPLETE	1	545283	Mk II
	GROUPED INSTRUMENT PANEL COMPLETE	1	551624	Mk III
	Ammeter	1	536034	Saloon
	Fuel gauge	1	536033@	Saloon
	Fuel gauge	1	545284@@	Saloon
	Fuel gauge	1	536033	Mk III Saloon
	Thermometer gauge	1	536035	Saloon
	Bulb for grouped instrument panel	1	GLB987	Saloon
47	Instrument nacelle	2	353395	Coupe
48	Ammeter	1	531607	Coupe
49	Oil pressure gauge	1	537150	Coupe
50	Thermometer gauge	1	531610	Coupe
51	Fuel gauge	1	545285	Mk II Coupe
	Fuel gauge	1	555701	Mk III Coupe
	Screw (1/4" UNF x 3" long) } Fixing nacelles to instrument panel	2	78532	Coupe
	Shakeproof washer } Fixing nacelles to instrument panel	2	78114	Coupe
	Grub screw fixing instruments to nacelle	4	353396	Coupe
	Plastic panel plate, RH	1	536227@	
	Plastic panel plate, LH	1	536053@	
	Plastic panel plate, RH	1	545054@@	
	Plastic panel plate, LH	1	545055@@	
	Bulb for panel plate	4	GLB281	
	Switch for oil level	1	536072@	
	Switch, fuel reserve and oil level	1	545050@@	
52	Switch for fuel reserve	1	536070@	also Mk III
53	Switch for side and park light	1	532861	
	Switch for side light	1	536070	NADA
	Switch for windscreen wiper	1	536069@	
54	Switch, on/off, for windscreen wiper	1	545052@@	
55	Switch, slow/fast, for windscreen wiper	1	545051@@	

@Mk II cars with suffix letters'A' and 'B'   @@ Mk II cars with suffix letter'C'onwards
NADA indicates parts peculiar to cars exported to the North American dollar area

*Asterisk indicates a new part which has not been used on any previous model.

INSTRUMENT PANEL AND FITTINGS, MK II AND MK III MODELS

1721

MK II AND MK III

MK I AND MK IA

H702

Plate Ref.	1 2 3 4 Description	Qty.	Part No.	Remarks
56	Switch for panel light, rheostat	1	536215	Coupe
	Switch for panel light, rheostat	1	536071	Saloon
57	Switch for heater	1	536083	
	Switch for ignition and starter, less barrel lock	1	547107	Mk II
58	Switch for ignition and starter, less barrel lock	1	551508	Mk III
	Insulating sleeve for panel switch Lucar blades	4	519870	
59	Barrel lock for ignition and starter switch	1	606024	
60	Bezel for panel switches	6	537136	
	Protection disc for panel switch bezel	6	512281	
	Sealing washer for bezel, RH	1	510227	
	Sealing washer for bezel, LH	1	510226	
61	WARNING LIGHT, IGNITION	1	503331	
62	WARNING LIGHT, OIL	1	503332	
63	WARNING LIGHT, CHOKE	1	503334	
	Glass for warning light, ignition red	1	507798	
	Glass for warning light, oil, green	1	507800	
	Glass for warning light, choke, amber	1	507801	
	Bezel for warning light	3	276574	
	Bulb for warning light	3	GLB987	
	Bulbholder for warning light	3	507799	
	Bulb for flasher and main beam warning light in speedometer	2	GLB987	
	Bulbholder	2	507799	
	Spacing bracket for warning light (three holes)	1	536172	
	HANDBRAKE AND BRAKE FLUID LEVEL WARNING LIGHT	1	526393	Mk II
64	HANDBRAKE AND BRAKE FLUID LEVEL WARNING LIGHT	1	547429	Mk III
	Bulb for warning light	1	GLB281	
65	Lens for warning light	1	534219	
66	Trim rail for warning lamp	1	384092	
67	Screw (6 UNC x 5/16" long) } Fixing warning light to trim panel	2	78734	Mk III
68	Spring washer } Fixing warning light to trim panel	2	3072	
69	Plain washer } Fixing warning light to trim panel	2	3886	
	Bracket for headlamp switch	1	502316	RH Stg
	Bracket for headlamp switch	1	502315	LH Stg
	Spring washer } Fixing bracket	1	3073	
	Set screw (10 UNF x 5/16" long) } Fixing bracket	1	78307	
	Switch for headlamp	1	511311@	
	Switch for headlamp	1	547093@@	RH Stg
	Switch for headlamp	1	547094@@	LH Stg
	Switch for headlamp	1	547096@@	NADA
	Switch for headlamp	1	555579	Mk III NADA
	SWITCH FOR FLASHING INDICATOR AND HEADLAMP FLASHER	1	545056	

@ Mk II cars with suffix letters 'A' and 'B' @@ Mk II cars with suffix letter 'C' onwards
NADA indicates parts peculiar to cars exported to the North American dollar area

*Asterisk indicates a new part which has not been used on any previous model.

INSTRUMENT PANEL AND FITTINGS, MK II AND MK III MODELS

1723

MK II AND MK III

MK I AND MK IA

H702

Plate Ref.	1 2 3 4 Description	Qty.	Part No.	Remarks
	Contact set for indicator switch	1	600443	
70	SWITCH FOR FLASHING INDICATOR AND HEADLAMP FLASHER	1	551503	Late Mk III
	Switch for overdrive	1	536084@	RH Stg
	Switch for overdrive	1	536085@	LH Stg
	Switch for overdrive	1	547091@@	RH Stg Mk II
	Switch for overdrive	1	547092@@	LH Stg Mk II
	Switch for overdrive	1	551628	RH Stg Mk III 4-speed
	Switch for overdrive	1	551629	LH Stg Mk III 4-speed
	Bracket for overdrive switch	1	548118	RH Stg
	Bracket for overdrive switch	1	548117	LH Stg
	Special set screw (Fixing bracket to shroud)	1	78307	
	Spring washer (Fixing bracket to shroud)	1	3073	
	Switch for intermediate gear lock	1	514834@	RH Stg Mk II Automatic
	Switch for intermediate gear lock	1	514835@	LH Stg Mk II Automatic
	Switch for intermediate gear lock	1	547089@@	RH Stg Mk II Automatic
	Switch for intermediate gear lock	1	547090@@	LH Stg Mk II Automatic
	Connector harness for instrument panel	1	532846@	Saloon
	Connector harness for instrument panel	1	536163@	Coupe
	Connector harness for instrument panel	1	541699@@	Saloon
	Connector harness for instrument panel	1	541700@@	Coupe
71	Connector harness for instrument panel	1	541700	Mk III saloon and coupe
	SPEEDOMETER COMPLETE, MPH	1	545236	Saloon 4-speed Mk II
	SPEEDOMETER COMPLETE, KPH	1	545237	Saloon 4-speed Mk II
72	SPEEDOMETER COMPLETE, MPH	1	533512@	Saloon and coupe Automatic Mk II
	SPEEDOMETER COMPLETE, KPH	1	533513@	Saloon and coupe Automatic Mk II
	SPEEDOMETER COMPLETE, MPH	1	545232@@	Saloon and coupe Automatic Mk II
	SPEEDOMETER COMPLETE, KPH	1	545233@@	Saloon and coupe Automatic Mk II
	SPEEDOMETER COMPLETE, MPH	1	533371@	Coupe 4-speed Mk II
	SPEEDOMETER COMPLETE, KPH	1	533372@	Coupe 4-speed Mk II
	SPEEDOMETER COMPLETE, MPH	1	545238@@	Coupe 4-speed Mk II
	SPEEDOMETER COMPLETE, KPH	1	545239@@	Coupe 4-speed Mk II
	SPEEDOMETER COMPLETE, MPH	1	555549	Saloon 4-speed Mk III
	SPEEDOMETER COMPLETE, KPH	1	555550	Saloon 4-speed Mk III
	SPEEDOMETER COMPLETE, MPH	1	555551	Coupe 4-speed Mk III
	SPEEDOMETER COMPLETE, KPH	1	555552	Coupe 4-speed Mk III
	SPEEDOMETER COMPLETE, MPH	1	555547	Saloon Automatic Mk III
	SPEEDOMETER COMPLETE, KPH	1	555548	Saloon Automatic Mk III
	SPEEDOMETER COMPLETE, MPH	1	555572	Coupe Automatic Mk III
	SPEEDOMETER COMPLETE, KPH	1	555573	Coupe Automatic Mk III
	Flexible trip control for speedometer	1	533302	
	Bulb for speedometer	1	GLB987	
	Felt washer for speedometer	1	241387	
	SPEEDOMETER CABLE	1	278983	4-speed
	Cable, inner	1	515665	4-speed
	Cable, outer	1	500549	4-speed
	Right-angled drive for speedometer at gearbox	1	277741	4-speed
	SPEEDOMETER CABLE	1	279268	Automatic
	Cable, inner	1	501352	Automatic
	Cable, outer	1	501353	Automatic
	Grommet for speedometer cable	1	69052	

@ Mk II cars with suffix letters 'A' and 'B'   @@ Mk II cars with suffix letter 'C' onwards

*Asterisk indicates a new part which has not been used on any previous model.

INSTRUMENT PANEL AND FITTINGS, MK II AND MK III MODELS

1725

MK II AND MK III

MK I AND MK IA

H702

COLLINS JONES

Plate Ref.	1 2 3 4 Description	Qty.	Part No.	Remarks
	Clip for speedometer cable at bell housing	1	512419	LH Stg Automatic
	Rubber grommet for clip	1	06860	LH Stg Automatic
	Cable cleat fixing speedometer cable to wiper leads at dash	1	240431	LH Stg Automatic
	Clip for speedometer cable at gearbox	1	50637	4- speed
	Bolt (¼" UNF x ⅜" long) } Fixing clip to gearbox	1	255206	4- speed
	Spring washer } Fixing clip to gearbox	1	3074	4- speed
	Nut (¼" UNF) } Fixing clip to gearbox	1	254810	4- speed
	Tachometer	1	531614@	Mk II
	Tachometer	1	545212@@	Mk II Coupe
73	Tachometer	1	551623	Mk III Coupe
	Plastic cover for steering column, RH	1	384583	4-speed RH and LH Stg also Automatic RH Stg only
	Plastic cover for steering column, RH	1	384584	Automatic LH Stg
	Plastic cover for steering column, LH	1	384585	4-speed RH and LH Stg
	Plastic cover for steering column, LH	1	384511	Automatic RH Stg
	Plastic cover for steering column, LH	1	384510	Automatic LH Stg
	Spring washer } Fixing covers together	4	3073	
	Set screw (10 UNF x 1⅜" long) } Fixing covers together	4	78312	
	Retaining plate for plastic covers	1	502326	Mk II 4-speed
	Retaining plate and switch mounting bracket	1	516382	RH Stg } Mk II Automatic
	Retaining plate and switch mounting bracket	1	516383	LH Stg } Mk II Automatic
	Retaining plate for plastic covers	1	502326	Mk III 4-speed and Automat
	Spring washer } Fixing covers to retainer	4	3073	
	Set screw (10 UNF x 5/16" long) } Fixing covers to retainer	4	78307	
	Finishing bead for plastic covers	1	359449	
	Cover plate with one hole and slot	1	359455	
	Cover plate with one hole	1	359456	
	Spring washer } Fixing cover plate	4	3073	2 off on Automatic
	Nut (10 UNF) } Fixing cover plate	4	257011	2 off on Automatic
	Gear change indicator plate	1	505190@	RH Stg — Automatic
	Gear change indicator plate	1	505191@	LH Stg — Automatic
	Gear change indicator plate	1	545258@@	Automatic
	Gear change indicator plate	1	545259@@	Automatic
	Gear change indicator plate	1	551468	RH Stg } Mk III — Automatic
	Gear change indicator plate	1	551469	LH Stg } Mk III — Automatic
	Bracket for indicator plate	1	505192	RH Stg — Automatic
	Bracket for indicator plate	1	505193	LH Stg — Automatic
	Screw (4 BA x ⅜" long) } Fixing bracket to indicator plate	2	77633	Automatic
	Spring washer } Fixing bracket to indicator plate	2	3072	Automatic
	Nut (4 BA) } Fixing bracket to indicator plate	2	4023	Automatic
	Bulbholder for indicator plate	1	505182	RH Stg — Automatic
	Bulbholder for indicator plate	1	505189	LH Stg — Automatic
	Bulb for gear change indicator	1	GLB280	Automatic

@ Mk II cars with suffix letters 'A' and 'B' @@ Mk II cars with suffix letter 'C' onwards

*Asterisk indicates a new part which has not been used on any previous model.

1727

INSTRUMENT UNIT CASING AND WARNING LIGHTS

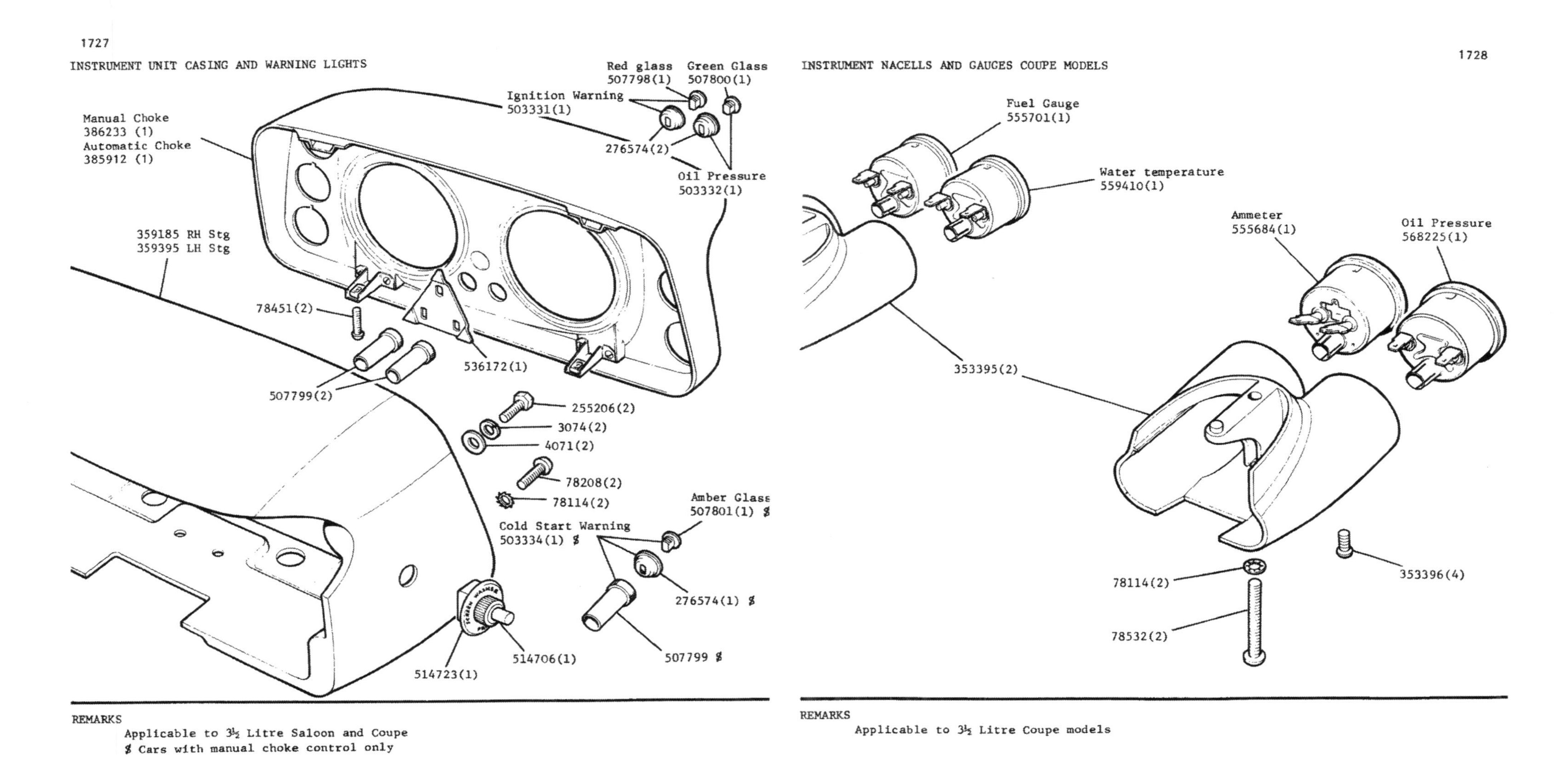

REMARKS
Applicable to 3½ Litre Saloon and Coupe
$ Cars with manual choke control only

1728

INSTRUMENT NACELLS AND GAUGES COUPE MODELS

REMARKS
Applicable to 3½ Litre Coupe models

1729

INSTRUMENTS AND SWITCHES SALOON

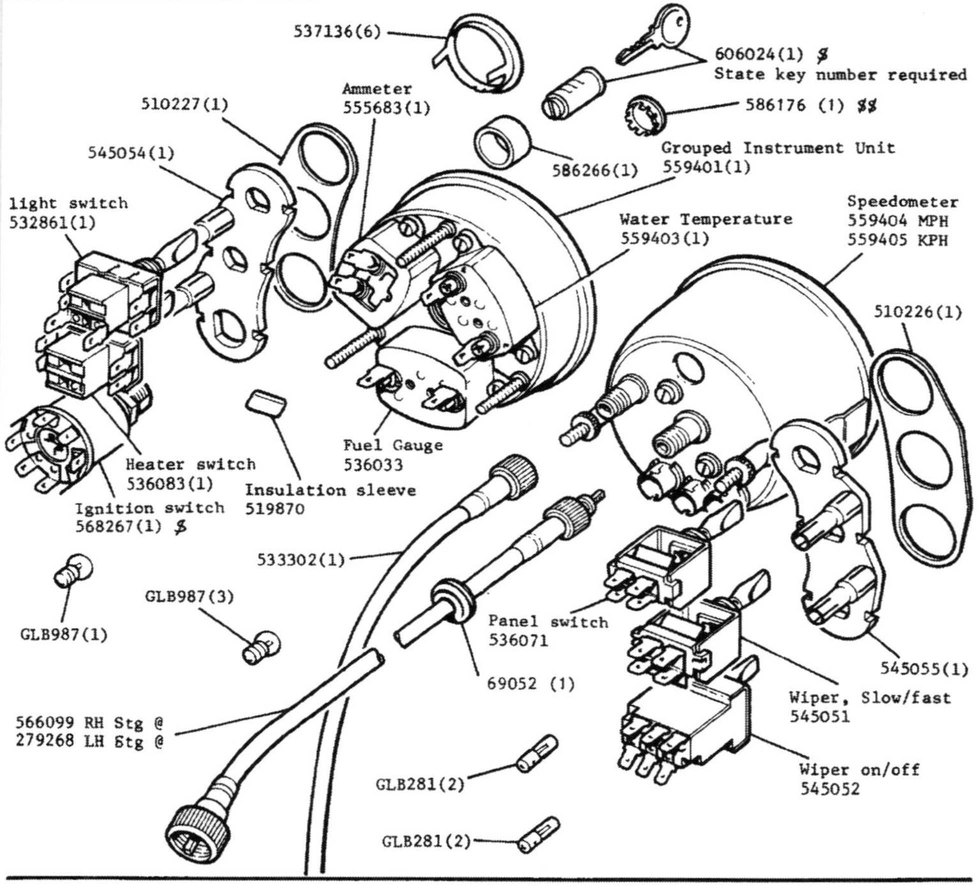

537136(6)
606024(1) $
State key number required
586176 (1) $$
510227(1)
Ammeter
555683(1)
545054(1)
586266(1)
Grouped Instrument Unit
559401(1)
Light switch
532861(1)
Water Temperature
559403(1)
Speedometer
559404 MPH
559405 KPH
510226(1)
Heater switch
536083(1)
Fuel Gauge
536033
Insulation sleeve
519870
Ignition switch
568267(1) $
533302(1)
GLB987(3)
GLB987(1)
Panel switch
536071
545055(1)
69052 (1)
Wiper, Slow/fast
545051
566099 RH Stg @
279268 LH Stg @
Wiper on/off
545052
GLB281(2)
GLB281(2)

REMARKS

Applicable to 3½ Litre Saloon models
@ 566099 Comprises inner cable 567840 and outer casing 567841
@ 279268 Comprises inner cable 501352 and outer casing 501353

$ Cars with ignition switch mounted on instrument unit.
$$ Cars with ignition switch mounted in steering column.

1730

INSTRUMENTS AND SWITCHES COUPE

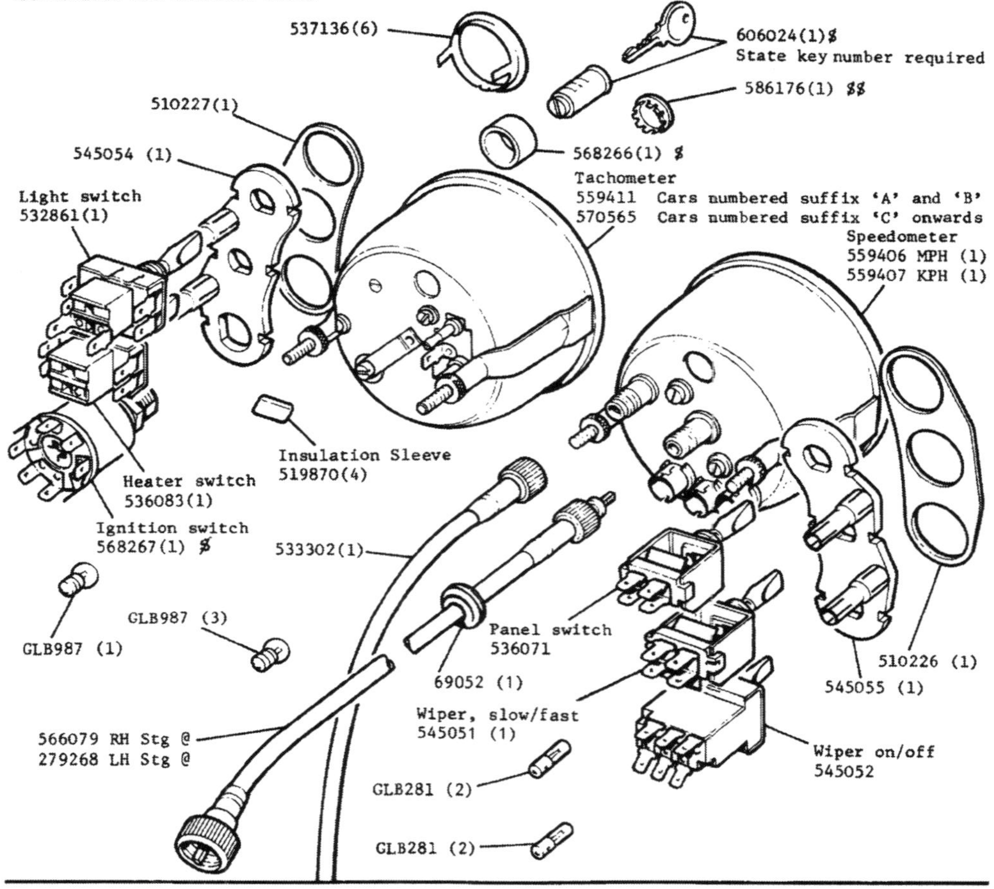

537136(6)
606024(1)$
State key number required
586176(1) $$
510227(1)
545054 (1)
568266(1) $
Light switch
532861(1)
Tachometer
559411 Cars numbered suffix 'A' and 'B'
570565 Cars numbered suffix 'C' onwards
Speedometer
559406 MPH (1)
559407 KPH (1)
Insulation Sleeve
519870(4)
Heater switch
536083(1)
Ignition switch
568267(1) $
533302(1)
GLB987 (3)
GLB987 (1)
Panel switch
536071
510226 (1)
545055 (1)
69052 (1)
Wiper, slow/fast
545051 (1)
566079 RH Stg @
279268 LH Stg @
Wiper on/off
545052
GLB281 (2)
GLB281 (2)

REMARKS

Applicable to 3½ Litre Coupe models

@ 566079 Comprises inner cable 567840 and outer casing 567841

@ 279268 Comprises inner cable 501352 and outer casing 501353

$ Cars with ignition switch mounted on instrument unit

$$ Cars with ignition switch mounted on steering column.

STEERING COLUMN COVERS AND CONTROLS

1731

THIEFPROOF COIL AND IGNITION SWITCH

1732

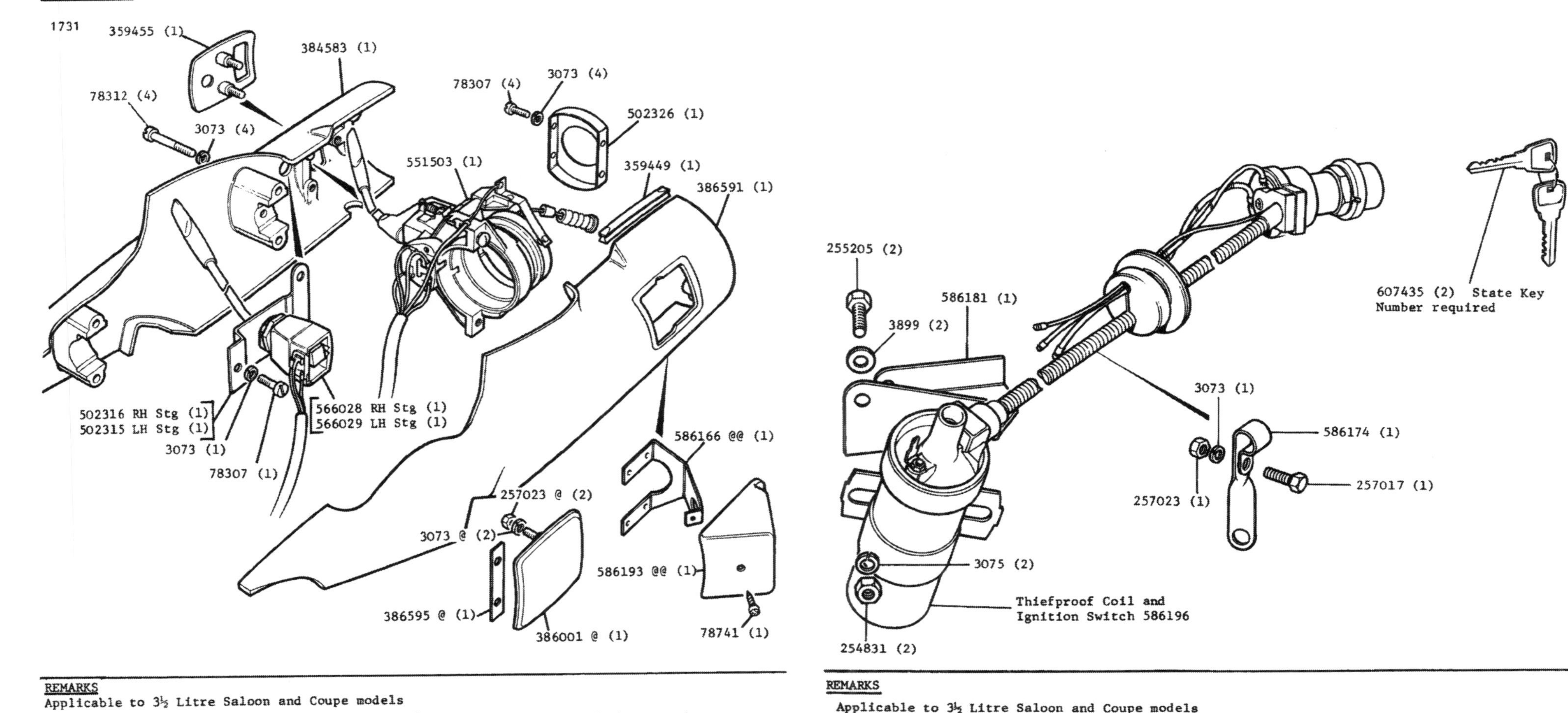

REMARKS

Applicable to 3½ Litre Saloon and Coupe models

@ Cars with ignition switch mounted on instrument unit.

@@Cars with ignition switch mounted on steering column

REMARKS

Applicable to 3½ Litre Saloon and Coupe models Home Market only.

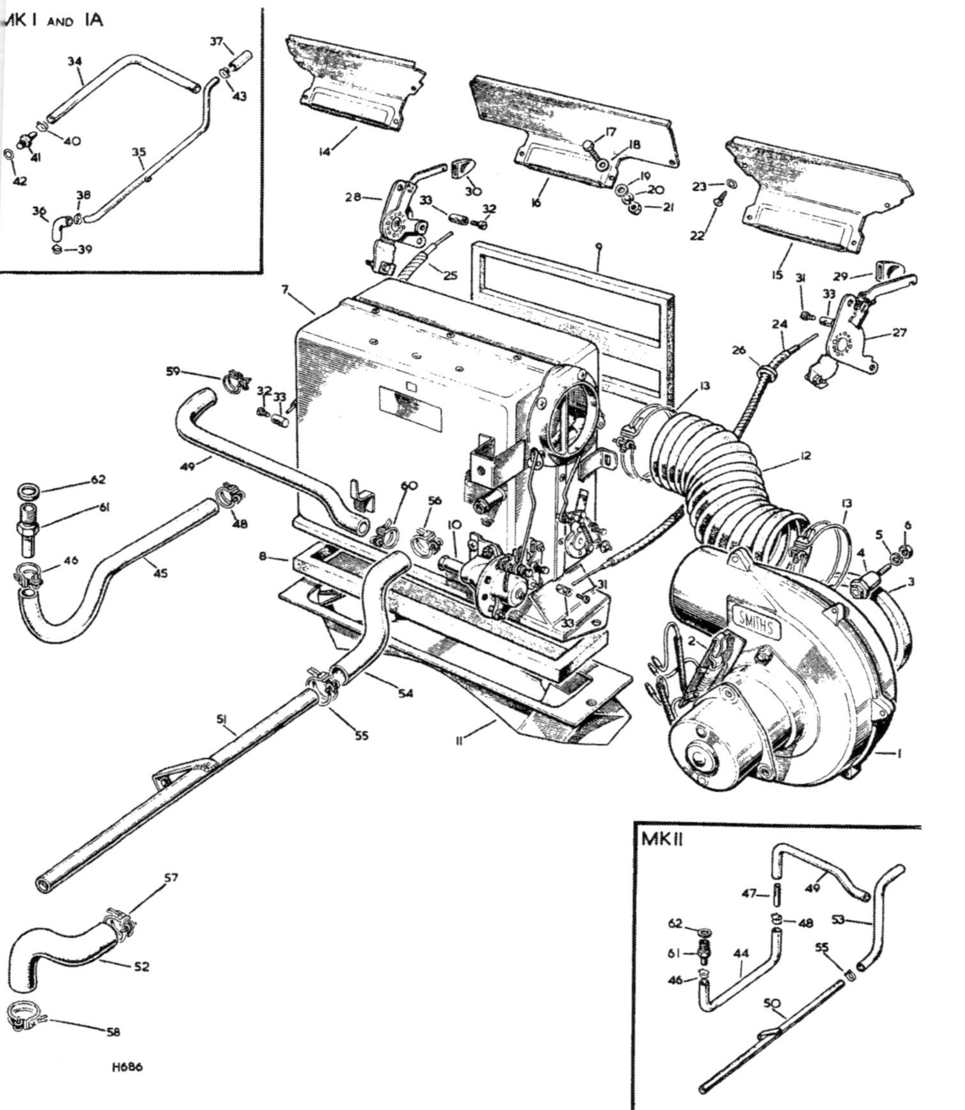

Plate Ref.	Description 1 2 3 4		Qty.	Part No.	Remarks
	BLOWER COMPLETE		1	356809	Mk I and Mk IA
1	BLOWER COMPLETE		1	359829	Mk II and Mk III
2	Resistance unit for blower		1	541898	
3	Rubber seal for blower to dash		1	352506	
4	Rubber mounting bush		3	356197	
5	Spring washer	Fixing blower	3	3074	
6	Nut (¼" UNF)		6	254810	
	HEATER COMPLETE		1	356808	Mk I
7	HEATER COMPLETE		1	384756	Mk IA, Mk II and Mk III
8	Rubber seal at heater base		1	352505	
9	Rubber seal, heater to dash		1	381168	
	Rubber seal, tap to heater		1	278217	
10	Tap for heater		1	356196	
11	Baffle		1	357454	
	Bolt (¼" UNF x ⅞" long)	Fixing baffle	2	255208	
	Plain washer		2	3900	
	Nut (¼" UNF)		2	254810	
	Set bolt (¼" UNF x ⅝" long)	Fixing heater to dash	2	255206	
	Plain washer		4	4071	
	Spring washer		4	3074	
12	Hose, blower to heater		1	386017	
13	Clip for hose		2	50347	
14	Demister nozzle, RH		1	352313	
15	Demister nozzle, LH		1	352314	
16	Demister nozzle, centre		1	352310	
17	Bolt (10 UNF x ½" long)	Fixing nozzle to dash ducting	6	257017	
18	Plain washer, thick		12	3816	
19	Plain washer, thin		18	4034	
20	Spring washer		6	3073	
21	Nut (10 UNF)		6	257023	
22	Drive screw	Fixing nozzle to dash drain channel	6	77704	
23	Plain washer		9	2874	
24	Control cable, heat		1	352479	
25	Control cable, demister		1	352480	
26	Grommet for control cables		2	312306	
	LEVER CONTROL AND SWITCH COMPLETE, HEAT		1	352477	Mk I and Mk IA
	Switch for lever control		1	600467	Mk I and Mk IA
27	Lever control complete, 'body heat'		1	359513	Mk II and Mk III
28	Lever control complete, 'demist'		1	352478	
	Spring washer	Fixing lever controls to facia frame	4	3073	
	Set bolt (10 UNF x ½" long)		4	257017	
	Control knob, 'heater'		1	352348	Mk I and Mk IA
29	Control knob, 'body heat'		1	359514	Mk II and Mk III
30	Control knob, 'demist'		1	352347	
	Special screw fixing knobs		2	250008	
31	Special screw, long, for cables		3	257307	
32	Special screw, short, for cables		1	257308	
33	Nipple for cables		4	356765	

*Asterisk indicates a new part which has not been used on any previous model.

Plate Ref.	1 2 3 4 Description	Qty.	Part No.	Remarks
34	Inlet hose for water heater	1	500626	Mk I
	Inlet hose for water heater	1	529471	Mk IA
35	Outlet pipe for water heater	1	500627	Mk I
36	Hose, outlet pipe to engine	1	274351	Mk I
37	Hose, outlet pipe to heater	1	513859	Mk I
38	Clip } For hose, outlet	1	GHC507	Mk I
39	Clip } pipe to engine	1	GHC507	Mk I
40	Clip for hose, except outlet pipe to engine	4	GHC507	Mk I
41	Adaptor for car heater inlet pipe	1	243535	Mk I
42	Joint washer for adaptor	1	243972	Mk I
43	Clip for outlet pipe	1	513860	Mk I
44	Hose, adaptor to inlet pipe	1	530441	Mk II
45	Hose, adaptor to tee piece	1	554963	Mk III
46	Clip, hose to inlet adaptor	1	GHC507	
47	Heater inlet pipe	1	529480	Mk II
48	Clip, hose to inlet pipe to tee piece	1	GHC507	
49	Hose, inlet pipe to heater	1	529479	
50	Outlet pipe to water heater	1	529466	Mk IA and Mk II
51	Outlet pipe to water heater	1	554960	Mk III
52	Hose, outlet pipe to engine	1	529468	Mk IA onwards
53	Hose, outlet pipe to heater	1	529467	Mk IA and Mk II
54	Hose, outlet pipe to heater	1	554916	Mk III
55	Clip } For hose, heater	1	GHC507	
56	Clip } to outlet pipe	1	GHC507	
57	Clip } For hose, outlet	1	603890	
58	Clip } pipe to engine	1	GHC507	
	Clip for hose, inlet pipe to cylinder head	1	GHC507	Mk IA
	Clip for hose, inlet pipe to heater connection	1	GHC507	Mk IA
59	Clip, hose to inlet pipe to valve unit	1	GHC507	Mk II and Mk III
60	Clip, hose to heater	1	GHC507	Mk II and Mk III
	Clip, inlet pipe to manifold	1	243321	Mk II and Mk III
	Set bolt (¼" UNC x ½" long) } Fixing clip	1	253004	Mk II and Mk III
	Spring washer } to manifold	1	3074	Mk II and Mk III
	Adaptor for heater inlet pipe	1	529475	Mk IA
61	Adaptor for heater inlet pipe	1	530115	Mk II and Mk III
	Joint washer for adaptor	1	530112	Mk IA
62	Joint washer for adaptor	1	530110	Mk II and Mk III

*Asterisk indicates a new part which has not been used on any previous model.

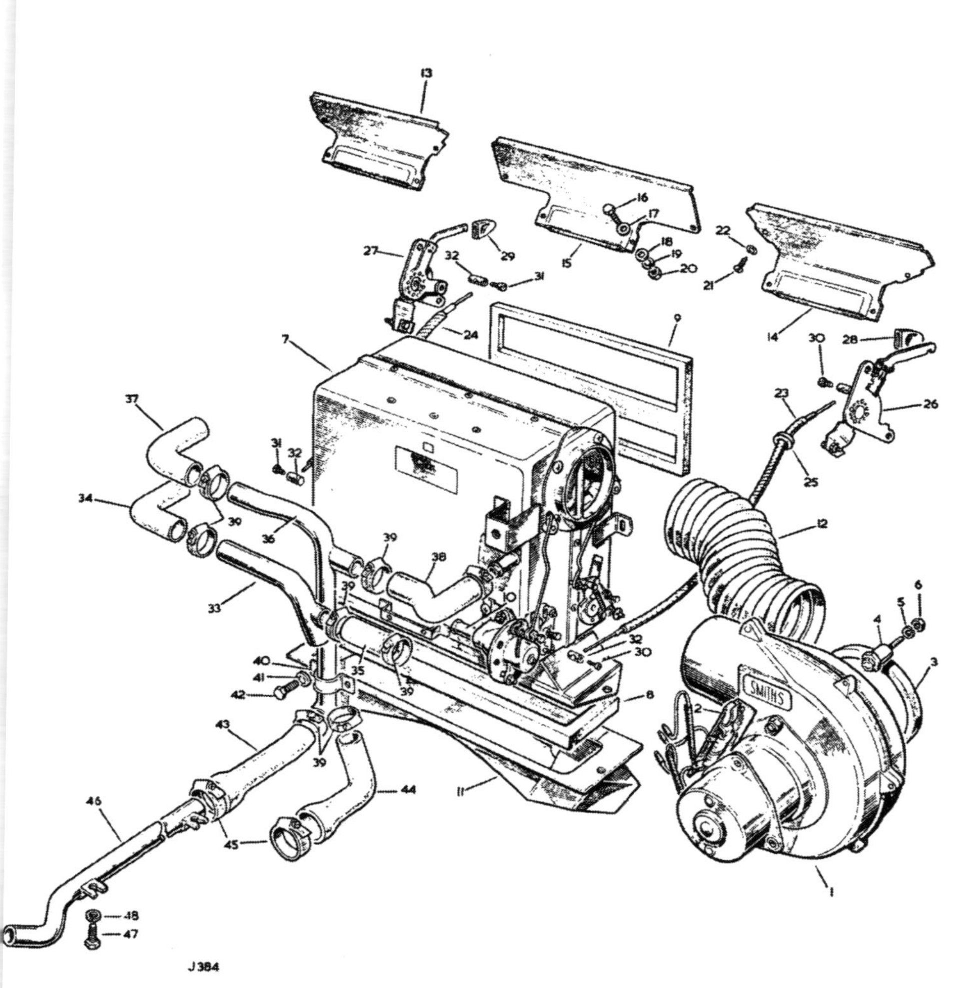

Plate Ref.	1 2 3 4 Description	Qty.	Part No.	Remarks
1	BLOWER COMPLETE	1	359829	
2	Resistance unit for blower	1	541898	
3	Rubber seal for blower to dash	1	352506	
4	Rubber mounting bush	3	356197	
5	Spring washer } Fixing blower	3	3074	
6	Nut (¼" UNF) } Fixing blower	6	254810	
7	HEATER COMPLETE	1	386012	
8	Rubber seal at heater base	1	352505	
9	Rubber seal, heater to dash	1	381168	
	Rubber seal, tap to heater	1	278217	
10	Tap for heater	1	356196	
	Insulating cover for heater unit	1	386301	
11	Baffle	1	357454	
	Bolt (¼" UNF x ⅞" long) } Fixing baffle	2	255208	
	Plain washer } Fixing baffle	2	3900	
	Nut (¼" UNF) } Fixing baffle	2	254810	
	Set bolt (¼" UNF x ⅝" long) } Fixing heater to dash	2	255206	
	Plain washer } Fixing heater to dash	4	4071	
	Spring washer } Fixing heater to dash	4	3074	
12	Hose, blower to heater	1	352434	
13	Demister nozzle, RH	1	352313	
14	Demister nozzle, LH	1	352314	
15	Demister nozzle, centre	1	352310	
16	Bolt (10 UNF x ½" long) } Fixing nozzle to dash ducting	6	257017	
17	Plain washer, thick } Fixing nozzle to dash ducting	12	3816	
18	Plain washer, thin } Fixing nozzle to dash ducting	18	4034	
19	Spring washer } Fixing nozzle to dash ducting	6	3073	
20	Nut (10 UNF) } Fixing nozzle to dash ducting	6	257023	
21	Drive screw } Fixing nozzle to dash drain channel	6	77704	
22	Plain washer } Fixing nozzle to dash drain channel	9	2874	
23	Control cable, heat	1	352479	
24	Control cable, demister	1	352480	
25	Grommet for control cables	2	312306	
26	Lever control complete 'body heat'	1	359513	
27	Lever control complete, 'demist'	1	352478	
	Spring washer } Fixing lever controls to facia frame	4	3073	
	Set bolt(10 UNFx½"long) } Fixing lever controls to facia frame	4	257017	
28	Control knob, 'body heat'	1	359514	
29	Control knob, 'demist'	1	352347	
	Special screw fixing knobs	2	250008	
30	Special screw, long for cables	3	257307	
31	Special screw, short, for cables	1	257308	
32	Nipple for cables	4	356765	
33	Pipe, engine inlet to diverter valve	1	562396	
34	Elbow, inlet pipe to diverter valve	1	578042*	
35	Hose, engine inlet pipe to heater unit	1	578043	

*Asterisk indicates a new part which has not been used on any previous model.

Plate Ref.	1 2 3 4 Description	Qty.	Part No.	Remarks
36	Pipe, engine outlet to divertor valve	1	562397	
37	Elbow, engine outlet pipe to divertor valve	1	578042*	
38	Hose, engine outlet pipe to heater unit	1	578079*	
39	Hose clip fixing heater hoses	10	517821	
40	Clip, pipe to stand-off bracket	1	56667	
41	Spring washer] Fixing clip	1	3073	
42	Bolt (10 UNF x ½" long)] to bracket	1	257017	
43	Hose, engine outlet	1	610812*	
44	Hose, engine inlet	1	610813*	Engines numbered up to suffix letter 'C'
	Hose, engine inlet	1	613591*	Engines numbered from suffix letter 'D'
45	Hose clip	2	GHC709	
46	Outlet pipe	1	603049	
47	Bolt (¼" UNC x ½" long)] Fixing outlet	2	253004	
48	Spring washer] pipe to engine	2	3074	
	Hose, outlet pipe to water pump	1	610810	
	Hose clip	2	GHC709	

*Asterisk indicates a new part which has not been used on any previous model.

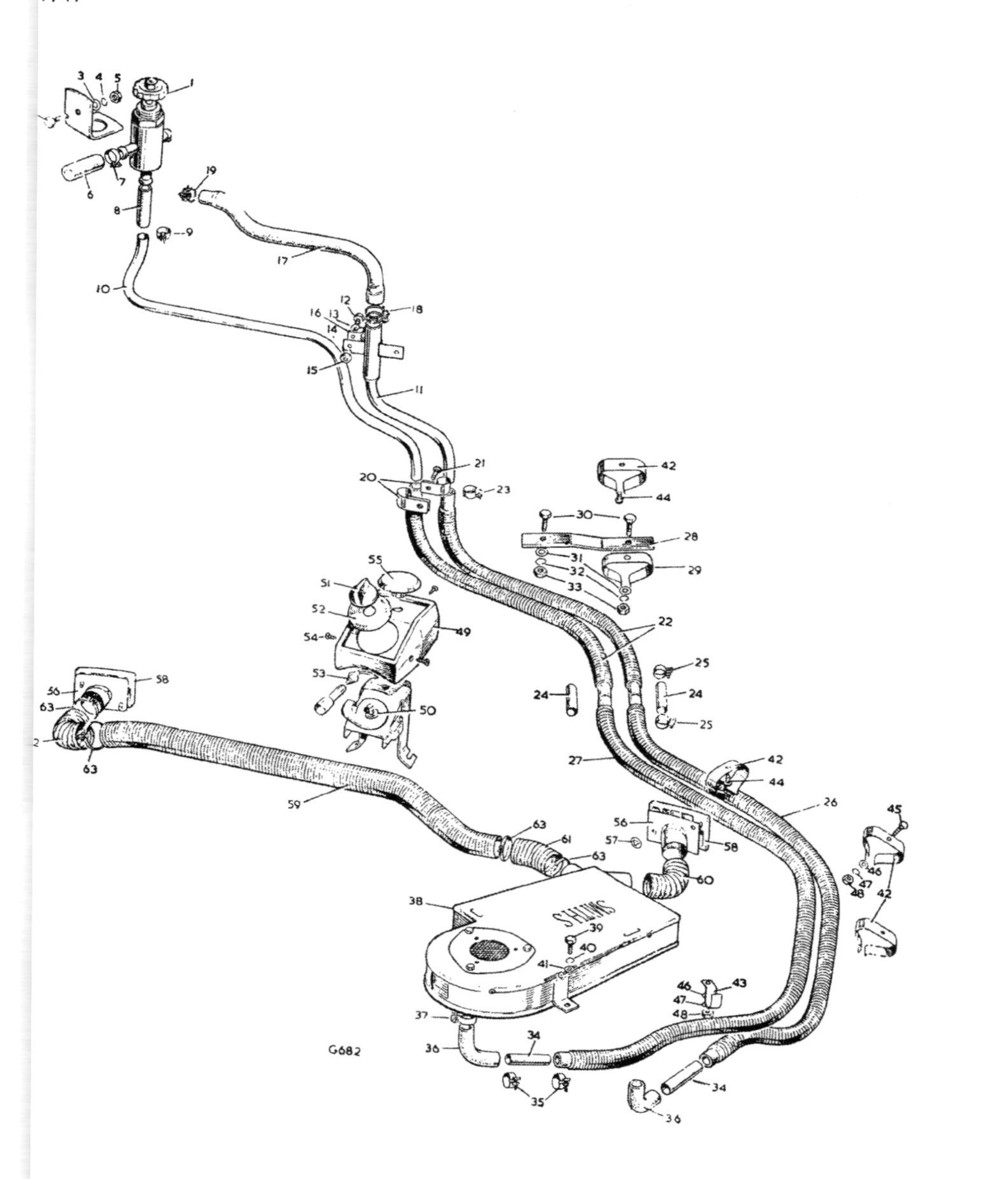

Plate Ref.	1 2 3 4 Description		Qty.	Part No.	Remarks
1	Valve unit for rear heater		1	557124	
2	Bolt (¼"UNF x ⅝"long)	Fixing valve unit to heater unit	1	255206	
3	Plain washer		1	3467	
4	Spring washer		1	3074	
5	Nut (¼" UNF)		1	254810	
6	Hose connecting valve unit to heater unit		1	557129	
7	Clip fixing hose		2	557130	
8	Hose, heater outlet pipe to valve unit		1	557129	
9	Clip for hose		2	557130	
10	Pipe to rear heater outlet insulated hose		1	557125	
11	Pipe and tee piece to rear heater insulated hose (inlet and heater inlet hose		1	557126	
12	Bolt (¼"UNF x ⅝"long)	Fixing pipe and tee piece to dash	1	255206	
	Bolt (¼"UNF x ¾"long)		1	255207	
13	Plain washer		2	3467	
14	Spring washer		2	3074	
15	Nut (¼" UNF)		2	254810	
16	Clip for heater pipe (outlet) (picks up at pipe and tee piece fixing)		1	219677	
17	Hose, heater inlet to inlet pipe		1	529479	
18	Clip, hose to inlet pipe		1	50307	
19	Clip, hose to heater		1	GHC507	
20	Clip for pipe	Fixing heater pipes, inlet and outlet, to dash	1	41908	
21	Drive screw		1	78137	
22	Insulated hose		2	557251	
23	Clip fixing hoses to pipes		2	GHC507	
24	Tube connecting insulated hoses		2	557120	
25	Clip, tubes to insulated hoses		4	GHC507	
26	Insulated hose, from rear heater inlet		1	557128	
27	Insulated hose, from rear heater outlet		1	557127	
28	Support bracket	For insulated hoses	1	557121	
29	Double clip		1	384760	
30	Bolt(¼" UNF x ⅝"long	Fixing support bracket and clip to sub-frame	2	255206	
31	Plain washer		2	3467	
32	Spring washer		2	3074	
33	Nut (¼"UNF)		2	254810	
34	Connector tube		2	557120	
35	Clip fixing insulated hose to connector tube and elbow		4	50304	
36	Elbow at rear heater		2	578046	
37	Clip fixing elbow to connector tube and rear heater		2	557130	
38	Heater unit for rear seat		1	384647	
39	Set screw (¼"UNF x ¾"long)	Fixing heater to rear seat pan	2	255207	
40	Spring washer		2	3074	
41	Plain washer		2	3931	
42	Double hose clip		5	384760	
43	Single hose clip		1	384761	

*Asterisk indicates a new part which has not been used on any previous model.

1743

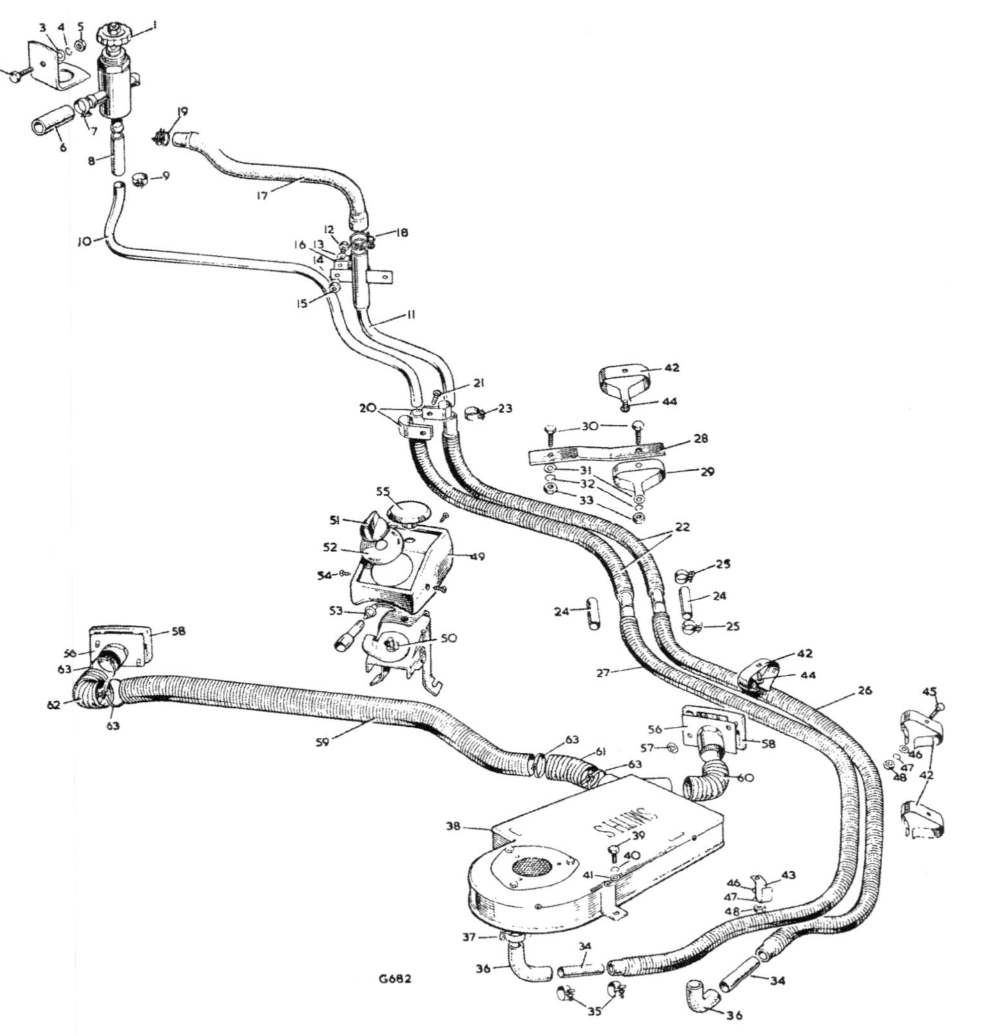

Plate Ref.	1 2 3 4 Description		Qty.	Part No.	Remarks
44	Drive screw	Fixing hose clips to floor	1	78654	
45	Screw(¼"UNF x ⅜"long)		5	78233	
46	Plain washer		6	3911	
47	Spring washer		5	3074	
48	Nut (¼")		5	254810	
49	Moulded cover for control box		1	555767	
	CARRIER BRACKET AND SWITCH ASSEMBLY FOR REAR HEATER		1	559173	
	Cover plate (stainless steel)		1	559175	
50	Switch for rear heater		1	559187	
51	Knob for switch		1	555770	
52	Escutcheon disc for switch		1	559176	
	Spacer for disc		1	559171	
	Diffuser plate		1	555764	
53	Bulb (12-volt, 2-watt) for switch		1	503352	
54	Screw fixing moulded cover		3	78719	
55	Blanking piece for cover plate		1	555773	
56	Escutcheon for rear heater air outlet		2	384683	
57	Retainer for mounting assembly and outlet grille		4	78714	
58	Outlet grille for rear heater		2	384648	
59	Air hose, long	For rear heater	1	384693	
60	Air hose, short		1	384702	
61	Connecting hose		4	384703	
62	Elbow, hose to escutcheon		1	384685	
63	Clip for rear heater hoses		4	384694	

*Asterisk indicates a new part which has not been used on any previous model.

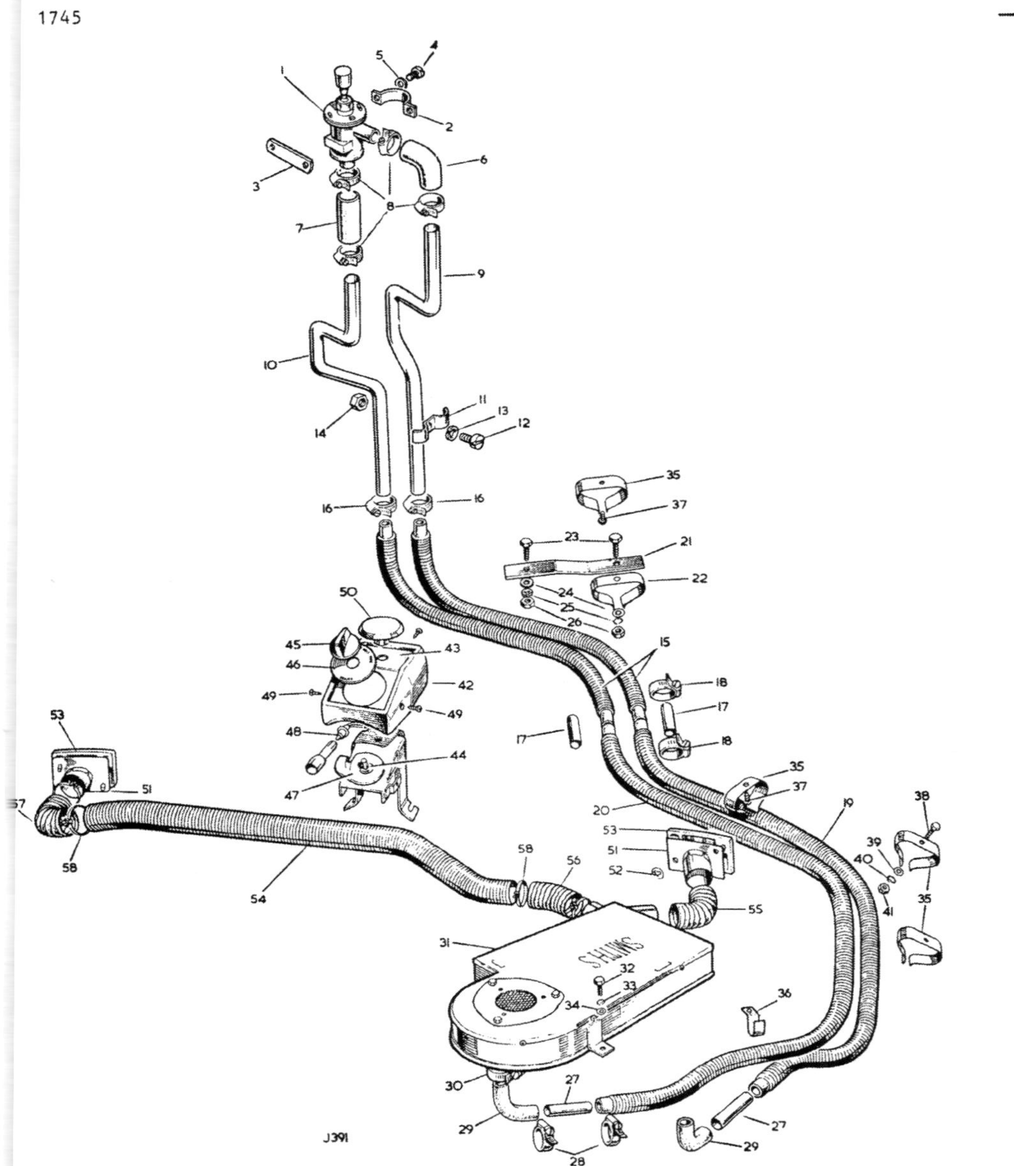

Plate Ref.	1 2 3 4 Description		Qty.	Part No.	Remarks
1	Divertor valve		1	562309	
2	Clamp	Fixing divertor valve to heater unit	1	562310	
3	Clamp plate		1	562311	
4	Bolt(¼"UNF x ⅝"long)		2	255206	
5	Spring washer		2	3074	
6	Elbow hose		1	578044*	
7	Connector hose		1	578043*	
8	Clip for hose		4	517821	
9	Heater pipe, inlet		1	562209	
10	Heater pipe, outlet		1	562210	
11	Clip	Fixing inlet and outlet pipes to dash	1	562211	
12	Bolt (10 UNF x ⅞"long)		1	257021	
13	Spring washer		1	3073	
14	Nut (10 UNF)		1	257023	
15	Insulated hose, front		2	578049*	
16	Clip fixing hoses to pipes		2	517821	
17	Tube connecting insulated hoses		2	557120	
18	Clip, tubes to insulated hoses		4	517821	
19	Insulated hose, from rear heater inlet		1	578047*	
20	Insulated hose, from rear heater outlet		1	578048*	
21	Support bracket	For insulated hoses	1	557121	
22	Double clip		1	384760	
23	Bolt(¼"UNF x ⅝" long)	Fixing support bracket and clip to sub-frame	2	255206	
24	Plain washer		2	3467	
25	Spring washer		2	3074	
26	Nut (¼"UNF)		2	254810	
27	Connector tube		2	557120	
28	Clip fixing insulated hose to connector tube and elbow		4	517821	
29	Elbow at rear heater		2	578046*	
30	Clip fixing elbow to connector tube and rear heater		2	517821	
31	Heater unit for rear seat		1	384647	
32	Set screw(¼" UNF x ¾"long)	Fixing heater to rear seat pan	2	255207	
33	Spring washer		2	3074	
34	Plain washer		2	3931	
35	Double hose clip		5	384760	
36	Single hose clip		1	384761	
37	Drive screw	Fixing hose clips to floor	1	78654	
38	Screw(¼"UNF x ⅝"long)		5	78232	
39	Plain washer		6	3911	
40	Spring washer		5	3074	
41	Nut (¼")		5	254810	
42	Moulded cover for control box		1	555767	

*Asterisk indicates a new part which has not been used on any previous model.

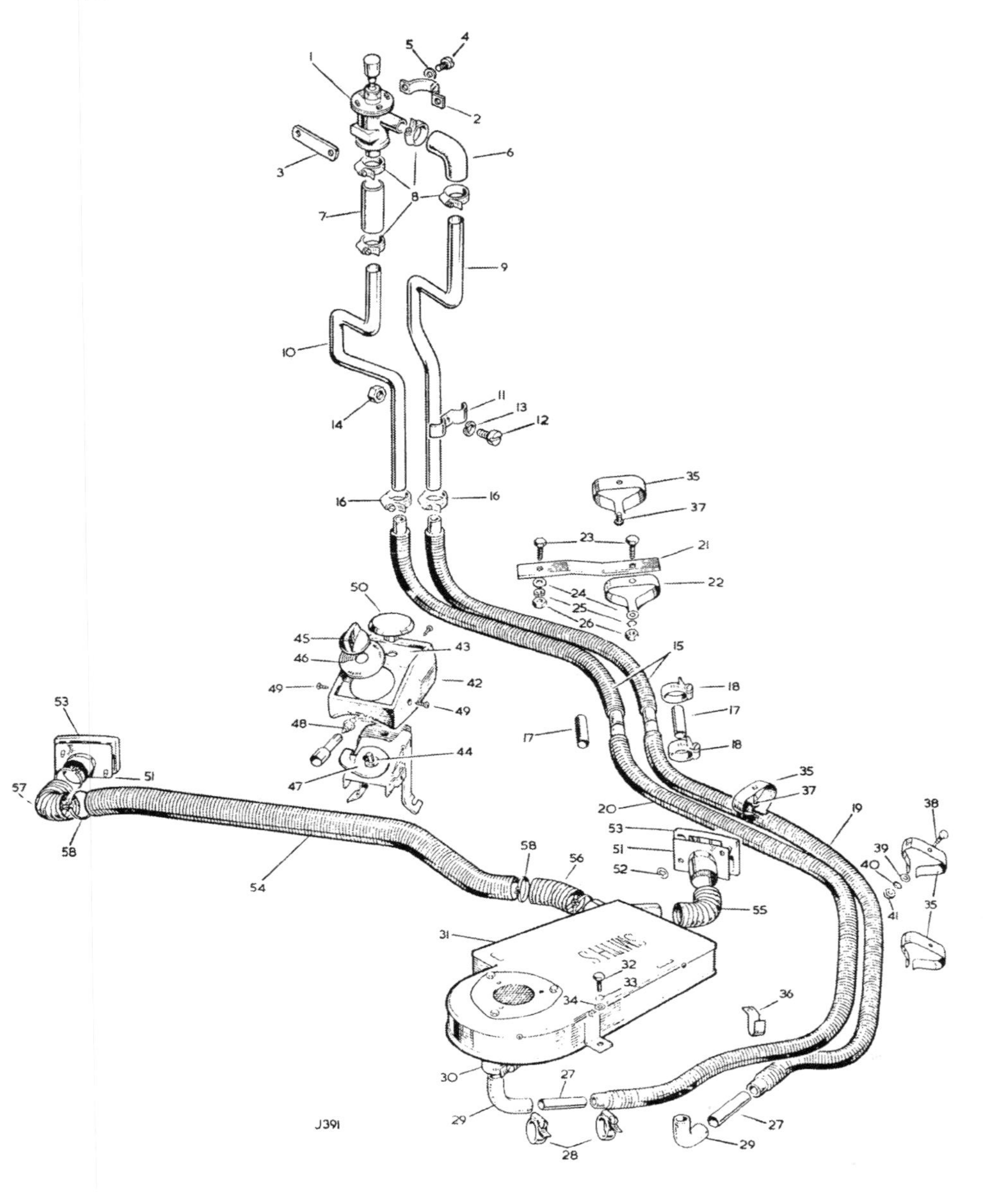

Plate Ref.	1 2 3 4	Description	Qty.	Part No.	Remarks
		CARRIER BRACKET AND SWITCH ASSEMBLY FOR REAR HEATER	1	559395	
43		Cover plate (stainless steel)	1	559175	
44		Switch for rear heater	1	559187	
45		Knob for switch	1	555770	
46		Escutcheon disc for switch	1	559176	
		Spacer for disc	1	559171	
47		Diffuser plate	1	555764	
48		Bulb (12-volt, 2-watt) for switch	1	GLB281	
49		Screw fixing moulded cover	3	78719	
50		Blanking piece for cover plate	1	555773	
		Harness for switch	1	559341	
51		Escutcheon for rear heater air outlet	2	384683	
52		Retainer for mounting assembly and outlet grille	4	78714	
53		Outlet grille for rear heater	2	384648	
54		Air hose, long — For rear heater	1	384693	
55		Air hose, short — For rear heater	1	384702	
56		Connecting hose — For rear heater	4	384703	
57		Elbow, hose to escutcheon — For rear heater	1	384685	
58		Clip for rear heater hoses — For rear heater	4	384694	

*Asterisk indicates a new part which has not been used on any previous model.

Plate Ref.	1 2 3 4 Description	Qty.	Part No.	Remarks
1	Air intake grille	1	352014	Cars with washer jets in intake grille
	Air intake grille complete with chrome screws	1	383661	Cars with washer jets on bonnet panel
	Drive screw] Fixing grille to dash	6	78394	
	Spire nut] Fixing grille to dash	6	313484	
	Drive screw] Fixing grille to dash	2	78747	
2	Rubber grommet] Fixing grille to dash	3	352135	
3	Filter for air intake grille	1	352599	
4	Sealing rubber for air intake lid	2	352171	
	Lid, RH] For air intake	1	351216	
5	Lid, LH] For air intake	1	351217	
	Spring washer] Fixing lid to body	4	3073	
	Nut(10 UNF)] Fixing lid to body	4	257023	
	COLD AIR CONTROL MECHANISM, RH	1	352156	
	COLD AIR CONTROL MECHANISM, LH	1	352157	
	Control mechanism, RH	1	352158	
6	Control mechanism, LH	1	352159	
	Bracket for RH mechanism	2	352160	
7	Bracket for LH mechanism	2	352161	
	Plain washer	2	3840	
	Split pin	2	2388	
	Set bolt(10 UNF x ½" long)] Fixing mechanism to dash	8	257017	
	Plain washer] Fixing mechanism to dash	8	3685	
	Spring washer] Fixing mechanism to dash	8	3073	
8	Joint pin Fixing mechanism to lid	2	352168	
	Split pin Fixing mechanism to lid	2	2388	
9	Return spring for control	2	356635	
10	Cold air controls	2	359139	
11	Retainer] Fixing cable to lever	2	356225	
12	Bolt (6 UNC x 5/16" long)] Fixing cable to lever	2	257194	
13	Support bracket for cold air control	2	352169	
	Set bolt (10 UNF x ½" long)] Fixing bracket to dash	4	257017	
	Plain washer] Fixing bracket to dash	4	3685	
	Spring washer] Fixing bracket to dash	4	3073	
	Spindle for butterfly ventilator control, RH	1	352586@	
14	Spindle for butterfly ventilator control, LH	1	352587@	
15	Plate with studs	2	352444@	
16	Sealing rubber] For ventilator control butterfly	2	352432@	
17	Plate with holes] For ventilator control butterfly	2	352537@	
	Plain washer] Fixing plates, sealing rubber and spindle in dash air tube	4	3816@	
	Spring washer] Fixing plates, sealing rubber and spindle in dash air tube	4	3073	
	Nut (10 UNF)] Fixing plates, sealing rubber and spindle in dash air tube	4	257023@	
	Split pin for butterfly stop	4	2426@	
	Face-level ventilator tube, RH	1	350979@	
18	Face-level ventilator tube, LH	1	350980@	
	Pop rivet fixing ventilator tube to dash tube	4	78257@	
	Face-level ventilator, RH	1	352395@	
19	Face-level ventilator, LH	1	352396@	
20	Bracket fixing ventilator to dash	2	352921@	
	Pop rivet fixing bracket to dash	4	78257@	] Alternatives
	Drive screw fixing bracket to dash	4	77704@	] Check before ordering

@Cars numbered up to 625001352, 626000341, 628000176, 630000748, 631000192, 633000443

*Asterisk indicates a new part which has not been used or any previous model.

1751

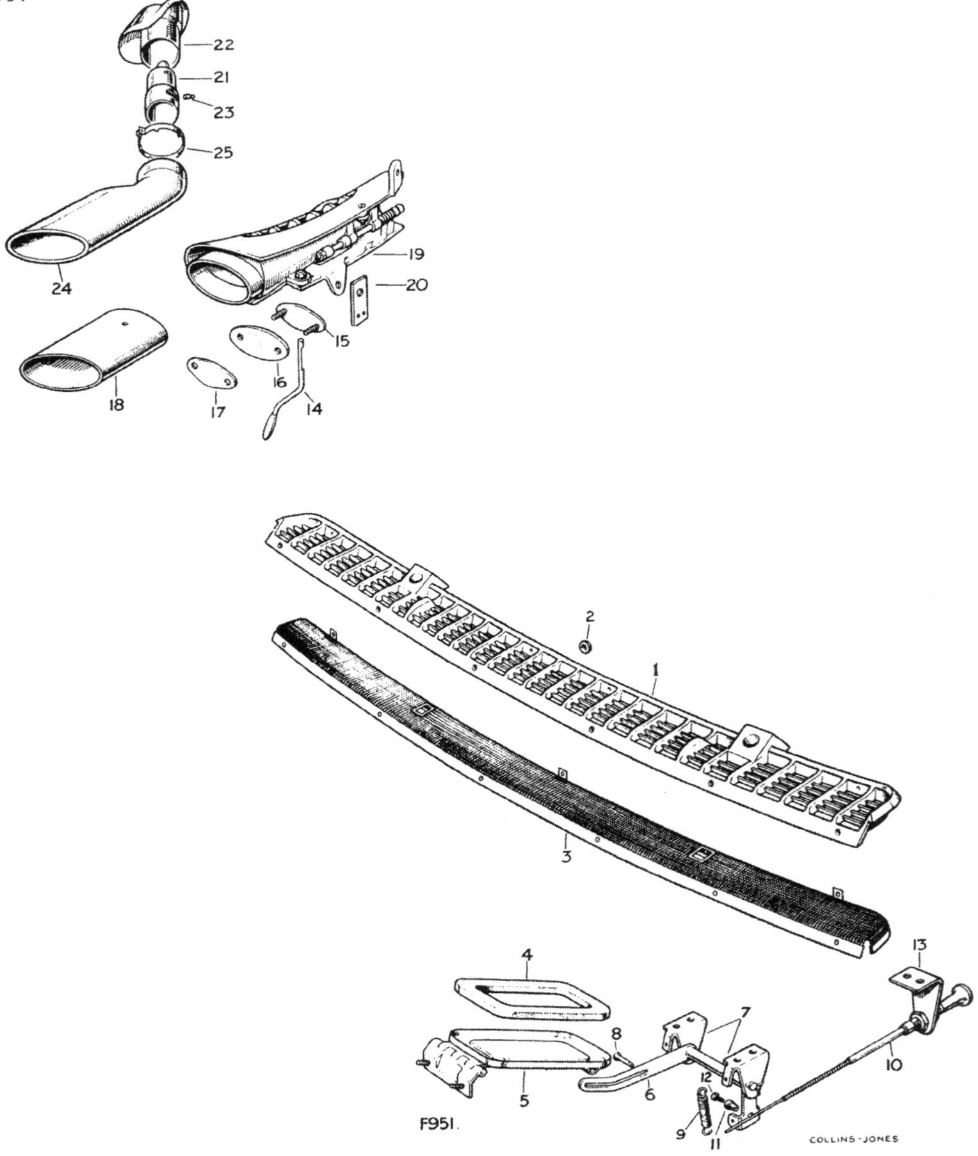

F951.

COLLINS-JONES

FRESH AIR CONTROLS

Plate Ref.	1 2 3 4 Description	Qty.	Part No.	Remarks
	Spire nut — Fixing ventilator to body	2	78267@	
	Drive screw — Fixing ventilator to body	4	78153@	
	Plain washer — Fixing ventilator to body	2	3851@	
	FACE-LEVEL VENTILATOR ASSEMBLY, RH	1	359445@@	Mk I, Mk IA and Mk II
	FACE-LEVEL VENTILATOR ASSEMBLY, LH	1	359446@@	Mk I, Mk IA and Mk II
	FACE-LEVEL VENTILATOR ASSEMBLY	2	384238	Mk III 3 Litre & 3½ Litre
21	Butterfly and housing for face-level ventilator	2	356379@@	
	Escutcheon for ventilator, RH	1	359447@@	Mk I, Mk IA and Mk II
22	Escutcheon for ventilator, LH	1	359448@@	Mk I, Mk IA and Mk II
	Escutcheon for ventilator	2	384239	Mk III 3 Litre & 3½ Litre
23	Special rivet fixing escutcheon	2	356633@@	
24	Rubber boot — For ventilator	2	356380@@	
25	Spring clip — For ventilator	2	356381@@	

@Cars numbered up to 625001352, 626000431, 628000175, 630000748, 631000192, 633000443
@@Cars numbered from 625001353, 626000431, 628000176, 630000748, 631000193, 633000444 onwards

*Asterisk indicates a new part which has not been used on any previous model.

1753

H641

FRONT GRILLE AND MOTIFS, MK I, MK IA AND MK II MODELS

1754

Plate Ref.	1 2 3 4 DESCRIPTION	Qty	Part No.	REMARKS
	FRONT END GRILLE COMPLETE, less name plate	1	357901	
1	Outer frame assembly for grille	1	357405††	
	Side motif, RH	1	356168†	
2	Side motif, RH	1	383723††	
	Side motif, LH	1	356169†	
3	Side motif, LH	1	383724††	
	Top motif	1	356170†	
4	Top motif	1	383721††	
	Bottom motif	1	356171†	
5	Bottom motif	1	383722††	
6	Centre motif	1	352086	
	Top rubber	2	352082†	
	Bottom rubber	2	352083†	
	Retainer for bottom rubber	2	352084†	
7	Badge for grille, plastic	1	352090	
8	Securing stud plate for centre motif	1	352346	
	Plain washer } Fixing plastic badge and centre motif to grille	2	3902	
	Spring washer } Fixing plastic badge and centre motif to grille	2	3073	
	Nut (10 UNF) } Fixing plastic badge and centre motif to grille	2	257023	
	Rib, centre LH	1	352072†	
	Rib, centre RH	1	352073†	
	Rib, 2nd and 3rd from centre LH	2	352528†	
	Rib, 2nd and 3rd from centre RH	2	352529†	
	Rib, 4th to 7th from centre LH	4	313118†	
	Rib, 4th to 7th from centre RH	4	313119†	
	Rib, 8th to 10th from centre LH	3	313114†	
	Rib, 8th to 10th from centre RH	3	313115†	
	Rib, 11th to 15th from centre LH	5	352070†	
	Rib, 11th to 15th from centre RH	5	352071†	
	Rib, 16th and 17th from centre LH	2	352068†	
	Rib, 16th and 17th from centre RH	2	352069†	
	Rib, 18th from centre LH	1	352066†	
	Rib, 18th from centre RH	1	352067†	
	Rib, 19th from centre LH	1	352064†	
	Rib, 19th from centre RH	1	352065†	
	Rib, 20th from centre LH and RH	2	352076†	
	Rib, 21st from centre LH and RH	2	352075†	
	Rib, 22nd from centre LH and RH	2	352074†	
9	Rib frame assembly, RH	1	357370††	
10	Rib frame assembly, LH	1	357371††	
	Transverse rib, top, upper	1	352078†	
11	Transverse rib, top, upper	1	357430††	
	Transverse rib, top, lower	1	352080†	
12	Transverse rib, top, lower	1	357432††	
	Transverse rib, bottom, upper	1	352081†	
13	Transverse rib, bottom, upper	1	357433††	
	Transverse rib, bottom, lower	1	352079†	
14	Transverse rib, bottom, lower	1	357431††	
15	Clip securing transverse ribs	8	381239††	

* Asterisk indicates a new part which has not been used on any previous Rover model

† Mk I, Mk IA and Mk II cars numbered up to 77001955, 77502430, 73500384, 74000297, 77100225, 77600499 73600016, 74100020, 77300382, 77800369, 73800077, 74300046

†† Mk II cars numbered from 77001956, 77502431, 73500385, 74000298, 77100226, 77600500 73600017, 74100021, 77300383, 77800370, 73800078, 74300047 onwards

FRONT GRILLE AND MOTIFS, MK I, MK IA AND MK II MODELS

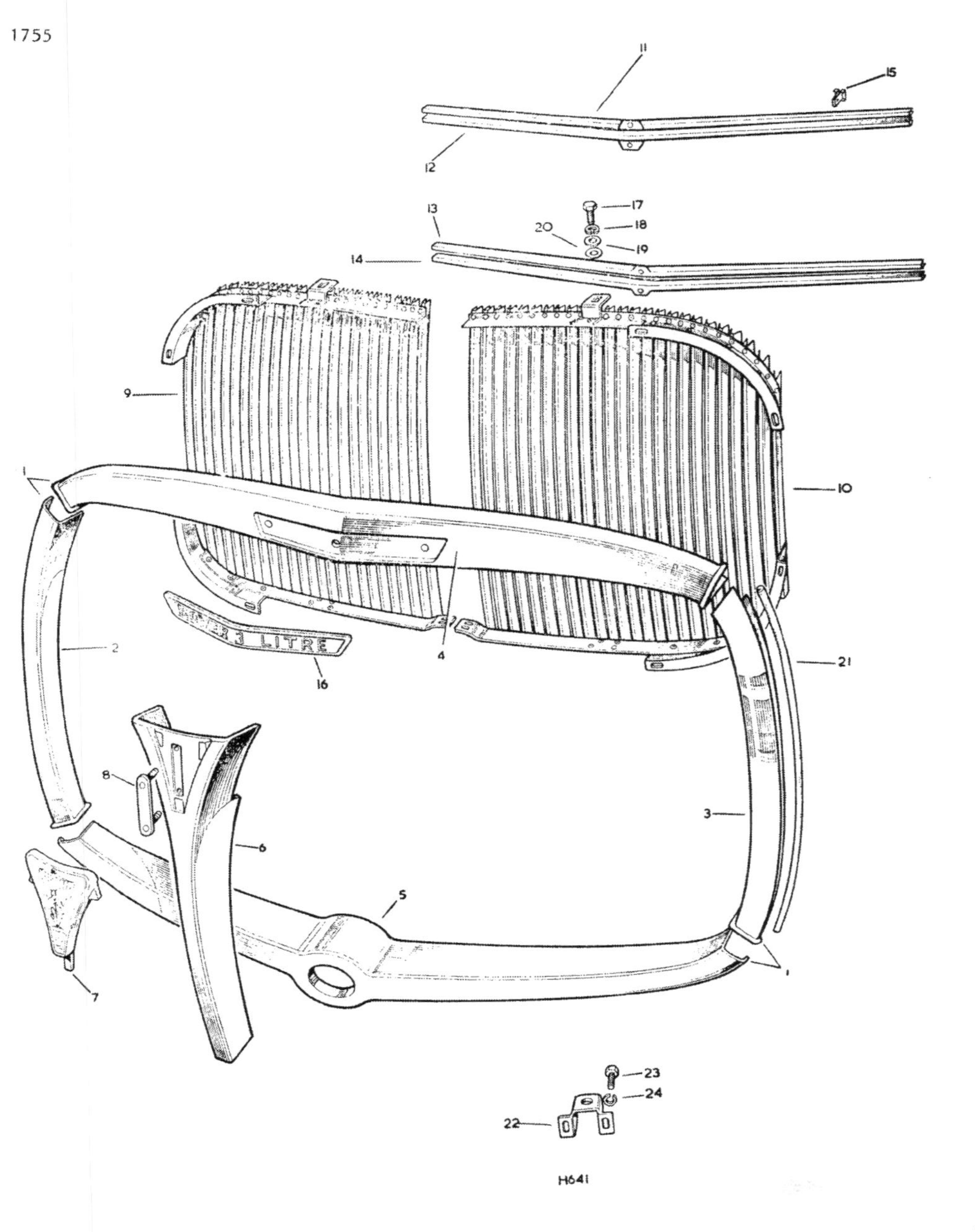

Plate Ref.	1 2 3 4	Description		Qty.	Part No.	Remarks
16		Name plate		1	357900	
17		Set bolt ($\frac{1}{4}$" UNF x $\frac{5}{8}$" long)	Fixing grille to body and bracket	4	255206	
18		Spring washer		4	3074	
19		Shim washer at top		As reqd	3524	
20		Plain washer		4	3900	
21		Plastic beading, radiator grille to front wing		2	386485	
22		Bracket fixing grille to body		2	553886	
23		Set bolt ($\frac{1}{4}$" UNF x $\frac{1}{2}$" long)	Fixing bracket to body	4	255204	
24		Spring washer		4	3074	

*Asterisk indicates a new part which has not been used on any previous model.

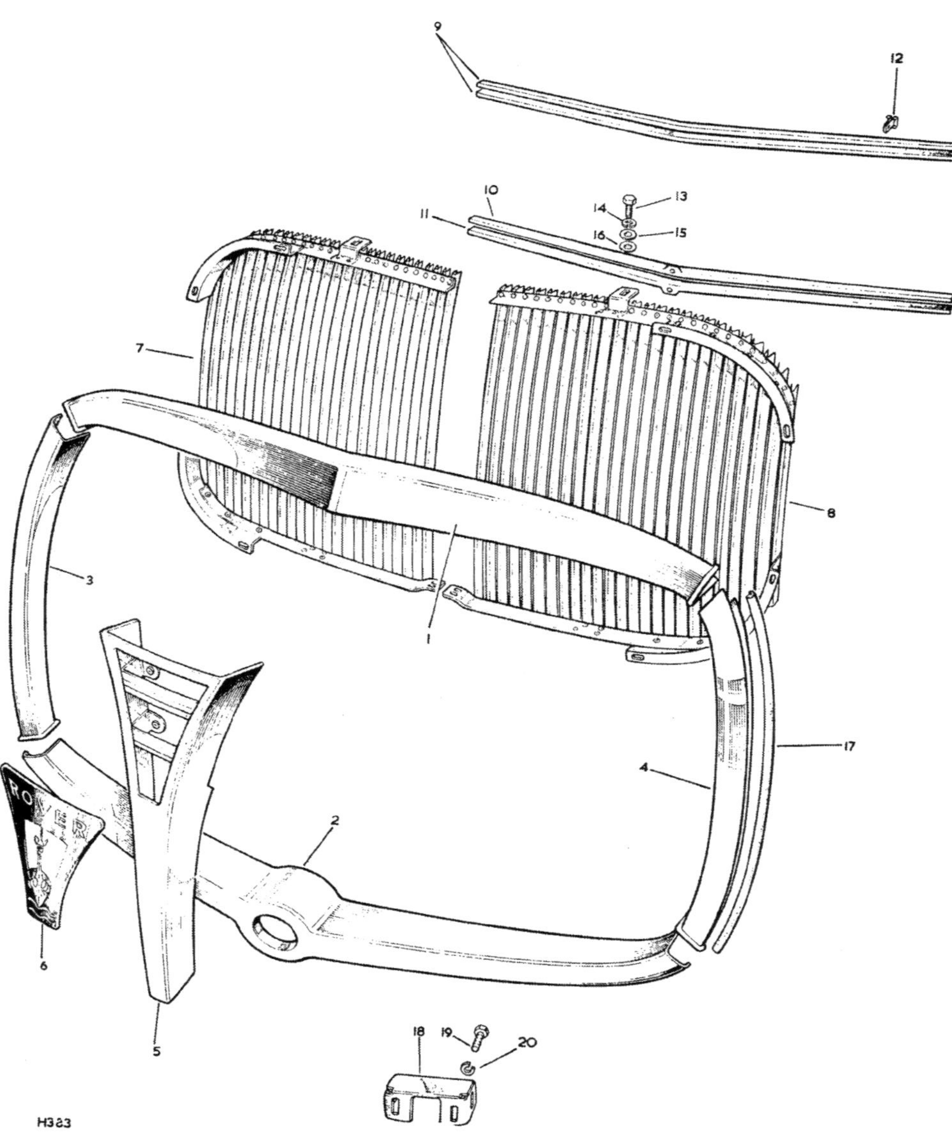

Plate Ref.	1 2 3 4 DESCRIPTION		Qty	Part No.	REMARKS
	FRONT END GRILLE COMPLETE		1	384918	
	OUTER FRAME ASSEMBLY		1	384939	
1	Top motif		1	384941	
2	Bottom motif		1	383722	
3	Side motif, RH		1	383723	
4	Side motif, LH		1	383724	
5	CENTRE MOTIF ASSEMBLY		1	384994	
	Gasket for badge		1	384991	
6	Badge for centre motif		1	384989	
7	Rib frame assembly, RH		1	384944	
8	Rib frame assembly, LH		1	384945	
9	Transverse rib, top		2	384938	
10	Transverse rib, bottom, upper		1	357433	
11	Transverse rib, bottom, lower		1	357431	
12	Clip securing transverse ribs		10	381239	
13	Set bolt (¼″ UNF x ⅝″ long)	Fixing grille to body	4	255206	
14	Spring washer		4	3074	
15	Plain washer		4	3900	
16	Shim washer (top)		4	3524	
17	Plastic beading, grille to front wing		2	357342	
18	Bracket, grille to surround		2	553886	
19	Set bolt (¼″ UNF x ½″ long)	Fixing bracket to grille surround	4	255204	
20	Spring washer		4	3074	

* Asterisk indicates a new part which has not been used on any previous Rover model

FRONT GRILLE AND MOTIFS 3½ LITRE

1759

255206 (2)
3074 (2)
3900 (2)
3524 (6)
387415 (1)
383723 (1)
Rib Frame Assembly RH 384944
384938 (4)
381239 (12)
Rib Frame Assembly LH 384945
386256 (1)
Outer Frame assembly 385982
386268 (1)
385984 (1)
553886 (2)
386485 (2)
255204 (4)
3074 (4)
383724 (1)
3900 (2)
3074 (2)
255208 (2)

REMARKS

Applicable to 3½ Litre Saloon and Coupe models
Front Grille Complete 385987

This page is intentionally left bank

1760

F 579

Plate Ref.	1 2 3 4 Description		Qty.	Part No.	Remarks	
	Front wing top and side, RH		1	351048		
1	Front wing top and side, LH		1	351049		
2	Support bracket		2	351185		
	Set bolt (1/4" UNF x 3/4" long)	Fixing wings to body	6	255207		
	Bolt (1/4" UNF x 1/2" long)	Fixing wings to body	12	255204		
	Set bolt (1/4" UNF x 1/2" long)	Fixing wings to body	8	255204		
	Set bolt (1/4" UNF x 5/8" long)	Fixing wings to body	2	255206		
	Plain washer	Fixing wings to body	40	3840		
	Shakeproof washer	Fixing wings to body	28	70884		
	Nut (1/4" UNF)	Fixing wings to body	12	254810		
3	Sealing rubber for front wing		2	352185		
4	Moulding for front wing		2	352413	Brass chrome plated	Alternatives. Check before ordering. Mk I, Mk IA and Mk II
	Moulding for front wing, RH		1	356353	Stainless steel	Alternatives. Check before ordering. Mk I, Mk IA and Mk II
	Moulding for front wing, LH		1	356354	Stainless steel	Alternatives. Check before ordering. Mk I, Mk IA and Mk II
	Clip	Fixing moulding to front wing	10	315974	12 off with stainless steel moulding	Mk I, Mk IA and Mk II
	Pop rivet	Fixing moulding to front wing	10	78257	12 off with stainless steel moulding	Mk I, Mk IA and Mk II
	Moulding for front wing, RH		1	384824	Mk III	
	Moulding for front wing, LH		1	384825	Mk III	
	Clip	Fixing moulding to front wing	12	384832	Mk III	
	Pop rivet	Fixing moulding to front wing	12	78704	Mk III	
	Finisher for front wing moulding, RH		1	352326	Brass chrome plated	Alternatives. Check before ordering
5	Finisher for front wing moulding, LH		1	352327	Brass chrome plated	Alternatives. Check before ordering
	Finisher for front wing moulding, RH		1	383279	Stainless steel screw fixing	Alternatives. Check before ordering
	Finisher for front wing moulding, LH		1	383280	Stainless steel screw fixing	Alternatives. Check before ordering
	Screw (6 UNC x 1/2" long)	Fixing finisher to wing	2	78299	For screw-type fixing finishers	Mk I, Mk IA and Mk II
	Plain washer	Fixing finisher to wing	4	4203	For screw-type fixing finishers	Mk I, Mk IA and Mk II
	Spring washer	Fixing finisher to wing	4	3072	For screw-type fixing finishers	Mk I, Mk IA and Mk II
	Nut (6 UNC)	Fixing finisher to wing	4	257203	For screw-type fixing finishers	Mk I, Mk IA and Mk II
	Polythene washer	Fixing finishers to wing	2	383281	For stud fixing finishers	Mk I, Mk IA and Mk II
	Brass nut (6 UNC)	Fixing finishers to wing	2	247239	For stud fixing finishers	Mk I, Mk IA and Mk II
	Motif, 'Mk III', for front wings		2	384821	Mk III	
	Friction bush fixing motif		4	359526	Mk III	
	Name plate, 'Rover', for front wings		2	352350	NADA	
	Nut (4 UNC)	Fixing name plate to front wings	8	257299	NADA	
	Plain washer	Fixing name plate to front wings	8	4032	NADA	
	Spring washer	Fixing name plate to front wings	8	3071	NADA	
6	Bonnet panel complete		1	381354		
7	Insulation pad for bonnet		1	355809		
8	Retaining wire for pad		2	355810		
9	Centre retaining wire for pad		1	355997		
	Anti-rattle tube		1	355999		
	Clip fixing retaining wire		3	355811		

*Asterisk indicates a new part which has not been used on any previous model.

F 579

Plate Ref.	Description (1 2 3 4)		Qty.	Part No.	Remarks
	Hinge for bonnet, RH		1	350948	
10	Hinge for bonnet, LH		1	350949	
	Bolt (5/16" UNF x 3/8" long)	Fixing hinge to body and bonnet	8	255024	
	Special set bolt	Fixing hinge to body and bonnet	4	78352	
	Plain washer	Fixing hinge to body and bonnet	14	3830	
	Shakeproof washer	Fixing hinge to body and bonnet	12	70822	
	Nut (5/16" UNF)	Fixing hinge to body and bonnet	2	254831	
	Protection strip for bonnet hinges		2	383490	
11	Spring for bonnet hinge		2	350972	
12	Lock for bonnet		1	350951	
	Set bolt (1/4" UNF x 1/2" long)	Fixing lock to bonnet	2	255204	
	Plain washer	Fixing lock to bonnet	2	3840	
	Shakeproof washer	Fixing lock to bonnet	2	70884	
13	Striker plate for bonnet lock		1	350343	
14	Slider catch for bonnet lock		1	350989	
15	Support for slider catch		1	350943	
	Nut (1/4" UNF)	Fixing striker plate, slider catch and support to body	4	254810	
	Plain washer	Fixing striker plate, slider catch and support to body	2	3840	
	Shakeproof washer	Fixing striker plate, slider catch and support to body	2	70884	
16	Spring for slider catch		1	356589	
17	Safety catch for bonnet		1	350945	
18	Spring	For safety catch	1	350946	
19	Pin	For safety catch	1	350941	
20	Split pin	For safety catch	1	2388	
21	Sealing rubber for rear of bonnet		1	352190	
22	Rubber buffer for bonnet side		8	311278	
23	BONNET LAMP		2	570827	Coupe
24	Gasket for lens		2	600457	Coupe
25	Bulb for bonnet lamp		2	GLB209	Coupe
26	Lens for bonnet lamp		2	600456	Coupe
27	Screw fixing lamp to underside of bonnet		4	78129	Coupe
28	Bracket for bonnet light switch at RH body bolt adjacent to hinge		1	555713	Coupe
29	Switch for bonnet light		1	541670	Coupe
	Drive screw	Fixing switch to bracket	1	78483	Coupe
	Fan disc washer	Fixing switch to bracket	1	536993	Coupe
30	Bracket for bonnet lock control		1	352154	
	Bolt (10 UNF x 1/2" long)	Fixing bracket to parcel shelf and channel	4	257017	
	Plain washer	Fixing bracket to parcel shelf and channel	6	3816	
	Spring washer	Fixing bracket to parcel shelf and channel	4	3073	
	Nut (10 UNF)	Fixing bracket to parcel shelf and channel	2	257023	
31	Control cable complete for bonnet catch		1	352508@	

@ Cars numbered up to 625900195, 626900004, 628900001, 630900067, 631900002, 633900002

*Asterisk indicates a new part which has not been used on any previous model.

1764

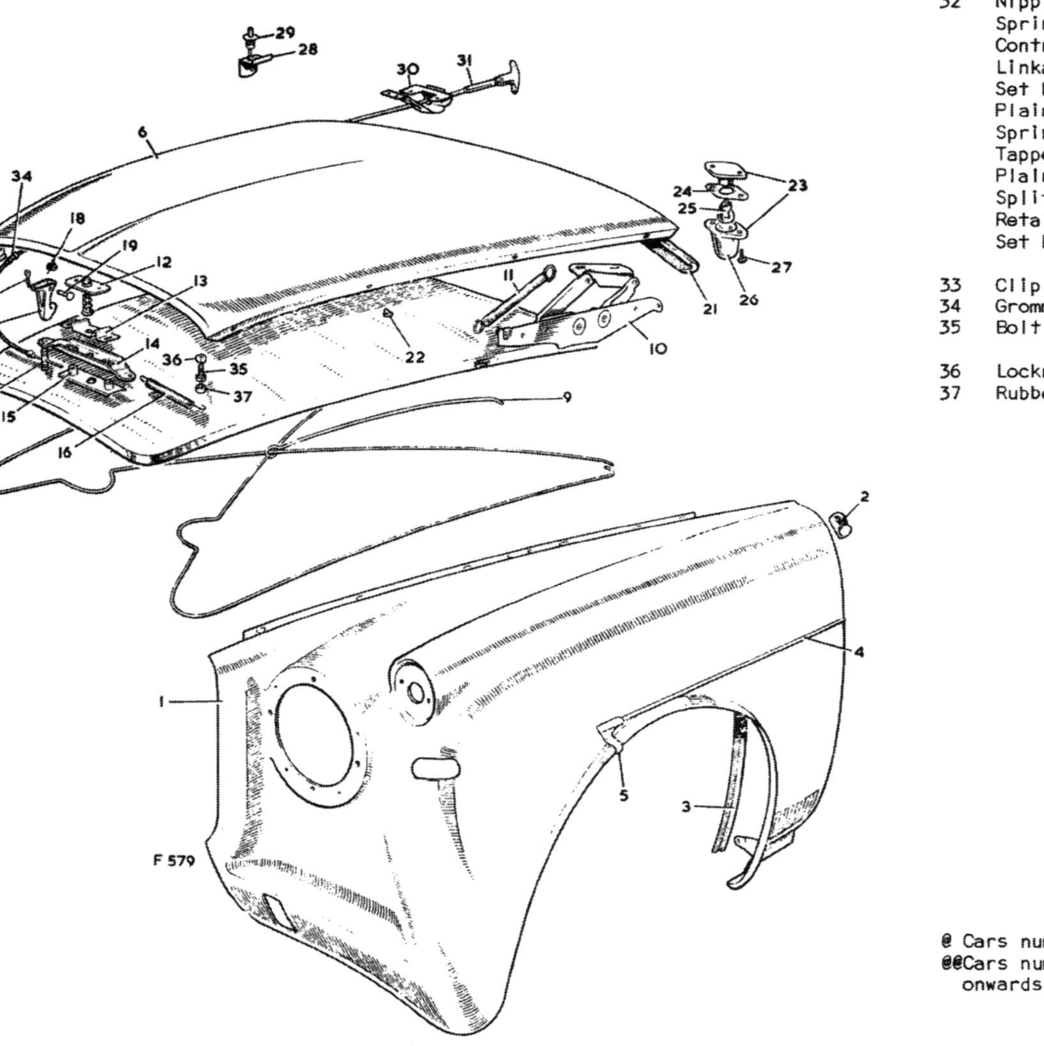

Plate Ref.	1 2 3 4 Description	Qty.	Part No.	Remarks
32	Nipple for control cable	1	352600@	
	Spring washer for nipple	1	3073@	
	Control cable complete for bonnet catch	1	352940@@	
	Linkage complete for bonnet lock	1	352928@@	
	Set bolt (8 UNC x $\frac{3}{8}$" long) ⎤ Fixing linkage	2	257123@@	
	Plain washer ⎥ to body	2	3902@@	
	Spring washer ⎥ at grille	2	3073@@	
	Tapped plate ⎦ surround	1	356215@@	
	Plain washer ⎤ Fixing linkage to	1	3902@@	
	Split pin ⎦ bonnet lock slider catch	1	3858@@	
	Retainer for bonnet control cable	1	356225@@	
	Set bolt (6 UNC x 5/16" long) fixing control cable to retainer	1	257194@@	
33	Clip ⎤ For bonnet	2	214228	
34	Grommet ⎦ control cable	2	214229	
35	Bolt (5/16" UNF x $1\frac{1}{4}$" long) for bonnet support buffer	2	255234	
36	Locknut (5/16" UNF) for support bolts	2	254831	
37	Rubber buffer for bonnet bolts	2	310591	

@ Cars numbered up to 625900195, 626900004, 628900001, 630900067, 631900002, 633900002
@@ Cars numbered from 625900196, 626900005, 628900002, 630900068, 631900003, 633900003 onwards

*Asterisk indicates a new part which has not been used on any previous model.

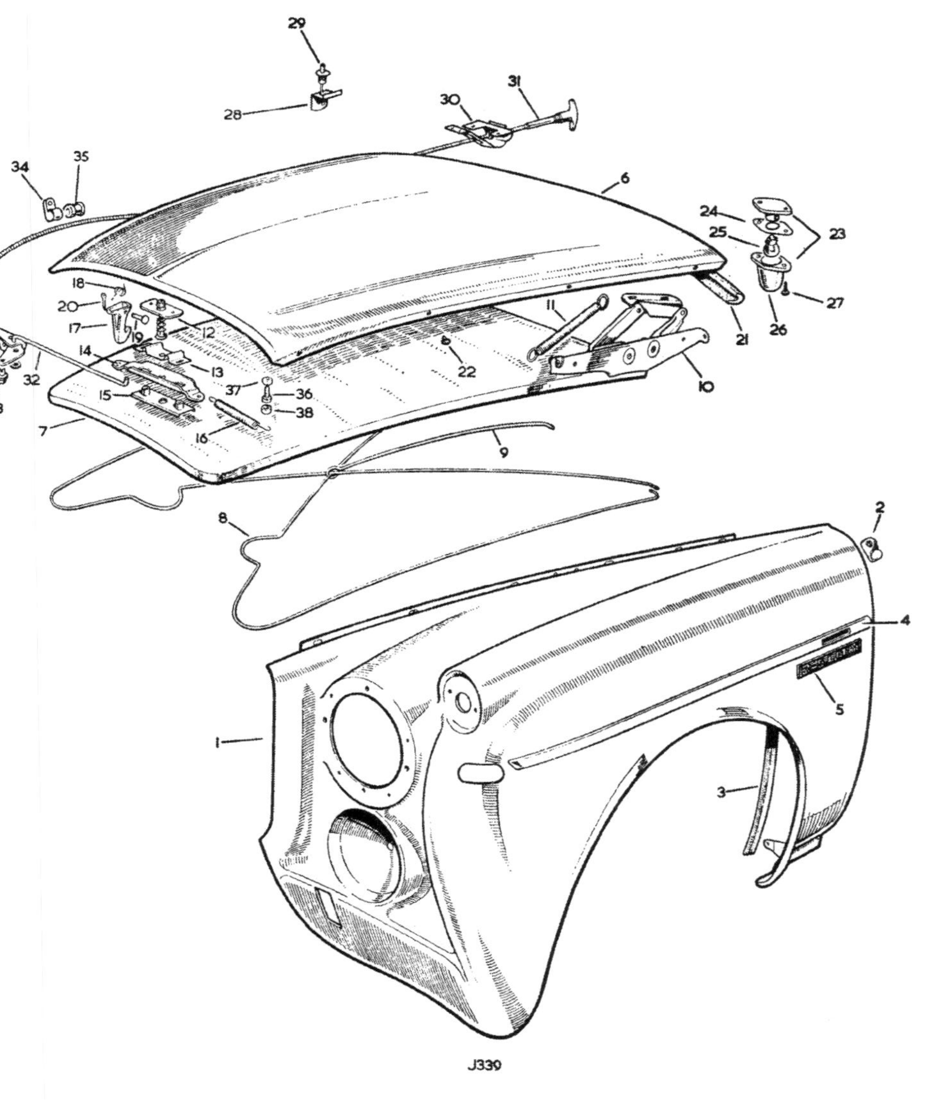

Plate Ref.	1 2 3 4	Description	Qty.	Part No.	Remarks
		Front wing top and side, RH	1	387083	
1		Front wing top and side, LH	1	387084	
2		Support bracket } Fixing wings to body	2	351185	
		Set bolt (¼" UNF x ¾" long) } Fixing wings to body	6	255207	
		Bolt (¼" UNF x ½" long) } Fixing wings to body	12	255204	
		Set bolt (¼" UNF x ½" long) } Fixing wings to body	8	255204	
		Set bolt (¼" UNF x ⅜" long) } Fixing wings to body	2	255206	
		Plain washer } Fixing wings to body	40	3840	
		Shakeproof washer } Fixing wings to body	28	70884	
		Nut (¼" UNF) } Fixing wings to body	12	254810	
3		Sealing rubber for front wing	2	352185	
		Moulding for front wing, RH	1	386271	Except Italy
4		Moulding for front wing, LH	1	386272	Except Italy
		Moulding for front wing, RH	1	384824	Italy
		Moulding for front wing, LH	1	384825	Italy
		Clip } Fixing moulding to front wing	12	384832	
		Pop rivet } Fixing moulding to front wing	12	78704	
5		Motif, '3.5- LITRE', for front wing	2	386009	
		Friction bush fixing motif	As reqd	359526	
6		Bonnet panel complete	1	381354	
7		Insulation pad for bonnet	1	355809	
8		Retaining wire for pad	2	355810	
9		Centre retaining wire for pad	1	355997	
		Anti-rattle tube	1	355999	
		Clip fixing retaining wire	3	355811	
		Hinge for bonnet, RH	1	350948	
10		Hinge for bonnet, LH	1	350949	
		Bolt (5/16" UNF x ⅝" long) } Fixing hinge to body and bonnet	8	255024	
		Special set bolt } Fixing hinge to body and bonnet	4	78352	
		Plain washer } Fixing hinge to body and bonnet	14	3830	
		Shakeproof washer } Fixing hinge to body and bonnet	12	70822	
		Nut(5/16" UNF) } Fixing hinge to body and bonnet	2	254831	
		Protection strip for bonnet hinges	2	383490	
11		Spring for bonnet hinge	2	350972	
12		Lock for bonnet	1	350951	
		Set bolt (¼" UNF x ½" long) } Fixing lock to bonnet	2	255204	
		Plain washer } Fixing lock to bonnet	2	3840	
		Shakeproof washer } Fixing lock to bonnet	2	70884	
13		Striker plate for bonnet lock	1	350343	
14		Slider catch for bonnet lock	1	350989	
15		Support for slider catch	1	350943	
		Nut (¼" UNF) } Fixing striker plate, slider catch and support to body	4	254810	
		Plain washer } Fixing striker plate, slider catch and support to body	2	3840	
		Shakeproof washer } Fixing striker plate, slider catch and support to body	2	70884	
16		Spring for slider catch	1	356589	
17		Safety catch for bonnet	1	350945	
18		Spring } For safety catch	1	350946	
19		Pin } For safety catch	1	350941	
20		Split pin } For safety catch	1	2388	
21		Sealing rubber for rear of bonnet	1	352190	
22		Rubber buffer for bonnet side	8	311278	

*Asterisk indicates a new part which has not been used on any previous model.

Plate Ref.	1 2 3 4 Description	Qty.	Part No.	Remarks
23	BONNET LAMP ASSEMBLY	1	570827	Coupe
24	Gasket for lens	2	600457	Coupe
25	Bulb for bonnet lamp (Alternatives)	2	GLB209	Coupe
	Bulb for bonnet lamp (Alternatives)	2	570829	Coupe
26	Lens for bonnet lamp	2	600456	Coupe
27	Screw fixing lamp to underside of bonnet	4	78129	Coupe
28	Bracket for bonnet light switch at RH body bolt adjacent to hinge	1	555713	Coupe
29	Switch for bonnet light	1	559061	Coupe
	Drive screw (Fixing switch to bracket)	1	78483	Coupe
	Fan disc washer (Fixing switch to bracket)	1	536993	
30	Bracket for bonnet lock control	1	352154	
	Bolt (10 UNF x ½" long) (Fixing bracket to parcel shelf and channel)	4	257017	
	Plain washer (Fixing bracket to parcel shelf and channel)	6	3816	
	Spring washer (Fixing bracket to parcel shelf and channel)	4	3073	
	Nut (10 UNF) (Fixing bracket to parcel shelf and channel)	2	257023	
31	Control cable complete for bonnet catch	1	352940	
32	Linkage complete for bonnet lock	1	352928	
	Set bolt (8 UNC x ⅜" long) (Fixing linkage to body at grille surround)	2	257123	
	Plain washer (Fixing linkage to body at grille surround)	2	3902	
	Spring washer (Fixing linkage to body at grille surround)	2	3073	
	Tapped plate (Fixing linkage to body at grille surround)	1	356215	
	Plain washer (Fixing linkage to bonnet lock slider catch)	1	3902	
	Split pin (Fixing linkage to bonnet lock slider catch)	1	3858	
33	Retainer for bonnet control cable	1	356225	
	Set bolt (6 UNC x 5/16" long)fixing control cable to retainer	1	257194	
34	Clip (For bonnet control cable)	2	214228	
35	Grommet (For bonnet control cable)	2	214229	
36	Bolt (5/16" UNF x 1¾" long)for bonnet support buffer	2	255234	
37	Locknut (5/16" UNF)for support bolts	2	254831	
38	Rubber buffer for bonnet bolts	2	310591	

*Asterisk indicates a new part which has not been used on any previous model.

Plate Ref. 1 2 3 4	Description	Qty.	Part No.	Remarks
	Rear wing top and side, RH	1	351050	
1	Rear wing top and side, LH	1	351051	
2	Beading, rear wing to body, top	2	356761	
3	Beading, rear wing to body, lower	2	352202	
	Set bolt (5/16" UNF x 3/4" long)	18	255226	
	Bolt (5/16" UNF x 3/4" long)	4	255226	
	Bolt (1/4" UNF x 1/2" long)	2	255204	
	Plain washer, large	26	3830	
	Plain washer, small — Fixing wings and beading	4	3840	
	Shakeproof washer, large	22	70822	
	Shakeproof washer, small	2	70884	
	Nut (5/16" UNF)	4	254831	
	Nut (1/4" UNF)	2	254810	
	Beading clip	2	356764	
4	Moulding for rear wing	2	352297	Brass chrome plated — Alternative. Check before ordering — MkI, MkIA and MkII
	Moulding for rear wing	2	352965	Stainless steel
	Finisher for rear wing moulding, RH	1	352299	Brass chrome plated
5	Finisher for rear wing moulding, LH	1	352300	Brass chrome plated
	Finisher for rear wing moulding, RH	1	352966	Stainless steel
	Finisher for rear wing moulding, LH	1	352967	Stainless steel
	Plain washer	2	4203	
	Spring washer — Fixing finisher to rear wing	2	3072	
	Nut (6 UNC)	2	257203	
	Clip — Fixing moulding to rear wing	14	315974	
	Pop rivet	14	78257	
	Moulding for rear wing, RH	1	384830	Mk III
	Moulding for rear wing, LH	1	384831	Mk III
	Clip fixing moulding to wing	14	384832	Mk III
	Pop rivet	14	78704	Mk III
	Motif for rear wing	6	384811	Mk III
6	Boot lid complete	1	357109@	
	Boot lid complete	1	351642@@	
	Hinge for boot lid, RH	1	350977	
7	Hinge for boot lid, LH	1	350978	
	Special set bolt, boot lid	8	78353	
	Special set screw, body — Fixing hinges to boot lid and body	8	78355	
	Plain washer	16	3830	
	Shakeproof washer	16	70822	
8	Lock for boot lid	1	350960@	
	Special set bolt	2	257017@	
	Plain washer — Fixing lock to boot lid	2	3685@	
	Shakeproof washer	2	74236	
9	Striker for boot lid lock	1	350655@	
	Plain washer	4	3830@	
	Shakeproof washer — Fixing striker to boot floor	2	70822@	
	Nut (5/16" UNF)	4	254810@	
10	Boot lid handle, less barrel lock	1	358185	
	Barrel lock for boot lid handle, 'FS' series	1	600873	State key number when ordering
11	Packing washer for handle	1	352991@	

@Cars numbered 625100706, 626100155, 628100055, 630100438, 631100208, 633100017

@@Cars numbered625100707, 626100156, 628100056, 630100439, 631100209, 633100018 onwards

*Asterisk indicates a new part which has not been used on any previous model.

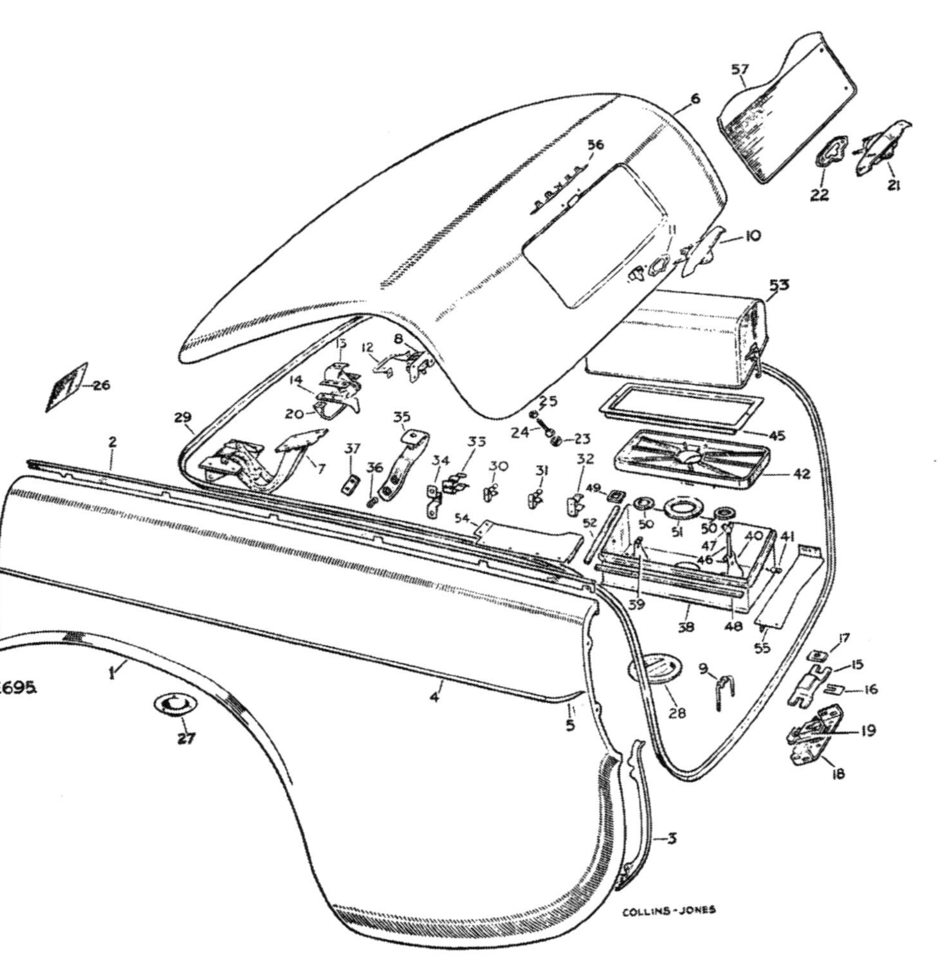

Plate Ref.	1 2 3 4 DESCRIPTION	Qty	Part No.	REMARKS
12	Securing bracket } Fixing handle to boot lid	1	352595†	
	Plain washer	2	3685†	
	Spring washer	2	3073†	
	Nut (10 UNF)	2	257023†	
13	Lock for boot lid	1	357344††	
	Screw (10 UNF x 5/8" long) } Fixing lock to boot lid	1	257019††	
	Screw (10 UNF x 1/2" long)	3	257017††	
	Plain washer	4	2225††	
	Spring washer	4	3073††	
14	Dovetail } For boot lid lock	1	356835††	
15	Striker plate	1	312018††	
16	Shim } Fixing stop bracket dovetail and striker plate	As reqd	312908††	
17	Washer plate	2	356836††	
	Spring washer	2	3074††	
	Plain washer	2	3831††	
	Nut (1/4" UNF)	2	254810††	
18	Mounting bracket, dovetail striker to boot lid	1	356837††	
	Screw (10 UNF) } Fixing mounting bracket	4	257019††	
	Plain washer	8	3886††	
	Spring washer	4	3073††	
	Nut (10 UNF)	4	257023††	
19	Dovetail striker, boot lid	1	356838††	
	Spring washer } Fixing dovetail striker	2	3074††	
	Nut (1/4" UNF)	2	254810††	
20	Securing bracket	1	356937††	
21	Boot lid handle, less barrel lock	1	357111††	
22	Sealing washer	1	356834††	
	Bolt (10 UNF x 1 3/4" long) } Fixing handle and lock to boot lid	1	257301††	
	Bolt (10 UNF x 2 1/8" long)	1	257302††	
	Distance tube	2	356905††	
	Plain washer	3	3902††	
	Spring washer	4	3073††	
	Nut (10 UNF)	2	257023††	
23	Rubber buffer for boot lid stop	2	352989	1959
	Rubber buffer for boot lid stop	2	310591	1960 onwards
24	Bolt (1/4" UNF x 3/4" long) } Fixing rubber buffer	2	73858	1959
	Bolt (1/4" UNF x 1 1/4" long)	2	255211	1960 onwards
25	Locknut (1/4" UNF)	2	254860	
26	Cover for shock absorber fixings	2	352378	
	Drive screw fixing covers	8	78140	
27	Cover for petrol tank drainage hole	1	350719	
	Spring washer } Fixing cover to boot floor	3	3073	
	Set bolt (10 UNF x 3/8" long)	3	257015	
28	Grommet covering spare wheel valve	1	305016	
29	Sealing rubber for boot aperture	1	357269	
30	Spring clip for jack	1	315031	
31	Spring clip for starting handle and wheelbrace	5	508035	
32	Spring clip for pump and wheelbrace	4	72085	
33	Clip for jack	1	316609	
34	Support bracket for clip	1	356384	
35	Strap retaining jack	1	356385	

* Asterisk indicates a new part which has not been used on any previous Rover model
† Cars numbered up to 625100706, 626100155, 628100055, 630100438, 631100208, 633100017
†† Cars numbered from 625100707, 626100156, 628100056, 630100439, 631100209, 633100018 onwards

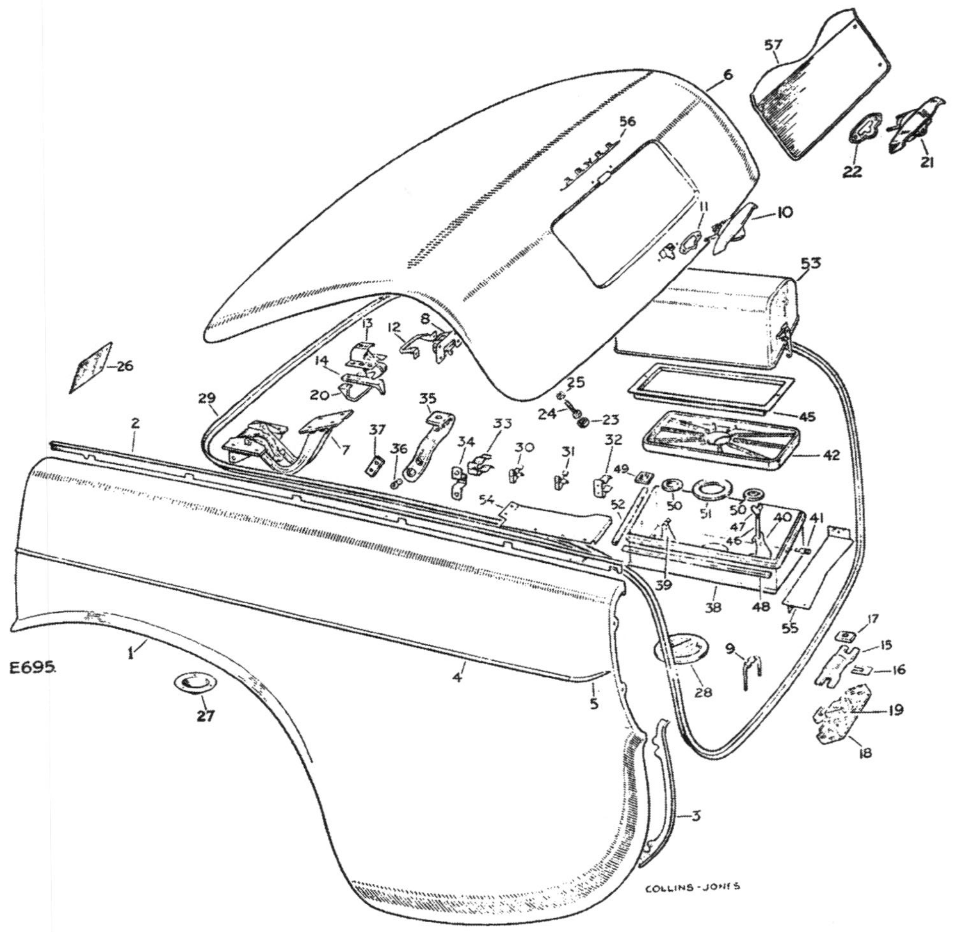

Plate Ref.	1 2 3 4 DESCRIPTION		Qty	Part No.	REMARKS
	Woodscrew	Fixing strap	1	20147	
	Plain washer	Fixing strap	1	3557	
36	Stud for strap		1	320345	
	Woodscrew fixing stud		1	20150	
37	Strap retaining wheel cover tool		1	357364	Mk IA, Mk II and Mk III
	Bolt (10 UNF x ½" long) fixing strap		2	257019	Mk IA, Mk II and Mk III
	Bolt (10 UNF x ⅝" long)	Fixing clips, tool stowage panel to rear wheel arch and strap to tool stowage panel	20	257019	17 off on 1960 models
	Plain washer, small	Fixing clips, tool stowage panel to rear wheel arch and strap to tool stowage panel	16	3902	17 off on 1960 models
	Plain washer, large	Fixing clips, tool stowage panel to rear wheel arch and strap to tool stowage panel	3	3852	17 off on 1960 models
	Spring washer	Fixing clips, tool stowage panel to rear wheel arch and strap to tool stowage panel	18	3073	17 off on 1960 models
	Nut (10 UNF)	Fixing clips, tool stowage panel to rear wheel arch and strap to tool stowage panel	18	257023	17 off on 1960 models
	Drive screw	Fixing tool stowage panel to body	1	69310	
	Plain washer	Fixing tool stowage panel to body	1	3902	
	Bolt (10 UNF x 1⅞" long)	Fixing jack support blocks to tool stowage panel	2	253871	
	Bolt (10 UNF x 1½" long)	Fixing jack support blocks to tool stowage panel	1	253869	
38	Base for battery box		1	512835†	
	Base for battery box		1	542367††	
39	Retaining bracket for battery, front		1	502864†	
	Retaining bracket for battery, front		1	542365††	
40	Retaining bracket for battery, rear		1	502863†	
	Retaining bracket for battery, rear		1	542364††	
	Spring washer	Fixing bracket to base	2	3074	
	Plain washer	Fixing bracket to base	2	2224	
	Nut (¼" UNF)	Fixing bracket to base	2	254810	
	Bolt (5/16" UNF x ⅝" long)	Fixing bracket and base to boot floor	4	255225	
	Plain washer	Fixing bracket and base to boot floor	8	2550	
	Spring washer	Fixing bracket and base to boot floor	4	3075	
	Nut (5/16" UNF)	Fixing bracket and base to boot floor	4	254811	
	Bolt (¼" UNF x 7/16" long)	Fixing earthing eyelet to boot floor	1	255205	
	Special earthing washer	Fixing earthing eyelet to boot floor	2	510170	
	Nut (¼" UNF)	Fixing earthing eyelet to boot floor	1	254810	
41	Pin for battery cover catch		2	502859†	
	Pin for battery cover catch		2	542362††	
42	Fibre glass tray for battery		1	515378†	
	Fibre glass tray for battery		1	542366††	
45	Clamping frame for battery		1	525276	
46	Special bolt	Fixing battery clamp	2	502860	
	Plain washer	Fixing battery clamp	2	3840	
47	Wing nut	Fixing battery clamp	2	252180	
48	Seal for battery box, long		2	524923	
49	Sealing (square) for earthing tag, battery box base to body floor		1	525519	
50	Sealing ring, small, battery box base to floor ..		2	524921†	
51	Sealing ring, large, battery box base to floor ..		1	524922	
52	Seal for battery box, short		2	524924	
53	Top cover complete for battery		1	542374	
54	FILLER PIECE FOR BOOT FLOOR, FRONT RH ..		1	354942	
	Plastic beading strip for cover		1	524930	
55	Filler piece for boot floor, rear RH		1	354948	
	Plain washer	Fixing filler pieces to boot floor	10	3816	
	Drive screw	Fixing filler pieces to boot floor	10	77704	

* Asterisk indicates a new part which has not been used on any previous Rover model
† Mk I, Mk IA and Mk II cars with suffix letters 'A' and 'B'
†† Mk II cars with suffix letter 'C' onwards

Plate Ref.	1 2 3 4 DESCRIPTION	Qty	Part No.	REMARKS
56	Name plate, 'Rover'	1	352350	Mk I and Mk IA
	Plain washer (Fixing name plate)	4	4032	Mk I and Mk IA
	Spring washer (Fixing name plate)	4	3071	Mk I and Mk IA
	Nut (4 UNC) (Fixing name plate)	4	257299	Mk I and Mk IA
	Single letter 'R' (For name plate on boot lid)	2	359238	Mk II
	Single letter 'O' (For name plate on boot lid)	1	359239	Mk II
	Single letter 'V' (For name plate on boot lid)	1	359240	Mk II
	Single letter 'E' (For name plate on boot lid)	1	359241	Mk II
	Motif, '3', for boot lid	1	359593	Mk II
	Motif, 'Litre', for boot lid	1	359592	Mk II
	Motif, 'Automatic', for boot lid	1	359174	Automatic; Mk II
	Motif, 'Coupé', for boot lid	1	359549	Coupé; Mk II
	Motif, 'Automatic', for boot lid	1	367571	Mk III
	Motif, 'Mk III', for boot lid	1	384821	Mk III
	Plastic fixings for name plate and motifs	As reqd	359526	
57	Mounting plate for rear number plate	1	357665	Except NADA and Australia
	Mounting plate for rear number plate	1	383469	Mk II and Mk III NADA and Australia
	Screw (8 UNC x ⅝" long) (Fixing mounting plate to boot lid)	4	78378	
	Plain washer, large (Fixing mounting plate to boot lid)	4	2983	
	Plain washer, small (Fixing mounting plate to boot lid)	4	4034	
	Spring washer (Fixing mounting plate to boot lid)	4	3073	
	Nut (8 UNC) (Fixing mounting plate to boot lid)	4	257143	

* Asterisk indicates a new part which has not been used on any previous Rover model
NADA indicates parts peculiar to cars exported to the North American dollar area

Plate Ref.	1 2 3 4 Description		Qty.	Part No.	Remarks
	Rear wing top and side, RH		1	351050	
1	Rear wing top and side, LH		1	351051	
2	Beading, rear wing to body, top		2	356761	
3	Beading, rear wing to body, lower		2	352202	
	Set bolt (5/16" UNF x 3/4" long)	Fixing wings and beading	18	255226	
	Bolt (5/16" UNF x 3/4" long)	Fixing wings and beading	4	255226	
	Bolt (1/4" UNF x 1/2" long)	Fixing wings and beading	2	255204	
	Plain washer, large	Fixing wings and beading	26	3830	
	Plain washer, small	Fixing wings and beading	4	3840	
	Shakeproof washer, large	Fixing wings and beading	22	70822	
	Shakeproof washer, small	Fixing wings and beading	2	70884	
	Nut (5/16" UNF)	Fixing wings and beading	4	254831	
	Nut 1/4" UNF)	Fixing wings and beading	2	254810	
	Beading clip	Fixing wings and beading	2	356764	
	Moulding for rear wing, RH		1	386006	Except Italy
4	Moulding for rear wing, LH		1	386007	Except Italy
	Moulding for rear wing RH		1	386316*	Italy
	Moulding for rear wing LH		1	386317*	Italy
	Clip fixing moulding to wing		14	384832	
	Pop rivet		14	78704	
5	Boot lid complete		1	351642	
	Hinge for boot lid, RH		1	350977	
6	Hinge for boot lid, LH		1	350978	
	Special set bolt, boot lid	Fixing hinges to boot lid and body	8	78353	
	Special set screw, body	Fixing hinges to boot lid and body	8	78355	
	Plain washer	Fixing hinges to boot lid and body	16	3830	
	Shakeproof washer	Fixing hinges to boot lid and body	16	70822	
7	Lock for boot lid		1	357344	
	Screw (10 UNF x 3/8" long)	Fixing lock to boot lid	1	257019	
	Screw (10 UNF x 1/2" long)	Fixing lock to boot lid	3	257017	
	Plain washer	Fixing lock to boot lid	4	2225	
	Spring washer	Fixing lock to boot lid	4	3073	
8	Dovetail	For boot lid lock	1	356835	
9	Striker plate	For boot lid lock	1	312018	
10	Shim		As reqd	312908	
11	Washer plate	Fixing stop bracket dovetail and striker plate	2	356836	
	Spring washer	Fixing stop bracket dovetail and striker plate	2	3074	
	Plain washer	Fixing stop bracket dovetail and striker plate	2	3831	
	Nut (1/4" UNF)	Fixing stop bracket dovetail and striker plate	2	254810	
12	Mounting bracket, dovetail striker to boot lid		1	356837	
	Screw (10 UNF)	Fixing mounting bracket	4	257019	
	Plain washer	Fixing mounting bracket	8	3886	
	Spring washer	Fixing mounting bracket	4	3073	
	Nut (10 UNF)	Fixing mounting bracket	4	257023	
13	Dovetail striker, boot lid		1	356838	
	Spring washer	Fixing dovetail striker	2	3074	
	Nut (1/4" UNF)	Fixing dovetail striker	1	254810	
14	Securing bracket		1	356937	

*Asterisk indicates a new part which has not been used on any previous model.

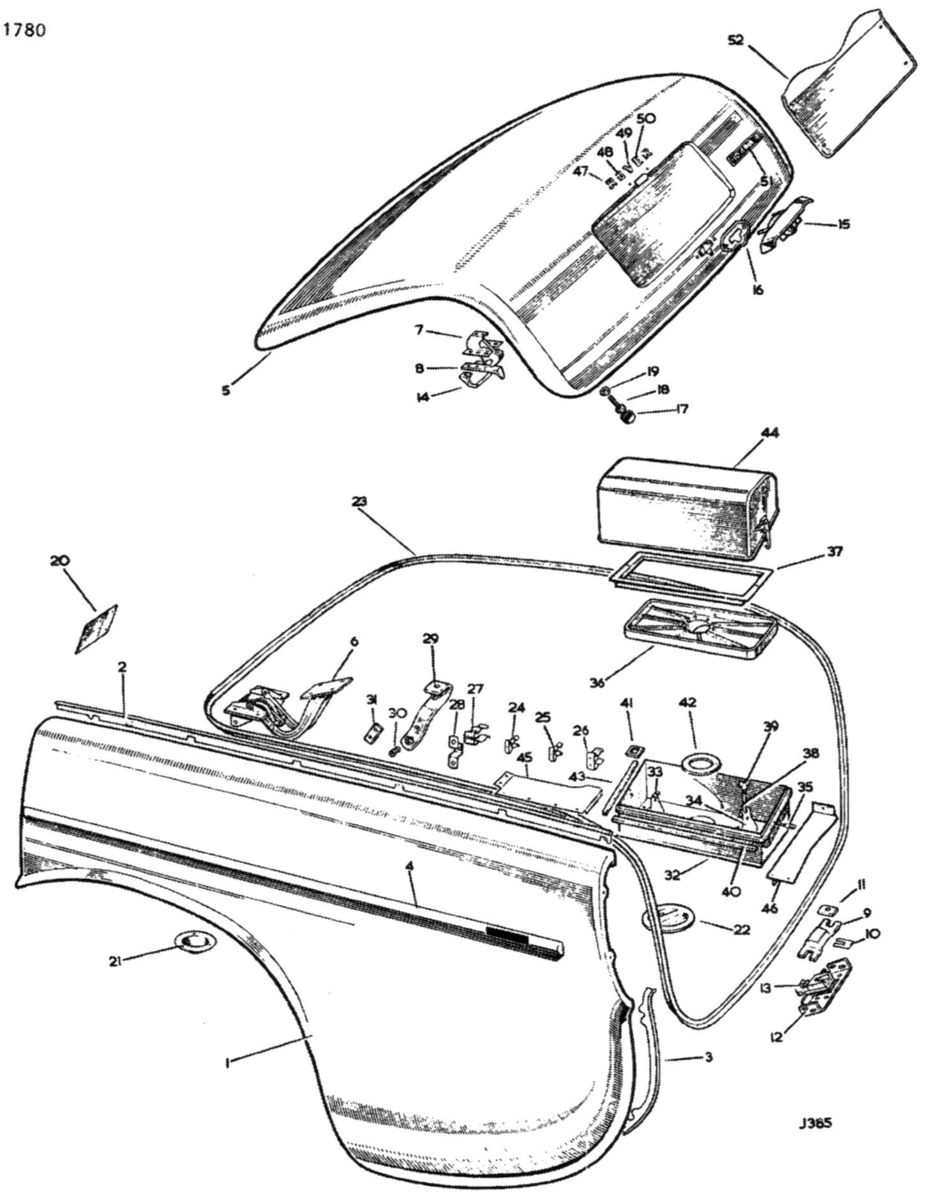

Plate Ref.	1 2 3 4 DESCRIPTION	Qty	Part No.	REMARKS
15	Boot lid handle, less barrel lock	1	357111	
	Barrel lock for handle, FS series	1	600873	
16	Sealing washer	1	356834	
	Bolt (10 UNF x $1\frac{3}{4}$" long) } Fixing handle and lock to boot lid	1	257301	
	Bolt (10 UNF x $2\frac{1}{8}$" long) } Fixing handle and lock to boot lid	1	257302	
	Distance tube } Fixing handle and lock to boot lid	2	356905	
	Plain washer } Fixing handle and lock to boot lid	3	3902	
	Spring washer } Fixing handle and lock to boot lid	4	3073	
	Nut (10 UNF) } Fixing handle and lock to boot lid	2	257023	
17	Rubber buffer for boot lid stop	2	310591	
18	Bolt ($\frac{1}{4}$" UNF x $1\frac{1}{4}$" long) } Fixing rubber buffer	2	255211	
19	Locknut ($\frac{1}{4}$" UNF) } Fixing rubber buffer	2	254860	
20	Cover for shock absorber fixings	2	352378	
	Drive screw fixing covers	8	78140	
21	Cover for fuel tank drainage hole	1	350719	
	Spring washer } Fixing cover to boot floor	3	3073	
	Set bolt (10 UNF x $\frac{3}{8}$" long) } Fixing cover to boot floor	3	257015	
22	Grommet covering spare wheel valve	1	305016	
23	Sealing rubber for boot aperture	1	357269	
24	Spring clip for jack	1	315031	
25	Spring clip for starting handle and wheelbrace ..	5	508035	
26	Spring clip for pump and wheelbrace	4	72085	
27	Clip for jack	1	316609	
28	Support bracket for clip	1	356384	
29	Strap retaining jack	1	356385	
	Woodscrew } Fixing strap	1	20147	
	Plain washer } Fixing strap	1	3557	
30	Stud for strap	1	320345	
	Woodscrew fixing stud	1	20150	
31	Strap retaining wheel cover tool	1	357364	
	Bolt (10 UNF x $\frac{7}{8}$" long) fixing strap	2	257019	
	Bolt (10 UNF x $\frac{5}{8}$" long) } Fixing clips, tool stowage panel to rear wheel arch and strap to tool stowage panel	20	257019	
	Plain washer, small } Fixing clips, tool stowage panel to rear wheel arch and strap to tool stowage panel	16	3902	
	Plain washer, large } Fixing clips, tool stowage panel to rear wheel arch and strap to tool stowage panel	3	3852	
	Spring washer } Fixing clips, tool stowage panel to rear wheel arch and strap to tool stowage panel	18	3073	
	Nut (10 UNF) } Fixing clips, tool stowage panel to rear wheel arch and strap to tool stowage panel	18	257023	
	Drive screw } Fixing tool stowage panel to body	1	69310	
	Plain washer } Fixing tool stowage panel to body	1	3902	
	Bolt (10 UNF x $1\frac{7}{8}$" long) } Fixing jack support blocks to tool stowage panel	2	253871	
	Bolt (10 UNF x $1\frac{1}{2}$" long) } Fixing jack support blocks to tool stowage panel	1	253869	
32	Base for battery box	1	542367	
33	Retaining bracket for battery, front	1	542365	
34	Retaining bracket for battery, rear	1	542364	
	Spring washer } Fixing bracket to base	2	3074	
	Plain washer } Fixing bracket to base	2	2224	
	Nut ($\frac{1}{4}$" UNF) } Fixing bracket to base	2	254810	

* Asterisk indicates a new part which has not been used on any previous Rover model

Plate Ref.	1 2 3 4 Description		Qty.	Part No.	Remarks
	Bolt (5/16" UNF x ⅝" long)	Fixing bracket and base to boot floor	4	255225	
	Plain washer		8	2550	
	Spring washer		4	3075	
	Nut (5/16" UNF)		4	254831	
	Bolt (¼" UNF x 9/16"long)	Fixing earthing eyelet to boot floor	1	255205	
	Special earthing washer		2	510170	
	Nut (¼" UNF)		1	254810	
35	Pin for battery cover catch		2	542362	
36	Fibre glass tray for battery		1	542366	
37	Clamping frame for battery		1	525276	
38	Special bolt	Fixing battery clamp	2	570120	
	Plain washer		2	3840	
39	Wing nut		2	252180	
40	Seal for battery box, long		2	524923	
41	Seal (square) for earthing tag, battery box base to body floor		1	525519	
42	Sealing ring, large, battery box base to floor		1	524922	
43	Seal for battery box, short		2	524924	
44	Top cover complete for battery		1	542374	
45	FILLER PIECE FOR BOOT FLOOR, FRONT RH		1	354942	
	Plastic beading strip for cover		1	524930	
46	Filler piece for boot floor, rear RH		1	354948	
	Plain washer	Fixing filler pieces to boot floor	10	3816	
	Drive screw		10	77704	
47	Letter 'R'	For name plate on boot lid	2	368558	
48	Letter 'O'		1	368559	
49	Letter 'V'		1	368560	
50	Letter 'E'		1	368561	
51	3.5. LITRE motif for boot lid		1	386009	
	Plastic fixings for name plate and motifs		As reqd	359526	
52	Mounting plate for rear number plate		1	357665	Except Canada
	Mounting plate for rear number plate		1	383469	Canada
	Screw (8 UNC x ⅝" long)	Fixing mounting plate to boot lid	4	78378	
	Plain washer, large		4	2983	
	Plain washer, small		4	4034	
	Spring washer		4	3073	
	Nut (8 UNC)		4	257143	

*Asterisk indicates a new part which has not been used on any previous model.

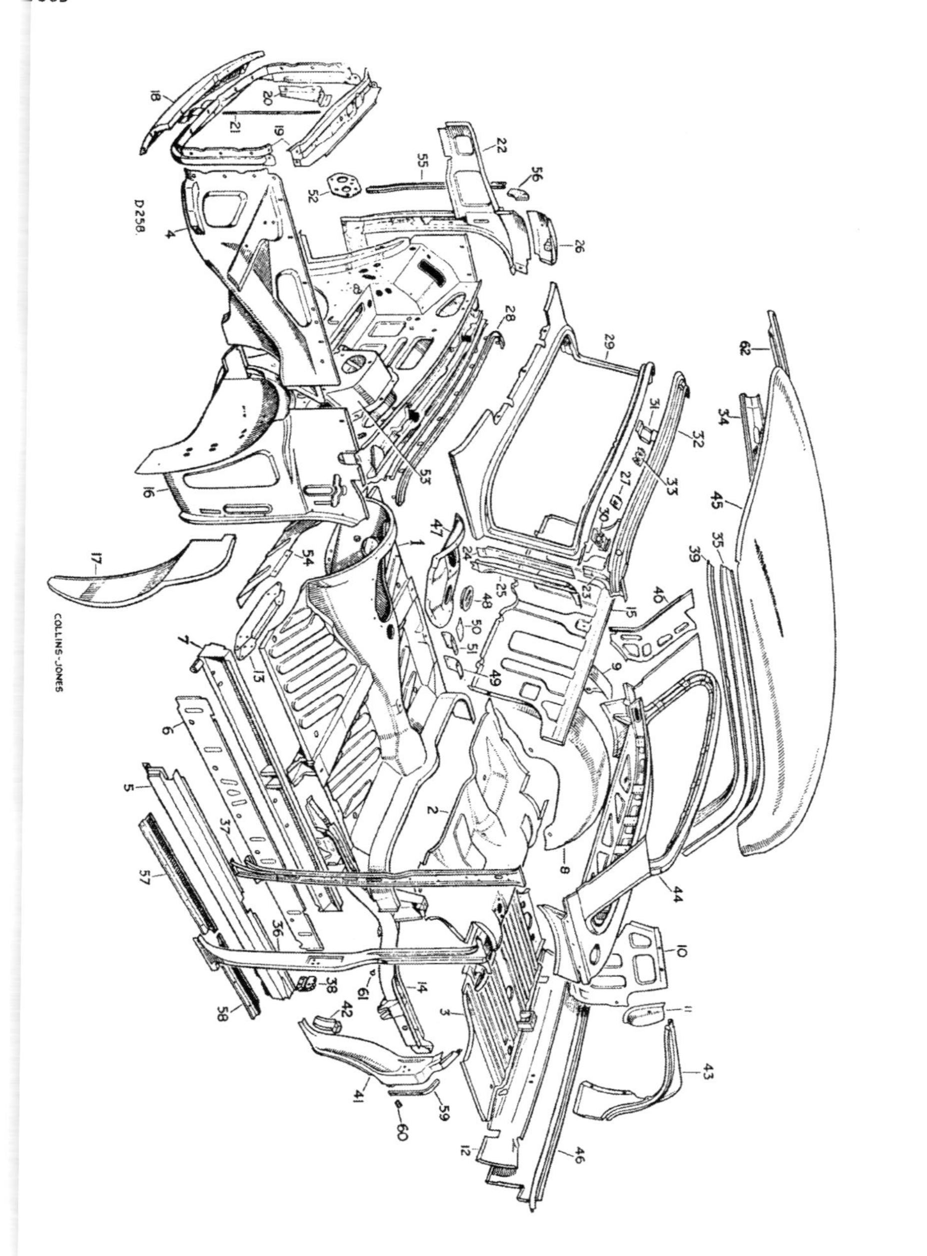

Plate Ref.	1 2 3 4 Description	Qty.	Part No.	Remarks
	BODY COMPLETE less doors, bonnet, boot lid and wings	1	387082	Saloon
	BODY COMPLETE less doors, bonnet, boot lid and wings	1	387085	Coupe
1	Floor complete	1	352634	3 litre
	Floor complete	1	387086	3½ Litre
	Mounting bracket for striker on boot floor	1	351766	
	Rivet fixing mounting bracket	12	78257	
	Cross-member, intermediate, RH outer	1	351718	
	Cross-member, intermediate, LH outer	1	351719	
2	Rear seat panel	1	350066	
3	Boot floor panel	1	352632	
	Front valance complete, RH	1	351088	
4	Front valance complete, LH	1	351089	
	Outer sill panel, RH	1	350120	Saloon
5	Outer sill panel, LH	1	350121	Saloon
	Outer sill panel, RH	1	351273	Coupe
	Outer sill panel, LH	1	351274	Coupe
	Spacer panel for sill, RH	1	350752	Saloon
6	Spacer panel for sill, LH	1	350753	Saloon
	Spacer panel for sill, RH	1	351780	Coupe
	Spacer panel for sill, LH	1	351781	Coupe
	Inner sill panel, RH	1	351734	Saloon
7	Inner sill panel, LH	1	351735	Saloon
	Inner sill panel, RH	1	351784	Coupe
	Inner sill panel, LH	1	351785	Coupe
	Jacking tube and bracket assembly, RH front	1	351736	
	Jacking tube and bracket assembly, LH front	1	351737	
	Jacking tube and bracket assembly, RH rear	1	351738	
	Jacking tube and bracket assembly, LH rear	1	351739	
8	Wheelarch panel, rear, inner RH	1	350094	
	Wheelarch panel, rear, inner LH	1	350095	
9	Wheelarch panel, rear, outer RH	1	350190	
	Wheelarch panel, rear outer LH	1	350191	
10	Rear valance panel, RH	1	350710	
	Rear valance panel, LH	1	350711	
11	Rear lamp cover, RH	1	350708	
	Rear lamp cover, LH	1	350709	
12	Reinforcement for body, rear, lower	1	351720	
13	Intermediate cross-member, centre	1	351723	
	Bolt (⅜" UNF x 3¼" long) – Fixing cross-member to body	6	256050	
	Plain washer – Fixing cross-member to body	12	2204	
	Spring washer – Fixing cross-member to body	6	3076	
	Nut (⅜" UNF) – Fixing cross-member to body	6	254812	
	Longitudinal rear channel complete, RH	1	351778	
14	Longitudinal rear channel complete, LH	1	351779	
15	Rear squab support panel	1	351704	
16	Shroud complete	1	351027	Saloon
	Shroud complete	1	351776	Coupe
	Panel, dash to front wing, RH	1	351142	
17	Panel, dash to front wing, LH	1	351143	
18	Lower front body panel	1	351176	3 litre
	Lower front body panel	1	356757	3½ litre

*Asterisk indicates a new part which has not been used on any previous model.

Plate Ref.	1 2 3 4 DESCRIPTION		Qty	Part No.	REMARKS
	Set bolt (¼" UNF x ½" long)	Fixing lower front panel to body	6	255204	
	Bolt (¼" UNF x ½" long)		4	255204	
	Plain washer		14	3840	
	Shakeproof washer		10	70884	
	Nut (¼" UNF)		4	254810	
19	Grille surround complete		1	351717	
	Bolt (¼" UNF x ⅝" long)	Fixing grille surround to body	2	255206	
	Plain washer		2	3840	
	Spring washer		2	3074	
	Nut (¼" UNF)		2	254810	
	Bolt (¼" UNF x ⅝" long)	Fixing bonnet locking platform to body	2	255206	
	Plain washer		2	3840	
	Spring washer		2	3074	
	Nut (¼" UNF)		2	254810	
20	Side gusset for grille surround, RH		1	351748	
	Side gusset for grille surround, LH		1	351749	
21	Stay, bonnet locking platform to lower front panel		2	350908	
	Plain washer	Fixing stay to lower front panel	4	3840	
	Shakeproof washer		4	70884	
	Nut (¼" UNF)		4	254810	
	'A' post, RH		1	351130	Saloon
22	'A' post, LH		1	351131	
	'A' post RH assembly		1	351672	Coupé
	'A' post LH assembly		1	351673	
	Side header to 'A' post panel, RH		1	351228	Saloon
23	Side header to 'A' post panel, LH		1	351229	
	Side header to 'A' post panel, RH		1	351490	Coupé
	Side header to 'A' post panel, LH		1	351491	
	Spacer panel, 'A' post RH		1	350310	Saloon
24	Spacer panel, 'A' post LH		1	350311	
	Spacer panel, 'A' post RH		1	351504	Coupé
	Spacer panel, 'A' post LH		1	351505	
	'A' post facing panel, RH		1	350176	Saloon
25	'A' post facing panel, LH		1	350177	
	'A' post facing panel, RH		1	350626	Coupé
	'A' post facing panel, LH		1	350627	
	Windscreen to dash panel, RH		1	381261	
26	Windscreen to dash panel, LH		1	381262	
27	Strap for windscreen glazing attachment		6	351177	
	Drain channel for windscreen		1	351219	
29	Windscreen panel assembly		1	351096	Saloon
	Windscreen panel assembly		1	350596	Coupé
	Visor mounting bracket header, RH		1	351168	Saloon
30	Visor mounting bracket header, LH		1	351169	
	Visor mounting bracket header, RH		1	351528	Coupé
	Visor mounting bracket header, LH		1	351529	
31	Mounting bracket for visor hook		2	350185	Saloon
	Mounting bracket for visor hook and interior mirror		1	351577	Coupé
32	Front header panel		1	350213	Saloon
	Front header panel assembly		1	350605	Coupé
33	Plate fixing mirror header		1	351214	Saloon 1959–60
	Plate fixing mirror header		1	351868	1961 onwards, saloon models
34	Side header panel, RH		1	350148	Saloon
	Side header panel, LH		1	350149	

* Asterisk indicates a new part which has not been used on any previous Rover model

Plate Ref.	1 2 3 4 Description		Qty.	Part No.	Remarks
	Side header panel, RH		1	351406	Coupe
	Side header panel, LH		1	351407	
	Cant rail, RH		1	350138	Saloon
35	Cant rail, LH		1	350139	
	Cant rail, RH		1	351404	Coupe
	Cant rail, LH		1	351405	
	'BC' post complete, RH		1	351100	Saloon
36	'BC' post complete, LH		1	351101	
	'BC' post complete, RH		1	387457*	Coupe
	'BC' post complete, LH		1	387458*	
	Moulding, stainless steel, for 'BC' post, RH		1	353391	
	Moulding, stainless steel, for 'BC' post, LH		1	353392	
	Drive screw fixing mouldings		2	78406	
	'BC' post closing plate RH		1	351102	Saloon
37	'BC' post closing plate, LH		1	351103	
	Striker for door lock, RH	10 UNF fixing	2	350656	Saloon Alternatives. Check before ordering
38	Striker for door lock, LH		2	350657	
	Striker for door lock, RH	¼" UNF 3-point fixing	2	356585	
	Striker for door lock, LH		2	356586	
	Striker for door lock, RH	¼" UNF 4-point fixing	2	356583	
	Striker for door lock, LH		2	356584	
	Tapping plate (10 UNF)	For striker plate	4	350140	
	Tapping plate (¼" UNF)		4	356587	
	Special screw (10 UNF)	Fixing striker to body	12	78356	
	Special screw (¼" UNF)		12	78591	
	Striker for door lock, RH		2	368546	Coupe
	Striker for door lock, LH		2	368547	
	Tapping plate for striker		4	351984	
	Shim, .050"	For striker	As reqd	383444	
	Shim, .080"		As reqd	353493	
	Screw (¼" UNF) fixing striker		12	78591	
	Shim for rear door striker		2	386065*	Saloon
	Packing for top hinge 'BC' post rear door, thick		2	386066*	
	Packing for top hinge 'BC' post rear door thin		2	386067*	
	Drip moulding, RH		1	350124	Saloon
39	Drip moulding, LH		1	350125	
	Drip moulding, RH		1	351408	Coupe
	Drip moulding, LH		1	351409	
40	Reinforcement 'D' post to backlight, RH		1	350336	Saloon
	Reinforcement 'D' post to backlight, LH		1	350337	
	Reinforcement 'D' post to backlight, RH		1	351512	Coupe
	Reinforcement 'D' post to backlight, LH		1	351513	

$ 3 Litre and 3½ Litre Coupe cars numbered up to:-
84004502, 84100295, 84300226, 84503374, 84600227, 84800209

$$3½ Litre Coupe cars numbered from:-
84004503, 84100296, 84300227, 84503375, 84600228, 84800210 onwards

*Asterisk indicates a new part which has not been used on any previous model.

Plate Ref.	1 2 3 4 Description	Qty.	Part No.	Remarks
	'D' post, RH	1	351118	
41	'D' post, LH	1	351119	Saloon
	'D' post assembly, RH	1	350600	Coupe
	'D' post assembly, LH	1	350601	
	'D' post to wheelarch panel, lower, RH	1	350128	
42	'D' post to wheelarch panel, lower, LH	1	350129	Saloon
43	Tonneau panel, RH	1	350078	
	Tonneau panel, LH	1	350079	
44	Backlight aperture complete	1	352633	Saloon Mk I, Mk IA
	Backlight aperture complete	1	380507	Coupe and Mk II
	Backlight aperture complete	1	385131	Saloon Mk III 3 Litre
	Backlight aperture complete	1	385133	Coupe 3½ Litre
	Location bracket for rear parcel shelf	2	384157	Mk III 3 litre and
	Pop rivet fixing bracket to parcel shelf	4	78631	3½ litre
	Support bracket, outer lower rear squab	2	384427	
	Pop rivet fixing brackets to squab panel	8	78248	
	Spacer bracket for rear squab	4	384582	
	Pop rivet fixing spacer bracket	8	78631	Mk III 3 litre saloon and
	Location spigot for rear cushion	4	384345	3½ litre saloon
	Pop rivet fixing spigot to heelboard	12	78248	
	Reinforcement bracket for spigot	2	384471	
	Blanking plate for speaker aperture	1	384194	
	Drive screw } Fixing blanking plate	3	78417	Mk III 3 Litre and 3½ Litre
	Nylon clinch nut } to rear parcel tray	3	364435	
45	Roof panel	1	350164	Saloon
	Roof panel	1	351303	Coupe
	Drip moulding, rear quarter, RH	1	353281	
	Drip moulding, rear quarter, LH	1	353282	
	Drive screw fixing moulding	4	78140	
	Texture plate, rear quarter, RH	1	383385	Coupe
	Texture plate, rear quarter, LH	1	383386	
	Drive screw	4	78140	
	Drive screw (stainless steel)	4	78406	
	Motif(Viking ship) for texture plate	2	359937	
46	Rear lower outer panel	1	358499	
47	Cover panel for gearbox	1	351215@	Mk I and Mk IA
	Cover panel for gearbox	1	357233@@	4-speed
	Cover panel for gearbox	1	359233	Mk II and Mk III 3 Litre 4-speed
	Cover panel for gearbox	1	350238%	Automatic
	Cover panel for gearbox	1	359234%%	
	Cover panel for gearbox	1	387410*	For use when 3½ litre body shell is supplied for Mk II and Mk III 3 litre Automatic models
	Cover panel for gearbox	1	385754	3½ Litre

@ Cars numbered up to 625101045, 626100262, 628100109, 630100840, 631100353, 63310078

@@Cars numbered from 625101046, 626100263, 628100110, 630100841, 631100354,633100079 and Mk IA models

% Mk I, Mk IA and Mk II cars numbered up to 77500949, 77600307, 77800115

%%Mk II cars numbered from 77500950, 77600308, 77800116 onwards

*Asterisk indicates a new part which has not been used on any previous model.

D258

COLLINS-JONES

Plate Ref.	1 2 3 4 Description		Qty.	Part No.	Remarks
48	Rubber grommet for gearbox cover panel		2	305016	
	Set bolt (10 UNF x ¾" long)	Fixing cover	12	257019	Mk I and Mk IA
	Plain washer	Fixing cover	12	3557	Mk I and Mk IA
	Spring washer	Fixing cover	12	3073	Mk I and Mk IA
	Special set bolt(10UNFx⅞" long)	Fixing cover	2	357475@@	Mk I and Mk IA
	Set bolt (¼" UNF x ¾" long) and washer	Fixing cover plate to tunnel	2	359583	
	Set bolt (¼" UNF x ⅞" long)	Fixing cover plate to tunnel	17	255206	
	Spring washer	Fixing cover plate to tunnel	17	3074	Mk II, Mk III 3 Litre and 3½ Litre
	Plain washer	Fixing cover plate to tunnel	17	3821	Mk II, Mk III 3 Litre and 3½ Litre
	Rubber seal, front, for cover panel		1	359934	
	Seal, front, for gearbox cover		1	385774	3½ Litre
	Rubber seal, top, for cover panel		1	359935	
	Seal,top, for gearbox cover		1	385775	3½ Litre
	Instruction plate, gearbox and overdrive oil capacities		1	383286	Mk II and Mk III 3 Litre 4-speed
	Rivet fixing plate to gearbox cover		4	78257	Mk II and Mk III 3 Litre 4-speed
49	Cover plate	For gearbox cover panel	1	352435	Mk I and Mk IA
50	Cover plate	For gearbox cover panel	1	352592	Mk I and Mk IA
51	Cover plate	For gearbox cover panel	1	352591	Mk I and Mk IA
	Cover plate for gearbox cover, RH side		1	356971@@	Mk II and Mk III 4-speed
	Drive screw fixing cover plates		11	77949	
	Cover panel for inspection hole in gearbox cover		1	385842	3½ Litre
	Seal for cover panel at side gearbox cover		2	385843	3½ Litre
	Seal for cover panel at front gearbox cover		1	385844	3½ Litre
	Seal for cover panel at rear gearbox cover		1	385845	3½ Litre
52	Cover plate for foot control holes		1	511817	4-speed
	Cover plate for foot control holes		1	511818	Automatic RH Stg
	Cover plate for foot control holes		1	512189	Automatic LH Stg
	Plain washer	For cover plate	2	3662	4-speed
	Plain washer	For cover plate	2	3985	Automatic. 1 off on 4-speed
	Drive screw fixing cover plates		1	70886	
	Cover plate for dash, LH		1	502405	
	Bolt (¼"UNF x ½" long)	Fixing cover plate to dash, LH	3	255204	
	Spring washer	Fixing cover plate to dash, LH	3	3074	
	Nut (¼" UNF)	Fixing cover plate to dash, LH	3	254810	
	Cover panel for dash top, RH		1	352015	Mk I
	Cover panel for dash top, RH		1	357502	Mk IA
	Cover panel for dash top, RH		1	383297	Mk II
	Cover panel for dash top, RH		1	384450	Mk III 3 Litre
	Cover panel for dash top, RH		1	385616	3½ Litre
53	Cover panel for dash top, LH		1	352016	
	Screw (10 UNF x ¾" long)	Fixing cover panels to dash	18	78272	
	Plain washer	Fixing cover panels to dash	22	3685	
	Spring washer	Fixing cover panels to dash	18	3073	
	Nut (10 UNF)	Fixing cover panels to dash	4	257023	

@@ Cars numbered from 625101046, 626100263, 628100110, 630100841, 631100354, 63310079 and Mk IA models

*Asterisk indicates a new part which has not been used on any previous model.

Plate Ref.	1 2 3 4	Description	Qty.	Part No.	Remarks
		Water shield for cover panel	1	352638	Mk I, Mk IA and Mk II
		Water shield foor cover panel	1	384446	Mk III 3 litre and 3½ litre
		Drive screw fixing water shield to dash	5	77704	
54		Passenger foot rest panel	1	357516	RH Stg
		Passenger foot rest panel	1	357515	LH Stg
		Drive screw } Fixing panel to dash	2	78160	
		Bolt (¼" UNF x ½" long) } Fixing panel to dash	1	255206	
		Bolt (¼" UNF x 1⅛" long) } Fixing panel to dash	1	256200	
		Plain washer } Fixing panel to dash	4	3900	
		Spring washer } Fixing panel to dash	2	3074	
		Nut (¼" UNF) } Fixing panel to dash	2	254810	
		Screw (¼" UNF x ⅝" long) } Fixing panel to floor	2	78208	
		Plain washer } Fixing panel to floor	2	3900	
		Spring washer } Fixing panel to floor	2	3074	
		Nut (¼" UNF) } Fixing panel to floor	2	254810	
		Clip fixing cold start control cable to foot rest panel	1	524423	
		Grommet for clip	1	214229	
		Distance tube between clip and panel	1	502809	
55		Sealing rubber for 'A' post to front wing	2	352471	
56		Sealing rubber for 'A' post to front wing, upper	2	352487	
		Sill tread plate, RH front	1	352302	Mk I, Mk IA and Mk II } Saloon
		Sill tread plate, RH front	1	384009	Mk III } Saloon
57		Sill tread plate, LH front	1	352303	Mk I } Saloon
		Sill tread plate, LH front	1	357520	Mk IA and Mk II } Saloon
		Sill tread plate, LH front	1	384010	Mk III } Saloon
		Drive screw fixing front tread plate to sill	16	78288	Mk I, Mk IA and Mk II
		Drive screw fixing tread plates	12	78395	Mk III 3 litre and 3½ litre
		Sill tread plate, RH front	1	353494	Coupe
		Sill tread plate, LH front	1	353495	Coupe
		Drive screw fixing front tread plates	10	78452	Coupe
		Sill tread plate, RH rear	1	352304	Mk I, Mk IA and Mk II } Saloon
		Sill tread plate, RH rear	1	384011	Mk III 3 litre and 3½ litre } Saloon
58		Sill tread plate, LH rear	1	352305	Mk I, Mk IA and Mk II } Saloon
		Sill tread plate, LH rear	1	384012	Mk III 3 Litre and 3½ Litre } Saloon
		Drive screw fixing rear tread plates to sill	16	78288	Mk I, Mk IA and Mk II } Saloon
		Drive screw fixing rear tread plates to sill	6	78395	Mk III 3 Litre and 3½ Litre
		Attachment plate for sill tread plate	20	384014	Mk III 3 Litre saloon and 3½ Litre saloon
		Sill tread plate, RH rear	1	357797	Coupe
		Sill tread plate, LH rear	1	357798	Coupe
		Drive screw fixing rear tread plate	6	78395	Coupe
		Exterior stainless steel moulding for sill, RH	1	386299	3½ Litre
		Exterior stainless steel moulding for sill, LH	1	386300	3½ Litre
		Clip fixing moulding to sill	24	384832	3½ Litre
		Pop rivet fixing clip	24	78704	3½ Litre
		Plug, spring steel 1"dia, rear lower body panel paint drain holes	8	386607*	3½ Litre
		Plug, spring steel 13/16" dia. for sill inner pannel paint drain holes	14	352561	3½ Litre
		Plug, polythene, ⅛" dia. for rear parcel tray panel paint drain holes	4	364617	3½ Litre

*Asterisk indicates a new part which has not been used on any previous model.

D258

COLLINS-JONES

BODY PANELS AND FITTINGS

Plate Ref.	1 2 3 4 DESCRIPTION		Qty	Part No.	REMARKS	
59	Sealing rubber, rear wing to 'D' post		2	352466		
60	Sealing rubber, rear wing to 'D' post, bottom		2	352467		
61	Rubber buffer for 'BC' post		2	311278		
62	Beading for drip moulding, RH		1	352207	Brass chrome plated	Saloon Alternatives Check before ordering
	Beading for drip moulding, LH		1	352208	Brass chrome plated	Saloon Alternatives Check before ordering
	Beading for drip moulding, RH		1	352948	Stainless steel	Saloon Alternatives Check before ordering
	Beading for drip moulding, LH		1	352949	Stainless steel	Saloon Alternatives Check before ordering
	Drive screw fixing beading		20	78295	Saloon	
	Beading for drip moulding, RH		1	353261	Coupé	
	Beading for drip moulding, LH		1	353262	Coupé	
	Drive screw fixing beading		24	78406	Coupé	
	Tacking strip for cant rail		2	354867		
	Tacking strip for 'A' post, lower		2	355844		
	Tacking strip for windscreen, side		2	354869		
	Tacking strip for 'BC' post		4	314974		
	Rubber drainage tube for windscreen		2	352628		
	Rubber drainage tube for backlight		2	352629		
	Spring cover for paint drain holes		30	352561		
	Bolt ($\frac{5}{16}$" UNF x $\frac{1}{2}$" long)	For redundant holes in floor	9	255223		
	Plain washer	For redundant holes in floor	12	3830		
	Spring washer	For redundant holes in floor	10	3075		
	Nut ($\frac{5}{16}$" UNF)	For redundant holes in floor	6	254811		
	Bolt ($\frac{1}{4}$" UNF x $\frac{1}{2}$" long)	For redundant holes in propeller shaft tunnel and wheelarches	As reqd	255204		
	Plain washer	For redundant holes in propeller shaft tunnel and wheelarches	As reqd	3900		
	Spring washer	For redundant holes in propeller shaft tunnel and wheelarches	As reqd	3074		
	Nut ($\frac{1}{4}$" UNF)	For redundant holes in propeller shaft tunnel and wheelarches	As reqd	254810		
	Rubber plug	For redundant accelerator and handbrake holes in dash	1	507842		
	Rubber plug	For redundant accelerator and handbrake holes in dash	1	265683		
	Rubber plug	For redundant accelerator and handbrake holes in dash	2	73198		
	Drive screw fixing blanking plate for redundant holes in dash and floor		As reqd	78155		

* Asterisk indicates a new part which has not been used on any previous Rover model

E106.

COLLINS-JONES

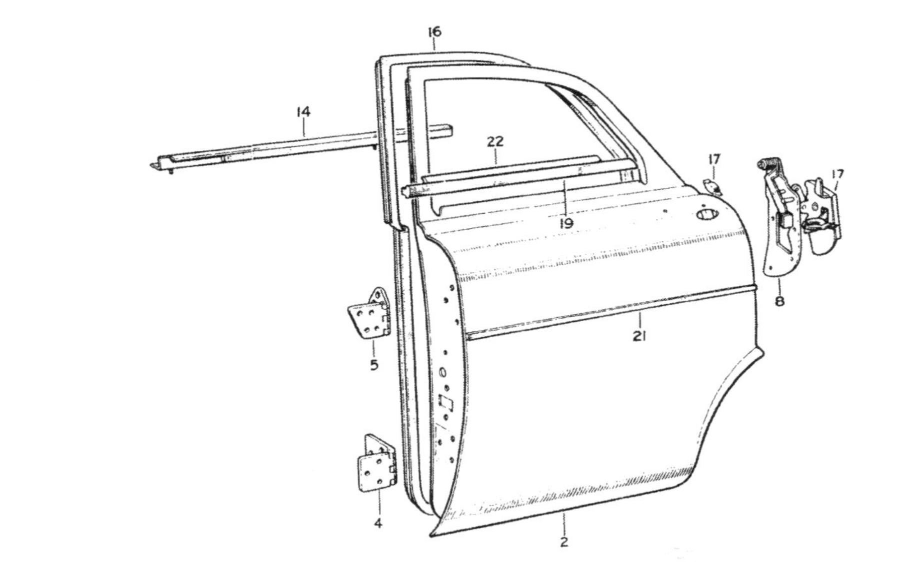

Plate Ref.	1 2 3 4 DESCRIPTION	Qty	Part No.	REMARKS
	DOOR ASSEMBLY, FRONT RH	1	356187†	Mk I
	DOOR ASSEMBLY, FRONT LH	1	356188†	Mk I
	DOOR ASSEMBLY, FRONT RH	1	356572††	Mk I
	DOOR ASSEMBLY, FRONT LH	1	356573††	Mk I
	DOOR ASSEMBLY, FRONT RH	1	358160	MKIA, MKII, MKIII 3 Litre and 3½ Litre
1	DOOR ASSEMBLY, FRONT LH	1	358161	MKIA, MKII, MKIII 3 Litre and 3½ Litre
	DOOR ASSEMBLY, REAR RH	1	356189†	
	DOOR ASSEMBLY, REAR LH	1	356190†	
	DOOR ASSEMBLY, REAR RH	1	358158††	
2	DOOR ASSEMBLY, REAR LH	1	358159††	
	Panel, outside front door, RH	1	350036	
	Panel, outside front door, LH	1	350037	
	Panel, outside rear door, RH	1	350042	
	Panel, outside rear door, LH	1	350043	
3	Hinge for front door	4	350963	
4	Hinge, lower, rear door	2	350965	
	Hinge, upper, RH rear door	1	350969	
5	Hinge, upper, LH rear door	1	350970	
	Special set screw fixing hinges to doors and body	50	350668	
	Lock complete for front door, RH	1	350938	
6	Lock complete for front door, LH	1	350939	
	Lock complete for rear door, RH	1	350934	
7	Lock complete for rear door, LH	1	350935	
	Lever for rear door lock, RH	1	350930	
8	Lever for rear door lock, LH	1	350931	
	Special screw fixing lock to door	16	78350	
	Packing piece for front door upper hinge ..	As reqd	356227	
	Door check arm, front doors	2	350957	
9	Door check arm, rear doors	2	350991	
	Seal for door check arm	4	351992	
10	Retainer for door check arm	4	350956	
11	Packing for door check retainer	4	352096	
	Set screw (¼" UNF x ⅝" long) } Fixing retainer and packing to door	8	255206	
	Shakeproof washer } Fixing retainer and packing to door	8	70884	
12	Special pin } Fixing check arm to body	4	352922	
	Split pin } Fixing check arm to body	4	2422	
	Support for outside panel on front door, RH ..	1	351996	
13	Support for outside panel on front door, LH ..	1	351997	
	Support for outside panel on rear door, RH ..	1	350983	
14	Support for outside panel on rear door, LH ..	1	350984	
	Packing piece for support, front door	2	356859	
	Packing piece for support, rear door..	2	356860	
	Plain washer } Fixing supports to inner door panels	8	3900	
	Spring washer } Fixing supports to inner door panels	8	3073	
	Nut (¼" UNF) } Fixing supports to inner door panels	8	254810	
	Grommet in door for check strap lubrication hole ..	4	356360	
	Sealing rubber for front door, RH	1	359953	
15	Sealing rubber for front door, LH	1	359954	
	Sealing rubber for rear door, RH	1	356788	
16	Sealing rubber for rear door, LH	1	356789	

* Asterisk indicates a new part which has not been used on any previous Rover model
† Cars numbered up to 625000979, 626000332, 628000110, 630000459, 631000115, 633000310
†† Cars numbered from 625000980, 626000333, 628000111, 630000460, 631000116, 633000311 onwards

1821

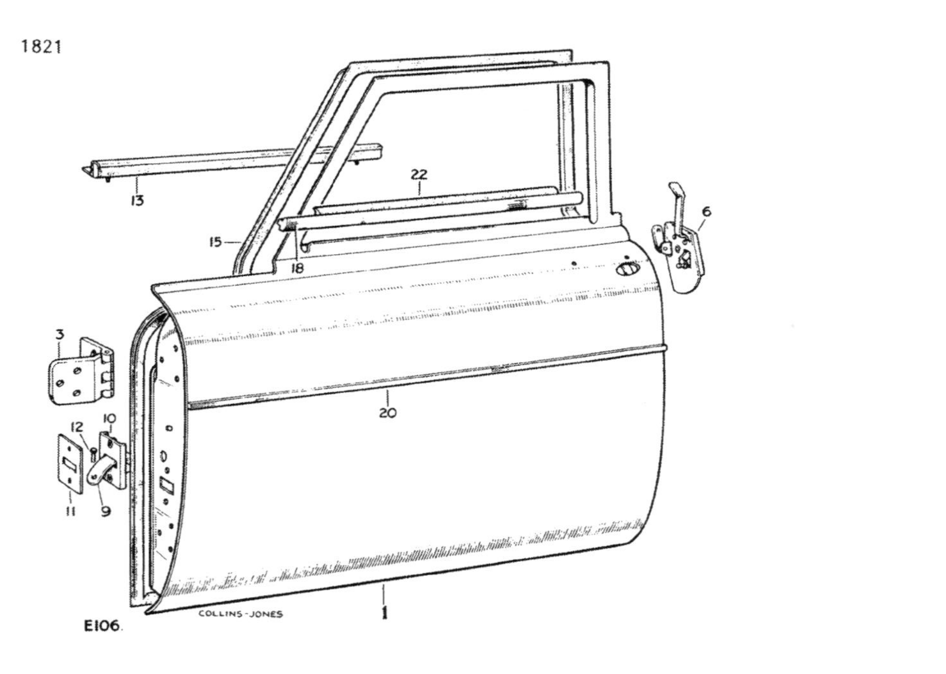

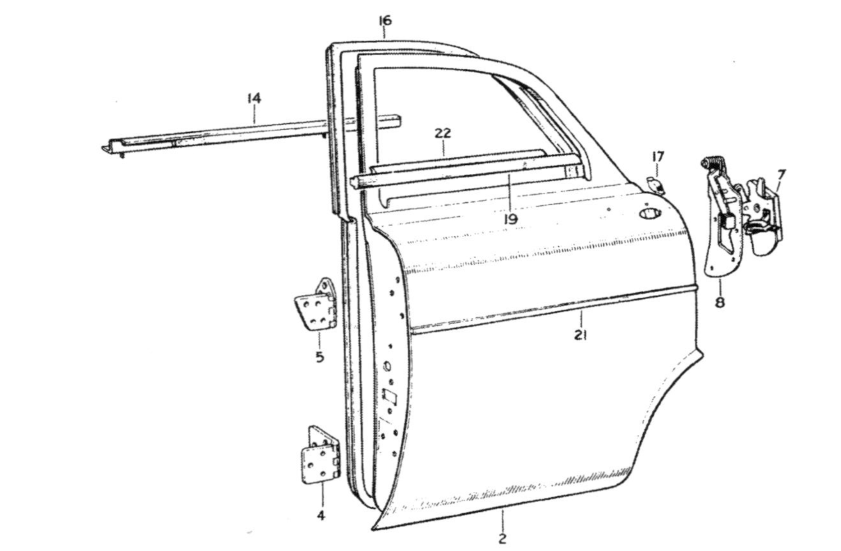

DOOR PANELS AND HINGES, SALOON MODELS

Plate Ref.	1 2 3 4 DESCRIPTION	Qty	Part No.	REMARKS		
	Rubber pad for rear door shut pillar, RH	1	352110			
17	Rubber pad for rear door shut pillar, LH	1	352111			
	Sealing rubber, rear door to wheelarch, RH	1	356767			
	Sealing rubber, rear door to wheelarch, LH	1	356768			
	Moulding for front door waist, RH	1	356361	Brass chrome plated Clip-on fixing	Mk I models Alternatives Check before ordering	
18	Moulding for front door waist, LH	1	356362			
	Moulding for front door waist, RH	1	352968	Stainless steel Clip-on fixing		
	Moulding for front door waist, LH	1	352969			
	Moulding for front door waist, RH	1	356857	Mk IA and Mk II models		
	Moulding for front door waist, LH	1	356858			
	Moulding for rear door waist, RH	1	356363	Brass chrome plated Clip-on fixing	Alternatives Check before ordering	
19	Moulding for rear door waist, LH	1	356364			
	Moulding for rear door waist, RH	1	352970	Stainless steel Clip-on fixing		
	Moulding for rear door waist, LH	1	352971			
	Clip fixing moulding to front doors	16	357201			
	Clip fixing moulding to rear doors	4	352451	For slide-on mouldings		
	Rivet fixing clip to front door panels	4	78252			
	Pop rivet fixing clip to front door panels	4	78257	For clip-on mouldings		
	Clip fixing moulding ends	8	356359			
	Pop rivet fixing clip to rear door panels	4	78257			
20	Moulding for front door	2	352328	Brass chrome plated	Alternatives Check before ordering	Mk I, Mk IA and Mk II
	Moulding for front door	2	352963	Stainless steel		
21	Moulding for rear door	2	352298	Brass chrome plated		
	Moulding for rear door	2	352964	Stainless steel		
	Clip } Fixing moulding	22	315974			
	Pop rivet } to door	22	78257			
	Moulding for front door, RH	1	384826	MKIII 3 Litre and 3½ Litre		
	Moulding for front door, LH	1	384827			
	Moulding for rear door, RH	1	384828			
	Moulding for rear door, LH	1	384829			
	Clip } Fixing moulding	22	384832			
	Pop rivet } to door	22	78704			
22	Weather seal for front and rear door waist	4	352211	Mk I		
	Weather seal for front door waist	2	356845	MKIA, MKII, MKIII 3 Litre and 3½ Litre		
	Weather seal for rear door waist	2	352211			

* Asterisk indicates a new part which has not been used on any previous Rover model

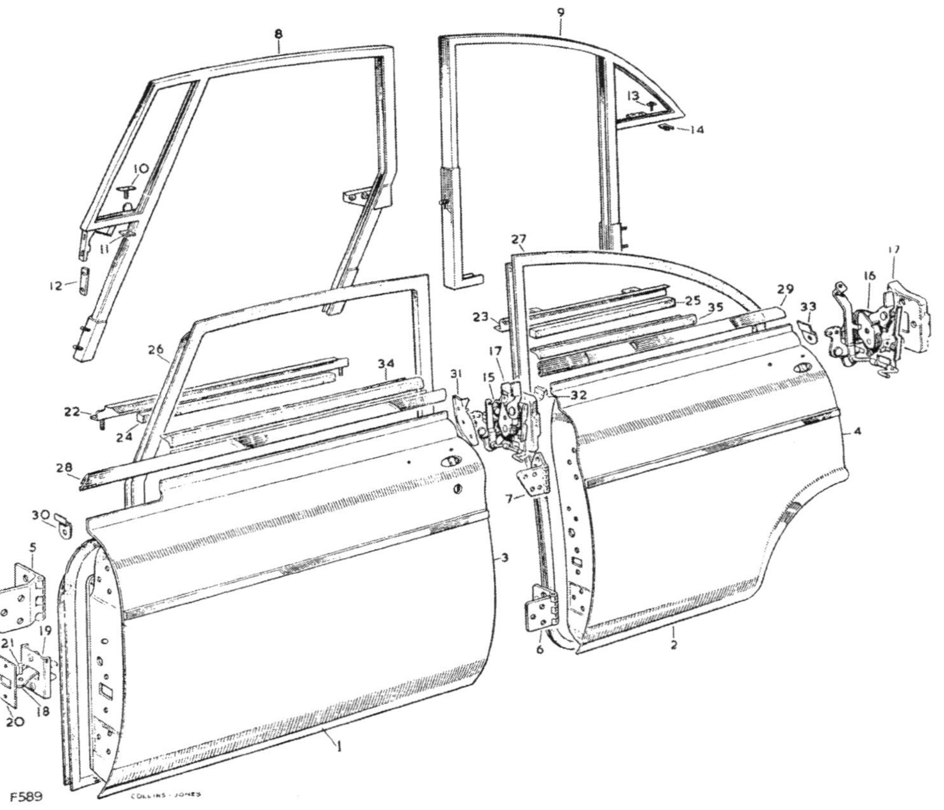

Plate Ref.	1 2 3 4 DESCRIPTION	Qty	Part No.	REMARKS
	DOOR ASSEMBLY, FRONT RH	1	380503	
1	DOOR ASSEMBLY, FRONT LH	1	380504	
	DOOR ASSEMBLY, REAR RH	1	380505	
2	DOOR ASSEMBLY, REAR LH	1	380506	
	Panel, outside front door, RH	1	351936	
3	Panel, outside front door, LH	1	351937	
	Panel, outside rear door, RH	1	351940	
4	Panel, outside rear door, LH	1	351941	
5	Hinge for front door	4	351986	
6	Hinge, lower, rear door	2	351998	
	Hinge, upper, RH rear door	1	350969	
7	Hinge, upper, LH rear door	1	350970	
	Special set screw fixing hinges to doors and body	50	350668	
	Door frame, upper, front door RH	1	353425	
8	Door frame, upper, front door LH	1	353426	
	Door frame, upper, rear door RH	1	353403	
9	Door frame, upper, rear door LH	1	353404	
	Plastic end finisher for rear of front door, RH	1	383346	
	Plastic end finisher for rear of front door, LH	1	383347	
10	Stud plate } Fixing frame to waist front on front doors	2	353422	
	Plain washer	2	3900	
	Spring washer	2	3074	
	Nut (¼" UNF)	2	254810	
11	Shim between frame and door at waist	12	312908	
	Set screw (¼" UNF x ⅜" long) } Fixing upper frames to front doors at waist, rear	4	78616	
	Shakeproof washer	4	78114	
	Bolt (10 UNF x ⅜" long) } Fixing frame to door at waist, rear of front doors	4	257019	
	Plain washer	4	3840	
	Spring washer	4	3074	
	Shim washer	8	3840	
12	Tapped plate } Fixing upper frame to front door waist	2	357857	
	Set screw (10 UNF x ½" long)	2	78276	
	Set screw (chrome) (10 UNF x ½" long)	2	78513	
	Plain washer (chrome)	2	4458	
	Plain washer } Upper frames to front door waist, front	2	3902	
	Spring washer	4	3073	
	Plain washer } Fixing upper frames to mounting brackets on front and rear doors	12	3900	
	Spring washer	12	3074	
	Nut (¼" UNF)	12	254810	
13	Stud plate } Fixing upper frame to rear door waist, rear	6	353422	
	Plain washer	6	3900	
	Spring washer	6	3074	
	Nut (¼" UNF)	6	254810	
14	Shim between frame and door waist	20	312908	
	Plain washer } Fixing upper frame to rear door, front	2	3900	
	Spring washer	2	3074	
	Nut (¼" UNF)	2	254810	
	Screw (10 UNF x $\frac{7}{16}$" long)	2	257004	
	Plain washer	2	3885	
	Spring washer	2	3073	

* Asterisk indicates a new part which has not been used on any previous Rover model

DOOR PANELS AND HINGES, COUPÉ MODELS

1825

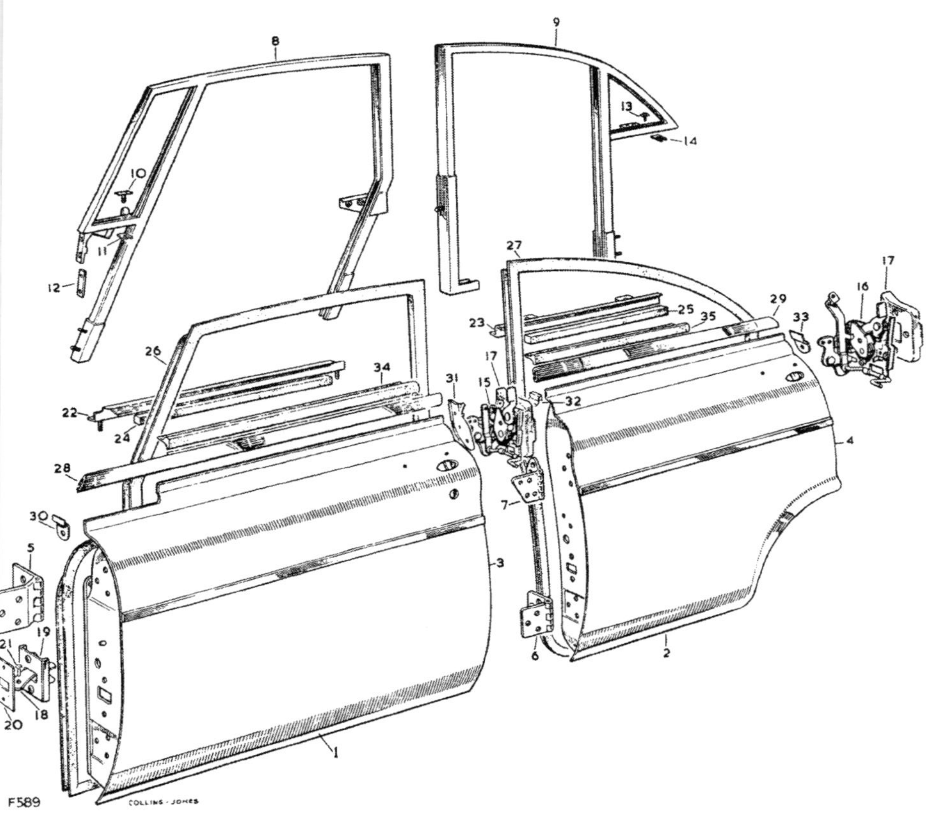

Plate Ref.	1 2 3 4 Description		Qty.	Part No.	Remarks
	Lock complete for front door, RH		1	385990	
15	Lock complete for front door, LH		1	385991	
	Lock complete for rear door, RH		1	368360	
16	Lock complete for rear door, LH		1	368361	
	Cover plate for door lock, RH		2	357912	
17	Cover plate for door lock, LH		2	357913	
	Special screw fixing lock to door		12	78583	
	Packing piece for front door upper hinge		As reqd	356227	
18	Door check arm, front doors		2	350957	
	Door check arm, rear doors		2	350991	
	Seal for check arm		4	351992	
19	Retainer for check arm		4	350956	
20	Packing for door check retainer		4	352096	
	Set screw(¼" UNF x ⅜" long)	Fixing retainer and packing to door	8	255206	
	Shakeproof washer		8	70884	
21	Special pin	Fixing check arm to body	4	356221	
	Split pin		4	2422	
22	Angle support for outside panel, front door		2	351979	
	Angle support for outside panel, rear door RH		1	351980	
23	Angle support for outside panel, rear door LH		1	351981	
24	Packing for support, front doors		2	356859	
25	Packing for support, rear doors		2	356860	
	Plain washer	Fixing supports to inner front door panel	4	3900	
	Spring washer		4	3074	
	Set bolt(¼" UNF x ½" long)		4	255204	
	Plain washer	Fixing supports to inner rear door panel	4	3900	
	Spring washer		4	3074	
	Nut (¼" UNF)		4	254810	
	Grommets in door for check strap lubrication hole		4	356360	
	Sealing rubber for front door, RH		1	385540	
26	Sealing rubber for front door, LH		1	385541	
	Sealing rubber for rear door, RH		1	383412	
27	Sealing rubber for rear door, LH		1	383413	
	Moulding for front door waist, RH		1	353397	
28	Moulding for front door waist, LH		1	353398	
	Moulding for rear door waist, RH		1	353399	
29	Moulding for rear door waist, LH		1	353400	
	Clip fixing moulding to doors		20	357201	
	End finisher, front, for front door, RH		1	359230	
30	End finisher, front, for front door, LH		1	359231	
	End finisher, rear, for front door, RH		1	357625	
31	End finisher, rear, for front door, LH		1	357626	
32	End finisher, front, for rear doors		2	359225	
	End finisher, rear, for rear door, RH		1	359226	
33	End finisher, rear, for rear door, LH		1	359227	
	Drive screw fixing finisher		8	78396	
34	Weather seal for front door waist		2	383509	
35	Weather seal for rear door waist		2	383508	
	Moulding for front door	Stainless steel	2	352963	MKIA and MKII 3 Litre
	Moulding for rear door		2	352964	
	Clip	Fixing moulding to door	22	315974	
	Pop rivet		22	78257	
	Moulding for front door, RH	Stainless steel	1	384826	MKIII 3 Litre and 3½ Litre
	Moulding for front door, LH		1	384827	
	Moulding for rear door, RH		1	384828	
	Moulding for rear door, LH		1	384829	
	Clip	Fixing moulding to door	22	384832	
	Pop rivet		22	78704	

*Asterisk indicates a new part which has not been used on any previous model.

DOOR HANDLES AND WINDOW REGULATORS, SALOON MODELS

Plate Ref. 1 2 3 4	Description	Qty.	Part No.	Remarks
	Control for door handle, RH front	1	384530	
1	Control for door handle, LH front	1	384531	
	Leather washer, remote control to door lock	2	356976	
	Spire nut (Fixing remote control to door and lock)	6	313484	
	Special locking washer (Fixing remote control to door and lock)	2	78363	
	Drive screw (Fixing remote control to door and lock)	6	78153	
	Control for door handle, RH rear	1	352535	
2	Control for door handle, LH rear	1	352536	
	Leather washer, remote control to door lock	2	356976	
3	Set screw (10 UNF x ⅜" long) (Fixing remote control to door and lock)	6	78301	
4	Plain washer (Fixing remote control to door and lock)	6	3885	
5	Shakeproof washer (Fixing remote control to door and lock)	6	311373	
6	Special locking washer (Fixing remote control to door and lock)	2	78363	
	Connecting link for front door lock, RH	1	352612	
7	Connecting link for front door lock, LH	1	352613	
	Retaining clip fixing connecting link to lock	2	352610	Mk I and Mk IA
8	Retaining clip fixing connecting link to lock	2	359879	Mk II and Mk III 3 Litre and 3½ Litre
	OUTSIDE DOOR HANDLE, RH LOCKING (Less barrel lock)	1	380707	
9	OUTSIDE DOOR HANDLE, LH LOCKING (Less barrel lock)	1	380708	
	Push-button mechanism (less barrel lock) for front door handle, RH	1	356410@	
	Push-button mechanism (less barrel lock) for front door handle, LH	1	356411@	
	Push-button mechanism (less barrel lock) for front door handle, RH	1	385117@@	
10	Push-button mechanism(less barrel lock) for front door handle, LH	1	385118@@	
11	Set bolt (For push-button mechanism)	2	356412	
12	Locknut (For push-button mechanism)	2	356413	
13	Set screw fixing push-button mechanism to door handle	4	356199	
	Barrel lock for door handle, RH	1	358174	
14	Barrel lock for door handle, LH	1	358173	
	Outside door handle, RH, plain	1	383455	
15	Outside door handle, LH, plain	1	383456	
	Washer for handle, large (Fixing door handles)	4	352416	
16	Washer for handle, large (Fixing door handles)	4	383457@@	
17	Washer for handle, small (Fixing door handles)	4	352417	
	Shakeproof washer (Fixing door handles)	8	74236	
18	Starlock washer (Fixing door handles)	4	78363@@	
19	Screw (10 UNF x ¾" long) (Fixing door handles)	8	257015	
	INSIDE DOOR HANDLE COMPLETE, chrome escutcheon	4	352625	Alternatives. Check before ordering
	INSIDE DOOR HANDLE COMPLETE, black plastic escutcheon	4	367173	Alternatives. Check before ordering
20	Inside door handle	4	352420	

@ MkI, Mk IA and Mk II cars with suffix letters 'A' and 'B'
@@ MkII cars with suffix letter 'C' onwards and Mk III 3 litre and 3½ litre

*Asterisk indicates a new part which has not been used on any previous model.

DOOR HANDLES AND WINDOW REGULATORS, SALOON MODELS

829

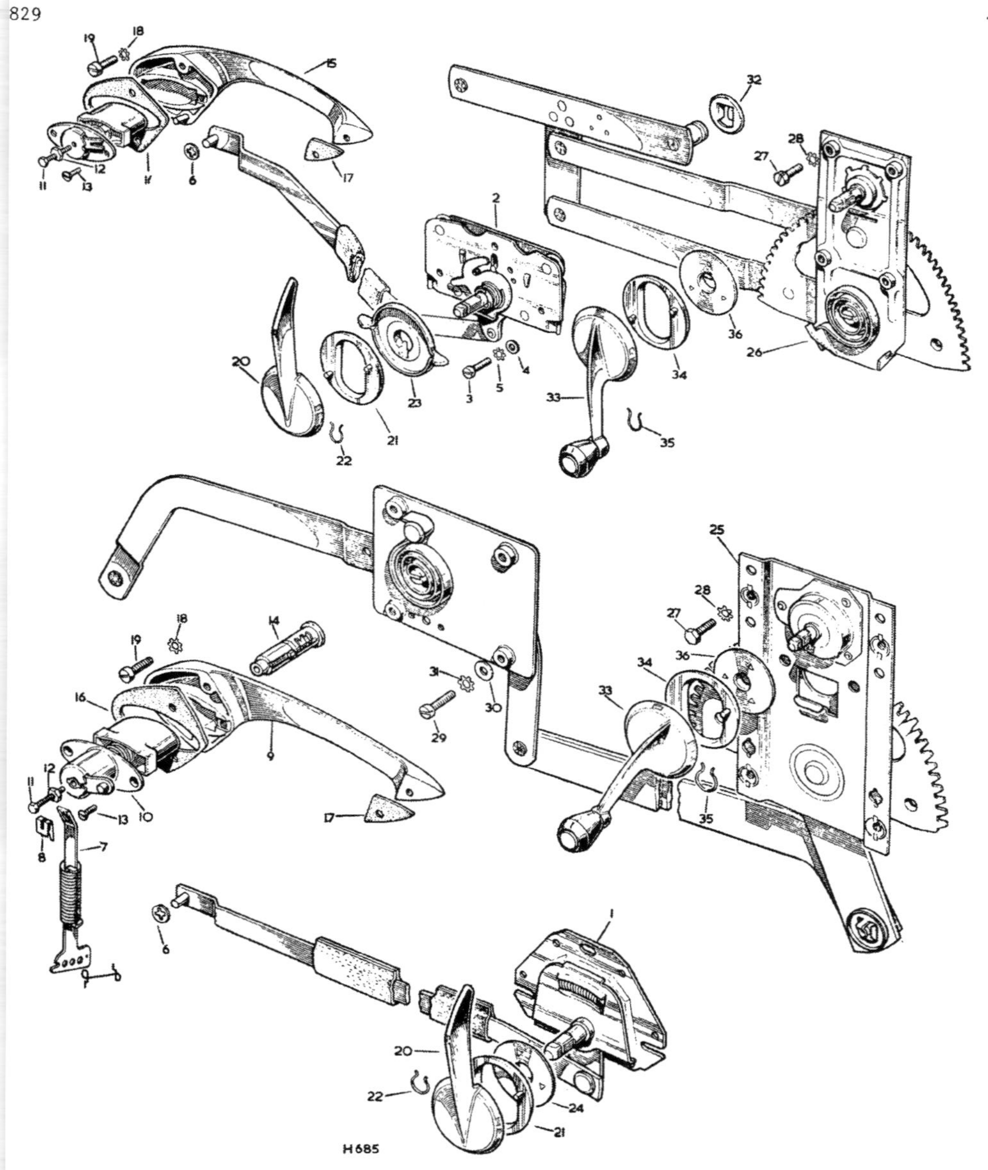

Plate Ref.	1 2 3 4 Description		Qty.	Part No.	Remarks
21	Escutcheon, black plastic	Fixing inside door handles	4	367164	
22	Spring clip		4	352423	
23	Escutcheon, rear door, for safety catch		2	352611	
24	Wearing plate for front door inside handle		2	352422	
	WINDOW REGULATOR, RH FRONT DOOR		1	352103	Mk I 3 Litre
	WINDOW REGULATOR, LH FRONT DOOR		1	352104	
	WINDOW REGULATOR, RH FRONT DOOR		1	358886	Mk IA, Mk II and Mk III 3 litre and 3½ litre
25	WINDOW REGULATOR, LH FRONT DOOR		1	358887	
	Handle end of regulator, RH		1	356193	Mk I 3 Litre only
	Handle end of regulator, LH		1	356194	
	Stud retainer fixing slide end to handle end		2	352636	
	Window regulator, RH rear door		1	352105	
26	Window regulator, LH rear door		1	352106	
	Slide for front window regulator		2	352012	Mk I 3 litre only
	Set screw (¼" UNF x 7/16" long)	Fixing slide to door	4	78251	
	Plain washer		4	3467	
	Shakeproof washer		4	70884	
27	Set screw (10 UNF x ⅜" long)	Fixing window regulators	16	78301	
28	Shakeproof washer		16	311373	
29	Set bolt (¼" UNF x 7/16" long)	Fixing front door regulator at lifitng arm end	8	78251	
30	Plain washer		8	3467	
31	Shakeproof washer		8	70884	
32	Stud retainer fixing rear door regulator arm to glass channel		4	352636	
33	WINDOW REGULATOR HANDLE COMPLETE, chrome knob		4	352624	Alternatives
	WINDOW REGULATOR HANDLE COMPLETE, black plastic kbob		4	367172	
34	Escutcheon, black plastic	Fixing regulator handles	4	367163	
35	Spring clip		4	352423	
36	Wearing plate		4	352422	

*Asterisk indicates a new part which has not been used on any previous model.

DOOR HANDLES AND WINDOW REGULATORS, COUPÉ MODELS

Plate Ref.	1 2 3 4 Description	Qty.	Part No.	Remarks
	Control for door handle, RH, front	1	357599	
1	Control for door handle, LH, front	1	357600	
2	Circlip } Fixing remote control to lock	4	353356	
3	Special washer } Fixing remote control to lock	4	353368	
	Screw (10 UNF x ½" long) } Fixing controls to inner panels	12	78276	
	Plain washer } Fixing controls to inner panels	12	2678	
	Spring washer } Fixing controls to inner panels	12	74236	
	Upper link } For RH front door lock	1	383557	
	Lower link } For RH front door lock	1	359882	
4	Upper link } For LH front door lock	1	383557	
5	Lower link } For LH front door lock	1	359882	
	Connecting link for rear doors	2	353384	
6	Retaining clip fixing link to handle	4	359879	
7	Spring for link on front doors	2	353383	
	Plain washer } Fixing link to lock	4	3902	
	Starlock washer } Fixing link to lock	4	78256	
8	Spring clip fixing link to private lock	2	359879	
	Pad for remote control link, front doors	2	359153	
	Pad for remote control link, rear doors	2	359830	
	Outside door handle, RH, front	1	353375@	
9	Outside door handle, LH, front	1	353376@	
	Outside door handle, RH, front	1	383449@@	
	Outside door handle, LH, front	1	383450@@	
	Push-button mechanism for door handle, RH, front	1	385119	
	Push-button mechanism for door handle, LH, front	1	385120	
	Outside door handle, RH, rear	1	353377@	
10	Outside door handle, LH, rear	1	353378@	
	Outside door handle, RH, rear	1	383451@@	
	Outside door handle, LH, rear	1	383452@@	
11	Seating washer, large } For handles	4	352416@	
	Seating washer, large } For handles	4	383457@@	
12	Seating washer, small } For handles	4	352417	
	Shakeproof washer } Fixing outside door handle	8	74236	
	Starlock washer } Fixing outside door handle	4	78363	
	Set screw (10 UNF x ⅜" long) } Fixing outside door handle	8	257015	
	Private lock for RH front door	1	368324	
13	Private lock for LH front door	1	368325	
14	Retaining clip fixing private lock	2	353488	
	INSIDE DOOR HANDLE COMPLETE, chrome escutcheon	4	352625	} Alternatives. Check before ordering
	INSIDE DOOR HANDLE COMPLETE, black plastic escutcheon	4	367173	} Alternatives. Check before ordering
15	Inside door handle	4	352420	
16	Escutcheon, black plastic } Fixing inside door handles	4	367164	
17	Spring clip } Fixing inside door handles	4	352423	

@ Mk II 3 Litre coupe cars with suffix letters 'A' and 'B'

@@ Mk II 3 Litre coupe cars with suffix letter 'C' onwards, Mk III and 3½ Litre

*Asterisk indicates a new part which has not been used on any previous model.

DOOR HANDLES AND WINDOW REGULATORS, COUPÉ MODELS

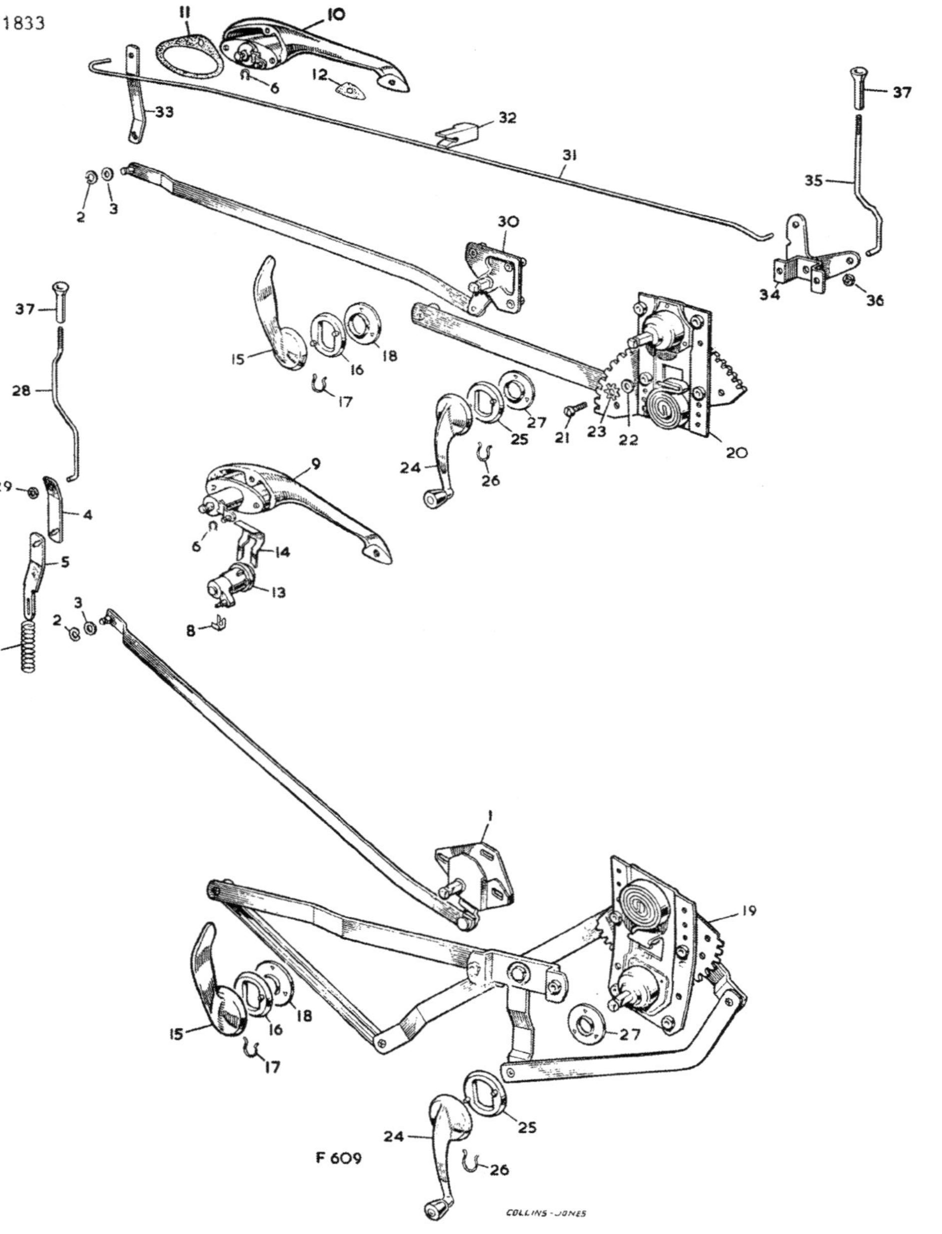

F 609

Plate Ref.	1 2 3 4 Description		Qty.	Part No.	Remarks
18	Wearing plate for inside handle		4	352422	
	Window regulator, RH, front door		1	357604	
19	Window regulator, RH, rear door		1	357567	
20	Window regulator, LH, rear door		1	357568	
21	Set screw (¼" UNF x 7/16" long)	Fixing regulators	20	78251	
22	Plain washer	Fixing regulators	20	3840	
23	Shakeproof washer	Fixing regulators	20	70884	
24	WINDOW REGULATOR HANDLE COMPLETE		4	367172	
25	Escutcheon, black plastic	Fixing regulator handles	4	367164	
26	Spring washer	Fixing regulator handles	4	352423	
27	Wearing plate	Fixing regulator handles	4	352422	
	Sill control rod, RH, front door		1	385713	
28	Sill control rod, LH, front door		1	385714	
29	Starlock washer fixing control rod to lock		4	78446	Early models
	Linkage clip fixing control rod to front door lock		2	368702	
	Control for door handle, RH, rear		1	357569	
30	Control for door handle, LH, rear		1	357570	
31	Sill control rod for rear door		2	385809	
32	Clip for sill control rod		2	359982	
33	Connecting link for rear doors		2	353384	
	Bell crank complete for control rod, RH		1	385988	
34	Bell crank complete for control rod, LH		1	385989	
	Screw (10 UNF x ⅜" long)	Fixing bell crank to doors	4	78301	
	Plain washer	Fixing bell crank to doors	4	4034	
	Spring washer	Fixing bell crank to doors	4	3073	
	Sill control rod, locking, RH, rear door		1	385711	
35	Sill control rod, locking, LH, rear door		1	385712	
36	Starlock washer fixing control rod to bell crank and locks		8	78446	
	Linkage clip fixing control rods and bell crank and rear door locks		6	368702	
37	Knob for sill lock		4	371200	

*Asterisk indicates a new part which has not been used on any previous model.

WINDSCREEN AND BACK LIGHT GLASS AND MOULDINGS, SALOON MODELS

1835

F614

Plate Ref.	1 2 3 4 Description	Qty.	Part No.	Remarks
1	Glass for windscreen, safety zone type	1	357641	RH Stg
	Glass for windscreen, safety zone type	1	357642	LH Stg
	Glass for windscreen, laminated, clear	1	385995	America.Dollar area, France, Algiers
	Glass for windscreen, laminated, Sundym tinted	1	385996*	
2	Rubber moulding for windscreen glass	1	352351	
	Exterior moulding for windscreen top, RH	1	352354	
	Exterior moulding for windscreen top, LH	1	352355	
	Exterior moulding for windscreen side, RH	1	352358	
	Exterior moulding for windscreen side, LH	1	352359	Brass chrome plated
	Exterior moulding for windscreen bottom, RH	1	352356	Alternative to stainless
	Exterior moulding for windscreen bottom, LH	1	352357	steel
	Exterior moulding for windscreen bottom,corner,RH	1	352360	Check before ordering
	Exterior moulding for windscreen bottom,corner,LH	1	352361	
	Cover plate for windscreen moulding top joint	1	352363	
	Cover plate for windscreen moulding bottom joint	1	352945	
	Exterior moulding for windscreen top, RH	1	352950	
3	Exterior moulding for windscreen top, LH	1	352951	
	Exterior moulding for windscreen side, RH	1	352954	
4	Exterior moulding for windscreen side, LH	1	352955	Stainless steel
	Exterior moulding for windscreen bottom, RH	1	352952	Alternative to brass
5	Exterior moulding for windscreen bottom, LH	1	352953	Check before ordering
	Exterior moulding for windscreen bottom corner,RH	1	356977	
6	Exterior moulding for windscreen bottom,corner,LH	1	356978	
7	Cover plate for windscreen moulding top joint	1	352959	
8	Cover plate for windscreen moulding bottom joint	1	352945	
	Drive screw fixing moulding at side and corner	8	78404	
9	Clip for mouldings	22	352938	
10	Support plate for clip	30	384770	
	Drive screw fixing clip	34	78417	
11	Fixing plate for moulding	4	352919	
	Drive screw fixing plate	4	77704	
	INTERIOR WINDSCREEN MOULDING,BOTTOM RH	1	356300	3-piece moulding
	INTERIOR WINDSCREEN MOULDING,BOTTOM LH	1	356301	Alternative to 2-piece type
	INTERIOR WINDSCREEN MOULDING,BOTTOM CENTRE	1	356303	Check before ordering
	INTERIOR WINDSCREEN MOULDING,BOTTOM RH	1	355677	2-piece moulding
	INTERIOR WINDSCREEN MOULDING,BOTTOM LH	1	355678	Alternative to 3-piece type Check before ordering
	INTERIOR WINDSCREEN MOULDING, BOTTOM RH	1	383688	Mk II 3 Litre saloon
	INTERIOR WINDSCREEN MOULDING, BOTTOM LH	1	383689	Mk II 3 Litre saloon
	Stud plate for centre moulding	2	356233	For 3-piece moulding
	Stud plate for interior moulding	4	356752	For 2-piece moulding
	Woodscrew fixing stud plate	8	78286	

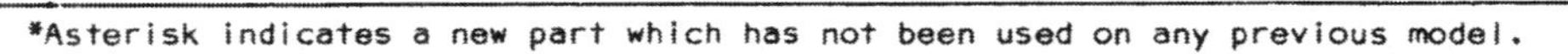
*Asterisk indicates a new part which has not been used on any previous model.

WINDSCREEN AND BACK LIGHT GLASS AND MOULDINGS, SALOON MODELS

1837

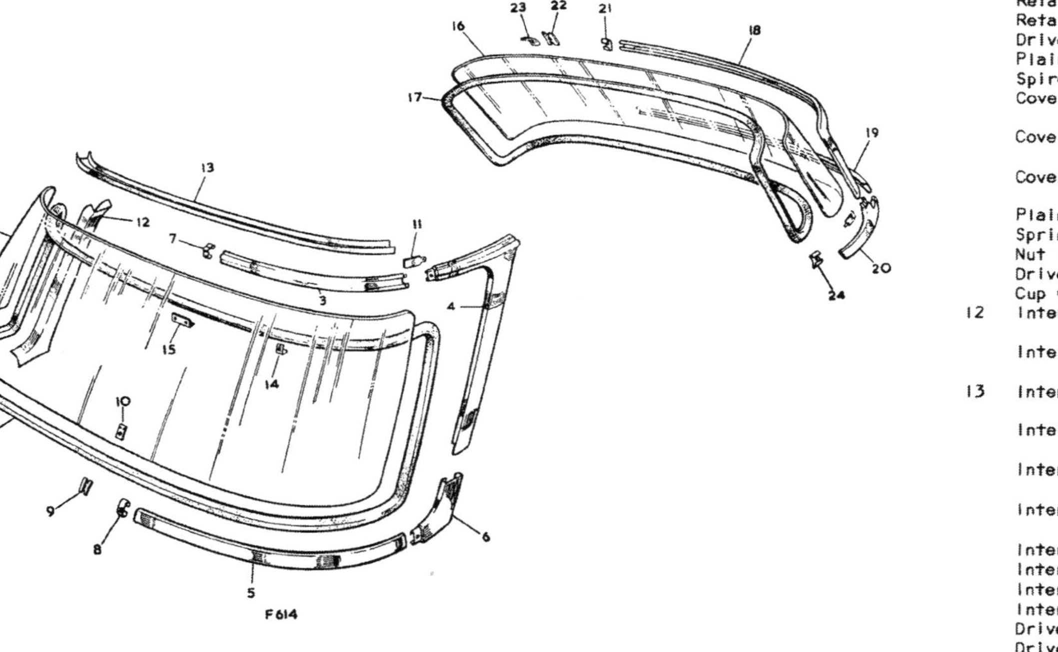

Plate Ref.	1 2 3 4	Description	Qty.	Part No.	Remarks
		INTERIOR WINDSCREEN MOULDING, BOTTOM RH	1	385715	Mk III 3 litre and 3½ litre saloon
		INTERIOR WINDSCREEN MOULDING, BOTTOM LH	1	385716	
		INTERIOR WINDSCREEN MOULDING, BOTTOM CENTRE	1	385717	
		'D' washer – Fixing mouldings to facia rail	4	383460	
		Spring washer – Fixing mouldings to facia rail	4	3073	
		Nut (10 UNF) – Fixing mouldings to facia rail	4	257011	
		Fixing plate for outer moulding	2	356299	For 3-piece moulding
		Woodscrew fixing plate	4	78286	
		Beading at end of centre moulding	2	356298	
		Woodscrew fixing beading	4	65556	
		Plain washer – Fixing centre moulding to body	2	2983	
		Nut (10 UNF) – Fixing centre moulding to body	2	257011	
		Retaining plate for bottom moulding	1	356640	For 2-piece moulding Mk I, Mk IA and Mk II 3 litre
		Retaining clip for cover plate	1	356641	
		Drive screw – Fixing retaining plate	1	78400	
		Plain washer – Fixing retaining plate	1	2678	
		Spire nut – Fixing retaining plate	1	313484	
		Cover plate for windscreen interior moulding joint	1	355666	Mk I and Mk IA – For 2 piece moulding
		Cover plate trimmed for windscreen interior moulding	1	381157	Mk II 3 litre Alternatives – For 2 piece moulding
		Cover plate, plastic, for windscreen interior moulding	1	381240	
		Plain washer – Fixing cover plate	4	2983	
		Spring washer – Fixing cover plate	4	3073	
		Nut (10 UNF) – Fixing cover plate	4	257011	
		Drive screw – Fixing moulding to facia rail	6	311248	
		Cup washer – Fixing moulding to facia rail	6	73961	
12		Interior windscreen moulding, side RH – 4-piece moulding 1 at each side 2 across the top	1	355367	Mk I and Mk IA 3 litre Quote car number and colour required
		Interior windscreen moulding side LH	1	355368	
13		Interior windscreen moulding, top RH	1	355371	
		Interior windscreen moulding, top LH	1	355372	
		Interior windscreen moulding, side RH – Side only, no top moulding	1	355673	
		Interior windscreen moulding, side LH	1	355674	
		Interior windscreen moulding, side RH	1	381162	Mk II 3 litre saloon
		Interior windscreen moulding, side LH	1	381163	
		Interior windscreen moulding, side RH	1	384285	Mk III 3 litre and 3½ litre saloon
		Interior windscreen moulding, side LH	1	384286	
		Drive screw – Fixing moulding to windscreen	2	311948	
		Drive screw – Fixing moulding to windscreen	2	78672	
		Cup washer – Fixing moulding to windscreen	4	73961	

*Asterisk indicates a new part which has not been used on any previous model.

F614 COLLINS JONES

Plate Ref.	1 2 3 4 Description	Qty.	Part No.	Remarks
	Joint plate complete for top moulding	1	355385	For use with 4-piece moulding
	Drive screw } Fixing moulding to bracket	4	311248	For use with 4-piece moulding
	Cup washer } Fixing moulding to bracket	4	73961	For use with 4-piece moulding
14	Bracket for windscreen top moulding	4	352583	For use with 4-piece moulding
15	Bracket for mirror	1	352582	For use with 4-piece moulding
	Set bolt (10 UNF x ½" long) } Fixing bracket to body	6	257017	For use with 4-piece moulding
	Plain washer } Fixing bracket to body	6	3557	For use with 4-piece moulding
	Spring washer } Fixing bracket to body	6	3073	For use with 4-piece moulding
16	Glass for back light	1	352001	
	Glass for back light, Sundym tinted	1	352935	
17	Rubber moulding for back light glass	1	352097@	
	Rubber moulding for back light glass	1	384391@@	
	Exterior moulding for back light, top RH	1	352941	
18	Exterior moulding for back light, top LH	1	352942	
	Exterior moulding for back light, bottom RH	1	352943	
19	Exterior moulding for back light, bottom LH	1	352944	
	Exterior moulding for back light, side RH	1	352946	
20	Exterior moulding for back light, side LH	1	352947	
21	Cover plate for back light moulding joints	2	352945	
22	Clip, top	22	352938	
23	Support plate for clip For back light mouldings	26	384770	
24	Clip, side	2	352994	
	Drive screw fixing top clip	26	78417	
	Clip, top corner, for back light mouldings	2	356307	For brass chrome plated mouldings
	Clip, top corner, for back light moulding	2	356996	For stainless steel mouldings
	Drive screw fixing clip	2	78417	
	Drive screw } Fixing side clip	2	78140	
	Plain washer } Fixing side clip	2	2905	
	Fixing plate for moulding	2	352919	
	Drive screw fixing plate	2	77704	

@ 3 litre cars numbered up to 625000879, 626000332, 628000110 630000459, 631000310, 633000115

@@ 3 litre cars numbered from 625000880, 626000333, 628000111,630000460, 631000311, 633000116 onwards

*Asterisk indicates a new part which has not been used on any previous model.

1841

F594

Plate Ref.	1 2 3 4 Description	Qty.	Part No.	Remarks
1	Glass for windscreen, zone toughened	1	353473	RH Stg
	Glass for windscreen, zone toughened	1	353474	LH Stg
	Glass for windscreen, laminated, clear	1	385992	NADA
	Glass for windscreen, laminated, Sundym tinted	1	385993*	
2	Rubber moulding for windscreen	1	383762	
3	Exterior moulding for windscreen, top	1	353257	
	Exterior moulding, side RH	1	353255	
4	Exterior moulding, side LH	1	353256	
5	Exterior moulding for windscreen, bottom RH	1	352952	
6	Exterior moulding for windscreen, bottom LH	1	352953	
	Exterior moulding for windscreen, bottom corner RH	1	356977	
7	Exterior moulding for windscreen, bottom corner LH	1	356978	
8	Cover plate for windscreen moulding, bottom joint	1	352945	
9	Securing plate for moulding at sides	2	386425	
	Drive screw (½" long) } Fixing moulding	2	78417	
	Drive screw (⅜" long) } Fixing moulding	8	78404	
	Clip for moulding } Fixing exterior mouldings	23	352938	
10	Support plate for clip } Fixing exterior mouldings	35	352939	
	Drive screw fixing clip } Fixing exterior mouldings	37	78417	
11	Interior windscreen 'A' post, RH	1	358641	
	Interior windscreen 'A' post. LH	1	358642	
	Drive screw fixing moulding	6	78459	
12	Edge strip, trimmed, for 'A' post, RH	1	383321	
	Edge strip, trimmed, for 'A' post, LH	1	383322	
	Drive screw } Fixing edge strip to 'A' post	4	78495	
	Drive screw } Fixing edge strip to 'A' post	2	77704	
13	Interior windscreen moulding, bottom RH	1	383725	Mk II 3 Litre
	Interior windscreen moulding, bottom LH	1	383726	Mk II 3 Litre
	Interior windscreen moulding, bottom RH	1	385715	Mk III 3 litre and 3½ litre
	Interior windscreen moulding, bottom LH	1	385716	Mk III 3 litre and 3½ litre
	Interior windscreen moulding, centre	1	385717	Mk III 3 litre and 3½ litre
	'D' washer } Fixing moulding to facia rail	4	383460	
	Spring washer } Fixing moulding to facia rail	4	3073	
	Nut (10 UNF) } Fixing moulding to facia rail	4	257011	
	Cover for joint, bottom windscreen moulding	1	381157	
	Retaining plate } Fixing cover plate	1	356640	
	Clip } Fixing cover plate	1	356641	
	Drive screw } Fixing retaining plate	1	78400	
	Plain washer } Fixing retaining plate	1	3685	
	Spire nut } Fixing retaining plate	1	313484	
14	Front header panel	1	358538	
15	Glass for back light	1	353264	
	Glass for back light, Sundym tinted	1	385753*	
16	Rubber moulding for back light glass	1	353265	
	Exterior moulding for back light, upper RH	1	353271	
17	Exterior moulding for back light, upper LH	1	353272	
18	Cover plate for joint, upper mouldings	1	386512*	
19	Clip fixing upper back light moulding	10	353273	
20	Fixing plate for upper back light moulding	1	353274	
21	Exterior moulding for back light, lower, centre	1	353275	
	Exterior moulding for back light, side RH	1	353276	
22	Exterior moulding for back light, side LH	1	353277	

NADA indicates parts peculiar to cars exported to the North American dollar area

*Asterisk indicates a new part which has not been used on any previous model.

1843

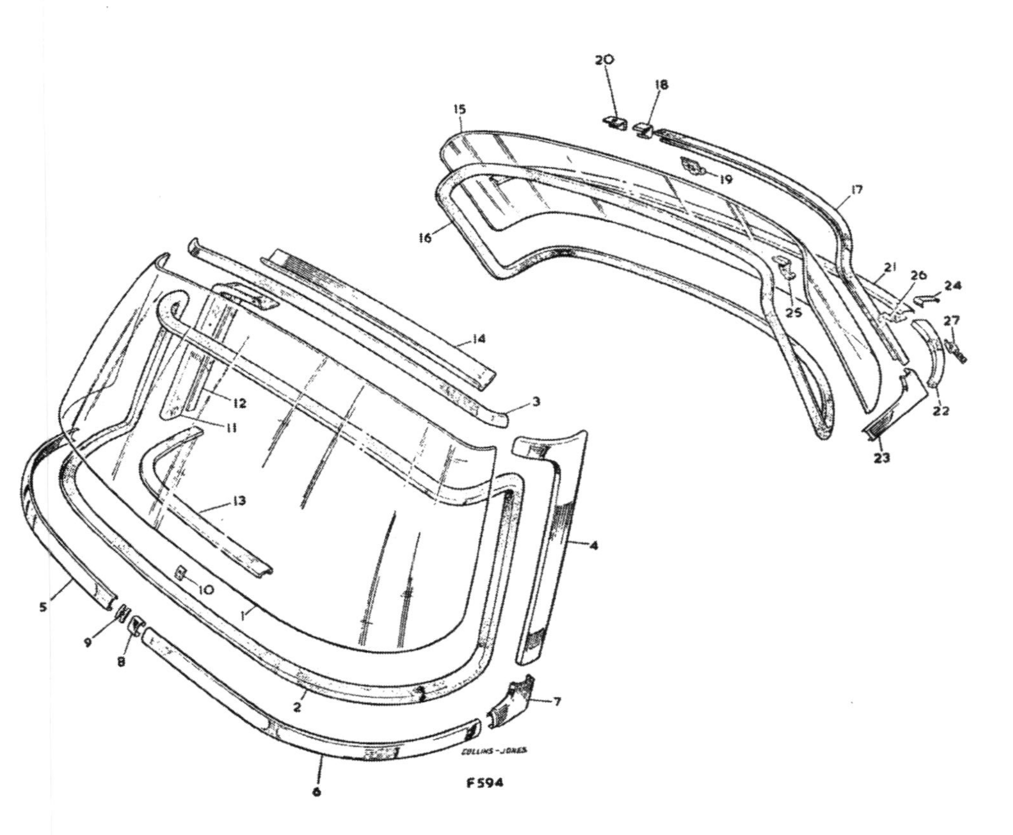

WINDSCREEN AND BACK LIGHT GLASS AND MOULDINGS, COUPÉ MODELS

Plate Ref.	1 2 3 4 DESCRIPTION	Qty	Part No.	REMARKS
	Exterior moulding, rear quarter and back light, RH ..	1	353283	
23	Exterior moulding, rear quarter and back light, LH ..	1	353284	
	Stud plate } Fixing rear quarter and back light mouldings	2	353287	
	Pad for stud plate	2	383687	
	Prestik washer	2	352911	
	Nut (10 UNF)	2	257023	
	Plain washer	2	3851	
	Clip	2	352994	
	Drive screw	2	78140	
24	Cover plate for joints, lower mouldings	2	353289	
25	Clip fixing lower, centre, back light mouldings ..	5	353278	
26	Fixing plate, lower, centre, back light moulding ..	2	353279	
	Packing washer for fixing plate	4	2678	
27	Clip fixing lower, side, back light moulding ..	2	353280	
	Drive screw } Fixing moulding clips and plates ..	21	78155	
	Drive screw	2	78140	
	Plain washer	2	4287	

* Asterisk indicates a new part which has not been used on any previous Rover model

1846

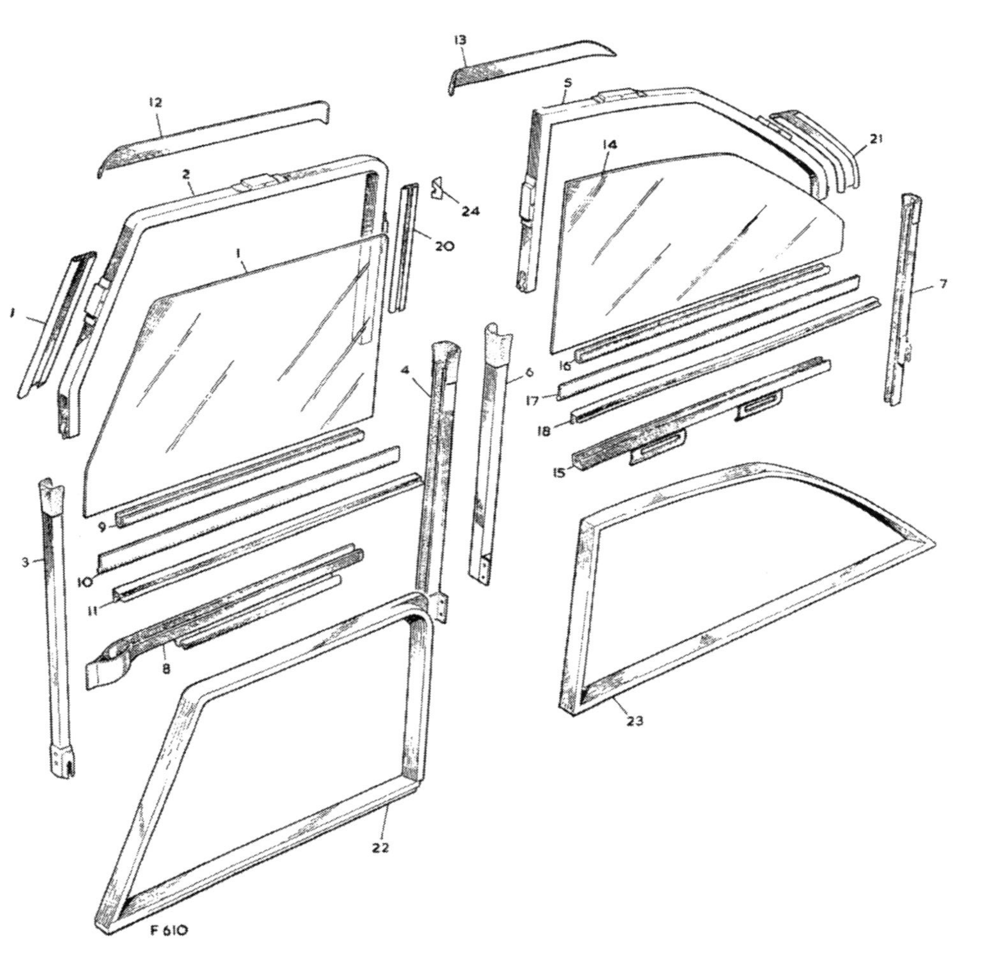

DOOR GLASS AND MOULDINGS, MK I SALOON MODELS

1847

Plate Ref.	1 2 3 4 DESCRIPTION	Qty	Part No.	REMARKS
1	Glass for front door	2	356915	
2	Channel complete, RH front door, upper	1	352030	
	Channel complete, LH front door, upper	1	352031	
3	Channel complete, RH front door, lower front	1	357408	
	Channel complete, LH front door, lower front	1	357409	
4	Channel complete, RH front door, lower rear	1	357406	
	Channel complete, LH front door, lower rear	1	357407	
5	Channel complete, RH rear door, upper	1	352092	
	Channel complete, LH rear door, upper	1	352093	
6	Channel complete, RH rear door, lower front	1	357412	
	Channel complete, LH rear door, lower front	1	357413	
7	Channel complete, RH rear door, lower rear	1	357410	
	Channel complete, LH rear door, lower rear	1	357411	
	Drive screw fixing upper channels	16	78126	
	Set screw (10 UNF x $\frac{3}{4}$" long) } Fixing lower channels	4	78384	
	Bolt (10 UNF x $\frac{3}{4}$" long) } Fixing lower channels	4	257020	
	Special bolt } Fixing lower channels	4	352594	
	Plain washer } Fixing lower channels	4	3685	
	Shakeproof washer } Fixing lower channels	4	74236	
	Nut (10 UNF) } Fixing lower channels	4	257011	
	Packing piece for lower glass channel, bottom	As reqd	352593	
8	Lifting channel for glass, RH front door	1	352004	
	Lifting channel for glass, LH front door	1	352005	
9	Glazing strip, front door	2	352023	
10	Draught strip for door glass	2	352125	
11	Draught strip support for front door	2	356746	
	Rivet fixing strips to support	12	77904	
	Drive screw } Fixing supports to door	8	77903	
	Shakeproof washer } Fixing supports to door	8	78090	
12	Front door louvre, RH	1	352040	
	Front door louvre, LH	1	352041	
13	Rear door louvre, RH	1	352044	
	Rear door louvre, LH	1	352045	
14	Glass for rear door	2	356914	
15	Lifting channel for glass, rear door, RH	1	352006	
	Lifting channel for glass, rear door, LH	1	352007	
16	Glazing strip, rear door	2	352024	
17	Draught strip for door glass, rear door	2	352124	
18	Draught strip support, rear door	2	356747	
	Rivet fixing strips to support	10	77904	
	Drive screw } Fixing supports to door	8	77903	
	Shakeproof washer } Fixing supports to door	8	78090	
	Support bracket complete for door glass	4	357230	
	Rivet } Fixing bracket to door	8	77904	
	Pop rivet } Fixing bracket to door	8	78257	

* Asterisk indicates a new part which has not been used on any previous Rover model

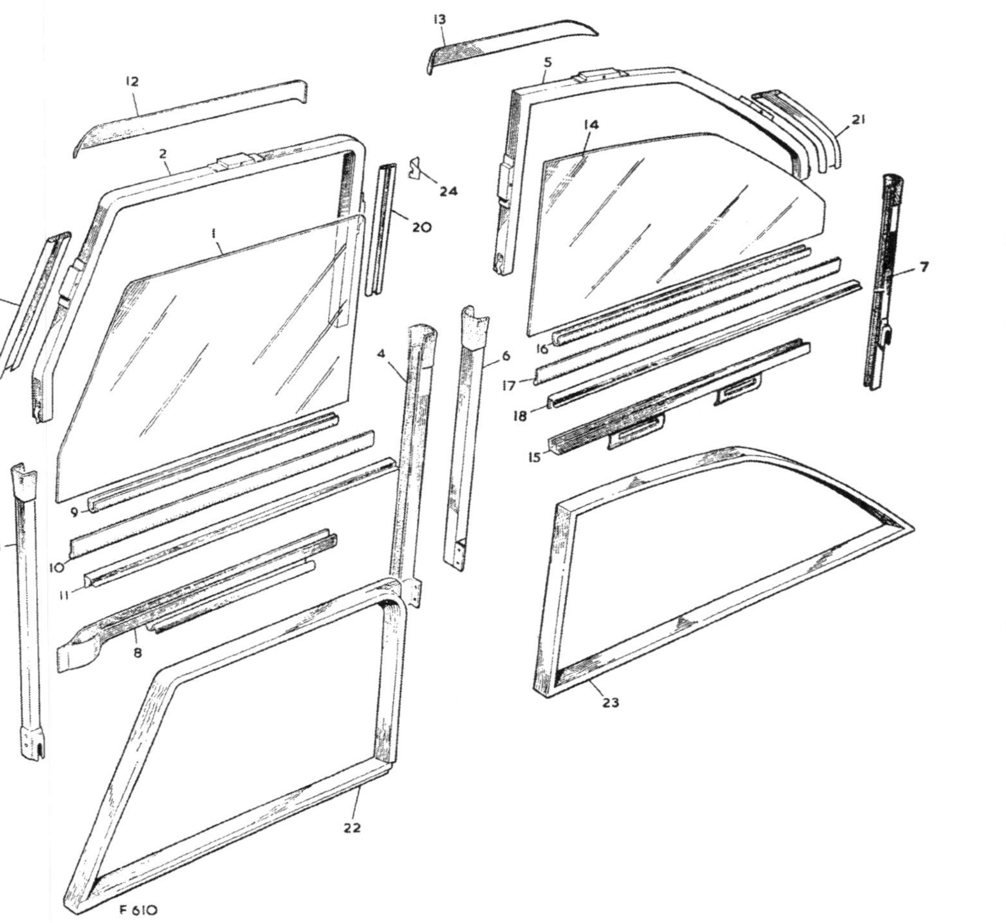

Plate Ref.	1 2 3 4 DESCRIPTION	Qty	Part No.	REMARKS
19	Filler strip for window opening at 'A' post	2	356278	
20	Filler strip for window opening at 'B' and 'C' posts	4	356280	
21	Filler strip for window opening at 'D' post, RH ..	1	356275	
	Filler strip for window opening at 'D' post, LH ..	1	356276	
	Drive screw, short, fixing strips to door	14	77903	For drive screw fixing type
	Drive screw, long, fixing strips to rear door ..	2	78160	For drive screw fixing type
	Clip fixing filler strips to door	16	356274	For clip fixing type
	Interior wood moulding, bottom, RH front door window opening	1	354632†	
	Interior wood moulding, bottom, LH front door window opening	1	354633†	
	Interior wood moulding, bottom, RH rear door window opening	1	354634†	
	Interior wood moulding, bottom, LH rear door window opening	1	354635†	
	Drive screw fixing bottom moulding to door ..	16	77704†	
22	Interior trim moulding, sides and top, RH front window	1	354640†	Quote car number and colour required
	Interior trim moulding, sides and top, LH front window	1	354641†	Quote car number and colour required
23	Interior trim moulding, sides and top, RH rear window	1	354642†	Quote car number and colour required
	Interior trim moulding, sides and top, LH rear window	1	354643†	Quote car number and colour required
	Interior wood moulding complete, RH front door window	1	355450††	
	Interior wood moulding complete, LH front door window	1	355451††	
	Interior wood moulding complete, RH rear door window	1	355478††	
	Interior wood moulding complete, LH rear door window	1	355479††	
	Attachment bracket for interior mouldings complete	8	355454	
	Woodscrew fixing bracket	16	20001	
24	Support bracket for side and top moulding ..	24	352457†	
	Pop rivet fixing bracket to door	48	78390†	
	Drive screw } Fixing moulding to bracket	24	78327†	
	Cup washer } Fixing moulding to bracket	24	74142†	

* Asterisk indicates a new part which has not been used on any previous Rover model
† Cars numbered up to 625001352, 626000431, 628000175, 630000748, 631000192, 633000443
†† Cars numbered from 625001353, 626000432, 628000176, 630000749, 631000193, 633000444 onwards

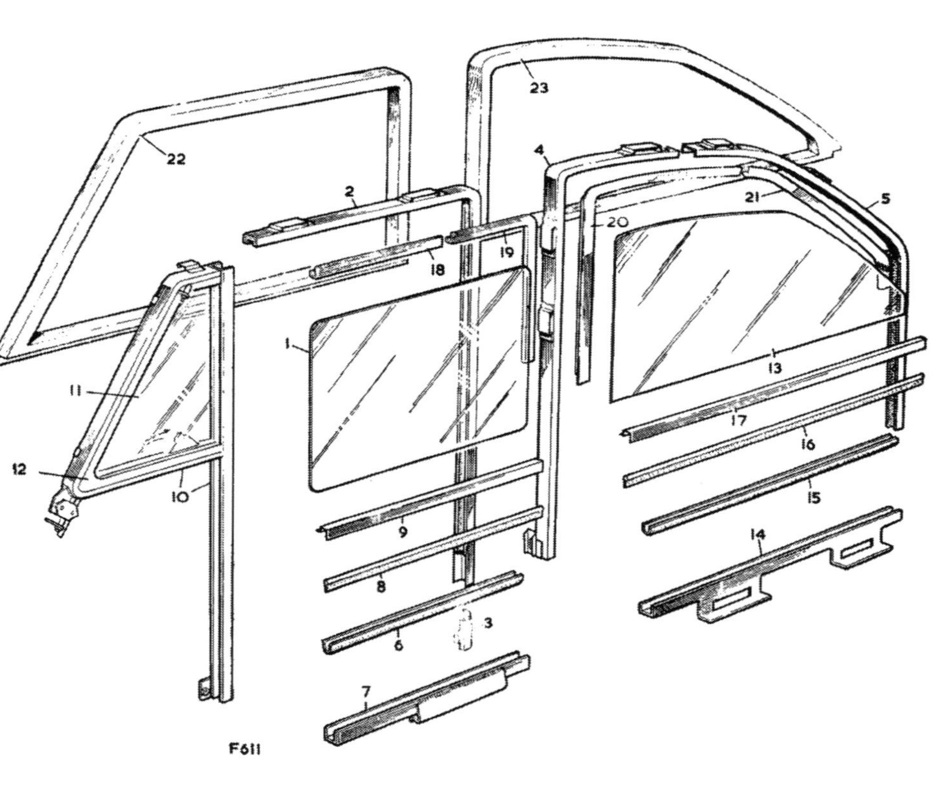

Plate Ref.	1 2 3 4 Description		Qty.	Part No.	Remarks
1	Glass for front door		2	356961	
	Glass for front door, Sundym tinted		2	356850*	
	Channel complete, front door, RH		1	386495	
2	Channel complete, front door, LH		1	386496	
	Support bracket, RH, for glass run channel		1	357365	
3	Support bracket, LH, for glass run channel		1	357366	
	Shim for front door channel, felt		4	312908	
	Channel complete, front, RH rear door		1	386490	
4	Channel complete, front, LH rear door		1	386491	
	Channel complete, rear, RH rear door		1	386492	
5	Channel complete, rear, LH rear door		1	386493	
	Drive screw	Fixing channels to doors	18	64358	
	Set screw (10 UNF x ¾" long), front	Fixing channels to doors	2	78384	
	Set bolt (10 UNF x ¾" long), rear	Fixing channels to doors	2	257020	
	Packing washer to rear door channel bottom		8	312908	
6	Glazing strip, front door		2	357453	
	Lifting channel for glass, RH front door		1	356878	
7	Lifting channel for glass, LH front door		1	357879	
8	Draught strip for front door glass		2	356846	
9	Draught strip support for front doors		2	356847	
	Rivet fixing strips to support		8	77904	
	Drive screw	Fixing supports to door	6	77903	
	Shakeproof washer	Fixing supports to door	6	78090	
	VENT WINDOW COMPLETE, FRONT DOOR RH		1	356887	
10	VENT WINDOW COMPLETE, FRONT DOOR LH		1	356888	
	Channel for door glass		2	386486	
	Vent glass complete, RH		1	356890	
	Vent glass complete, LH		1	356891	
11	Vent glass only		2	356889	
	Vent glass only, Sundym tinted		2	356898*	
	Catch pin and washers for vent window, RH		1	316917	
	Catch pin and washers for vent window, LH		1	316918	
	Rubber moulding for vent glass, RH		1	356892	
12	Rubber moulding for vent glass, LH		1	356893	
	Drive screw, short, fixing vent to door, upper		4	78174	
	Drive screw, long, fixing vent to door, upper		6	78175	
	Set bolt (¼" UNF x ½" long)	Fixing vent to lower door	2	255204	
	Spring washer	Fixing vent to lower door	2	3074	
	Plain washer	Fixing vent to lower door	2	3831	
13	Glass for rear door		2	356914	
	Glass for rear door, Sundym tinted		2	352936*	
	Lifting channel for glass, rear door, RH		1	352006	
14	Lifting channel for glass, rear door, LH		1	352007	
15	Glazing strip, rear door		2	352024	
16	Draught strip for door glass, rear door		2	352124	
17	Draught strip support, rear door		2	356747	
	Rivet fixing strips to support		10	77904	
	Drive screw	Fixing supports to door	8	77903	
	Shakeproof washer	Fixing supports to door	8	78090	
	Support bracket complete for door glass		2	357230	
	Rivet	Fixing bracket to door	8	77904	
	Pop rivet	Fixing bracket to door	8	78257	

*Asterisk indicates a new part which has not been used on any previous model.

Plate Ref.	1 2 3 4	Description	Qty.	Part No.	Remarks
18		Filler strip, window opening, front door top	2	356280	
		Filler strip, window opening at 'C' post	2	356280	
		Filler strip, window opening, front door, RH 'B' post	1	357459	
19		Filler strip, window opening, front door, LH 'B' post	1	357460	
		Filler strip, window opening at rear door, RH front	1	357473	
20		Filler strip, window opening at rear door, LH front	1	357474	
		Filler strip, window opening, rear door, RH rear	1	357463	
21		Filler strip, window opening, rear door, LH rear	1	357464	
		Clip fixing filler strips to door	26	356274	
		Interior wood moulding complete, RH front door	1	356872	Mk IA 3 litre
		Interior wood moulding complete, LH front door	1	356873	Mk IA 3 litre
		Interior wood moulding complete, RH rear door	1	356940	Mk IA 3 litre
		Interior wood moulding complete, LH rear door	1	356941	Mk IA 3 litre
		Interior wood moulding complete, RH — Front doors — Walnut mouldings	1	358957	Mk II, Mk III 3 litre and 3½ litre saloons numbered up to 84000891A, 84300009A
22		Interior wood moulding complete, LH — Front doors — Walnut mouldings	1	358958	Mk II, Mk III 3 litre and 3½ litre saloons numbered up to 84000891A, 84300009A
		Interior wood moulding complete, RH — Rear doors — Walnut mouldings	1	358960	Mk II, Mk III 3 litre and 3½ litre saloons numbered up to 84000891A, 84300009A
23		Interior wood moulding complete, LH — Rear doors — Walnut mouldings	1	358961	Mk II, Mk III 3 litre and 3½ litre saloons numbered up to 84000891A, 84300009A
		Interior wood moulding complete RH front door	1	385898*	3½ Litre Saloons numbered 84000892A 84300010A onwards
		Interior wood moulding complete LH front door	1	385899*	3½ Litre Saloons numbered 84000892A 84300010A onwards
		Interior wood moulding complete RH rear door	1	385886*	3½ Litre Saloons numbered 84000892A 84300010A onwards
		Interior wood moulding complete LH rear door	1	385887*	3½ Litre Saloons numbered 84000892A 84300010A onwards
		Attachment bracket for mouldings	8	355454	
		Woodscrew fixing bracket	16	20001	

*Asterisk indicates a new part which has not been used on any previous model.

DOOR GLASS AND FITTINGS, COUPÉ MODELS

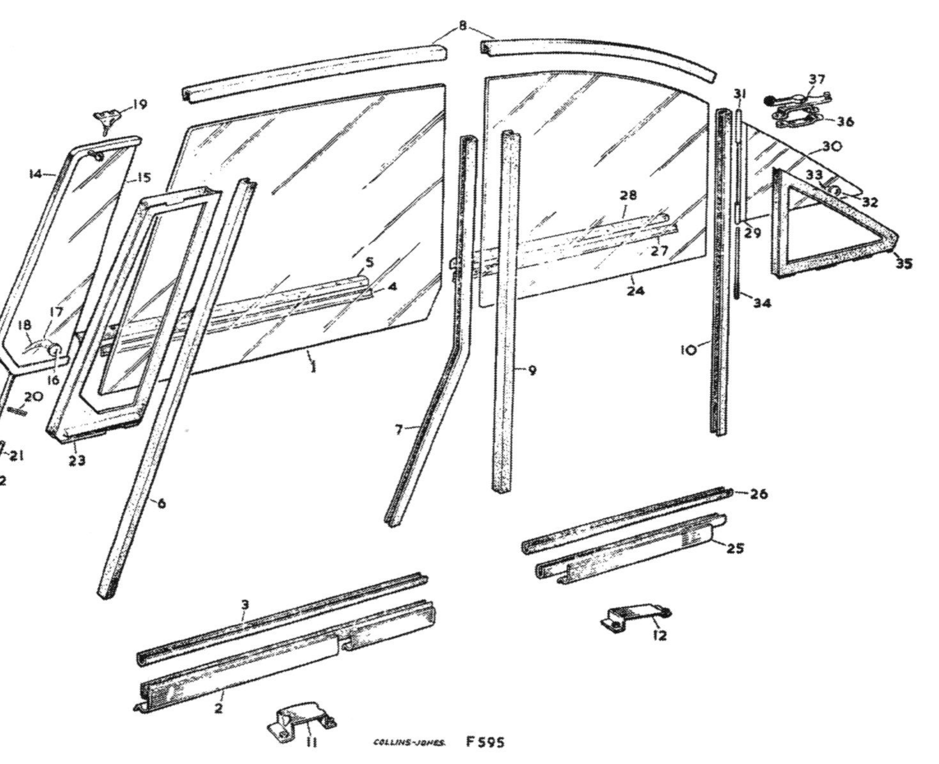

Plate Ref.	1 2 3 4 Description	Qty.	Part No.	Remarks
1	Glass for front door	2	353448	
	Glass for front door, Sundym tinted	2	353446*	
	Lifting channel for glass, RH front door	1	357578	
2	Lifting channel for glass, LH front door	1	357579	
3	Glazing strip, front door	2	357592	
4	Draught strip for front door glass	2	357576	
5	Draught strip support for front doors	2	357575	
	Rivet fixing strips to support	10	77904	
6	Channel complete, front door, front	2	353436	
7	Channel complete, front door, rear	2	353437	
8	Channel complete, top, front and rear doors	4	353435	
9	Channel complete, rear door, front	2	353438	
10	Channel complete, rear door, rear	2	353439	
	Bracket for glass stop on front door, RH	1	359144	
11	Bracket for glass stop on front door, LH	1	359145	
	Bracket for glass stop on rear door, RH	1	359140	
12	Bracket for glass stop on rear door, LH	1	359141	
	Bolt ($\frac{1}{4}$" UNF x $\frac{1}{2}$" long) } Fixing glass stop	8	255204	
	Spring washer } brackets to front and	8	3074	
	Plain washer } rear door panels	8	3931	
	VENT WINDOW COMPLETE, FRONT DOOR RH	1	383719	
13	VENT WINDOW COMPLETE, FRONT DOOR LH	1	383720	
	Frame for vent window, RH	1	357930	
14	Frame for vent window, LH	1	357931	
15	Glass for vent window	2	383527	
	Glass for vent window, Sundym tinted	2	383528*	
	Boss and peg for vent window, RH	1	367166	
16	Boss and peg for vent window, LH	1	367167	
	Rubber bush for catch	2	357819	
	Spacing washer	2	357938	
	Locking ring	2	357939	
	Waved washer for catch	2	357940	
	Mills pin for catch	2	357941	
	Spring for push-button	2	357942	
17	Push-button for catch	2	357943	
	Handle for vent window, RH	1	357944	
18	Handle for vent window, LH	1	357945	
	Pivot bracket for vent window, top RH	1	357946	
19	Pivot bracket for vent window, top LH	1	357947	
	Screw (6 UNC x $\frac{3}{8}$" long) fixing pivot bracket to door frame, top	4	78456	
20	Pin for vent window pivot, bottom	2	383518	
	Locating bush for pivot, bottom RH	1	359202	
21	Locating bush for pivot, bottom LH	1	359203	
22	Spring for locating bush	2	509536	
	Plain washer } Fixing	4	4148	
	Locknut (5/16" UNF) } pivot pin	4	254851	
	Sealing rubber, vent window to door, RH	1	383535	
23	Sealing rubber, vent window to door, LH	1	383536	
24	Glass for rear doors	2	353457	
	Glass for rear doors, Sundym tinted	2	353455*	
	Lifting channel for glass, RH rear door	1	357545	
25	Lifting channel for glass, LH rear door	1	357546	
26	Glazing strip, rear doors	2	357544	

*Asterisk indicates a new part which has not been used on any previous model.

1858

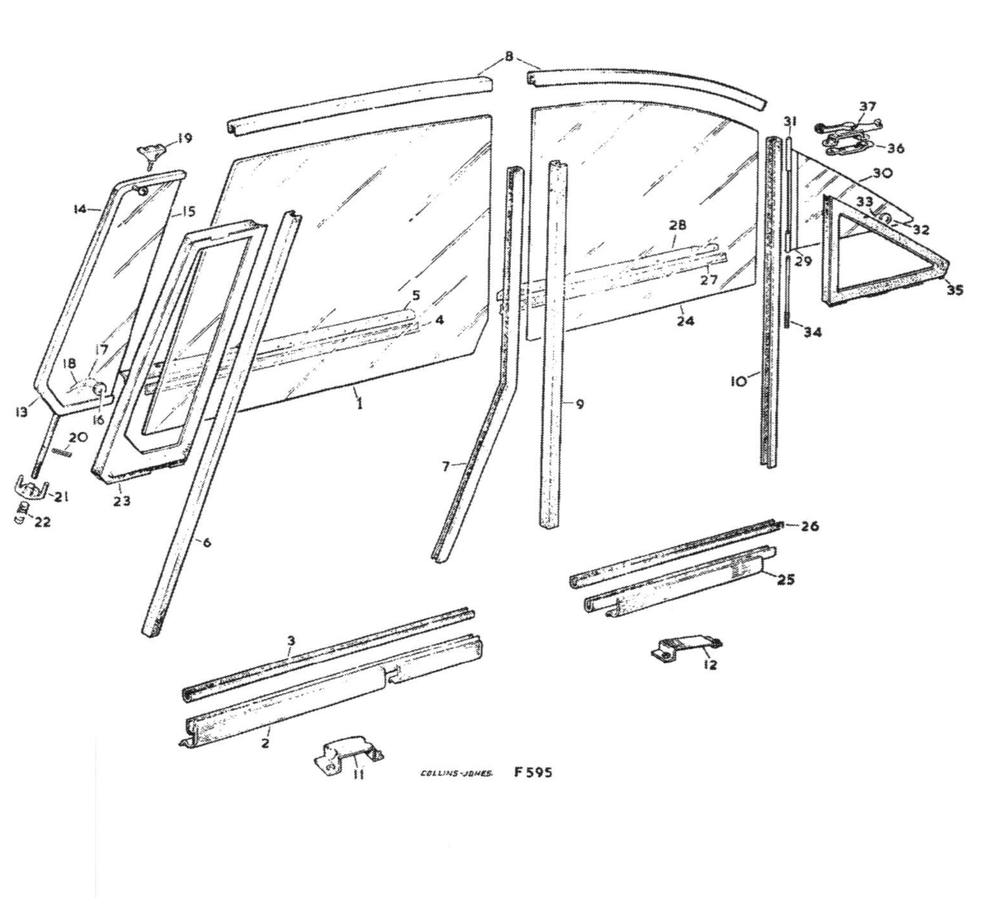

Plate Ref.	1 2 3 4 Description	Qty.	Part No.	Remarks
27	Draught strip for rear door glass	2	357565	
28	Draught strip support for rear doors	2	357564	
	Rivet fixing strips to support	8	77904	
	Drive screw ⎤ Fixing supports	16	77903	
	Shakeproof washer ⎦ to doors	16	78090	
	VENT WINDOW COMPLETE, REAR DOOR, RH	1	357807	
29	VENT WINDOW COMPLETE, REAR DOOR, LH	1	357808	
30	Glass for vent window	2	357815	
	Glass for vent window, Sundym tinted	2	357816*	
31	Pivot assembly	2	357811	
	Rubber bush for catch	2	357819	
32	Boss for catch	2	357817	
33	Slide for catch	2	357818	
	Screw (¼" UNF x 5/16" long) fixing catch to vent window	2	78454	
34	Pivot pin, vent window lower to rear doors	2	357869	
	Grub screw locking pin	2	250013	
	Sealing rubber, vent window to door, RH	1	357863	
35	Sealing rubber, vent window to door, LH	1	357864	
36	Support bracket for vent window catch	2	357963	
	Plain washer ⎤ Fixing support	4	4032	
	Spring washer ⎬ bracket to	4	3072	
	Screw (6 UNC x ½" long) ⎦ door frame	4	78457	
	Plain washer between support bracket and door frame	8	4032	
	Catch for vent window, RH rear door	1	369604	
37	Catch for vent window, LH rear door	1	369605	
	Screw (6 UNC x ⅜" long) fixing catch to bracket	4	78456	
	Fibre washer for catch operating screw	4	4422	
	Screw (10 UNF x ½" long) ⎤ Fixing seal	8	78276	
	Plain washer ⎬ retainer to	8	3867	
	Spring washer ⎥ mounting angle	8	3073	
	Tapped plate ⎦	2	383382	

*Asterisk indicates a new part which has not been used on any previous model.

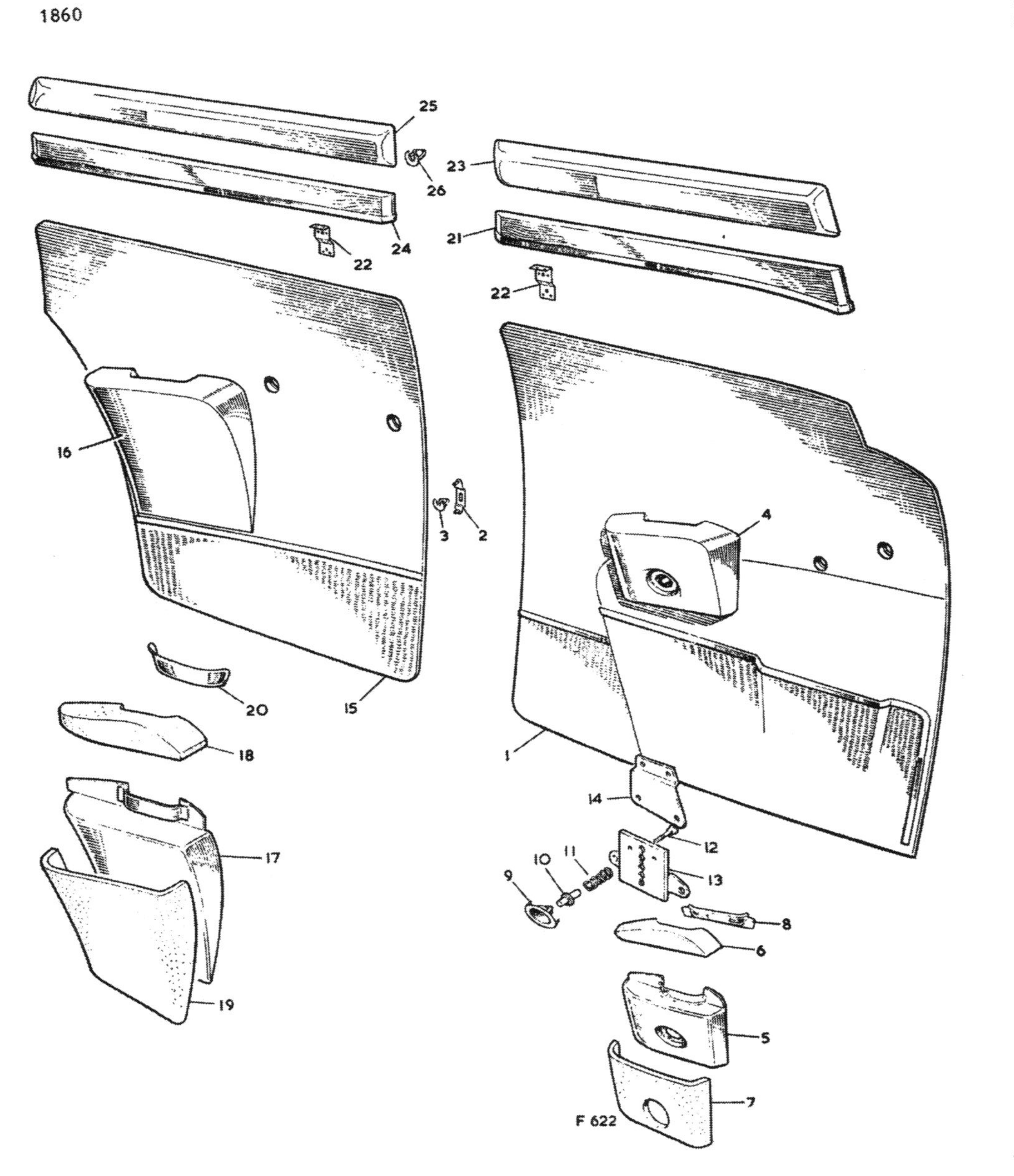

Plate Ref.	1 2 3 4 DESCRIPTION	Qty	Part No.	REMARKS
	DOOR CASING COMPLETE, RH FRONT	1	355627	Quote car number and
1	DOOR CASING COMPLETE, LH FRONT	1	355628	colour required
2	Clip } Fixing	13	314164	
3	Clip } door casing	6	73037	
	SIDE ARMREST COMPLETE, RH FRONT	1	354179	Quote car number and
4	SIDE ARMREST COMPLETE, LH FRONT	1	354180	colour required
	Panel for armrest, RH	1	354181	
5	Panel for armrest, LH	1	354182	
	Topping for armrest, top RH	1	354210	
6	Topping for armrest, top LH	1	354211	
7	Topping for armrest, front	2	354241	
8	Door pull escutcheon	2	354208	
	Drive screw fixing escutcheon on finisher to panel	20	77906	
9	Escutcheon for push-button	2	354197	
10	Push-button, front portion	2	354194	
11	Spring for push-button	2	354196	
12	Push-button, rear portion	2	354195	
	Adjustment plate for armrest, RH	1	355840	
13	Adjustment plate for armrest, LH	1	355841	
	Bracket, RH } For armrest	1	354212†	
	Bracket, RH } For armrest	1	355036††	
	Bracket, LH } For armrest	1	354213†	
14	Bracket, LH } For armrest	1	355037††	
	Screw ($\frac{1}{4}$" UNF x $\frac{1}{2}$" long) } Fixing bracket to front door	4	78230	
	Plain washer } Fixing bracket to front door	4	3831	
	Spring washer } Fixing bracket to front door	4	3074	
	Nut ($\frac{1}{4}$" UNF) } Fixing bracket to front door	4	254810	
	Set screw (10 UNF x $\frac{3}{8}$" long) at armrest bracket	4	78308	
	DOOR CASING COMPLETE, RH REAR	1	355658	Quote car number and colour required
15	DOOR CASING COMPLETE, LH REAR	1	355663	
	SIDE ARMREST COMPLETE, RH REAR	1	354216	
16	SIDE ARMREST COMPLETE, LH REAR	1	354217	
	Panel for armrest, RH	1	354167	
17	Panel for armrest, LH	1	354168	
	Topping for armrest, top RH	1	354218	
18	Topping for armrest, top LH	1	354219	
19	Topping for armrest, front	2	354226	
	Trim clip, short, for door casing	28	73037	20 off from ††
	Trim clip, long, for door casing	10	72602††	
	Drive screw } Fixing armrest	8	77704	
	Plain washer } Fixing armrest	8	4034	
20	Escutcheon for rear door pull	2	354176	
	Drive screw } Fixing escutcheon to armrest and door	2	77923	
	Cup washer } Fixing escutcheon to armrest and door	2	354177	
	Cap for screw cup } Fixing escutcheon to armrest and door	2	354178	
	WAIST ROLL AND WOOD CAPPING COMPLETE, RH FRONT DOOR	1	355154†	
	WAIST ROLL AND WOOD CAPPING COMPLETE, LH FRONT DOOR	1	355155†	
	Wood capping for RH front door	1	354017†	
	Wood capping for LH front door	1	354018†	

* Asterisk indicates a new part which has not been used on any previous Rover model
† Cars numbered up to 625001352, 626000431, 628000175, 630000748, 631000192, 633000443
†† Cars numbered from 625001353, 626000432, 628000176, 630000749, 631000193, 633000444 onwards

1862

Plate Ref.	1 2 3 4 DESCRIPTION	Qty	Part No.	REMARKS
	Woodscrew fixing capping to roll	8	20003†	
	Woodscrew } Fixing beading	10	78286†	
	Woodscrew } to capping	8	65556†	
	Clip for waist roll	4	73037†	
	Wood moulding for RH front door waist	1	355761††	
21	Wood moulding for LH front door waist	1	355762††	
22	Retaining bracket for waist moulding	16	359697††	
	Woodscrew fixing bracket to moulding	32	20115††	
	Drive screw fixing moulding to door	24	77707††	
	Woodscrew fixing beading to capping	10	22780††	
	WAIST ROLL COMPLETE, RH FRONT DOOR, CHARCOAL GREY	1	380691††	
23	WAIST ROLL COMPLETE, LH FRONT DOOR, CHARCOAL GREY	1	380692††	
	Clip for waist roll	8	73037††	
	Drive screw fixing front and rear door waist roll to door	8	78126†	
	End capping for front door, RH front	1	354103†	
	End capping for front door, RH rear	1	354105†	
	End capping for front door, LH front	1	354104†	
	End capping for front door, LH rear	1	354106†	
	End capping for 'A' post waist, RH	1	352484†	
	End capping for 'A' post waist, LH	1	352485†	
	Drive screw fixing end capping to 'A' post	6	78327†	
	WAIST ROLL AND WOOD CAPPING COMPLETE, RH REAR DOOR	1	355156†	
	WAIST ROLL AND WOOD CAPPING COMPLETE, LH REAR DOOR	1	355157†	
	Wood capping for RH rear door	1	354027†	
	Wood capping for LH rear door	1	354028†	
	Woodscrew fixing capping to roll	8	20003†	
	Woodscrew } Fixing beading	10	78286†	
	Woodscrew } to capping	8	65556†	
	Clip for waist roll	6	73037†	
24	Wood moulding for rear door waist	2	355758††	
	Drive screw fixing moulding to door	8	78175††	
	Woodscrew fixing beading to moulding	10	22780††	
	WAIST ROLL COMPLETE, RH REAR DOOR, CHARCOAL GREY	1	380693††	
25	WAIST ROLL COMPLETE, LH REAR DOOR, CHARCOAL GREY	1	380694††	
26	Clip for waist roll	8	73037††	
	End capping, chrome, for rear door, RH front	1	354107†	
	End capping, chrome, for rear door, RH rear	1	354109†	
	End capping, chrome, for rear door, LH front	1	354108†	
	End capping, chrome, for rear door, LH rear	1	354110†	
	Drive screw fixing end capping to door	24	78327†	

* Asterisk indicates a new part which has not been used on any previous Rover model
† Cars numbered up to 625001352, 626000431, 628000175, 630000748, 631000192, 633000443
†† Cars numbered from 625001353, 626000432, 628000176, 630000749, 631000193, 633000444 onwards

864

F601

Plate Ref.	1 2 3 4 Description	Qty.	Part No.	Remarks
	DOOR CASING COMPLETE, RH FRONT	1	359707	Saloon
	DOOR CASING COMPLETE, LH FRONT	1	359705	State colour required
	DOOR CASING COMPLETE, RH FRONT	1	358709	Coupe
1	DOOR CASING COMPLETE, LH FRONT	1	358710	State colour required
2	Staples } Fixing	14	314164	Saloon
3	Trim clips } door casing	14	73037	(20 off on coupe)
	Packing strip(plastic), front, 27" long	2	381078	
	Packing strip(plastic), rear, 25" long	2	381079	Saloon
	Trim fastener fixing strips	4	68087	
4	ARMREST ASSEMBLY FOR FRONT DOORS	2	358646	State colour required
5	Outer panel for armrest	2	358894	
6	Padding for face	2	359006	
7	Padding for top	2	359005	
8	Finisher for finger lift	2	359524	
	Mounting bracket for armrest, RH	1	359586	Saloon
	Mounting bracket for armrest, LH	1	359587	
9	Mounting bracket for armrest	2	358664	Coupe
	Screw (¼" UNF x 7/16" long) } Fixing brackets	6	78251	
	Plain washer } to doors	6	3831	
	Spring washer	6	3074	
	Nut (¼" UNF)	6	254810	
10	BACK PANEL FOR ARMREST	2	358647	
	Nylon slide, RH	1	358660	
	Nylon slide, LH	1	359976	
	Screw (¼" UNF x 1" long) } Fixing back panel	8	78589	
	Spring washer } to bracket	8	3074	
	Spire nut	8	357788	
	Plain washer	4	2217	
	Drive fastener for armrest	4	359977	
	DOOR CASING COMPLETE, RH REAR	1	359691	Saloon. Except Mk III
	DOOR CASING COMPLETE, LH REAR	1	359692	3 Litre and 3½ Litre State colour required
	DOOR CASING COMPLETE, RH REAR	1	384001	Mk III 3 litre and
	DOOR CASING COMPLETE, LH REAR	1	384002	3½ Litre saloon State colour required
	DOOR CASING COMPLETE, RH REAR	1	359066	Coupe
11	DOOR CASING COMPLETE, LH REAR	1	359067	State colour required
	Staples } Fixing door	15	314164	
	Trim clips } casing	18	73037	
	Trim clip for door casing	12	72602	
	ARMREST ASSEMBLY, RH REAR	1	359079	Except Mk III 3 litre and
12	ARMREST ASSEMBLY, LH REAR	1	359080	3½ litre saloon State colour required
	ARMREST ASSEMBLY, RH REAR	1	384003	Mk III 3 litre and 3½ litre
	ARMREST ASSEMBLY, LH REAR	1	384004	saloon State colour required
	Outer panel for armrest, RH	1	359081	Except Mk III 3 litre and
13	Outer panel for armrest, LH	1	359082	3½ Litre saloon
	Outer panel for armrest, RH	1	384033	Mk III 3 litre and 3½ Litre
	Outer panel for armrest, LH	1	384034	saloon
	Padding, top, armrest, RH	1	359098	Except Mk III 3 litre and
14	Padding, top, armrest, LH	1	359099	3½ litre saloon
	Padding, top, armrest, RH	1	384027	Mk III 3 litre and 3½
	Padding, top, armrest, LH	1	384028	litre saloon

*Asterisk indicates a new part which has not been used on any previous model.

F601

COLLINS-JONES

Plate Ref.	1 2 3 4 Description	Qty.	Part No.	Remarks
15	Padding, face, armrest	2	359100	
	Finger pull for armrest, RH	1	359086	Except Mk III 3 Litre and 3½ Litre saloon
16	Finger pull for armrest, LH	1	359087	
	Ashtray complete for armrest	2	384015	Mk III 3 litre and 3½ litre saloon
	Finger pull in armrest, RH	1	384021	
	Finger pull in armrest, LH	1	384022	
	Self-tapping screw, pull to panel	4	78498	
	Screw (6 UNC) } Fixing ashtray to armrest	4	78686	
	Shakeproof washer }	4	78249	
	Plain washer }	4	4203	
	Nut (6 UNC) }	4	257191	
	Drive screw } Fixing armrest to casings	8	78417	
	Plain washer }	8	4034	
	Plastic nut for armrest	8	359319	
	Wood moulding for RH front door waist } Cherrywood veneered mouldings	1	383380	3 litre and 3½ litre saloon Cars numbered up to 84000891A 84300008A
	Wood moulding for LH front door waist }	1	383381	
	Wood moulding for RH front door waist	1	385900*	3½ Litre saloon Cars numbered from:- 84000892A, 84300009A
	Wood moulding for LH front door waist	1	385901*	
	Retaining bracket for waist mouldings	16	359697	Saloon
	Woodscrew(⅜" long) } Fixing bracket to moulding	48	20001	
	Woodscrew (½" long) }	32	20115	
	WAIST ROLL COMPLETE FOR RH FRONT DOOR	1	356933	
	WAIST ROLL COMPLETE FOR LH FRONT DOOR	1	356934	
	Clip for waist roll	8	73037	
	WOOD MOULDING FOR RH FRONT DOOR WAIST	1	359033	3 Litre and 3½ Litre coupe Cars numbered up to 84500610A
17	WOOD MOULDING FOR LH FRONT DOOR WAIST	1	359034	
	WOOD MOULDING FOR RH FRONT DOOR WAIST	1	385904*	3½ Litre Coupe Cars numbered from 84500611A
	WOOD MOULDING FOR LH FRONT DOOR WAIST	1	385905*	
18	Bracket fixing moulding	6	359046	Coupe
	Woodscrew fixing bracket to moulding	24	20003	
19	Retaining clip	4	383475	
	Woodscrew } Fixing retaining clip	4	20114	
	Plain washer }	4	3886	
	Beading for RH front door moulding	1	359043	
20	Beading for LH front door moulding	1	359044	
21	Plastic finisher for sill locking knob	2	359869	
	Plastic end finisher, rear of RH front door moulding	1	383346	
	Plastic end finisher, rear of LH front door moulding	1	383347	
	Wood moulding for RH rear door waist } Cherrywood veneered mouldings	1	383378	3 litre and 3½ litre saloon Cars numbered up to 84000891A, 84300008A
	Wood moulding for LH rear door waist }	1	383379	

*Asterisk indicates a new part which has not been used on any previous model.

DOOR CASINGS AND ARMRESTS, MK II, MK III 3 LITRE AND 3½ LITRE SALOON AND COUPE MODELS

Plate Ref.	1 2 3 4 Description	Qty.	Part No.	Remarks
	Wood moulding for RH rear door waist	1	385888*	3½ Litre saloon
	Wood moulding for LH rear door waist	1	385889*	Cars numbered from 84000892A, 84300009A
	Retaining bracket	16	359697	
	WAIST ROLL COMPLETE FOR RH REAR DOOR	1	355631	Saloon
	WAIST ROLL COMPLETE FOR LH REAR DOOR	1	355632	
	Clip for waist roll	8	73037	
	WOOD MOULDING FOR RH REAR DOOR WAIST	1	359047	3 litre and 3½ litre
22	WOOD MOULDING FOR LH REAR DOOR WAIST	1	359048	Coupe Cars numbered up to 84500610A
	WOOD MOULDING FOR RH REAR DOOR WAIST	1	385892*	3½ Litre coupe
	WOOD MOULDING FOR LH REAR DOOR WAIST	1	385893*	Cars numbered from:- 84500611A
23	Bracket fixing moulding	6	359046	Coupe
	Woodscrew fixing bracket to moulding	24	20003	
24	Retaining clip	4	383475	
	Woodscrew fixing retaining clip	4	20114	
	Plain washer	4	3886	
25	Beading for moulding, rear doors	2	359057	
26	Plastic finisher for sill locking knob	2	359869	

*Asterisk indicates a new part which has not been used on any previous model.

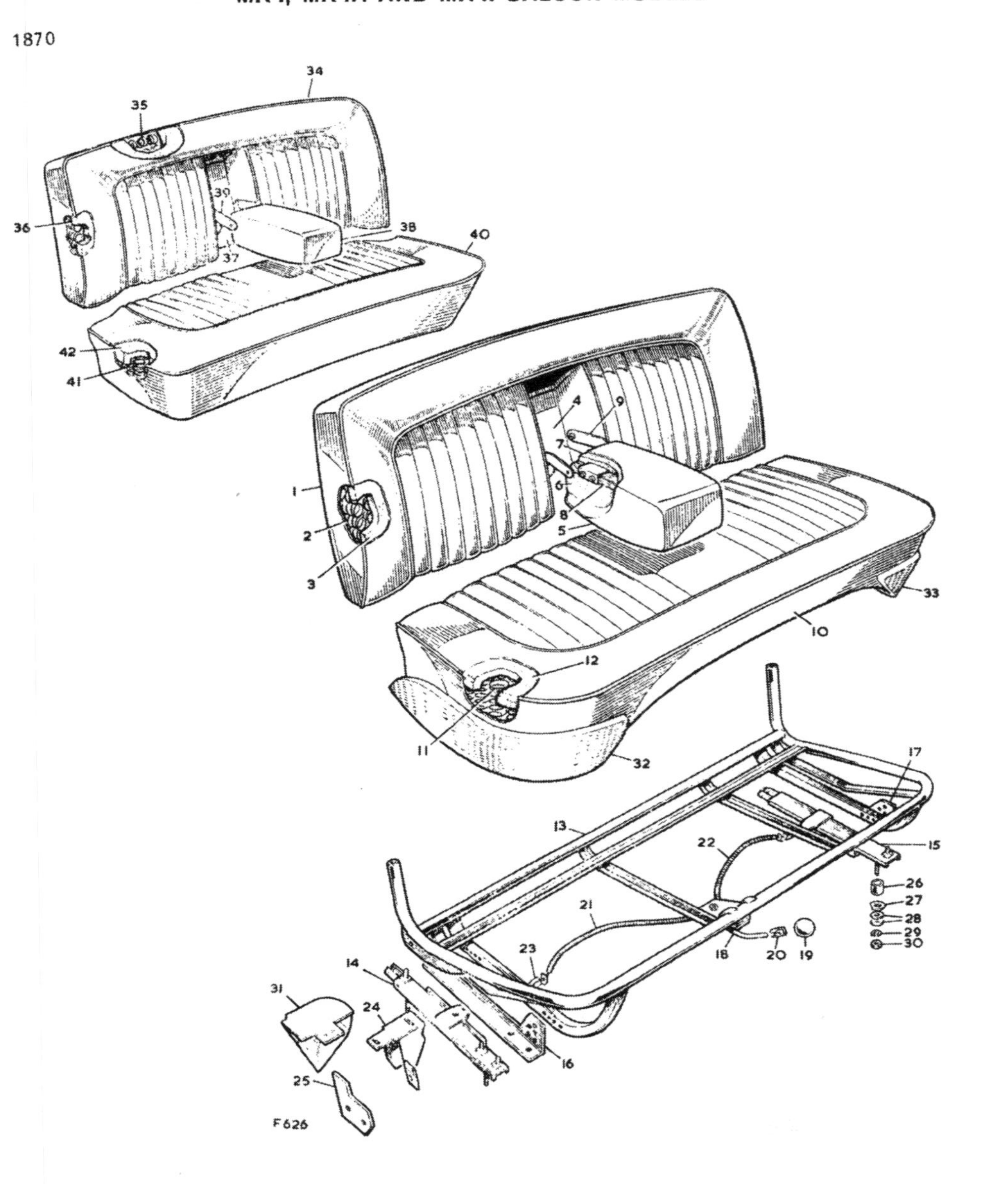

Plate Ref.	1 2 3 4 DESCRIPTION	Qty	Part No.	REMARKS
1	FRONT SQUAB COMPLETE	1	355554	Quote car number and colour required
	FRAME FOR SQUAB	1	354034†	Mk I and
	FRAME FOR SQUAB	1	358974††	Mk IA
	FRAME FOR SQUAB	1	359348	Mk II
	Spring case for squab	1	355401	Mk I and Mk IA
2	Spring case for squab	1	359349	Mk II
	Side plate for squab frame	2	354261	Mk I and Mk IA
	Side panel for squab frame	2	381081	Mk II
	Screw (10 UNF x 1" long) } Fixing side plate to squab frame	4	78325	
	Plain washer	4	3903	
	Nut (10 UNF)	4	257023	
	Flute padding for squab, plastic foam ..	As reqd	355984	
	Topping for top roll	1	354319†	Mk I and Mk IA
	Topping, RH, for flute panel	1	354318†	
	Topping, LH, for flute panel	1	354326†	
	Topping for side roll, RH	1	354320†	
	Topping for side roll, LH	1	354321†	
	Topping for top roll	1	355410††	
	Topping, RH, for flute panel	1	355408††	
	Topping, LH, for flute panel	1	355409††	
	Topping, RH, for surround panel	1	355406††	
	Topping, LH, for surround panel	1	355407††	
	Topping, RH, for flute panel	1	359351	Mk II
	Topping, LH, for flute panel	1	359352	
3	Topping, top and sides	1	359353	
	Clip } Fixing cover	20	331071	
	Clip	40	315960	
	Clip fixing squab to seat frame at rear	12	315264	
	Back panel for front squab	1	355542	Quote car number and colour required
	Clip fixing back panel to squab frame	11	354330	
4	Front armrest recess complete	1	355543	Quote car number and colour required
5	CENTRE ARMREST COMPLETE	1	315075	
	Topping for armrest, front	1	354039†	Mk I and Mk IA
	Topping for armrest, back	1	354038†	
	Tongue for armrest	1	354037†	
	Tongue for armrest	1	355905††	
	Link for centre armrest	2	354040†	
	Topping for armrest, front	1	315027††	
	Topping for armrest, back	1	315028††	
6	Topping for armrest, front	1	359376	Mk II
7	Topping for armrest, back	1	359375	
8	Tongue for armrest	1	359374	
	Link for centre armrest	2	355403††	Mk I and Mk IA
9	Link for armrest	2	359377	Mk II
	Special set screw } Fixing links to armrest	4	354822	
	Shakeproof washer	4	70884	
	Bolt ($\frac{1}{4}$" UNF x $\frac{3}{4}$" long) } Fixing links to squab frame	4	255207	Alternative to clip-type fixing
	Plain washer	4	3840	
	Shakeproof washer	4	70884	
	Nut ($\frac{1}{4}$" UNF)	4	254810	

* Asterisk indicates a new part which has not been used on any previous Rover model
† Cars numbered up to 625000979, 626000332, 628000110, 630000459, 631000115, 633000310
†† Cars numbered from 625000980, 626000333, 628000111, 630000460, 631000116, 633000311 onwards

REAR SEAT AND BENCH-TYPE FRONT SEAT MK I, MK IA AND MK II SALOON MODELS

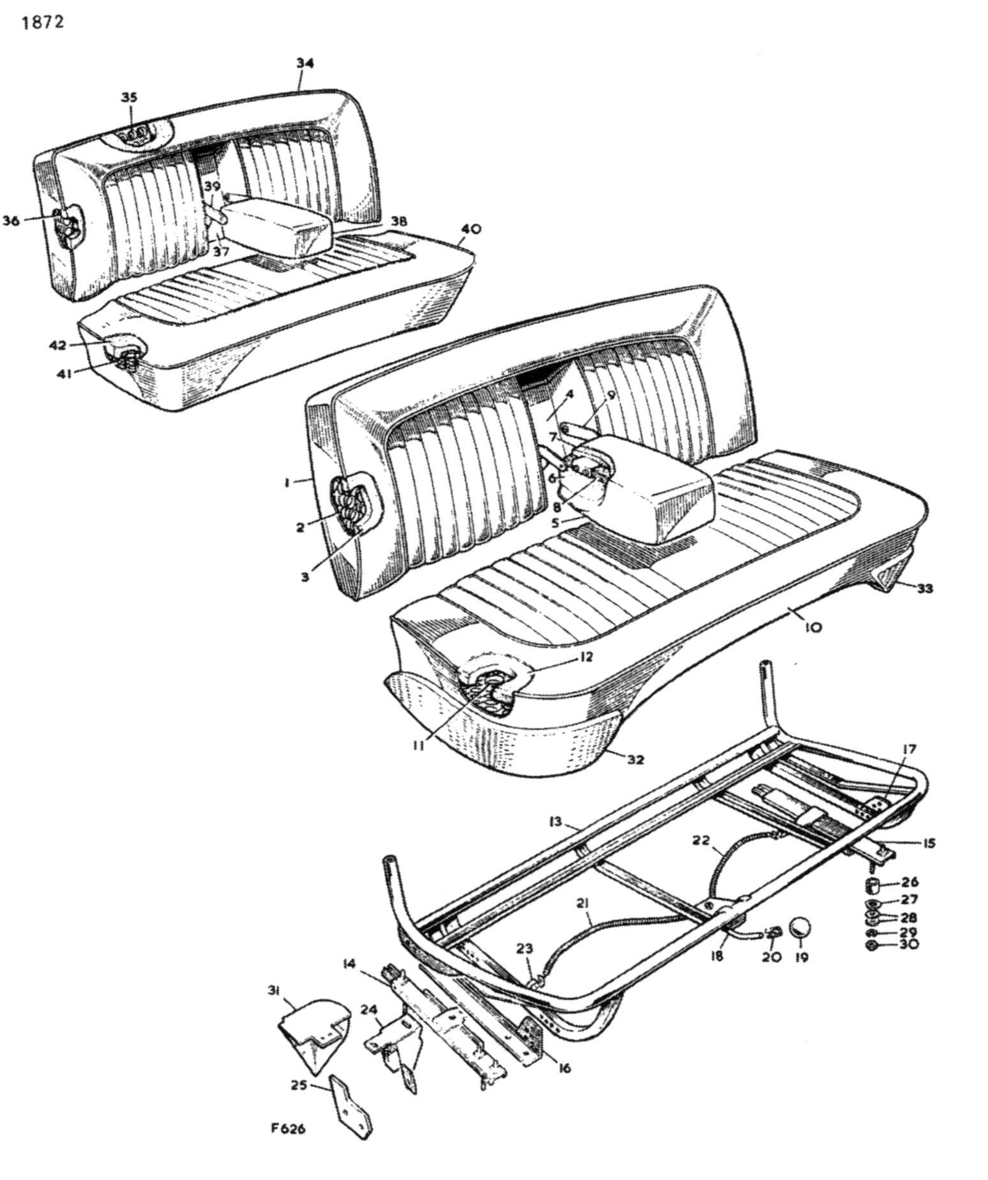

REAR SEAT AND BENCH-TYPE FRONT SEAT MK I, MK IA AND MK II SALOON MODELS

Plate Ref.	1 2 3 4 DESCRIPTION	Qty	Part No.	REMARKS
	Spire nut and clip	4	357788	Alternative to bolt and washer fixing
10	FRONT CUSHION COMPLETE	1	355861	Quote car number and colour required
	Spring case	1	354035	Mk I and Mk IA
11	Spring case	1	359331	Mk II
	Topping for side roll, RH	1	354316†	
	Topping for side roll, LH	1	354317†	
	Topping for flute panel	1	354314†	Mk I and Mk IA
	Topping for front roll	1	354315†	
	Topping for cushion surround	1	355411††	
	Topping for flute panel	1	355412††	
	Topping for flute panel	1	359714	Mk II
12	Topping for surround panel	1	359336	
	Clip for cover and pockets	52	331071	
	Front seat frame complete	1	355852	Mk I and Mk IA
13	Front seat frame complete	1	359383	Mk II
	Seat slide, RH	1	354000	Mk I and
	Seat slide, LH	1	354001	Mk IA
14	Seat slide, RH	1	357745	
15	Seat slide, LH	1	357746	Mk II
16	Mounting bracket, RH } Slides to	1	359384	
17	Mounting bracket, LH } seat frame	1	359385	
	Bolt (5/16" UNF x 7/8" long) } Fixing seat slides to seat frame	4	255227	
	Plain washer	4	3899	Mk I and Mk IA
	Spring washer	4	3075	
	Nut (5/16" UNF)	4	254811	
	Bolt (5/16" UNF x 1½" long) } Fixing mounting brackets to seat frame	3	256222	
	Plain washer	4	3830	
	Spring washer	4	3075	Mk II
	Nut (5/16" UNF)	4	254811	
	Spring washer } Fixing mounting brackets	6	3075	
	Nut (5/16" UNF) } to seat slides	6	254811	
	Front seat control	1	354049	Mk I and Mk IA
18	Front seat control	1	359017	Mk II
	Screw (5/16" UNF x 1¼" long) } Fixing seat slides to floor	2	78324	
	Bearing washer	4	311941	
	Packing washer	4	3793	
	Plain washer	4	311199	
	Spring washer	4	3075	
	Nut (5/16" UNF)	4	254811	Mk I and Mk IA
	Set bolt (¼" UNF x ¾" long) } Fixing handle to frame	1	255007	
	Plain washer	1	3840	
	Nut (¼" UNF)	1	254850	
	Knob for handle control	1	354309	
	Clip fixing knob	1	279955	
19	Knob for operating handle	1	359019	Mk II
20	Clip fixing knob	1	359020	
	Cable complete, long, for seat control	1	354800	Mk I and
	Cable complete, short, for seat control	1	354801	Mk IA

* Asterisk indicates a new part which has not been used on any previous Rover model
† Cars numbered up to 625000979, 626000332, 628000110, 630000459, 631000115, 633000310
†† Cars numbered from 625000980, 626000333, 628000111, 630000460, 631000116, 633000311 onwards

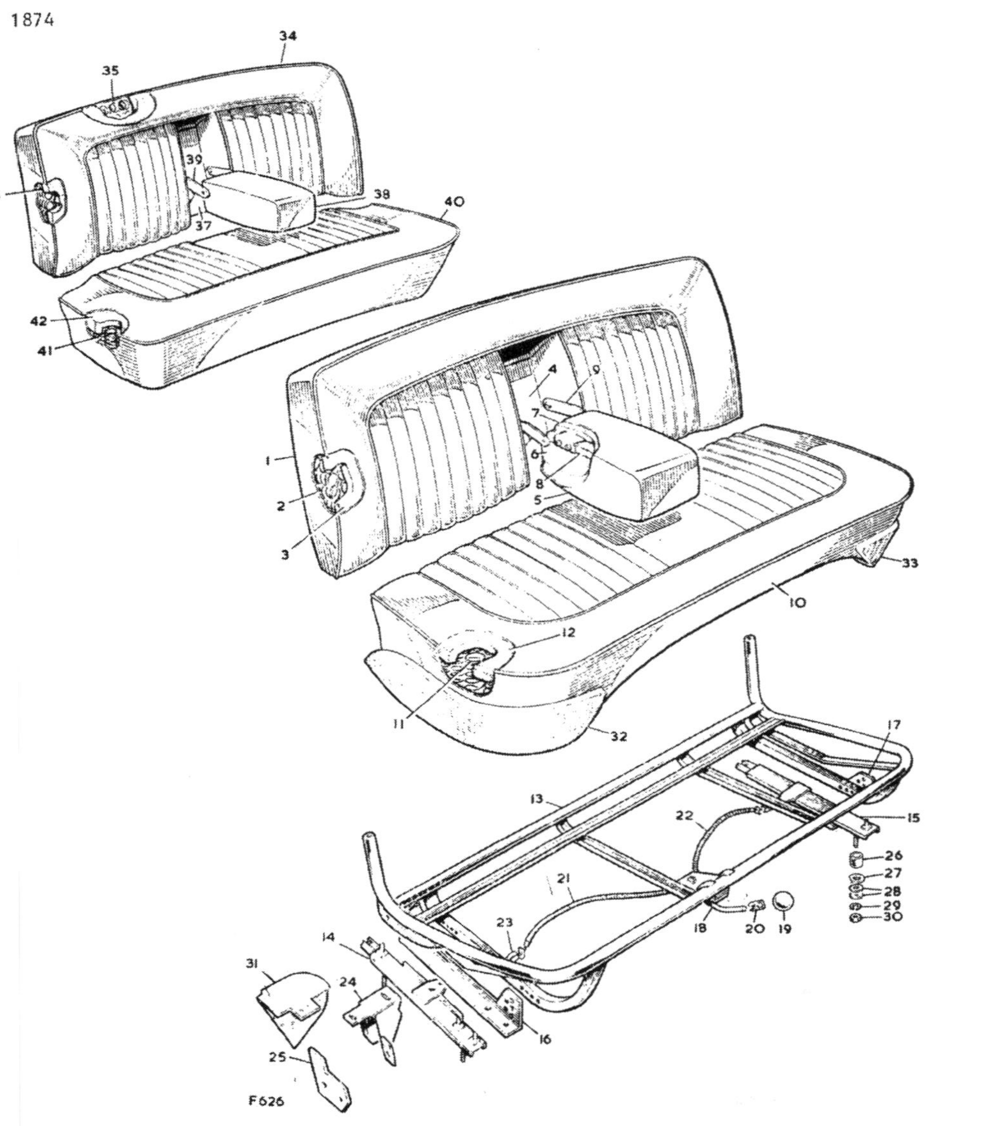

Plate Ref.	1 2 3 4 DESCRIPTION	Qty	Part No.	REMARKS
21	Cable complete, long, for seat control	1	359390	
22	Cable complete, short, for seat control	1	359391	Mk II
23	Tab washer for control cables	2	359014	
	Spindle } For retaining cable to seat slide	2	312689	
	Bush } For retaining cable to seat slide	2	312690	Mk I and Mk IA
	Nut (2 BA) } For retaining cable to seat slide	2	2247	
24	Support bracket for slide, outer	2	354853	
25	Reinforcement plate	2	359541	
	Bolt ($\frac{5}{16}$" UNF x $\frac{3}{4}$" long) } Fixing support bracket to floor	6	255226	
	Spring washer } Fixing support bracket to floor	6	3075	
	'D' washer } Fixing support bracket to floor	6	311199	
	Plain washer } Fixing support bracket to floor	6	3899	
	Nut ($\frac{5}{16}$" UNF) } Fixing support bracket to floor	6	254811	
26	Bearing washer } Fixing front and rear of outer slides to floor	4	311941	Mk II
27	'D' washer } Fixing front and rear of outer slides to floor	2	311199	
28	Plain washer } Fixing front and rear of outer slides to floor	2	3830	
29	Spring washer } Fixing front and rear of outer slides to floor	4	3075	
30	Nut ($\frac{5}{16}$" UNF) } Fixing front and rear of outer slides to floor	4	254811	
31	Cover complete, RH } For seat support brackets	1	355063	
	Cover complete, LH } For seat support brackets	1	355064	
	Drive screw fixing covers	8	77704	
32	Valance complete, RH	1	355394	Quote car number and
33	Valance complete, LH	1	355395	colour required
	Drive screw } Fixing valances	4	77704	
	Plain washer } Fixing valances	4	3816	
	Polythene cover for squab } For front bench seat	1	381110	
	Polythene cover for cushion } For front bench seat	1	381111	
34	REAR SQUAB COMPLETE	1	357349	Quote car number and colour required
	Spring case for rear squab	1	354079†	Mk I and
	Spring case for rear squab	1	355981††	Mk IA
35	Spring case for rear squab	1	359742	Mk II
	Topping for rear squab top rail	1	354808†	
	Topping for RH rear squab flute panel	1	354809†	
	Topping for LH rear squab flute panel	1	354810†	
	Topping for side roll, RH	1	354806†	
	Topping for side roll, LH	1	354807†	
	Topping for rear squab upper roll	1	355643††	Mk I and Mk IA
	Topping for RH rear squab flute panel	1	355416††	
	Topping for LH rear squab flute panel	1	355417††	
	Topping, RH, for surround panel	1	355414††	
	Topping, LH, for surround panel	1	355415††	
	End topping for rear squab	2	355321††	
	Topping for RH rear squab flute panel	1	359728	
	Topping for LH rear squab flute panel	1	359729	Mk II
36	Topping for surround panel	1	359727	
	Clip } For cover	26	331071	
	Clip } For cover	13	315960	
37	Rear armrest recess complete	1	355604	Quote car number and colour required
	Rivet fixing armrest recess	2	78118	

* Asterisk indicates a new part which has not been used on any previous Rover model
† Cars numbered up to 625000979, 626000332, 628000110, 630000459, 631000115, 633000310
†† Cars numbered from 625000980, 626000333, 628000111, 630000460, 631000116, 633000311 onwards

REAR SEAT AND BENCH-TYPE FRONT SEAT MK I, MK IA AND MK II SALOON MODELS

1876

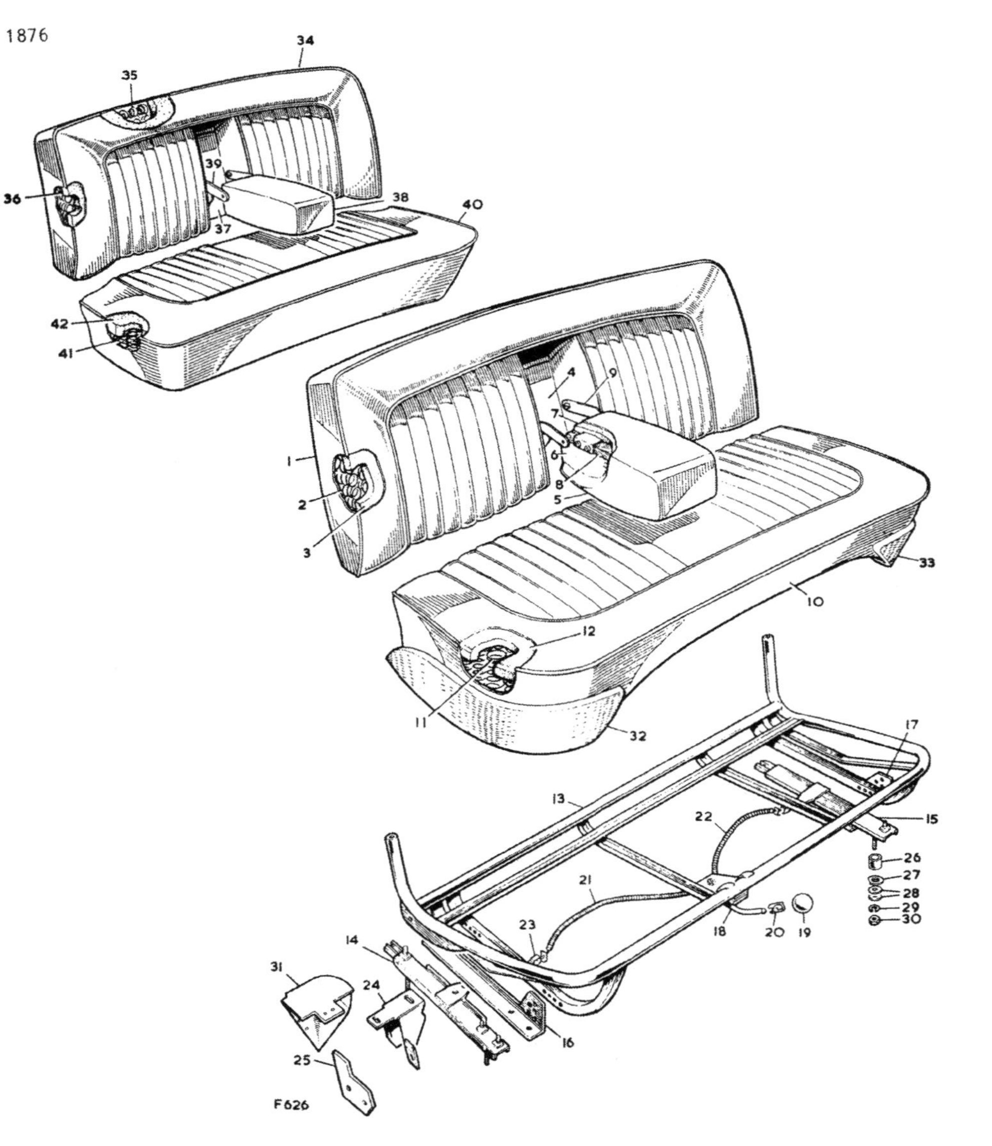

REAR SEAT AND BENCH-TYPE FRONT SEAT MK I, MK IA AND MK II SALOON MODELS

1877

Plate Ref.	1 2 3 4 DESCRIPTION	Qty	Part No.	REMARKS
38	CENTRE ARMREST COMPLETE	1	315075	Quote car number and colour required
	Topping for armrest, front	1	354251†	
	Topping for armrest, back	1	354252†	
	Tongue for armrest	1	354253†	Mk I and Mk IA
	Link for centre armrest	2	354254†	
	Topping for armrest, front	1	315028††	
	Topping for armrest, back	1	315027††	
	Topping for armrest, back	1	315028	Mk II
	Topping for armrest, front	1	359758	
	Tongue for armrest	1	355905††	
39	Link for centre armrest	2	381224††	
	Special set screw } Fixing links	4	310896	
	Shakeproof washer } to armrest	4	70884	
	Bolt (¼" UNF x ⅝" long) }	4	255206	
	Plain washer } Fixing links to	4	3840	Alternative to clip-type fixing
	Shakeproof washer } squab frame	4	70884	
	Nut (¼" UNF) }	4	254810	
	Spire nut and clip	4	357788	Alternative to nut and bolt fixing
	Drive screw fixing rear squab back panel to squab frame	8	78140	
	Set bolt (¼" UNF x 1¼" long) } Fixing	4	255211	
	Plain washer } squab	4	2552	
	Rubber washer } to body	4	355085	
	Drive screw fixing end wing to rear squab	2	77704	
40	REAR CUSHION COMPLETE	1	355860	Quote car number and colour required
	Spring case	1	354078	Mk I and Mk IA
41	Spring case	1	359741	Mk II
	Topping for front roll	1	354323†	
	Topping for side roll, RH	1	354324†	
	Topping for side roll, LH	1	354325†	Mk I and Mk IA
	Topping for rear cushion surround	1	355419††	
	Topping for flute panel	1	354322†	
	Topping for flute panel	1	355420††	
	Topping, rear flute panel	1	359714	Mk II
42	Topping for rear surround	1	359715	
	Clip for cover and pocket	48	331071	

* Asterisk indicates a new part which has not been used on any previous Rover model
† Cars numbered up to 625000979, 626000332, 628000110, 630000459, 631000115, 633000310
†† Cars numbered from 625000980, 626000333, 628000111, 630000460, 631000116, 633000311 onwards

REAR SEAT, MK II COUPÉ MODELS

F625

Plate Ref.	1 2 3 4 Description		Qty.	Part No.	Remarks
1	REAR SQUAB AND ARMREST ASSEMBLY		1	358515	Quote car number and colour required
	Rear squab frame complete		1	357838	
	Topping, top centre, rear squab		1	358516	
2	Topping, surround RH, rear squab		1	358517	
	Topping, surround LH, rear squab		1	358518	
	Topping, centre RH, rear squab		1	358519	
	Topping, centre LH, rear squab		1	358520	
3	CENTRE ARMREST RECESS TRIM COMPLETE		1	357998	Quote car number and colour required
4	CENTRE ARMREST ASSEMBLY REAR SQUAB		1	358524	Quote car number and colour required
5	Topping, back, for armrest		1	353646	
6	Topping, front, for armrest		1	353645	
7	Tongue for armrest		1	353647	
8	Link assembly for armrest		1	353651	
	Set screw(¼" UNF x ½" long)	Fixing links to armrest	4	354822	
	Shakeproof washer	Fixing links to armrest	4	70884	
	Bolt(¼" UNF x ¾" long)	Fixing armrest bracket to squab frame	4	255007	
	Plain washer	Fixing armrest bracket to squab frame	4	3840	
	Spire nut and clip	Fixing armrest bracket to squab frame	4	357788	
9	Bracket fixing squab to body		4	381018	
	Bolt (¼" UNF x 1¼" long)	Fixing bracket to squab frame	8	255011	
	Plain washer	Fixing bracket to squab frame	8	2552	
	Nut (¼" UNF)	Fixing rear squab to body	4	254810	
	Plain washer	Fixing rear squab to body	4	2552	
	Rubber washer	Fixing rear squab to body	4	355085	
	Bolt (¼" UNF x ¾" long)	Fixing rear squab armrest bracket to body	2	255207	
	Plain washer	Fixing rear squab armrest bracket to body	2	2552	
	Spring washer	Fixing rear squab armrest bracket to body	2	3074	
10	REAR CUSHION ASSEMBLY, RH		1	353565	Quote car number and colour required
11	Main latex cushion for RH seat		1	353562	
12	Surround for cushion (latex)for RH seat		1	353542	
13	REAR CUSHION ASSEMBLY, LH		1	353566	Quote car number and colour required
	Main latex cushion for LH seat		1	353563	
	Surround for cushion(latex)for LH seat		1	353543	
14	CENTRE PAD BETWEEN REAR CUSHIONS		1	357974	Quote car number and colour required
	Centre pad (latex)		1	353585	
	Drive screw	Fixing centre pad to body	8	77704	
	Plain washer	Fixing centre pad to body	8	4203	
15	Plinth for ashtray, rear seat		1	359157	
	Drive screw fixing plinth to body		4	77704	
16	ASHTRAY AND CIGAR LIGHTER HOLDER COMPLETE		1	383686	
17	Bowl for ashtray		1	359122	
18	Chrome top for ashtray		1	359123	
	Screw(10 UNF x ¾" long)	Fixing ashtray assembly to plinth	2	78384	
	Plain washer	Fixing ashtray assembly to plinth	2	3851	
19	CIGAR LIGHTER COMPLETE FOR REAR SEAT		1	565850	
	Element for cigar lighter		1	600280	
	Bulb for cigar lighter		1	530054	
	Knob for cigar lighter		1	600593	
	Flange for knob		1	600602	

*Asterisk indicates a new part which has not been used on any previous model.

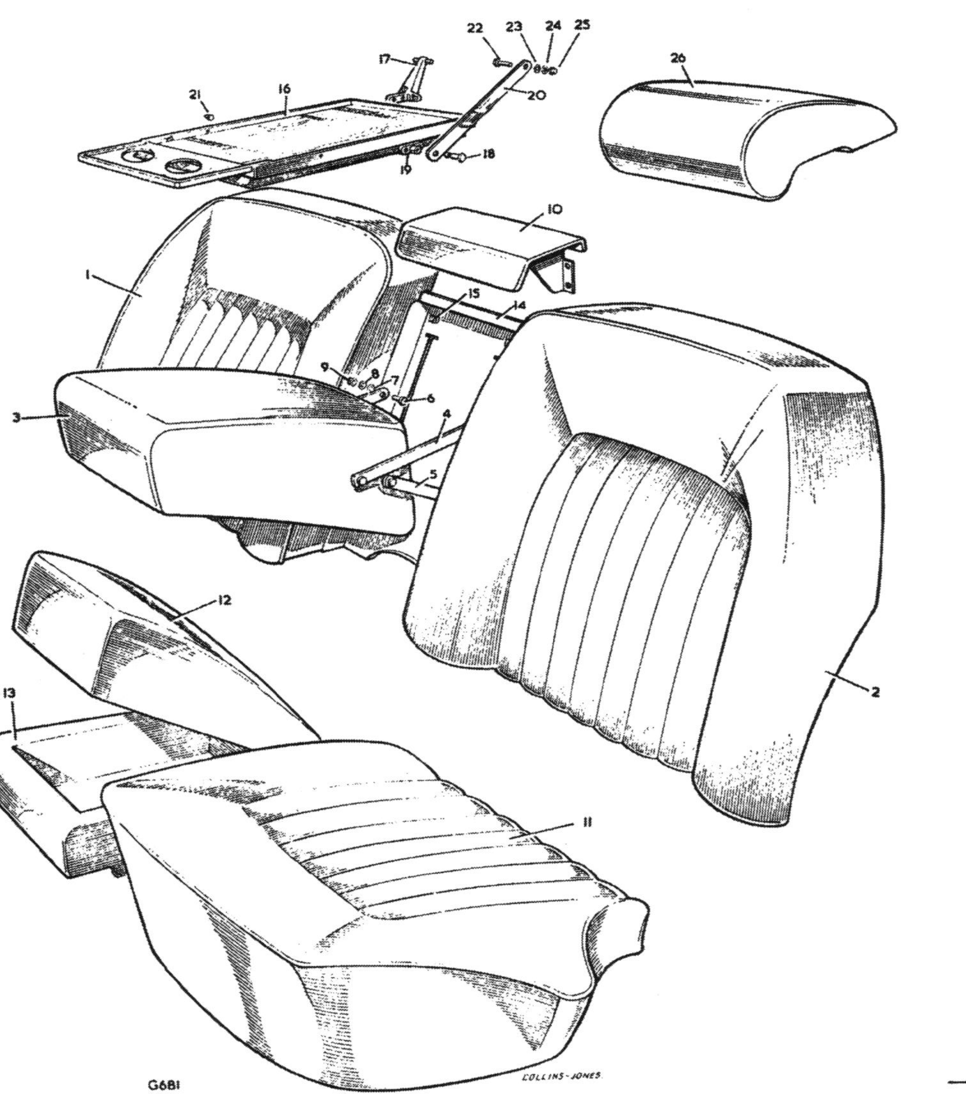

Plate Ref.	1 2 3 4 Description	Qty.	Part No.	Remarks
1	Rear squab assembly, RH	1	383911	Quote car number
2	Rear squab assembly, LH	1	383912	and colour required
3	Centre armrest assembly	1	383913	
4	Top links for centre armrest	2	383833	
5	Bottom links for centre armrest	2	383834	
	Shakeproof washer } Fixing links	4	70884	
	Set screw (¼" UNF x ½" long) } to armrest	4	354822	
6	Shaft	4	365096	
7	Plain washer } Fixing links to	2	3840	
8	Waved washer } squab frame	2	353368	
9	Retaining clip	4	78573	
10	Rear squab spacer assembly	1	384344	Quote car number and colour required
	Drive screw fixing spacer to frame	4	78549	
	Drive screw } Fixing squab	8	78549	
	Plain washer } to frame	8	4203	
	Rear cushion assembly, RH	1	383880	Quote car number and colour
11	Rear cushion assembly, LH	1	383881	required
12	SPACER, TOP ASSEMBLY (between cushions)	1	384053	
	Clip fixing cover to panel	12	331071	
13	TRINKET BOX ASSEMBLY FOR REAR SEAT	1	384054	Quote car number and colour required
	Hinge for trinket box	2	384062	
	Drive screw fixing hinge	6	78483	
	Bolt (10 UNF x ½" long) } Fixing	2	257017	
	Spring washer } trinket box	2	3073	
	Plain washer } to spacer top	2	3851	
	Drive screw } Fixing trinket box	4	77704	
	Plain washer } to body	4	4203	
14	Guide panel for picnic tray	1	383829	
	Buffer for panel	1	234429	
	Drive screw fixing panel to squab frame	6	78549	
15	Anti-rattle clip for guide panel	2	385515	
	Retaining clip for anti-rattle clip	2	385517	
	PICNIC TRAY COMPLETE FOR REAR SQUAB	1	384245	
16	Picnic tray	1	384159	
17	Guide bracket, RH	1	383981	
	Guide bracket, LH	1	383982	
18	Pivot shaft	2	383965	
19	Nylon bush	2	384882	
20	Link for picnic tray	2	383948	
21	Friction plug for flap	2	384251	
	Handle for picnic tray	1	385514	
22	Shaft	2	365096	
23	Washer } Fixing picnic tray	2	3840	
24	Waved washer } link to squab frame	2	353368	
25	Retaining clip	2	78573	
	Bolt (¼" UNF x 1¼" long) } Fixing	4	255211	
	Plain washer } rear squab	4	2552	
	Sealing washer } to body	4	355085	
	Headrest complete, RH	1	384847	Optional equipment. Quote
26	Headrest complete, LH	1	384848	car number and colour required

*Asterisk indicates a new part which has not been used on any previous model.

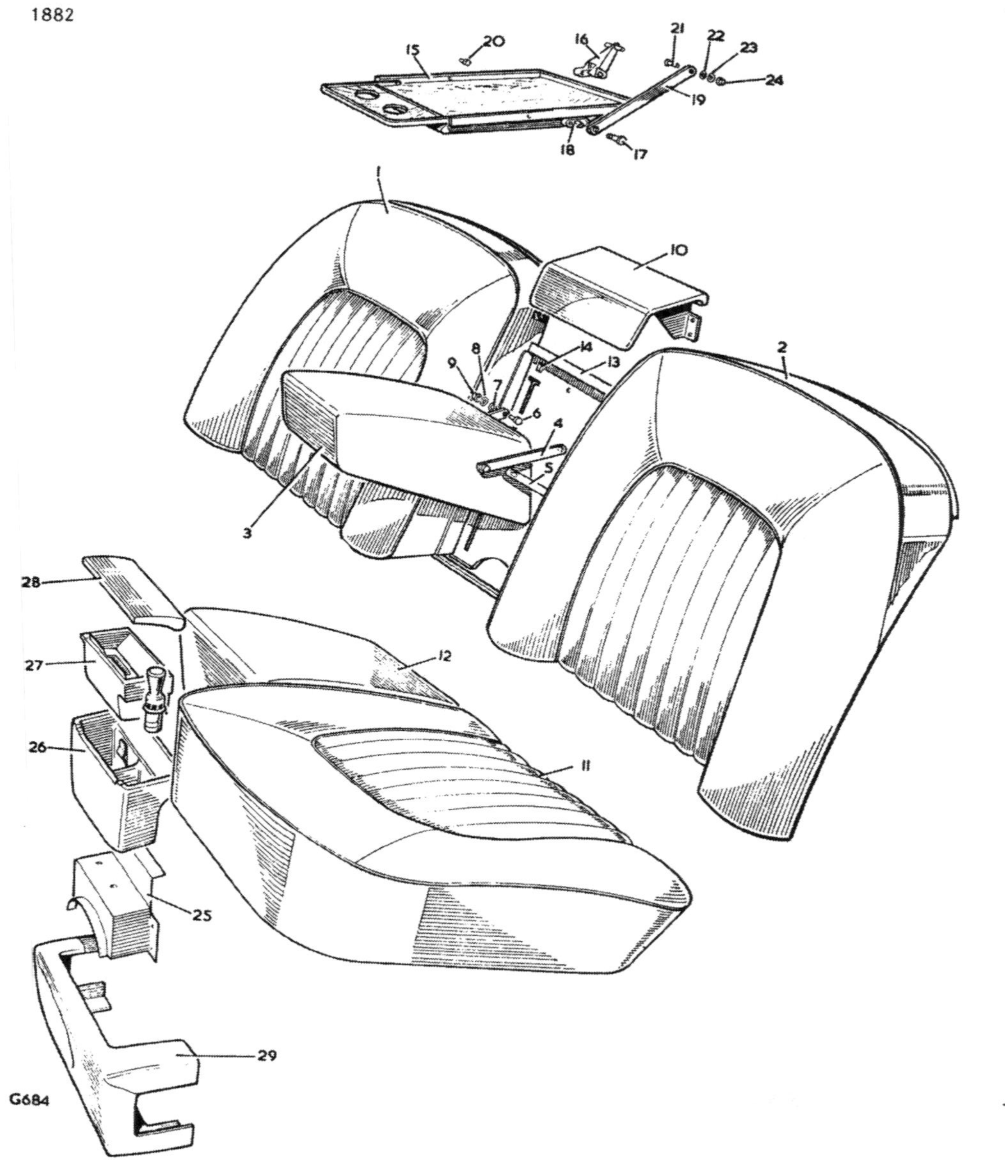

Plate Ref.	1 2 3 4 Description	Qty.	Part No.	Remarks
1	Rear squab assembly, RH	1	384103	Quote car number and colour
2	Rear squab assembly, LH	1	384104	
3	Centre armrest assembly	1	384726	
4	Top links for centre armrest	2	383833	
5	Bottom links for centre armrest	2	383834	
	Shakeproof washer] Fixing links to armrest	4	70884	
	Set screw (¼" UNF x ½" long)] Fixing links to armrest	4	354822	
6	Shaft	4	365096	
7	Plain washer] Fixing links to squab frame	2	3840	
8	Waved washer] Fixing links to squab frame	2	353368	
9	Retaining clip] Fixing links to squab frame	4	78573	
10	Rear squab spacer assembly	1	383919	Quote car number and colour required
	Drive screw fixing spacer to frame	4	78549	
	Drive screw] Fixing squab to frame	8	78549	
	Plain washer] Fixing squab to frame	8	4203	
	Rear cushion assembly, RH	1	384560	Quote car number and colour required
11	Rear cushion assembly, LH	1	384561	
12	Spacer assembly for rear seat (between cushions)	1	383995	
	Drive screw] Fixing spacer to body	4	77704	
	Plain washer] Fixing spacer to body	4	4203	
13	Guide panel for picnic tray	1	383829	
	Buffer for panel	1	234429	
	Drive screw fixing panel to squab frame	6	78549	
14	Anti-rattle clip for guide panel	2	385515	
	Retaining clip for anti-rattle clip	2	385517	
	PICNIC TRAY COMPLETE FOR REAR SQUAB	1	384245	
15	Picnic tray	1	384159	
16	Guide bracket, RH	1	383981	
	Guide bracket, LH	1	383982	
17	Pivot shaft	2	383965	
18	Nylon bush	2	384882	
19	Link for picnic tray	2	383948	
20	Friction plug for flap	2	384251	
	Handle for picnic tray	1	385514	
21	Shaft	2	365096	
22	Plain washer] Fixing picnic tray link to squab frame	2	3840	
23	Waved washer] Fixing picnic tray link to squab frame	2	353368	
24	Retaining clip] Fixing picnic tray link to squab frame	2	78573	
	Bolt (¼" UNF x 1¼" long)] Fixing rear squab to body	4	255211	
	Plain washer] Fixing rear squab to body	4	2552	
	Sealing washer] Fixing rear squab to body	4	355085	
25	Plinth for ashtray, rear seat	1	359157	
	Drive screw fixing plinth to body	4	77704	
26	ASHTRAY AND CIGAR LIGHTER HOLDER COMPLETE	1	383686	

*Asterisk indicates a new part which has not been used on any previous model.

REAR SEATS AND PICNIC TRAY, MKIII 3 LITRE AND 3½ LITRE COUPE MODELS

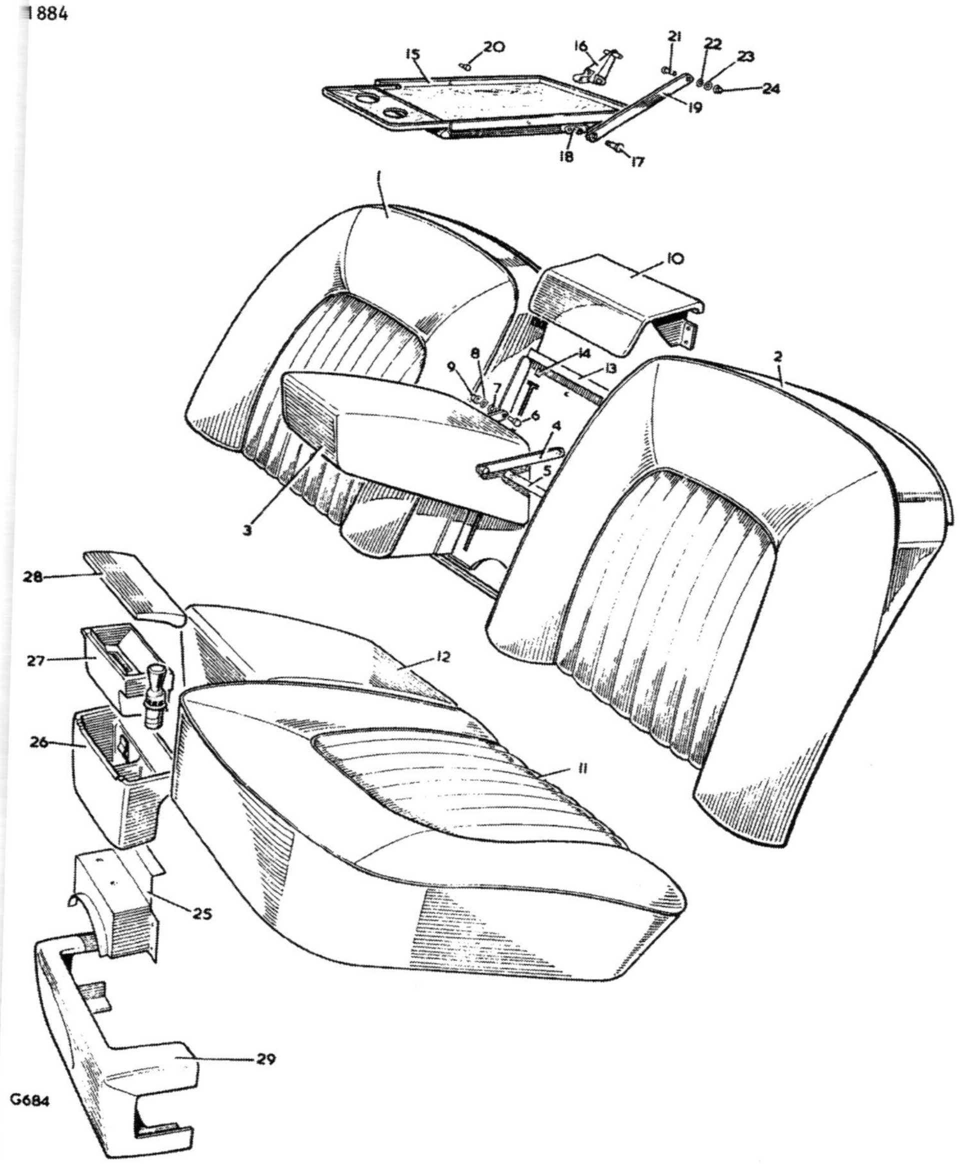

REAR SEATS AND PICNIC TRAY, MK III 3 LITRE AND 3½ LITRE COUPE MODELS

Plate Ref.	1 2 3 4	Description	Qty.	Part No.	Remarks
27		Bowl for ashtray	1	359122	
28		Chrome top for ashtray	1	359123	
		Screw (10 UNF x ¾" long) } Fixing ashtray	2	78384	
		Plain washer } assembly to plinth	2	3851	
		CIGAR LIGHTER COMPLETE FOR REAR SEAT	1	565850	
		Element for cigar lighter	1	600280	
		Bulb for cigar lighter	1	530054	
		Knob for cigar lighter	1	600593	
		Flange for knob	1	600602	
29		Cover panel for ashtray (trimmed)	1	384005	Quote car number and colour required

*Asterisk indicates a new part which has not been used on any previous model.

Plate Ref.	1 2 3 4 Description	Qty.	Part No.	Remarks
1	SCUTTLE TRIM PANEL COMPLETE, RH	1	357509	Quote car number and
	SCUTTLE TRIM PANEL COMPLETE, LH	1	357510	colour required
	Special 'Z' clip for scuttle trim panel	4	354877	
2	Finisher for RH scuttle trim panel	1	355774	Mk I, Mk IA and
	Finisher for LH scuttle trim panel	1	355775	Mk II saloon
	Finisher for scuttle trim panel, RH	1	384019	Mk III 3 litre and 3½ litre
	Finisher for scuttle trim panel, LH	1	384020	saloon
	Finisher for scuttle trim panel, RH	1	358522	
	Finisher for scuttle trim panel, LH	1	358523	Coupe
	Fixing plate, 'A' post lower kick plate	2	383273	
	Drive screw fixing plate	4	78140	
	Retaining plate for LH trim panel	1	355776	
	Drive screw (3/8" long) Fixing finisher to trim panel	2	78288	
	Drive screw (1" long) Fixing finisher to trim panel	1	78672	
	Drive screw (3/8" long) Fixing finisher to trim panel	1	311248	
	Clip fixing scuttle trim panel	6	72602	
3	Canopy trim complete	1	355490	Saloon.
	'BC' POST TRIM COMPLETE, UPPER RH	1	383352	
4	'BC' POST TRIM COMPLETE, UPPER LH	1	383353	
	'BC' POST TRIM COMPLETE, UPPER, RH	1	383338	Coupe
	'BC' POST TRIM COMPLETE, UPPER, LH	1	383339	
	Clip for 'BC' post trim	2	72602	
	'BC'POST TRIM COMPLETE, LOWER RH	1	355238	3 litre saloon and coupe
5	'BC' POST TRIM COMPLETE, LOWER LH	1	355239	
	'BC' POST TRIM COMPLETE LOWER, RH	1	386514*	3½ litre saloon models with inertia reel safety harness. Quote car number and colour required
	'BC' POST TRIM COMPLETE, LOWER, LH	1	386515*	
	'BC' POST TRIM COMPLETE, LOWER	2	386518*	3½ litre coupe models with inertia reel safety harness
	Clip for 'BC' post trim	2	355224	
	Moulding, black plastic, for 'BC' post, upper	4	383320	Coupe
6	Back light trim complete, top	1	355495	Saloon. Quote car number and colour required
	REAR CENTRE HEADER PAD ASSEMBLY	1	383428	Coupe
	Rubber interior, upper — For back light trim	1	354986	@
	Rubber interior, lower — For back light trim	1	354985	@
	Drive screw fixing back light trim	2	78140	@
	CANT RAIL AND 'D' POST CASING COMPLETE, RH	1	355435	Saloon.
7	CANT RAIL AND 'D' POST CASING COMPLETE, LH	1	355436	
	CANT RAIL TRIM (BLACK), RH	1	357919	
	CANT RAIL TRIM (BLACK) LH	1	357920	Coupe
	Clip for rear back light, cant rail and 'D' post casing	33	730378@@	Saloon
	Drive screw fixing cant rail trim	12	78495	Coupe
8	ROOF TRIM COMPLETE	1	355699	Saloon. Quote car number and colour required.

@ Cars numbered up to 625000979, 626000332, 628000110, 630000459, 631000115, 633000310

@@ Cars numbered from 625000980, 626000333, 628000111, 630000460, 631000116, 633000311 onwards

*Asterisk indicates a new part which has not been used on any previous model.

BODY TRIMMING AND CARPETS

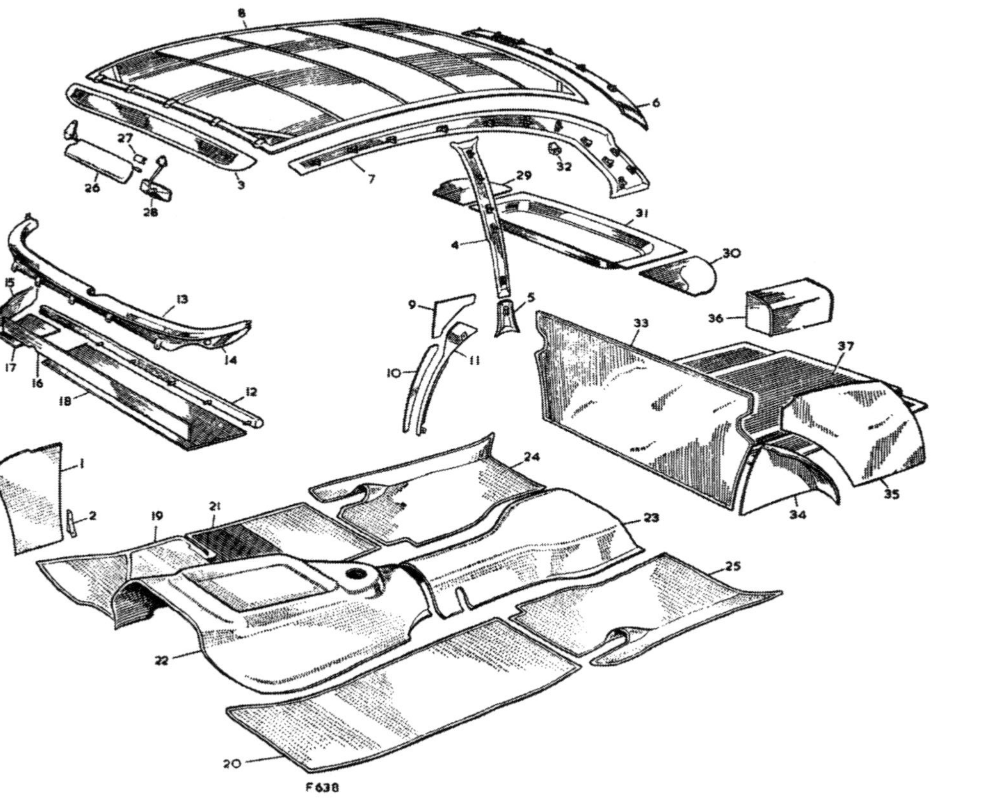

Plate Ref. 1 2 3 4	Description	Qty.	Part No.	Remarks
	Roof frame complete	1	354142@	Saloon
	Roof frame complete	1	355425@@	Saloon
	Listing rail, No. 1 (front)	1	355316@	Saloon
	Listing rail, No. 2	1	355317@	Saloon
	Listing rail, Nos. 3 and 4	2	354229@	Saloon
	Listing rail, No. 5 (rear)	1	354231@	Saloon
	Listing rail, Nos. 1 (front), 2 and 3	3	355426@@	Saloon
	Listing rail, No. 4 (rear)	1	355428@@	Saloon
	Rubber sleeve	8	310654	Saloon
	Clip fixing trim to frame	50	331071	Saloon
	Drive screw] Fixing roof trim	6	77704	Saloon
	Cup washer] Fixing roof trim	2	354177	Saloon
	Drive screw] Fixing roof trim	2	77707	Saloon
	Cap for screw cup] Fixing roof trim	2	354178	Saloon
	ROOF TRIM COMPLETE	1	358533	Quote car number and colour required; Coupe
	Roof frame complete	1	357880	Coupe
	Listing rail, No. 1	1	357915	Coupe
	Listing rail, No. 2	1	357916	Coupe
	Listing rail, No. 3	1	357917	Coupe
	Listing rail, No. 4	1	357918	Coupe
	Clip fixing trim to frame	44	331071	Coupe
	Plain washer] Fixing frame to roof	4	4203	Coupe
	Drive screw] Fixing frame to roof	4	77704	Coupe
	REAR QUARTER TRIM COMPLETE, RH TOP	1	354550@	Quote car number and colour required
	REAR QUARTER TRIM COMPLETE, LH TOP	1	354551@	Quote car number and colour required
	Rubber interior for rear top quarter trim	2	355194@	
	Drive screw fixing rear top quarter trim	4	77704@	
	REAR HEADER SIDE PANEL COMPLETE, RH	1	355498@@	Quote car number and colour required
	REAR HEADER SIDE PANEL COMPLETE, LH	1	355499@@	Quote car number and colour required
	Clip for rear side header panel	4	726028@	
	LOWER QUARTER TRIM COMPLETE, RH	1	354279@	Quote car number and colour required
	LOWER QUARTER TRIM COMPLETE, LH	1	354280@	Quote car number and colour required
	Clip for lower quarter trim	2	72602@	
	REAR QUARTER TRIM COMPLETE, RH	1	355707@@	Quote car number and colour required
	REAR QUARTER TRIM COMPLETE, LH	1	357539@@	Quote car number and colour required
	Clip for quarter trim	6	730378@	
	Windhose for doors	4	358157	Quote car number and colour required
	Drive screw fixing windhose retainer to floor at 'D' post	2	77892	
	Rear seat valance trim complete, one piece	1	354566	Quote car number and colour required
	Rear seat valance trim complete, RH	1	386398	Mk III 3 litre and 3½ litre saloon
	Rear seat valance trim complete, LH	1	386399	Mk III 3 litre and 3½ litre saloon

@ Cars numbered up to 625000979, 626000332, 628000110, 630000459, 631000310, 633000115
@@ Cars numbered from 625000980, 626000333, 628000111, 630000460, 631000311, 633000116 onwards

*Asterisk indicates a new part which has not been used on any previous model.

Plate Ref.	1 2 3 4 Description	Qty.	Part No.	Remarks
	Rear seat valance trim complete, RH	1	383790	Mk II coupe
	Rear seat valance trim complete, LH	1	383791	
	Rear seat valance trim compelte, RH	1	384767	Mk III 3 litre and 3½ litre coupe
	Rear seat valance trim complete, LH	1	384768	
	Drive screw fixing seat valance trim	12	77704	
9	Harness cover complete, upper RH	1	354559	Quote car number and colour required
10	Harness cover complete, lower RH	1	354561	
	Harness cover complete, upper LH	1	357533	
	Harness cover complete, lower LH	1	357535	
11	Wheelarch edge trim complete, RH	1	354875	
	Drive screw } Fixing wheelarch edge trim to body	6	311948	
	Cup washer }	6	73961	
	Crash rail trimmed assembly for parcel shelf	1	355723%	
	Crash rail trimmed assembly for parcel shelf (charcoal grey)	1	355891%%	
12	Crash rail trimmed assembly for parcel shelf (trim colours)	1	358784	Mk IA and Mk II. Quote car number and colour required
	CRASH RAIL TRIMMED ASSEMBLY FOR PARCEL SHELF	1	384270	Mk III 3 Litre and 3½ Litre Quote car number and colour required
	Beading for crash rail	1	384261	
	Drive screw fixing beading	7	78627	
	Plain washer } Fixing crash rail to parcel shelf	7	3902	
	Spring washer }	7	3073	
	Nut (10 UNF) }	7	257023	Mk I and Mk IA
	Nut (8 UNC) }	7	257131	Mk II
	Beading for crash rail padding	1	354164	
	Drive screw fixing beading	5	78126	
	FACIA CRASH RAIL TRIMMED ASSEMBLY	1	354507 $	
	FACIA CRASH RAIL TRIMMED ASSEMBLY	1	355980 $$	Mk I and Mk IA
13	FACIA CRASH RAIL TRIMMED ASSEMBLY	1	381151	Mk II
	FACIA CRASH RAIL TRIMMED ASSEMBLY	1	384236	Mk III 3 Litre and 3½ Litre
	Facia rail, steel	1	357205	Mk I and Mk IA
14	Facia rail, steel	1	381149	Mk II
	Facia rail, steel	1	384018	Mk III 3 Litre and 3½ Litre
	Pop rivet fixing facia rail centre to ends	4	78390	
	Clip fixing cover to rail	12	331071	
	Securing plate fixing facia rail to frame	4	352390	
	Drive screw fixing plate to rail	8	77707	
	Drive screw fixing facia rail ends to dash	4	78155@@	
	Tie bracket, RH } Fixing facia rail to dash	1	352568@	
	Tie bracket, LH }	1	352569@	
	Set bolt(10 UNF x ½" long) } Fixing tie bracket to dash	2	257017@	
	Plain washer }	2	3685@	
	Spring washer }	2	3073@	

@ Cars numbered up to 625000979, 626000332, 628000110, 630000459, 631000310, 633000115

@@ Cars numbered from 625000980, 626000333, 628000111, 630000460, 631000311, 633000116 onwards

$ Cars numbered up to 625001352, 626000431, 628000175, 630000748, 631000192, 633000443

$$ Cars numbered from 625001353, 626000432, 628000176, 630000749, 631000193, 633000444 onwards

% Cars numbered up to 625101045, 626100262, 628100109, 630100840, 631100353, 633100078

%% Cars numbered from 625101046, 626100263, 628100110, 630100841, 631100354, 633100079 onwards

*Asterisk indicates a new part which has not been used on any previous model.

F638

BODY TRIMMING AND CARPETS

Plate Ref.	1 2 3 4 Description	Qty.	Part No.	Remarks
	Plain washer } Fixing bracket to facia rail	2	3902	
	Set bolt (10 UNF x ½" long) } Fixing bracket to facia rail	2	257017	
	Drive screw fixing rail to facia frame	4	78439	
15	Parcel shelf side trim, RH	1	354511	Mk I, Mk IA and Mk II. Quote car number and colour required
	Parcel shelf side trim, LH	1	354512	
	Parcel shelf side trim, RH	1	354511	Mk III. 3 Litre and 3½ Litre Quote car number and colour required
	Parcel shelf side trim, LH	1	384522	
	Clip for side trim	4	355220	
16	Parcel shelf back trim	1	357525	RH Stg Mk I, Mk IA and Mk II Quote car number and colour required
	Parcel shelf back trim	1	357542	LH Stg
	Rivet fixing stiffening angle on back trim	7	311160	
17	Parcel shelf floor trim, RH	1	355032	RH Stg } Mk I, Mk IA and Mk II. Quote car number and colour required
18	Parcel shelf floor trim, LH	1	355033	RH Stg
	Parcel shelf floor trim, RH	1	355054	LH Stg
	Parcel shelf floor trim, LH	1	355053	LH Stg
	Parcel shelf floor trim, RH	1	384645	RH Stg } Mk III. 3 Litre and 3½ Litre. Quote car number and colour required
	Parcel shelf floor trim, LH	1	384643	RH Stg
	Parcel shelf floor trim, RH	1	384644	LH Stg
	Parcel shelf floor trim, LH	1	384646	LH Stg
	Parcel shelf, upper casing RH, small (dull black trim)	1	384549	Mk III 3 Litre and 3½ Litre
	Parcel shelf, upper casing LH, large (dull black trim)	1	384548	
	Trim clip fixing casing to facia frame	8	385741	
	Packing, front floor, RH	1	354910	
	Packing, front floor, LH	1	354911	
	Packing, tunnel	1	354913	
	Packing, gearbox cover	1	354912	4-speed
	Packing, gearbox cover	1	354937	Automatic
	Front carpet complete, RH	1	355259	RH Stg } Quote car number and colour required } Mk I and Mk IA
	Front carpet complete, RH	1	355158	LH Stg
	Rubber heel mat	1	354286	RH Stg
	Front carpet complete, LH	1	355271	RH Stg 4-speed } Mk I Quote car number and colour required
	Front carpet complete, LH	1	355265	RH Stg Automatic
	Front carpet complete, LH	1	355929	LH Stg 4-speed
	Front carpet complete, LH	1	355930	LH Stg Automatic
	Front carpet complete, LH	1	357440	RH Stg 4-speed and } Mk IA. Quote car number and colour required
	Front carpet complete, LH	1	357441	LH Stg Automatic
	Rubber heel mat	1	354976	LH Stg
	Rear carpet complete	1	355204	Bench-type seats } Quote car number and colour required } Mk I and Mk IA
	Rear carpet complete	1	355201	Bucket-type seats

*Asterisk indicates a new part which has not been used on any previous model.

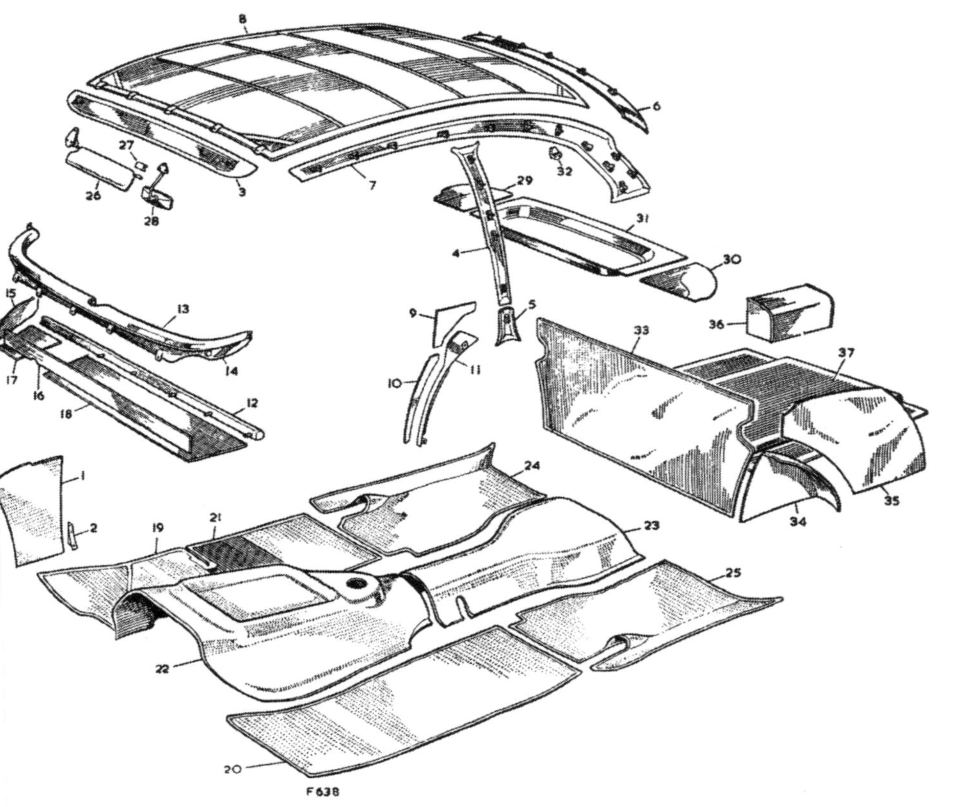

Plate Ref.	1 2 3 4 Description	Qty.	Part No.	Remarks		
	Front carpet, RH	1	381159	4-speed	Cars with suffix letters'A'and'B'	
	Front carpet, RH	1	359180	Automatic		
19	Front carpet, RH	1	359180	RH Stg 4-speed and Automatic with suffix letter'C'		Mk II 3Litre saloon State colour req-uired
	Front carpet, RH	1	359320	LH Stg		
20	Front carpet, LH	1	359181	RH Stg		
	Front carpet, LH	1	359321	LH Stg		
21	Rubber heel mat	1	354286	RH Stg		
	Rubber heel mat	1	354976	LH Stg		
	Gearbox and front tunnel carpet, cut carpet (not moulded)	1	359847	Automatic		
	Gearbox and front tunnel carpet, moulded carpet	1	380484	RH Stg	Auto-matic	Alterna-tives
	Gearbox and front tunnel carpet, moulded carpet	1	380485	LH Stg		
22	Gearbox and front tunnel carpet, moulded carpet	1	380486	RH Stg	4-speed with bucket seats	
	Gearbox and front tunnel carpet, moulded carpet	1	380487	LH Stg		
23	Tunnel carpet, rear	1	380490	With bucket seats		
	Tunnel carpet, rear	1	380491	With bench seat		
	Rear carpet complete, RH	1	359303	Automatic with bench seat		
	Rear carpet complete, LH	1	359304			
24	Rear carpet complete, RH	1	358918	With bucket seats		
25	Rear carpet complete, LH	1	358966			
	Front carpet complete, RH	1	359180	RH Stg		Mk II 3 Litre coupe Quote car number and colour required
	Front carpet complete, LH	1	359181	RH Stg		
	Front carpet complete, RH	1	359320	LH Stg		
	Front carpet complete, LH	1	359321	LH Stg		
	Rubber heel mat	1	354286	RH Stg		
	Rubber heel mat	1	354976	LH Stg		
	Gearbox and front tunnel cover carpet	1	380486	RH Stg	4-speed	
	Gearbox and front tunnel cover carpet	1	380487	LH Stg		
	Gearbox and front tunnel cover carpet	1	380488	RH Stg	Auto-matic	
	Gearbox and front tunnel cover carpet	1	380489	LH Stg		
	Tunnel cover carpet, rear	1	380509			
	Rear carpet complete, RH	1	359281			
	Rear carpet complete, LH	1	359282			
	Front carpet, RH	1	385542	RH Stg Mk III 3 Litre		Quote car number anc colour required
	Front carpet, RH	1	386312	RH Stg 3½Litre		
	Front carpet, RH	1	385593	LH Stg		
	Front carpet, LH	1	385543	RH Stg Mk III 3 Litre		
	Front carpet, LH	1	386344	LH Stg 3½ Litre		
	Front carpet, LH	1	385594	LH Stg		
	Rubber heel mat	1	354286	RH Stg		
	Rubber heel mat	1	354976	LH Stg		
	Gearbox and front tunnel carpet	1	380486	RH Stg	Mk III 3 litre 4-speed	
	Gearbox and front tunnel carpet	1	380487	LH Stg		
	Gearbox and front tunnel carpet	1	380488	RH Stg	Mk III 3 Litre Automatic	
	Gearbox and front tunnel carpet	1	380489	LH Stg		

*Asterisk indicates a new part which has not been used on any previous model.

BODY TRIMMING AND CARPETS

1915

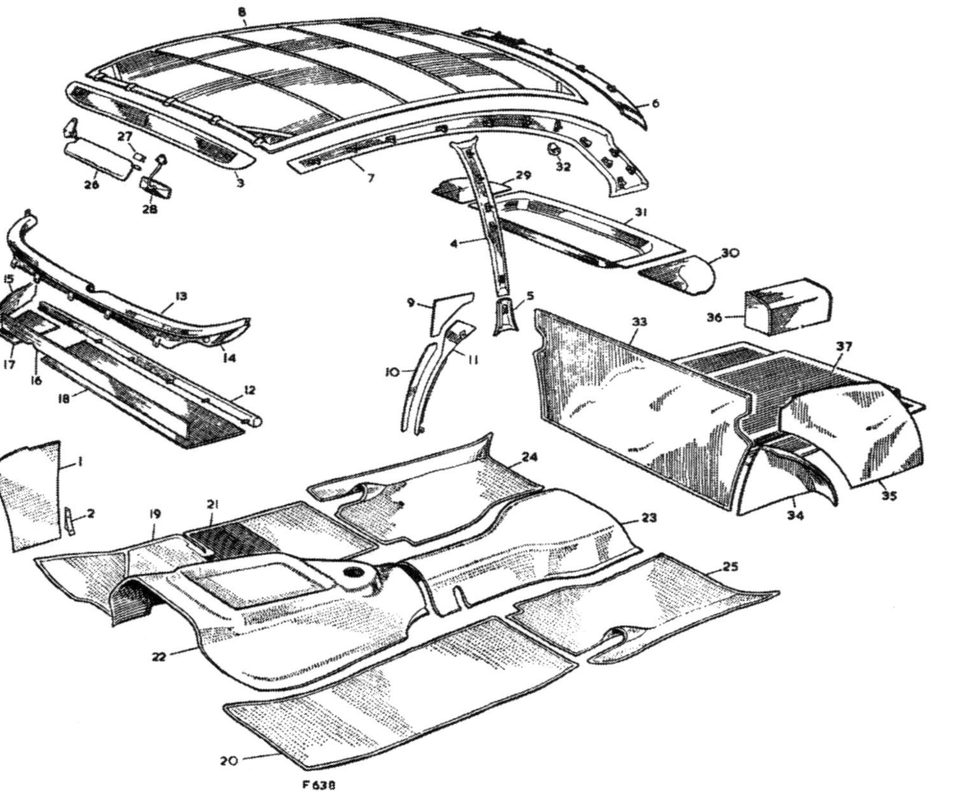

Plate Ref. 1 2 3 4	Description	Qty.	Part No.	Remarks
	Front gearbox cover carpet	1	386310	RH Stg
	Front gearbox cover carpet	1	386311	LH Stg
	Tunnel carpet	1	387168	RH Stg] Saloon] 3½ Litre State colour required
	Tunnel carpet	1	387169	LH Stg]
	Tunnel carpet	1	387170	RH Stg] Coupe
	Tunnel carpet	1	387171	LH Stg]
	Rear tunnel carpet	1	380981	Saloon] Mk III 3 litre
	Rear tunnel carpet	1	380982	Coupe] Quote car number and colour required
	Rear carpet, RH	1	385544	] Saloon
	Rear carpet, LH	1	385545	]
	Rear carpet, RH	1	385600	] Coupe
	Rear carpet, LH	1	385601	]
	Carpet clip	As reqd	77210	
	Drive screw fixing carpet clip	As reqd	77704	
	Carpet clip, front carpet to dash	2	355847	
	Carpet clip and bracket assembly	1	355845	LH Stg
	Drive screw fixing clip	4	78160	2 on LH Stg cars
	Carpet protection plate	4	359917	Mk II] Alternatives. Cars with safety harness fixing points incorporated in the body shell
	Special screw for sealing fixing points for safety harness at tunnel floor	8	381176	Plastic screw]
	Special screw for sealing fixing points for safety harness at tunnel floor	8	383805	Steel screw]
26	Sun visor, padded, no mirror, biscuit	1	358885	] RH Stg]
	Sun visor, padded, no mirror, grey	1	358860	]
	Sun visor, padded, no mirror, biscuit	1	359409	] LH Stg] MkI, MkIA, Mk II and Mk III saloon models
	Sun visor, padded, no mirror, grey	1	359410	]
	SUn visor, padded, with mirror, biscuit	1	358883	] RH Stg] Padded visor with triangular fixing foot NB. Only biscuit-coloured sun visors are fitted Mk III
	Sun visor, padded, with mirror, grey	1	358858	]
	Sun visor, padded, with mirror, biscuit	1	358884	] LH Stg
	Sun visor, padded, with mirror, grey	1	358859	]
	Mirror for padded sun visor	1	358953	]
	Sun visor, padded, no mirror, biscuit	1	359115	] RH Stg]
	Sun visor, padded, no mirror, grey	1	359119	]
	Sun visor, padded, no mirror, biscuit	1	359116	] LH Stg] Mk II and Mk III coupe models Padded visor with square fixing foot NB. Only biscuit-coloured sun visors are fitted to Mk III
	Sun visor, padded, no mirror, grey	1	359120	]
	Sun visor, padded, with mirror, biscuit	1	358950	] RH Stg
	Sun visor, padded, with mirror, grey	1	359117	]
	Sun visor padded with mirror biscuit	1	358951	] LH Stg
	Sun visor, padded, with mirror, grey	1	359118	]
	Mirror for padded sun visor	1	358953	]

*Asterisk indicates a new part which has not been used on any previous model.

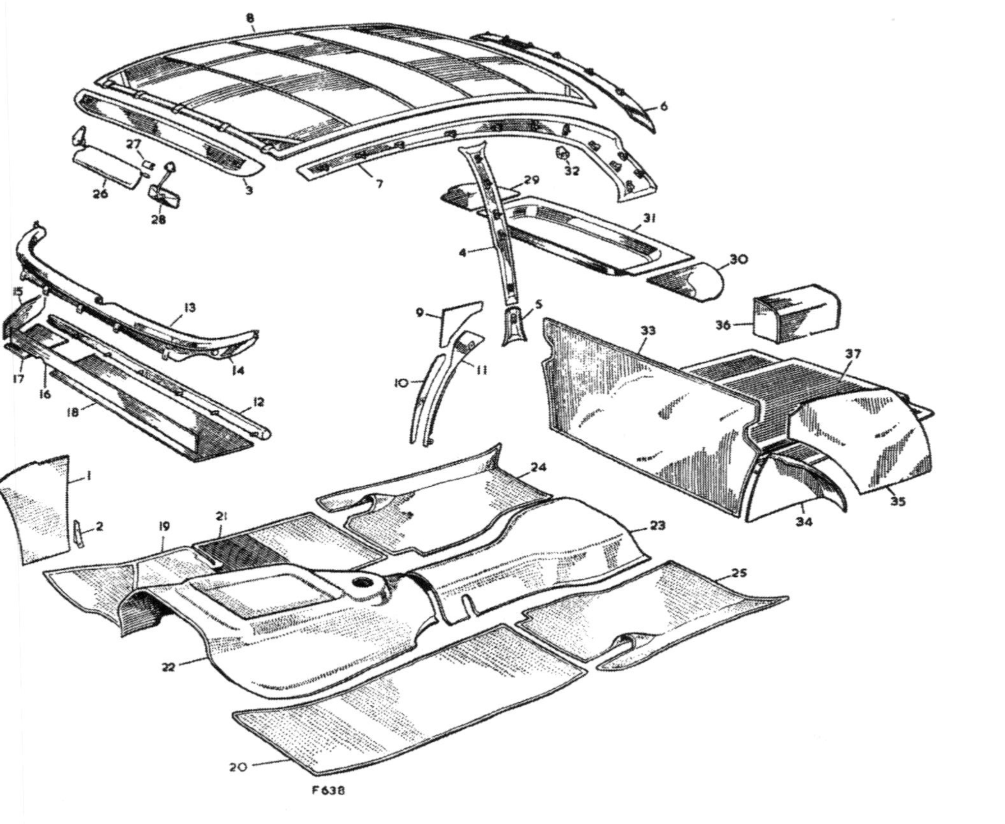

Plate Ref.	1 2 3 4 Description	Qty.	Part No.	Remarks
	Sun visor, padded, no mirror, oatmeal	1	386160 RH Stg	Saloon — Late Mk III 3 Litre and 3½ Litre
	Sun visor, padded, no mirror, oatmeal	1	386162 LH Stg	Saloon — Late Mk III 3 Litre and 3½ Litre
	Sun visor, padded, with mirror, oatmeal	1	386161 RH Stg	Saloon — Late Mk III 3 Litre and 3½ Litre
	Sun visor, padded, with mirror, oatmeal	1	386163 LH Stg	Saloon — Late Mk III 3 Litre and 3½ Litre
	Sun visor, padded, no mirror, oatmeal	1	386164 RH Stg	Coupe — Late Mk III 3 Litre and 3½ Litre
	Sun visor, padded, no mirror, oatmeal	1	386166 LH Stg	Coupe — Late Mk III 3 Litre and 3½ Litre
	Sun visor, padded, with mirror, oatmeal	1	386165 RH Stg	Coupe — Late Mk III 3 Litre and 3½ Litre
	Sun visor, padded, with mirror, oatmeal	1	386167 LH Stg	Coupe — Late Mk III 3 Litre and 3½ Litre
	Special screw (⅜" long) } Fixing sun visor	6	78308	Alternative
	Drive screw (1" long) } Fixing sun visor	6	313785	Alternative
	Hook for sun visor	2	313482@	
	Drive screw fixing hook	4	311948@	
27	Hook for sun visor	2	355497@@	
	Drive screw fixing hook	4	311248@@	
28	DRIVING MIRROR	1	386482	
	Glass only	1	380699	Except NADA
	DRIVING MIRROR, ADJUSTABLE, ANTI-DAZZLE	1	383764	NADA
	Drive screw fixing mirror	3	313785@@	
	Ashtray in front seat	1	315096@	
	Back plate for ashtray	1	312103@	
	Drive screw fixing ashtray	2	77704@	
	Ashtray in front seat	2	314719@@	
	Back plate for ashtray	2	317199@@	
	Drive screw fixing ashtray	6	74141@@	
29	Rear parcel shelf, RH	1	355716	Mk I and Mk IA saloon. Quote car number and colour required
30	Rear parcel shelf, LH	1	355717	Mk I and Mk IA saloon. Quote car number and colour required
31	Rear parcel shelf, centre	1	356735	Mk I and Mk IA saloon. Quote car number and colour required
	Rear parcel shelf, centre panel	1	359546	Mk II 3 litre saloon
	Rear parcel shelf assembly	1	384109	Mk III 3 litre and 3½ litre saloon. Quote car number and colour required
	Centre finisher for parcel shelf	1	384117	Mk III 3 litre and 3½ litre saloon. Quote car number and colour required
	End finisher, RH, for parcel shelf	1	384122	Mk III 3 litre and 3½ litre saloon. Quote car number and colour required
	End finisher, LH, for parcel shelf	1	384123	Mk III 3 litre and 3½ litre saloon. Quote car number and colour required
	Rear parcel shelf trimmed assembly	1	381031	Mk II 3 litre coupe. Quote car number and colour required
	Stud fixing rear speaker on parcel shelf	4	381140	Mk II 3 litre coupe. Quote car number and colour required
	Support bracket, rear support angle to body	7	357756	Mk II 3 litre coupe. Quote car number and colour required
	Drive screw	7	77704	Mk II 3 litre coupe. Quote car number and colour required
	Plain washer	7	4203	Mk II 3 litre coupe. Quote car number and colour required
	Drive screw fixing front support angle to body	6	77704	Mk II 3 litre coupe. Quote car number and colour required
	Rear parcel tray assembly	1	384109	Mk III 3 litre and 3½ litre coupe. Quote car number and colour required
	Centre finisher for parcel tray	1	384139	Mk III 3 litre and 3½ litre coupe. Quote car number and colour required
	End finisher, RH, for parcel tray	1	384144	Mk III 3 litre and 3½ litre coupe. Quote car number and colour required
	End finisher, LH, for parcel tray	1	384145	Mk III 3 litre and 3½ litre coupe. Quote car number and colour required
	Plastic edge finisher, rear, for parcel shelf	1	355818@@	Saloon
	Plastic edge finisher, front, for parcel shelf	1	355819@@	Saloon
	Rubber seal, parcel shelf to rear quarter	2	317024	Saloon
	Coat hook, steel	2	356648@@	
	Drive screw fixing coat hook	4	311948@@	
	Mounting bracket for coat hanger	2	383760	Alternative coat hangers. Check before ordering
	Drive screw fixing bracket to body	4	78140	Alternative coat hangers. Check before ordering
	Coat hanger, male portion	2	359279	Alternative coat hangers. Check before ordering
	Coat hanger, female portion	2	359280	Alternative coat hangers. Check before ordering

@ Cars numbered up to 625000979, 626000332, 628000110, 630000459, 631000310, 633000115

@@ Cars numbered from 625000980, 626000333, 628000111, 630000460, 631000311, 633000116 onwards

NADA indicates parts peculiar to cars exported in the North American dollar area

*Asterisk indicates a new part which has not been used on any previous model.

1919

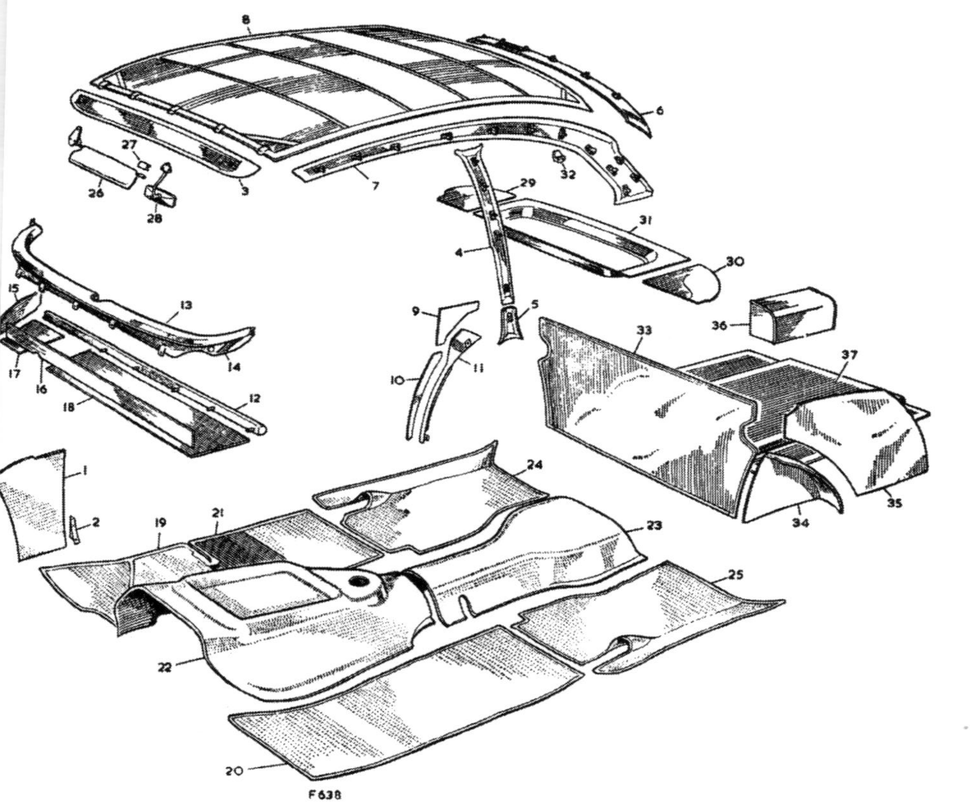

Plate Ref.	1 2 3 4	Description	Qty.	Part No.	Remarks
		Loose ring for coat hanger	2	359889	Alternative coat hangers
32		Coat hanger, one-piece plastic	2	383326	Check before ordering
		Drive screw fixing coat hanger	4	311948	
		Licence holder	1	381119	
		Boot trim complete, front	1	355858	Mk I 3 Litre
		Boot trim complete, side RH	1	354728	Mk I 3 Litre
		Boot trim complete, front	1	357527	Mk IA 3 litre
		Boot trim complete, side RH	1	357493	Mk IA 3 litre
33		Boot trim complete, front	1	355858	Mk II and Mk III 3 Litre
		Boot trim complete, side RH	1	354728	Mk II and Mk III 3 Litre
		Boot trim complete front	1	386302	3½ Litre
		Boot trim complete, side RH	1	386245	3½ Litre
		Plain washer } Fixing front and side trim	5	3816	
		Drive screw } Fixing front and side trim	5	77704	
		Boot trim complete for wheelarch, RH	1	354737	
34		Boot trim complete for wheelarch, LH	1	354741	
35		Boot trim compelte for tools	1	354745	
		Retainer for tools trim	1	354951	
36		Boot trim complete for battery cover	1	354731	
		Boot trim complete for single petrol pump	1	504429	Mk I only
37		Rubber mat for boot floor	1	356884	

*Asterisk indicates a new part which has not been used on any previous model.

CONSOLE UNIT

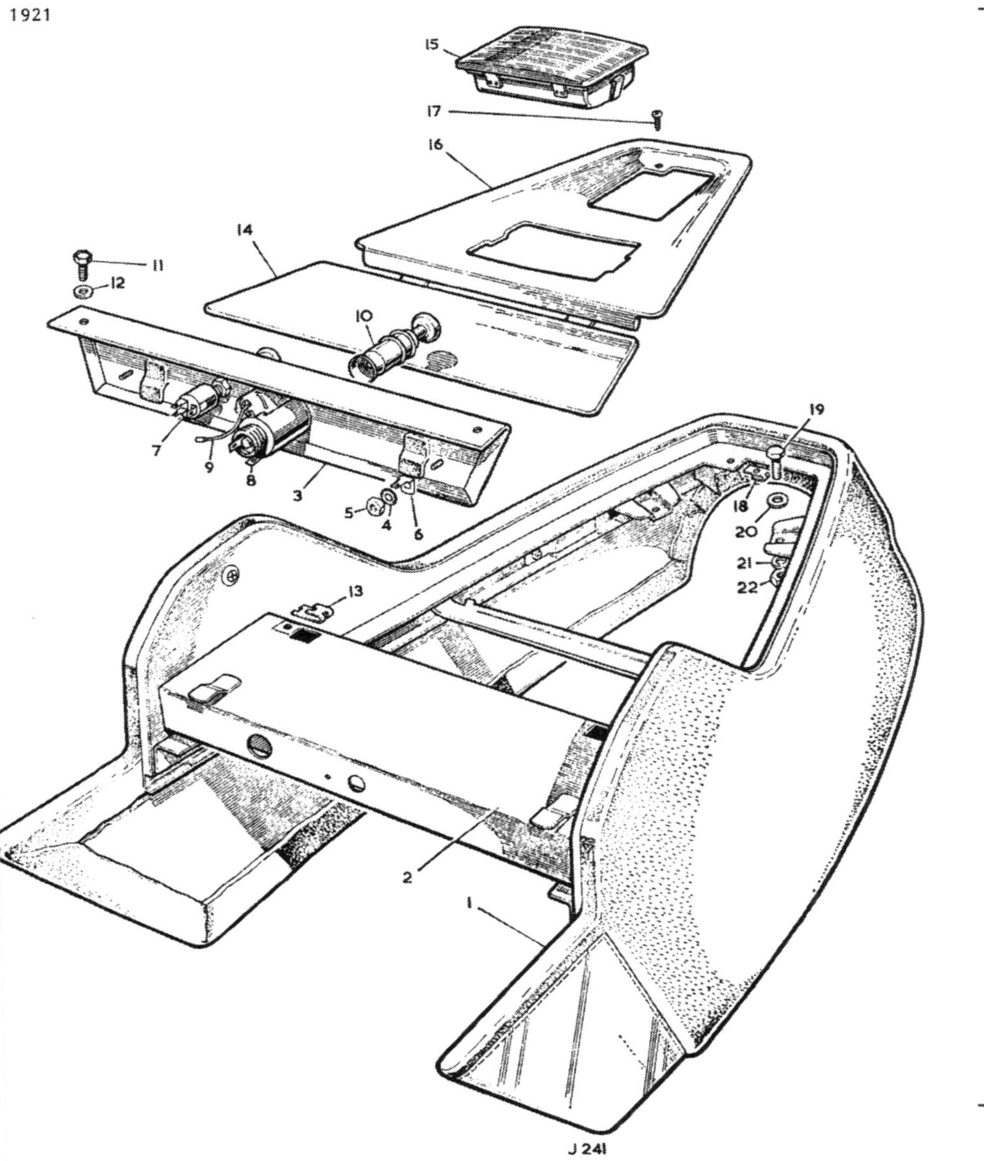

CONSOLE UNIT 3½ LITRE

Plate Ref.	1 2 3 4 Description		Qty.	Part No.	Remarks	
1	Casing trimmed assembly		1	386172	RH Stg	Quote car number and colour required
	Casing trimmed assembly		1	386291	LH Stg	
2	Switch panel housing trimmed assembly		1	386052		
3	Switch panel trimmed assembly		1	386303		
4	Fan disc washer	Fixing switch panel	2	513282		
5	Nut (10 UNF)		2	257023		
6	Lucar blade		1	547402		
7	Switch for fog lamps		1	551277		
8	CIGAR LIGHTER ASSEMBLY		1	565850		
9	Bulb		1	530054		
10	Element		1	600280		
	Knob		1	600281		
	Flange for knob		1	600282		
11	Special bolt	Fixing switch panel assembly	2	78787		
12	Plain washer		2	4586		
13	Spire fix nut		2	78786		
	Drive screw	Fixing switch panel housing to console unit	2	78778		
	Drive screw		3	78777		
	Drive screw		2	78784		
	Cup washer		2	78780		
	Plain washer		5	3998		
14	Tray trimmed assembly		1	386182	Quote car number and colour required	
15	Ash tray		1	364275		
16	Gear lever cover panel trimmed assembly		1	386032	Quote car number and colour required	
17	Drive screw	Fixing cover panel to console unit	1	311248		
18	Spire nut		1	314394		
19	Bolt (¼" UNF x ⅜" long)	Fixing console unit to body	2	255206		
20	Plain washer		2	3900		
21	Spring washer		2	3074		
22	Nut (¼" UNF)		2	254810		

*Asterisk indicates a new part which has not been used on any previous model.

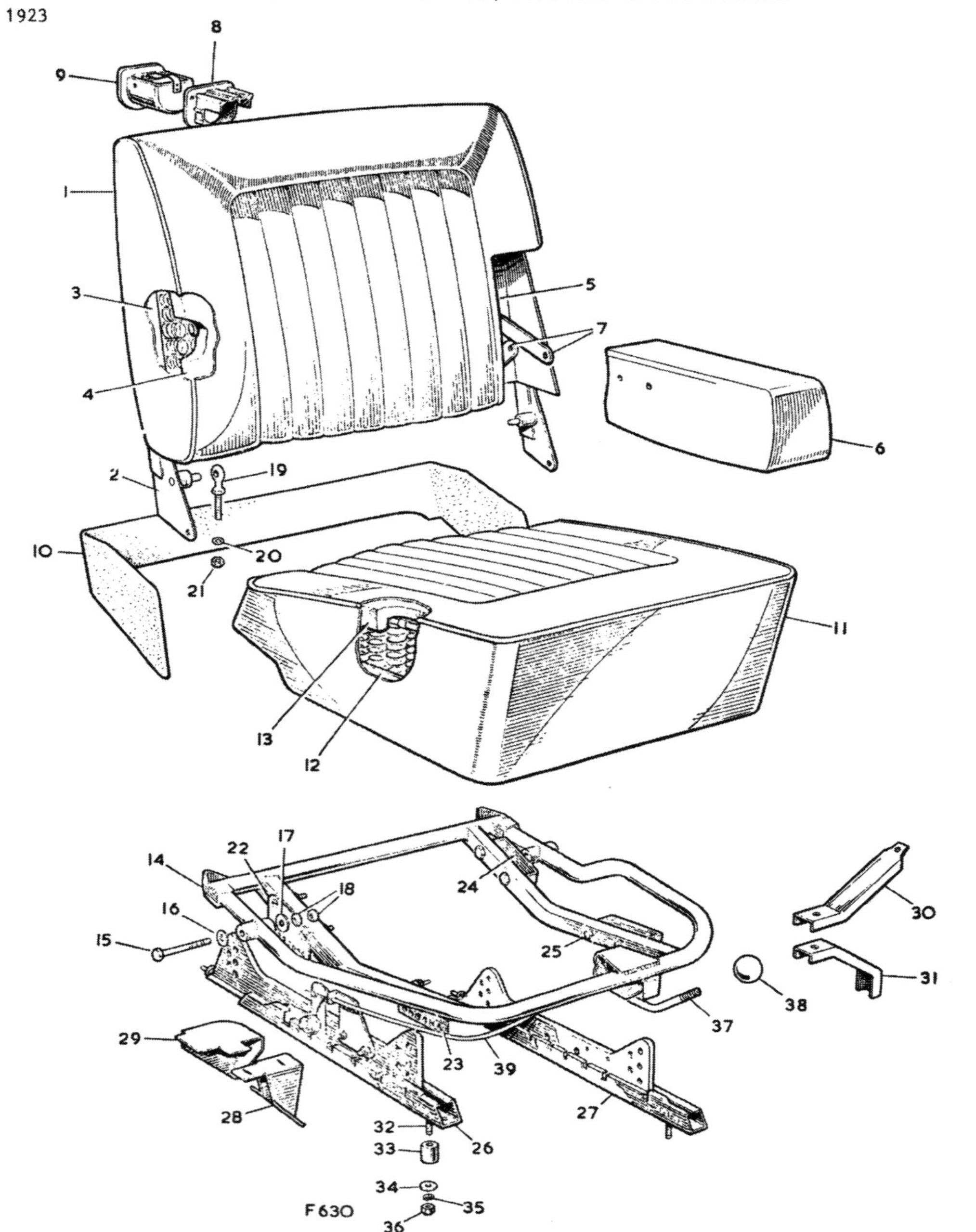

FRONT BUCKET-TYPE SEATS, MK I AND MK IA MODELS

Plate Ref.	1 2 3 4 DESCRIPTION	Qty	Part No.	REMARKS
1	BUCKET SEAT SQUAB COMPLETE, RH	1	355580	Quote car number and colour required
	BUCKET SEAT SQUAB COMPLETE, LH	1	355581	
2	Squab frame for bucket seat, RH	1	317217	
	Squab frame for bucket seat, LH	1	317218	
3	Spring case for bucket seat squab, RH	1	314413	
	Spring case for bucket seat squab, LH	1	314414	
	Topping, centre, for bucket seat squab, RH	1	354823†	
	Topping, centre, for bucket seat squab, LH	1	354824†	
	Topping for top panel	1	354821†	
	Topping for top roll	1	354821†	
	Topping for RH roll	1	354813†	
	Topping for RH roll	1	354812†	
	Topping for LH roll	1	354811†	
	Topping for LH roll	1	354814†	
	Topping for flute panel	2	355423††	
4	Topping, RH, for surround	1	355421††	
	Topping, LH, for surround	1	355422††	
	Flute padding for squab, plastic foam	As reqd	355984	
	Clip } Fixing cover to frame	24	315960	
	Clip	6	331071	
5	Armrest recess complete	2	314644	Quote car number and colour required
	Drive screw fixing armrest recess to frame	4	77906	
6	ARMREST COMPLETE, RH	1	354718	Quote car number and colour required
	ARMREST COMPLETE, LH	1	354719	
	Tongue bracket for armrest	2	354849	
	Washer plate } Fixing tongue bracket	2	314508	
	Screw (10 UNF x 1⅛" long)	4	78335	
	Spring washer	4	3073	
	Nut (10 UNF)	4	257023	
	Woodscrew	2	20153	
	Topping for RH armrest, front	1	354825	
	Topping for LH armrest, front	1	354826	
	Topping for armrest, rear	2	314498	
7	Links for armrest	2	314434	
	Special set screw } Fixing links to armrest	4	354822	
	Shakeproof washer	4	70884	
	Screw (¼" UNF x ⅝" long) } Fixing links to squab	3	78232	
	Plain washer	3	3840	
	Spring washer	3	3074	
	Nut (¼" UNF)	3	254810	
	BACK PANEL FOR SQUAB, RH, TRIMMED	1	315570	Quote car number and colour required
	BACK PANEL FOR SQUAB, LH, TRIMMED	1	315571	
	Clip for back panel	3	314514	
	Clip fixing panel to frame	22	72602	
	Pocket assembly, trimmed	2	315578	Quote car number and colour required
	Drive screw fixing pocket to frame	10	77707	
	Finger plate for pocket	2	315009	
	Drive screw fixing plate to frame	4	77707	
8	Ashtray holder	2	317199	
	Drive screw fixing holder	6	74141	
9	Bowl for ashtray	2	314719	

* Asterisk indicates a new part which has not been used on any previous Rover model
† Cars numbered up to 625000979, 626000332, 628000110, 630000459, 631000310, 633000115
†† Cars numbered from 625000980, 626000333, 628000111, 630000460, 631000311, 633000116 onwards

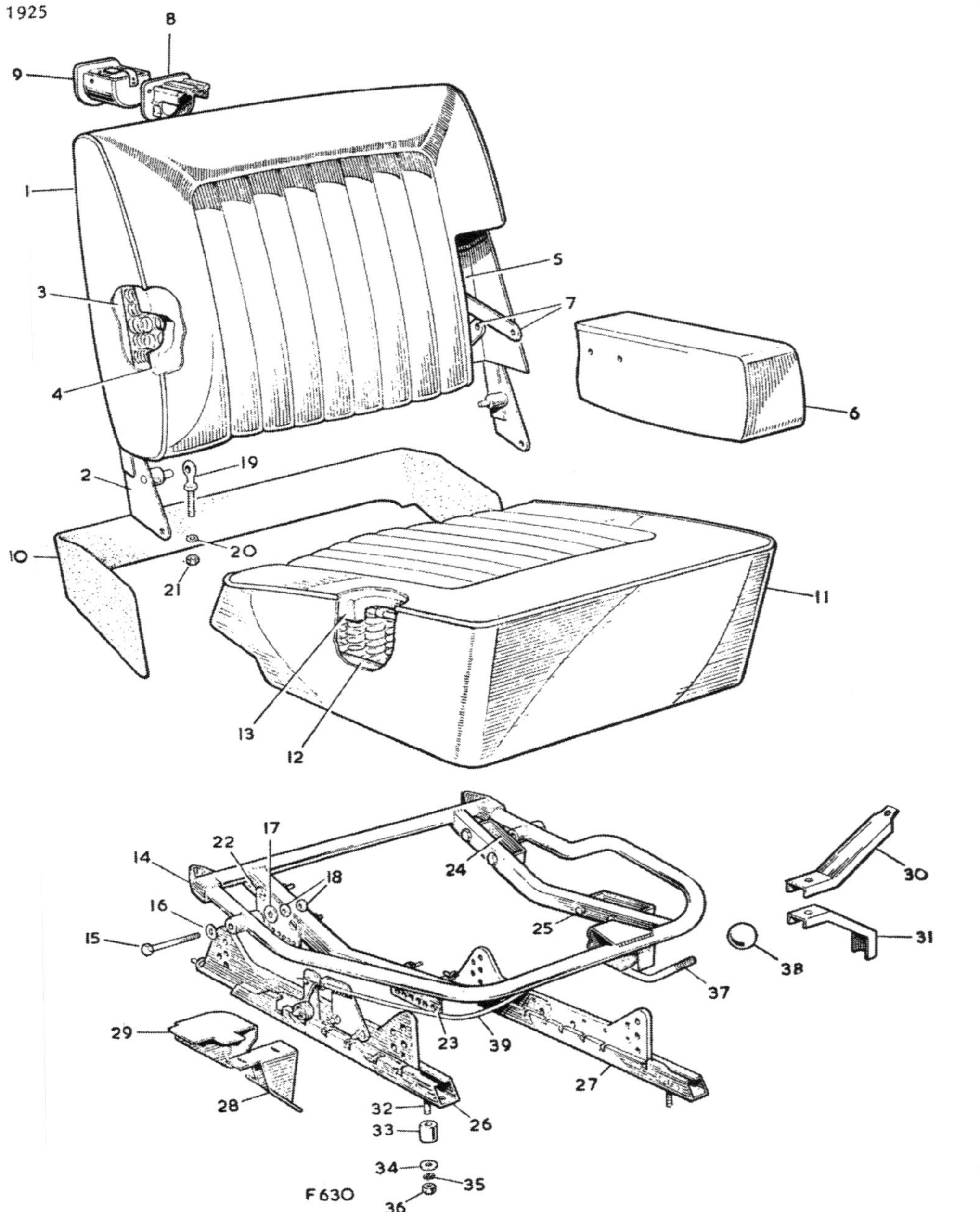

Plate Ref.	1 2 3 4 DESCRIPTION		Qty	Part No.	REMARKS
10	Rear valance for bucket seat, RH, trimmed		1	355948	Quote car number and colour required
	Rear valance for bucket seat, LH, trimmed		1	355949	
	Drive screw fixing valance to frame		8	78185	
11	CUSHION FOR BUCKET SEAT, RH, TRIMMED		1	355867	Quote car number and colour required
	CUSHION FOR BUCKET SEAT, LH, TRIMMED		1	355868	
12	Spring case for bucket seat, RH		1	354819	
	Spring case for bucket seat, LH		1	354820	
	Topping for bucket seat cushion, centre		2	354816†	
	Topping, side RH, for bucket seat cushion, RH		1	354817†	
	Topping, side RH, for bucket seat cushion, LH		1	354818†	
	Topping for bucket seat flute panel		2	315609††	
	Topping, side LH, for bucket seat cushion, RH		1	354818	
	Topping, side LH, for bucket seat cushion, LH		1	354817	
	Topping for bucket seat cushion front roll		2	354815†	
13	Topping for bucket seat cushion surround		2	355424††	
	Flute padding for cushion, plastic foam		As reqd	355983	
	Centre panel padding for cushion, plastic foam		As reqd	355985	
14	Seat frame for bucket seat, RH		1	357335	
	Seat frame for bucket seat, LH		1	357336	
15	Pivot bolt ($\frac{5}{16}$" UNF x $2\frac{1}{4}$" long)	Fixing squab to seat frame	4	256027	
16	Plain washer, large		4	2719	
17	Plain washer, small		4	3830	
18	Locknut ($\frac{5}{16}$" UNF)		8	254851	
19	Eyebolt for seat adjustment		4	355872	
20	Plain washer	Fixing eyebolt to seat frame	16	3830	
21	Nut ($\frac{5}{16}$" UNF)		4	254811	
22	Support plate, frame to slide, rear RH		1	355075	
	Support plate, frame to slide, rear LH		1	355076	
23	Support plate, frame to slide, front RH		1	355077	
	Support plate, frame to slide, front LH		1	355078	
24	Support bracket, seat to slide, RH rear		1	355047	
	Support bracket, seat to slide, LH rear		1	355048	
25	Support bracket, seat to slide, RH front		1	355045	
	Support bracket, seat to slide, LH front		1	355046	
	Bolt ($\frac{1}{4}$" UNF x $1\frac{1}{2}$" long)	Fixing support and support plate to seat frame	8	256003	
	Screw ($\frac{1}{4}$" UNF x $1\frac{3}{8}$" long)		8	78344	
	Plain washer		16	3840	
	Spring washer		16	3074	
	Nut ($\frac{1}{4}$" UNF)		16	254810	
	Bolt ($\frac{5}{16}$" UNF x $\frac{7}{8}$" long)	Fixing seat slide to support and support plate	8	255227	
	Plain washer		8	3830	
	Spring washer		8	3075	
	Nut ($\frac{5}{16}$" UNF)		8	254811	
26	Seat slide complete, RH, with catch		1	354000	
	Seat slide complete, LH, with catch		1	354001	
27	Seat slide complete, without catch		2	354851	
28	Support bracket for seat slide, outer		2	354853	
	Bolt ($\frac{5}{16}$" UNF x $\frac{3}{4}$" long)	Fixing support bracket to floor	6	255226	
	Plain washer		6	311199	
	Spring washer		6	3075	
	Nut ($\frac{5}{16}$" UNF)		6	254811	

* Asterisk indicates a new part which has not been used on any previous Rover model
† Cars numbered up to 625000979, 626000332, 628000110, 630000459, 631000115, 633000310
†† Cars numbered from 625000980, 626000333, 628000111, 630000460, 631000116, 633000311 onwards

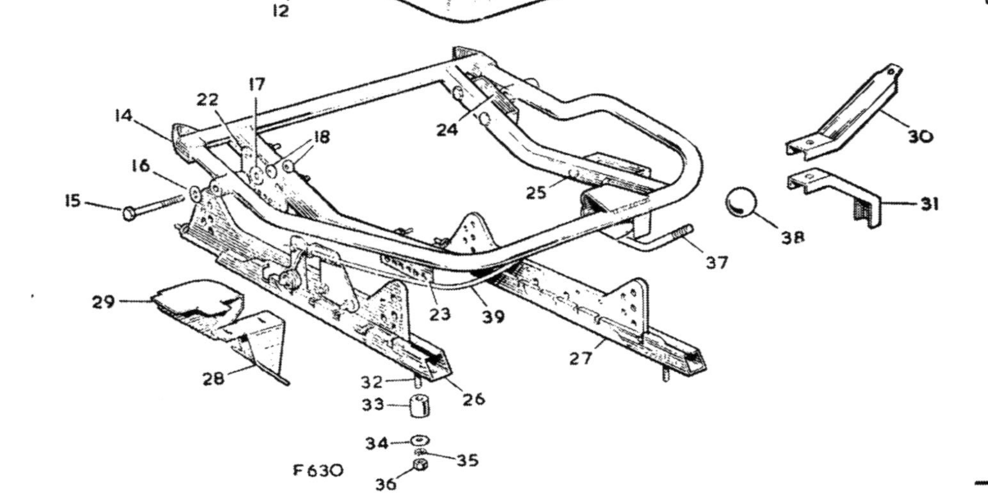

Plate Ref.	1 2 3 4 DESCRIPTION	Qty	Part No.	REMARKS
29	Cover for seat slide, RH outer	1	355063	
	Cover for seat slide, LH outer	1	355064	
	Drive screw fixing cover	8	77704	
	Cover for seat support, inner RH	1	355069	
	Cover for seat support, inner LH	1	355070	
	Drive screw fixing cover	8	77704	
30	Support bracket, top, for seat slide, inner	2	354854	
31	Support bracket, bottom, for seat slide, inner	2	354855	
	Set bolt ($\frac{5}{16}$" UNF x $\frac{1}{2}$" long) } Fixing support brackets to propeller shaft tunnel	4	255223	
	Spring washer } Fixing support brackets to propeller shaft tunnel	4	3075	
32	Screw ($\frac{5}{16}$" UNF x $1\frac{1}{4}$" long) } Fixing outer seat slides to floor and supports	2	78339	
33	Bearing washer } Fixing outer seat slides to floor and supports	4	311941	
34	Plain washer } Fixing outer seat slides to floor and supports	4	311199	
35	Spring washer } Fixing outer seat slides to floor and supports	4	3075	
36	Nut ($\frac{5}{16}$" UNF) } Fixing outer seat slides to floor and supports	4	254811	
	Screw ($\frac{5}{16}$" UNF x $\frac{3}{4}$" long) } Fixing seat slides to floor, inner	4	78343	
	Plain washer, small } Fixing seat slides to floor, inner	4	2218	
	Plain washer, large } Fixing seat slides to floor, inner	4	3299	
	Spring washer } Fixing seat slides to floor, inner	4	3075	
	Nut ($\frac{5}{16}$" UNF) } Fixing seat slides to floor, inner	4	254811	
	Bolt ($\frac{5}{16}$" UNF x $\frac{1}{2}$" long) } For redundant seat slide holes in floor	2	255223	
	Plain washer } For redundant seat slide holes in floor	4	3830	
	Spring washer } For redundant seat slide holes in floor	2	3075	
	Nut ($\frac{5}{16}$" UNF) } For redundant seat slide holes in floor	2	254811	
37	Operating handle, RH	1	315010	
	Operating handle, LH	1	315011	
	Set bolt ($\frac{1}{4}$" BSF x $\frac{1}{2}$" long) } Fixing operating handle to seat frame	2	237139	
	Shim } Fixing operating handle to seat frame	As reqd	313713	
	Shakeproof washer } Fixing operating handle to seat frame	2	78114	
	Plain washer } Fixing operating handle to seat frame	2	3665	
38	Knob for operating handle	2	313380	
39	Cable complete for seat slide	2	355060	
	Spindle } For retaining cable to seat slide	2	312689	
	Bush } For retaining cable to seat slide	2	312690	
	Nut (2 BA) } For retaining cable to seat slide	2	2247	
	Rear carpet complete	1	355201	Quote car number and colour required
	Carpet clip	As reqd	77210	
	Drive screw fixing carpet clip	As reqd	77704	
	Polythene cover for squab and cushion	4	381104	

* Asterisk indicates a new part which has not been used on any previous Rover model

FRONT BUCKET-TYPE SEATS, MK II SALOON MODELS

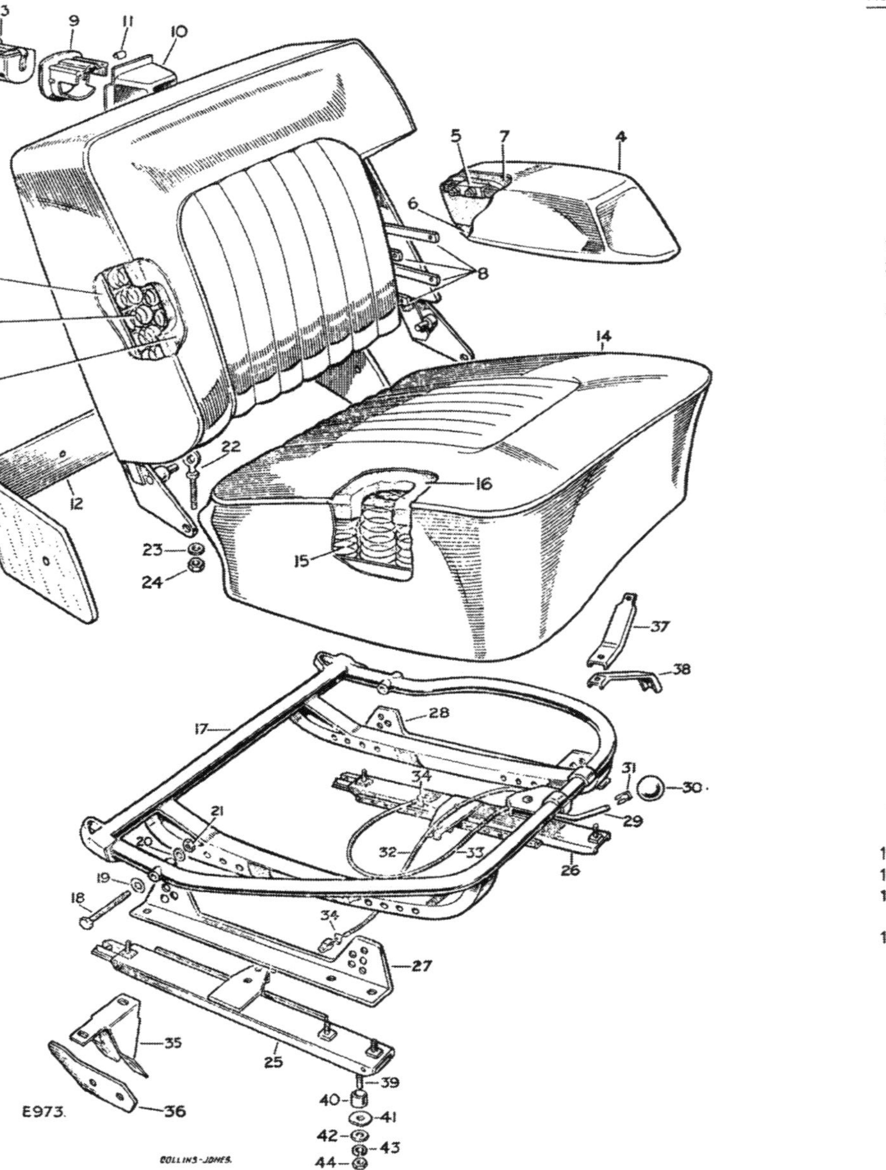

FRONT BUCKET-TYPE SEATS, MK II SALOON MODELS

Plate Ref.	1 2 3 4 DESCRIPTION	Qty	Part No.	REMARKS
	BUCKET SEATS COMPLETE ASSEMBLY ..	1	380263	Comprises two bucket seats with cushions and all fixings. State colour required
	BUCKET SEAT COMPLETE, RH (less cushion) ..	1	359594	Comprises one bucket seat, RH, with slides and fixings (less cushion). State colour required
	BUCKET SEAT COMPLETE, LH (less cushion) ..	1	359595	Comprises one bucket seat, LH, with slides and fixings (less cushion). State colour required
	BUCKET SEAT SQUAB COMPLETE, RH ..	1	359596	State colour required
1	Squab frame for bucket seat, RH	1	359484	
2	Spring case for squab, RH	1	359504	
	Topping for flute panel	1	359508	
3	Topping for surround panel	1	359506	
	Clip } Fixing cover	24	315960	
	Clip } to squab frame	6	331071	
	Armrest recess complete, RH	1	357733	State colour required
	Drive screw fixing recess to frame..	2	77903	
4	ARMREST COMPLETE, RH	1	359600	State colour required
5	Tongue bracket for armrest	1	359537	
6	Topping for armrest, front	1	359535	
7	Topping for armrest, rear	1	357738	
8	Armrest links for bucket seats, RH	1	359538	
	Special set screw } Fixing links	4	354822	
	Shakeproof washer } to armrest	4	70884	
	Screw (¼" UNF x ⅝" long) } Fixing links to squab frame	3	78232	
	Plain washer }	3	3840	
	Spring washer }	3	3074	
	Nut (¼" UNF) }	3	254810	
	BACK PANEL FOR SQUAB, RH	1	358833	State colour required
	Clips for back panel	3	314514	
	Clip fixing panel to frame	11	72602	
	Pocket assembly, trimmed	1	315578	State colour required
	Drive screw fixing pocket to frame..	5	77707	
	Finger plate for pocket	1	315509	
	Drive screw fixing plate to frame	2	77707	
9	Ashtray holder	1	317199	
	Drive screw fixing holder to frame	3	74141	
10	Ashtray shield	1	317014	
11	Spring clip for ashtray shield	1	315264	
12	Valance assembly for squab, RH	1	355948	State colour required
	Drive screw fixing valance to frame	4	78185	
13	Bowl for ashtray	1	314719	
	BUCKET SEAT SQUAB COMPLETE, LH ..	1	359597	State colour required
	Squab frame for bucket seat, LH	1	359485	
	Spring case for squab, LH	1	359505	
	Topping for flute panel	1	359508	
	Topping for surround panel	1	359507	

* Asterisk indicates a new part which has not been used on any previous Rover model

E973

Plate Ref.	1 2 3 4 DESCRIPTION	Qty	Part No.	REMARKS
	Clip } Fixing cover	24	315960	
	Clip } to squab frame	6	331071	
	Armrest recess complete, LH	1	357734	State colour required
	Drive screw fixing recess to frame..	2	77906	
	ARMREST COMPLETE, LH	1	359601	State colour required
	Tongue bracket for armrest	1	359537	
	Topping for armrest, front	1	359536	
	Topping for armrest, rear	1	357738	
	Armrest links for bucket seat, LH	1	359539	
	Special set screw } Fixing links	4	354822	
	Shakeproof washer } to armrest	4	70884	
	Screw (¼" UNF x ¾" long) } Fixing links to squab frame	3	78232	
	Plain washer } Fixing links to squab frame	3	3840	
	Spring washer } Fixing links to squab frame	3	3074	
	Nut (¼" UNF) } Fixing links to squab frame	3	254810	
	BACK PANEL FOR SQUAB, LH	1	358834	State colour required
	Clips for back panel	3	314514	
	Clip fixing panel to frame	11	72602	
	Pocket assembly, trimmed	1	315578	State colour required
	Drive screw fixing pocket to frame..	5	77707	
	Finger plate for pocket	1	315509	
	Drive screw fixing plate to frame	2	77707	
	Ashtray holder	1	317199	
	Drive screw fixing holder to frame	3	74141	
	Ashtray shield	1	317014	
	Spring clip for ashtray shield	1	315264	
	Valance assembly for squab, LH	1	355949	State colour required
	Drive screw fixing valance to frame	4	78185	
	Bowl for ashtray	1	314719	
14	CUSHION COMPLETE, RH	1	359598	} State colour required
	CUSHION COMPLETE, LH	1	359599	} State colour required
15	Spring case for cushion, RH	1	354819	
	Spring case for cushion, LH	1	354820	
16	Topping for flute panel	2	357685	
	Topping for surround panel	2	357686	
	Seat frame for bucket seat, RH	1	359529	
17	Seat frame for bucket seat, LH	1	359530	
18	Pivot bolt } Fixing hinge plate to seat frame	4	256027	
19	Friction washer } Fixing hinge plate to seat frame	4	2719	
20	Plain washer } Fixing hinge plate to seat frame	4	3830	
21	Thin nut (5/16" UNF) } Fixing hinge plate to seat frame	8	254851	
22	Eye bolt, seat adjusting	4	355872	
23	Plain washer } Eye bolt to	16	3830	
24	Nut (5/16" UNF) } cushion frame	8	254811	
25	Seat slide, RH	2	357745	
26	Seat slide, LH	2	357746	
27	Mounting bracket, RH } Seat slide	2	359384	
28	Mounting bracket, LH } to frame	2	359385	

* Asterisk indicates a new part which has not been used on any previous Rover model

Plate Ref.	1 2 3 4 DESCRIPTION		Qty	Part No.	REMARKS
	Bolt ($\frac{5}{16}$" UNF x $1\frac{1}{2}$" long)	Fixing mounting brackets to cushion frame	8	256222	
	Plain washer		8	3830	
	Spring washer		8	3075	
	Nut ($\frac{5}{16}$" UNF)		8	254811	
	Spring washer	Fixing mounting brackets to seat slides	12	3075	
	Nut ($\frac{5}{16}$" UNF)		12	254811	
29	Operating handle complete		2	359017	
30	Knob for handle		2	359019	
31	Clip fixing knob		2	359020	
	Bolt ($\frac{1}{4}$" UNF x $\frac{3}{4}$" long)	Fixing operating handles to seat frames	2	255007	
	Plain washer		2	2213	
	Locknut ($\frac{1}{4}$" UNF)		2	254850	
32	Control cable ($21\frac{1}{2}$" long)	For seat slides	2	359390	
33	Control cable (16" long)		2	359556	
34	Tab washer for control cables		4	359014	
35	Support bracket for seat slide, outer		2	354853	
36	Reinforcement plate for support bracket		2	359541	
	Bolt ($\frac{5}{16}$" UNF x $\frac{3}{4}$" long)	Fixing support bracket to floor	6	255226	
	Spring washer		6	3075	
	'D' washer		6	311199	
	Nut ($\frac{5}{16}$" UNF)		6	254811	
	Seat support cover assembly, RH		1	359798	State colour required
	Seat support cover assembly, LH		1	359799	
37	Bracket, top inner, for seat slide		2	358629	
38	Bracket, lower inner, for seat slide		2	358630	
	Set bolt ($\frac{5}{16}$" UNF x $\frac{1}{2}$" long)	Fixing bracket to tunnel	4	255223	
	Spring washer		4	3075	
39	Screw ($\frac{5}{16}$" UNF x $1\frac{1}{2}$" long)	Fixing front and rear of outer slides and front and rear of inner slides	4	78595	
40	Bearing washer		4	311941	
41	'D' washer		8	311199	
42	Plain washer		4	3830	
43	Spring washer		8	3075	
44	Nut ($\frac{5}{16}$" UNF)		8	254811	
	'D' washer, packing, to suit floor		As reqd 8	311199	
	Cover complete, RH outer support bracket		1	355063	
	Cover complete, LH outer support bracket		1	355064	
	Drive screw fixing covers		8	77704	
	Polythene cover for squab and cushion		4	381104	

* Asterisk indicates a new part which has not been used on any previous Rover model

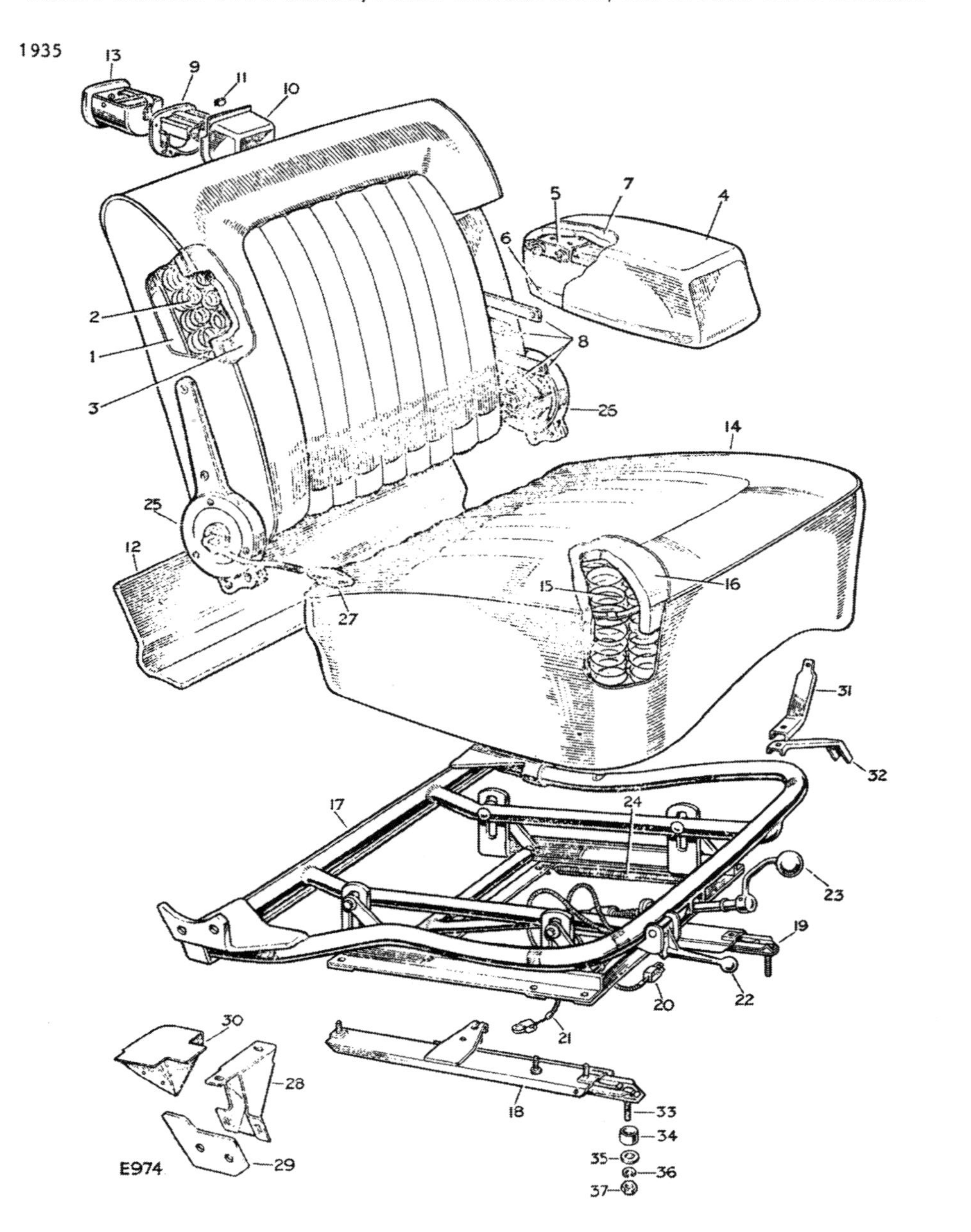

FRONT BUCKET-TYPE SEATS, FULLY ADJUSTABLE, MK IA AND MK II MODELS

Plate Ref.	1 2 3 4 DESCRIPTION	Qty	Part No.	REMARKS
	BUCKET SEAT SQUAB COMPLETE, RH ..	1	358850	Coupé } Quote car number and colour required
	BUCKET SEAT SQUAB COMPLETE, LH ..	1	358851	
	BUCKET SEAT SQUAB COMPLETE, RH ..	1	358621	Mk IA saloon
	BUCKET SEAT SQUAB COMPLETE, LH ..	1	358622	
	BUCKET SEAT SQUAB COMPLETE, RH ..	1	357687	Mk II saloon
	BUCKET SEAT SQUAB COMPLETE, LH ..	1	357688	
	Squab frame, RH	1	358565	Mk IA saloon
	Squab frame, LH	1	358566	
1	Squab frame, RH	1	357805	Mk II saloon and coupé
	Squab frame, LH	1	357806	
	Spring case for squab, RH	1	358623	Mk IA saloon
	Spring case for squab, LH	1	358624	
2	Spring case for squab, RH	1	357747	Mk II saloon and coupé
	Spring case for squab, LH	1	357748	
	Topping for flute panel	2	358627	Mk IA saloon
	Topping for flute panel	2	359188	Mk II saloon and coupé
	Topping for surround panel, RH squab	1	358625	Mk IA saloon
	Topping for surround panel, LH squab	1	358626	
3	Topping for surround panel, RH squab	1	359186	Mk II saloon and coupé
	Topping for surround panel, LH squab	1	359187	
	Clip } Fixing cover	24	315960	
	Clip } to frame	6	331071	
	Armrest recess complete	2	358544	Mk IA saloon. Quote car number and colour required
	Armrest recess complete, RH seat	1	359743	Mk II saloon and coupé
	Armrest recess complete, LH seat	1	359744	Quote car number and colour required
	Drive screw fixing recess to frame..	4	77906	
	ARMREST COMPLETE, RH	1	358546	Mk IA saloon
	ARMREST COMPLETE, LH	1	358547	Quote car number and colour required
4	ARMREST COMPLETE, RH	1	359550	Mk II saloon and coupé
	ARMREST COMPLETE, LH	1	359551	Quote car number and colour required
	Tongue bracket for armrest	1	358548	Mk IA saloon
	Washer plate } Fixing tongue to bracket	1	358552	
	Screw } Fixing tongue to bracket	2	78335	
	Spring washer } Fixing tongue to bracket	2	257023	
	Woodscrew } Fixing tongue to bracket	1	20153	
5	Tongue bracket	2	357721	Mk II saloon and coupé
	Topping for armrest, front RH	1	358553	Mk IA saloon
	Topping for armrest, front LH	1	358554	
	Topping for armrest, rear RH	2	358555	
6	Topping for armrest, front RH	1	357736	Mk II saloon and coupé
	Topping for armrest, front LH	1	357737	
7	Topping for armrest, rear	2	357738	
	Armrest links for bucket seat, RH	1	358557	Mk IA saloon
	Armrest links for bucket seat, LH	1	358558	
8	Armrest link and bracket	2	357731	Mk II saloon and coupé
	Set screw (¼" UNF x ½" long) } Fixing link to armrest	8	354822	
	Shakeproof washer } Fixing link to armrest	8	70884	
	Screw (¼" UNF x ⅝" long) } Fixing link to squab frame	6	78232	
	Plain washer } Fixing link to squab frame	6	3840	
	Spring washer } Fixing link to squab frame	6	3074	
	Nut (¼" UNF) } Fixing link to squab frame	6	254810	
	BACK PANEL COMPLETE FOR SQUAB, RH ..	1	358854	Coupé. Quote car number and colour required
	BACK PANEL COMPLETE FOR SQUAB, LH ..	1	358855	

* Asterisk indicates a new part which has not been used on any previous Rover model

FRONT BUCKET-TYPE SEATS, FULLY ADJUSTABLE, MK IA AND MK II MODELS

E974.

COLLINS-JONES

FRONT BUCKE'‑TYPE SEATS, FULLY ADJUSTABLE, MK IA AND MK II MODELS

Plate Ref.	1 2 3 4 DESCRIPTION	Qty	Part No.	REMARKS
	BACK PANEL COMPLETE FOR SQUAB, RH	1	358561	Saloon. Quote car number and colour required
	BACK PANEL COMPLETE FOR SQUAB, LH	1	358562	Saloon. Quote car number and colour required
	Clip for back panel	6	314514	
	Clip fixing back panel to frame	22	72602	
	Pocket assembly, trimmed for squab	2	315578	Quote car number and colour required
	Drive screw fixing pocket to frame	10	77707	
	Finger plate for pocket	2	315509	
	Drive screw fixing plate to frame	4	77707	
9	Ashtray holder	2	317199	Saloon
	Drive screw fixing holder	6	74141	Saloon
10	Shield for ashtray	2	317014	Saloon
11	Spring clip for shield	2	315264	Saloon
12	Valance assembly for bucket seat	2	358540	Quote car number and colour required
	Drive screw fixing valance	12	78185	Saloon
13	Bowl for ashtray	2	314719	Saloon
	CUSHION FOR BUCKET SEAT, RH	1	358597	Mk IA saloon. Quote car number and colour required
	CUSHION FOR BUCKET SEAT, LH	1	358598	Mk IA saloon. Quote car number and colour required
14	CUSHION FOR BUCKET SEAT, RH	1	353588	Mk II saloon and coupé. Quote car number and colour required
	CUSHION FOR BUCKET SEAT, LH	1	353589	Mk II saloon and coupé. Quote car number and colour required
15	Spring case for cushion, RH	1	357683	
	Spring case for cushion, LH	1	357684	
	Topping for cushion flute panel	2	315609	Mk IA saloon
	Topping for bucket seat surround	2	355424	Mk IA saloon
16	Topping for cushion flute panel	2	357685	Mk II saloon and coupé
	Topping for cushion surround panel	2	357686	Mk II saloon and coupé
	SEAT FRAME FOR BUCKET SEAT, RH	1	357801	
17	SEAT FRAME FOR BUCKET SEAT, LH	1	357802	
18	Seat slide, RH	2	357745	
19	Seat slide, LH	2	357746	
20	Cable, long (For seat adjustment)	2	358681	
21	Cable, short (For seat adjustment)	2	358682	
22	Seat slide adjustment handle	2	358683	
23	Raise and fall handle complete	2	358639	
24	Spring for seat frame	4	380492	
25	'Lyback' mechanism with handle, RH (For RH seat)	1	358633	Mechanism sliding fit to actuating rod. Alternatives Check before ordering
26	'Lyback' mechanism without handle, LH (For RH seat)	1	358636	Mechanism sliding fit to actuating rod. Alternatives Check before ordering
	'Lyback' mechanism with handle, LH (For LH seat)	1	358634	Mechanism sliding fit to actuating rod. Alternatives Check before ordering
	'Lyback' mechanism without handle, RH (For LH seat)	1	358635	Mechanism sliding fit to actuating rod. Alternatives Check before ordering
	'Lyback' mechanism with handle (For RH seat)	1	383269	Mechanism bolted to actuating rod. Alternatives Check before ordering
	'Lyback' mechanism without handle (For RH seat)	1	383272	Mechanism bolted to actuating rod. Alternatives Check before ordering
	'Lyback' mechanism with handle (For LH seat)	1	383270	Mechanism bolted to actuating rod. Alternatives Check before ordering
	'Lyback' mechanism without handle (For LH seat)	1	383271	Mechanism bolted to actuating rod. Alternatives Check before ordering
	Bolt (3/8" UNF x 3/4" long) Fixing 'Lyback' mechanism to cushion frame	8	358556	
	Nut (3/8" UNF) Fixing 'Lyback' mechanism to cushion frame	8	78465	
	Screw (1/4" UNF x 7/8" long) fixing mechanism to squab frame	8	78402	
27	Knob for operating handle	2	359021	
	Screw (5/16" UNF x 1/2" long) (For holes in floor)	2	255023	
	Plain washer (For holes in floor)	4	3830	
	Spring washer (For holes in floor)	2	3075	
	Nut (5/16" UNF) (For holes in floor)	2	254811	

* Asterisk indicates a new part which has not been used on any previous Rover model

FRONT BUCKET-TYPE SEATS, FULLY ADJUSTABLE, MK IA AND MK II MODELS

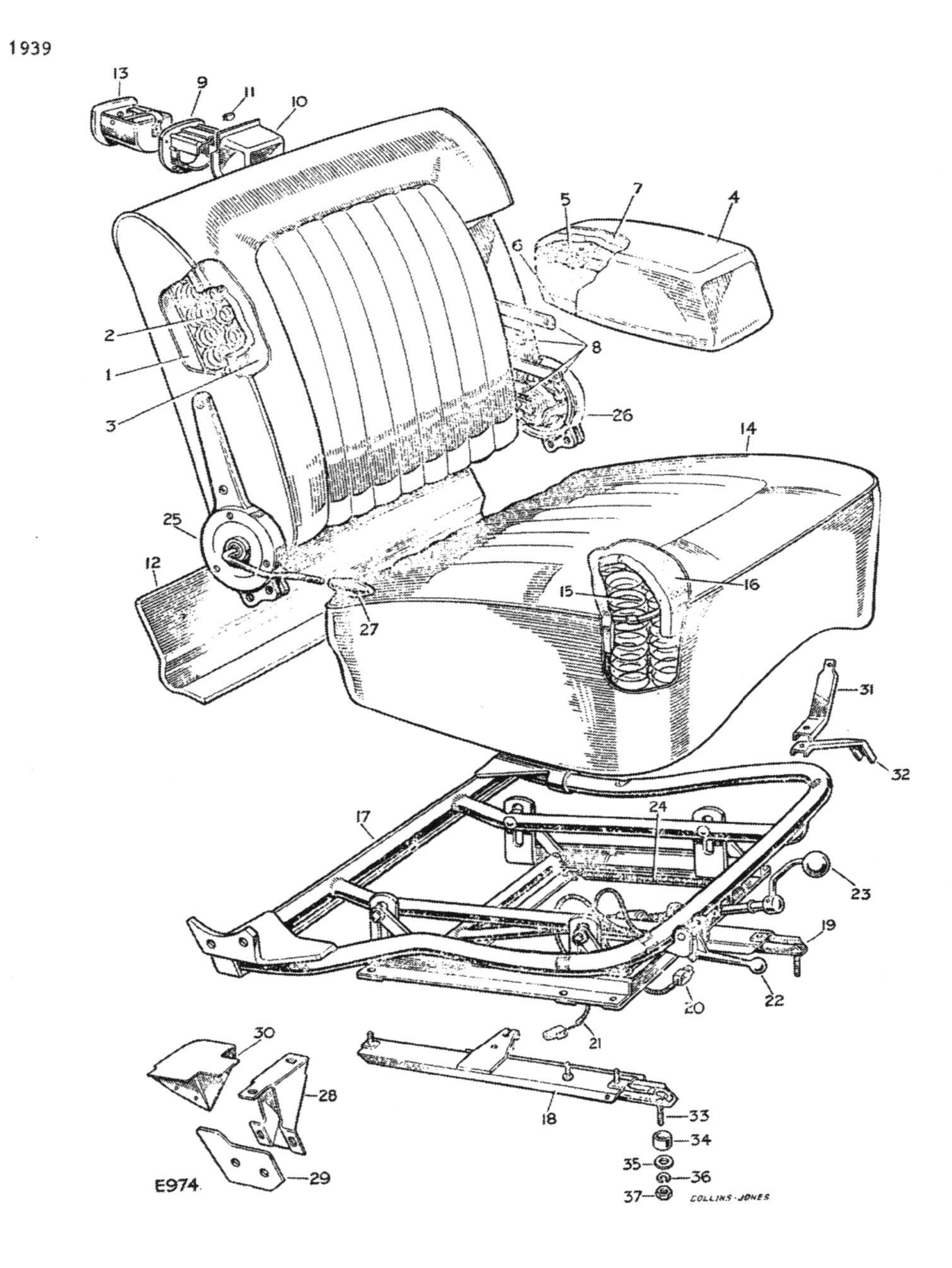

FRONT BUCKET-TYPE SEATS, FULLY ADJUSTABLE, MK IA AND MK II MODELS

Plate Ref.	1 2 3 4 DESCRIPTION	Qty	Part No.	REMARKS
28	Support bracket for seat slide, outer	2	354853	
29	Reinforcement plate for support bracket	2	359541	
30	Cover for support bracket, outer	2	355063	
	Bolt ($\frac{5}{16}$" UNF x $\frac{3}{4}$" long) } Fixing support bracket to floor	6	255226	
	Spring washer } Fixing support bracket to floor	6	3075	
	'D' washer } Fixing support bracket to floor	6	311199	
	Nut ($\frac{5}{16}$" UNF) } Fixing support bracket to floor	6	254811	
31	Bracket, top, inner, for seat slide	2	358629	
32	Bracket, lower, inner, for seat slide	2	358630	
	Set bolt ($\frac{5}{16}$" UNF x $\frac{1}{2}$" long) } Fixing bracket to tunnel	4	255223	
	Spring washer } Fixing bracket to tunnel	4	3075	
33	Screw ($\frac{5}{16}$" UNF x $1\frac{1}{4}$" long) } Fixing front and rear of outer slides and front and rear of inner slides	4	78472	
34	Bearing washer } Fixing front and rear of outer slides and front and rear of inner slides	4	311941	
	'D' washer } Fixing front and rear of outer slides and front and rear of inner slides	8	311199	
35	Plain washer } Fixing front and rear of outer slides and front and rear of inner slides	4	3830	
36	Spring washer } Fixing front and rear of outer slides and front and rear of inner slides	8	3075	
37	Nut ($\frac{5}{16}$" UNF) } Fixing front and rear of outer slides and front and rear of inner slides	8	254811	
	'D' washer, packing, to suit floor	As reqd 8	311199	
	Polythene cover for squab and cushion	4	381108	

* Asterisk indicates a new part which has not been used on any previous Rover model

FRONT BUCKET TYPE SEATS, FULLY ADJUSTABLE
MKIII 3 LITRE AND 3½ LITRE SALOON AND COUPE MODELS

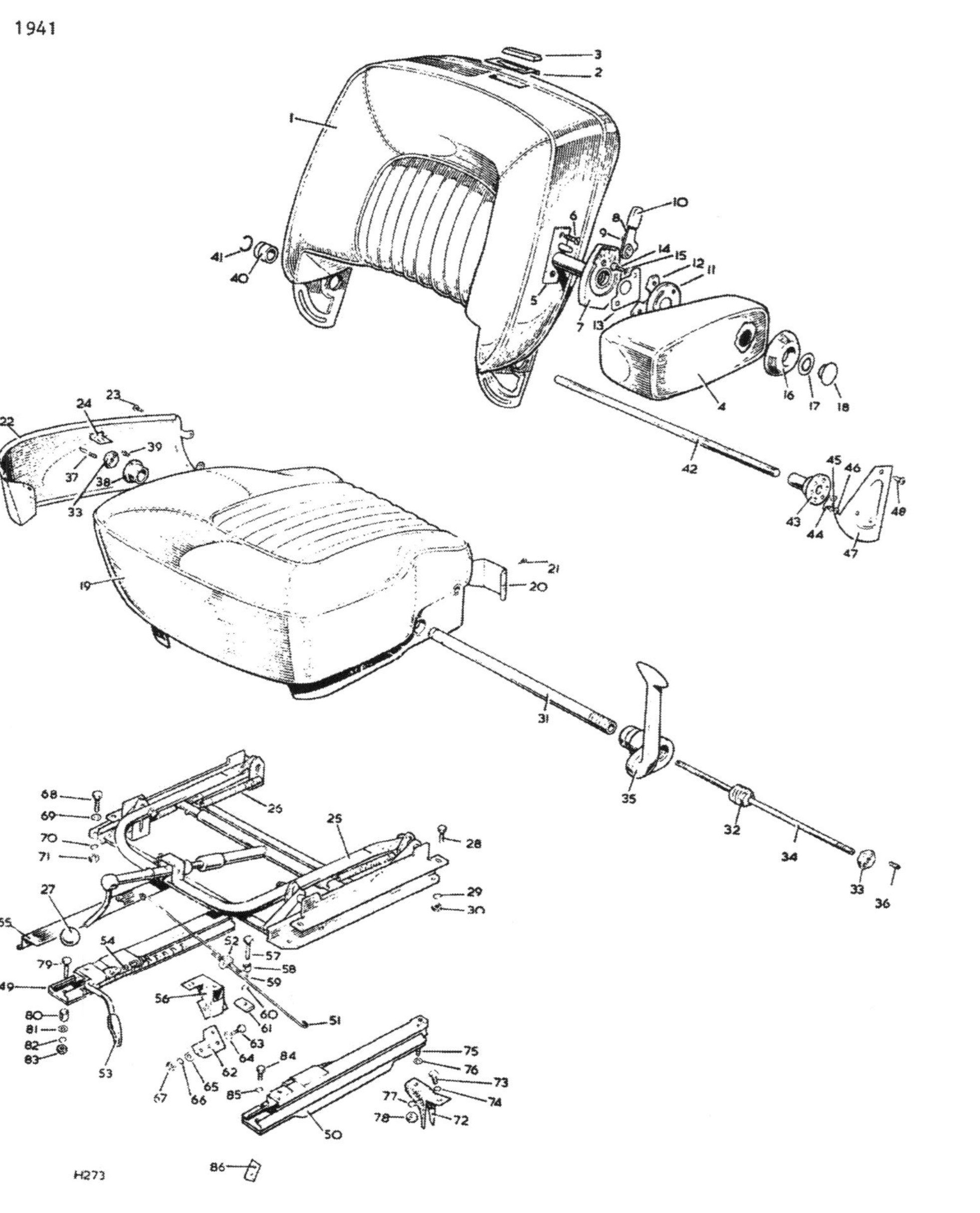

FRONT BUCKET-TYPE SEATS, FULLY ADJUSTABLE
MK III 3 LITRE AND 3½ LITRE SALOON AND COUPE MODELS

Plate Ref.	Description (1 2 3 4)	Qty.	Part No.	Remarks
	FRONT BUCKET SEAT COMPLETE, RH	1	386406	Quote car number and colour required
	FRONT BUCKET SEAT COMPLETE, LH	1	386407	
1	SQUAB ASSEMBLY	2	384890	
	Back panel trimmed assembly	2	384900	
	Drive screw — Fixing back panel to squab	8	78544	
	Nylon clinch nut — Fixing back panel to squab	8	364436	
2	Escutcheon at headrest aperture	2	384899	
3	Blanking plug for escutcheon	2	384910	
	Drive screw fixing escutcheon to squab	4	78527	
	Trim clip fixing hide to frame	16	331071	
4	Centre armrest trimmed assembly for driver's seat only	1	384774	Quote car number and colour required
5	Base plate for armrest	1	385861	
6	Screw (¼" UNF x 1" long)fixing base plate to seat	2	78592	
7	Plastic cover for adjusting mechanism	1	384796	
8	CAM LEVER ASSEMBLY	1	384788	RH Stg
	CAM LEVER ASSEMBLY	1	385708	LH Stg
9	Locating spring	1	384791	
	Screw (6 UNC x 5/16" long) — Fixing spring to lever	1	257194	Early models only
	Shakeproof washer — Fixing spring to lever	1	78538	
	Plain washer — Fixing spring to lever	1	2874	
10	Knob for lever	1	384792	
11	Cover plate (chrome)	1	384783	
12	Armrest stop	1	384785	
13	Cam retaining plate	1	384784	
14	Screw (10 UNF x ⅜" long) — Fixing armrest stop	3	257008	
15	Shakeproof washer — Fixing armrest stop	3	71082	
16	Escutcheon for armrest	1	384419	
17	Spire fastener fixing base plate to armrest	1	384797	Early Mk III 3 Litre
18	Plug for base plate spindle	1	384423	
	Plain washer — Fixing escutcheon to base plate	1	4575	Late Mk III 3 Litre and 3½ Litre
	Special spring washer — Fixing escutcheon to base plate	1	385825	
	Cover disc — Fixing escutcheon to base plate	1	385826	
	Self-locking screw — Fixing escutcheon to base plate	1	78746	
19	Cushion assembly	2	383851	Quote car number and colour required
20	Support leg finisher trim	2	384210	
21	Drive screw fixing finisher to cushion	4	78747	
22	Stainless steel finisher, lower, RH	1	383991	
	Stainless steel finisher, lower, LH	1	383992	
23	Drive screw fixing finisher to seat	4	78575	
24	Retaining bracket for lower finisher	2	383994	
	Drive screw fixing bracket to seat	4	78485	
25	RISE AND FALL MECHANISM COMPLETE	2	383984	
26	Spring for rise and fall mechanism	2	385180	
27	Knob for rise and fall mechanism	2	385106	
28	Bolt (5/16" UNF x ¾" long) — Fixing rise and fall mechanism to seat	8	255226	
29	Spring washer — Fixing rise and fall mechanism to seat	8	3075	
30	Nut (5/16" UNF) — Fixing rise and fall mechanism to seat	8	254831	
31	Compression tube assembly, RH seat	1	368298	
	Compression tube assembly, LH seat	1	368299	

*Asterisk indicates a new part which has not been used on any previous model.

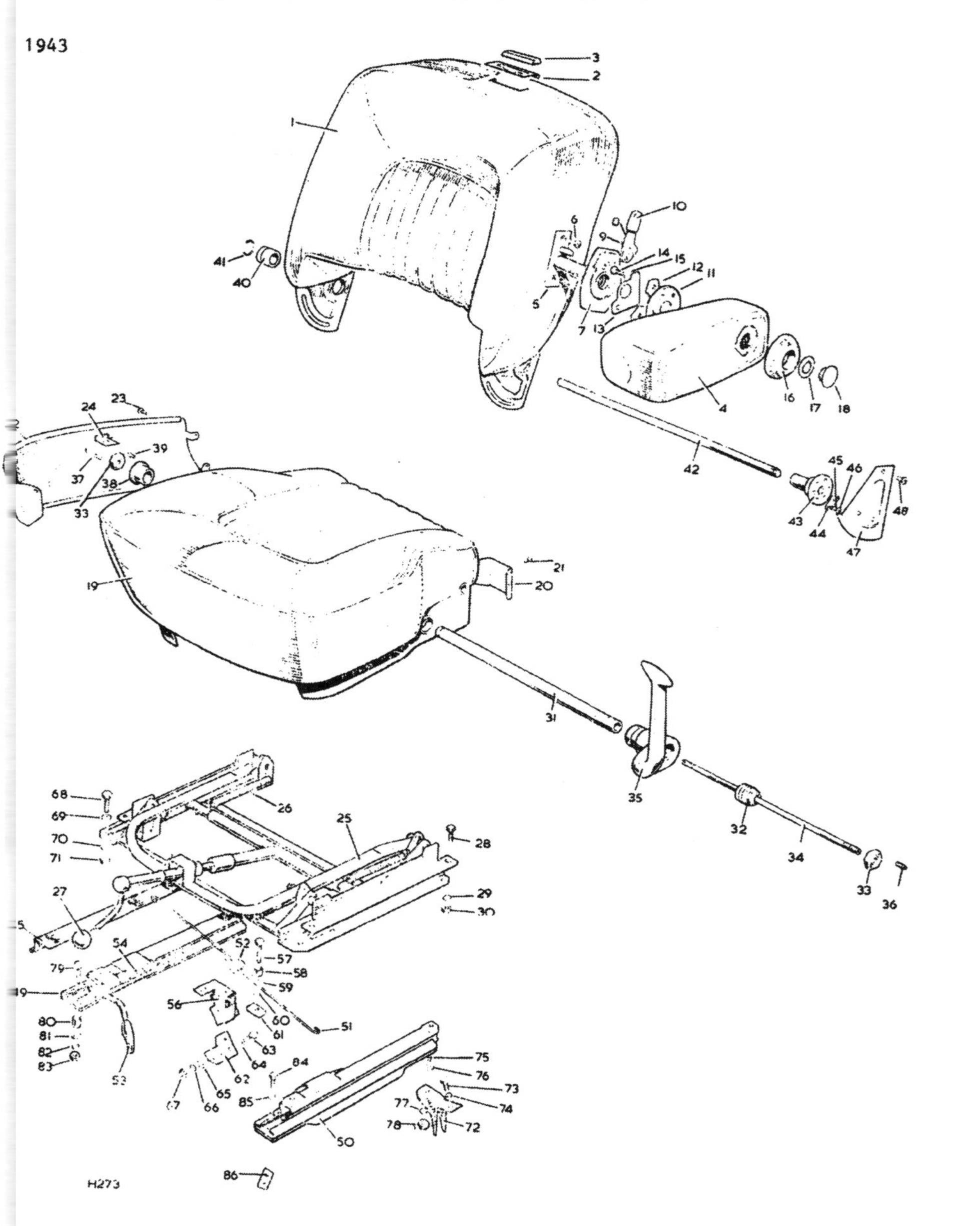

Plate Ref.	1 2 3 4 Description	Qty.	Part No.	Remarks
32	Inner locking pad for RH seat	1	365151	
	Inner locking pad for LH seat	1	365152	
33	Outer locking pad	4	367904	
34	Tie bar for locking seat	2	362531	
35	Seat locking lever, RH seat	1	384336	
	Seat locking lever, LH seat	1	384337	
36	Locking pad dowel, short	2	362534	
37	Locking pad dowel, long	2	362533	
38	Compression tube bearing	2	370654	
39	Drive screw fixing bearing to squab frame	2	78485	
40	Bearing, fixed pivot to squab	2	362519	
41	Circlip fixing bearing to squab	2	362520	
42	Torsion bar for squab support	2	384709	
43	Loose pivot, torsion bar to squab	2	384710	
44	Retainer for torsion bar	2	362518	
45	Spring washer } Fixing pivot and retainer to squab	3	3073	
46	Screw (10 UNF x ½" long) } Fixing pivot and retainer to squab	3	78593	
47	Cover plate, stainless steel, RH } For pivot legs	2	383942	Early models
	Cover plate, stainless steel, LH } For pivot legs	2	383943	Early models
48	Drive screw fixing cover plates to legs	6	78575	Early models
	Cover plate stainless steel RH } For pivot legs	2	386557*	Late models
	Cover plate stainless steel LH } For pivot legs	2	386558*	Late models
	Drive screw short } Fixing cover plates to legs	8	78575	Late models
	Drive screw long } Fixing cover plates to legs	4	78713	Late models
	Nylon spacer	8	370711	When fitted
49	Seat slide, RH	2	384100	
50	Seat slide, LH	2	384101	
51	TIE ROD FOR SEAT SLIDE	2	384099	
52	Socket screw (¼" UNF x ¾" long) for tie rod	4	78737	
53	Finger lift for seat slide	2	384168	
54	Screw (4 UNC x ⅜" long) fixing finger lift to slide	2	78702	
55	Cover, stainless steel, for outer seat slide	2	383875	
56	Mounting bracket for outer slide, RH	1	383938	
	Mounting bracket for outer slide, LH	1	383939	
57	Bolt (5/16" UNF x 1½" long) } Fixing seat slides, outer rear, to mounting brackets	2	256222	
58	Bearing washer } Fixing seat slides, outer rear, to mounting brackets	2	311941	
59	Plain washer } Fixing seat slides, outer rear, to mounting brackets	2	3830	
60	Spring washer } Fixing seat slides, outer rear, to mounting brackets	2	3075	
61	Tapping plate } Fixing seat slides, outer rear, to mounting brackets	2	384636	
	Screw (5/16" UNF x ¾" long) fixing tapped plate to floor	2	78329	
62	Reinforcement plate	2	383940	
63	Bolt (5/16" UNF x ¾" long) } Fixing mounting bracket and reinforcing plate to floor	10	255226	
64	Plain washer } Fixing mounting bracket and reinforcing plate to floor	10	3899	
65	'D' washer } Fixing mounting bracket and reinforcing plate to floor	10	311199	
66	Spring washer } Fixing mounting bracket and reinforcing plate to floor	10	3075	
67	Nut (5/16" UNF) } Fixing mounting bracket and reinforcing plate to floor	10	254831	
68	Bolt (5/16" UNF x 1" long) } Fixing rise and fall mechanism to seat slides	8	255028	
69	Plain washer } Fixing rise and fall mechanism to seat slides	8	3830	
70	Spring washer } Fixing rise and fall mechanism to seat slides	8	3075	
71	Nut (5/16" UNF) } Fixing rise and fall mechanism to seat slides	8	254831	

*Asterisk indicates a new part which has not been used on any previous model.

FRONT BUCKET TYPE SEATS, FULLY ADJUSTABLE
MKIII 3 LITRE AND 3½ LITRE SALOON AND COUPE MODELS

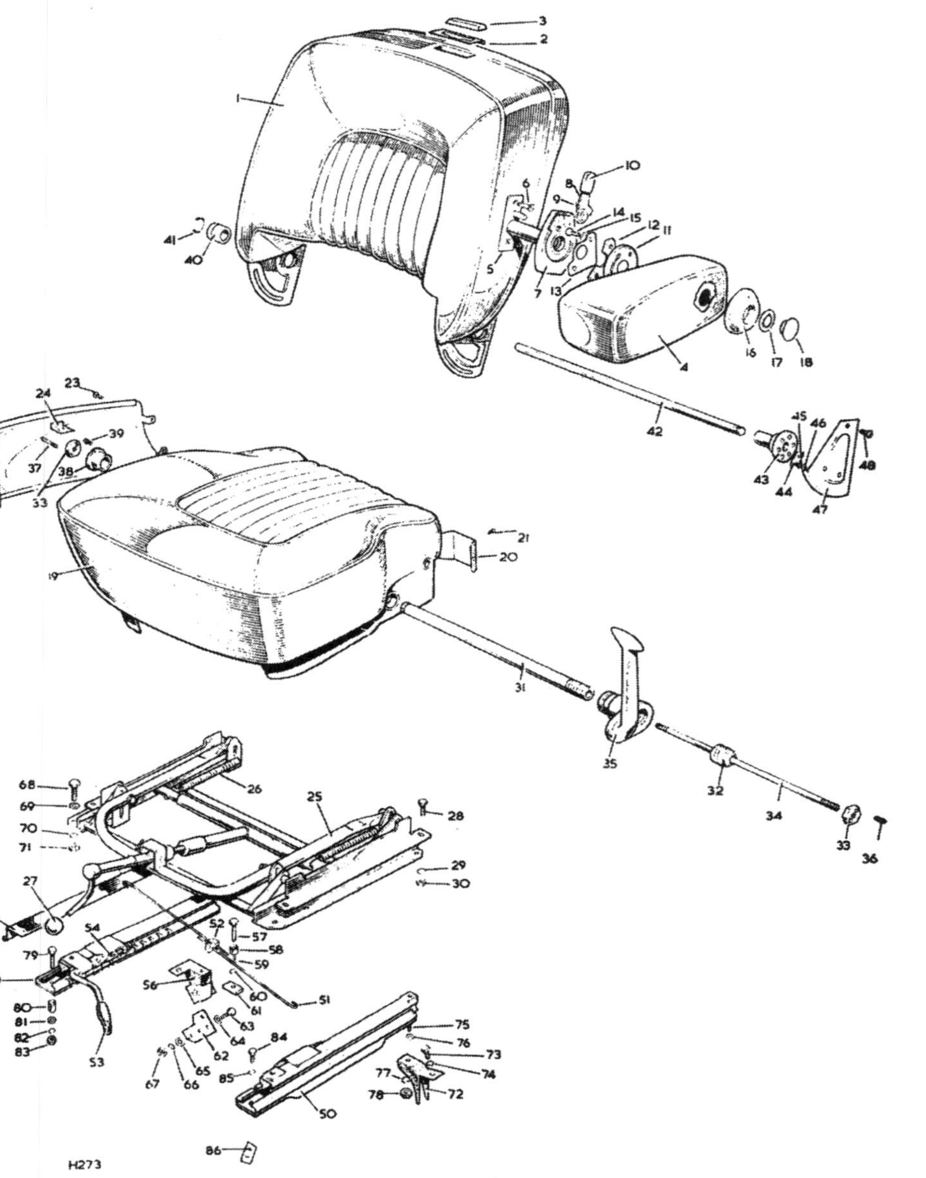

Plate Ref.	1 2 3 4	Description		Qty.	Part No.	Remarks
72		Mounting bracket trim assembly, RH, for seat slide, inner		1	384533	3 Litre Quote car number and colour required
		Mounting bracket trim assembly, LH, for seat slide, inner		1	384534	
		Mounting bracket trim assembly, RH for seat slide, inner		1	385909	3½ Litre
		Mounting bracket trim assembly, LH for seat slide, inner		1	385910	
73		Bolt (5/16" UNF x ⅝" long)	Fixing bracket to tunnel	4	255225	
74		Spring washer		4	3075	
75		Screw (5/16" UNF x 1" long)	Fixing seat slides, inner, rear, to mounting brackets	2	255228	
76		'D' washer		2	311199	
77		Spring washer		2	3075	
78		Nut (5/16" UNF)		2	254831	
79		Bolt (5/16" UNF x 1½" long)	Fixing seat slides, outer front, to body	2	256222	
80		Bearing washer		2	311941	
81		'D' washer		2	311199	
82		Spring washer		2	3075	
83		Nut (5/16" UNF)		2	254831	
84		Bolt (5/16" UNF x 1" long)	Fixing seat slides, inner front, to body	2	255228	
85		Spring washer		2	3075	
86		Tapped plate		2	385521	
		'D' washer, packing, to suite floor		8	311199	
		Rubber grommet for holes in floor		8	312886	
		Polythene cover for front cushion		2	384837	
		Polythene cover for front squab		2	384838	

*Asterisk indicates a new part which has not been used on any previous model.

H707

OPTIONAL EQUIPMENT, REAR BENCH SEAT ASSEMBLY, MK III 3 LITRE AND 3½ LITRE SALOON MODELS

Plate Ref.	1 2 3 4 Description		Qty.	Part No.	Remarks
1	REAR BENCH SEAT SQUAB ASSEMBLY		1	385624	State colour required
2	CENTRE ARMREST ASSEMBLY		1	385663	State colour required
3	Tongue for armrest		1	385530	
4	Top link	For centre armrest	2	383833	
5	Bottom link	For centre armrest	2	383834	
6	Shakeproof washer	Fixing links for armrest	4	70884	
7	Set screw(¼" UNF x½" long)	Fixing links for armrest	4	354822	
8	Shaft	Fixing links to squab frame	4	365096	
9	Waved washer	Fixing links to squab frame	4	353368	
10	Plain washer	Fixing links to squab frame	4	3911	
11	Retaining clip	Fixing links to squab frame	4	78573	
12	Recess panel for armrest aperture		1	385671	State colour required
13	Fastener fixing recess panel		2	315960	
14	CUSHION ASSEMBLY		1	385676	State colour required
15	Spring case for cushion		1	385677	
16	Bolt (¼" UNF x 1¼" long)	Fixing rear squab to body	4	255211	
17	Plain washer	Fixing rear squab to body	4	2552	
18	Sealing washer	Fixing rear squab to body	4	355085	

*Asterisk indicates a new part which has not been used on any previous model.

OPTIONAL EQUIPMENT, FRONT SEAT HEADREST

H275

OPTIONAL EQUIPMENT, FRONT SEAT HEADREST Mk III 3 LITRE AND 3½ LITRE

Plate Ref.	1 2 3 4 Description	Qty.	Part No.	Remarks
	FRONT SEAT HEADREST KIT COMPLETE	1	385122%	
	FRONT SEAT HEADREST KIT COMPLETE	1	387456*%%	
1	HEADREST, trimmed	1	384849%	
	HEADREST, TRIMMED	1	386499*%%	
	Release button	1	384870	
	Ferrule for release button	1	385949	
2	Base plate, trimmed, for headrest	1	384854	
3	Drive screw fixing base plate to headrest	4	78527	
	Distance piece, pillar stop	1	367932%%	
	Drive screw fixing pillar stop to headrest retainer	1	78544%%	
4	Retainer for headrest in squab	1	386260	
5	Nylon guide for headrest in squab	1	386264	
	Screw (10 UNF x 1⅜" long) } Fixing retainer and guide to seat squab	2	78477	
	Spire fix } Fixing retainer and guide to seat squab	2	78723	
	Drive screw } Fixing retainer and guide to seat squab	2	78727	
	Drive screw } Fixing retainer and guide to seat squab	2	78527	
6	Spring } For headrest retainer	1	384887	
	Pop rivet } For headrest retainer	2	78257	
7	LAMP FOR HEADREST	1	559180	
8	Bulb, 10-watt for headrest lamp	1	559186	
9	Screw (6UNC x 5/16" long) } Fixing lamp to base plate	1	78738	This headrest kit is suit-
10	Fan disc washer } Fixing lamp to base plate	1	536993	able for either the driver's
11	Nut (6 UNC) } Fixing lamp to base plate	1	257191	or passenger's seat
12	Housing for lamp	1	559181	State colour required
13	Screw (6 UNC x 1" long) } Fixing lamp housing to headrest	2	78739	@@ NOTE: Only one harness
14	Fan disc washer } Fixing lamp housing to headrest	2	536993	555787 is required whether
15	Nut (6 UNC) } Fixing lamp housing to headrest	2	257191	one or two headrests are
16	Switch for lamp	1	541652	fitted
	Connector block, lead to harness	1	555974	
17	Harness, switch and lamp to plug	1	555971	
	Cable clip (pick-up at headrest slide fixing)	1	4557	
18	Harness, plug socket to tunnel	1	555785	
19	Grommet for bottom edge of squab	1	520012	
20	Cable clip	4	555833	
21	Connector block, seat harness to tunnel harness	1	555975	
22	HARNESS, FEED PICK-UP FROM CIGAR LIGHTER	1	555787@@	
	In-line fuse holder	1	600546	
	Fuse, 10 amp	1	541567	
	'Crimp-on' nipples	4	536937	
	Cable cleat	1	240429	

% 3 Litre and 3½ Litre cars numbered with suffix letter 'A'
%% 3½Litre cars numbered with suffix letter 'B' onwards

*Asterisk indicates a new part which has not been used on any previous model.

OPTIONAL EQUIPMENT, FOG AND LONG-RANGE DRIVING LAMPS

1952

D383

Plate Ref.	Description (1 2 3 4)		Qty.	Part No.	Remarks
	FOGLAMP COMPLETE ASSEMBLY		1	533881	Mk I
	LONG RANGE DRIVING LAMP COMPLETE ASSEMBLY		1	601709	Mk IA, Mk II and Mk III
1	FOGLAMP COMPLETE SFT type		1	274782	
	FOGLAMP COMPLETE LUCAS 'NOVA'		1	606846*	
	LONG RANGE DRIVING LAMP LUCAS 'NOVA'		1	606847*	
2	Light unit for foglamp		1	276810	
	Light unit for long range driving lamp		1	276812	
	Bulb for foglamp		1	238838	
	Bulb for long range driving lamp		1	519960	
4	Bracket for foglamp		1	500621	
	Distance piece for foglamp bracket		1	532641	Mk IA, Mk II and Mk III
5	Plinth for foglamp		1	500623	
6	Clamp plate		1	500622	
7	Bolt (5/16" UNF x 1⅛" long)	Fixing bracket to front bumper bar	2	255029	Mk I
	Bolt (5/16" UNF x 1¾" long)	Fixing bracket to front bumper bar	2	255234	Mk IA, Mk II and Mk III
8	Plain washer	Fixing bracket to front bumper bar	2	2550	
9	Spring washer		2	3075	
10	Harness for foglamp		1	512324	
	Lead, fuse to harness		1	512325	RH Stg
	Lead, fuse to harness		1	512326	LH Stg
	Earthing lead		1	512328	
	Cable clip		1	50637	LH Stg
	Cable clip		5	237749	
	Cable cleat		6	240431	
11	Switch for foglamp, under instrument unit		1	512253	
12	Switch for foglamp on steering column shroud		1	511311	
13	Mounting bracket, lower, for switch		1	502316	RH Stg — When overdrive is fitted or one switch only is required
	Mounting bracket, lower, for switch		1	502315	LH Stg — When overdrive is fitted or one switch only is required
	Mounting bracket, lower, for switch		1	502316	RH Stg — When no overdrive is fitted and two switches are required
	Mounting bracket, upper, for switch		1	502318	When no overdrive is fitted and two switches are required
	Mounting bracket, lower, for switch		1	502315	LH Stg — When no overdrive is fitted and two switches are required
	Mounting bracket, upper, for switch		1	502317	When no overdrive is fitted and two switches are required
14	Screw (10 UNF x 5/16" long)	Fixing mounting bracket	1	78307	2 off when two switches are fitted
15	Spring washer	Fixing mounting bracket	1	3073	
16	Cover plate for shroud with one hole		1	352496	When one switch is required
	Cover plate for shroud with two holes		1	352497	When two switches are required

FOGLAMP SWITCHES ON STEERING COLUMN SHROUD INSTEAD OF UNDER INSTRUMENT UNIT

These may be fitted, if required, to all models except those with automatic transmission. Models with overdrive may be fitted with one switch on the steering column shroud only. It is recommended that in cases where two foglamps are required and overdrive is fitted both switches be positioned under the instrument unit for ease of operation.

The parts listed above, items 12 to 16 inclusive, which must be ordered separately if required, cover switches and fixings for mounting on the steering column shroud. These replace the switch, part number 512253, which is for mounting under the instrument unit.

*Asterisk indicates a new part which has not been used on any previous model.

Plate Ref. 1 2 3 4	Description	Qty.	Part No.	Remarks
	Cylinder liner, shaped	6	516498	3 Litre only
	Electric immersion heater for cylinder block	1	511302	3 Litre only
	Rear hub sealing modification kit	1	520808	
	Front shock absorber, heavy duty	2	536595	
	Rear shock absorber, heavy duty	2	536596	
	Power steering conversion kit	1	524122	RH Stg } 1961 and Mk IA cars numbered up to 72500551, 72600090, 72800081, 73000766, 73100150, 73300152
	Power steering conversion kit (Cars with single-bearing inner column)	1	524123	LH Stg
	Power steering conversion kit	1	536418	RH Stg } Mk IA cars numbered from 72500552, 72600091, 72800082, 73000767, 73100151, 73300153 onwards.
	Power steering conversion kit (Cars with double-bearing inner column)	1	536419	LH Stg
	Power steering conversion kit	1	536885	RH Stg } Mk II models with suffix letter 'A'
	Power steering conversion kit	1	536886	LH Stg
	Power steering conversion kit	1	600011	RH Stg } Mk II models with suffix letter 'B' onwards
	Power steering conversion kit	1	600012	LH Stg
				(Cars fitted with steering damper) Please note that there are no conversion kits available for models prior to 1961)
	Steering wheel extension kit, giving 2¼" extension	1	525885	
	Continental conversion lens masks for headlamps, pair	1	601975	
	Chrome exhaust tailpipe finisher complete	1	601536	3 Litre only
	BADGE BAR ASSEMBLY	1	316564	
	Badge bar	1	518242	
	PVC washer	2	529152	
	Set screw	2	255226	
	Spring washer	2	3075	
	Plain washer	2	2249	
	Mudflap kit, front set	1	380730	
	Mudflap kit, rear set	1	380728	

*Asterisk indicates a new part which has not been used on any previous model.

Plate Ref. 1 2 3 4	Description	Qty.	Part No.	Remarks
	Exterior driving mirror kit 'Solar' door fixing type	1	605627	
	Tow bar assembly	1	605695*	
	ELECTRICAL KIT FOR TOW BAR	1	605000	3 Litre
	Flasher unit, Bosch	1	563201	
	Drive screw } Fixing flasher unit	1	77704	
	Plain washer	2	4035	
	Socket, seven-pin, for trailer connection	1	563217	
	Bracket for socket	1	559470	
	Set bolt (¼" UNF x ¾" long) } Fixing bracket to tow bar	2	255207	
	Spring washer	2	3074	
	Nut (¼" UNF)	2	254810	
	Screw (10 UNF) } Fixing socket to bracket	3	78756	
	Spring washer	3	3073	
	Nut (10 UNF)	3	257023	
	Plug, seven-pin, for trailer connection	1	563216	
	Cable for trailer plug	1	560501	
	Bracket for warning light	1	559458	
	Screw (10 UNF x ½" long) } Fixing bracket to underside of instrument panel	1	78718	
	Plain washer	1	3557	
	Shakeproof washer	1	74236	
	Nut (10 UNF)	1	257023	
	Flasher warning light	1	559472	
	Brake lamp relay	1	551531	
	Drive screw fixing relay to RH valance	2	78494	
	Lead for warning light	1	559439	
	Conversion harness	1	559463	
	Cable clip	3	50639	
	Screw (10 UNF x ½" long) } Fixing cable clip to tow bar	3	78593	
	Spring washer	3	3073	
	Nut (10 UNF)	3	257023	
	ELECTRICAL KIT FOR TOW BAR	1	605693*	3½ Litre
	Flasher unit, Bosch	1	573081	
	Drive screw } Fixing flasher unit	1	78522	
	Plain washer	2	3816	
	Socket, 7 pin for trailer connection	1	563217	
	Bracket for socket	1	559470	
	Set bolt (¼" UNF x ¾" long) } Fixing bracket to tow bar	2	255207	
	Spring washer	2	3074	
	Nut (¼" UNF)	2	254810	
	Screw (10 UNF) } Fixing socket to bracket	3	78756	
	Spring washer	3	3073	
	Nut (10 UNF)	3	257023	
	Fixing plate for flasher unit	1	573082	
	Drive screw } Fixing flasher unit to fixing plate	1	78770	
	Fan disc washer	1	513282	

*Asterisk indicates a new part which has not been used on any previous model.

Plate Ref. 1 2 3 4 Description		Qty.	Part No.	Remarks	
Plug, 7 pin for trailer connection		1	563216	3½ Litre	
Cable for trailer plug		1	560501		
Bracket for warning light		1	559458		
Screw (10 UNF x ½" long)	Fixing bracket to underside of instrument panel	1	78718		
Plain washer		1	3557		
Shakeproof washer		1	74236		
Nut (10 UNF)		1	257023		
Flasher warning light		1	559472		
Lead for warning light		1	559439		
Conversion harness		1	567846*		
Cable clip		3	50639		
Screw (10 UNF x ½" long)	Fixing cable clip to tow bar	3	78593		
Spring washer		3	3073		
Nut (10 UNF)		3	257023		
Fitting instructions		1	605694		
Glass for back light, heated, with leads		1	386390	Saloon	3 Litre and early 3½ Litre
Glass for back light, heated, with leads		1	386392	Coupe	
Glass for back light, heated, with leads		1	386605*	Saloon	Late 3½ Litre models
Glass for back light, heated, with leads		1	386608*	Coupe	
Bracket for switch		1	359259		
Set screw (10 UNF x ½" long)	Fixing bracket to facia frame	2	257005		
Disc washer		2	513282		
Plain washer		2	3685		
Switch for heated back light		1	563275		
Bulb for switch		1	503352		
Earth clip for back light cable		1	236366		
Cable, heater junction to back light		1	536008	Mk IA 3 Litre	
Cable, heater switch to back light switch		1	536061	Mk II 3 Litre	
Cable, ignition switch to back light switch		1	547483	Mk III 3 Litre	
Cable, ignition switch to back light switch		1	570678	3½ Litre	
Cable, back light to switch		1	536003	Mk IA 3 Litre	
Cable, back light to switch		1	536935	Mk II 3 Litre	
Cable, back light to switch		1	551636	Mk II 3 Litre	
Cable, back light to switch		1	565866	3½ Litre	
Cable, back light to switch		1	586390*	3½ Litre with "Hot line" type back light	
Earth lead for back light		1	573194		
'Lucrimp' 3-way connector for switch lead		1	536937		
Drive screw	Fixing back light earth lead to body	1	78137		
Disc washer		1	513282		
Roof rack complete		1	358248		
Safety harness kit, front seat, RH	For bench seat or standard bucket seats	1	358253	Saloon	Mk II 3 Litre cars with suffix letters 'A' and 'B' (Cars without anchor points incorporated in the body shell)
Safety harness kit, front seat, LH		1	358188		
Safety harness kit, adjustable bucket seats		2	358255		
Safety harness kit, rear seat		3	358190		
Safety harness kit, front seats		2	380695	Coupe	
Safety harness kit, rear seats		2	358190		
Safety harness kit for front seat (bench or bucket type seats)		2	386248	Mk II with suffix letter 'C' and Mk III 3 Litre and 3½ Litre (Cars with anchor points incorporated in the body shell)	
Safety harness kit for rear seat		?	386545		

*Asterisk indicates a new part which has not been used on any previous model.

Plate Ref. 1 2 3 4 Description		Qty.	Part No.	Remarks	
Anchorage kit for front safety harness		2	605097	Alternative types Check before ordering Late 3½ Litre models	
Safety harness kit, inertia reel, front seat, RH		1	387463		
Safety harness kit, inertia reel, front seat, LH		1	387464		
Safety harness kit, Britax inertia reel, one-handed operaton, front seat, RH		1	389444*		
Safety harness kit, Britax inertia reel, one-handed operation, front seat, LH		1	389445*		
Safety harness kit, rear seat, press button release type		2	386545		
Floor mat, driver's and passenger's, front		1	356062	RH Stg	Rubber link type
Floor mat, driver's and passenger's, front		1	356063	LH Stg	
Floor mat, rear		1	356064		
Fitted floor mats, front set, Charcoal Grey, rubber		1	380947	RH Stg	
Fitted floor mats, front set, Charcoal Grey, rubber		1	385115	LH Stg	
Fitted floor mats, rear set, Charcoal Grey, rubber		1	380948		
Fitted floor mats, front set, Beige, nylon fur		1	380949	RH Stg	
Fitted floor mats, front set, Beige, nylon fur		1	385116	LH Stg	
Fitted floor mats, rear set, Beige, nylon fur		1	380950		
Pillar pull, RH	Cashmere Beige	1	354312	Pillar pulls with coat hook	Alter-natives Check before ordering
Pillar pull, LH		1	354313		
Pillar pull, RH	Biscuit	1	355927		
Pillar pull, LH		1	355928		
Pillar pull, RH	Tan	1	355934		
Pillar pull, LH		1	355935		
Pillar pull, RH	Light Grey	1	354220		
Pillar pull, LH		1	354221		
Pillar pull, RH	Biscuit	1	316696	Pillar pulls without coat hook	
Pillar pull, LH		1	316697		
Pillar pull, RH	Silver Grey	1	316350		
Pillar pull, LH		1	316351		
Pillar pull, RH	Oatmeal trim	1	386168		
Pillar pull, LH		1	386169		
Drive screw fixing pillar pull		4	70167		
REAR SEAT HEADREST KIT		1	385129	Mk III 3 Litre and 3½ Litre saloon only. Quote car number and colour required.	
Rear headrest, RH		1	384847		
Rear headrest, LH		1	384848		

*Asterisk indicates a new part which has not been used on any previous model.

Plate Ref.	1 2 3 4	Description	Qty.	Part No.	Remarks
		Brake wheel cylinder overhaul kit, front	1	275744	Drum brakes
		Disc brake overhaul kit, front	1	513673@	Disc
		Disc brake overhaul kit	1	531939@@	brakes
		Brake wheel cylinder overhaul kit, rear	1	271511	For wheel cylinder 268464/5
		Brake wheel cylinder overhaul kit, rear	1	513684	For wheel cylinder 501256/7
		Brake wheel cylinder overhaul kit, rear	1	605222	3½ Litre
		Brake master cylinder overhaul kit	1	502333	3 Litre
		Brake master cylinder overhaul kit	1	605164	3½ Litre
		Brake adjuster overhaul kit	1	518844	Drum brakes
		Clutch master cylinder overhaul kit	1	601611	
		Clutch slave cylinder overhaul kit	1	512511	Mk I and Mk IA 3 Litre
		Clutch slave cylinder overhaul kit	1	541809	Mk II and Mk III 3 Litre
		Repair kit for servo unit	1	278509	For servicing servo unit 274187
		Repair kit for servo unit	1	514088	For servicing servo unit 510287
		Repair kit for servo unit	1	535674	For servicing servo unit 532553
		Repair kit for servo unit	1	600554	For servicing servo unit 545900
		Piston conversion kit for servo unit	1	535673	This kit, together with overhaul kit 535674, may be used to convert early servos to the latest type piston
		Major repair kit } For brake servo unit	1	601331	} 3½ Litre
		Air control valve kit } For brake servo unit	1	601907	} 3½ Litre
		Minor repair kit } For fuel pump	1	605167	} 3½ Litre
		Major repair kit } For fuel pump	1	605168	} 3½ Litre
		Decarbonising gasket kit	1	GEG163	Mk I and Mk IA
		Decarbonising gasket kit	1	GEG164	Mk II and Mk III
		Decarbonising gasket kit	1	GEG165	3½ Litre
		Engine overhaul gasket kit	1	600384	Mk I and Mk IA
		Engine overhaul gasket kit	1	600382	Mk II and Mk III
		Engine overhaul gasket kit	1	605124	3½ Litre
		Water pump overhaul kit	1	503031	3 Litre
		Joint washer kit for gearbox	1	601305	3 Litre 4-speed
		Seal kit for power steering pump	1	605175	
		Seal kit for power steering unit	1	542276	Mk IA and Mk II 3 Litre
		Seal kit for power steering unit	1	607151*	Mk III 3 Litre and 3½ Litre

@ Suitable for cars numbered up to 72500084a, 72600022a, 72800012a, 73000124a, 73100020a 73300023a

@@ Suitable for cars numbered from 72500085a, 72600023a, 72800013a, 73000125a, 73100021a 73300024a onwards

*Asterisk indicates a new part which has not been used on any previous model.

Plate Ref.	1 2 3 4	Description	Qty.	Part No.	Remarks
		Special nut	1	600555	} The nylon petrol pipes may be repaired by using these three items
		Nipple for nylon pipe	1	600556	} The nylon petrol pipes may be repaired by using these three items
		Adaptor for nylon pipe	1	600557	} The nylon petrol pipes may be repaired by using these three items
		Brake fluid, 1 quart tin	As reqd	262079	
		'Barseal' sealing pellet for radiator	1	601314	
		White brake grease tube	1	514577	
		Red brake rubber grease, tube	1	514578	
		Cleaning fluid for brake master cylinder parts, ½ pint tin	1	535642	
		Bostik 1775, adhesive, 1¾ oz tube	1	601736	
		Hylomar sealing compound, 4 oz tube	1	534244	
		Anti-freeze mixture, Ethylene Glycol	As reqd	13644	3 Litre
		Anti-freeze mixture	1	605529	} 3½ Litre
		Plastigage strip, box of 24	1	605238	} 3½ Litre
		Coolant inhibitor, 18 oz	1	605765	} 3½ Litre
		Thread sealant/lubricant, EC 776, 1 pint	1	605764	} 3½ Litre

*Asterisk indicates a new part which has not been used on any previous model.

Plate Ref.	1 2 3 4	Description	Qty.	Part No.	Remarks
		Wheel brace	1	50217	3 Litre
		Wheel brace	1	570467	3½ Litre
		Wheel cover removal tool	1	534469	Mk IA and Mk II 3 Litre
		Lifting jack	1	503018	1959
		Lifting jack	1	513508	1960 onwards
		Lifting jack	1	545953	Late 3½ Litre models
		Plastic buffer for jack	1	578626*	Late 3½ Litre models
		Tyre pump	1	523638	
		Conne tion for tyre pump	1	524959	
		Starting handle	1	230694	
		Combination pliers	1	278957	3 Litre
		Pliers	1	2703	3½ Litre
		Screwdriver	1	234654	Mk I, Mk IA and Mk II 3 Litre
		Screwdriver dual purpose	1	565770	Mk III and 3½ Litre
		Spanner (⅜" x 11/16" AF)	1	277217	
		Spanner (⅜" x 9/16" AF)	1	276397	
		Spanner (½" x 7/16" AF)	1	276396	
		Adjustable spanner	1	279527	Early models
		Spanner (5/16" x ⅜" AF)	1	549840	Late models alternative to adjustable spanner
		Spanner (2 BA x 4 BA)	1	572503	Late models alternative to adjustable spanner
		Sparking plug spanner (4¼" long)	1	276322	3 Litre
		Sparking plug spanner (3" long)	1	276323	3 Litre
		Sparking plug spanner	1	565434	3½ Litre
		Tommy bar	1	1403	3 Litre
		Tommy bar	1	565443	3½ Litre
		Tyre pressure gauge	1	562019	
		Luggage strap	1	352601	

*Asterisk indicates a new part which has not been used on any previous model.

Plate Ref.	1 2 3 4	Description	Qty.	Part No.	Remarks
		Felted carpet, Rush Green, 40" wide	As reqd	92018	
		Felted carpet, Red, 40" wide	As reqd	92020	
		Felted carpet, Mid-Blue, 40" wide	As reqd	92017	
		Felted carpet, Silver Grey, 40" wide	As reqd	92021	
		Felted carpet, Charcoal Grey, 40" wide	As reqd	92022	
		Felted carpet, Beige, 40" wide	As reqd	92019	
		Carpet binding lace, Rush green	As reqd	92032	
		Carpet binding lace, Red	As reqd	92034	
		Carpet binding lace, Mid-Blue	As reqd	92031	
		Carpet binding lace, Silver Grey	As reqd	92035	
		Carpet binding lace, Charcoal Grey	As reqd	92036	
		Carpet binding lace, Beige	As reqd	92281	
		Velvet, Rush Green, 48" wide	As reqd	92024	
		Velvet, Red, 48" wide	As reqd	92026	
		Velvet, Mid-Blue, 48" wide	As reqd	92023	
		Velvet, Silver Grey, 48" wide	As reqd	92027	
		Velvet, Charcoal Grey, 48" wide	As reqd	92028	
		Velvet, Beige, 48" wide	As reqd	92025	
		Leathercloth, Rush Green, 50" wide	As reqd	92007	
		Leathercloth, Red, 50" wide	As reqd	92009	
		Leathercloth, Mid-Blue, 50" wide	As reqd	92006	
		Leathercloth, Silver Grey, 50" wide	As reqd	92010	
		Leathercloth, Charcoal Grey, 50" wide	As reqd	92011	
		Leathercloth, Beige, 50" wide	As reqd	92008	
		Leathercloth, Black	As reqd	92175	
		Hide, Rush Green	As reqd	92132	
		Hide, Red	As reqd	92058	
		Hide, Mid-Blue	As reqd	92055	
		Hide, Silver Grey	As reqd	92059	
		Hide, Charcoal Grey	As reqd	92135	
		Hide, Beige	As reqd	92057	
		Woollen cloth, Rush Green, 50" wide	As reqd	92067	
		Woollen cloth, Red, 50" wide	As reqd	92069	
		Woollen cloth, Mide-Blue, 50" wide	As reqd	92066	
		Woollen cloth, Silver Grey, 50" wide	As reqd	92070	
		Woollen cloth, Beige, 50" wide	As reqd	92068	
		Terylene thread, Silver Grey	As reqd	92041	
		Terylene thread, Beige	As reqd	92039	
		Nylon thread, Charcoal Grey	As reqd	92334	
		Headlining PVC, Light Grey, 50" wide	As reqd	92029	
		Headlining PVC, Beige 50" wide	As reqd	92030	
		Grey felt, 72" wide	As reqd	91414	
		Natural felt, 54" wide	As reqd	91890	
		Thin plain cloth, 38" wide	As reqd	91829	
		Black cloth, 54" wide	As reqd	91459	
		Stout plain cloth, 73" wide	As reqd	91956	
		Knitted back leathercloth, Rush Green	As reqd	92073	
		Knitted back leathercloth, Red	As reqd	92075	
		Knitted back leathercloth, Mid-Blue	As reqd	92072	
		Knitted back leathercloth, Silver Grey	As reqd	92076	
		Knitted back leathercloth, Beige	As reqd	92074	
		Knitted back leathercloth, Charcoal Grey	As reqd	92081	

*Asterisk indicates a new part which has not been used on any previous model.

Plate Ref. 1 2 3 4	DESCRIPTION	Qty	Part No.	REMARKS
	Linen thread, No. 18, Rush Green	As reqd	92050	
	Linen thread, No. 18, Red	As reqd	92052	
	Linen thread, No. 18, Mid-Blue	As reqd	92049	
	Linen thread, No. 18, Silver Grey	As reqd	92053	
	Linen thread, No. 18, Beige	As reqd	92051	1959–60 models
	Linen thread, No. 25, Rush Green	As reqd	92044	
	Linen thread, No. 25, Mid-Blue	As reqd	92043	
	Linen thread, No. 25, Silver Grey	As reqd	92047	
	Linen thread, No. 25, Beige	As reqd	92045	
	Nylon thread, No. 25, Red	As reqd	92333	
	1961 Mk I AND Mk IA MODELS			
	Felted carpet, Rush Green, 40" wide	As reqd	92018	
	Felted carpet, Red, 40" wide	As reqd	92109	
	Felted carpet, Blue, 40" wide	As reqd	92110	
	Felted carpet, Dark Grey, 40" wide	As reqd	92112	
	Felted carpet, Tan, 40" wide	As reqd	92111	
	Felted carpet, Charcoal Grey, 40" wide	As reqd	92022	
	Carpet binding lace, Rush Green	As reqd	92032	
	Carpet binding lace, Red	As reqd	92122	
	Carpet binding lace, Blue	As reqd	92123	
	Carpet binding lace, Dark Grey	As reqd	92173	
	Carpet binding lace, Tan	As reqd	92124	
	Carpet binding lace, Charcoal Grey	As reqd	92036	
	Velvet, Rush Green, 48" wide	As reqd	92024	
	Velvet, Red, 48" wide	As reqd	92113	
	Velvet, Blue, 48" wide	As reqd	92114	
	Velvet, Dark Grey, 48" wide	As reqd	92116	
	Velvet, Tan, 48" wide	As reqd	92115	
	Velvet, Charcoal Grey, 48" wide	As reqd	92028	
	Leathercloth, Rush Green, 50" wide	As reqd	92007	
	Leathercloth, Red, 50" wide	As reqd	92105	
	Leathercloth, Blue, 50" wide	As reqd	92106	
	Leathercloth, Light Grey, 50" wide	As reqd	92108	
	Leathercloth, Tan, 50" wide	As reqd	92107	
	Leathercloth, Charcoal Grey, 50" wide	As reqd	92011	
	Leathercloth, Black, 50" wide	As reqd	91934	
	Hide, Rush Green	As reqd	92132	
	Hide, Red	As reqd	92101	
	Hide, Blue	As reqd	92102	
	Hide, Light Grey	As reqd	92104	
	Hide, Tan	As reqd	92103	
	Hide, Charcoal Grey	As reqd	92135	
	Headlining PVC, Light Grey, 50" wide	As reqd	92029	
	Headlining PVC, Biscuit, 50" wide	As reqd	92117	
	Grey felt, 72" wide	As reqd	91414	
	Natural felt, 54" wide	As reqd	91890	
	Thin plain cloth, 38" wide	As reqd	91829	
	Black cloth, 54" wide	As reqd	91459	
	Stout plain cloth, 73" wide	As reqd	91956	
	Nylon thread, White	As reqd	92336	
	Nylon thread, Biscuit	As reqd	92337	
	Nylon thread, Charcoal Grey	As reqd	92334	

* Asterisk indicates a new part which has not been used on any previous Rover model

Plate Ref. 1 2 3 4	DESCRIPTION	Qty	Part No.	REMARKS
	Knitted back leathercloth, Rush Green	As reqd	92073	
	Knitted back leathercloth, Red	As reqd	92128	
	Knitted back leathercloth, Blue	As reqd	92126	
	Knitted back leathercloth, Grey	As reqd	92129	
	Knitted back leathercloth, Tan	As reqd	92127	
	Knitted back leathercloth, Charcoal Grey	As reqd	92081	1961 Mk I and Mk IA models
	Linen thread, No. 25, Rush Green	As reqd	92044	
	Nylon thread, No. 25, Red	As reqd	92333	
	Linen thread, No. 25, Blue	As reqd	92043	
	Linen thread, No. 25, Grey	As reqd	92047	
	Linen thread, No. 25, Tan	As reqd	92134	
	Seaming cord for seat trim	As reqd	91171	
	Black flexible plastic for door trim panel	As reqd	355387	
	Twine	As reqd	91160	
	Spring wire	As reqd	81698	
	Sheet wadding, 36" wide opened	As reqd	91341	
	Horse hair, black	As reqd	91295	
	Acetate wadding, 37" wide	As reqd	91980	
	Elastic, black, 1" wide	As reqd	91977	
	Fine black canvas, 54" wide	As reqd	91291	
	Webbing, black	As reqd	91888	
	Black felt, 50" wide	As reqd	91819	
	Thin-faced felt, 72" wide	As reqd	91415	
	Thin-faced felt	As reqd	91945	
	Black rubbered cloth, 60" wide	As reqd	91290	
	Felted plastic, brown, 52" wide	As reqd	91935	All Mk I and Mk IA models
	Felted plastic, black, 52" wide	As reqd	91965	
	Plywood, ⅛" thick	As reqd	85156	
	Plywood, ⅜" thick	As reqd	85158	
	Flut-wad, 3" wide	As reqd	91955	
	Flut-wad, 5" wide	As reqd	91954	
	Flut-wad, 7" wide	As reqd	92071	
	Sealing strip, 1/16" x ⅜" wide	As reqd	13279	
	Sealing strip, 1/16" x ½" wide	As reqd	13296	
	Sealing strip, 1/16" x 1¼" wide	As reqd	13388	
	Bittac pitch compound, ½ gallon tin	As reqd	262095	
	Bittac pitch compound, 1 gallon tin	As reqd	262096	
	Flintkote sheet	As reqd	82344	
	Compressed paper board, ⅜" x 26" x 20"	As reqd	91818	
	Waterproof paper, 54" wide	As reqd	13205	
	Cord	As reqd	91845	

* Asterisk indicates a new part which has not been used on any previous Rover model

Plate Ref.	1 2 3 4	Description	Qty.	Part No.	Remarks
		Felted carpet, Green, 40" wide	As reqd	92018	
		Felted carpet, Red, 40" wide	As reqd	92109	
		Felted carpet, Blue, 40" wide	As reqd	92110	
		Felted carpet, Tan, 40" wide	As reqd	92111	
		Felted carpet, Grey, 40" wide	As reqd	92188	
		Felted carpet, Dark Fawn, 40" wide	As reqd	92189	
		Felted carpet, Toledo Red, 40" wide	As reqd	92412	
		Felted carpet, Mortlake Brown, 40" wide	As reqd	92409	
		Felted carpet, Birch Grey 40" wide	As reqd	92435	
		Plain carpet, Birch Grey	As reqd	92376	
		Carpet binding lace, Green	As reqd	92032	
		Carpet binding lace, Red	As reqd	92122	
		Carpet binding lace, Blue	As reqd	92123	
		Carpet binding lace, Tan	As reqd	92124	
		Carpet binding lace, Grey	As reqd	92173	
		Carpet binding lace, Dark Fawn	As reqd	92174	
		Carpet binding lace, Toledo Red	As reqd	92227	
		Carpet binding lace, Mortlake Brown	As reqd	92413	
		Carpet binding lace, Birch Grey	As reqd	92282	
		Velvet, Green, 48" wide	As reqd	92024	
		Velvet, Red, 48" wide	As reqd	92113	
		Velvet, Blue, 48" wide	As reqd	92114	
		Velvet, Tan, 48" wide	As reqd	92115	
		Velvet, Grey, 48" wide	As reqd	92190	
		Velvet, Stone, 48" wide	As reqd	92191	
		Velvet, Toledo Red, 48" wide	As reqd	92414	
		Velvet Buckskin, 48" wide	As reqd	92415	
		Velvet, Buffalo, 48" wide	As reqd	92416	
		Velvet, Sandalwood, 48" wide	As reqd	92417	
		Velvet, Mulberry	As reqd	92452	
		Velvet, Black	AS reqd	92456	
		Velvet, Saddle Tan	As reqd	92453	
		Leathercloth, Green, 50" wide	As reqd	92180	
		Leathercloth Red, 50" wide	As reqd	92105	
		Leathercloth, Blue, 50" wide	As reqd	92106	
		Leathercloth, Tan, 50" wide	As reqd	92181	
		Leathercloth, Grey, 50" wide	As reqd	92182	
		Leathercloth, Stone, 50" wide	As reqd	92183	
		Leathercloth, Dull Black, 50" wide	As reqd	92175	
		Leathercloth, Toledo Red, 50" wide	As reqd	92228	
		Leathercloth, Buckskin, 50" wide	As reqd	92391	
		Leathercloth, Buffalo, 50" wide	As reqd	92392	
		Leathercloth, Sandalwood, 50" wide	As reqd	92393	
		Leathercloth, Mulberry	As reqd	92439	
		Leathercloth, Saddle Tan	As reqd	92436	
		Leathercloth, Ebony	As reqd	92175	

*Asterisk indicates a new part which has not been used on any previous model.

Plate Ref.	1 2 3 4	Description	Qty.	Part No.	Remarks
		Hide, Green	As reqd	92169	
		Hide, Red	As reqd	92101	
		Hide, Blue	As reqd	92102	
		Hide, Tan	As reqd	92170	
		Hide, Grey	As reqd	92171	
		Hide, Stone	As reqd	92172	
		Hide, Toledo Red	As reqd	92234	
		Hide, Buckskin	As reqd	92406	
		Hide, Buffalo	As reqd	92407	
		Hide, Sandalwood	As reqd	92408	
		Hide, Mulberry	As reqd	92437	
		Hide, Saddle Tan	As reqd	92438	
		Hide, Ebony	As reqd	92454	
		Headlining PVC, Light Grey, 50" wide	As reqd	92029	
		Headlining PVC, Biscuit, 50" wide	As reqd	92117	
		PVC Headlining, Oatmeal	As reqd	92462	
		Grey felt, 72" wide	As reqd	91414	
		Natural felt, 54" wide	As reqd	91890	
		Thin plain cloth, 38" wide	As reqd	91829	
		Black cloth, 54" wide	As reqd	91459	
		Stout plain cloth, 60" wide, blue stripe	As reqd	92215	
		Knitted back leathercloth, Green	As reqd	92184	
		Knitted back leathercloth, Red	As reqd	92128	
		Knitted back leathercloth, Blue	As reqd	92126	
		Knitted back leathercloth, Tan	As reqd	92185	
		Knitted back leathercloth, Grey	As reqd	92186	
		Knitted back leathercloth, Stone	As reqd	92187	
		Knitted back leathercloth, Black	As reqd	92216	
		Knitted back leathercloth, Toledo Red	As reqd	92230	
		Knitted back leathercloth, Buckskin	As reqd	92394	
		Knitted back leathercloth, Buffalo	As reqd	92395	
		Knitted back leathercloth, Sandalwood	As reqd	92396	
		Knitted back leathercloth, Mulberry	As reqd	92440	
		Knitted back leathercloth, Saddle Tan	As reqd	92441	
		Knitted back leathercloth Ebony	As reqd	92276	
		Unsupported PVC, Biscuit	As reqd	92193	
		Unsupported PVC, Oatmeal	As reqd	92239	
		Nylon thread, Green	As reqd	92331	
		Nylon thread, Red	As reqd	92333	
		Nylon thread, Blue	As reqd	92330	
		Nylon thread, Tan	As reqd	92332	
		Nylon thread, Grey	As reqd	92334	
		Nylon thread, Stone	As reqd	92335	
		Nylon thread, White	As reqd	92336	
		Nylon thread, Biscuit	As reqd	92337	
		PVC bonded polyether, Toledo Red	As reqd	92245	
		PVC bonded polyether, Buckskin	As reqd	92397	
		PVC bonded polyether, Buffalo	As reqd	92398	
		PVC bonded polyether, Sandalwood	As reqd	92399	
		PVC bonded polyether, Mulberry	As reqd	92442	
		PVC bonded polyether, Saddle Tan	As reqd	92443	
		PVC bonded polyether, Ebony	As reqd	92268	

*Asterisk indicates a new part which has not been used on any previous model.

TRIMMING RAW MATERIALS, MK II AND MK III 3 LITRE AND 3½ LITRE MODELS

1967

Plate Ref. 1 2 3 4	Description	Qty.	Part No.	Remarks
	Linen thread, Green	As reqd	92195	
	Linen thread, Red	As reqd	92196	
	Linen thread, Blue	As reqd	92194	
	Linen thread, Tan	As reqd	92198	
	Linen thread, Grey	As reqd	92047	
	Linen thread, Stone	As reqd	92197	
	Linen thread, Black	As reqd	91257	
	Linen thread, Toledo Red	As reqd	92237	
	Linen thread, Buckskin	As reqd	92197	
	Linen thread, Buffalo	As reqd	92418	
	Linen thread, Sandalwood	AS reqd	92198	
	Linen thread, Fawn	As reqd	92286	
	Nylon thread, Buckskin	As reqd	92420	
	Nylon thread, Buffalo	As reqd	92421	
	Nylon thread, Sandalwood	As reqd	92419	
	Seaming cord for seat trim	As reqd	91171	
	Black flexible plastic for door trim panel	As reqd	355387	
	Twine	As reqd	91160	
	Spring wire	As reqd	81698	
	Polyprepylene rod ⅛" dim for seat piping	As reqd	92479*	
	Sheet wadding, 36" wide opened	As reqd	92496	
	Horse hair, black	As reqd	91295	
	Acetate wadding, 37" wide	As reqd	91980	
	Elastic, black, 1" wide	As reqd	91977	
	Fine black canvas, 54" wide	As reqd	91291	
	Webbing, black	As reqd	91888	
	Black felt, 50" wide	As reqd	91819	
	Thin-faced felt, 72" wide	As reqd	91415	
	Thin-faced felt	As reqd	91945	
	Black rubbered cloth, 60" wide	As reqd	91290	
	Felted plastic, brown, 52" wide	As reqd	91935	
	Felted plastic, black, 52" wide	As reqd	91965	
	Plywood, ⅛" thick	As reqd	85156	
	Plywood, ¾" thick	As reqd	85158	
	Flut-wad, 3" wide	As reqd	91955	
	Flut-wad, 5" wide	As reqd	91954	
	Flut-wad, 7" wide	As reqd	92071	
	Sealing strip, 1/16" x ⅜" wide	As reqd	13279	
	Sealing strip, 1/16" x ½" wide	As reqd	13296	
	Sealing strip, 1/16" x 1¼" wide	As reqd	13388	
	Bittac pitch compound, ½ gallon tin	As reqd	262095	
	Bittac pitch compound, 1 gallon tin	As reqd	262096	
	Flintkote sheet	As reqd	82344	
	Compressed paper board ⅛" x 26" x 20"	As reqd	91818	
	Waterproof paper, 54" wide	As reqd	13205	
	Cord	As reqd	91845	

*Asterisk indicates a new part which has not been used on any previous model.

1968

CELLULOSE AND FINISHING MATERIALS

Plate Ref. 1 2 3 4	DESCRIPTION		Qty	Part No.	REMARKS
	ACP 'Deoxidine' cleaner, ½ gallon tin		As reqd	261883	
	Cellulose primer, 1 pint tin		As reqd	261884	
	Cellulose filler, 1 pint tin		As reqd	261885	
	Thinner, 1 pint tin		As reqd	261906	
	Thinner, 1 gallon tin		As reqd	261909	
	Half-hour air-drying enamel, 1 pint tin	Black	As reqd	261886	1959–60 models
	Half-hour air-drying enamel, 1 gallon tin		As reqd	261901	
	Half-hour air-drying enamel, 1 pint tin	Dover White	As reqd	507219	
	Half-hour air-drying enamel, 1 gallon tin		As reqd	507220	
	Half-hour air-drying enamel, 1 pint tin	Dark Blue	As reqd	507221	
	Half-hour air-drying enamel, 1 gallon tin		As reqd	507222	
	Half-hour air-drying enamel, 1 pint tin	Smoke Grey	As reqd	502016	
	Half-hour air-drying enamel, 1 gallon tin		As reqd	502020	
	Half-hour air-drying enamel, 1 pint tin	Dove Grey	As reqd	522914	
	Half-hour air-drying enamel, 1 gallon tin		As reqd	522915	
	Half-hour air-drying enamel, 1 pint tin	Light Grey	As reqd	507217	
	Half-hour air-drying enamel, 1 gallon tin		As reqd	507218	
	Half-hour air-drying enamel, 1 pint tin	Shadow Green	As reqd	507223	
	Half-hour air-drying enamel, 1 gallon tin		As reqd	507224	
	Half-hour air-drying enamel, 1 pint tin	Rush Green	As reqd	507225	
	Half-hour air-drying enamel, 1 gallon tin		As reqd	507226	
	Half-hour air-drying enamel, 1 pint tin	Heather Brown	As reqd	507227	
	Half-hour air-drying enamel, 1 gallon tin		As reqd	507228	
	Half-hour air-drying enamel, 1 pint tin	Light Brown	As reqd	507229	
	Half-hour air-drying enamel, 1 gallon tin		As reqd	507230	
	Half-hour air-drying enamel, 1 pint tin	Ivory	As reqd	357112	1961 and Mk IA models
	Half-hour air-drying enamel, 1 gallon tin		As reqd	357113	
	Half-hour air-drying enamel, 1 pint tin	Storm	As reqd	357114	
	Half-hour air-drying enamel, 1 gallon tin	Grey	As reqd	357115	
	Half-hour air-drying enamel, 1 pint tin	Medium	As reqd	357116	
	Half-hour air-drying enamel, 1 gallon tin	Grey	As reqd	357117	
	Half-hour air-drying enamel, 1 pint tin	Royal	As reqd	357118	
	Half-hour air-drying enamel, 1 gallon tin	Blue	As reqd	357119	
	Half-hour air-drying enamel, 1 pint tin	Slate	As reqd	357120	
	Half-hour air-drying enamel, 1 gallon tin	Grey	As reqd	357121	
	Half-hour air-drying enamel, 1 pint tin	Norse	As reqd	357122	
	Half-hour air-drying enamel, 1 gallon tin	Blue	As reqd	357123	
	Half-hour air-drying enamel, 1 pint tin	Black	As reqd	261886	
	Half-hour air-drying enamel, 1 gallon tin		As reqd	261901	
	Half-hour air-drying enamel, 1 pint tin	Smoke	As reqd	502016	
	Half-hour air-drying enamel, 1 gallon tin	Grey	As reqd	502020	
	Half-hour air-drying enamel, 1 pint tin	Shadow	As reqd	507223	
	Half-hour air-drying enamel, 1 gallon tin	Green	As reqd	507224	
	Half-hour air-drying enamel, 1 pint tin	Rush	As reqd	507225	
	Half-hour air-drying enamel, 1 gallon tin	Green	As reqd	507226	
	Half-hour air-drying enamel, 1 pint tin	Pine	As reqd	358309	
	Half-hour air-drying enamel, 1 gallon tin	Green	As reqd	358310	
	Half-hour air-drying enamel, 1 pint tin	Light	As reqd	358311	
	Half-hour air-drying enamel, 1 gallon tin	Navy	As reqd	358312	
	Half-hour air-drying enamel, 1 pint tin	Burgundy	As reqd	358313	
	Half-hour air-drying enamel, 1 gallon tin		As reqd	358314	

* Asterisk indicates a new part which has not been used on any previous Rover model

Plate Ref.	1 2 3 4	Description		Qty.	Part No.	Remarks
		Half-hour air-drying enamel, 1 pint tin		As reqd	261886	
		Half-hour air-drying enamel, 1 gallon tin	Black	As reqd	261901	
		Half-hour air-drying enamel, 1 pint tin	Pine	As reqd	358309	
		Half-hour air-drying enamel, 1 gallon tin	Green	As reqd	358310	
		Half-hour air-drying enamel, 1 pint tin	Light	As reqd	358311	
		Half-hour air-drying enamel, 1 gallon tin	Navy	As reqd	358312	
		Half-hour air-drying enamel, 1 pint tin	Burgandy	As reqd	358313	
		Half-hour air drying enamel, 1 gallon tin		As reqd	358314	
		Half-hour air-drying enamel, 1 pint tin	White	As reqd	381049	
		Half-hour air-drying enamel, 1 gallon tin		As reqd	380205	
		Half-hour air drying enamel, 1 pint tin	Steel	As reqd	381050	
		Half-hour air drying enamel, 1 gallon tin	Blue	As reqd	380206	
		Half-hour air-drying enamel, 1 pint tin	Charcoal	As reqd	381051	
		Half-hour air drying enamel, 1 gallon tin	Grey	As reqd	380207	
		Half-hour air-drying enamel, 1 pint tin	Marine	As reqd	381052	
		Half-hour air-drying enamel, 1 gallon tin	Grey	As reqd	380208	
		Half-hour air-drying enamel, 1 pint tin	Juniper	As reqd	381053	
		Half-hour air-drying enamel, 1 gallon tin	Green	As reqd	380209	
		Half-hour air-drying enamel, 1 pint tin	Stone	As reqd	381054	
		Half-hour air-drying enamel, 1 gallon tin	Grey	As reqd	380210	
		Half-hour air-drying enamel, 1 pint tin	Arden	As reqd	366504	
		Half-hour air-drying enamel, 1 gallon tin	Green	As reqd	366505	
		Half-hour air-drying enamel, 1 pint tin	Admiralty	As reqd	385146	
		Half-hour air-drying enamel, 1 gallon tin	Blue	As reqd	385147	
		Half-hour air-drying enamel, 1 pint tin	Bordeaux	As reqd	385148	
		Half-hour air-drying enamel, 1 gallon tin	Red	As reqd	385149	
		Half-hour air-drying enamel, 1 pint tin	Silver	As reqd	385150	
		Half-hour air-drying enamel, 1 gallon tin	Birch	As reqd	385151	
		Half-hour air-drying enamel, 1 pint tin	Burnt	As reqd	366548	
		Half-hour air drying enamel, 1 gallon tin	Grey	As reqd	366549	
		Half-hour air-drying enamel, 1 pint tin	Zircon	As reqd	366506	3½ Litre only
		Half-hour air-drying enamel, 1 gallon tin	Blue	As reqd	366507	
		Half-hour air-drying enamel, 1 pint tin,	Matt Black	As reqd	387088	For road wheels 3½ Litre only
		Polypropylene paint, 1 pint tin,	Black	As reqd	387414	For lower sill panels 3½ Litre only

*Asterisk indicates a new part which has not been used on any previous model.

Plate Ref.	1 2 3 4	Description	Qty.	Part No.	Remarks
		Touch-up paint 2 oz tin, Black, for body repairs	As read	385178	
		Touch-up paint 2 oz tin, Dove Grey, for body repairs	As reqd	387167	
		Touch-up paint 2 oz tin, Smoke Grey, for body repairs	As reqd	385190	
		Touch-up pencil, Light Grey, for body repairs	As reqd	356286	
		Touch-up pencil, Dover White, for body repairs	As reqd	356287	1959-60 3 Litre
		Touch-up pencil, Dark Blue, for body repairs	As reqd	356288	
		Touch-up paint, 2 oz tin Shodow Green, for body repairs	As reqd	385189	
		Touch-up paint, 2 oz tin, Rush Green, for body repairs	As reqd	385188	
		Touch-up pencil, Heather Brown, for body repairs	As reqd	356291	
		Touch-up pencil, Light Brown, for body repairs	As reqd	356292	
		Touch-up paint, 2 oz tin, Black, for body repairs	As reqd	385178	
		Touch-up paint, 2 oz tin, Rush Green, for body repairs	As reqd	385188	
		Touch-up paint, 2 oz tin, Shadow Green, for body repairs	As reqd	385189	
		Touch-up paint, 2 oz tin, Smoke Grey for body repairs	As reqd	385190	
		Touch-up paint, 2 oz tin, Ivory, for body repairs	As reqd	385191	
		Touch-up paint, 2 oz tin, Storm Grey, for body repairs	As reqd	385192	
		Touch-up paint, 2 oz tin, Medium Grey, for body repairs	As reqd	385193	1961 and Mk IA 3 Litre
		Touch-up paint, 2 oz tin, Royal Blue, for body repairs	As reqd	385194	
		Touch-up paint, 2 oz tin, Slate Grey, for body repairs	As reqd	385195	
		Touch-up paint, 2 oz tin, Norse Blue for body repairs	As reqd	385196	
		Touch-up paint, 2 oz tin, Pine Green, for body repairs	As reqd	385181	
		Touch-up paint, 2 oz tin, Light Navy, for body repairs	As reqd	385182	
		Touch-up paint, 2 oz tin, Burgundy, for body repairs	As reqd	385183	

*Asterisk indicates a new part which has not been used on any previous model.

Plate Ref.	1 2 3 4	Description	Qty.	Part No.	Remarks
		Touch-up paint, 2 oz tin, Black, for body repairs	As reqd	385178	
		Touch-up paint, 2 oz tin, Pine Green, for body repairs	As reqd	385181	
		Touch-up paint, 2 oz tin, Light Navy, for body repairs	As reqd	385182	
		Touch-up paint, 2 oz tin, Burgundy, for body repairs	As reqd	385183	
		Touch-up paint, 2 oz tin, Juniper Green	As reqd	385172	Mk II 3 Litre
		Touch-up paint, 2 oz tin, Steel Blue	As reqd	385184	
		Touch-up paint, 2 oz tin, White	As reqd	385173	
		Touch-up paint, 2 oz tin, Stone Grey	As reqd	385185	
		Touch-up paint, 2 oz tin, Marine Grey	As reqd	385186	
		Touch-up paint, 2 oz tin, Charcoal Grey	As reqd	385187	
		Touch-up paint, 2 oz tin, Arden Green	1	385171	
		Touch-up paint, 2 oz tin, Admirality Blue	1	385174	
		Touch-up paint, 2 oz tin, Bordeaux Red	1	385175	
		Touch-up paint, 2 oz tin, Burnt Grey	1	385176	Mk III 3 Litre and 3½ Litre
		Touch-up paint, 2 oz tin, White	1	385173	
		Touch-up paint, 2 oz tin, Silver Birch	1	385177	
		Touch-up paint, 2 oz tin, Juniper Green	1	385172	
		Touch-up paint, 2 oz tin, Black	1	385178	
		Touch-up paint, 2 oz tin Zirion Blue	1	366591	
		Touch-up paint, 2 oz tin, Matt Black, for road wheels	1	387087	
		Touch-up paint, 2 oz tin, Polypropylene, Black for lower sill panels	1	387413	

*Asterisk indicates a new part which has not been used on any previous model.

This page is intentionally left bank

Printed in Great Britain
by Amazon

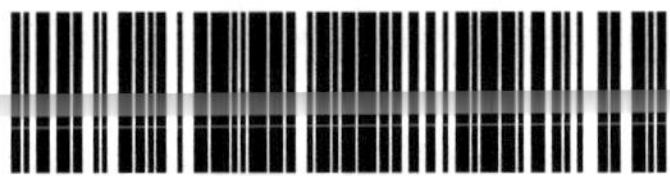